3 Oct. 1978

To Len,

with many thanks +
all best wishes,

Lawrence Rugelman [illegible]
Gerard Cohen [illegible]

Waves in the Ocean

FURTHER TITLES IN THIS SERIES

1 J.L. MERO
THE MINERAL RESOURCES OF THE SEA
2 L.M. FOMIN
THE DYNAMIC METHOD IN OCEANOGRAPHY
3 E.J.F. WOOD
MICROBIOLOGY OF OCEANS AND ESTUARIES
4 G. NEUMANN
OCEAN CURRENTS
5 N.G. JERLOV
OPTICAL OCEANOGRAPHY
6 V. VACQUIER
GEOMAGNETISM IN MARINE GEOLOGY
7 W.J. WALLACE
THE DEVELOPMENT OF THE CHLORINITY/SALINITY CONCEPT IN OCEANOGRAPHY
8 E. LISITZIN
SEA-LEVEL CHANGES
9 R.H. PARKER
THE STUDY OF BENTHIC COMMUNITIES
10 J.C.J. NIHOUL
MODELLING OF MARINE SYSTEMS
11 O.I. MAMAYEV
TEMPERATURE—SALINITY ANALYSIS OF WORLD OCEAN WATERS
12 E.J. FERGUSON WOOD and R.E. JOHANNES
TROPICAL MARINE POLLUTION
13 E. STEEMANN NIELSEN
MARINE PHOTOSYNTHESIS
14 N.G. JERLOV
MARINE OPTICS
15 G.P. GLASBY
MARINE MANGANESE DEPOSITS
16 V.M. KAMENKOVICH
FUNDAMENTALS OF OCEAN DYNAMICS
17 R.A. GEYER
SUBMERSIBLES AND THEIR USE IN OCEANOGRAPHY AND OCEAN ENGINEERING
18 J.W. CARUTHERS
FUNDAMENTALS OF MARINE ACOUSTICS
19 J.C.J. NIHOUL
BOTTOM TURBULENCE
21 C.C. VON DER BORCH (Editor)
SYNTHESIS OF DEEP-SEA DRILLING RESULTS IN THE INDIAN OCEAN
22 P. DEHLINGER
MARINE GRAVITY

Elsevier Oceanography Series 20

Waves in the Ocean

PAUL H. LeBLOND

Professor of Oceanography and Physics,
The University of British Columbia,
Vancouver, British Columbia, Canada

and

LAWRENCE A. MYSAK

Professor of Mathematics and Oceanography,
The University of British Columbia,
Vancouver, British Columbia, Canada

ELSEVIER SCIENTIFIC PUBLISHING COMPANY
Amsterdam – Oxford – New York 1978

ELSEVIER SCIENTIFIC PUBLISHING COMPANY
335 Jan van Galenstraat
P.O. Box 211, Amsterdam, The Netherlands

Distributors for the United States and Canada:

ELSEVIER NORTH-HOLLAND INC.
52, Vanderbilt Avenue
New York, N.Y. 10017

Library of Congress Cataloging in Publication Data

LeBlond, Paul H
Waves in the ocean.

(Elsevier oceanography series ; 20)
Bibliography: p.
Includes indexes.
I. Ocean waves. I. Mysak, Lawrence A., joint author. II. Title.
GC211.2.L43 551.4'7022 77-18815
ISBN 0-444-41602-1

ISBN: 0-444-41602-1 (VOL. 20)
ISBN: 0-444-41623-4 (Series)

Printed in Great Britain by Page Bros (Norwich) Ltd, Norwich

PREFACE

The study of ocean waves once occupied a significant position within the mainstream of physics, Newton, Bernoulli, Laplace, Stokes, Kelvin, Poincaré and Taylor, to enumerate but a few of the great physicists and mathematicians of yesteryears, preoccupied themselves with the theory of ocean waves. Modern physics has moved on to a deeper level of reality in its search for the ultimate clockwork of nature. Nevertheless, fluid dynamics and the study of waves remains central to meteorology, oceanography, geophysics and cosmology, not to mention its innumerable engineering applications. As more effort is devoted to the study of the ocean and to the role played by the fluctuations of its properties on the weather, on fisheries and on littoral geomorphology, it remains clear that an understanding of ocean waves is essential for a proper grasp of ocean dynamics.

The past decade has witnessed a bloom of new discoveries and interpretations of ocean waves. Recent advances in modern oceanography are in great part attributable to technological progress in instrument design and in data handling. The development of compact and reliable sensors has allowed sampling of the ocean at rates and on scales hitherto inaccessible. High-speed computers have made possible great refinements in the analysis of data and have provided theoreticians with revolutionary new tools. It is not surprising that the study of ocean waves is evolving so rapidly, in a scientific milieu where the feedback loop between observation and theory is becoming ever tighter.

Recent results on waves in the ocean are scattered in hundreds of research articles, a situation which makes it difficult for anyone to initiate himself to the field. Available textbooks are either out of date or narrowly specialized to some particular aspect of the subject. Although it is impossible to write a definitive account of a rapidly evolving field, such as that of ocean waves, we have attempted to present a broad review of this field, ranging from capillary to planetary waves, with emphasis on modern developments. This text is primarily intended for graduate students, whom we encourage to test their comprehension by working out the exercises given at the ends of most sections. We also hope that as a broad review of the field, it will prove useful to its practitioners. The presentation proceeds from the simple to the complex; it is eclectic rather than comprehensive, particular emphasis being given to recent topics at the expense of those for which exhaustive references are available. Although a staggering number of works are compiled in the bibliography, that list of references is by no means complete: only some of those papers which we deemed most relevant have been cited.

Throughout the writing of this book, we have received encouragement and assistance from many friends and colleagues. Several of them have carefully read various parts of the book before publication. We would especially like to thank Dr. Len Todd, Dr. Richard E. Thomson and Mr. Daniel Wright for their help. Dr. Todd carefully dissected a preliminary draft of the first half of the book, and Dr. Thomson thoroughly read several chapters of the penultimate version. Both provided us with many useful and illuminating comments. Mr. Wright painstakingly worked through two versions of the complete manuscript and, in so doing, considerably helped to eliminate many errors and obscurities. We are also grateful to the following colleagues who read and criticized individual chapters: Drs. V.T. Buchwald, C.J.R. Garrett, P. Gent, A.E. Gill, R.H.J. Grimshaw, M.S. Howe, B.A. Hughes, J.C. McWilliams, J.W. Miles, G.T. Needler, S. Pond and R.W. Stewart. We should also like to thank Maryse Ellis, who efficiently typed the bulk of the final manuscript, and Sylvie Lacerte, who drafted most of the figures. Finally, we wish to thank our wives, Josée and Mary, for patiently listening to an unending chatter about the book during the last three years.

We dedicate this book to our parents who, in our earlier years, taught us a way of life which we shall always respect.

PAUL H. LeBLOND
LAWRENCE A. MYSAK

Vancouver, February 1977.

CONTENTS

GLOSSARY OF COMMONLY USED SYMBOLS

Roman letters

a	Wave amplitude
A	Complex amplitude
$B(k)$	Bond number [Section 11]
$B(z)$	Burger number [Section 20]
$\boldsymbol{c}, c$	Phase velocity and its magnitude
$\boldsymbol{c}_g, c_g$	Group velocity and its magnitude
c_v	Specific heat at constant volume
E	Energy density
$\langle E \rangle$	Energy density averaged over the phase
f	Coriolis parameter
$\boldsymbol{g}, g$	Acceleration of gravity and its magnitude
g'	Reduced gravity for a two-layer fluid
h_n	Equivalent depth
H	Depth of the ocean
H_1, H_2	Upper and lower layer depths in a two-layer fluid
J	The Jacobian operator
$\boldsymbol{k}, k$	Wavenumber vector and its magnitude
$\boldsymbol{k}_h$	Horizontal component of the wavenumber vector
L	Wavelength; lengthscale; in Section 6, Lagrangian density
$\langle M \rangle$	Average wave momentum
$\boldsymbol{n}$	Outward normal to the fluid boundary
N	Brunt-Väisälä frequency
p	Pressure
p_a	Atmospheric pressure
p_0	Hydrostatic pressure
p'	Perturbation pressure
q	Barotropic potential vorticity
q_s	Baroclinic potential vorticity
$\boldsymbol{r}$	Position vector measured from the centre of the Earth
r_e	Barotropic Rossby radius of deformation
r_i	Internal Rossby radius of deformation
r_{i2}	Internal Rossby radius of deformation for a two-layer fluid
R	The radius of the Earth
Ro	The Rossby number
$R(\omega)$	Autocovariance function
S	Salinity; phase function
$S(k, \omega)$	Power spectrum
t	Time

T	Temperature; wave period; transmission coefficient
T_{ij}	Radiation stress tensor
$\boldsymbol{u}$	Velocity vector, with components u, v, w
$\boldsymbol{U}$	Mean current vector with components U, V, W
$\boldsymbol{x}$	Position vector with components x, y, z
Z	Vertical dependence of the vertical component of velocity

Greek letters

α	Bottom slope
β	Derivative of the Coriolis parameter with respect to latitude
Γ	Circulation
δ	Relative density difference in a two-layer fluid
δ_{ij}	Kronecker's symbol
$\delta(x)$	Dirac's delta function
ϵ	A small parameter
ζ	Relative vorticity
η	Surface displacement
θ_h	Angle made by a horizontal vector with the easterly direction
$\boldsymbol{\kappa}$	Rotated or nondimensional wavenumber vector
λ	Azimuthal angle in spherical polar coordinates; parameter describing the properties of the medium of propagation
μ	Molecular viscosity
ν	Kinematic viscosity (molecular)
ξ	Vertical displacement in the fluid
$\Pi(z)$	Vertical dependence of the perturbation pressure
ρ	Mass density
$\rho_0(z)$	Mean density of the basic state
ρ_*	A reference density
τ_{ij}	Stress
ϕ	Latitude; velocity potential
ψ	Stream function
ω	Angular frequency
Ω	Angular velocity of the Earth

Special symbols

$\mathcal{L}$	$\partial_{tt} + f^2$
∇_h	The horizontal component of the gradient
∇_h^2	$\partial_{xx} + \partial_{yy}$
$\langle \cdot \rangle$	An average over the phase
$P\{\cdot\}$	The expectation value
$\overline{(\ )}$	Average over a wavelength

CHAPTER 1

INTRODUCTION

1. A BRIEF ORIENTATION

Dynamic oceanography may be defined as the study of the ocean's response to external forces acting on its boundaries, and to internal forces working directly in its midst or arising in answer to external forces. At the sea surface the driving forces are mainly meteorological and consist of the atmospheric pressure gradient, the wind stress and the buoyancy forces resulting from radiation, evaporation and precipitation. At the ocean floor and on continental margins, localized seismic disturbances occur intermittently; the whole ocean bed itself also undergoes small-amplitude, low-frequency oscillations because of the Earth tide. The interior of the ocean is acted upon by body forces due to the Earth's gravity and rotation, by tidal forces and by nonconservative forces such as turbulent stresses.

Because of the time-varying nature of most of these forces, the overall response of the ocean is generally time-dependent. Further, because of the wide range of frequencies associated with these different forces, the oceanic response will extend over an equally wide frequency range. The frequency spectrum associated with wave motions ranges from capillary waves, with periods of less than a second, to low-frequency baroclinic planetary waves with periods of up to several years.

Our study of ocean waves will be based on an idealization of the ocean as a linear, nondissipative mechanical system. A particular wave motion, i.e. a travelling perturbation of an equilbrium state, may be adequately represented by a linear model provided the ratio of particle speed to phase speed is small compared to unity. This criterion, which implies that the wave amplitude is small compared to the wavelength, holds for most (but not all) oceanic wave phenomena. In a nondissipative system all waves are undamped, an idealization which is tenable only if the decay time of a wave greatly exceeds its period. Again, most ocean waves are sufficiently durable that their properties may be usefully described by a nondissipative theory.

In a linear wave model most modes are uncoupled and can be classified and studied independently. In the first four chapters of this book, we shall examine the free oscillations of the ocean in situations of increasing geometrical complexity. The assumption of linearity will occasionally be relaxed, however, and the principal modifications of the linear solutions by nonlinear effects will be explored briefly.

In the real ocean, waves interact with each other and with the mean flow and stratification. They grow due to the action of external forces or through processes of internal instability, and they decay as a result of molecular and turbulent friction and diffusion. Because of all these processes, the oceanic response cannot be represented by a discrete spectrum of the undamped normal modes of a bounded ocean. It must be written in terms of a continuous spectrum of modes whose amplitude and form change in time and space because of nonlinear and diffusive interactions and the external forces. Therefore, after a chapter of transition concerned with spectral, statistical and stochastic methods, the second half of the book continues with a study of wave interactions, stability of mean flows and wave generation and decay.

In keeping with the attitude that limnology is but a specialized aspect of oceanography, we remind the reader that most of the wave phenomena described here are common to both lakes and oceans. An intimate similarity also exists between many oceanic and atmospheric waves, the latter being reviewed in the recent book by Gossard and Hooke (1975). Only wave motions are discussed in this book; other time-dependent phenomena, of a convective or diffusive nature, are discussed extensively in the book by Turner (1973). Finally, for a succinct and systematic review of oceanic variability, we refer the reader to a recent book by Monin et al. (1977).

2. A GLANCE AT OCEANIC WAVES

The ocean basins are filled with a slightly compressible and electrically conducting liquid lying on the surface of a weakly magnetized rotating sphere. The kinds of forces acting in such a medium are well known, and one can therefore directly infer what types of waves it can support.

Compressibility allows the existence of sound waves. Electrical conductivity in the presence of a magnetic field leads to the possibility of Alfvén waves and also splits the basic sound wave into a fast and a slow wave (Cabannes, 1970; Chapter 2). The Earth's magnetic field is so weak, however, that the associated electromagnetic restoring forces are insignificant compared to elastic and other restoring forces and are customarily neglected in oceanography. Gravity waves arise through the restoring action of gravity on water particles displaced from equilibrium levels, such as a free surface or an internal geopotential surface in a stratified fluid. In addition, at any surface of contact between two different fluids (such as air and water), surface tension acts as a restoring force and gives rise to short, high-frequency capillary waves. Rotation introduces the Coriolis force, acting at right angles to the velocity vector, which allows the existence of inertial or gyroscopic waves. Finally, variations of the equilibrium potential vorticity (Section 4) due to changes in depth or latitude give rise to very slow, large-scale oscillations called planetary or Rossby waves.

These five basic types of oceanic waves (sound, capillary, gravity, inertial and planetary) generally occur together, with the five basic restoring forces all acting simultaneously to produce more complicated mixed types of oscillations. The relative importance of each restoring force in any particular case depends on the properties of the medium, the geometry of the container and the frequency and wavelength of the oscillations. All these waves are not merely theoretical possibilities, but are found to exist in the ocean, as may be seen from a glance at some observed data.

An exhaustive survey of wave data would confuse rather than illuminate the subject at this early stage. The observations presented here should convince the reader of the scope, ubiquity and diversity of oceanic waves and familiarize him with some methods of data presentation.

The rather short wind-generated gravity waves, associated with seasickness, shore erosion or artistic seascapes, are by far the best known and most extensively studied ocean waves. The casual observer is, however, usually unaware of the existence of other types of waves, less conspicuous perhaps, but equally significant in oceanic dynamics.

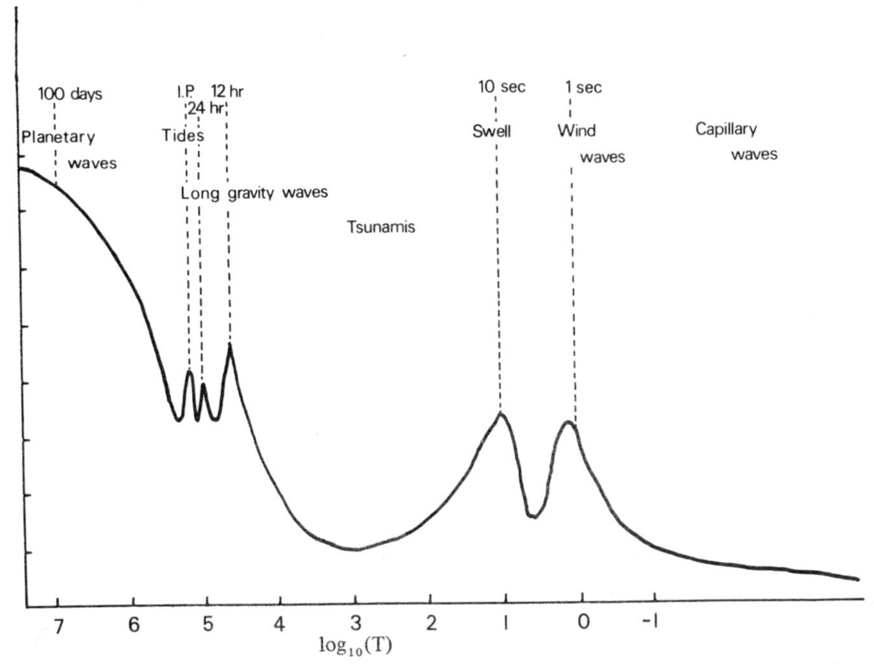

Fig. 2.1. Schematic energy spectrum of oceanic variability, showing the different types of waves occurring in the ocean. I.P. denotes the inertial period and is defined as $\pi/\Omega|\sin\phi|$, where Ω = magnitude of the Earth's rotation vector and ϕ is the geographic latitude (Section 3). In this picture I.P. = 35 hours, corresponding to a latitude of ± 20°. The relative amplitudes of the various parts of the spectrum do not necessarily reflect actual conditions.

Surface waves (travelling oscillations of sea level) occupy an extraordinarily broad range of wavelengths and periods, as evidenced by their contribution to the oceanic energy spectrum (Fig. 2.1.). Passing from shorter to longer periods in Fig. 2.1, one first observes capillary waves, dominated by surface tension effects, followed by a broad band of gravity waves, mostly wind-induced. Long-period gravity waves arise in response to meteorological forcing or as a consequence of earthquakes. The tides are another type of forced gravity wave. At very long periods, gravity loses its dominant dynamic role to differential rotation effects and surface waves become "planetary" waves, and manifest themselves as slowly drifting, large-scale current systems.

The interior of the ocean is no quieter than its upper surface. Although hidden from direct view, this internal agitation is readily discovered by measurements and manifests itself in oscillations of temperature and salinity and of the associated currents (Fig. 2.2.).

Some types of wave motions show up as variations in horizontal currents, rather than as oscillations about equilibrium levels. An ingenious method of illustrating such variability consists of constructing *progressive vector diagrams*, an example of which is shown and explained in Fig. 2.3.

Waves are of course space–time structures which may be detected and represented through their spatial as well as their temporal variability. In order to establish unambiguously the wave-nature of a phenomenon, data are necessary *both* in time and space.

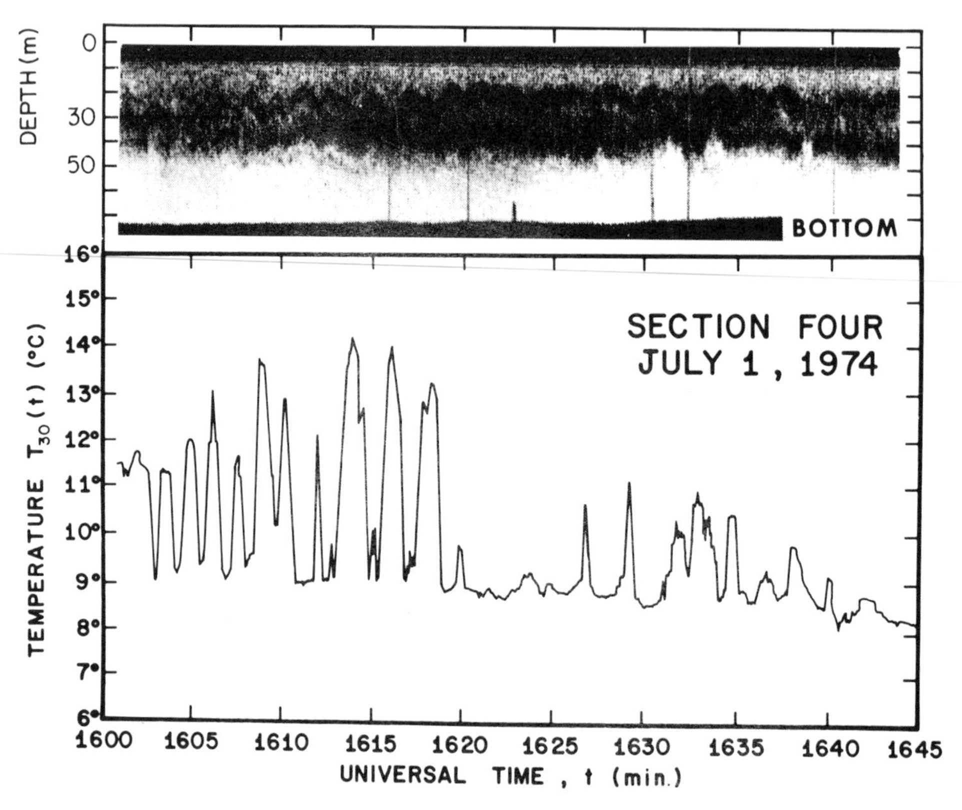

Fig. 2.2 Above: internal wave packet in 80 m of water off New Jersey, as observed by a 20 kHz acoustic echo-sounder. Below: the temperature record of the same wave packet, obtained from a towed thermistor at 30 m depth. A 135° turn was executed shortly after 1625 UT. (From Apel et al., 1976.)

Examples of wavy spatial variations are shown in the satellite photograph of Fig. 2.4 and in the aerial photo of Fig. 2.5.

3. THE BASIC EQUATIONS

A serious study of the dynamics of the ocean is based on the mathematical description of the time-independent motion of a thin layer of stratified fluid lying on the surface of the rotating Earth. These motions are governed by conservation laws for mass and momentum, an equation of state and the laws of thermodynamics. We shall use an Eulerian (as opposed to Lagrangian) description of the fluid so that the particle velocity $\boldsymbol{u}$, pressure p, density ρ, temperature T and salinity S are treated as functions of the position vector $\boldsymbol{r}$ (measured outward from the Earth's center) and the time t. All positions will be referred to a right-handed orthogonal coordinate system which is uniformly rotating with the Earth's angular velocity $\boldsymbol{\Omega}$, the latter having a magnitude of

$$\Omega = |\boldsymbol{\Omega}| = 7.29 \cdot 10^{-5} \text{ rad/s}.$$

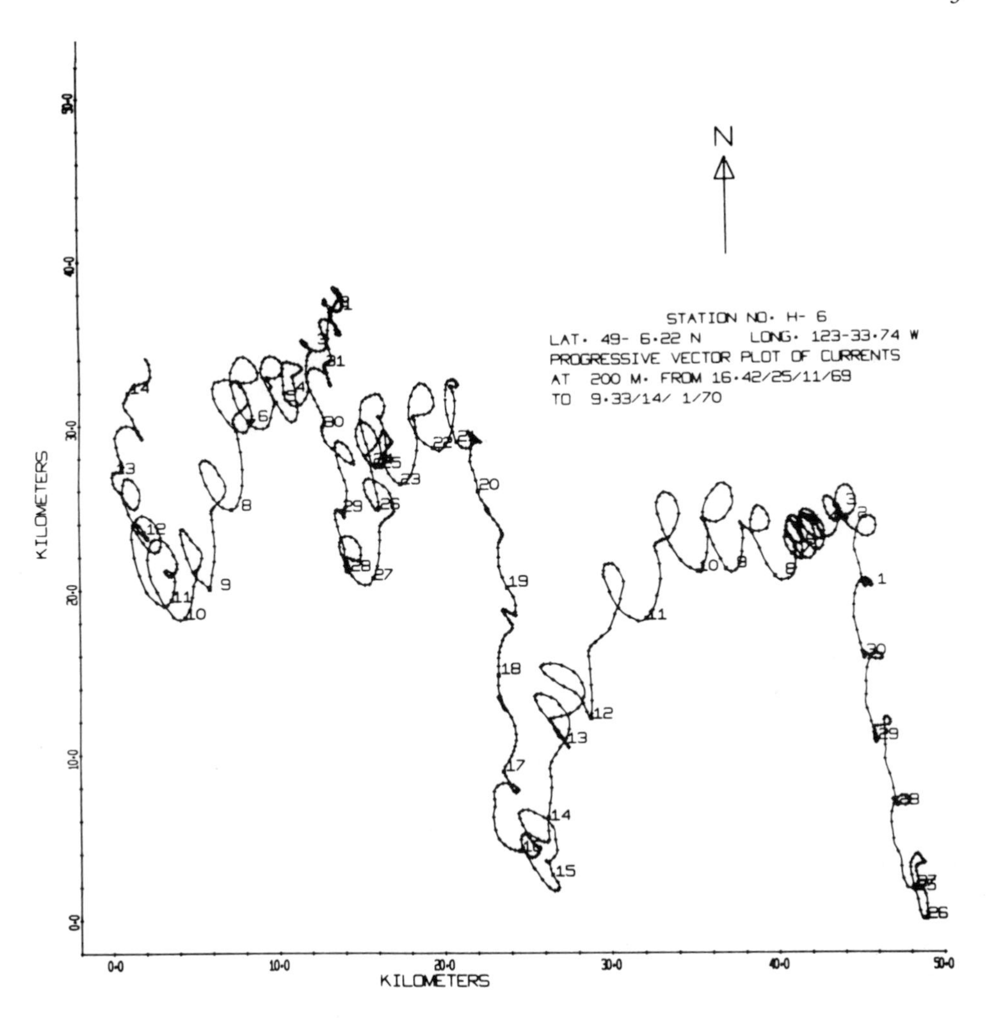

Fig. 2.3. A progressive vector diagram constructed from successive cumulative values of north–south and east–west components of current velocity from records obtained at 10-minute intervals over a 50-day period during November 25, 1969 through January 14, 1970 in the Strait of Georgia, British Columbia. The plotted positions correspond to the horizontal displacement of the water that would occur if the motion in the entire neighboring area of the location of the instrument was the same as at this location. The record starts in the lower right-hand corner (dates 23 and 25 are nearly superimposed). (From Tabata and Stickland, 1972.)

The velocity $\boldsymbol{u}$ in this rotating frame of reference is related to the inertial (i.e. nonrotating) velocity $\boldsymbol{u}_{\text{inert}}$ by the equation

$$\boldsymbol{u}_{\text{inert}} = \boldsymbol{u} + \boldsymbol{\Omega} \times \boldsymbol{r}.$$

If R denotes the mean radius of the Earth (measured from the Earth's center to the undisturbed sea surface) and z denotes the distance measured vertically upward from the undisturbed sea surface, then

$$r = |\boldsymbol{r}| = R + z. \tag{3.1}$$

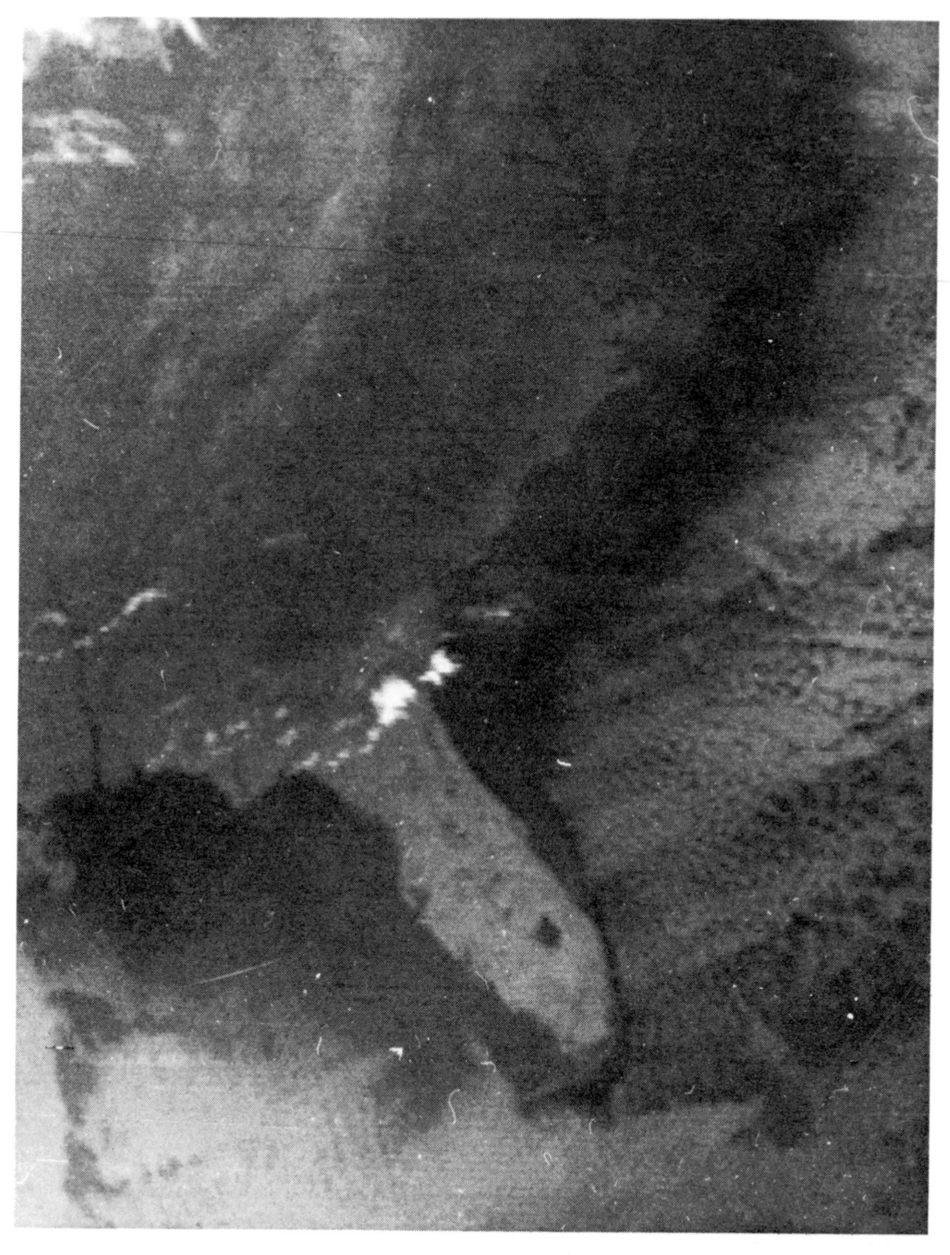

Fig. 2.4. Satellite photograph of the eastern North American coast, showing wavy, eddy-like structures on the coastal side of the Gulf Stream. (From National Environment Satellite Service, Washington, D.C.)

Fig. 2.5. Surface manifestations of internal waves travelling on a shallow pycnocline (a sharp density interface) in the Strait of Georgia, British Columbia. The waves are propagating towards the left (northwestward). As a scale reference, note the barge in the upper central part of the picture – it is about 70 m long.

The position of the undisturbed sea surface is thus given by the equation $z = 0$.

The conservation of mass is expressed by

$$\frac{\partial \rho}{\partial t} + \nabla \cdot (\rho \boldsymbol{u}) = 0, \tag{3.2a}$$

$$\text{or} \quad \frac{D\rho}{Dt} + \rho \nabla \cdot \boldsymbol{u} = 0, \tag{3.2b}$$

$$\text{where} \quad \frac{D}{Dt} = \frac{\partial}{\partial t} + \boldsymbol{u} \cdot \nabla$$

denotes the derivative following the motion. Momentum conservation is expressed as

$$\rho \frac{D\boldsymbol{u}}{Dt} + \rho 2 \boldsymbol{\Omega} \times \boldsymbol{u} = \rho[\boldsymbol{g} - \boldsymbol{\Omega} \times (\boldsymbol{\Omega} \times \boldsymbol{r})] - \nabla p + \boldsymbol{F}, \tag{3.3}$$

where $\boldsymbol{g} = -g\hat{z}$ denotes the gravitational acceleration ($g = 9.81$ m/s^2 and $\hat{z}$ denotes a unit vector in the vertical direction) and $\boldsymbol{F}$ represents the sum of all the other forces per unit volume acting on the fluid, including the tide-producing forces as well as molecular and turbulent friction. The magnitude of the ratio of the centrifugal term, $\rho \boldsymbol{\Omega} \times (\boldsymbol{\Omega} \times \boldsymbol{r})$, to the gravitational term, ρg, is less than $3 \cdot 10^{-3}$ throughout the ocean; henceforth the term $\rho \boldsymbol{\Omega} \times (\boldsymbol{\Omega} \times \boldsymbol{r})$ will be neglected, so that the effects of rotation on the ocean are manifested only through the Coriolis term $\rho 2 \boldsymbol{\Omega} \times \boldsymbol{u}$. The term $D\boldsymbol{u}/Dt$ formally takes the form

$$\frac{D\boldsymbol{u}}{Dt} = \frac{\partial \boldsymbol{u}}{\partial t} + \boldsymbol{u} \cdot \nabla \boldsymbol{u};$$

$\boldsymbol{u} \cdot \nabla \boldsymbol{u}$ is, however, useful only in Cartesian coordinates. The coordinate-free or invariant vector form, which holds for any orthogonal coordinate system, is given by

$$\boldsymbol{u} \cdot \nabla \boldsymbol{u} = (\nabla \times \boldsymbol{u}) \times \boldsymbol{u} + \tfrac{1}{2}\nabla(\boldsymbol{u} \cdot \boldsymbol{u}). \tag{3.4}$$

Sea water is a complex solution of many salts. Therefore, in addition to pressure and temperature, we require the salinity S in order to specify a thermodynamic state, where S (in parts per thousand: ‰) is loosely defined as the mass in grams of all dissolved solids in one kilogram of sea water (for a more precise definition, see Fofonoff, 1962). Thus the salinity lumps all solutes together under one thermodynamic variable. The density of sea water is then given by an equation of state of the form

$$\rho = \rho(p, T, S). \tag{3.5}$$

This relation is nonlinear in p, T and S and has no simple analytical form (Mamayev, 1975). Values are often read from standard oceanographic tables (Lafond, 1951) or estimated through best-fit polynomials valid over a restricted range of S and T. Bryan and Cox (1972), for example, achieve better than about 1% accuracy with a quadratic form:

$$\rho = A + x_1(T - B) + x_2(S - C) + x_3(T - B)^2.$$

The coefficients x_1, x_2, x_3, A, B and C are determined by fitting to accurate data and are functions of pressure. The same authors also present and discuss a nine-term polynomial fit. All we need to know for the present purpose is that in general $(\partial\rho/\partial S)_{p,T} > 0$, $(\partial\rho/\partial p)_{S,T} > 0$, and that, except below the temperature of maximum density, $(\partial\rho/\partial T)_{S,p} < 0$. The temperature of maximum density lies above the freezing temperature at atmospheric pressure only if the salinity is less than 24.69‰ (McLellan, 1965; Chapter 3); this is of great relevance in limnology, but rarely so in oceanography. In the open ocean the salinity ranges from 29 to 36‰.

To close the above system of equations (3.2), (3.3) and (3.5) for the seven unknown functions $\boldsymbol{u}, p, \rho, T, S$ (where $\boldsymbol{u}$ has three components), we require two more conservation equations for T and S. Conservation of internal energy is expressed by the equation (Batchelor, 1967)

$$\frac{\mathrm{D}}{\mathrm{D}t}(\rho c_v T) = \nabla \cdot (k_T \nabla T) + Q_T, \tag{3.6}$$

where c_v denotes the specific heat at constant volume, k_T the thermal conductivity and Q_T represents all sources and sinks of heat. In particular Q_T includes heating due to compression and cooling due to expansion, the mechanical production of heat due to shearing motions and all heat transfer terms through the sea surface (solar heating, evaporative cooling, sensible heat flux to the atmosphere and long-wave radiation into space). Conservation of salt is expressed by an equation analogous to (3.6):

$$\mathrm{D}S/\mathrm{D}t = \nabla \cdot (K_S \nabla S) + Q_S, \tag{3.7}$$

where K_S denotes the coefficient of molecular diffusion of salt and Q_S includes all sources and sinks of salt due to such phenomena as ice melting and formation, precipitation and evaporation.

As mentioned in Section 2, the waves which can propagate through the interior of the ocean fall into two broad categories: high-frequency acoustic waves which arise from the slight compressibility of sea water, and the much lower frequency internal, inertial and planetary waves which exist because of the gravitational and rotational forces. The propagation of sound waves in the ocean is governed by the usual linear wave equation with phase speed $c_s \simeq 1.5 \cdot 10^3$ m/s. Ocean acoustics is generally concerned with a band of frequencies lying between 1 Hz and 100 kHz (Tolstoy and Clay, 1966). Thus the periods of acoustic waves lie in the range $T_s = 1$ to 10^{-5} s, with a wavelength $L_s (= c_s T_s)$ ranging from a few kilometers to about one centimeter. Internal, inertial and planetary waves, on the other hand, span the period range of minutes to months with corresponding wavelengths of tens of meters to hundreds of kilometers. Thus the dispersion curves of the latter waves are widely separated from that of sound waves in the period–wavenumber plane and little interaction between them is expected. Nevertheless, even though the overall dynamic balance of the ocean is not affected by sound waves, they are useful as sensors to measure or detect oceanic properties and other ocean wave processes (Tolstoy and Clay, 1966; Porter et al., 1974; Baer and Jacobson, 1974; Mooers, 1975a). Henceforth we shall filter out the acoustic propagation mode in our governing equations by assuming that our model ocean is incompressible, which means that

$$(\partial\rho/\partial p)_\eta = 0, \tag{3.8}$$

where here η denotes the entropy. Since in reality $(\partial\rho/\partial p)_\eta = 1/c_s^2$, the assumption (3.8) implies that in our model ocean the local sound speed is infinite. Further, for the most part of this book we shall assume that the ocean is nondiffusive ($K_T = 0 = K_S$, where

$K_T = k_T/\rho c_v$ is the thermal diffusivity). Then since neither compression nor diffusion takes place, the density cannot change along a particle path:

$$\mathrm{D}\rho/\mathrm{D}t = 0. \tag{3.9}$$

In view of (3.9), (3.2b) reduces to the "continuity equation":

$$\nabla \cdot \boldsymbol{u} = 0. \tag{3.10}$$

Thus for an incompressible, nondiffusive fluid, the momentum equation (3.3) together with (3.9) and (3.10) form a closed set of equations for $\boldsymbol{u}$, p and ρ. On the other hand, to study the effects of diffusion in an incompressible model ocean, the equation set (3.3), (3.5)–(3.7) and (3.10) would be used to find $\boldsymbol{u}$, p, ρ, T and S.

In this book our attention will be focused on the first set of equations, i.e. (3.3), (3.9) and (3.10). Most of the presentation will be concerned with free waves obeying the unforced and nondissipative momentum equation, which is known as the Euler equation for a rotating fluid:

$$\rho\frac{\mathrm{D}\boldsymbol{u}}{\mathrm{D}t} + \rho 2\,\boldsymbol{\Omega}\times\boldsymbol{u} + \nabla p - \rho \boldsymbol{g} = \boldsymbol{0}. \tag{3.11}$$

This equation together with (3.9) and (3.10) will be referred to as the *adiabatic equations*, since they include neither friction nor heat or salt diffusion or sources.

The equations (3.9)–(3.11) represent a homogeneous, nonlinear system of partial differential equations of evolution type, i.e. a set of equations which describe the time rate of change of $\boldsymbol{u}$, p and ρ in terms of the spatial derivatives of these quantities. Because of the time dependence of the equations and of the finite dimensions of the ocean, we are confronted with solving a nonlinear, initial boundary-value problem. At $t = 0$ say, initial conditions for $\boldsymbol{u}$, p and ρ must be prescribed, and at a rigid boundary, the normal component of velocity must vanish:

$$\boldsymbol{u}\cdot\boldsymbol{n} = 0, \tag{3.12}$$

where $\boldsymbol{n}$ is a vector normal to the boundary. In some limiting cases, however, we can consider problems which are periodic in space and time, in which case no initial conditions are required.

At the disturbed sea surface $z = \eta(x, y, t)$, we require continuity of stress and displacement. In the absence of molecular and turbulent viscosity and surface tension, these conditions are expressed as

$$p_{\mathrm{ocean}} = p_{\mathrm{atmosphere}} \quad \text{at } z = \eta, \tag{3.13a}$$

$$\text{and} \quad \frac{\mathrm{D}}{\mathrm{D}t}(z-\eta) = 0 \quad \text{at } z = \eta. \tag{3.13b}$$

Since $\mathrm{D}z/\mathrm{D}t = w$, this second condition may also be written as

$$w = \mathrm{D}\eta/\mathrm{D}t \quad \text{at } z = \eta, \tag{3.14}$$

where w is the component of velocity in the vertical direction.

Exercises Section 3

1. Show that for constant ρ the centrifugal term in (3.3) can be written as the gradient of a scalar and hence incorporated into the pressure term.

2. Consider a simplified equation of state of the form $\rho = \rho_*(1 - \alpha T + \beta S)$ and the following diffusion equations for heat and salt:

$$\mathrm{D}T/\mathrm{D}t = K_T \nabla^2 T, \quad \mathrm{D}S/\mathrm{D}t = K_S \nabla^2 S,$$

where ρ_*, α, β, K_T and K_S are constants. Show that if $\mathrm{D}\rho/\mathrm{D}t = 0$ is to hold, the function $F = -\alpha K_T T + \beta K_S S$ must be harmonic, i.e. $\nabla^2 F = 0$. What is the meaning of this condition?

3. Consider a uniformly rotating, nonhomogeneous fluid in which the rotation vector $\boldsymbol{\Omega}$ is anti-parallel to gravity:

$$\boldsymbol{g} = -g\hat{\boldsymbol{z}} \quad \text{and} \quad \boldsymbol{\Omega} = \Omega\hat{\boldsymbol{z}}.$$

Show that for this configuration, and in the presence of the centrifugal force, there is no static solution ($\boldsymbol{u} \equiv 0$) of

$$\rho\frac{\mathrm{D}\boldsymbol{u}}{\mathrm{D}t} + \rho 2\boldsymbol{\Omega} \times \boldsymbol{u} = \rho[\boldsymbol{g} - \boldsymbol{\Omega} \times (\boldsymbol{\Omega} \times \boldsymbol{r})] - \nabla p$$

which is compatible with the linearized form of the equation of state and the diffusion equations used in Exercise 3.2. (Greenspan, 1968.) Hint: Use cylindrical polar coordinates.

4. In equation (3.12), we state that $\boldsymbol{u} \cdot \boldsymbol{n} = 0$ at a solid boundary. In a viscous fluid, the non-slip boundary condition $\boldsymbol{u} \cdot \boldsymbol{s} = 0$ must also hold, where $\boldsymbol{s}$ is a unit vector along the boundary. Show that when this latter condition is satisfied,

$$\boldsymbol{\zeta} \cdot \boldsymbol{n} = 0,$$

where $\boldsymbol{\zeta} = \nabla \times \boldsymbol{u}$.

5. Show that in the absence of heat or salt diffusion, equation (3.9) implies that $\partial\rho/\partial t = 0$ on a solid boundary.

4. PROPERTIES AND CONSEQUENCES OF THE ADIABATIC EQUATIONS

To the best of our knowledge, the questions of existence, uniqueness and stability of solutions to the basic evolution system (3.9)–(3.11), with appropriate initial and boundary conditions are only partly resolved. Howard and Siegmann (1969) have shown that the solutions of the linearized form of the system are indeed unique and have established existence proofs for special cases. A number of questions pertaining to the stability and existence of periodic solutions of the nonstratified Navier-Stokes equations [(3.10) and (3.11) with ρ = constant and $\boldsymbol{\Omega} = 0$ but with the addition of the usual viscous terms)] have been studied by Iooss (1972). Marchuk (1972) and Kordzadze (1974) establish the uniqueness of solutions of another specialized form of the evolution equations, the so-called long wave equations (see Chapter 3), under a broad range of initial and boundary values and in the presence of certain terms modelling turbulent diffusion.

Uniqueness thus appears easier to prove than existence or stability, and we shall have occasion to remark on these properties in those special cases which have been studied in more detail. Such questions are of course difficult to resolve for nonlinear systems of equations, and the situation is complicated here by the presence of a moving boundary at $z = \eta$, the position of which must be determined in the course of the analysis.

Nevertheless, from the adiabatic system (3.9)–(3.11) one can derive equations describing the behaviour of quantities such as the vorticity, circulation and energy of the fluid, as well as some general conservation laws.

The vorticity consists of two distinct parts in a rotating coordinate frame: a relative vorticity, $\boldsymbol{\zeta} = \nabla \times \boldsymbol{u}$, and a planetary vorticity $2\boldsymbol{\Omega}$, which together make up the absolute vorticity $\boldsymbol{\zeta} + 2\boldsymbol{\Omega}$. A vorticity equation is obtained by taking the curl of (3.11) (after dividing by ρ), with (3.4) used for $\boldsymbol{u} \cdot \nabla \boldsymbol{u}$:

$$\frac{\partial}{\partial t}(\boldsymbol{\zeta} + 2\boldsymbol{\Omega}) + \nabla \times [(\boldsymbol{\zeta} + 2\boldsymbol{\Omega}) \times \boldsymbol{u}] = \frac{1}{\rho^2}\nabla\rho \times \nabla p, \tag{4.1}$$

or in a more usual form,

$$\frac{\mathrm{D}}{\mathrm{D}t}(\boldsymbol{\zeta} + 2\boldsymbol{\Omega}) = [(\boldsymbol{\zeta} + 2\boldsymbol{\Omega}) \cdot \nabla]\boldsymbol{u} + \frac{1}{\rho^2}\nabla\rho \times p. \tag{4.2}$$

The absolute vorticity may change along a path line due to vortex stretching [the first term on the right-hand side of (4.2)], and because of baroclinic torques (the second term). A flow in which isopycnals and isobars are not parallel, so that $\nabla\rho \times \nabla p \neq \boldsymbol{0}$, is said to be *baroclinic*; in a *barotropic* flow, $\nabla\rho \times \nabla p = \boldsymbol{0}$. The nonlinear advection terms which make up part of the total derivative $\mathrm{D}/\mathrm{D}t$ in (4.2) only redistribute vorticity from one place to another; when integrated over a closed volume, their contribution vanishes, as does that of the vortex stretching terms.

The circulation Γ relative to the rotating coordinate system is defined as

$$\Gamma = \oint_C \boldsymbol{u} \cdot \mathrm{d}\boldsymbol{l} = \iint_S \boldsymbol{\zeta} \cdot \boldsymbol{n}\, \mathrm{d}A, \tag{4.3}$$

where C is a closed material contour made up of fluid particles, $\mathrm{d}\boldsymbol{l}$ is an element of path length, and S is a surface bounded by C and with outward normal $\boldsymbol{n}$. We shall now exploit the relation

$$\frac{\mathrm{d}}{\mathrm{d}t}\iint_S \boldsymbol{R}\cdot\boldsymbol{n}\,\mathrm{d}A = \iint_S \frac{\mathrm{D}\boldsymbol{R}}{\mathrm{D}t}\cdot\boldsymbol{n}\,\mathrm{d}A - \iint_S [\boldsymbol{R}\cdot\nabla\boldsymbol{u}]\cdot\boldsymbol{n}\,\mathrm{d}A,$$

which holds for an arbitrary vector $\boldsymbol{R}$ and results from taking into account the changes in position and shape of the moving surface element $\boldsymbol{n}\mathrm{d}A$ (Batchelor, 1967; Section 3.1). Integrating (4.2) over S and using the above relation, we find the appropriate generalization of Kelvin's theorem for a stratified rotating fluid:

$$\frac{\mathrm{d}}{\mathrm{d}t}[\Gamma + \iint_S 2\boldsymbol{\Omega}\cdot\boldsymbol{n}\,\mathrm{d}A] = \iint_S \frac{1}{\rho^2}(\nabla\rho\times\nabla p)\cdot\boldsymbol{n}\,\mathrm{d}A. \tag{4.4}$$

The local, or relative circulation Γ varies because of baroclinic effects (the right-hand side); it is also subject to changes associated with variations of the planetary circulation $\iint_S 2\boldsymbol{\Omega}\cdot\boldsymbol{n}\,\mathrm{d}A$.

Even though neither vorticity nor circulation are locally conserved in a rotating stratified fluid, another quantity, the potential vorticity, is readily shown to be invariant along a path line. Taking the scalar product of (4.1) with $\nabla\rho$, we obtain

$$\nabla\rho\cdot\frac{\partial}{\partial t}(\boldsymbol{\zeta}+2\boldsymbol{\Omega}) + \nabla\rho\cdot\nabla\times[(\boldsymbol{\zeta}+2\boldsymbol{\Omega})\times\boldsymbol{u}] = 0 \tag{4.5}$$

or, using the identity $\nabla\cdot(\boldsymbol{A}\times\boldsymbol{B}) = \boldsymbol{B}\cdot(\nabla\times\boldsymbol{A}) - \boldsymbol{A}\cdot(\nabla\times\boldsymbol{B})$,

$$\nabla\rho\cdot\frac{\partial}{\partial t}(\boldsymbol{\zeta}+2\boldsymbol{\Omega}) - \nabla\cdot\{\nabla\rho\times[(\boldsymbol{\zeta}+2\boldsymbol{\Omega})\times\boldsymbol{u}]\} = 0. \tag{4.6}$$

Using the formula for $\boldsymbol{A}\times(\boldsymbol{B}\times\boldsymbol{C})$ and the Lagrangian invariance of ρ, i.e. (3.9), one readily establishes that

$$\nabla\rho\times[(\boldsymbol{\zeta}+2\boldsymbol{\Omega})\times\boldsymbol{u}] = -(\boldsymbol{\zeta}+2\boldsymbol{\Omega})\frac{\partial\rho}{\partial t} - [(\boldsymbol{\zeta}+2\boldsymbol{\Omega})\cdot\nabla\rho]\boldsymbol{u}, \tag{4.7}$$

which, upon substitution into (4.6), yields, provided $\nabla\cdot\boldsymbol{\Omega} = 0$,

$$\frac{\mathrm{D}}{\mathrm{D}t}[(\boldsymbol{\zeta}+2\boldsymbol{\Omega})\cdot\nabla\rho] = 0. \tag{4.8}$$

This result shows that $(\boldsymbol{\zeta}+2\boldsymbol{\Omega})\cdot\nabla\rho$, which is defined as the *potential vorticity*, is conserved along a path line. It is a special case of Ertel's theorem (see Greenspan, 1968; Section 1.6). A further specialization to shallow fluids will be obtained later (Chapter 3) and found particularly useful in discussing long-period planetary waves. For a detailed discussion of the balance and redistribution of potential vorticity in the ocean, we refer the reader to Thomson and Stewart (1977).

An adiabatic system must be energy conserving. Taking the product of the momentum equation (3.11) with $\boldsymbol{u}$, we find

$$\rho\frac{\mathrm{D}}{\mathrm{D}t}\left(\frac{u^2}{2}\right) + \nabla\cdot(\boldsymbol{u}p) - \rho\boldsymbol{g}\cdot\boldsymbol{u} = 0, \tag{4.9}$$

where $\boldsymbol{u}^2 = \boldsymbol{u}\cdot\boldsymbol{u}$. Let us define $\boldsymbol{x}$ as a position vector measured from a fixed point on the undistorted sea surface ($z = 0$) (see Fig. 5.1). Then $\boldsymbol{u} = \mathrm{D}\boldsymbol{x}/\mathrm{D}t$ and, with the help of $\mathrm{D}\rho/\mathrm{D}t = 0$, (4.9) may be rewritten as

$$\frac{\mathrm{D}}{\mathrm{D}t}\left[\frac{\rho u^2}{2} - \rho \boldsymbol{g} \cdot \boldsymbol{x}\right] + \nabla \cdot (\boldsymbol{u}p) = 0, \tag{4.10}$$

which is recognized as a conservation equation for the energy density: $\frac{1}{2}\rho u^2$ is the kinetic energy component and $-\rho \boldsymbol{g} \cdot \boldsymbol{x}$ the work done against gravity. The vector $\boldsymbol{u}p$ is the energy flux, but is not uniquely defined, as any nondivergent vector may be added to it. According to (4.10) energy is not conserved locally, but only on a global scale. When (4.10) is integrated over a fixed volume of fluid, we obtain

$$\frac{\partial}{\partial t}\iiint_V \left(\frac{\rho u^2}{2} - \rho \boldsymbol{g} \cdot \boldsymbol{x}\right) \mathrm{d}V = -\iint_S p\boldsymbol{u} \cdot \boldsymbol{n}\, \mathrm{d}A; \tag{4.11}$$

where V is the volume of the fluid, S is the surface of V and $\boldsymbol{n}$ is the outward normal to S. The right-hand side of (4.11) vanishes on solid bounding walls ($\boldsymbol{u} \cdot \boldsymbol{n} = 0$) or on a free surface with zero mean velocity on which p is uniform. Under these conditions, the total energy is invariant. The energy balance in a more general, dissipative system will be examined later (Chapter 8).

Exercises Section 4

1. Given the existence of a quantity S (such as the salinity) which remains invariant along a path line, prove the general form of Ertel's theorem in an inviscid fluid:

$$\frac{\mathrm{D}}{\mathrm{D}t}\left[\frac{(\boldsymbol{\zeta} + 2\boldsymbol{\Omega}) \cdot \nabla S}{\rho}\right] = -\frac{1}{\rho^3}\nabla S \cdot (\nabla p \times \nabla \rho). \tag{4.12}$$

Show that the right-hand side vanishes when $S = S(p, \rho)$.

2. Show that the steady state Euler equation (3.11) has a first integral in the form of a Bernoulli function $B_1 = u^2/2 + \int \mathrm{d}p/\rho + gz$, where B_1 is constant along a streamline.

3. In an irrotational flow, the vorticity vanishes everywhere, and one can write the velocity vector as $\boldsymbol{u} = \nabla\phi$, where ϕ is a scalar potential. Show that

$$\nabla\left[\frac{\partial\phi}{\partial t} + \frac{u^2}{2} + \frac{p}{\rho_*} + gz\right] = 0, \tag{4.13}$$

where ρ_* is a uniform density. Equation (4.13) then implies the existence of another Bernoulli function $B_2(t)$, given by

$$B_2(t) = \frac{\partial\phi}{\partial t} + \frac{u^2}{2} + \frac{p}{\rho_*} + gz.$$

5. APPROXIMATIONS

Some "traditional" approximations commonly used in dynamic oceanography deserve careful explanation since they are often confusing to the novice. In this section we introduce three useful approximations of the system (3.9)–(3.11) which are used, either separately or in some combination, in order to find time-dependent solutions of these equations. They are usually known as the Boussinesq, the linear and the β-plane approximations. Subsequently we shall see that each of these approximations imposes important physical and geometrical restrictions on the solutions of the appropriate governing equations.

The Boussinesq and linear approximations

To discuss these approximations, it is convenient to introduce the hydrostatic equilibrium state described by

$$\boldsymbol{u} \equiv \boldsymbol{0}, \quad p = p_0(z), \quad \rho = \rho_0(z). \tag{5.1}$$

Equations (3.9) and (3.10) are then identically satisfied and (3.11) reduces to

$$\partial p_0/\partial z = -\rho_0 g.^{\dagger} \tag{5.2}$$

Thus for any given density distribution $\rho_0(z)$, (5.2) can be integrated to find p_0. An important quantity characterizing the static state (5.1) and (5.2) is the Brunt-Väisälä frequency $N(z)$, defined by

$$N^2 = -\frac{g}{\rho_0}\frac{\partial \rho_0}{\partial z}. \tag{5.3}$$

For *stable* static equilibrium in an incompressible fluid, $N^2 > 0$ for all z, or equivalently, $\partial\rho_0/\partial z < 0$ for all z. The quantity N, which is also known as the stability or buoyancy frequency (see Turner, 1973), is the natural (angular) frequency of oscillation associated with small-amplitude, simple-harmonic motion of a neutrally buoyant element of fluid moving up and down along a line parallel to $\boldsymbol{g}$. The period $2\pi/N$ varies in the ocean from a few minutes in the thermocline to several hours in the deep ocean. In a compressible fluid, the right-hand side of (5.3) contains the extra term $-g^2/c_s^2$, where c_s is the sound speed in the ocean (see Phillips, 1966; Section 2.4). The only circumstance where compressibility becomes important in determining the stability of natural waters is in lakes, at temperatures near that of maximum density (Osborn and LeBlond, 1974), and in the deep ocean where $\partial\rho_0/\partial z$ is very small.

To describe motions which represent departures from the static state (5.1), (5.2), we introduce the perturbation pressure p' and density ρ' defined by the equations

$$p = p_0 + p', \quad \rho = \rho_0 + \rho'. \tag{5.4}$$

Substituting (5.4) into (3.9)–(3.11), we find that (3.10) remains unchanged whereas (3.9) and (3.11) take the forms

---

† For future notational convenience, in which subscripts will be used to denote partial derivatives, we use partial rather than ordinary derivatives in these equations.

$$\frac{D\rho'}{Dt} + w\frac{\partial \rho_0}{\partial z} = 0, \tag{5.5}$$

$$\rho\frac{D\boldsymbol{u}}{Dt} + \rho 2\boldsymbol{\Omega} \times \boldsymbol{u} + \nabla p' - \rho'\boldsymbol{g} = \boldsymbol{0}. \tag{5.6}$$

Thus for motion in a stratified, incompressible fluid, it is only the density perturbation field that is important in the gravity term of (5.6). The fluid is now acted upon by a reduced gravitational (or buoyancy) acceleration $g' = \rho' g/\rho$ and a modified pressure $p' = p - p_0$. The buoyancy force $-\rho'\boldsymbol{g}$ pulls back any particle displaced from its original equilibrium level. This internal restoring force gives rise to internal wave oscillations in a stably stratified fluid.

From (5.4) we can write

$$\rho = \rho_0(1 + \rho'/\rho_0).$$

Since $\rho'/\rho_0 \ll 1$ uniformly throughout the ocean [in fact $\rho'/\rho_0 = 0(10^{-3})$], the density perturbation produces a very small correction to the inertia and Coriolis accelerations in (5.6). However, as mentioned above, the density perturbation is important in the buoyancy term. The approximation first introduced by Boussinesq in 1903 is based on these observations, and in mathematical terms consists of replacing ρ in (5.6) by some constant, reference value ρ_*, say. For practical purposes we can think of ρ_* as a depth-averaged value of $\rho_0(z)$. We shall see in Chapters 2 and 3 that for the Boussinesq approximation to hold, we must have $N^2H/g \ll 1$, where H denotes a vertical length scale. In a compressible fluid, on the other hand, other restrictions are necessary [for example, that all particle velocities must be much less in magnitude than the sound speed: see Turner (1973, Section 1.3) for more details] but they will not be discussed here. We will casually refer to a fluid in which the Boussinesq approximation has been made as a *Boussinesq fluid*. When using a Boussinesq fluid, it is also customary to use a slightly different definition of N^2:

$$N^2 = -\frac{g}{\rho_*}\frac{\partial \rho_0}{\partial z}. \tag{5.3'}$$

Since $\rho_0 = \rho_*[1 + 0(10^{-3})]$ for the oceans, this definition hardly affects the numerical value of N^2 as given by (5.3) for a non-Boussinesq fluid.

The equations which represent the linearization of (3.9)–(3.11) about the static state (5.1), (5.2) are obtained by neglecting the products of all perturbation quantities $\boldsymbol{u}, p', \rho'$. Thus (3.10), being linear in $\boldsymbol{u}$, remains unchanged whereas (3.9) and (3.11) reduce to

$$\frac{\partial \rho'}{\partial t} + w\frac{\partial \rho_0}{\partial z} = 0, \tag{5.7}$$

$$\rho_0\frac{\partial \boldsymbol{u}}{\partial t} + \rho_0 2\boldsymbol{\Omega} \times \boldsymbol{u} + \nabla p' - \rho'\boldsymbol{g} = \boldsymbol{0}. \tag{5.8}$$

Loosely speaking, linearization is valid for wave motions of infinitesimal amplitude, which implies that the particle speed is much smaller than the phase speed. However, for any one particular class of wave motions implied by (3.10), (5.7) and (5.8), the validity of the linearization will depend strongly on the time, length and velocity scales associated with the motions under consideration. As an example we note that for wave motions in which rotation is important and the time scale is $0(\Omega^{-1})$, the neglect of the nonlinear terms, $\boldsymbol{u} \cdot \nabla \rho$ and $\boldsymbol{u} \cdot \nabla \boldsymbol{u}$ in (3.9) and (3.11), respectively, is valid provided the Rossby number Ro is small, viz.,

$$Ro \equiv U/\Omega L \ll 1,$$

where L and U are the characteristic length and velocity scales (see Exercise 5.2). It is important to realize that a condition of the type $Ro \ll 1$ based on scaling arguments is only a *local* condition on the validity of linearization. It is very difficult in practice (in fact usually impossible) to obtain a global condition, i.e., one that is uniformly valid for a wide range of the time, length and velocity scales. To obtain such a condition we would first have to find the exact solution of the nonlinear equations and then compare this with the exact solution of the linearized equations.

The above two approximations can be combined to give the linearized Boussinesq equations, which consist of (3.10), (5.7) and

$$\frac{\partial \boldsymbol{u}}{\partial t} + 2\boldsymbol{\Omega} \times \boldsymbol{u} + \frac{1}{\rho_*}\nabla p' - \boldsymbol{g}'_* = \boldsymbol{0}, \tag{5.9}$$

where $\boldsymbol{g}'_* = (\rho'/\rho_*)\boldsymbol{g}$.

The β-plane approximation

Because we are primarily interested in oceanic motions of limited horizontal extent and because of the mathematical problems involved in dealing with spherical polar coordinates (e.g. see Miles, 1974c) it is desirable to introduce a Cartesian metric centered at some reference latitude and longitude that locally approximates the spherical metric in this chosen neighborhood of the Earth's surface. In this metric the effects of sphericity are retained by approximating f, the local vertical (or radial) component of $2\boldsymbol{\Omega}$, with a linear function of y, a latitudinal coordinate which is measured positive northward from the reference latitude (see below). This approximation is known as the β-plane approximation and was first introduced by Rossby (1939) in connection with his study of zonal circulation in the atmosphere. The symbol β has been traditionally used to denote $\partial f/\partial y$.

It is important to mention here that the β-plane approximation consists solely of a set of geometric approximations; it does not involve any assumptions regarding the relative magnitudes of the various dynamic terms in the basic equations. Since this point has not been emphasized in the standard references on the β-plane approximation (e.g. see Veronis, 1963a, b and 1973) and because of the fundamental importance of the so-called β-plane equations, we feel that it is important to present a thorough account of this approximation.

Let λ, ϕ, r denote a set of spherical polar coordinates and let u, v, w, denote the corresponding velocity components in the direction of increasing azimuth, latitude and radius respectively (see Fig. 5.1). In terms of these coordinates the line element is given by

$$ds^2 = r^2 \cos^2\phi \, d\lambda^2 + r^2 \, d\phi^2 + dr^2; \tag{5.10}$$

hence the metric coefficients are $h_1 \equiv h_\lambda = r \cos\phi$, $h_2 \equiv h_\phi = r$ and $h_3 \equiv h_r = 1$. The

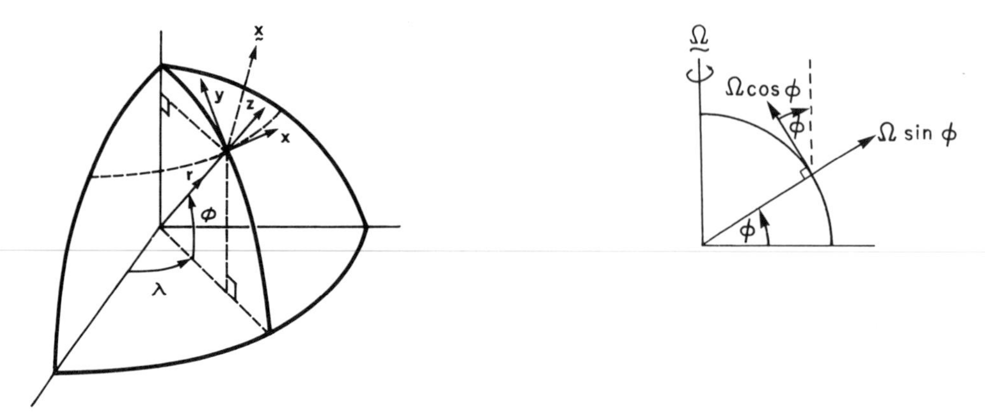

Fig. 5.1. The spherical polar coordinate system and the Cartesian coordinate system used in the β-plane approximation.

Fig. 5.2. The resolution of **Ω** into vertical (radial) and horizontal (tangential) components.

general expression for $\nabla \cdot \boldsymbol{F}$ in orthogonal curvilinear coordinates thus implies that the continuity equation (3.10) takes the form

$$\frac{1}{r\cos\phi}\frac{\partial u}{\partial\lambda}+\frac{1}{r\cos\phi}\frac{\partial}{\partial\phi}(v\cos\phi)+\frac{1}{r^2}\frac{\partial}{\partial r}(r^2w) = 0. \tag{5.11}$$

Upon resolving **Ω** into radial (vertical) and latitudinal (tangential) components (see Fig. 5.2), we find that

$$\begin{aligned}2\boldsymbol{\Omega}\times\boldsymbol{u} &= 2\Omega(0,\cos\phi,\sin\phi)\times(u,v,w)\\ &= 2\Omega(-v\sin\phi+w\cos\phi,\, u\sin\phi,\, -u\cos\phi).\end{aligned} \tag{5.12}$$

Thus, from (3.4), (3.11) and (5.12) it follows that the azimuthal component of the momentum equation takes the form

$$\begin{aligned}&\frac{\partial u}{\partial t}+\frac{u}{r\cos\phi}\frac{\partial u}{\partial\lambda}+\frac{v}{r\cos\phi}\frac{\partial}{\partial\phi}(u\cos\phi)+\frac{w}{r}\frac{\partial}{\partial r}(ru)\\ &\quad -2\Omega v\sin\phi+2\Omega w\cos\phi+\frac{1}{\rho r\cos\phi}\frac{\partial p}{\partial\lambda} = 0.\end{aligned} \tag{5.13a}$$

Analogous momentum equations can be written down for v and w:

$$\frac{\partial v}{\partial t}+\frac{u}{r\cos\phi}\frac{\partial v}{\partial\lambda}+\frac{v}{r}\frac{\partial v}{\partial\phi}+\frac{w}{r}\frac{\partial}{\partial r}(rv)+\frac{u^2}{r}\tan\phi+2\Omega u\sin\phi+\frac{1}{\rho r}\frac{\partial p}{\partial\phi} = 0; \tag{5.13b}$$

$$\frac{\partial w}{\partial t}+\frac{u}{r\cos\phi}\frac{\partial w}{\partial\lambda}+\frac{v}{r}\frac{\partial w}{\partial\phi}+w\frac{\partial w}{\partial r}-\frac{u^2+v^2}{r}-2\Omega u\cos\phi+\frac{1}{\rho}\frac{\partial p}{\partial r} = 0. \tag{5.13c}$$

In spherical polar coordinates, the density equation (3.9) takes the following form:

$$\frac{\partial\rho}{\partial t}+\frac{u}{r\cos\phi}\frac{\partial\rho}{\partial\lambda}+\frac{v}{r}\frac{\partial\rho}{\partial\phi}+w\frac{\partial\rho}{\partial r} = 0. \tag{5.14}$$

Let x, y, z be a set of curvilinear coordinates defined by

$$x = (R\cos\phi_0)\lambda, \quad y = R(\phi-\phi_0), \quad z = r-R, \tag{5.15}$$

where ϕ_0 is a reference latitude. The origin of this right-handed system lies on the undisturbed sea surface ($r = R$) at latitude ϕ_0 and longitude $\lambda = 0$. Henceforth the velocity components u, v, w will be identified with the x (eastward), y (northward), z (vertically upward) directions. Under the transformation (5.15), (5.11) becomes

$$R\cos\phi_0 u_x - \sin\left(\phi_0 + \frac{y}{R}\right)v + \cos\left(\phi_0 + \frac{y}{R}\right)[Rv_y + 2w + (R+z)w_z] = 0, \tag{5.16}$$

where the partial derivatives are denoted by subscripts. It is now convenient to introduce the nondimensional coordinates x', y', z' and velocity components u', v', w' by the relations

$$x' = x/L, \quad y' = y/L, \quad z' = z/H,$$
$$u' = u/U, \quad v' = v/U, \quad w' = w/W, \tag{5.17}$$

where L and U are characteristic horizontal length and velocity scales and H and W are characteristic vertical length and velocity scales. Using (5.17) in (5.16) and then dropping the primes, we readily obtain, after some rearrangement of the terms, the continuity equation (3.10) in the form

$$\cos\phi_0 u_x + \cos\left(\phi_0 + \frac{L}{R}y\right)v_y + \left(1 + \frac{H}{R}z\right)\cos\left(\phi_0 + \frac{L}{R}y\right)\frac{WL}{UH}w_z$$
$$- \sin\left(\phi_0 + \frac{L}{R}y\right)\frac{L}{R}v + 2\cos\left(\phi_0 + \frac{L}{R}y\right)\frac{WL}{UR}w = 0. \tag{5.18}$$

Let us now assume that we have chosen our scales correctly in the normalization (5.17) and that the quantities u_x, v_y, w_z, v, w, y and z in (5.18) are all of order unity. Since we shall be dealing with motions whose lateral scale L will necessarily be limited by the presence of continental boundaries and hence will generally be smaller than R, the quantity $(L/R)y$ will be small compared to unity and $\cos\phi$ and $\sin\phi$ can be expanded in a convergent Taylor series about the reference latitude $y = 0$:

$$\cos\left(\phi_0 + \frac{L}{R}y\right) = \cos\phi_0\left\{1 - \tan\phi_0\frac{L}{R}y - \frac{1}{2}\left(\frac{L}{R}\right)^2 y^2 + 0\left[\left(\frac{L}{R}\right)^3\right]\right\},$$
$$\sin\left(\phi_0 + \frac{L}{R}y\right) = \cos\phi_0\left\{\tan\phi_0 + \frac{L}{R}y - \frac{\tan\phi_0}{2}\left(\frac{L}{R}\right)^2 y^2 + 0\left[\left(\frac{L}{R}\right)^3\right]\right\}. \tag{5.19}$$

From (5.18) and (5.19) it is clear that to zeroth order in L/R and H/R the combined first three terms in (5.18) reduce to

$$\cos\phi_0\left(u_x + v_y + \frac{WL}{UH}w_z\right). \tag{5.20}$$

A convenient scale for the vertical velocity is †

$$W = HU/L; \tag{5.21}$$

† This choice of scale is not unique; a different choice will be made in Section 18.

Using this choice of scale, together with the expansions (5.19), the continuity equation (5.18) now becomes

$$u_x + \left[v_y + \left(1 + \frac{H}{R}z\right)w_z\right]\left\{1 - \tan\phi_0 \frac{L}{R}y - \frac{1}{2}\left(\frac{L}{R}\right)^2 y^2 + 0\left[\left(\frac{L}{R}\right)^3\right]\right\}$$
$$- v\left\{\tan\phi_0 \frac{L}{R} + \left(\frac{L}{R}\right)^2 y + 0\left[\left(\frac{L}{R}\right)^3\right]\right\} + 2w\frac{H}{R}\left\{1 - \tan\phi_0 \frac{L}{R}y + 0\left[\left(\frac{L}{R}\right)^2\right]\right\} = 0. \quad (5.22)$$

If we now assume that

$$H/R \ll 1, \quad (5.23)$$

$$(L/R)^2 \ll 1, \quad (5.24)$$

$$\tan\phi_0(L/R) \ll 1, \quad (5.25)$$

equation (5.22) can be approximated by the Cartesian form

$$u_x + v_y + w_z = 0. \quad (5.26)$$

From (5.17) and (5.21) it follows that (5.26) holds also for the dimensional variables. The inequalities (5.23)–(5.25) are the fundamental approximations included under the label β-plane approximation; if the line element (5.10) is transformed into x, y, z coordinates and these approximations are invoked, we readily find that

$$ds^2 = dx^2 + dy^2 + dz^2.$$

Thus, although according to (5.15) the coordinates x and y are orthogonal curvilinear coordinates lying on the spherical surface $z = 0$, under the approximations (5.23)–(5.25) x and y behave as ordinary Cartesian coordinates that can be thought of as spanning a plane (the so-called β-plane) that is tangent to the surface $z = 0$ at the reference latitude and longitude.

Approximation (5.23) means that we are dealing with a thin layer of fluid and are neglecting the radial distortion in moving from one value of z to another. Since $R = 6.4 \cdot 10^6$ m and $H = 0(5 \cdot 10^3\,\mathrm{m})$ for the deepest ocean basins, $H/R = 0(10^{-3})$, which indeed concurs with (5.23). Approximation (5.24) states that the horizontal scale of the motion must be appreciably less than the Earth's radius. If $L \leqslant 0(10^6\,\mathrm{m})$ then $(L/R)^2 = 0(10^{-2})$, which also agrees well with (5.24). The final approximation is the most stringent: for scales $L \leqslant 0(10^6\,\mathrm{m})$, $L/R = 0(10^{-1})$, and hence (5.25) will then be valid provided ϕ_0 corresponds to mid- or low latitudes, so that $\tan\phi_0 \lesssim 1$. Indeed (5.25) rules out the application of the resulting β-plane equations [see (5.30)–(5.32) below] to the study of motions at high latitudes or in polar seas. For these cases, a different approach involving cylindrical coordinates must be used (see Exercise 5.3).

Under the approximations (5.23)–(5.25) it is easy to see that the nonlinear inertial terms in (5.13) reduce to the Cartesian form

$$uu_x + vu_y + wu_z, \quad (5.27)$$

the pressure term simplifies to

$$\frac{1}{\rho}\frac{\partial p}{\partial x}, \quad (5.28)$$

and the Coriolis terms take the form

$$2\Omega(-v\sin\phi + w\cos\phi)$$

$$= 2\Omega U\left[-v'\left(\sin\phi_0 + \cos\phi_0\frac{L}{R}y' + \ldots\right) + w'\frac{H}{L}\cos\phi_0\left(1 - \tan\phi_0\frac{L}{R}y' + \ldots\right)\right]$$

$$\simeq 2\Omega U\left[-v'\left(\sin\phi_0 + \cos\phi_0\frac{L}{R}y'\right) + w'\frac{H}{L}\cos\phi_0\right]$$

$$= 2\Omega\left[-v\left(\sin\phi_0 + \cos\phi_0\frac{y}{R}\right) + w\cos\phi_0\right]. \qquad (5.29)$$

The momentum equations for v and w and (5.14) can also be readily simplified in the same manner. For the sake of completeness we now write here the full *β-plane equations* (in terms of *dimensional* variables) as implied by the three approximations (5.23)–(5.25):

$$u_t + uu_x + vu_y + wu_z - fv + \tilde{f}w + \frac{1}{\rho}p_x = 0, \qquad (5.30a)$$

$$v_t + uv_x + vv_y + wv_z + fu + \frac{1}{\rho}p_y = 0, \qquad (5.30b)$$

$$w_t + uw_x + vw_y + ww_z - (u^2 + v^2)/R - \tilde{f}u + \frac{1}{\rho}p_z = -g, \qquad (5.30c)$$

$$u_x + v_y + w_z = 0, \qquad (5.31)$$

$$\rho_t + u\rho_x + v\rho_y + w\rho_z = 0. \qquad (5.32)$$

In (5.30) f is the linear approximation to the local vertical component of $2\mathbf{\Omega}$ and is given by

$$f = f_0 + \beta y, \text{ where } \quad f_0 = 2\Omega\sin\phi_0, \beta = 2\Omega\cos\phi_0/R, \qquad (5.33)$$

and $\tilde{f}$ is the approximate local horizontal component of $2\,\mathbf{\Omega}$ and is given by

$$\tilde{f} = 2\Omega\cos\phi_0.$$

The quantity f is known as the *Coriolis parameter*. At mid-latitudes ($\phi_0 \simeq \pm 45°$), $f_0 \simeq \pm 10^{-4}$ rad/s and $\beta \simeq 1.6 \cdot 10^{-11}$ m^{-1} rad/s.

It can be fairly easily shown that the β-plane equations (5.30)–(5.32) also ensure that the potential vorticity remains conserved (Ertel's Theorem) as stated in its coordinate-free form in Section 4, provided we also impose the condition $L/R \ll 1$. For this conservation law to hold, it is crucial that $\tilde{f}$ be constant. In some β-plane approximations, $\tilde{f}$ is also allowed to vary with y and then the above theorems do not hold (see Grimshaw, 1975c). Since the energy equation (4.10) is independent of $\mathbf{\Omega}$, this discussion does not apply to that equation and hence the conservation of energy, correct to $0(L/R)$, is also ensured by the β-plane equations.

For smaller scale motions such as internal waves in a rotating system, it is natural to set L equal to H. From (5.29) it thus follows that for $\phi_0 \neq 0$ f can be approximated by f_0, and hence $f_0 v$ is comparable in magnitude to $\tilde{f}w$. That is to say, sphericity effects can be neglected, but it is important to retain both components of the rotation vector. With $L = H$, we find $[(u^2 + v^2)/R]/\boldsymbol{u} \cdot \nabla w = 0(H/R)$ and hence the term $-(u^2 + v^2)/R$ can be neglected in (5.30c). The set of equations with the β-term neglected are often referred to as *f-plane equations*, in which f is a constant. The f-plane equations are also useful for larger scale, nonequatorial motions for which $L > H$, so long as we have $L \cos \phi_0 / R \ll 1$, the condition needed in order to neglect the β term in f [see (5.33)].

Thus in summary three different sets of equations are required to deal with mid-latitude motions of varying horizontal scales. For very large scales in which $L = 0(R)$ the original equations (3.9)–(3.11) expressed in spherical polar coordinates must be used. For intermediate scales [$L/R = 0(10^{-1})$], the β-plane equations (5.30)–(5.32) are suitable, whereas for shorter scales still, the f-plane equations [(5.30)–(5.32) with f constant] are adequate.

It is interesting to note that for a β-plane centered at the equator ($\phi_0 = 0$), $f_0 = 0$, and βy, of course, also vanishes at $y = 0$. Hence very close to the equator, the $\tilde{f}w$ term in (5.30) can be important and therefore should not be neglected in studying equatorial dynamics. According to Veronis (1963b) the $\tilde{f}w$ term can be neglected, however, if the oceanic regions under consideration do not lie within a degree or so of the equator.

Finally, we return to a discussion of large-scale motions in mid-latitude ocean basins for which L is large compared to H and introduce a final set of approximations which will yield what we shall call the *hydrostatic β-plane equations*. If we assume that

$$(H/L) \cot \phi_0 \ll 1, \tag{5.34}$$

$$HR/L^2 \ll 1, \tag{5.35}$$

then consideration of (5.29) implies that $\tilde{f}w$ can be neglected in (5.30). Equation (5.34) clearly rules out equatorial ocean basins, and (5.35) restricts our attention to large-scale motions. Lastly, if we also assume the hydrostatic approximation, i.e. neglect all the vertical acceleration terms in (5.30c), these three approximations applied to (5.30) give

$$\frac{\mathrm{D}u}{\mathrm{D}t} - fv + \frac{1}{\rho} p_x = 0, \tag{5.36a}$$

$$\frac{\mathrm{D}v}{\mathrm{D}t} + fu + \frac{1}{\rho} p_y = 0, \tag{5.36b}$$

$$\frac{1}{\rho} p_z = -g, \tag{5.36c}$$

where $\mathrm{D}/\mathrm{D}t = \partial_t + u\partial_x + v\partial_y + w\partial_z$. We shall call equations (5.31), (5.32) and (5.36) the *hydrostatic β-plane equations*. In analogy with the f-plane equations, putting $f =$ constant in (5.36) reduces these equations to the *hydrostatic f-plane equations*. The justification for making the hydrostatic approximation depends very much on the length, time and velocity scales of the phenomena under consideration and will be dealt with separately in Chapter 3.

Exercises Section 5

1. Consider equations (3.10), (5.5) and (5.6) with $u = v = 0$, $\boldsymbol{\Omega} = \boldsymbol{0}$ and p', ρ', w as functions of t alone. Show that the resulting equations imply that

$$\xi_{tt} + N^2 \xi = 0,$$

where $\xi = \int w \, dt$ is the vertical displacement of an isopycnal (surface of constant density) in the fluid. Interpret this result.

2(a). If Ω^{-1}, L and U denote typical time, length and velocity scales, show that the neglect of the nonlinear terms in (5.5) and (5.6) is valid provided the Rossby number is small, viz.,

$$Ro \equiv U/\Omega L \ll 1.$$

2(b). If $\boldsymbol{\Omega} = \boldsymbol{0}$ and T now denote a typical time scale, determine the criterion for the linearization of (5.5) and (5.6).

(Ans. $U \ll L/T$.)

3. Consider the set of spherical polar coordinates θ, λ, r, where θ is now the co-latitude ($\pi/2 -$ latitude), and λ and r are as defined in Fig. 5.1. Let ρ, λ, z be the set of curvilinear coordinates defined by

$$\rho = R\theta, \quad \lambda = \lambda, \quad z = r - R, \tag{5.37}$$

which has its origin at the north pole, $\theta = 0$. Find the set of approximations analogous to (5.23)–(5.25) that transforms $\nabla \cdot \boldsymbol{u} = 0$ written in terms of θ, λ, r into the *cylindrical* form

$$\frac{1}{\rho}\frac{\partial}{\partial \rho}(\rho u) + \frac{1}{\rho}\frac{\partial v}{\partial \lambda} + \frac{\partial w}{\partial z} = 0,$$

where u, v, w are the velocity components in the directions of increasing ρ, λ, z. Under these approximations the coordinates defined by (5.37) behave like cylindrical coordinates with ρ, λ spanning a plane tangent at the pole (LeBlond, 1964).

4. Show that the application of the approximations (5.23)–(5.25) to the inertial terms in (5.13) does indeed lead to (5.27).

5. Show that (5.35) implies that the term $-(u^2 + v^2)/R$ is large compared to $\boldsymbol{u} \cdot \nabla w$ in the third equation of (5.30).

6. Use the normalization (5.17) and (5.21) together with $t' = t/T$ to determine the condition(s) under which the hydrostatic approximation holds. Distinguish between the cases $T \gg \Omega^{-1}$, $T = 0(\Omega^{-1})$ and $T \ll \Omega^{-1}$. How does your answer change when (5.6) is used as the basic momentum equation, rather than (3.11)? (In the second case the hydrostatic approximation means that p' and ρ' also satisfy $p'_z = -\rho' g$.)

7. Consider the *Mercator* coordinates λ, μ, r, where r is the radius, λ is the longitude and μ is related to the latitude by the relations

$$\operatorname{sech} \mu = \cos \phi, \quad \tanh \mu = \sin \phi, \quad \sinh \mu = \tan \phi.$$

Show that the metric distance is given by

$$ds^2 = m^2(d\lambda^2 + d\mu^2) + dr^2,$$

where $m = r \cos \phi$. Find $\nabla \cdot \boldsymbol{u}$ in terms of the coordinates λ, μ, r and find the approximations that transform this divergence into Cartesian form.

6. GENERAL PROPERTIES OF PLANE AND NEARLY-PLANE WAVES

Waves occurring in different physical systems have so much in common that many of their properties may be described without reference to the underlying dynamical system. It will be useful to introduce here some of these general wave properties before discussing specific types of waves. We will thus familiarize ourselves at this stage with general basic concepts of linear wave theory such as phase and group velocities, energy density and wave action, as well as with some techniques and approximations which will be used in further chapters. The reader may also consult an excellent review by Bretherton (1971) on the general linearized theory of wave propagation.

Kinematics: ray theory

A propagating wave is a travelling disturbance about an equilibrium state; it may be imagined as a moving surface or front, on which some physical quantity (such as temperature, or pressure) retains a coherent contrast with respect to its surroundings. From another, less geometrical point of view, wave motion may be viewed as a mechanism by which energy and information may be transferred from point to point without commensurate transport of mass. The very simplest wave is a plane wave, written as

$$\phi(\boldsymbol{x}, t) = \mathrm{Re}\{A \exp\,[i(\boldsymbol{k} \cdot \boldsymbol{x} - \omega t)]\}. \tag{6.1}$$

The amplitude of the wave, A, is a constant, and the phase, $\boldsymbol{k} \cdot \boldsymbol{x} - \omega t$, varies in time at a fixed point with the *angular frequency* ω (for convenience, taken as positive) and in space at a fixed time with the *wavenumber vector* $\boldsymbol{k}$. The period T of the wave is the time interval between the passage of successive crests; it is related to ω by $T = 2\pi/\omega$. The distance between successive crests is the wavelength $L = 2\pi/k$, where $k \equiv |\boldsymbol{k}|$. A crest moves, with speed $c = L/T = \omega/k$, which is known as the *phase speed*. At any instant t_0, a wave front is defined by $\boldsymbol{k} \cdot \boldsymbol{x} - \omega t_0 = \text{constant}$, which is clearly the equation of a family of parallel planes with normal vector $\boldsymbol{k}$. As time increases, these planes move with the phase speed in the direction $\boldsymbol{k}$: the phase velocity $\boldsymbol{c}$ is therefore parallel to $\boldsymbol{k}$. An expression for $\boldsymbol{c}$ in terms of ω and $\boldsymbol{k}$ is given in (6.7). The description of the wave is completed by a *dispersion relation* $\omega = \omega(\boldsymbol{k})$ arising from the equations describing the dynamics of the system.

A slightly more general wave form is given by

$$\phi(\boldsymbol{x}, t) = A(\boldsymbol{x}, t) \exp\,[iS(\boldsymbol{x}, t)], \tag{6.2}$$

where, as in (6.1), the real part of the right-hand side is implied. We shall henceforth routinely omit the notation $\mathrm{Re}\{\cdot\}$ for the real part of complex quantities, except in cases where some confusion might arise from this omission. The amplitude $A(\boldsymbol{x}, t)$ of the wave form (6.2) is assumed to be slowly varying in space and time compared to the variations in the phase function $S(\boldsymbol{x}, t)$.

Such slowly modulated waves are called nearly-plane waves, and their behaviour can be described through the simplified geometrical techniques of ray theory, first developed in optics and often referred to as "geometrical optics" or as the WKB approximation.

For a plane wave (6.1), the wavenumber and the frequency are given respectively by the gradient and the negative of the time derivative of the phase function. By analogy, the local values of the slowly varying wavenumber and frequency of the nearly-plane wave may be defined by

$$\boldsymbol{k} = \nabla S, \quad \omega = -\frac{\partial S}{\partial t}. \tag{6.3}$$

These definitions immediately imply that

$$\frac{\partial \boldsymbol{k}}{\partial t} + \nabla \omega = \boldsymbol{0}, \tag{6.4}$$

a relation often called the conservation of crests equation, which can be interpreted as follows. The vector $\boldsymbol{k}$ may be thought of as a directional density of crests in space, and the frequency ω as a flux of crests past a fixed point. Under the slowly varying assumptions of ray theory, crests are neither destroyed nor created and their total number must be conserved. The crest density and flux may then be related through a conservation equation of the form (6.4).

The rate of progression of a surface of constant phase $S(\boldsymbol{x}, t) = \text{constant}$ is found by letting $\mathrm{d}S(\boldsymbol{x}, t) = 0$, which gives

$$\frac{\partial S}{\partial t}\mathrm{d}t + \nabla S \cdot \mathrm{d}\boldsymbol{x} = 0. \tag{6.5}$$

Using (6.3), and (6.5) the *phase velocity* $\boldsymbol{c} = \mathrm{d}\boldsymbol{x}/\mathrm{d}t$ on $S = \text{constant}$ is related to the frequency by

$$\omega = \boldsymbol{k} \cdot \boldsymbol{c}. \tag{6.6}$$

Since $\boldsymbol{c}$ and $\boldsymbol{k}$ are parallel and $c = |\boldsymbol{c}| = \omega/k$, it follows from (6.6) that the phase velocity vector is given by

$$\boldsymbol{c} = \omega \boldsymbol{k}/k^2. \tag{6.7}$$

The underlying dynamics again yield a dispersion relation between ω and $\boldsymbol{k}$:

$$\omega(\boldsymbol{x}, t) = \sigma[k(\boldsymbol{x}, t); \lambda(\boldsymbol{x}, t)]. \tag{6.8}$$

Here ω varies not only through $\boldsymbol{k}(\boldsymbol{x}, t)$, but, for constant $\boldsymbol{k}$, also through variations in the properties of the medium itself, as expressed through the parameter $\lambda(\boldsymbol{x}, t)$. Substituting the dispersion relation (6.8) into the conservation of crests equation (6.4) we find (in tensor notation)

$$\frac{\partial k_i}{\partial t} + \frac{\partial \sigma}{\partial k_j}\frac{\partial k_j}{\partial x_i} + \frac{\partial \sigma}{\partial \lambda}\frac{\partial \lambda}{\partial x_i} = 0. \tag{6.9}$$

Since (6.3) implies that $\nabla \times \boldsymbol{k} = \boldsymbol{0}$, we have $\partial k_j/\partial x_i = \partial k_i/\partial x_j$. If we then define the *group velocity* $\boldsymbol{c}_g$ by the relation

$$c_{gi} = \partial \sigma/\partial k_i, \tag{6.10}$$

equation (6.9) may be rewritten (upon reverting to vector notation)

$$\frac{\partial \boldsymbol{k}}{\partial t} + \boldsymbol{c}_g \cdot \nabla \boldsymbol{k} = -\frac{\partial \sigma}{\partial \lambda}\nabla \lambda. \tag{6.11}$$

Similarly, by differentiating (6.8) with respect to t and using (6.4), we obtain

$$\frac{\partial \omega}{\partial t} + \boldsymbol{c}_g \cdot \nabla \omega = \frac{\partial \sigma}{\partial \lambda} \frac{\partial \lambda}{\partial t}. \qquad (6.12)$$

Equations (6.11) and (6.12) have the same characteristic curves, which are obtained by integrating the relationships

$$\mathrm{d}\boldsymbol{x}/\mathrm{d}t = \boldsymbol{c}_g = \partial \sigma / \partial \boldsymbol{k} \qquad (6.13)$$

(see Courant and Hilbert, vol. II, 1962; Chapter 1, Section 5). These characteristic curves are called *rays* in this wave context and their significance is best illustrated by choosing them as a kind of natural coordinate system for describing the waves. Integration of (6.13) gives

$$\boldsymbol{x} - \int \boldsymbol{c}_g \, \mathrm{d}t = \boldsymbol{\alpha}$$

where $\boldsymbol{\alpha}$ is a constant vector which varies from one ray to the next. The time derivative along a ray, i.e. keeping $\boldsymbol{\alpha}$ fixed, is then

$$\left(\frac{\mathrm{d}}{\mathrm{d}t}\right)_{\boldsymbol{\alpha}} = \frac{\partial}{\partial t} + \boldsymbol{c}_g \cdot \nabla. \qquad (6.14)$$

Coming back to (6.11) and (6.12), it is clear that in a completely homogeneous and time-independent medium, $\boldsymbol{k}$ and ω are both conserved along a ray. Any variations of $\boldsymbol{k}$ or ω along a ray result from spatial or temporal inhomogeneities, as given by the right-hand sides of these equations. Given some initial and boundary conditions and an additional statement on the variation of wave amplitude along a ray, one can construct the wave solution over the rest of the space–time domain in which the ray theory is applicable. The validity of ray theory is related to that of the conservation of crests equation (6.4), which breaks down when rays intersect or overlap (at caustic surfaces). At a less catastrophic stage, ray theory begins to lose its applicability when the amplitude $A(\boldsymbol{x}, t)$ varies appreciably over space or time scales comparable to the wavelength or the period respectively. More specific analytical criteria of validity depend on the exact form of the dynamical wave equation considered. When applicable, ray theory proves to be an extremely powerful tool, simplifying the solution of wave problems, usually stated in terms of partial differential equations, to ordinary integration along rays (see Exercise 6.1).

In practice, ray theory is applied as follows:

(1) From the dynamical equations and knowledge of the variability of the medium of propagation, a dispersion relation of the form (6.8) is derived for nearly-plane waves.

(2) Starting from known values of $\boldsymbol{k}$ and ω along a curve which is not itself a characteristic, it is possible to calculate, through (6.13), the directions of the rays emanating from that curve.

(3) The wave front is then allowed to move forward at a speed $\boldsymbol{c}_g$ (in general, a function of position) for a short time step along the characteristics.

(4) New values of $\boldsymbol{k}$ and ω are then evaluated from (6.11) and (6.12) and used to recalculate the directions of the rays and the new values of group velocity using (6.13). The procedure is continued, time step by time step. This technique has been used

extensively to describe the refraction of surface gravity waves in harbours, for example, and readily lends itself to automatic numerical computation (Skovgaard et al., 1975, 1976).

Ray theory may be summarized by equation (6.13), which describes the ray paths themselves, and (6.11) and (6.12) which specify the variations of $\boldsymbol{k}$ and ω along these paths. Using the time derivative along a ray, as defined by (6.14), these equations become

$$\frac{\mathrm{d}x_i}{\mathrm{d}t} = \frac{\partial\sigma}{\partial k_i}, \tag{6.15a}$$

$$\frac{\mathrm{d}k_i}{\mathrm{d}t} = -\frac{\partial\sigma}{\partial\lambda}\frac{\partial\lambda}{\partial x_i}, \tag{6.15b}$$

$$\frac{\mathrm{d}\omega}{\mathrm{d}t} = \frac{\partial\sigma}{\partial\lambda}\frac{\partial\lambda}{\partial t}. \tag{6.15c}$$

In addition, the relation (6.8) must be given for the function $\sigma(\boldsymbol{k}, \lambda)$. The first two of the above equations are identical in form to Hamilton's equations of classical particle dynamics (Goldstein, 1959; Chapter 7), with σ playing the role of a Hamiltonian and k_i the generalized momenta. This fundamental analogy between wave and particle mechanics opens the door to the well-known duality exploited in quantum mechanics. Equations (6.15a)–(6.15c) will be referred to as the Hamiltonian ray equations, or simply as the ray equations.

The above results may be generalized to wave propagation in a moving medium. Let $\boldsymbol{U}(\boldsymbol{x}, t)$ be an ambient current which varies over length and time scales well in excess of the wavelength and period. As the waves are advected by the current, the frequency observed by a stationary observer becomes

$$\omega = \boldsymbol{k} \cdot [\boldsymbol{c}_0 + \boldsymbol{U}], \tag{6.16}$$

where $\boldsymbol{c}_0$ is the phase velocity measured relative to the moving medium. The frequency ω_0, as measured by an observer travelling with velocity $\boldsymbol{U}$, is given by a dispersion relation of the form (6.8); the subscript 0 will be used to distinguish between quantities measured by a moving observer from those measured in a stationary frame of reference.

The wavenumber $\boldsymbol{k}$ is unaffected by a translation of coordinates and need not be burdened by a subscript in the moving frame. The frequency $\omega_0(\omega_0 = kc_0)$ is commonly referred to as the *Doppler-shifted* frequency.

Letting

$$\omega = \sigma_0[\boldsymbol{k}(\boldsymbol{x}, t); \lambda'(\boldsymbol{x}, t)] + \boldsymbol{k} \cdot \boldsymbol{U}, \tag{6.17}$$

the ray paths are now given by

$$\frac{\mathrm{d}x_i}{\mathrm{d}t} = \frac{\partial\sigma_0}{\partial k_i} + U_i. \tag{6.18}$$

Further, (6.15b) and (6.15c) become

$$\frac{\mathrm{d}k_i}{\mathrm{d}t} = -\frac{\partial\sigma_0}{\partial\lambda'}\frac{\partial\lambda'}{\partial x_i} - k_j\frac{\partial U_j}{\partial x_i}, \tag{6.19a}$$

$$\frac{d\omega}{dt} = \frac{\partial \sigma_0}{\partial \lambda'}\frac{\partial \lambda'}{\partial t} + k_j \frac{\partial U_j}{\partial t}, \tag{6.19b}$$

where differentiation along a ray is now expressed as

$$\frac{d}{dt} = \frac{\partial}{\partial t} + (\mathbf{c}_{g0} + \mathbf{U}) \cdot \nabla. \tag{6.20}$$

Stationary phase: ray theory as an asymptotic formulation

It has already been noted that ray theory is a valid representation of wave propagation so long as the waves described are nearly plane, their amplitude, wavenumber and frequency varying only slowly in time and space. When is such a situation likely to prevail? There spring to mind many simple cases, based on optics, but equally valid for other types of waves, where a geometrical ray theory is inadequate: in diffraction phenomena, in the vicinity of complex sources, etc. It will presently be seen that ray theory is essentially an asymptotic approximation to the representation of dispersive wave fields.

For simplicity, we will consider only the one-dimensional case. Let us assume that at $t = 0$ an initial disturbance of the form $F(x, 0)$ is created in a medium which can sustain simple plane waves of the form (6.1). These plane waves may be used as Fourier components to construct a solution of the form

$$F(x, t) = \int_{-\infty}^{\infty} G(k) \exp\,[i(kx - \omega(k)t]\,dk, \tag{6.21}$$

where $\quad G(k) = \dfrac{1}{2\pi}\displaystyle\int_{-\infty}^{\infty} F(x, 0) \exp\,(-ikx)\,dx,$

in order to satisfy the initial condition. For large x and t the integral (6.21) may be estimated by the method of stationary phase. This technique is now standard (Carrier et al., 1966, Chapter 6); the reader may also turn to Section 50 for a derivation of the technique. Qualitatively speaking, if the phase $(kx - \omega t)$ is a much more rapidly varying function of k than the amplitude $G(k)$, the main contribution to the integral will be from the vicinity of a point of stationary phase where

$$\frac{d}{dk}\,[kx - \omega(k)t] = 0. \tag{6.22}$$

Anywhere else, the phase is varying so rapidly compared to the amplitude that the oscillatory behaviour of the integrand is self-cancelling when integrated. Provided $\omega''(k) \neq 0$ and $G(k)$ is slowly varying, the first term in the asymptotic expansion of (6.21) as $t \to \infty$ is

$$F(x, t) \simeq G(k_s) \left(\frac{2\pi}{t|\omega''(k_s)|}\right)^{1/2} \exp\left[i\left\{k_s x - \omega(k_s)t - \frac{\pi}{4}\,\mathrm{sgn}\,\omega''(k_s)\right\}\right], \tag{6.23}$$

where $k_s(x, t)$ is the solution of (6.22). We immediately notice that the paths of stationary phase are identical to the rays defined in (6.13). The asymptotic solution has the form of a slowly varying plane wave and the phase propagates locally at the phase speed appropriate to the wavenumber k_s. The position at which the solution (6.23) is found in space as given by (6.22), and hence the overall propagation properties, are however determined by the group speed, $c_g(k_s)$. This fact leads us to suspect that this latter speed may play yet another role beyond that revealed by ray theory: if the whole solution $F(x, t)$ propagates at this speed, we may associate the group velocity with the rate of energy propagation.

The energy density of the waves described by (6.23) is proportional to the amplitude squared, viz.:

$$A^2 = G(k)G^*(k)2\pi/t|\omega''(k)|,$$

where G^* denotes the complex conjugate of G, and $k \equiv k_s$. The amount of energy $\hat{E}$ between two points x_1 and x_2 is equal to

$$\hat{E}(t) = \int_{x_1}^{x_2} \alpha(k) \frac{2\pi G(k)G^*(k)}{t|\omega''(k)|} \, dx, \tag{6.24}$$

where $\alpha(k)$ is some proportionality factor, relating the square of the amplitude to the energy.

In a coordinate system travelling with the group velocity, i.e. along the rays $x = \omega'(k)t$, (6.24) becomes, for $\omega''(k) > 0$,

$$\hat{E}(t) = 2\pi \int_{k_1}^{k_2} \alpha(k)G(k)G^*(k) \, dk,$$

where k_1 and k_2 are defined through $x_1 = \omega'(k_1)t$, $x_2 = \omega'(k_2)t$.

In a uniform medium, k_1 and k_2 are invariant along a ray [see (6.15b)] and $\hat{E}(t)$ is no longer a function of time. Therefore, the energy content in an interval in x-space whose ends travel at the group velocity is invariant, and the energy may be said to propagate at the group velocity. Differentiation of (6.24) with respect to time gives

$$\frac{\partial \hat{E}}{\partial t} = \int_{x_1}^{x_2} \frac{\partial}{\partial t}(\alpha A^2) \, dx + \alpha(k_2)\omega'(k_2)A^2(x_2) - \alpha(k_1)\omega'(k_1)A^2(x_1) = 0,$$

which can hold for any x_1, x_2 only if

$$\frac{\partial}{\partial t}(\alpha A^2) + \frac{\partial}{\partial x}(c_g \alpha A^2) = 0. \tag{6.25}$$

The quantity αA^2 is the amount of energy per unit length in the direction of propagation over a unit area with normal in that same direction, i.e. the spatial energy density $\langle E \rangle$. The flux of energy density is $c_g \alpha A^2$, i.e., the energy travels at the group velocity. In addition, in a homogeneous medium, $\partial c_g / \partial x = 0$ and (6.25) implies that the energy density remains invariant along a ray.

A variational approach

The variational approach introduced by Whitham (1965; 1974), and discussed further by Bretherton and Garrett (1968) and Garrett (1968), provides an elegant method of extending the energy balance equation (6.25) to more general situations. These variational techniques put nonlinear problems on the same footing as linear ones and allow an unprecedented degree of generality in treating slowly varying wave fields.

The equations of fluid dynamics or any one of their specialized forms [such as the adiabatic equations (3.9)–(3.11)] may be derived from a variational (Hamilton's) principle of the form

$$\delta \iint L(Z, Z_x, Z_t; \boldsymbol{x}, t)\mathrm{d}\boldsymbol{x}\, \mathrm{d}t = 0, \tag{6.26}$$

where $\boldsymbol{Z}$ stands for the dependent variables $(\boldsymbol{u}, p, \rho)$, derivatives are denoted by subscripts, L is the Lagrangian density and integration is over all $\boldsymbol{x}$ and t. For an account of the calculus of variations and its application to continuum physics, the reader should consult Gelfand and Fomin (1963) and Yourgrau and Mandelstam (1968). The dynamical equations are the Euler-Lagrange equations implied by (6.26):

$$\frac{\partial L}{\partial Z_i} - \frac{\partial}{\partial t}\left[\frac{\partial L}{\partial(\partial Z_i/\partial t)}\right] - \frac{\partial}{\partial x_j}\left[\frac{\partial L}{\partial(\partial Z_i/\partial x_j)}\right] = 0. \tag{6.27}$$

One should note that L is not uniquely defined, and although it may be written in the form $L = T - V$, with T the kinetic and V the potential energy densities, that is not the only possible nor necessarily the most desirable formulation.

Solutions of (6.27) may consist of nonlinear nearly-plane waves, with slowly varying amplitude, frequency and wavenumber. Whitham (1965) assumed that the slow variation of a wave train could be derived from an averaged variational principle

$$\delta \iint \mathcal{L}(\omega, \boldsymbol{k}, A; \boldsymbol{x}, t)\mathrm{d}\boldsymbol{x}\, \mathrm{d}t = 0, \tag{6.28}$$

where $\mathcal{L}$ is the time average of L over a period:

$$\mathcal{L}(\omega, \boldsymbol{k}, A; \boldsymbol{x}, t) = \frac{1}{T}\int_0^T L\, \mathrm{d}t. \tag{6.29}$$

The averaging process eliminates the dependence on the short time and space scales [i.e. $0(1/\omega)$, $0(1/k)$ respectively] characterizing the local phase variation in favour of the slow variations of the wave parameters ω, $\boldsymbol{k}$ and A. The behaviour of slowly varying properties is then given by the Euler-Lagrange equations (6.27) for the averaged Lagrangian $\mathcal{L}$. Recalling that ω and $\boldsymbol{k}$ are related to the phase function S through (6.3), we note that there are really only two dependent variables in $\mathcal{L}$: S and A. Thus

$$\mathcal{L}(\omega, \boldsymbol{k}, A; \boldsymbol{x}, t) = \mathcal{L}(S_t, \nabla S, A; \boldsymbol{x}, t)$$

and the Euler equations of (6.28) may be written directly from (6.27) by replacing L by $\mathcal{L}$ and letting $Z_1 = A, Z_2 = S$:

$$\mathcal{L}_A = 0, \tag{6.30}$$

$$\frac{\partial}{\partial t}\mathcal{L}_\omega - \frac{\partial}{\partial x_i}\mathcal{L}_{k_i} = 0. \tag{6.31}$$

Subscripts are used to denote differentiation with respect to the explicit dependence of $\mathcal{L}$ on the dependent variables (A, ω, $\boldsymbol{k}$) and independent variables ($\boldsymbol{x}$, t), whereas the conventional derivatives include complete differentiation with respect to one independent variable, keeping the others fixed. For example,

$$\frac{\partial \mathcal{L}}{\partial t} = \mathcal{L}_\omega \frac{\partial \omega}{\partial t} + \mathcal{L}_{k_i} \frac{\partial k_i}{\partial t} + \mathcal{L}_A \frac{\partial A}{\partial t} + \mathcal{L}_t. \tag{6.32}$$

The first Euler-Lagrange equation (6.30) expresses a relation between ω, $\boldsymbol{k}$ and A, a generalization to nonlinear waves of the dispersion relation (6.8). This relation may be written in the implicit form

$$G[\omega, \boldsymbol{k}, A; \lambda(\boldsymbol{x}, t)] = 0. \tag{6.33}$$

For small-amplitude waves, the energy density, and hence also $\mathcal{L}$, is proportional to A^2, and the relation between frequency and wavenumber is independent of amplitude, so that in that case $\mathcal{L}$ must be of the form

$$\mathcal{L} = G[\omega, \boldsymbol{k}; \lambda(\boldsymbol{x}, t)]A^2. \tag{6.34}$$

The nonlinear dispersion relation (6.33) then reduces to

$$G[(\omega, \boldsymbol{k}; \lambda(\boldsymbol{x}, t)] = 0, \tag{6.35}$$

an implicit form equivalent to (6.8). We also note that (6.34) and (6.35) imply that

$$\mathcal{L} \equiv 0 \tag{6.36}$$

for linear waves, which implies that $\mathcal{T} = \mathcal{V}$, i.e. energy equipartition on the average. Let us now restrict our attention to small-amplitude waves, for which (6.34) and (6.36) hold. The second Euler-Lagrange equation (6.31) has the form of a conservation equation for a quantity with density $\mathcal{L}_\omega$ and flux $-\mathcal{L}_{k_i}$. In a frame of reference with respect to which the fluid is locally at rest, the group velocity $\boldsymbol{c}_g$ may be written from (6.34) and (6.35) as

$$c_{gi} = \frac{\partial \sigma}{\partial k_i} = -\frac{\mathcal{L}_{k_i}}{\mathcal{L}_\omega}. \tag{6.37}$$

Equation (6.31) may then be recast as

$$\frac{\partial}{\partial t} \mathcal{L}_\omega + \frac{\partial}{\partial x_i}(c_{gi}\mathcal{L}_\omega) = 0, \tag{6.38}$$

which is a conservation equation for the (as yet unidentified) quantity $\mathcal{L}_\omega$. For linear waves $\mathcal{L}_\omega$ (from 6.34) is proportional to A^2. Using (6.35) to express ω as a function of $\boldsymbol{k}$ and λ, we may write

$$\mathcal{L}_\omega = f(\boldsymbol{k}, \lambda)A^2. \tag{6.39}$$

The quantity $\mathcal{L}_\omega$ is clearly related to the energy density of the waves; it is not however identical to the energy density, as $f(\boldsymbol{k}, \lambda)$ is not, in general, the same function of $\boldsymbol{k}$ as the $\alpha(\boldsymbol{k})$ appearing in (6.24) (see Exercise 6.6).

For any dynamical system described by a variational principle of the form (6.28), Noether's theorem shows that there exists a conservation equation corresponding to any group of transformations with respect to which the variational principle is invariant (Gelfand and Fomin, 1963; p. 177). Invariance with respect to time implies energy conservation; translational invariance in space implies momentum conservation. In view of (6.4), the pair (6.30) and (6.31) and the notation illustrated in (6.32), we can verify that

$$\frac{\partial}{\partial t}[\omega\mathcal{L}_\omega - \mathcal{L}] - \frac{\partial}{\partial x_i}[\omega\mathcal{L}_{k_i}] = -\mathcal{L}_t. \tag{6.40}$$

If $\mathcal{L}$ is invariant with respect to the time coordinate, i.e. if $\mathcal{L}$ does not depend explicitly on t, $\mathcal{L}_t = 0$ and we identify the homogeneous form of (6.40) with the energy conservation equation. This result also directly follows from Noether's theorem. For small-amplitude waves, $\mathcal{L} \equiv 0$, and

$$\omega \mathcal{L}_\omega = \langle E \rangle, \tag{6.41}$$

where $\langle E \rangle$ is the average energy density of the waves. With $\boldsymbol{c}_g$ given by (6.37) for a medium at rest and $\mathcal{L}_t = 0$, (6.40) reduces to the simple energy conservation equation

$$\frac{\partial \langle E \rangle}{\partial t} + \frac{\partial}{\partial x_i}(c_{gi}\langle E \rangle) = 0. \tag{6.42}$$

On the other hand, (6.38) is not subject to conditions of time invariance. In a frame of reference at rest with respect to the fluid, and in which (6.37) holds, the second Euler-Lagrange equation (6.31) becomes, for small-amplitude waves ($\mathcal{L} \equiv 0$ and $\omega \mathcal{L}_\omega = \langle E \rangle$)

$$\frac{\partial}{\partial t}\frac{\langle E \rangle}{\omega} + \frac{\partial}{\partial x_i}\left(c_{gi}\frac{\langle E \rangle}{\omega}\right) = 0. \tag{6.43}$$

The quantity $\langle E \rangle/\omega$ is well known from classical mechanics as an "adiabatic" invariant, i.e. a quantity which does not change when a physical system is subjected to externally determined variations of one of its parameters (Landau and Lifshitz, 1960; p. 154). In the study of wave propagation, the quantity $\langle E \rangle/\omega$ is called the *wave-action density*. As we shall see, conservation of wave action is more fundamental than that of energy, since it applies to moving systems as well as to stationary ones. It is only in a time-invariant system, where $\mathrm{d}\omega/\mathrm{d}t = 0$, that (6.43) reduces to (6.42).

A frame of reference in which the fluid is locally at rest may be moving with a slowly varying velocity $\boldsymbol{U}(\boldsymbol{x}, t)$ with respect to an inertial coordinate system. Thus (6.43) also applies within a moving system and may be rewritten (Garrett, 1968),

$$\frac{\partial}{\partial t}\frac{\langle E_0 \rangle}{\omega_0} + \nabla \cdot \left(\boldsymbol{c}_{g0}\frac{\langle E_0 \rangle}{\omega_0}\right) = 0, \tag{6.44}$$

where the subscript 0, as mentioned earlier, refers to quantities observed within the moving system. Equation (6.44) provides us with a generalization of the energy balance, expressed by (6.42) for a system at rest, to a moving system:

$$\frac{\partial \langle E_0 \rangle}{\partial t} + \nabla \cdot (\boldsymbol{c}_{g0}\langle E_0 \rangle) - \frac{\langle E_0 \rangle}{\omega_0}\frac{\mathrm{d}\omega_0}{\mathrm{d}t} = 0, \tag{6.45}$$

where $\mathrm{d}/\mathrm{d}t$ is the derivative along a ray (in the advected coordinates in which the fluid is at rest) as given by (6.14). If there were no interaction between the waves and the ambient current, the energy of a group would be conserved and (6.45) would reduce to (6.42), or (6.25) in a one-dimensional case. The last term in (6.45) must then represent the interaction between the waves and the flow, in terms of a rate of energy exchange between the two.

With $\omega_0 = \omega - \boldsymbol{k} \cdot \boldsymbol{U}$ and using (6.19a) and (6.19b), we find

$$\frac{\mathrm{d}\omega_0}{\mathrm{d}t} = \left(\frac{\partial \lambda'}{\partial t} + U_j\frac{\partial \lambda'}{\partial x_j}\right)\frac{\partial \sigma_0}{\partial \lambda'} - k_i(c_{g0})_j\frac{\partial U_i}{\partial x_j}. \tag{6.46}$$

It is usually possible to relate the variations of the basic state, as expressed by the derivatives of λ', to the current shear by an equation of the form (Exercise 6.7)

$$\frac{1}{\lambda'}\left(\frac{\partial\lambda'}{\partial t} + U_j\frac{\partial\lambda'}{\partial x_j}\right) + \Lambda_{ij}\frac{\partial U_i}{\partial x_j} = 0, \tag{6.47}$$

where Λ_{ij} is a tensor independent of $\boldsymbol{U}$ and its derivatives. The energy equation in a moving fluid, (6.45), may then be recast as

$$\frac{\partial\langle E_0\rangle}{\partial t} + \nabla\cdot(\boldsymbol{c}_{g0}\langle E_0\rangle) + T_{ij}\frac{\partial U_i}{\partial x_j} = 0, \tag{6.48}$$

where $T_{ij}(\partial U_i/\partial x_j)$ is the rate of energy exchange per unit volume along a ray (in the moving coordinate system) and may be equated to the rate of working of the interaction stress tensor T_{ij}, given by

$$T_{ij} = \frac{E_0}{\omega_0}\left[\lambda'\frac{\partial\sigma_0}{\partial\lambda'}\Lambda_{ij} + (c_{g0})_j k_i\right], \tag{6.49}$$

against the spatial rate of strain of the ambient flow. The interaction stress T_{ij} will later be identified, by examples, with the mean momentum flux tensor or radiation tensor of the waves.

The elegance and directness of variational techniques are well illustrated by the facility with which the governing equations for average wave properties are inferred from the Euler-Lagrange equations (6.30), (6.31). These techniques are also of great value in their applicability to nonlinear waves. Some of the most recent work using these methods is included in a book edited by Leibovich and Seebass (1974).

Exercises Section 6

1. Let the solution of the wave equation

$$\phi_{tt} - [c(\boldsymbol{x}/L)]^2\nabla^2\phi = 0$$

be of the form

$$A(\boldsymbol{x}/L)\exp[i(k_0 S(\boldsymbol{x}) - \omega t],$$

where L is the length scale of variations of c and A. Show that by substituting the assumed solution into the wave equation one obtains, for large $k_0 L$, the following relations for the phase and amplitude variation:

$$(\nabla S)^2 = n^2 = (k/k_0)^2,$$

$$(\nabla \ln A)\cdot\nabla S = -\tfrac{1}{2}\nabla^2 S,$$

where $k = \omega/c$. The function S is the "eikonal" functional of geometrical optics, and it is clear from its definition that $S(\boldsymbol{x})$ = constant gives the instantaneous position of a wave front. The quantity n is the local index of refraction with respect to a uniform medium k_0 (Pearson, 1966).

2. One-dimensional surface waves in deep water have the dispersion relation

$$\sigma(\boldsymbol{k}) = g|\boldsymbol{k}|.$$

Use the ray equation (6.15a) to estimate arrival time as a function of frequency for waves generated at a distant concentrated storm. Show that the arrival time increases linearly with frequency and that the distance from the generating area may be estimated from the slope of the arrival time curve (Snodgrass, et al., 1966). A similar technique is used to estimate the distances of pulsars (Lyne and Rickett, 1968).

3. Use the dispersion relation given in Problem 6.2 to find the behaviour of $F(x, t)$, as given by (6.23). Interpret the results for a fixed time as a function of position as well as for a fixed position as a function of time.

4. Verify (6.40) using (6.4), (6.30), (6.31) and the expansion (6.32).

5. Derive the "wave-momentum" equation

$$\frac{\partial}{\partial t}(k_i \mathcal{L}_\omega) + \frac{\partial}{\partial x_j}[-k_i \mathcal{L}_{k_j} + \mathcal{L}\delta_{ij}] = \mathcal{L}_{x_i} \tag{6.50}$$

from the same premises as (6.40). Note that the momentum density for linear waves is

$$k_i \mathcal{L}_\omega = k_i \langle E \rangle / \omega, \tag{6.51}$$

i.e. a vector in the direction of phase propagation with magnitude $\langle E \rangle / c$.

6(a). Given the one-dimensional Klein-Gordon equation

$$\phi_{tt} - \alpha^2 \phi_{xx} + \beta^2 \phi = 0, \tag{6.52}$$

construct a Lagrangian density $L(\phi, \phi_x, \phi_t)$ which, when substituted in (6.27), yields (6.52). This function is the difference between the kinetic and potential energy densities.

6(b). Letting $\phi(x, t)$ be a plane-wave solution of the form (6.1), construct an average Lagrangian function $\mathcal{L}$ for small-amplitude waves satisfying (6.29).

6(c). Show that $\langle E \rangle = \omega \mathcal{L}_\omega$, and derive the dispersion relation, the energy equation and the wave-action equation from (6.30), (6.42) and (6.43).

7. Consider wave propagation in a fluid of variable depth H where $\sigma_0 = \sigma_0(\mathbf{k}, H)$. In the presence of a two-dimensional horizontal current $\mathbf{U} = (U, V, 0)$ which satisfies the continuity equation

$$\frac{\partial}{\partial x_i}(U_i H) + \frac{\partial H}{\partial t} = 0,$$

show that the tensor Λ_{ij} in (6.49) is simply δ_{ij}.

CHAPTER 2

FREE WAVES: SHORT WAVELENGTHS

7. INTRODUCTION

The wave motions that we discuss in this book arise as perturbations of stable states under the four restoring effects of surface tension, gravity, the Coriolis force and the restoring torque associated with the conservation of potential vorticity. All these effects act simultaneously, in proportions which depend on the length and period of the waves, on the degree of density stratification and on the geometrical configuration of the basin in which the waves propagate. Many different types of dynamic balances are possible, and it is not surprising to find that the nomenclature of oceanic waves is cluttered with a great number of special types of waves, named either after their discoverer or after some dynamic or geometrical peculiarity.

In the following three chapters, we guide the reader through the labyrinth of wave-types along a path which takes us through a series of situations of increasing geometrical complexity (Fig. 7.1). In each case, the simplest waveforms are discussed in order to get as clear a picture as possible of the basic dynamics of each particular situation. More specifically, we shall consider primarily free plane-wave solutions of the linearized adiabatic equations (3.9)–(3.11).

We start in Section 8 with a study of plane-wave propagation in an unbounded stratified rotating fluid, in which gravity and the Coriolis force combine to produce internal-gyroscopic waves. The reflection of such waves from plane rigid or free boundaries is then discussed in Section 9. In Section 10, the real ocean is modelled as a waveguide in which the waves described in the preceding sections bounce back and forth between a rigid flat bottom and a free upper surface. The general properties of the vertical modal structure of these waves are presented in that section. Our path then bifurcates: the rest of Chapter 2 is devoted to the relatively short, high-frequency gravity and capillary waves, in which the role of the Coriolis force may be neglected. The linear theory of short waves is developed in Section 11, followed by sections on nonlinear effects (12) and average transport properties (13). The second branch of the path takes us to Chapter 3, entirely devoted to a study of waves of large spatial scales and long periods that may be described by the hydrostatic f-plane or β-plane equations. The two streams of study rejoin in Chapter 4, where extra geometrical complexity is introduced by adding lateral boundaries to the ocean.

8. SIMPLE GYROSCOPIC AND INTERNAL WAVES

Our knowledge of oceanic wave motion must ultimately be based on an understanding of the simplest possible types of waves which may exist in a stratified rotating fluid. We thus begin this chapter with an account of the properties of plane waves in an unbounded

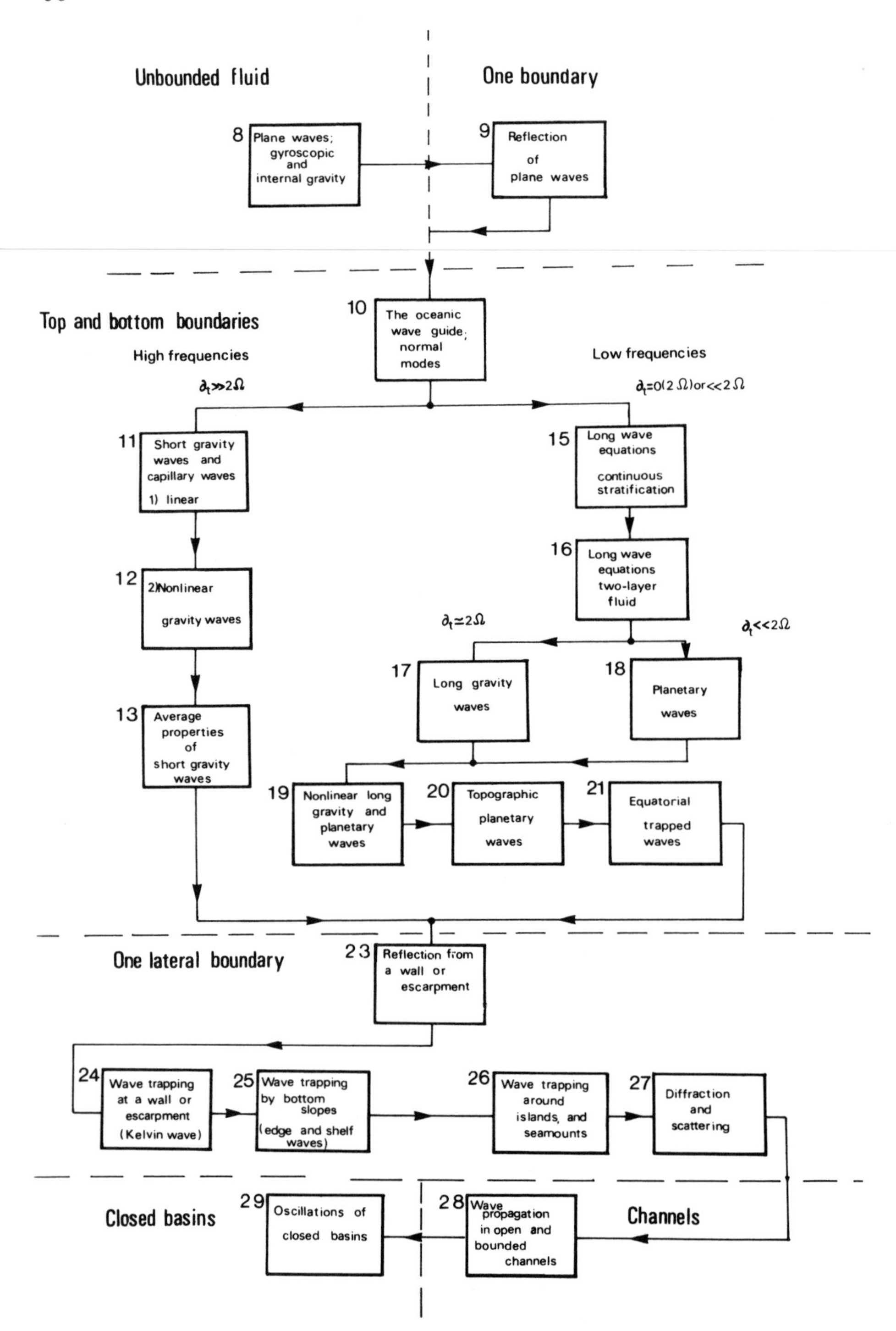

Fig. 7.1. A schematic guide to the material presented in Chapters 2, 3 and 4. The numbers refer to section numbers in the book.

fluid endowed with uniform density stratification and rotation rate. The results which we obtain will be a key piece in resolving the puzzle of time-dependent oceanic motions and will later be used as building blocks to construct more complex types of waves.

We base the analysis on the linearized equations (5.7) and (5.8) and on the continuity equation (3.10), rewritten here for convenience:

$$\rho_0(\partial_t + 2\boldsymbol{\Omega}x)\boldsymbol{u} = -\nabla p + \rho \boldsymbol{g}, \tag{8.1}$$

$$\rho_t + w\rho_{0z} = 0, \tag{8.2}$$

$$\nabla \cdot \boldsymbol{u} = 0. \tag{8.3}$$

Notice that in (8.1) and (8.2) the perturbation pressure and density appear without primes. By a long series of manipulations, it is possible to eliminate from (8.1)–(8.3) all variables but the pressure and obtain an equation which is valid for arbitrary rotation rate and stratification (Needler and LeBlond, 1973). Although mathematically satisfying, this exercise reveals little of the properties of waves in a rotating stratified fluid. Let us rather proceed from the simple to the more complicated by isolating the effects of rotation and stratification, examining first the properties of those waves which are found when only one of these effects is present.

Gyroscopic waves

In a rotating but unstratified fluid, where $\rho_0 = \rho_*$ and $\rho \equiv 0$, the buoyancy force vanishes in (8.1): any waves that arise must owe their existence to the presence of the Coriolis force. Such waves are called *gyroscopic waves* and have been described by Kelvin (1880a), Bjerknes et al. (1933), and Görtler (1944) among others. A more recent account of their properties is found in Greenspan (1968, chapter 4). Gyroscopic waves have also commonly been referred to as *inertial waves,* a designation which we shall avoid, as it is also used to refer to a limiting case of gyroscopic waves at the "inertial" frequency $\omega = 2\Omega \sin \phi$.

In the absence of stratification, (8.2) is identically satisfied and (8.1) becomes

$$[\partial_t + 2\boldsymbol{\Omega}x]\boldsymbol{u} = -\nabla(p/\rho_*). \tag{8.4}$$

Taking the curl of (8.4) to eliminate the pressure, we recover a linearized form of the vorticity equation (4.2):

$$\partial_t(\nabla \times \boldsymbol{u}) - (2\boldsymbol{\Omega} \cdot \nabla)\boldsymbol{u} = 0. \tag{8.5}$$

We seek plane-wave solutions of (8.5) and (8.3) of the form

$$\begin{pmatrix} \boldsymbol{u} \\ p \end{pmatrix} = \begin{pmatrix} \boldsymbol{U} \\ \Pi \end{pmatrix} \exp\left[i(\boldsymbol{k} \cdot \boldsymbol{x} - \omega t)\right], \tag{8.6}$$

where $\boldsymbol{U}$ and Π are amplitude constants and $\omega > 0$. Substituting (8.6) into (8.3) and (8.5), we obtain

$$\boldsymbol{k} \cdot \boldsymbol{U} = 0, \tag{8.7}$$

$$i\omega(\boldsymbol{U} \times \boldsymbol{k}) = (2\boldsymbol{\Omega} \cdot \boldsymbol{k})\boldsymbol{U}. \tag{8.8}$$

From (8.7), the waves are transverse: all displacements occur in a plane normal to the propagation vector. In view of (8.7),

$$|\boldsymbol{U} \times \boldsymbol{k}| = Uk,$$

where $U = |\boldsymbol{U}|$. Taking magnitudes of both sides of (8.8), we thus find that

$$\omega = |2\,\boldsymbol{\Omega} \cdot \boldsymbol{k}|/k. \tag{8.9}$$

This is the dispersion relation for gyroscopic waves. In terms of the angle between the $\boldsymbol{\Omega}$ and $\boldsymbol{k}$ vectors (Fig. 8.1) it may be rewritten as

$$\omega = 2\Omega|\cos\alpha|, \tag{8.10}$$

where $0 \leqslant \alpha \leqslant \pi$. Wave propagation is possible only in the frequency band $0 \leqslant \omega \leqslant 2\Omega$. Gyroscopic waves are highly anisotropic: the direction of propagation is determined by the frequency. For a given ω in the allowed band, all wavenumbers lie on a conical surface of half angle α (or $\pi - \alpha$ if $\alpha > \pi/2$) centered on the $\boldsymbol{\Omega}$ -axis (Fig. 8.1). Phase propagation also takes place along that conical surface with a velocity

$$\boldsymbol{c} = \frac{|2\boldsymbol{\Omega} \cdot \boldsymbol{k}|}{k^3}\boldsymbol{k}.$$

The group velocity is most simply derived using tensor notation. Using (8.9) we find

$$c_{gi} = \frac{\partial}{\partial k_i}\left(\pm\frac{2\Omega_j k_j}{k}\right) = \pm 2\left(\frac{\Omega_i}{k} - \frac{\Omega_j k_j}{k^3}k_i\right) \tag{8.11}$$

or, reverting to vector notation

$$\begin{aligned} \boldsymbol{c}_g &= \pm\frac{2}{k^3}[\boldsymbol{\Omega}(\boldsymbol{k} \cdot \boldsymbol{k}) - \boldsymbol{k}(\boldsymbol{k} \cdot \boldsymbol{\Omega})] \\ &= \pm\frac{2}{k^3}[\boldsymbol{k} \times (\boldsymbol{\Omega} \times \boldsymbol{k})], \end{aligned} \tag{8.12}$$

where $\pm$ corresponds to $\boldsymbol{\Omega} \cdot \boldsymbol{k} \lessgtr 0$. Clearly the group velocity is perpendicular to the phase velocity: although individual waves appear to go in the direction $\boldsymbol{k}$ because of their phase propagation, the energy, and hence the whole pattern of a wave group really moves at right angles to that direction. This rather unfamiliar property is also readily deduced from a geometrical argument. Since $\boldsymbol{c}_g$ is the gradient of ω in $\boldsymbol{k}$-space, it must be in a direction normal to lines of constant ω, i.e. in this case, normal to the lines $\alpha =$ constant. But $\boldsymbol{k}$ and $\boldsymbol{c}$ are vectors originating from the origin in $\boldsymbol{k}$-space (and also in physical space) and hence lie along $\alpha =$ constant. So, $\boldsymbol{k}$ and $\boldsymbol{c}_g$ are normal to each other; their relative directions are shown in Figs. 8.1 and 8.2. The group velocity is towards the $\boldsymbol{\Omega}$ -axis, in the direction of increasing ω (i.e. increasing $|\cos\alpha|$).

The particle motion associated with the wave may be inferred from the momentum equation (8.4). Let us define a right-handed Cartesian system by the unit vectors $(\hat{\boldsymbol{a}}, \hat{\boldsymbol{k}}, \hat{\boldsymbol{b}})$, where $\hat{\boldsymbol{k}}$ and $\hat{\boldsymbol{a}}$ are in the directions of the wavenumber and of $\boldsymbol{c}_g$ respectively, and $\hat{\boldsymbol{b}}$ is perpendicular to both (Fig. 8.2). The velocity $\boldsymbol{u}$ may then be written as

$$u = u_a\hat{\boldsymbol{a}} + u_b\hat{\boldsymbol{b}},$$

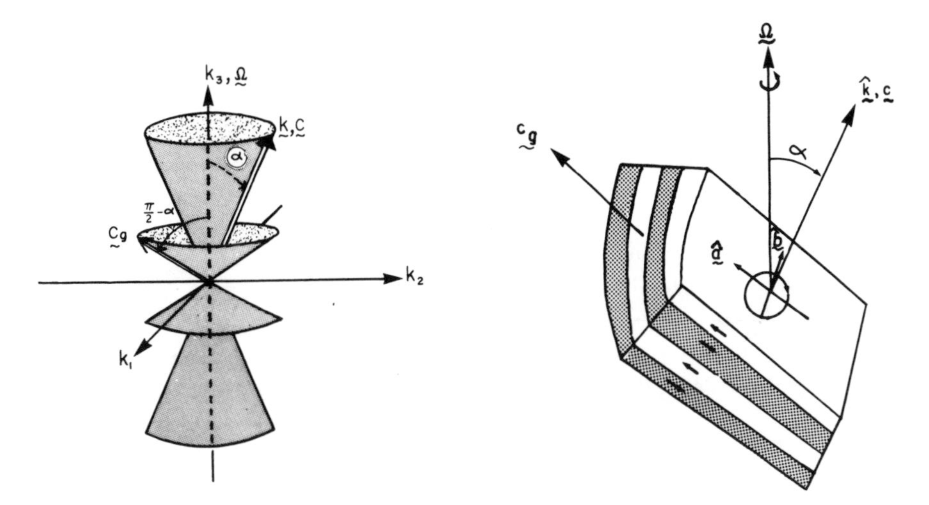

Fig. 8.1. Propagation diagram for gyroscopic waves, with the rotation vector taken as $\mathbf{\Omega} = (0, 0, \omega)$. The frequency ω is constant on the surface of the cone, on which $|\cos\alpha| =$ constant. The wavenumber $\boldsymbol{k}$ and phase velocity $\boldsymbol{c}$ lie on the surface of this cone; one wavenumber is shown (with $k_3 > 0$). The group velocity is normal to $\boldsymbol{k}$ and for any chosen $\boldsymbol{k}$ lying on the inner cone, $\boldsymbol{c}_g$ lies on the outer cone, on which $\sin\alpha =$ constant; this outer cone is a ray-surface which contains all the rays emanating from the origin at a given frequency. As ω decreases from its maximum value at $\alpha = 0$, the inner cone opens out and gradually flattens towards the (k_1, k_2)-plane, while the ray-surface closes in on the $\mathbf{\Omega}$-axis.

Fig. 8.2. A slice of the ray cone of Fig. 8.1 (a ray-surface of finite thickness), showing the direction of orbital motion and the alternation and progression of the phase through the shell, in the $\boldsymbol{k}$-direction. The coordinate system defined by the unit vectors $(\hat{\boldsymbol{a}}, \hat{\boldsymbol{k}}, \hat{\boldsymbol{b}})$ is also indicated.

and the rotation vector as

$$\mathbf{\Omega} = \Omega(\cos\alpha\hat{\boldsymbol{k}} + \cos\alpha'\hat{\boldsymbol{a}}),$$

where α' is the angle which $\boldsymbol{c}_g$ makes with $\mathbf{\Omega}$ $[\alpha' = \{\pi - (\operatorname{sgn} k_3)\pi/2\} - \alpha$, and $0 \leqslant \alpha' \leqslant \pi]$. The three components of the momentum equation (8.4), in the directions $\hat{\boldsymbol{k}}, \hat{\boldsymbol{a}}, \hat{\boldsymbol{b}}$ are then

$$\partial_t u_a = -2\Omega\cos\alpha\, u_b, \tag{8.13a}$$

$$\partial_t u_b = 2\Omega\cos\alpha\, u_a, \tag{8.13b}$$

$$2\Omega\cos\alpha' u_b = \hat{\boldsymbol{k}}\cdot\nabla(p/\rho_*). \tag{8.13c}$$

Since $\nabla(p/\rho_*) \propto \hat{\boldsymbol{k}}$ by (8.6), the pressure appears only in (8.13c).

The momentum balance in a plane normal to $\hat{\boldsymbol{k}}$ is entirely between the local rate of change of velocity and the Coriolis acceleration. From (8.13) and (8.10),

$$|u_a/u_b| = 1$$

and the motion is circularly polarized. The orbital motion is traced in a direction opposite to the sense of rotation of the fluid as a whole (as given by $\mathbf{\Omega}$) (see Exercise 8.1). The direction of orbital motion is determined by the fact that the Coriolis acceleration must always point forwards to the centre of the orbital circles in order to close the orbits. We also note that it is only the component of $\mathbf{\Omega}$ normal to $\boldsymbol{u}$ which is relevant to the orbital motion and which determines its frequency. The Coriolis acceleration due to the other component of $\mathbf{\Omega}$ is always kept balanced [see (8.13c)] by the pressure forces which are necessary to keep the orbital motion in a plane normal to $\hat{\boldsymbol{k}}$. The geometry of the orbital motion in relation to phase and group propagation is sketched in Fig. 8.2.

The spreading of a disturbance from a source is best visualized for a monochromatic perturbation of frequency ω_0. Oscillations at ω_0 impart a motion of the same frequency to the adjacent fluid, forcing the particle displacements in planes on which harmonic motion is possible at that frequency. The wave spreads at the group velocity on a double thin conical shell at an angle α' about the $\mathbf{\Omega}$ vector, with α' as given above. The thickness of the conical shell depends on the size of the wave source (as well as on viscosity and on the inevitably finite bandwidth in frequency forcing, which we do not take into account here). The phase propagates through the conical shell, as shown in Fig. 8.2, with the crests appearing on the outside surface, and disappearing on the inside surface.

Two limiting cases are of special interest. When $\alpha = 0$, the wavenumber is purely vertical and $\omega = 2\Omega$. The orbital motion all lies in the plane normal to $\mathbf{\Omega}$, and the group velocity vanishes. The *inertial oscillations* which are frequently observed in the ocean (Fig. 2.3) are an example of this limiting case. In the other limit, $\alpha = \pi/2$, $\omega = 0$ and the motion is entirely in a plane containing the $\mathbf{\Omega}$-axis. From (8.4) and (8.5) it follows that $\partial_t = 0$,

$$(2\mathbf{\Omega}\cdot\nabla)(\boldsymbol{u},p) = 0. \tag{8.14}$$

Hence the steady response is also uniform along $\mathbf{\Omega}$. If there is any velocity in that direction, it can only be due to the steady translation of a source. We have thus arrived, somewhat circuitously, at the *Taylor-Proudman* theorem, of which (8.14) is a concise statement (Taylor, 1921a; Greenspan, 1968; section 1.3). The condition (8.14) forbids any gradients of pressure or velocity to exist along the rotation vector in an unaccelerated, inviscid flow. This constraint leads to the existence of zones of stagnation (relative to the steadily moving source) which have been called *Taylor columns*. Such structures have been observed in laboratory experiments (Greenspan, 1968, section 1.1).

The time-dependent response to an impulsively started object has been studied by Bretherton (1967). The qualitative behaviour is most simply imagined in terms of waves whose group velocity becomes ever closer to the $\mathbf{\Omega}$-axis as time progresses, until a Taylor column appears as the asymptotic steady-state solution.

Internal gravity waves

In a stratified but nonrotating fluid ($\mathbf{\Omega} \equiv \mathbf{0}$) the only restoring force is that due to buoyancy. Oscillations about the equilibrium state will be called *internal gravity waves,* indicating both their dynamic origin and the fact that they occur within the interior of the fluid and not on an upper free surface of density discontinuity, as the more familiar surface gravity waves which will be explored later (Section 11).

Starting again from the equations (8.1)–(8.3), it is now convenient to split the velocity vector into two parts:

$$\boldsymbol{u} = \boldsymbol{u}_h + w\hat{\boldsymbol{z}} \tag{8.15}$$

where $\boldsymbol{u}_h$ is a two-dimensional vector in a plane normal to the direction of the gravity vector $\boldsymbol{g}$ and w is the velocity component anti-parallel to $\boldsymbol{g}$ and in the direction of the vertical unit vector $\hat{z}$. The basic equations (8.1)–(8.3) then become, with $\boldsymbol{\Omega} = \boldsymbol{0}$,

$$\rho_0 \partial_t \boldsymbol{u}_h = -\nabla_h p, \tag{8.16}$$

$$\rho_0 w_t = -p_z - \rho g, \tag{8.17}$$

$$\rho_t + w\rho_{0z} = 0, \tag{8.18}$$

$$\nabla_h \cdot \boldsymbol{u}_h = -w_z, \tag{8.19}$$

where ∇_h is the gradient operator normal to $\boldsymbol{g}$. Differentiating (8.17) with respect to time and substitution of ρ_t gives

$$\rho_0(w_{tt} + N^2 w) = -p_{zt}, \tag{8.20}$$

where $N^2 = -g\rho_{0z}/\rho_0$ is the stability frequency as defined in (5.3). Eliminating $\boldsymbol{u}_h$ between (8.16) and (8.19) yields another equation relating w and p:

$$\rho_0 w_{zt} = \nabla_h^2 p. \tag{8.21}$$

Finally, elimination of p from (8.20) and (8.21) yields an equation for w alone:

$$\nabla^2 w_{tt} + N^2 \nabla_h^2 w - \frac{N^2}{g} w_{ztt} = 0. \tag{8.22}$$

Let us now assume for simplicity that $dN^2/dz = 0$. Then (8.22) has constant coefficients and we may assume there exist plane wave solutions of the form

$$\begin{bmatrix} \boldsymbol{u}_h \\ w \\ p \end{bmatrix} = \begin{bmatrix} U_h \\ W \\ \Pi \end{bmatrix} \exp\left[i(\boldsymbol{k} \cdot \boldsymbol{x} - \omega t)\right]. \tag{8.23}$$

Substitution of (8.23) into (8.22) yields the dispersion relation

$$k^2 - \frac{N^2}{\omega^2}(k_1^2 + k_2^2) + ik_3\frac{N^2}{g} = 0, \tag{8.24}$$

which must hold for a nontrivial solution. The presence of an imaginary coefficient in (8.24) indicates that at least one of the components of $\boldsymbol{k}$ must be complex. Complex wavenumbers are usually associated with exponential growth or decay and are difficult to reconcile with the constraints of an adiabatic system. The problem is, however, only apparent. Consider the mean density field $\rho_0(z)$; the condition N^2 = constant means that

$$\rho_0(z) = \rho_0(0)\exp(-N^2 z/g). \tag{8.25}$$

In the absence of energy sources, the energy density of waves travelling through a uniformly stratified medium should remain uniform. Assuming equipartition [see below, equation (8.65) and following comments], the energy density is twice the kinetic energy density for linear waves:

$$E = \rho_0(z)|\boldsymbol{u}|^2.$$

If E is to be independent of z, $|\boldsymbol{u}|$ must be proportional to $\exp(N^2 z/2g)$. On substituting

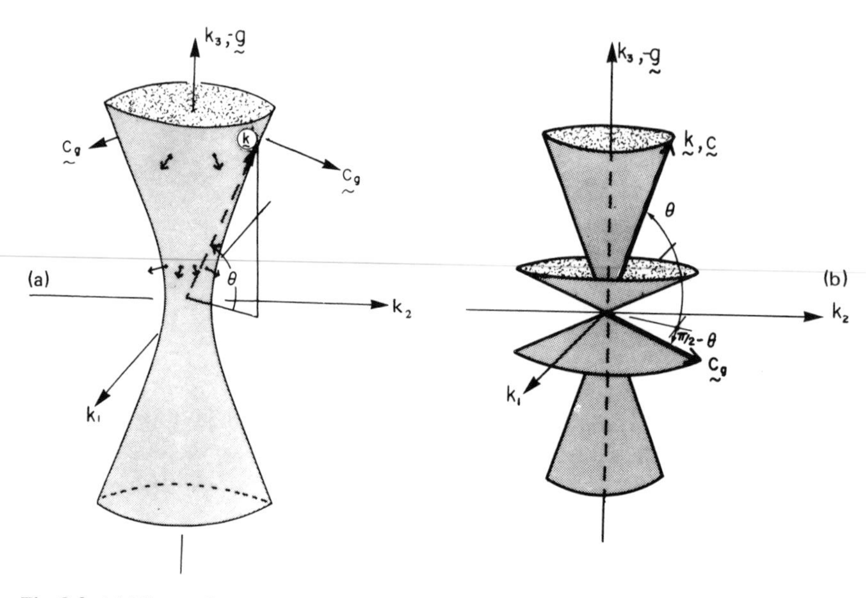

Fig. 8.3. (a) The surface of constant ω for pure internal gravity waves. The hyperboloid gradually opens up towards the (k_1, k_2)-plane as the frequency increases towards N. The group velocity is outwards, as shown by the arrows sticking out of the hyperboloid. (b) Propagation diagram for internal gravity waves under the Boussinesq approximation. The hyperboloid of Fig. 8.3a has been pinched at the waist and has degenerated into a cone. Compare this situation with that shown in Fig. 8.1 for gyroscopic waves and note the difference in the direction of the group velocity between the two cases.

$k_3 = \mathrm{Re}(k_3) + i\,\mathrm{Im}\,(k_3)$ into (8.24) and separating into real and imaginary parts, we indeed find that

$$\mathrm{Im}\,(k_3) = -N^2/2g, \tag{8.26}$$

as well as the following real dispersion relation:

$$(k_1^2 + k_2^2)\left(\frac{N^2}{\omega^2} - 1\right) = [\mathrm{Re}\,(k_3)]^2 + \left(\frac{N^2}{2g}\right)^2. \tag{8.27}$$

Writing $k_1^2 + k_2^2 = k_h^2$ and $k_h^2 + [\mathrm{Re}(k_3)]^2 = \boldsymbol{k}^2$, the dispersion relation may be rewritten explicitly for ω^2:

$$\omega^2 = N^2 k_h^2/[k^2 + (N^2/2g)^2]. \tag{8.28}$$

The frequency ranges from $\omega = 0$ for upward phase propagation ($k_h = 0$) to $\omega = N$ for horizontal phase propagation ($k_3 = 0$) at high wavenumbers ($k_h \gg N^2/2g$). A surface of constant ω in wavenumber space lies on a hyperboloid of revolution as shown in Fig. 8.3a. The group velocity is in the direction of the normal to that surface in the direction of increasing ω, i.e. away from the $\boldsymbol{g}$-axis.

For short waves, such that $k \gg N^2/2g$, (8.28) may be simplified to

$$\omega = N \cos\theta, \tag{8.29}$$

where $|\theta| \leqslant \pi/2$ is the angle between the horizontal and the wavenumber vector, as shown in Fig. 8.3. The short-wave approximation invoked to pass from (8.28) to (8.29) is the Boussinesq approximation mentioned in Section 5, and which states that variations in density have a negligible influence in the inertia terms. Had we used this approximation ab initio, replacing ρ_0 by ρ_* in (8.20) and (8.21), the term in N^2/g in (8.22) would not have appeared in this equation and the dispersion relation (8.29) (in which $N^2 = -g\rho_{0z}/\rho_*$) would have followed immediately. The Boussinesq approximation thus reveals itself as a short-wavelength approximation, valid when the wavenumber k greatly exceeds N^2/g, the inverse of the scale height of the stratification. As N is smaller than 10^{-2} rad s^{-1} over most of the ocean, the approximation should be valid for $k \gg 10^{-6}$ m^{-1}, i.e. for wavelengths much shorter than 6000 km. Volosov (1974) has made a formal study of the influence and applicability of this approximation.

Under the Boussinesq approximation, the hyperboloidal surface of constant ω shown in Fig. 8.3a reduces to the conical surface shown in Fig. 8.3b. The phase velocity, under (8.29), is

$$\boldsymbol{c} = \frac{N\cos\theta}{k^2}\boldsymbol{k} \tag{8.30}$$

and the group velocity is

$$\boldsymbol{c}_g = \frac{Nk_3^2}{k^3 k_h}\left[k_1, k_2, \left(\frac{-k_h^2}{k_3^2}\right)k_3\right]. \tag{8.31}$$

Clearly, $\boldsymbol{c}_g \cdot \boldsymbol{c} = 0$ and the energy propagates normally to the phase, as for gyroscopic waves, with the difference that the direction of $\boldsymbol{c}_g$ now is from the inside toward the outside of the surface of constant ω, as shown in Fig. 8.3b. This situation implies that phase and energy propagation are in opposite vertical directions: whenever the phase has an upward component of propagation ($k_3 > 0$), the vertical component of energy propagation is downward. The relation between rays and wave fronts may be seen in Fig. 8.4 showing photographs from the experiments performed by Mowbray and Rarity (1967a). The phase structure across rays, shown in the fifth panel of Fig. 8.4, has been analyzed by Thomas and Stevenson (1972), who developed a similarity solution including the influence of friction to describe the variation across a beam of waves.

From (8.3) or (8.19) plane internal gravity waves are transverse waves: particles of fluid oscillate in a plane normal to $\boldsymbol{k}$. The details of the motion are again most clearly elucidated in a local right-handed Cartesian system defined by unit vectors $(\hat{\boldsymbol{k}}, \hat{\boldsymbol{a}}, \hat{\boldsymbol{b}})$, similar to that used in looking at gyroscopic waves, with $\hat{\boldsymbol{k}}$ along the wavenumber vector and $\hat{\boldsymbol{a}}$ in the direction of $\boldsymbol{c}_g$. Writing

$$\boldsymbol{u} = u_a\hat{\boldsymbol{a}} + u_b\hat{\boldsymbol{b}}, \tag{8.32a}$$

$$\boldsymbol{g} = -g(\sin\theta\hat{\boldsymbol{k}} + \sin\theta'\hat{\boldsymbol{a}}), \tag{8.32b}$$

where θ' is the angle which the group velocity makes with the horizontal $[\theta' = \theta - \pi/2(\operatorname{sgn} k_3), |\theta'| \leqslant \pi/2]$.

The momentum equations (8.1) with $\boldsymbol{\Omega} = \boldsymbol{0}$ become

$$\rho_0\partial_t u_a = -g\sin\theta', \tag{8.33a}$$

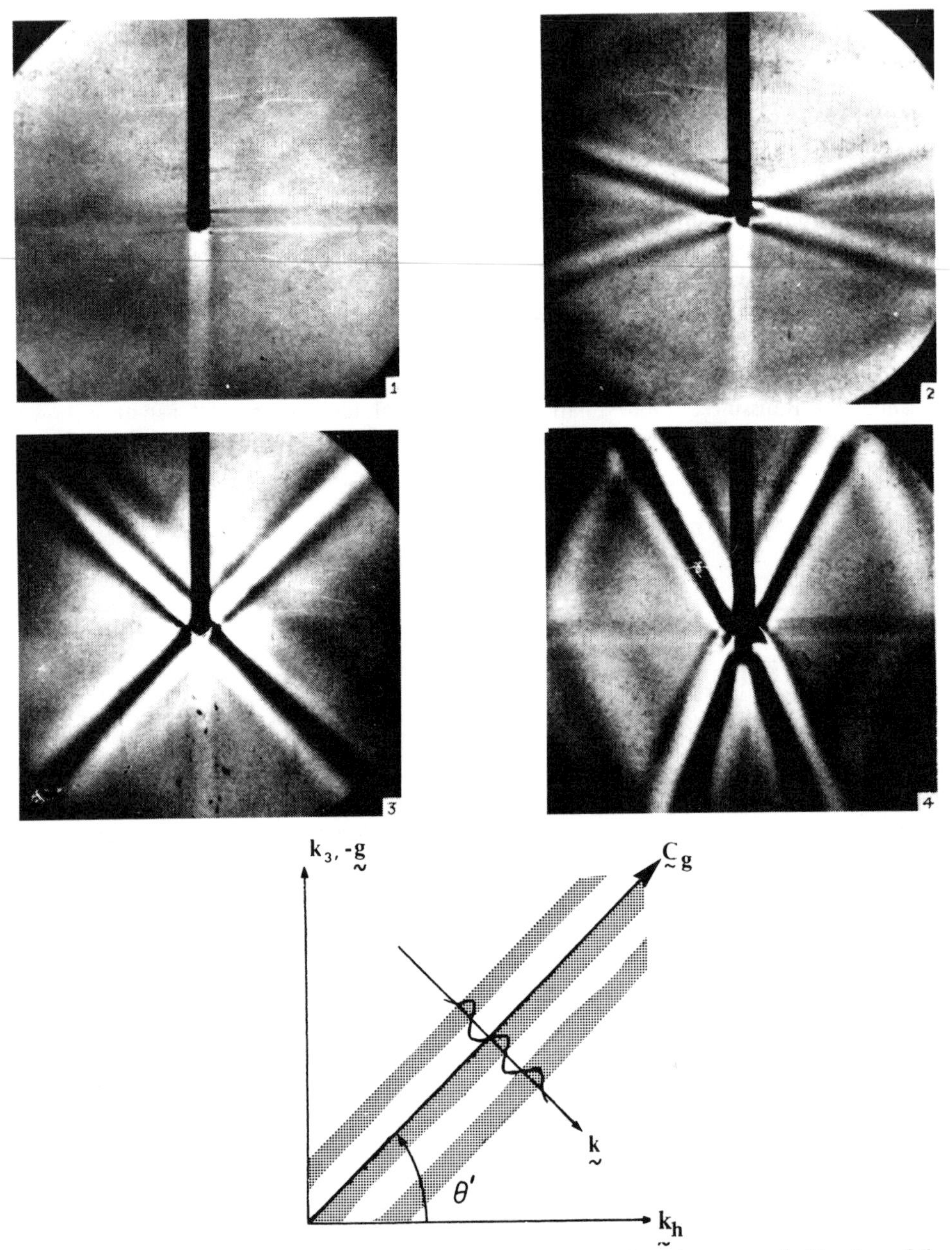

Fig. 8.4. Schlieren pictures of the phase configuration of internal gravity waves. A side view of the undisturbed fluid is shown in (1) and patterns corresponding to increasing forcing frequencies appear in the following sequence: (2) $\omega/N = 0.318$; (3) $\omega/N = 0.615$; (4) $\omega/N = 0.900$.

The X-shaped figures seen in these pictures are rays, emanating from a source attached to the bottom of the rigid vertical bar visible as a dark line. The alternation of light and dark bands along these rays represent wave crests, propagating across the ray as sketched in the bottom panel and in a manner similar, except for the direction of propagation, to that illustrated in Fig. 8.2. (From Mowbray and Rarity, 1967.)

$$\rho_0 \partial_t u_b = 0, \tag{8.33b}$$

$$-\rho g \sin\theta' = \hat{\boldsymbol{k}} \cdot \nabla p. \tag{8.33c}$$

From (8.32b), $u_b \equiv 0$ for plane waves, which are then linearly polarized in a plane which includes $\boldsymbol{g}$ and $\hat{\boldsymbol{k}}$, with the oscillation taking place normal to $\hat{\boldsymbol{k}}$. From (8.33a), the restoring force on particle displacements is that component of gravity which is normal to $\hat{\boldsymbol{k}}$. The other component of the buoyancy force, along $\hat{\boldsymbol{k}}$, is balanced by a pressure gradient sufficient to maintain the transverse nature of the wave. There is obviously a profound similarity between internal gravity waves and gyroscopic waves, as evidenced by the similarity of the dispersion relations (8.29) and (8.10) and of the momentum equations (8.33) and (8.13). The analogy between the behaviour of stratified and rotating fluids has been thoroughly discussed by Veronis (1970).

The limiting cases of the dispersion relation (8.29) are of special interest. For $\theta = 0$, the phase propagation is horizontal; particle displacements are strictly vertical (along $\boldsymbol{g}$) and the group velocity vanishes. The frequency becomes equal to the Brunt-Väisälä frequency N, which is then seen to characterize infinitesimal vertical oscillations of fluid particles, and hence the stability of the fluid. For $N > 0$, buoyancy acts as a restoring force and the fluid is stably stratified; when $N < 0$, the stratification is unstable and convection is initiated. In the other limit, $\theta \to \pm\pi/2$ and $\omega = 0$, describing steady flow. In the steady state, and with $\boldsymbol{\Omega} = \boldsymbol{0}$, (8.1) implies

$$\boldsymbol{g} \times \nabla p = \boldsymbol{0} \tag{8.34}$$

which we recognize as the analogue in stratified fluids of (8.14), the Taylor-Proudman theorem of rotating systems. In the present case, gradients perpendicular to $\boldsymbol{g}$ must vanish and the phenomenon analogous to the Taylor column is called blocking: a horizontally moving object is accompanied by a column of fluid perpendicular to $\boldsymbol{g}$ and moving with the object.

Plane waves in a rotating stratified fluid

The oscillations of a rotating stratified fluid share the properties of gyroscopic and internal gravity waves. For simplicity of presentation, we adopt the Boussinesq approximation and restrict our attention to a uniform exponential stratification (N^2 = constant). The basic equations (8.1) and (8.2) are then

$$\rho_* [\partial_t + 2\boldsymbol{\Omega} \mathrm{x}] \boldsymbol{u} = -\nabla p + \rho \boldsymbol{g}, \tag{8.35}$$

$$\rho_t - (\rho_* N^2/g) w = 0. \tag{8.36}$$

The rotation vector $\boldsymbol{\Omega}$ is not assumed to be collinear with $\boldsymbol{g}$. Eliminating the density perturbation ρ, between (8.35) and (8.36), we obtain

$$[\partial_t + 2\boldsymbol{\Omega} \mathrm{x}] \boldsymbol{u}_t = -\nabla(p/\rho_*)_t + \boldsymbol{g}(N^2 w/g). \tag{8.37}$$

The pressure term is in turn eliminated by taking the curl of (8.37); with $\boldsymbol{\Omega}$ and N^2 independent of position, we find

$$\nabla \times \boldsymbol{u}_{tt} - 2(\boldsymbol{\Omega} \cdot \nabla) \boldsymbol{u}_t - (N^2/g) \nabla \times (\boldsymbol{g} w) = \boldsymbol{0}. \tag{8.38}$$

Plane wave solutions of the form

$$\boldsymbol{u} = \boldsymbol{U} \exp [i(\boldsymbol{k} \cdot \boldsymbol{x} - \omega t)], \tag{8.39}$$

where $\boldsymbol{U} = (U, V, W)$, will be transverse, satisfying (8.7) $(\boldsymbol{k} \cdot \boldsymbol{U} = 0)$. Substituting (8.39) into (8.38) yields the algebraic vector equation

$$\omega^2 \boldsymbol{k} \times \boldsymbol{U} - 2i(\boldsymbol{\Omega} \cdot \boldsymbol{k})\omega \boldsymbol{U} + (N^2 W/g)(\boldsymbol{k} \times \boldsymbol{g}) = 0. \tag{8.40}$$

It is now convenient to pass directly to the right-handed coordinate system defined by the unit vectors $(\hat{\boldsymbol{k}}, \hat{\boldsymbol{a}}, \hat{\boldsymbol{b}})$, identical to that used in the study of internal gravity waves. With $\boldsymbol{U}$ and $\boldsymbol{g}$ in the form as given in (8.32) and

$$W = -U_a \sin \theta', \tag{8.41}$$

where θ' is defined following (8.32b), equation (8.40) has the components

$$\omega^2 k U_b + 2i\omega(\boldsymbol{\Omega} \cdot \boldsymbol{k}) U_a = 0, \tag{8.42a}$$

$$\omega^2 k U_a - 2i\omega(\boldsymbol{\Omega} \cdot \boldsymbol{k}) U_b - kN^2 U_a \cos^2\theta = 0. \tag{8.42b}$$

in the directions $\hat{\boldsymbol{a}}$ and $\hat{\boldsymbol{b}}$ respectively. Eliminating U_a or U_b from the above pair of equations, we find

$$\omega^2 [k^2(\omega^2 - N^2 \cos^2 \theta) - 4(\boldsymbol{\Omega} \cdot \boldsymbol{k})^2] = 0. \tag{8.43}$$

Excluding the steady state case $\omega = 0$, we find the dispersion relation for mixed gyroscopic–internal gravity waves:

$$\omega^2 = N^2 \cos^2 \theta + 4\Omega^2 \cos^2 \alpha, \tag{8.44}$$

where the angle α, as defined earlier (Fig. 8.1), is the angle between the wavenumber and the rotation vector. The dispersion relations for gyroscopic waves (8.10) and internal gravity waves (8.29) are special cases of (8.44). If one considers plane wave motion in a stratified rotating fluid as a simple harmonic motion resulting from the superposition of the two restoring forces $N^2 \cos^2 \theta$ and $4\Omega^2 \cos^2 \alpha$, the additive result (8.44) follows immediately.

The surface of constant ω may again be shown to be a conical surface through the origin of wavenumber space, similar to those shown in Fig. 8.1 for gyroscopic waves and in Fig. 8.3b for internal gravity waves. The orientation of the axis of the cone depends, however, on the angle between the rotation vector and the vertical. The components of the Earth's rotation vector in the coordinate system (x, y, z) introduced in Section 5 (see Figs. 5.1 and 5.2) are

$$\boldsymbol{\Omega} = (0, \Omega \cos \phi, \Omega \sin \phi),$$

where ϕ is the latitude. We abbreviate by letting

$$f = 2\Omega \sin \phi, \quad \tilde{f} = 2\Omega \cos \phi. \tag{8.45}$$

Recalling that θ is the angle between the wavenumber vector and the horizontal plane, we may now rewrite (8.43) as

$$k_1^2(N^2 - \omega^2) + k_2^2(N^2 + \tilde{f}^2 - \omega^2) + 2k_2 k_3 f\tilde{f} = k_3^2(\omega^2 - f^2). \tag{8.46}$$

Let us introduce rotated wavenumber coordinates $(\kappa_1, \kappa_2, \kappa_3)$, defined by

$$\begin{pmatrix} \kappa_1 \\ \kappa_2 \\ \kappa_3 \end{pmatrix} = \begin{pmatrix} 1 & 0 & 0 \\ 0 & \cos \nu & \sin \nu \\ 0 & -\sin \nu & \cos \nu \end{pmatrix} \begin{pmatrix} k_1 \\ k_2 \\ k_3 \end{pmatrix}, \tag{8.47}$$

where the angle of rotation ν about the x_1-axis is given by

$$\tan 2\nu = \sin 2\phi \bigg/ \left(\frac{N^2}{4\Omega^2} + \cos 2\phi \right). \tag{8.48}$$

In terms of the components of $\mathbf{\Omega}$, (8.46) takes the more symmetric form

$$(N^2 - \omega^2)\kappa_1^2 + (\omega_a^2 - \omega^2)\kappa_2^2 = (\omega^2 - \omega_b^2)\kappa_3^2, \tag{8.49}$$

where $\omega_a^2 = (\tilde{f}^2 + N^2)\cos^2 \nu + f\tilde{f} \sin 2\nu + f^2 \sin^2 \nu$, (8.50a)

$$\omega_b^2 = f^2 \cos^2 \nu - f\tilde{f} \sin 2\nu + (\tilde{f}^2 + N^2)\sin^2 \nu. \tag{8.50b}$$

It may be shown (Exercise 8.5) that ω_a and ω_b are the maximum and minimum frequencies compatible with real wavenumber components in (8.49). Thus,

$$\omega_b < \omega < \omega_a. \tag{8.51}$$

We note that N, f and $\tilde{f}$ also lie between ω_b and ω_a. We are now in a position to examine the form of the surface $\omega =$ constant in the $\boldsymbol{\kappa}$-plane over the frequency range (8.51). With reference to Fig. 8.5a, we first observe that in the range $\omega_b < \omega < N < \omega_a$, $\omega =$ constant on the surface of a cone of elliptical cross-section centered about the κ_3-axis. This cone opens up and the ellipticity of its cross-section increases with frequency (see Fig. 8.5b) until at $\omega = N$ the surface of constant ω degenerates to a pair of planes (Fig. 8.5c). Finally, for $\omega_b < N < \omega < \omega_a$, $\omega =$ constant lies in the surface of a cone centered about the κ_2-axis (Fig. 8.5d) which gradually shrinks towards that axis as $\omega \to \omega_a$.

The group velocity is everywhere in the direction of the normal to the $\omega =$ constant surface (in the direction of increasing ω), and may be deduced from the geometry of that surface. In the rotated coordinate system $(\kappa_1, \kappa_2, \kappa_3)$, we find by differentiation of (8.49) with respect to κ_i that

$$\boldsymbol{c}_g = \frac{1}{\omega\kappa^2} [\kappa_1(N^2 - \omega^2), \kappa_2(\omega_a^2 - \omega^2), \kappa_3(\omega_b^2 - \omega^2)]. \tag{8.52}$$

The derivation of the group velocity vector referred to the original unrotated coordinate system proceeds readily from (8.46) and is left as an exercise for the reader. The particle motion is again most clearly elucidated with reference to the local coordinate system defined by the unit vectors $(\hat{\boldsymbol{k}}, \hat{\boldsymbol{a}}, \hat{\boldsymbol{b}})$ used earlier. From (8.42a), the motion is elliptically polarized in a plane normal to $\hat{\boldsymbol{k}}$, the degree of polarization depending on the frequency and, through it, on the direction of propagation.

Except for a superficial wind-mixed layer, the upper layers of the ocean are strongly stratified in the sense that for all latitudes ϕ

$$N^2 \gg 4\Omega^2 \cos 2\phi. \tag{8.53}$$

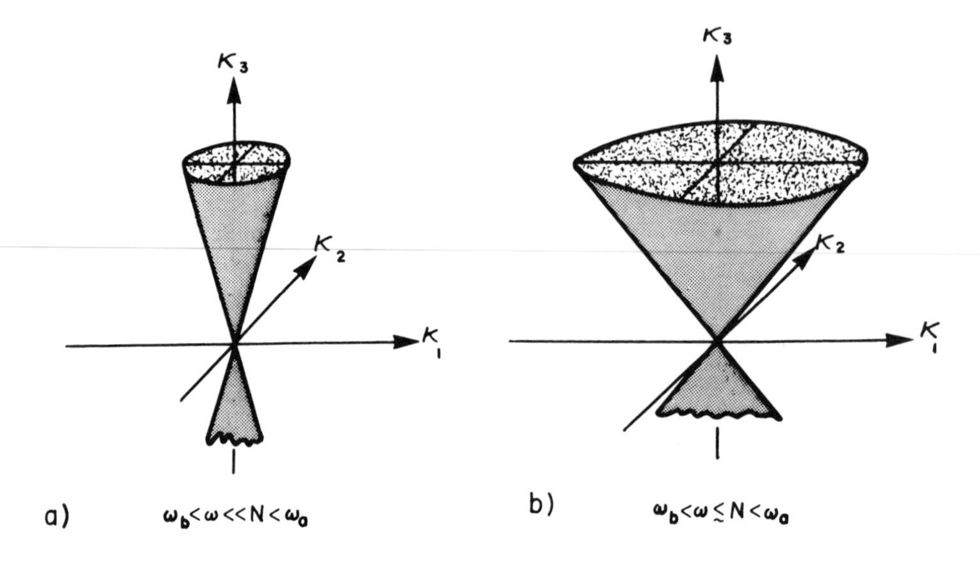

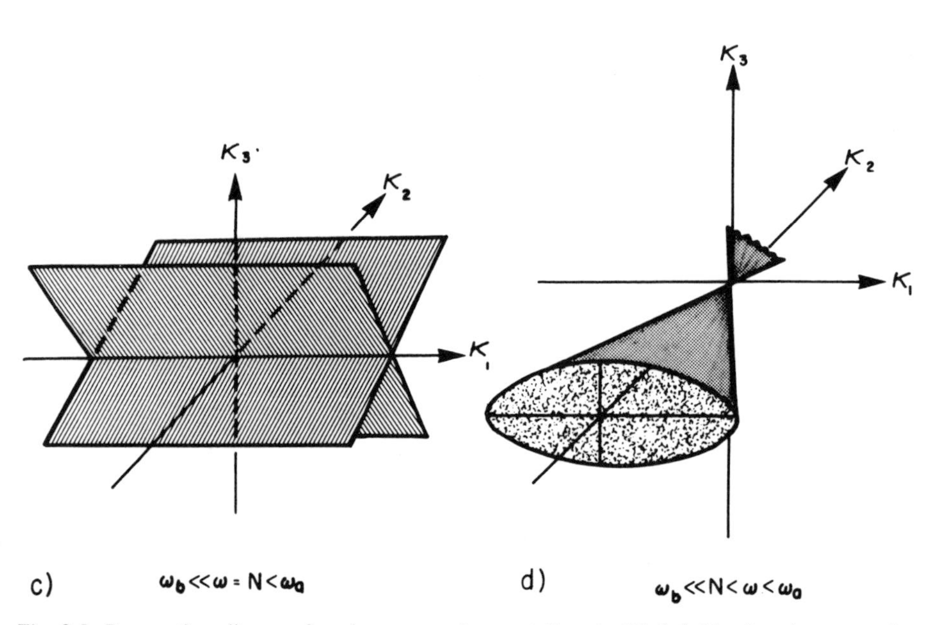

Fig. 8.5. Propagation diagram for plane waves in a rotating stratified fluid referred to rotated axes (κ_1, κ_2, κ_3) as defined by (8.47). The surfaces illustrated are those on which ω = constant, where ω is related to the components of $\boldsymbol{\kappa}$ by (8.49). The frequency lies between the limits $\omega_b < \omega < \omega_a$, where ω_a and ω_b are given by (8.50).

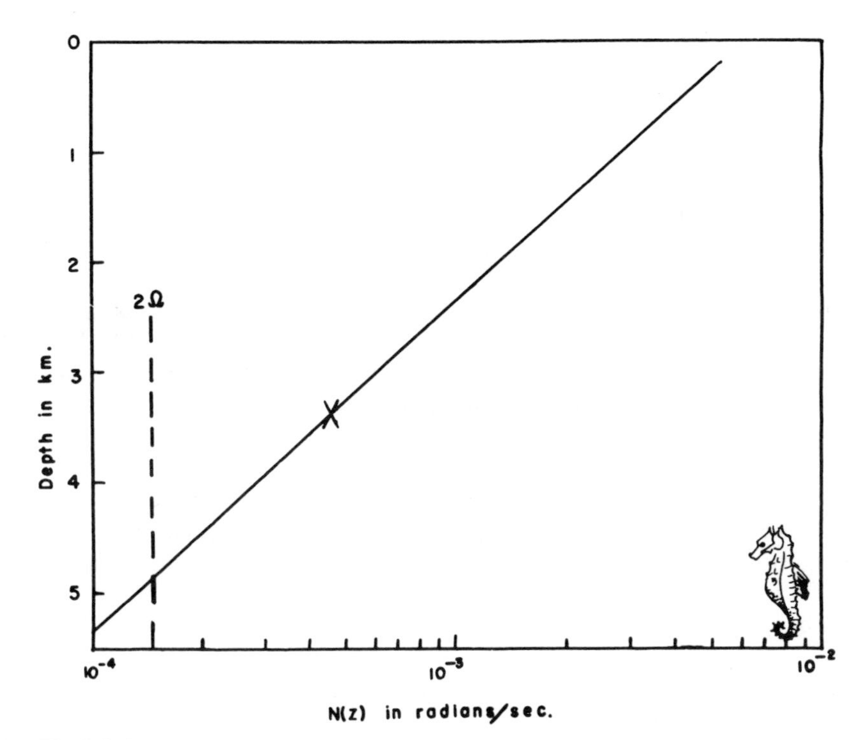

Fig. 8.6. Garrett and Munk's (1972b) best fit to the Brunt-Väisälä frequency in the deep ocean. The heavy line obeys the relation $N(z) = N(-200)\exp[(z + 200)/1300]$, with the depth in metres and $N(-200) = 5.23 \cdot 10^{-3}$ rad s^{-1}. The depth at which $N^2/4\Omega^2 = 10$ is shown by the cross on the $N(z)$ line. Above this depth, the ocean is strongly stratified at all latitudes and the simplified dispersion relation (8.56) may be used. Note that the depth distribution of $N(z)$ shown here is applicable only for $z \leqslant -200$ m; near the sea surface, a wind-mixed layer is usually found, in which N is very small (cf. Fig. 16.1).

The typical stratification in the upper 5 km of the ocean, as shown in Fig. 8.6, shows that this inequality holds over the upper 3 km of the ocean. Under (8.53), the tilt angle ν becomes

$$\tan 2\nu \simeq (4\Omega^2/N^2)\sin 2\phi. \tag{8.54}$$

Thus, for $N^2 \gg 4\Omega^2$, ν varies from zero at the equator ($\phi = 0$), attains a maximum in absolute value at $\phi = \pm 45°$ and decreases again to zero at the poles. Note that even when N^2 is not much greater than $4\Omega^2$, (8.54) may still provide a good approximation to (8.48) near $\phi = \pm\pi/4$. With $N^2 \gg 4\Omega^2$, and in view of (8.54), $\nu = 0(4\Omega^2/N^2)$ and the maximum and minimum frequencies ω_a and ω_b are given by

$$\omega_a^2 = N^2[1 + 0(4\Omega^2/N^2)], \tag{8.55a}$$

$$\omega_b^2 = f^2[1 + 0(4\Omega^2/N^2)]. \tag{8.55b}$$

Further, for small ν, we find from (8.47) that

$$\kappa_2 = k_2[1 + 0(4\Omega^2/N^2)] \quad \text{and} \quad \kappa_3 = k_3[1 + 0(4\Omega^2/N^2)].$$

Neglecting terms of $0(4\Omega^2/N^2)$ in (8.49), the dispersion relation reduces to

$$\frac{k_1^2 + k_2^2}{k_3^2} = \frac{\omega^2 - f^2}{N^2 - \omega^2}, \tag{8.56}$$

which is simply (8.46) with $\tilde{f} \equiv 0$. For this simplified case, the surface of constant ω lies on a cone of circular cross-section. The simplified relation (8.56) applies at all depths only at the poles, where $\tilde{f} = 0$; at other latitudes, (8.56) is not applicable to the deep ocean.

The near-inertial oscillations of frequency $\omega \simeq f$ commonly observed in the ocean are usually detected by the horizontal circularly polarized currents associated with them. Leaman and Sanford (1975) have also analyzed the vertical phase variation of such currents for which $f \lesssim \omega \ll N$ and found a predominantly upward phase propagation, corresponding to downward energy propagation. Internal waves near the inertial frequency are thus seen to arise near the sea surface and propagate their energy downwards. The refraction of such waves due to depth variations in N and latitudinal variations in f will be discussed in Section 9. The wave patterns generated by a moving disturbance in a rotating stratified fluid have been examined by Redekopp (1975) (see Section 53).

The energetics of plane waves

Let us define the vertical displacement ξ of a particle from its equilibrium level through

$$w = \xi_t.$$

Integrating (8.2) with respect to time, we find that the density perturbation is directly proportional to the displacement:

$$\rho = \rho_0 N^2 \xi / g. \tag{8.57}$$

Taking the scalar product of (8.1) with $\boldsymbol{u}$, and substituting for w and ρ from (8.56) and (8.57) we obtain an energy conservation law

$$\partial_t [\tfrac{1}{2}\rho_0(\boldsymbol{u} \cdot \boldsymbol{u} + N^2\xi^2)] + \nabla \cdot (p\boldsymbol{u}) = 0, \tag{8.58}$$

which is a special case of (4.10) for small-amplitude motions in stratified fluid.

The quantities $\rho_0(\boldsymbol{u} \cdot \boldsymbol{u})/2$ and $\rho_0 N^2 \xi^2/2$ are respectively the kinetic and potential parts of the energy density of the wave motion. The energy flux $p\boldsymbol{u}$ is defined only up to an arbitrary nondivergent function. This means that although the average value of $p\boldsymbol{u}$ may not coincide in direction with the group velocity of plane waves, the two estimates of the energy flux, $p\boldsymbol{u}$ and $\langle E \rangle \boldsymbol{c}_g$ may be reconciled through the addition of the arbitrary flux vector $\boldsymbol{F}$ satisfying $\nabla \cdot \boldsymbol{F} = 0$.

It is the average rather than the fluctuating part of the energy density and flux which are of primary interest. In order to construct correct averages of complex quantities, we note that if

$$\phi = A\, e^{iS} \tag{8.59}$$

where $S = \boldsymbol{k} \cdot \boldsymbol{x} - \omega t$ is the phase function and A is a complex amplitude constant, then

$$\mathrm{Re}\,\{\phi\} = \tfrac{1}{2}(\phi + \phi^*) = \tfrac{1}{2}[A\,\mathrm{e}^{iS} + A^*\,\mathrm{e}^{-iS}].$$

Defining the average over a complete cycle of the phase S through the operation

$$\langle\psi\rangle = \frac{1}{2\pi}\int_0^{2\pi} \psi\,\mathrm{d}S, \tag{8.60}$$

it follows that the average value of the square of the real part of ϕ is given by

$$\langle(\mathrm{Re}\{\phi\})^2\rangle = \tfrac{1}{2}AA^*. \tag{8.61}$$

The average energy density $\langle E\rangle$ is then given by

$$\langle E\rangle = \frac{\rho_0}{4}\left[U\cdot U^* + \frac{N^2}{\omega^2}ww^*\right]. \tag{8.62}$$

Under the Boussinesq approximation, $\langle E\rangle$ is readily evaluated for plane waves from the velocity components in the $(\hat{\mathbf{k}}, \hat{\mathbf{a}}, \hat{\mathbf{b}})$ coordinate system, using (8.41) and (8.42):

$$\langle E\rangle = \rho_*\frac{U_aU_a^*}{4}\left[\left(1 + \frac{4\Omega^2\cos^2\alpha}{\omega^2}\right) + \frac{N^2\cos^2\theta}{\omega^2}\right]. \tag{8.63}$$

The term in the first bracket is proportional to the mean kinetic energy density; the second term to the mean gravitational potential energy density. From the dispersion relation (8.44),

$$\langle E\rangle = \rho_* U_a U_a^*/2. \tag{8.64}$$

The ratio of mean gravitational potential to mean kinetic energy $\langle E_P\rangle/\langle E_K\rangle$ is given by

$$\frac{\langle E_P\rangle}{\langle E_K\rangle} = \left(\frac{N^2\cos^2\theta}{\omega^2 + 4\Omega^2\cos^2\alpha}\right), \tag{8.65}$$

which is generally less than unity. Equipartition between gravitational potential energy and kinetic energy is not found in a rotating fluid. If, on the other hand, we interpret $\rho_* U_a U_a^* \Omega^2\cos^2\alpha/\omega^2$ as a rotational potential energy, associated with the stretching of the vortex lines of the uniform rotation field by small velocity perturbations, equipartition between the total potential energy and the kinetic energy holds.

The energy flux $\langle p\boldsymbol{u}\rangle$ averages to

$$\langle p\boldsymbol{u}\rangle = \tfrac{1}{4}(\Pi\boldsymbol{U}^* + \Pi^*\boldsymbol{U}), \tag{8.66}$$

where Π is the (complex) amplitude of the fluctuating pressure. The pressure may be evaluated in terms of the components of $\boldsymbol{U}$ from the $\boldsymbol{k}$-component of the momentum equation (8.37). The calculation of the energy flux is left as an exercise to the reader.

Let us take a closer look at the upper ocean approximation $\tilde{f} = 0$. The basic equations (8.1) and (8.2), may be manipulated to express the velocity components in terms of pressure gradients as

$$\rho_0(\partial_{tt} + f^2)u = -(\partial_{xt} + f\partial_y)p, \tag{8.67}$$

$$\rho_0(\partial_{tt} + f^2)v = -(\partial_{yt} - f\partial_x)p, \tag{8.68}$$

$$\rho_0(\partial_{tt} + N^2)w = -p_{zt}. \tag{8.69}$$

The amplitudes of pressure (Π) and velocity (U) fluctuations for plane waves are then related by

$$\rho_0 U = (\omega k_1 + ifk_2)\Pi/(\omega^2 - f^2), \tag{8.70}$$

$$\rho_0 V = (\omega k_2 - ifk_1)\Pi/(\omega^2 - f^2), \tag{8.71}$$

$$\rho_0 W = -\omega k_3 \Pi/(N^2 - \omega^2). \tag{8.72}$$

The total mean energy density, evaluated either from (8.62) or directly from the general case (8.64), is under the Boussinesq approximation:

$$\langle E \rangle = \frac{\rho_*}{2} \frac{k^2}{(k_1^2 + k_2^2)} WW^*. \tag{8.73}$$

Using (8.70)–(8.72) and (8.56), the ratio of vertical kinetic energy to horizontal kinetic energy becomes

$$\frac{WW^*}{UU^* + VV^*} = \frac{\omega^2(\omega^2 - f^2)}{(\omega^2 + f^2)(N^2 - \omega^2)}. \tag{8.74}$$

This ratio vanishes at $\omega = f$, at which frequency the motion is purely horizontal; in contrast, at $\omega = N$, the kinetic energy is all contained in the vertical motion. As inferred from (8.65), the ratio of gravitational potential energy to kinetic energy is less than unity; in the case $\tilde{f} = 0$,

$$\frac{\langle E_P \rangle}{\langle E_K \rangle} = \frac{N^2(\omega^2 - f^2)}{\omega^2(N^2 - f^2) + f^2(N^2 - \omega^2)}. \tag{8.75}$$

Observed ratios of $\langle E_P \rangle$ to total mean kinetic energy and to its horizontal component are shown in Fig. 8.7. Observations of this nature were discussed by Fofonoff (1969); a series

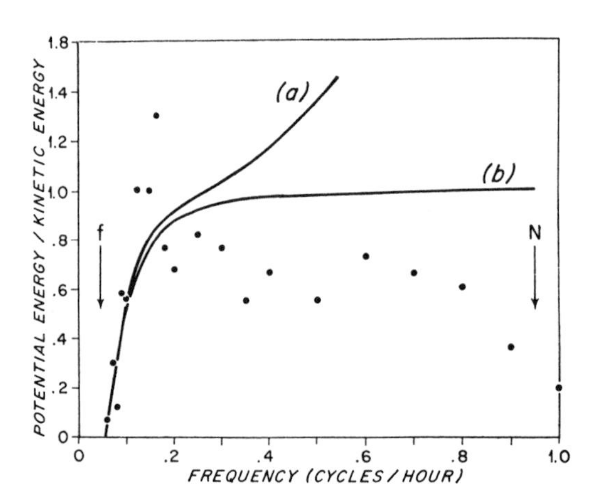

Fig. 8.7. The ratio of gravitational potential energy to horizontal kinetic energy (a) and to total kinetic energy (b) for internal waves, as deduced from (8.74) and (8.75). The points are estimates of the ratio of gravitational potential energy to horizontal kinetic energy based on current measurements by Voorhis (1968). (From Fofonoff, 1969.)

of consistency relations similar to (8.74) and (8.75), but relating all the variables of an internal wave field have been presented by Müller and Siedler (1976). A synthesis of the observations has been used by Garrett and Munk (1972b, 1975) to construct an empirical spectrum of oceanic internal waves; this spectrum will be discussed in Section 53.

From the relations (8.80) to (8.72) between U and Π, the energy flux, as given by (8.66), becomes

$$\langle pu \rangle = \frac{\rho_*(N^2 - \omega^2)WW^*}{2\omega(k_1^2 + k_2^2)} \left[k_1, k_2, \frac{-k_3(\omega^2 - f^2)}{(N^2 - \omega^2)} \right]. \tag{8.76}$$

Using (8.73) for $\langle E \rangle$ and (8.52), with $\tilde{f} = 0$, for $\boldsymbol{c}_g$, we readily establish that

$$\langle pu \rangle = \langle E \rangle \boldsymbol{c}_g, \tag{8.77}$$

i.e. the energy travels at the group velocity.

Exercises Section 8

1. Using the momentum equation (8.13) show that particle orbits in gyroscopic waves are always traced in a sense opposite to that of the rotation of the fluid as a whole.
2. Show, using (8.49) and (8.52) that $\boldsymbol{c} \cdot \boldsymbol{c}_g = 0$ for gyroscopic–internal gravity waves, under the Boussinesq approximation.
3. Calculate the frequency range and sketch the surfaces of constant ω, as done in Fig. 8.5, for the case $\nu = -\pi/4$, where ν is the rotation angle given by (8.48).
4. Obtain a dispersion relation for small-amplitude waves in a rotating stratified fluid without using the Boussinesq approximation.
5. Prove that ω_a and ω_b, as given in (8.50), are indeed the maximum and minimum frequencies for which the solution of (8.49) consists of real values of $\kappa_1, \kappa_2, \kappa_3$.
6. Using (8.66) calculate the energy flux, under the Boussinesq approximation for gyroscopic internal gravity waves when $f = 0$, $\tilde{f} \neq 0$, i.e. at the equator. Show that $\langle pu \rangle = \langle E \rangle \boldsymbol{c}_g$.
7. Show that plane gyroscopic–internal gravity waves are *exact* solutions of the non-linear adiabatic equations (3.9)–(3.11).
8. Calculate the co-spectra $\langle uv \rangle$, $\langle uw \rangle$, $\langle vw \rangle$, for plane waves using (8.70)–(8.72). For example, $\langle uv \rangle = k_1 k_2 \Pi\Pi^*/4(\omega^2 - f^2)\rho_0^2$.
9. Show that for small-amplitude waves in a Boussinesq fluid the following conservation law holds:

$$\frac{\partial}{\partial t}\left\{\frac{1}{2}\left[\frac{(\nabla \times \boldsymbol{u})^2}{N^2} + (\nabla\xi)^2\right]\right\} - \nabla \cdot \left[\xi \frac{\partial \boldsymbol{u}}{\partial z}\right] = 0,$$

where ξ is the vertical displacement of the fluid from its equilibrium level (Milder, 1976).

9. REFLECTION AND REFRACTION OF GYROSCOPIC–INTERNAL GRAVITY WAVES

In order to adapt the plane-wave solutions developed in the previous section to the constraints of a bounded ocean, it is necessary to examine how gyroscopic–internal gravity waves reflect from oceanic boundaries. Rays emanating from a source within the ocean eventually reach the bottom or the air–sea interface; how are these rays reflected?

Let us restrict our attention to a uniformly and strongly stratified ($N = \text{constant} \gg 2\Omega$) Boussinesq fluid, for which the plane-wave dispersion relation is given by (8.56). Only reflection from plane surfaces is considered here; scattering from rough surfaces may be examined by the stochastic methods of Section 34. The case of strong stratification is equivalent to taking $\tilde{f} \equiv 0$; in that case, there is complete horizontal symmetry and we can choose the reflecting surface as the plane

$$z - \alpha x = 0, \tag{9.1}$$

where $\alpha > 0$. The boundary contains the vector $\boldsymbol{s} = (1, 0, \alpha)$ and has its outward (into the fluid) normal along $\boldsymbol{n} = (-\alpha, 0, 1)$. The wavenumber $\boldsymbol{k}$ of the incident wave is restricted to lie entirely in the (x, z)-plane ($k_2 = 0$). This special choice for the orientation of the wavenumber vector simplifies the algebra while illustrating the relevant features of the reflection process. As will be seen later, the generalization to an arbitrary plane of incidence is straightforward. The case $\tilde{f} \neq 0$ for an ocean with horizontal boundaries will be considered in the next section; the more general case of sloping boundaries is left as an exercise to the reader.

When $k_2 = 0$, the dispersion relation (8.56) for the strongly stratified ocean reduces to

$$k_3^2/k_1^2 = (N^2 - \omega^2)/(\omega^2 - f^2) = R^2, \tag{9.2}$$

where we shall take $R > 0$: $k_3/k_1 = \pm R$. Since $\boldsymbol{c}$ and $\boldsymbol{c}_g$ are orthogonal, it follows that the slope of the rays (in the direction $\boldsymbol{c}_g$) has magnitude $1/R$.

Reflection by a rigid wall

Reflection by a rigid wall is most naturally described in terms of the velocity components. From (8.70)–(8.72),

$$U = -(k_3/k_1)W, \tag{9.3}$$

$$V = (ifk_3/\omega k_1)W, \tag{9.4}$$

$$\Pi = \rho_0[(\omega^2 - N^2)/\omega k_3]W. \tag{9.5}$$

The incident wave is a plane wave for which $\boldsymbol{c}_g$ points towards the reflecting boundary (Fig. 9.1). Using the subscript i for "incident", we write:

$$W_i = W_0 \exp\,[i(k_1 x + k_3 z - \omega t)]. \tag{9.6}$$

For the reflected wave $\boldsymbol{c}_g$ points away from the reflecting boundary. We write (with a subscript r for "reflected"):

$$W_r = AW_0 \exp\,[i(l_1 x + l_3 z - \sigma t)], \tag{9.7}$$

where A is the amplitude reflection coefficient. On the boundary (9.1), the normal velocity must vanish: $\boldsymbol{U} \cdot \boldsymbol{n} = 0$, which takes the form

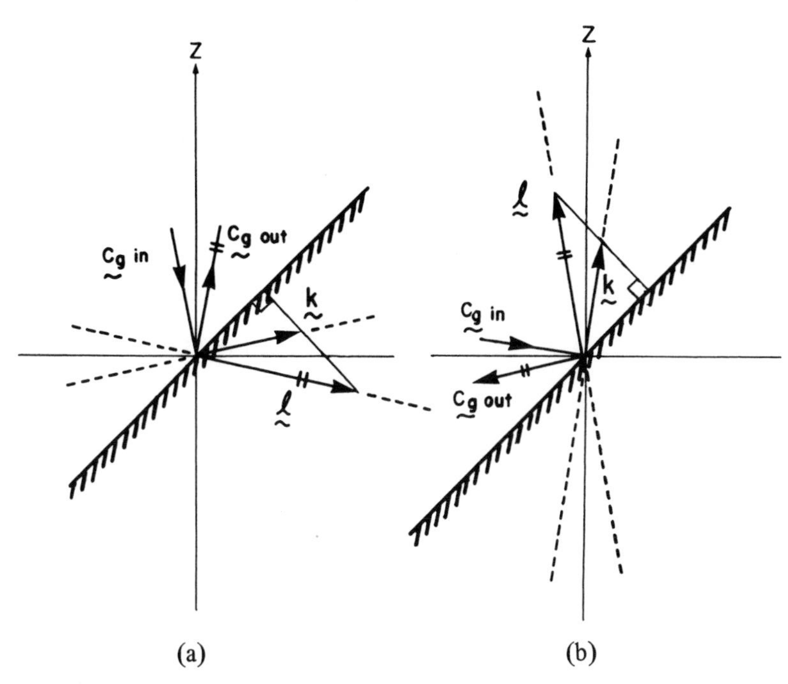

Fig. 9.1. The reflection of plane waves from a plane boundary in a strongly stratified rotating fluid. The reader should refer to Fig. 8.3b for a depiction of the relative directions of phase and group velocities. (a) The high-frequency transmissive case $\alpha R < 1$, for which the slope of the rays is greater than that of the bottom. (b) The low-frequency reflective case, where $\alpha R > 1$. The magnitude and the direction of the incident and reflected wavenumbers are shown, but only the directions of the incident and reflected group velocities are indicated.

$$W_i + W_r - \alpha(U_i + U_r) = 0. \tag{9.8}$$

Using (9.1) and (9.3), together with the assumed forms for W_i and W_r, this condition becomes

$$(1 + \alpha k_3/k_1) \exp\{i[(k_1 + \alpha k_3)x - \omega t]\} + A(1 + \alpha l_3/l_1) \exp\{i[(l_1 + \alpha l_3)x - \sigma t]\} = 0 \tag{9.9}$$

which must hold for all values of x and t. This is patently impossible unless

$$\sigma = \omega, \tag{9.10}$$

and $\quad l_1 + \alpha l_3 = k_1 + \alpha k_3, \qquad (9.11)$

i.e., $\quad \boldsymbol{l} \cdot \boldsymbol{s} = \boldsymbol{k} \cdot \boldsymbol{s}$.

The relation (9.10) merely equates the frequency of the reflected wave to that of the incident wave, a fact that we could have guessed ab initio. The second relation, (9.11), imposes a conservation of crests along the boundary by equating the incident and reflected wavenumber components parallel to the wall. Since $\omega = \sigma$, the reflected wave components l_1 and l_3 satisfy the same dispersion relation (9.2) obeyed by k_1 and k_3, a result which we may express as

$$k_3 = \pm Rk_1; \quad l_3 = \pm Rl_1. \tag{9.12}$$

How are the signs to be chosen in (9.12)? If like signs are selected in the two relations, substitution of (9.12) into (9.11) implies that $k_1 = l_1$ and hence, again from (9.11), $k_3 = l_3$; the reflected wave then propagates exactly in the same direction as the incident wave. This is surely not the correct solution and we must conclude that opposite signs must be picked:

$$k_3 = Rk_1; \quad l_3 = -Rl_1 \tag{9.13a}$$

$$\text{or} \quad k_3 = -Rk_1; \quad l_3 = Rl_1. \tag{9.13b}$$

This argument is confirmed by geometrical construction, as shown in Fig. 9.1. Given an incident wave with wavenumber $\boldsymbol{k} = (k_1, 0, k_3)$ and amplitude W_0, equations (9.9)–(9.11) and (9.13) imply that the reflected wave has wavenumber components $(l_1, 0, l_3)$ and amplitude AW_0, where

$$l_1 = \frac{(1 \pm \alpha R)}{(1 \mp \alpha R)} k_1, \tag{9.14a}$$

$$l_3 = -\frac{(1 \pm \alpha R)}{(1 \mp \alpha R)} k_3, \tag{9.14b}$$

$$A = -(1 \pm \alpha R)/(1 \mp \alpha R). \tag{9.15}$$

We notice that the reflection coefficient A is not, in general, of unit magnitude. As the magnitude of the wavenumber changes upon reflection, so does the magnitude of the group velocity; furthermore the beam width (as measured by the shortest distance between two parallel rays) is also altered, and the amplitude reflection coefficient cannot be equal to unity if the out-going energy flux is to be equal to the incident energy flux. The reader may convince himself that energy is indeed conserved by working out Exercise 9.1. From (9.15), it is only for a horizontal boundary ($\alpha = 0$) or for the limiting cases $R = 0$ ($\omega = N$) and $R = \infty$ ($\omega = f$) that $|A| = 1$. Neither the magnitude of the wavenumber vector nor the beam width are modified by reflection in that case.

When $\alpha R < 1$, i.e., when the magnitude of the bottom slope is less than that of the rays, l_1 and k_1 are of the same sign and l_3 and k_3 of opposite signs, as shown in Fig. 9.1a. In that case, the group velocity reflects about the vertical axis, the reflected and incident rays making equal angles about the vertical. Energy may propagate onto the reflecting plane from either the $+x$ or $-x$ side and the horizontal component of the energy flux has the same direction after reflection as it had before. The bottom is said to be horizontally transmissive for the upper frequency range for which

$$0 \leqslant \alpha R < 1, \text{i.e.} \; (\alpha^2 N^2 + f^2)/(1 + \alpha^2) < \omega^2 \leqslant N^2.$$

For $\alpha R > 1$ on the other hand, energy may be incident only from the $-x$ side, as shown in Fig. 9.1b. Reflection of the rays now takes place about a horizontal axis, and the direction of the x-component of the energy flux is reversed. For such slopes, i.e. for $1 < \alpha R \leqslant \infty$, the bottom is said to be horizontally reflective. This case covers the lower frequency range

$$f^2 \leqslant \omega^2 < (\alpha^2 N^2 + f^2)/(1 + \alpha^2).$$

The terms transmissive and reflective have been introduced with horizontal energy propagation in mind: as the ocean is thin (i.e. $H/L \ll 1$), this is usually the direction of interest. If we consider the behaviour of the vertical components of the rays, however, it is clear from Fig. 9.1 that the attribution of the qualifiers should be reversed.

The reflection process just described may seem a bit strange to anyone familiar only with the analogous phenomenon in optics or acoustics. In place of Snell's law on the equality of the angles made by the rays with respect to the normal of the reflecting surface, we find that this equality occurs for angles about the vertical (for $\alpha R < 1$) or about the horizontal direction (for $\alpha R > 1$). The wavenumber vectors reflect about the horizontal for $\alpha R < 1$ and about the vertical when $\alpha R > 1$, but may both point into the reflecting plane, as in Fig. 9.1a, or out of it as in Fig. 9.1b. Further, to satisfy (9.11), the magnitudes of the incident and reflected wavenumbers must, in general, be different. Hence the reflection process provides a mechanism for energy exchange between different scales of motion. The consequences of this exchange process have been examined by Phillips (1963a).

Reflection by the sea surface

At the air–sea interface $z = \eta(x, y, t)$, we apply the boundary conditions (3.13) and (3.14):

$$p_0(\eta) + p(x, y, \eta, t) = p_a, \tag{9.16a}$$

$$w(x, y, \eta, t) = \frac{D\eta}{Dt}, \tag{9.16b}$$

where p_0 is the hydrostatic part of the pressure and p the perturbation due to the wave motion. For free, unforced waves, the atmospheric pressure p_a is taken to be a constant. The conditions (9.16) are awkward to apply because of their nonlinearity and because they hold on a moving boundary, the position of which is unknown and must be found as part of the solution of the reflection problem. For small-amplitude waves of the type considered here, these difficulties are easily circumvented by expanding p and w about their values at the equilibrium position of the surface $z = 0$. Thus,

$$p_0(0) + \eta p_{0z}(0) + \ldots + p(x, y, 0, t) + \eta p_z(x, y, 0, t) + \ldots = p_a, \tag{9.17a}$$

$$w(x, y, 0, t) + \eta w_z(x, y, 0, t) + \ldots = \eta_t + \ldots. \tag{9.17b}$$

The nonlinear terms in $D\eta/Dt$ are dropped on the basis of the same linearizing assumption used to write (8.1)–(8.3), i.e., the phase speed is much greater than the particle speed. A further linearization of the expanded surface boundary conditions (9.17) involves a separate assumption on the smallness of the surface displacement. These matters will be considered in detail in the discussion of nonlinear effects in Section 12. Using (5.2) for the hydrostatic pressure gradient p_{0z}, (9.17) is linearized into

$$p_0(0) + p(x, y, 0, t) - \rho_0 g\eta = p_a, \tag{9.18a}$$

$$w(x, y, 0, t) = \eta_t. \tag{9.18b}$$

Eliminating η, these two conditions may be combined as

$$p_t - \rho_0 g w = 0 \text{ at } z = 0, \tag{9.19}$$

which may in turn be expressed solely in terms of the pressure through the use of (8.69). The result is

$$(\partial_{tt} + N^2)p + g p_z = 0 \text{ at } z = 0. \tag{9.20}$$

In a strongly stratified fluid, $\omega^2 < N^2$ (see 8.55a); furthermore, to remain consistent with the Boussinesq approximation $N^2 p \ll g p_z$. Hence, for gyroscopic–internal gravity waves in a Boussinesq fluid, (9.20) reduces to

$$p_z = 0, \tag{9.21}$$

which, from (8.69), is equivalent to $w = 0$. A free surface is indistinguishable from a rigid wall under these conditions. With the incident and reflected waves as given by (9.6) and (9.7), and a horizontal boundary ($\alpha = 0$ in 9.1), (9.14) and (9.15) reduce to

$$l_1 = k_1; \quad l_3 = -k_3; \quad A = -1. \tag{9.22}$$

A horizontal surface, be it rigid or free, is always horizontally transmissive, since $k_1 = l_1$. As $A = -1$, the waves suffer a phase change of 180° upon reflection.

For the more general case of propagation in a plane which does not coincide with the (x, z)-plane ($k_2 \neq 0$, in the above formulation) the dispersion relation (9.2) becomes

$$k_3^2/(k_1^2 + k_2^2) = R_2. \tag{9.23}$$

Any boundary condition on $z - \alpha x = 0$ now imposes the additional condition

$$k_2 = l_2. \tag{9.24}$$

Using (9.10), (9.11) and (9.24) as well as the dispersion relation (9.23) satisfied by both $\boldsymbol{l}$ and $\boldsymbol{k}$, it is then possible to solve for the components of $\boldsymbol{l}$ in terms of those of $\boldsymbol{k}$. The details are left as an exercise.

When $\tilde{f} \neq 0$, a similar procedure is followed, except that the general dispersion relation (8.46) must now be used to solve for k_3 and l_3. Reflection from horizontal boundaries for this case is treated in the next section.

The above reflection rules will be applicable to cases where N varies with z only provided the scale depth of the variation of N is well in excess of the vertical wavelength: $N_z \ll N k_3$. Furthermore, for variable $N(z)$ and $f(y)$, refraction occurs and the rays become curved. The solution between a pair of reflecting boundaries cannot be constructed without taking the curvature of the rays into account.

Refraction

The refraction of plane waves by inhomogeneities in f and N is most simply treated by the methods of ray theory presented in Section 6. Examples are given by Hughes (1964) and Kroll (1975). We limit our attention to two special cases which illustrate the influence of N and f variations separately.

Consider first waves emanating from a point at a depth $z = -d$ in a strongly stratified fluid in which $N(z)$ is a monotonic increasing function of z and $f =$ constant. Since the medium varies only in the z-direction, ω, k_1 and k_2 are invariant along a ray according

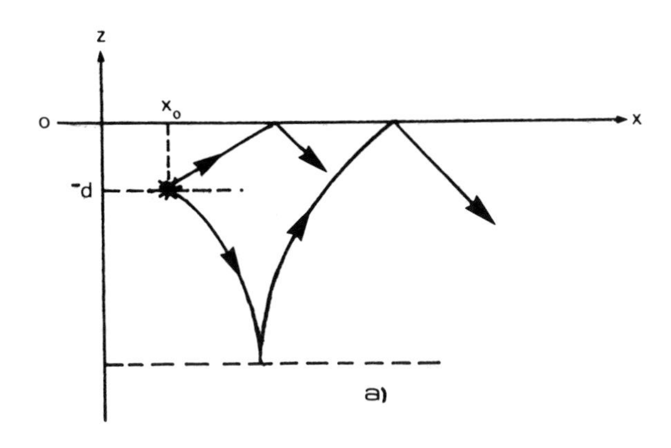

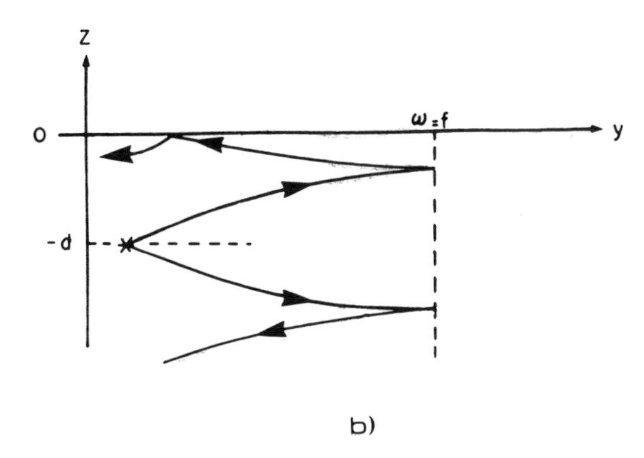

Fig. 9.2. Wave refraction in a nonhomogeneously stratified rotating fluid. (a) $f =$ constant, $N_z > 0$; only two rays are depicted, for simplicity. (b) $N =$ constant, $f = f_0 + \beta y$; the rays are turned back towards the equator at the critical latitude where $\omega = f$.

to (6.15). We choose $k_2 = 0$ for simplicity. The relevant dispersion relation is then (8.56):

$$k_1^2/k_3^2 = (\omega^2 - f^2)/(N^2 - \omega^2), \tag{9.25}$$

and the components of the group velocity are given by

$$\boldsymbol{c}_g = [k_1, 0, -k_3(k_1^2/k_3^2)](N^2 - \omega^2)/\omega k^2. \tag{9.26}$$

The direction of a ray is that of $\boldsymbol{c}_g$.

Consider a ray pointing down into the fluid (i.e. one with $k_3 > 0$). Along this ray, N^2 decreases and hence $(N^2 - \omega^2)$ decreases; from (9.25), k_3 decreases. From (9.26), the ray is refracted towards the vertical, as shown in Fig. 9.2a. No wave energy penetrates beyond the depth given by $N(z) = \omega$, where both components of the group velocity vanish, and where the wave is reflected back up.

Refraction in the β-plane, where $f = f_0 + \beta y$, and with $N =$ constant, may be analyzed in a similar fashion. From (6.15), it now follows that ω, k_1 and k_3 are invariant along a ray. Let us now choose $k_1 = 0$; then the dispersion relation (8.56) reduces to

$$k_2^2/k_3^2 = (\omega^2 - f^2)/(N^2 - \omega^2) \tag{9.27}$$

and the group velocity becomes

$$c_g = [0, k_2, -k_3(k_2^2/k_3^2)](N^2 - \omega^2)/\omega k^2. \tag{9.28}$$

Along a ray pointing north ($k_2 > 0$), f^2 increases and $(\omega^2 - f^2)$ decreases; from (9.27), k_2 decreases. According to (9.28), this ray is refracted towards the horizontal and, as shown in Fig. 9.2b, is turned back at a "critical latitude", where $\omega^2 = f^2$.

Applications of ray theory to wave propagation in the presence of topographic features such as seamounts, ridges or continental shelves may be found in the works of Magaard (1962), Sandstrom (1969) and Baines (1974). An active area of research using the methods of ray theory has been the study of generation of internal waves of tidal frequency at continental boundaries. Recent observations such as those of Wunsch and Hendry (1972) and of Gould and McKee (1973) have shown that internal tides have large amplitudes in the vicinity of continental slopes, and contribute a significant fraction of the horizontal velocity field at tidal frequencies. The existence of internal tides in the ocean has long been known (Fjeldstad, 1933) and it was quickly recognized (Zeilon, 1934) that such internal oscillations would readily be generated at continental shelves. The first theories elaborated by Rattray (1960) and Rattray et al. (1969) considered the interaction of the surface tide with a step-like shelf. With a flat bottom on each side of a depth discontinuity, the solution can either be written in terms of normal modes (see Section 10) on each side, or represented by a beam of rays emanating from the depth discontinuity, as shown in Fig. 9.3. The wave energy is all contained in the beam of rays and the shaded areas in Fig. 9.3 are shadow zones where, except for noise, no internal wave energy should be found. The theory has been refined to include dissipation (Prinsenberg et al., 1974) and sloping continental slopes (Prinsenberg and Rattray, 1975). However, efforts to measure the theoretically predicted beam patterns in the field have not been particularly successful (Barbee et al., 1975; Schott, 1977).

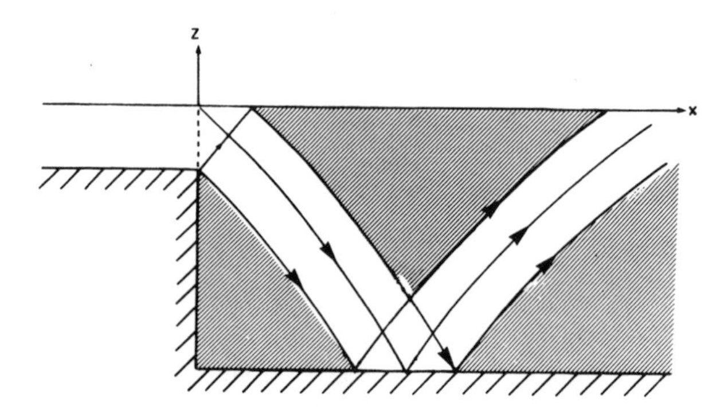

Fig. 9.3. Internal wave generation at a depth discontinuity. Rays penetrate the deep ocean and the shelf area (where the rays are not shown), carrying the energy of a perturbation occurring at the shelf break. The shaded areas are shadow zones, where no wave energy is to be expected.

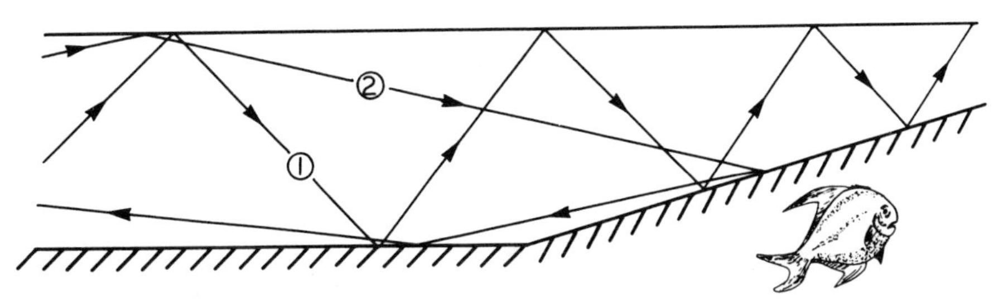

Fig. 9.4. Wave propagation in the presence of a bottom slope. The bottom is everywhere horizontally transmissive for ray (*1*) since the bottom slope is inferior to that of the rays. For ray (*2*), corresponding to a lower-frequency wave, the bottom is reflective and the wave is turned back.

Another application of the ray theory method bears on the problem of internal wave propagation up a sloping bottom. Rays sketched in Fig. 9.4 show how waves can propagate up a slope as long as the bottom is horizontally transmissive. Reflection back into deeper water occurs when the slope steepens and the bottom becomes horizontally reflective. As the orientation of the rays depends on frequency, it is evident that a bottom slope which is transmissive for a certain frequency may be reflective for another, lower frequency. These properties may be exploited in model studies to construct nonreflecting beaches (using a small enough, transmissive slope) or geometrical filters (a large slope section will act as a high-frequency pass filter for gyroscopic–internal gravity waves). The problem of internal wave propagation along a linear bottom slope has also been solved analytically by Wunsch (1968) for a fluid with a constant Brunt-Väisälä frequency. Solutions have also been found for a wide class of bottom profiles by Magaard (1962), Manton and Mysak (1971) and by Sandstrom (1976), using a functional equation method. A laboratory study of internal wave propagation over a slope in a nonrotating fluid by Cacchione and Wunsch (1974) shows that the linear theory outlined above is in good agreement with observations in the transmissive case $\alpha R < 1$. With $\alpha R \geqslant 1$, however, nonlinear and viscous effects become important and the linear ray theory fails to provide an adequate description of the experimental results.

Exercises Section 9

1. Show [using (8.70), for example] that the component of mean energy flux normal to a rigid reflecting boundary, as in Fig. 9.1, is the same in magnitude (but opposite in sign) for the out-going and in-going waves.

2. Show that when $\tilde{f} = 0$, and reflection off the plane $z - \alpha x = 0$, the reflected wavenumber components may be written in terms of the incident wavenumber components as:

$$l_1 = \frac{(1+\alpha^2R^2)k_1 + 2\alpha k_3}{(1-\alpha^2R^2)},$$

$$l_2 = k_2,$$

$$l_3 = -\frac{(1+\alpha^2R^2)k_3 + 2\alpha R^2 k_1}{(1-\alpha^2R^2)}.$$

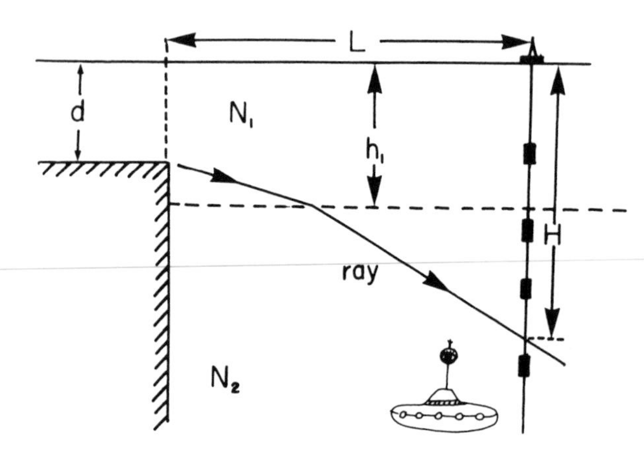

Fig. 9.5. The detection of an internal wave ray issuing from the edge of the continental shelf (cf. Exercise 9.4).

3. Let $\tilde{f} \neq 0$ and $k_1 = 0$; calculate the amplitude reflection coefficient A for plane gyroscopic–internal gravity waves incident on the surface $z - \alpha y = 0$.

4. Consider the geometrical situation shown in Fig. 9.5, where a shallow shelf (of depth $d = 100\,\text{m}$) is adjacent to a very deep basin. The oceanic stratification may be approximated by a two-layer system, with $N_1 = 2000^{1/2} f$ above a depth $h_1 = 200\,\text{m}$ and $N_2 = 10f$ below. The internal tides produced at the shelf break have a frequency $\omega = \sqrt{2}f$. A string of instruments is placed at a distance $L = 20\,\text{km}$ from the shelf break. At what depth H will the ray shown emanating from the shelf corner intercept the line of instruments? (Ans. 1769 m.) Suppose a strong storm comes along and enough upwelling occurs to bring about a 20% increase in N_1 near the shelf break: calculate the change in H. (Ans 91 m upwards.) Such variations in upper layer structure near the edge of the shelf present one of the difficulties inherent in predicting ray paths and internal wave structure off the shelf. (See for example Hayes and Halpern, 1976a.)

5. Consider a two-layer fluid system in rotation about an axis normal to a (plane) surface of density discontinuity. Given an incident plane wave on the upper side of the interface, construct the geometry of the refracted and reflected wavenumbers and group velocities and also obtain expressions for the amplitudes of these waves, (Scott, 1975.)

10. THE OCEAN OF UNIFORM DEPTH: NORMAL MODES

In an ocean of uniform depth H with small surface slopes, the top and bottom boundary conditions are applied on essentially parallel boundaries. This model ocean may then be treated as a waveguide where wave solutions satisfying both boundary conditions may be constructed by the superposition of simpler waves. In a uniformly stratified fluid (N = constant) on the f-plane (f = constant), this superposition may be effected in terms of a pair of plane waves which interfere at the top ($z = 0$) and at the bottom ($z = -H$) of the ocean in precisely the right fashion to satisfy the boundary conditions. When N varies with depth, the rays are curved and the interference pattern is not so readily constructed. In that case, it is simpler to use a normal mode representation by separating the vertical and horizontal structures of the wave field. Long waves in the presence of a variable f (the β-plane) will be treated in Chapter 3.

Uniform stratification: N = constant

Consider a uniformly (but not necessarily strongly) stratified Boussinesq fluid bounded by a rigid flat bottom at $z = -H$ and a free surface at $z = 0$. The time-dependent solution satisfying $w = 0$ at $z = -H$ and the condition (9.19) $[p_t - \rho_0 g w = 0]$ at $z = 0$ will be constructed by adding a pair of plane waves with wavenumbers $\boldsymbol{k}$ and $\boldsymbol{l}$ and frequencies ω and σ respectively. Each wave must satisfy the dispersion relation (8.46). In order to satisfy the surface and bottom boundary conditions everywhere on the horizontal bounding planes as well as for all values of the time, it is clear that the horizontal components of the wavenumbers and the frequencies of the two waves must be identical:

$$k_1 = l_1: \quad k_2 = l_2: \quad \omega = \sigma. \tag{10.1}$$

The vertical wavenumber components k_3 and l_3 must then be the two roots of (8.46) for given values of k_1 and k_2. Solving (8.46) for k_3, we find

$$k_3, l_3 = \delta \pm \gamma, \tag{10.2}$$

where $$\delta = f\tilde{f}k_2/(\omega^2 - f^2), \tag{10.3a}$$

$$\gamma = [(k_1^2 + k_2^2)\mu - k_1^2\omega^2\tilde{f}^2]^{1/2}/(\omega^2 - f^2), \tag{10.3b}$$

$$\mu = (\omega^2 - f^2)(N^2 - \omega^2) + \omega^2\tilde{f}^2. \tag{10.3c}$$

Each dependent variable describing the motion may be written in the form

$$\phi = \phi_0 \exp[i(k_1 x + k_2 y + \delta z - \omega t)][e^{i\gamma z} + A e^{-i\gamma z}]. \tag{10.4}$$

Since the bottom boundary condition necessitates

$$w(x, y, -H, t) = 0, \tag{10.5}$$

the constant A must take the value

$$A = -e^{-2i\gamma H},$$

when $\phi = w$, and we can write the vertical velocity due to the superposition of both plane waves as

$$w = w_0 \sin[\gamma(z+H)] \exp[i(k_1 x + k_2 y + \delta z - \omega t)]. \tag{10.6}$$

At the free surface, the pressure and the vertical velocity are related through (9.19); with a harmonic dependence $e^{-i\omega t}$, this condition becomes

$$i\omega p + \rho_* g w = 0. \tag{10.7}$$

In order to relate p and w within the body of the fluid, we proceed to eliminate u and v from (8.1) and (8.2) to obtain (the derivation is left as an exercise)

$$\rho_* [\tilde{f}\partial_{tt} + (\partial_{tt} + f^2)(\partial_{tt} + N^2)]w = -[\tilde{f}\partial_t(\partial_{xt} + f\partial_y) + (\partial_{tt} + f^2)\partial_{zt}]p. \tag{10.8}$$

For plane-wave solutions, (10.8) becomes

$$[(\omega^2 - f^2)\partial_z - \tilde{f}(\omega k_1 + ifk_2)]p = -(i\mu\rho_*/\omega)w, \tag{10.9}$$

where μ is defined by (10.3c). Let $\Pi(z)$ be the z-dependent part of the pressure perturbation. Substituting for w from (10.6), (10.9) takes the form of an ordinary differential equation for $\Pi(z)$ which is readily integrated to yield

$$\Pi(z) = \frac{-\mu\rho_* w_0 e^{i\delta z}}{2\omega(\omega^2 - f^2)} \left[\frac{e^{i\gamma(z+H)}}{(i\gamma + i\delta - s)} - \frac{e^{-i\gamma(z+H)}}{(-i\gamma + i\delta - s)} \right], \tag{10.10}$$

where $\quad s = \tilde{f}(\omega k_1 + ifk_2)/(\omega^2 - f^2)$.

Using the definitions of δ, γ, μ, and s, (10.10) may be rewritten more elegantly as

$$\Pi(z) = \frac{i(\omega^2 - f^2)\rho_* w_0 e^{i\delta z}}{\omega(k_1^2 + k_2^2)} \left\{ \frac{\tilde{f}\omega k_1}{(\omega^2 - f^2)} \sin[\gamma(z+H)] + \gamma \cos[\gamma(z+H)] \right\}. \tag{10.11}$$

Finally, substituting for p and w into (10.7), we obtain a dispersion relation between the frequency ω and the horizontal components of the wavenumber for the combined waves:

$$\frac{\tan \gamma H}{\gamma H} = \frac{(\omega^2 - f^2)}{[g(k_1^2 + k_2^2) - \tilde{f}\omega k_1]H}. \tag{10.12}$$

Since γ is a function of both ω and $\boldsymbol{k}_h = (k_1, k_2)$ (see 10.3b), the relation (10.12) between ω and $\boldsymbol{k}_h$ is transcendental, making it impossible to find an explicit relation of the form $\omega = \omega(\boldsymbol{k}_h)$.

The relation (10.12) was first obtained (without recourse to the Boussinesq approximation) by Saint-Guily (1970). A special case applicable to a homogeneous ocean ($N = 0$) was derived by Johns (1965b).

A number of features of the solution are easily perceived from the dispersion relation (10.12) and from the general form of the variables. We first notice that the superposition of the two waves still leaves a net vertical phase propagation in the interference pattern; lines of constant phase appear to issue from one of the boundaries and propagate with a vertical velocity component given by

$$c_3 = \omega\delta/(k_1^2 + k_2^2 + \delta^2). \tag{10.13}$$

There is always a phase difference of

$$\delta H = f\tilde{f}k_2 H/(\omega^2 - f^2) \tag{10.14}$$

between the two bounding planes. This phase difference vanishes when propagation is entirely in the x-direction (east–west) or when $\tilde{f}=0$ or when $f=0$.

The dispersion relation (10.12), relating the horizontal wavenumber to the frequency, is best solved graphically. Let

$$(k_1, k_2) \;=\; k_h(\cos\theta_h, \sin\theta_h), \tag{10.15}$$

where θ_h is the angle which the horizontal component of the wavenumber (of magnitude k_h) makes with the x-axis. From (10.3b),

$$k_h \;=\; \gamma(\omega^2 - f^2)/(\mu - \omega^2\tilde{f}^2\cos^2\theta_h)^{1/2}. \tag{10.16}$$

Eliminating k_1 and k_2 in (10.12) by means of (10.15) and then using (10.16) to replace k_h in favor of γ, the dispersion relation (10.12) may be recast as

$$\tan\gamma H \;=\; \frac{[\mu - \omega^2\tilde{f}^2\cos^2\theta_h]}{g(\omega^2 - f^2)} \cdot \frac{H}{(\gamma - \gamma^*)H}, \tag{10.17}$$

where $\quad \gamma^* \;=\; \tilde{f}\omega\cos\theta_h(\mu - \omega^2\tilde{f}^2\cos^2\theta_h)^{1/2}/g(\omega^2 - f^2).$ (10.18)

In the frequency range $\omega_b \leqslant \omega \leqslant \omega_a$, where ω_a and ω_b are given by (8.50), the plane-wave dispersion relation (8.46) has real wavenumber solutions; therefore $\mu - \omega^2\tilde{f}^2\cos^2\theta_h \geqslant 0$ so that (10.16) is real. Outside that range of frequencies, γ is imaginary and may be written $\gamma = i\Gamma$; (10.17) then takes the form

$$\tanh\Gamma H \;=\; \frac{(\omega^2\tilde{f}^2\cos^2\theta_h - \mu)}{g(\omega^2 - f^2)} \; \frac{H}{(\Gamma - \Gamma^*)H}, \tag{10.19}$$

where $\Gamma^* = i\gamma^*$. The roots of (10.17) and (10.19) may be found graphically by plotting both sides of these equations as functions of γH (or ΓH) for given values of ω and θ_h. Note that the presence of the angle θ_h in the dispersion relation (10.19) implies that wave propagation is not isotropic in the horizontal plane, but depends (through θ_h) on the direction of propagation. This feature is to be attributed to the presence of the $\tilde{f}$ component of rotation, which has been retained here. The special case $\theta_h = \pi/2$ (northward propagation: $k_1 = 0, k_2 > 0$) is particularly simple, and will be examined first.

For $\theta_h = \pi/2, \gamma^* = 0$, and (10.17) reduces to

$$\tan\gamma H \;=\; H\mu/g(\omega^2 - f^2)\gamma H. \tag{10.20}$$

The positions of the roots of this equation are shown in Fig. 10.1. For $\omega_b < f < \omega \leqslant \omega_a$, $\gamma > 0$ and the roots are found on the curve labelled ω_1 in Fig. 10.1; for the range $\omega_b \leqslant \omega < f < \omega_a$, $\gamma < 0$ and the roots lie along the curve labelled ω_2. Each root may be labelled by the number of extrema of $\sin[\gamma(z+H)]$, the amplitude of the vertical velocity w (cf. 10.6). From the continuity equation (8.3), $w_z = 0$ implies $\boldsymbol{k}_h \cdot \boldsymbol{u}_h = 0$, so that an extremum of w (where $w_z = 0$) coincides with a zero-crossing of the horizontal velocity. In both of the above frequency ranges, there exists an infinity of roots, corresponding to a series of possible vertical normal modes. For high mode numbers, the roots occur at

$$|\gamma|H \;\simeq\; |n|\pi, \;\; |n| \;\gtrsim\; 3. \tag{10.21}$$

We note that the mode $n = 0$, with eigenvalue $|\gamma_0| < \pi/2$, occurs only in the upper frequency range $\omega_b < f < \omega \leqslant \omega_a$.

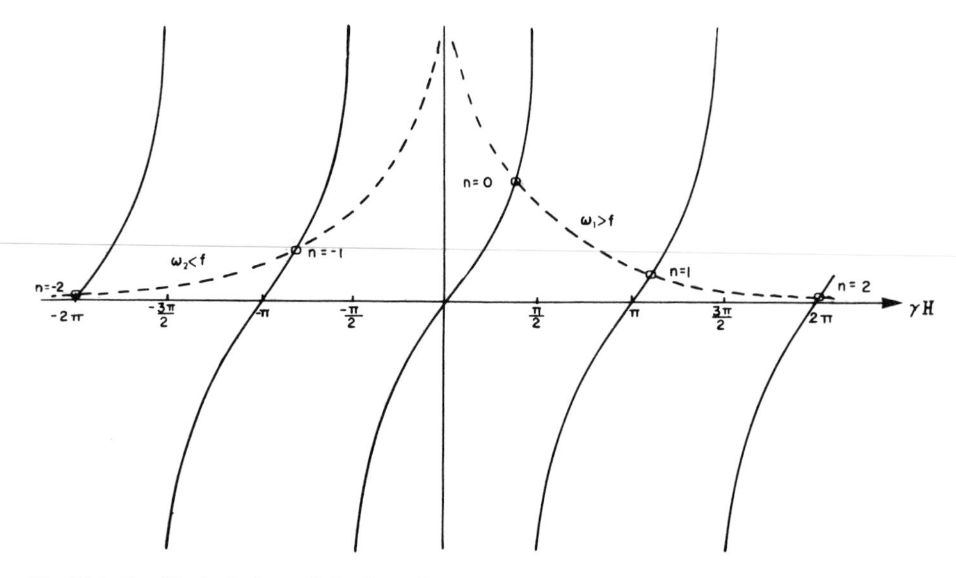

Fig. 10.1. Graphical solutions of the dispersion relation (10.20), valid in the frequency range $\omega_b < \omega < \omega_a$. The solid lines are the graph of tan γH; the broken curves that of $H\mu/g(\omega^2 - f^2)\gamma H$. The two solutions presented are for $\omega_b < f < \omega_1 < \omega_a (\gamma > 0)$ and for $\omega_b < \omega_2 < f < \omega_a (\gamma < 0)$. The roots γ_n occur at the intersections as indicated.

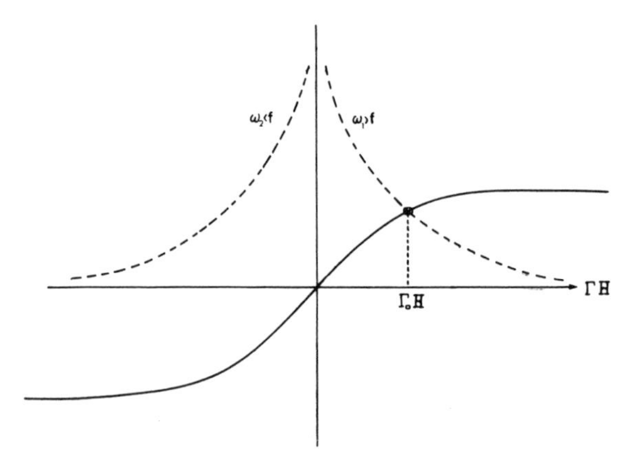

Fig. 10.2. Graphical solutions of the dispersion relation (10.22). The solid line shows tanh ΓH; the broken line, $-H\mu/g(\omega^2 - f^2)\Gamma H$, where $\mu < 0$ since ω lies outside the range $\omega_b \leqslant \omega \leqslant \omega_a$. The only root is found for $\omega > f\,(\Gamma > 0)$. The root Γ_0 is located at the intersection of the two curves. For $\omega < \omega_b < f$, there is no intersection and hence no solution.

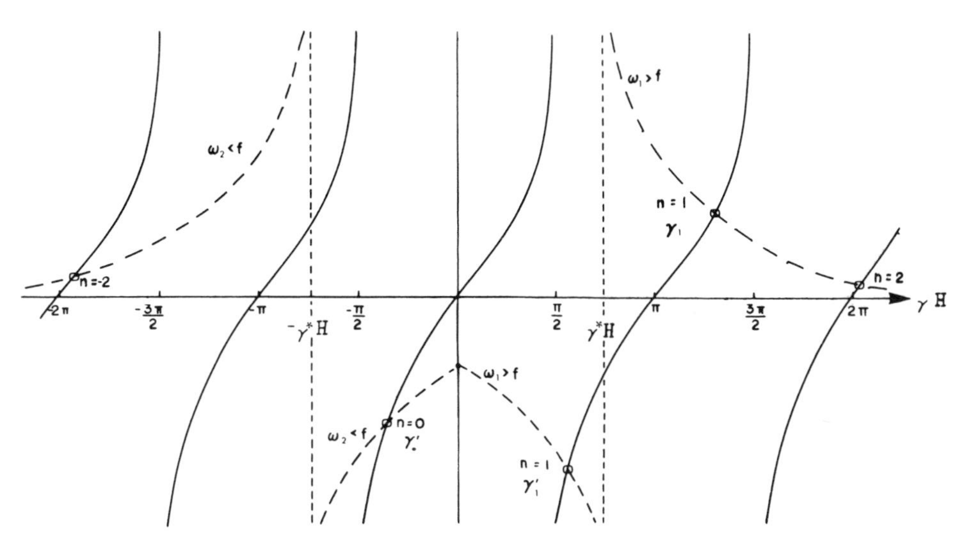

Fig. 10.3. Graphical solutions of (10.17) for the case $\tilde{f}\cos\theta_h > 0$ and $\frac{1}{2}\pi < |\gamma^*|H < \pi$. The broken line curve is now the right-hand side of (10.17) for two values of ω chosen on either side of f, as in Fig. 10.1.

Whenever $\omega > \omega_a$ or $\omega < \omega_b$, γ is imaginary, $\mu < 0$, and (still for $\theta_h = \pi/2$), (10.19) becomes

$$\tanh \Gamma H = -\mu/g(\omega^2 - f^2)\Gamma, \tag{10.22}$$

which has only one solution Γ_0 provided $\omega > f$ (Fig. 10.2). The amplitude of the vertical velocity is now of the form sinh $[\Gamma(z + H)]$ and has no extrema in $-H \leqslant z \leqslant 0$.

The horizontally propagating wave solutions in a simple flat-bottom oceanic wave-guide thus fall into two categories (for $\theta_h = \pi/2$). For $\omega > f$ there exists an $n = 0$ mode, for which the horizontal velocity has no zero-crossings in $-H \leqslant z \leqslant 0$; this mode is called the *barotropic* or *surface mode* since it is associated with the displacements of the free surface from its equilibrium position. Indeed, if the free surface is replaced by a rigid lid, on which $w(0) = 0$, (10.6) leads to the dispersion relation

$$\tan \gamma H = 0, \tag{10.23}$$

which has nontrivial solutions

$$\gamma H = \pm n\pi, \quad n = 1, 2, \ldots \tag{10.24}$$

only for real values of γ. The barotropic mode disappears (on the f-plane) for an ocean bounded by a rigid upper plane. The second category of solutions consist of a denumerable infinity of *internal* or *baroclinic modes*, $n = 1, 2, \ldots$, which exist only in the frequency range $\omega_b \leqslant \omega \leqslant \omega_a$. These waves result from the interference of a pair of plane waves with real wavenumber components. For high mode numbers ($n \gtrsim 3$), the influence of the free surface on the internal modes becomes negligible and the solutions of (10.20) become nearly identical with those of (10.23).

The general case $\theta_h \neq \pi/2$ is slightly more complicated but essentially of the same

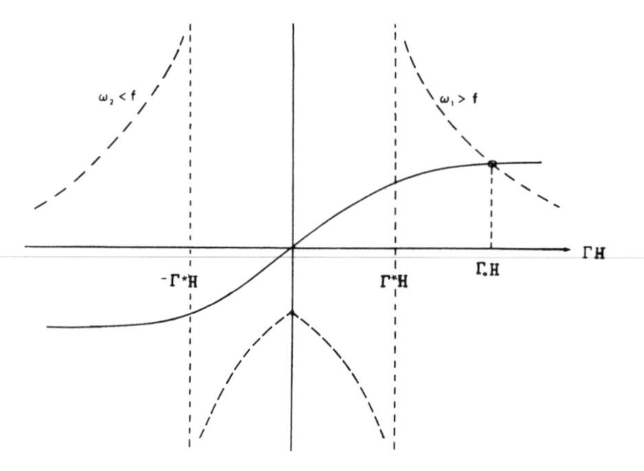

Fig. 10.4. Graphical solutions of (10.19) under the same conditions as in Fig. 10.3; this solution is to be compared to the $\Gamma^* = 0$ case in Fig. 10.2.

nature. Let $\tilde{f} \cos \theta_h > 0$; then, from (10.18), γ^* is real and positive in the range $\omega_b < f < \omega \leqslant \omega_a$, and real and negative in $\omega_b \leqslant \omega < f < \omega_a$. A graphical solution is shown in Fig. 10.3 for $\pi/2 < \gamma^* H < \pi$. For $\omega > f$, the barotropic root γ_0 is modified slightly to a baroclinic root $\gamma_1' > \pi/2$; the other roots $\gamma_1, \gamma_2, \ldots$ are essentially unchanged. For $\omega < f$, the first baroclinic root γ_1 is shifted to a barotropic root at $\gamma_0' \geqslant -\pi/2$; the roots $\gamma_2, \gamma_3, \ldots$ are not significantly modified. The presence of γ^* modifies the form of the eigenfunctions for those modes which have a root $|\gamma| < |\gamma^*|$, introducing an additional extremum in the vertical velocity profile for $\omega > f$, and removing an extremum for $\omega < f$. Outside the range of frequencies for which γ and γ^* are real (cf. Fig. 10.4), the single root γ_0 is larger than when $\gamma^* = 0$, but is still found only for $\omega > f$. The case $\tilde{f} \cos \theta_h < 0$ leads to similar modifications and is left as an exercise.

The horizontal propagation of perturbations from equilibrium in a uniformly stratified f-plane ocean with a flat bottom may then be represented by a superposition of vertical normal modes, the exact synthesis being determined by the initial conditions. For any given frequency ω, γ_n (and hence k_h) increases as the mode number n increases. The horizontal phase speed $|c_n| = \omega k_h$ decreases with increasing mode number, and the wave pattern disperses.

We have noted in our discussion of plane wave refraction in Section 9, that in a strongly stratified ocean rays are turned back towards the equator from a critical latitude where $\omega^2 = f^2$. Using (8.45), the critical latitude for the strongly stratified ocean is then given by

$$|\phi_c| = \sin^{-1}(\omega/2\Omega). \tag{10.25}$$

In the more general case of arbitrary stratification, the critical latitude may be found as that at which the northward (or southward) component of the group velocity vanishes. From the general dispersion relation (8.46), we find by differentiation that

$$\frac{\partial \omega}{\partial k_2} = \frac{k_2(N^2 + \tilde{f}^2 - \omega^2) + k_3 f\tilde{f}}{\omega k^2}.$$

Substituting into (8.46) for k_3 in terms of k_2 when $\partial\omega/\partial k_2 = 0$, we obtain

$$\frac{k_1^2}{k_2^2}\frac{(N^2-\omega^2)}{(N^2+\tilde{f}^2-\omega^2)} - 1 = \frac{(\omega^2-f^2)(N^2+\tilde{f}^2-\omega^2)}{f^2\tilde{f}^2}.$$

Let us consider the relatively simple case of purely poleward propagation ($k_1 = 0$). Substituting for f and $\tilde{f}$ in terms of the latitude from (8.45) into the above relation, with $k_1 = 0$, and solving for the critical latitude ϕ_c, we find that

$$|\phi_c| = \sin^{-1}\left[\frac{\omega^2}{4\Omega^2} - \frac{\omega^2}{N^2}\left(\frac{\omega^2}{4\Omega^2} - 1\right)\right]^{1/2}. \tag{10.26}$$

This result is a special case of that obtained for arbitrary k_1 by Hughes (1964). The reader may verify that the presence of a nonzero k_1 diminishes $|\phi_c|$. Note that for low frequencies, such that $\omega/N \ll 1$, (10.26) reduces to (10.25).

Nonuniform Stratification

The wave superposition method just presented is strongly restricted in its applicability to the real ocean by the basic assumption of a uniform Brunt-Väisälä frequency, necessary for the existence of plane-wave solutions. A simple separation of variables technique can be used to handle arbitrary $N(z)$, provided the horizontal rotation component is dropped. With $\tilde{f} = 0$, i.e. the ocean is assumed to be strongly stratified in the sense of (8.53), and upon eliminating the density perturbation from the vertical momentum equation, the basic linearized momentum and continuity equations (8.1)–(8.3) become, for a harmonic time dependence $e^{-i\omega t}$,

$$\rho_0(-i\omega u - fv) = -p_x, \tag{10.27}$$

$$\rho_0(-i\omega v + fu) = -p_y, \tag{10.28}$$

$$\rho_0(\omega^2 - N^2)w = -i\omega p_z, \tag{10.29}$$

$$u_x + v_y + w_z = 0. \tag{10.30}$$

Let us separate the horizontal and vertical dependences of the variables by letting

$$\rho_0(u, v) = \mathfrak{D}(z)[U^{(h)}(x,y), V^{(h)}(x,y)], \tag{10.31}$$

$$p = \mathfrak{D}(z)P(x,y), \tag{10.32}$$

$$w = \frac{i\omega}{g}Z(z)P(x,y). \tag{10.33}$$

The new variable $\mathfrak{D}(z)$ has the dimensions of density; it includes the vertical dependence of the horizontal velocity components. $P(x, y)$ is a specific pressure (pressure/density), and $Z(z)$ is a dimensionless amplitude function for the vertical dependence of w. Substituting (10.31)–(10.33) into (10.27)–(10.30) the governing equations become

$$-i\omega U^{(h)} - fV^{(h)} = -P_x, \tag{10.34}$$

$$-i\omega V^{(h)} + fU^{(h)} = -P_y, \tag{10.35}$$

$$\rho_0(\omega^2 - N^2)Z = -g\mathfrak{D}_z, \tag{10.36}$$

$$\frac{\mathfrak{D}}{\rho_0}(U_x^{(h)} + V_y^{(h)}) + \frac{i\omega}{g} Z_z P = 0. \tag{10.37}$$

Only the last equation (10.37) includes both horizontal and vertical dependences. Introducing a separation constant h_n, we may rewrite (10.37) as

$$\frac{g(U_x^{(h)} + V_y^{(h)})}{i\omega P} = \frac{-\rho_0 Z_z}{\mathfrak{D}} = \frac{1}{h_n}. \tag{10.38}$$

The equations (10.34)–(10.37) may then be split into two sets. For the horizontally dependent variables, with $\boldsymbol{U}^{(h)} = (U^{(h)}, V^{(h)}, 0)$ and $2\boldsymbol{\Omega} = (0, 0, f)$, (10.34), (10.35) and (10.38) become

$$i\omega \boldsymbol{U}^{(h)} - 2\boldsymbol{\Omega} \times \boldsymbol{U}^{(h)} = \nabla P, \tag{10.39}$$

$$\nabla \cdot \boldsymbol{U}^{(h)} = \frac{i\omega}{gh_n} P. \tag{10.40}$$

These equations are to be supplemented by appropriate boundary conditions and/or periodicity assumptions for $\boldsymbol{U}^{(h)}$ and P. The vertical dependence is determined by (10.36) and (10.38). Eliminating P, we find that

$$Z'' - \frac{N^2}{g} Z' + \frac{(N^2 - \omega^2)}{gh_n} Z = 0, \tag{10.41}$$

which is valid for any $N(z)$, and does not rely on the Boussinesq approximation. The primes denote differentiation with respect to z. In terms of Z, the bottom and surface boundary conditions (10.5) and (9.19) become

$$Z(-H) = 0, \tag{10.42a}$$

$$Z' - Z/h_n = 0 \quad \text{on } z = 0. \tag{10.42b}$$

The solution of the basic equations (10.27)–(10.30) can then be found by solving two boundary-value problems: one in the horizontal plane [equations (10.39) and (10.40), with lateral boundary conditions], the other in the vertical direction [equations (10.41), (10.42)]. The solutions of these two problems have been extensively discussed by Kamenkovich (1973). The procedure involved in obtaining actual solutions is most simply illustrated for an uncomplicated basin as follows. Let us say that we wish to find the frequencies of free oscillations of a stratified ocean basin on the f-plane, with a given $N(z)$, and of a shape which allows construction of separated solutions in the horizontal plane. A square basin provides a simple illustration; in that case, $(\boldsymbol{U}^{(h)}, P)$ have Fourier series representations

$$(U^{(h)}, V^{(h)}, P) = \sum_{k_1, k_2} (A, B, C)[\exp i(k_1 x + k_2 y)], \tag{10.43}$$

where k_1 and k_2 take on discrete values. Substituting into (10.39) and (10.40), we find a set of three homogeneous algebraic equations for the Fourier coefficients (A, B, C):

$$\begin{pmatrix} -i\omega, & -f, & ik_1 \\ f, & -i\omega, & ik_2 \\ ik_1, & ik_2, & -i\omega/gh_n \end{pmatrix} \begin{pmatrix} A \\ B \\ C \end{pmatrix} = \begin{pmatrix} 0 \\ 0 \\ 0 \end{pmatrix}. \tag{10.44}$$

Nontrivial solutions will be found only if the determinant vanishes, i.e. for

$$h_n = (\omega^2 - f^2)/g(k_1^2 + k_2^2). \tag{10.45}$$

Inserting this value of h_n into the vertical boundary value problem, one will then find that there are solutions which satisfy both the equation (10.41) and the boundary conditions (10.42) only for certain eigenfrequencies ω_n. The dependence of ω_n on the wavenumbers k_1 and k_2 through (10.45) is then the dispersion relation. For a closed basin, k_1 and k_2 are quantized and h_n will take on discrete values, $h_n(r, s)$ say, with r, s as bookkeeping indices running over integer values. The vertical problem (10.41), (10.42) leads to a series of possible vertical modes, i.e. values of $\omega = \omega_n(r, s)$ for each horizontal mode as specified by $h_n(r, s)$.

Let us now restrict our attention to plane waves (in the horizontal) in the absence of lateral boundaries. For N = constant and under Boussinesq's approximation, the dispersion relation arising from the vertical eigenvalue problem is easily shown to reduce to (10.12) under the special case $\tilde{f} = 0$. In that case, we already know that there exist a discrete spectrum of baroclinic (internal) modes for $f < \omega < N$ and a barotropic (surface) mode for the range $\omega > f$. This behaviour should, at least qualitatively, carry over to the case of nonuniform stratification. The problem which now arises is that of interpreting the meaning of a depth-dependent limiting frequency $N(z)$ for internal waves.

The surface wave may well be modified by the presence of stratification, but it does not owe its existence to the presence of the density structure. We will exclude it from consideration in our preliminary discussion of the wave properties of a nonuniformly stratified ocean by replacing the surface boundary condition (10.42b) by

$$Z(0) = 0. \tag{10.46}$$

Introducing a new variable ϕ defined by

$$\phi(z) = Z(z) \exp\left(-\frac{1}{2} \int^{z} \frac{N^2}{g}\, dz\right), \tag{10.47}$$

the vertical dependence equation (10.41) is reformulated as

$$\phi'' + \left[\frac{NN_z}{g} - \frac{1}{4}\frac{N^4}{g^2} + \frac{k_h^2(N^2 - \omega^2)}{(\omega^2 - f^2)}\right]\phi = 0, \tag{10.48}$$

where (10.45) has been used for h_n and $k_1^2 + k_2^2$ is abbreviated as k_h^2. This equation has the homogeneous boundary conditions

$$\phi(0) = \phi(-H) = 0. \tag{10.49}$$

A simple graphical argument (Fig. 10.5) shows that a function $\phi(z)$ governed by a differential equation of the form $\phi_{zz} + G(z)\phi = 0$, such as (10.48), cannot satisfy homogeneous boundary conditions, i.e., come back to its starting value, unless $G(z) > 0$. The curvature of the function $\phi(z)$ (which has the sign of ϕ_{zz}) must be of a sign opposite to that of the

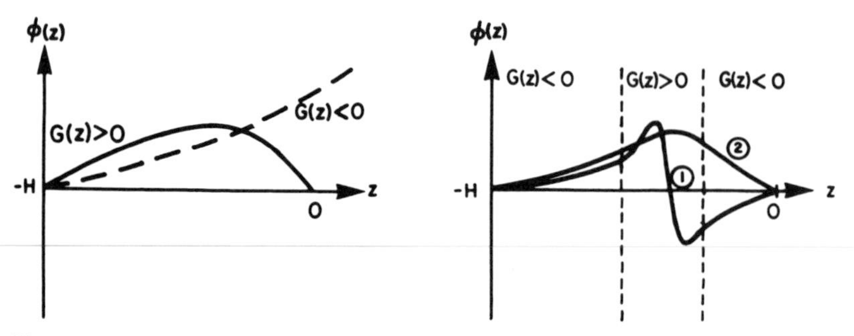

Fig. 10.5 (a) The behaviour of the function $\phi(z)$ obeying $\phi_{zz} + G(z)\phi = 0$ in the interval $[-H, 0]$, where $G(z)$ is of one sign over the whole interval. It is clear that the boundary conditions $\phi(0) = \phi(-H) = 0$ can be satisfied only if $G(z) > 0$. (b) Two possible solutions (first two modes for $\phi(z)$ obeying the same equation, with $G(z) > 0$ only over part of the interval.

function itself, as is the case for simple oscillatory functions such as sin z and cos z. This oscillatory behaviour need not prevail over the whole range of z: $\phi(z)$ may be turned back towards its original value in one (or more) oscillatory region(s), surrounded by nonoscillatory, or exponential-like depth ranges (Fig. 10.5b)

It thus follows that if ω^2 is larger than a certain maximum value, the coefficient of ϕ in (10.48), denoted by $G(z)$, will not be positive for any z in $[-H, 0]$ and that there will be no internal wave solutions of (10.49) with $k_h^2 > 0$ (corresponding to real wave propagation) satisfying (10.49). Evanescent modes, with $k_h^2 < 0$, merely give an exponential decay away from their source; they are not very interesting from the point of view of wave propagation, but they have to be taken into account to describe the complete field of motion in the vicinity of some disturbance oscillating over a wide frequency spectrum. If the Boussinesq approximation is used, the first two terms in the coefficient of ϕ drop out and $G(z)$ changes sign at $\omega = N(z)$ and at $\omega = f$.

We have already limited ourselves in writing (10.27)–(10.29) to the strongly stratified ocean where $N^2(z) \gg f^2$ at all depths. The function $G(z)$ is then positive somewhere in $-H \leqslant z \leqslant 0$ provided $f < \omega < N_{max}$, where N_{max} is the maximum value of $N(z)$. In the ranges of depths for which $\omega < N(z)$, $G(z) > 0$ and $\phi(z)$ has an oscillatory behaviour; outside these ranges, $G(z) < 0$ and the wave amplitude decays exponentially.

A strong maximum of $N(z)$ is often found near the ocean surface, near which the high N-value shown in Fig. 8.6 decreases sharply in a superficial (about 100 m deep) wind-mixed and seasonally-influenced layer. For frequencies approaching N_{max}, internal waves will be trapped in a narrow range of depths. An illustration of wave trapping by a pycnocline [sharp maximum in $N(z)$] is shown in Fig. 10.6.

The ultimate pycnocline is a density discontinuity, at which N becomes unbounded. Let such a discontinuity, between an upper layer of density ρ_1 and a deeper fluid of density $\rho_2 > \rho_1$, occur at a depth $z = -d$. The density profile is then given by

$$\rho_0(z) = \rho_2 - (\rho_2 - \rho_1)\Theta(z + d), \tag{10.50}$$

where $\Theta(z + d)$ is the unit step function. The stability frequency is then

$$N(z) = g(\rho_2 - \rho_1)\delta(z + d)/\rho_0, \tag{10.51}$$

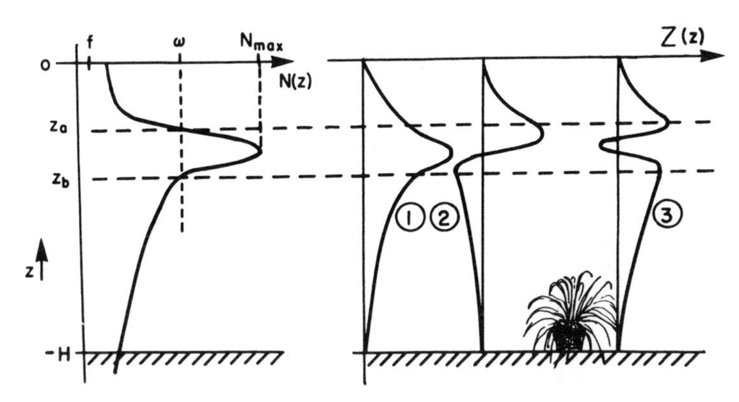

Fig. 10.6. Internal wave trapping in a thermocline. The relative values of f, ω and $N(z)$ are shown on the left. The solutions to $\phi_{zz} + G(z)\phi = 0$ have oscillatory behaviour only in $z_b < z < z_a$ and decay away from that layer. The vertical dependence $Z(z)$ of the vertical velocity w is shown on the right for the first three modes.

which is infinite at $z = -d$. There is then no upper limit to the frequency of internal oscillations at a density discontinuity. Such interfacial waves are similar to surface waves in that respect and in many others. After all, oscillations of a free surface may be considered as a limiting case of interfacial waves when the density of the upper layer goes to zero. Our simple pictures of oscillatory behaviour squeezed into a thin oscillatory region now fails, since the thickness of that region is now zero and there is no room left for any oscillatory vertical structure of the velocity field. The solution must be continued across the interface by using suitable matching conditions. One such condition is evidently that the vertical velocity must remain continuous across the interface: a gap is not allowed to appear between the two fluid regions; thus

$$Z_1 = Z_2, \tag{10.52}$$

where the subscript 1 pertains to $z > -d$ and subscript 2 to the lower layer. The other matching condition (which turns out to be continuity of pressure: see Exercise 10.4), is readily derived from (10.41). Recalling the definition of N^2, as given in (5.3), we may rewrite (10.41) as

$$(\rho_0 Z')' - \frac{\rho_0}{gh_n}(\omega^2 + g\rho_{0z}/\rho_0)Z = 0. \tag{10.53}$$

With ρ_0 as given by (10.50), this equation is now integrated over a thin layer including $z = -d$. Letting the thickness of this layer decrease to zero and using (10.52), we find

$$\rho_1 Z_1' - \rho_2 Z_2' + (\rho_2 - \rho_1)Z_1/h_n = 0. \tag{10.54}$$

Note that (10.54) reduces to the free surface boundary condition (10.42b) when $\rho_1 = 0$.

The simplest interfacial waves are oscillations of a single density discontinuity in an otherwise unstratified fluid. For a fluid without free surface there then exists only one wave mode: the interfacial wave. The reader is invited to examine the properties of this wave in more detail in Exercises 10.5 and 10.6. In the presence of two density steps two interfacial modes are possible, and in general for n discontinuities, there are n interfacial waves. In a fluid which is continuously stratified and also includes n density jumps, there exist in general a countable infinity of vertical modes associated with the continuous structure and n interfacial waves. In addition, if the surface is free, a surface barotropic mode is present.

General properties of normal mode solutions

This discussion has been rather qualitative. The whole theory of the possible solutions of the vertical boundary value problem (10.41), (10.42) may be stated and resolved on a formal basis using the Sturm-Liouville theory (a modern presentation of this theory is given in Swanson, 1968). The above presentation is partly based on strict results obtained by Yih (1965, Chapter 2) and Yanowitch (1962) for a nonrotating fluid; however, these results are readily carried over to the rotating case. The main results are briefly summarized below:

(1) The countably infinite spectrum of eigenvalues of (10.41) and (10.42), augmented by interfacial waves propagating on density discontinuities (including the surface mode) and a steady geostrophic current to take into account any nonzero time-averaged motion, form a complete set of functions, in terms of which any horizontally propagating disturbance in a fluid layer with parallel boundaries may be described.

(2) For a given frequency ω, the wavenumber k_h increases, and the phase speed $|\boldsymbol{c}|$ decreases with mode number. This has already been seen to be true for the special case $N =$ constant (Fig. 10.1).

(3) For a given wavelength, interfacial modes have greater speeds than the internal modes resulting from the continuous stratification. All other parameters being constant, the speed of an interfacial wave increases with the magnitude of the density jump across it.

(4) For all wave types, in any stratification, c^2 decreases and ω^2 increases as k_h^2 increases, so that waves are always normally dispersive, with $\partial c/\partial k_h < 0$ and $c_g < c$.

This last statement can be proved in an elementary fashion by extension of a method given by Yih (1974) for nonrotating fluids. Consider a certain internal wave mode, characterized by a horizontal wavenumber k_h, frequency ω and vertical velocity with vertical dependence given by $Z(z)$. Let us then perturb k_h^2, ω^2 and $Z(z)$ by small increments δk_h^2, $\delta\omega^2$ and δZ. From (10.45), the perturbation in the separation constant h_n is then

$$\delta h_n = (\delta\omega^2 - gh_n\delta k_h^2)/gk_h^2. \tag{10.55}$$

Substituting for the perturbed variable Z and the parameters k_h, ω^2 and h_n in (10.53) and its boundary conditions (10.42), we find that the perturbations must satisfy

$$[\rho_0(\delta Z)']' - \frac{\rho_0}{gh_n}(\omega^2 + g\rho_{0z}/\rho_0)\delta Z =$$
$$\rho_0 Z[h_n\delta\omega^2 - (\omega^2 + g\rho_{0z}/\rho_0)\delta h_n]/gh_n^2. \tag{10.56}$$

$$\delta Z = 0 \quad \text{on } z = -H, \tag{10.57a}$$

$$(\delta Z)' - \delta Z/h_n = -Z\delta h_n/h_n^2 \quad \text{on } z = 0. \tag{10.57b}$$

Multiplying (10.56) by Z and (10.53) by δZ, subtracting, and integrating over the fluid depth we obtain, after using the boundary conditions (10.42) and (10.57),

$$g\rho_0(0)Z^2(0)\delta h_n = \delta h_n \int_{-H}^{0} \rho_0 Z^2(\omega^2 + g\rho_{0z}/\rho_0)\,\mathrm{d}z - h_n \delta\omega^2 \int_{-H}^{0} \rho_0 Z^2\,\mathrm{d}z. \tag{10.58}$$

At the free surface, the density drops abruptly to zero; its derivative there may be written

$$\rho_{0z}(0) = -\rho_0(0)\delta(z), \tag{10.59}$$

where $\delta(z)$ is the Dirac delta function. Extending the limits of integration to a point $z = 0+$ just above the surface,

$$\int_{-H}^{0+} g\rho_{0z} Z^2\,\mathrm{d}z = \int_{-H}^{0} g\rho_{0z} Z^2\,\mathrm{d}z - g\rho_0(0)Z^2(0) \equiv I_1. \tag{10.60}$$

Defining also

$$I_2 \equiv \int_{-H}^{0} \rho_0 Z^2\,\mathrm{d}z, \tag{10.61}$$

the equation (10.58) takes the compact form

$$(I_1 + \omega^2 I_2)\delta h_n - h_n \delta\omega^2 I_2 = 0. \tag{10.62}$$

By definition, $I_2 \geqslant 0$. Multiplying (10.53) by Z and integrating as above, we also find that

$$(I_1 + \omega^2 I_2) = -gh_n \int_{-H}^{0} \rho_0 (Z')^2\,\mathrm{d}z < 0. \tag{10.63}$$

From (10.62) and (10.63), it follows that

$$\delta h_n / \delta\omega^2 < 0. \tag{10.64}$$

In combination with (10.55), this inequality also implies that

$$\delta\omega^2 / \delta k_h^2 > 0, \tag{10.65a}$$

which proves part of statement (4) above. From (10.45), (10.64) and the basic relation $\omega^2 = k_h^2 c^2$, it also follows directly that

$$\delta c^2 / \delta k_n^2 < 0, \tag{10.65b}$$

which, for c and k_h both positive, proves the other half of the statement.

Instead of continuing with a detailed discussion of wave properties in the whole frequency interval $\omega > f$, we now divide our attention between two special cases of great interest. In later sections of this chapter, we address ourselves to the high-frequency, high-wavenumber case, for which $\omega^2 \gg f^2$ and the influence of rotation may be neglected. These sections provide a basic coverage of the familiar high-frequency surface gravity waves. In Chapter 3, the low-frequency cases of gravity waves, for which $\omega \gtrsim f$, appropriate to tidal motions, and the β-plane oscillations with $\omega \ll f$ will be examined.

Exercises Section 10

1. Derive equation (10.8) from the momentum and mass conservation equations (8.1) and (8.2), assuming that $2\mathbf{\Omega} = (0, \tilde{f}, f)$.

2. Derive a dispersion relation equivalent to (10.12) for an ocean bounded by a rigid top. Note that the result could be obtained directly from (10.12) by letting $g \to \infty$.

3. Discuss the solutions for the rigid-topped ocean using graphical methods as in Fig. 10.1 and compare them with the solutions obtained above. Why is the barotropic mode lost when the upper surface is made rigid?

4. Show that the second matching condition (10.54) applicable across a density interface may be derived, in a fashion analogous to (9.19) for a free surface, by insisting that the pressure be continuous across the interface.

5. Show that the dispersion relation for interfacial waves in a fluid bounded by horizontal rigid plates at $z = 0, -H$, and unstratified except for a density discontinuity from $\rho_0 = \rho_1$ above $z = -d$ to $\rho_0 = \rho_2 > \rho_1$ below $z = -d$, is given by

$$\omega = \sqrt{\frac{g}{h_n}(\rho_2 - \rho_1)\{\rho_2 \coth\,[\omega(H-d)/\sqrt{gh_n}] + \rho_1 \coth\,[\omega d/\sqrt{gh_n}]\}^{-1}}, \tag{10.66}$$

where h_n is related to the horizontal wavenumber k_h via (10.45).

6. Using h_n as given in (10.45), consider the case of long waves ($k_h \to 0$) and reduce (10.66) to

$$\omega^2 - f^2 = \frac{g(\rho_2 - \rho_1)k_h^2 d(H-d)}{\rho_2 d + \rho_1(H-d)} \tag{10.67}$$

For $f \neq 0$, such waves are clearly dispersive. Letting $f^2 = 0$, show that given the total depth H and the densities ρ_1 and ρ_2, the maximum wave speed will be found for a discontinuity located at $z = -d$, where

$$d = \frac{\rho_1 H}{(\rho_2 - \rho_1)}\,[(\rho_2/\rho_1)^{1/2} - 1]. \tag{10.68}$$

Find d/H for fresh water ($\rho_1 = 1.000$) over salt water ($\rho_2 = 1.025$) and for fresh water over mercury ($\rho_2 = 13.6$).

7. Show that for a homogeneous fluid ($N = 0$) the vertical problem (10.41) and (10.42) has the solution

$$Z = A \sinh\,[\omega(z+H)/\sqrt{gh_n}] \tag{10.69}$$

with h_n given implicitly by

$$\omega = \sqrt{g/h_n}\tanh\,(\omega H/\sqrt{gh_n}). \tag{10.70}$$

In particular, for long waves ($\omega H/\sqrt{gh_n} \ll 1$), show that

$$h_n = H. \tag{10.71}$$

8. Show that for a rigid-topped Boussinesq fluid with constant N, the solution of (10.41) and (10.42) is

$$Z = A_n \sin\,(n\pi z/H), \quad n = 1, 2, \ldots, \tag{10.72a}$$

$$h_n = \frac{H(N^2 - \omega^2)}{g} \cdot \frac{H}{n^2\pi^2}. \tag{10.72b}$$

For typical oceanic conditions, show that $h_n \ll H$.

11. HIGH-FREQUENCY GRAVITY AND CAPILLARY WAVES OF SMALL AMPLITUDE

The study of high-frequency surface waves is one of the oldest and most developed topics of oceanographic fluid dynamics. The early developments of the theory, from the work of Stokes (1847) onwards are gathered in Lamb's treatise on hydrodynamics (1945). More modern accounts may be found in books by Stoker (1957), Kinsman (1965) and Phillips (1966). The thrust of the theory has long shifted to the study of nonlinear effects, which will be discussed briefly in the next section, and to the problem of wind-wave generation (Section 51). At this stage, we shall describe the basic properties of surface gravity waves of small amplitude (already introduced as the barotropic mode of a stratified fluid in Section 10) and introduce and discuss surface tension as a wave-producing restoring force.

Basic formulation

Let us first modify the free surface boundary condition to include the effect of surface tension. At surfaces of contact between two different fluids, or between different phases of the same material, sharp gradients of intermolecular forces arise and produce a tension force acting in the plane of the surface of contact. The origin and nature of surface tension has been reviewed recently by Brown (1974). Any curvature of a surface under tension will give rise to a restoring component of force directed towards the centre of curvature and tending to restore the surface to a plane configuration. The restoring force per unit area is a pressure p_s given by Laplace's formula

$$p_s = \sigma\left(\frac{1}{R_1} + \frac{1}{R_2}\right), \tag{11.1}$$

where σ is the surface-tension coefficient and R_1, R_2 are the principal radii of curvature at a given point on the surface. The coefficient σ is usually a decreasing function of temperature and of contamination of the surface; it has units of force/length and a typical value of $74 \cdot 10^{-3}$ N/m for an air/fresh-water interface at 10°C. The variation of σ with temperature, salinity, etc., may be found in standard physical tables. Many of the fascinating capillary effects which are produced by surface tension are discussed by Boys (1959) and illustrated in a film by Trefethen (1972). A deeper discussion of capillary effects will be found in Levich (1962).

At high frequencies ($\omega^2 \gg f^2 + \tilde{f}^2$) both components of rotation are negligible, and the symmetry of the problem in the horizontal plane allows us to choose the x-axis in the direction of wave propagation. For plane waves, $\partial_y \equiv 0$, and in the configuration of Fig. 11.1 the radii of curvature R_1 and R_2 become,

$$R_2 = \infty,$$

$$R_1 = -(1 + \eta_x^2)^{3/2}/\eta_{xx},$$

so that

$$p_s = -\sigma\eta_{xx}/(1 + \eta_x^2)^{3/2}. \tag{11.2}$$

The surface boundary condition (9.16a) is modified to

$$p_0(\eta) + p(x, \eta, t) = p_a + p_s, \tag{11.3}$$

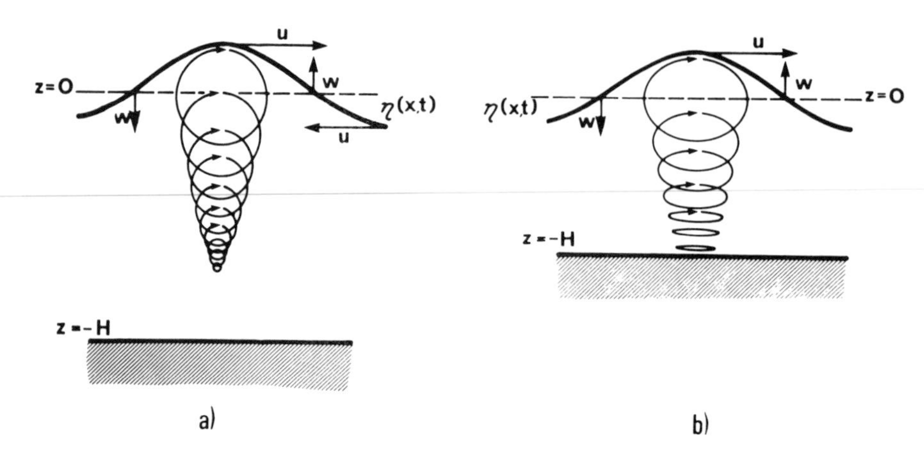

Fig. 11.1. The geometry of surface waves, showing the surface displacement $\eta(x, t)$ and the directions of the velocity components u and w at various points. On the left, the approximate particle orbits are shown for a wave in deep water ($\lambda \simeq H$), and on the right, for a wave in water of intermediate depth ($\lambda \simeq 2.5H$).

where p_0 is the hydrostatic pressure and p the perturbation due to the waves, as earlier. Expanding $p_0(\eta)$ and $p(x, \eta, t)$ about the equilibrium level $z = 0$ and using a linear approximation to (11.2), the modified form of (9.18a) is now

$$p_0(0) + p(x, 0, t) - \rho_0(0)g\eta = p_a - \sigma\eta_{xx}. \tag{11.4}$$

Eliminating the surface displacement using (9.18b), we obtain

$$p_t - \rho_0 g w + \sigma w_{xx} = 0 \quad \text{at } z = 0. \tag{11.5}$$

Substituting for w in terms of p from (8.69), we obtain

$$(\partial_{tt} + N^2)p + gp_z - \frac{\sigma}{\rho_0} p_{xxz} = 0 \quad \text{at } z = 0, \tag{11.6}$$

a condition which differs from (9.20) solely through the presence of the surface tension term. The boundary condition on a flat bottom is unchanged:

$$p_z = 0 \quad \text{at } z = -H. \tag{11.7}$$

For an unstratified ($N = 0$), nonrotating fluid, the momentum equation (8.1) reduces to

$$\rho_* \boldsymbol{u}_t = -\nabla p. \tag{11.8}$$

Taking the divergence of this equation and using the continuity equation (8.3), we find that p is a harmonic function:

$$\nabla^2 p = 0. \tag{11.9}$$

It is well known that Laplace's equation does not have wave solutions; it is the boundary condition (11.6) at the free surface, where the restoring force is acting, which allows the system (11.6), (11.7) and (11.9) to have free travelling-wave solutions.

An alternate, and more traditional, approach to the formulation of the problem of surface waves in an unstratified and nonrotating fluid is based on the irrotationality of the flow field under these conditions. For $\nabla\rho = 0$, $\boldsymbol{\Omega} = \boldsymbol{0}$ and in the absence of viscosity, the vorticity equation (4.2) reduces to

$$\mathrm{D}\boldsymbol{\zeta}/\mathrm{D}t = \boldsymbol{\zeta} \cdot \nabla \boldsymbol{u}. \tag{11.10}$$

If $\boldsymbol{\zeta} = \boldsymbol{0}$ at some initial time t_0, it then remains zero at $t > t_0$ (this result is known as Helmholtz's theorem). Thus,

$$\boldsymbol{\zeta} = \nabla \times \boldsymbol{u} = \boldsymbol{0},$$

and the velocity may be written in terms of a potential:

$$\boldsymbol{u} = \nabla\phi. \tag{11.11}$$

From the continuity equation, it then follows that ϕ is a harmonic function, i.e.,

$$\nabla^2\phi = 0. \tag{11.12}$$

The existence of a Bernoulli function for irrotational flow has already been demonstrated in (4.13). Thus,

$$\phi_t + \tfrac{1}{2}\boldsymbol{u} \cdot \boldsymbol{u} + (p_0 + p)/\rho_* + gz = \text{constant} \tag{11.13}$$

throughout the fluid. The boundary conditions on $z = 0, -H$ may be expressed in terms of ϕ and of its derivatives through (11.11), (11.13) and the kinematic boundary condition (9.18b) (Exercise 11.2). The introduction of a potential function is advantageous because it allows the problem to be formulated in terms of a scalar. We have chosen to express the problem in terms of another scalar, the perturbation pressure p, because the latter can also be used in a rotating or stratified fluid. A scalar velocity potential does not exist in a rotating or stratified fluid because the flow is rotational.

Plane waves of the form

$$p = \Pi(z)\, \mathrm{e}^{i(kx-\omega t)} \tag{11.14}$$

satisfy (11.7) and (11.9) provided

$$\Pi(z) = A \cosh\,[k(z + H)], \tag{11.15}$$

where A is an amplitude constant. Substituting p into the surface boundary condition (11.6) (with $N = 0$ and $\rho_0 = \rho_*$) gives the dispersion relation

$$\omega^2 = \left(gk + \frac{\sigma k^3}{\rho_*}\right) \tanh kH. \tag{11.16}$$

The relative importance of surface tension and gravity restoring forces on the waves is clearly recognized in (11.16), and may be characterized through the Bond (or Weber) number (Catchpole and Fulford, 1966), defined as

$$B(k) = \sigma k^2/g\rho_*. \tag{11.17}$$

For very short waves $[B(k) \gg 1]$ capillarity dominates; long waves $[B(k) \ll 1]$ on the

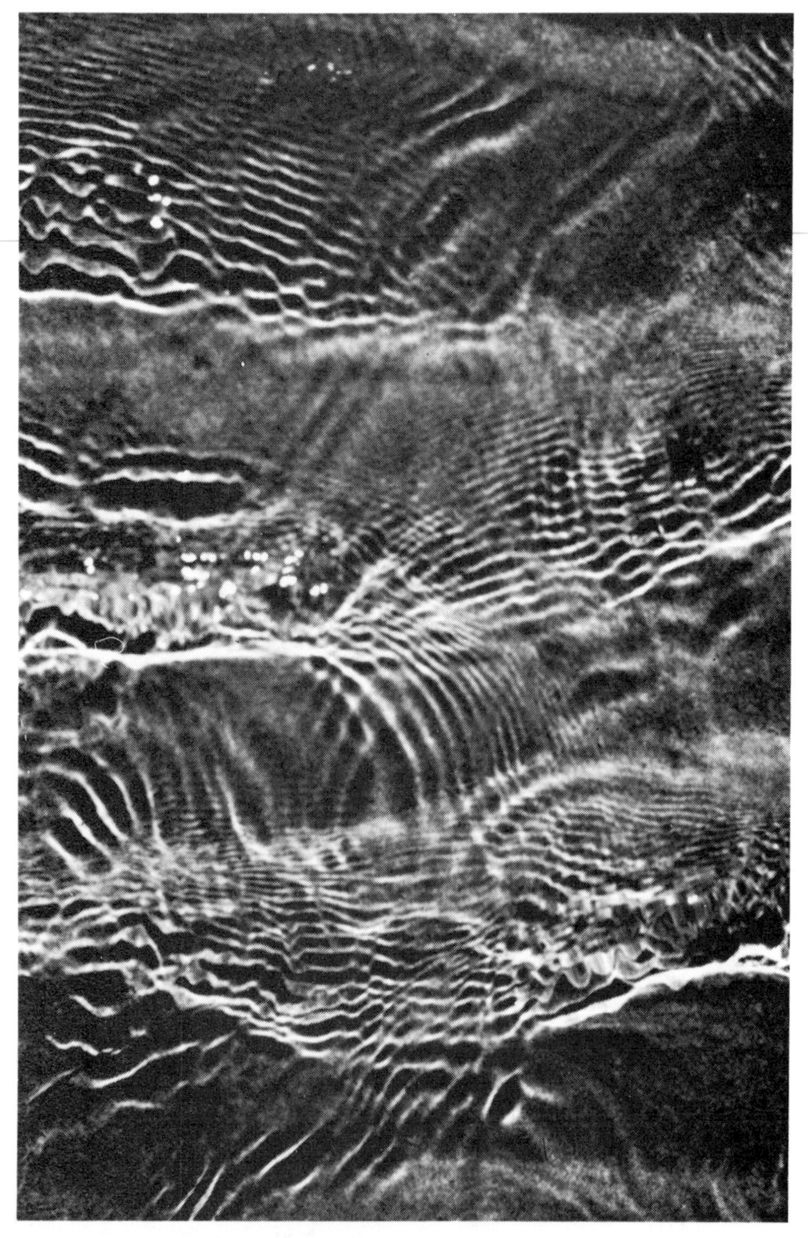

Fig. 11.2. Capillary waves generated at the forward face of steep gravity waves advancing towards the top of the picture. There are four gravity wave crests visible, the upper one at the edge of the top of the figure; the second crest from the bottom occupies only the left half of the picture. The water depth is about 5 cm and the wavelength $\lambda \simeq 20$ cm. The anomalous dispersion of capillary waves is clearly noticeable: short waves lie forward of longer ones. Frictional dissipation is also evident in the decreased amplitude and eventual disappearance of the shorter capillary waves.

other hand, are nearly purely gravitational. The two restoring forces are equal at $B = 1$; for an air–water interface at 10°C, this occurs at $k = 0.36 \cdot 10^3$ rad m^{-1} and a wavelength $\lambda = 2\pi/k = 17.26$ mm.

From (11.16) the phase speed is found to be

$$c^2 = \left(\frac{g}{k} + \frac{\sigma k}{\rho_*}\right) \tanh kH. \tag{11.18}$$

Since $c^2 = c^2(k)$, the waves are dispersive. Two types of dispersion occur, depending on the wavenumber. For large values of k (short waves), surface tension dominates and $\partial c^2/\partial k > 0$: capillary waves show anomalous dispersion, the short waves travelling faster than the long ones. This behaviour is visible in Fig. 11.2, and is in marked contrast with the dispersive properties of gravity waves (the $B \ll 1$ limit) discussed in the previous section. The phase speed is plotted as a function of k in Fig. 11.3, and has a minimum near $B = 1$. Water depths of interest usually are greatly in excess of 17.3 mm, so that we may use the deep water limit $kH \gg 1$ ($\tanh kH \simeq 1$) to estimate the value of the minimum velocity and the wavenumber at which it occurs. The minimum velocity c_{min} is found at $B = 1$ and is given by

$$c_{min} = (4g\sigma/\rho_*)^{1/4}. \tag{11.19}$$

With the numerical values appropriate to the air–water interface at 10°C, $c_{min} =$ 0.23 m s^{-1}. This value sets the lower limit for the speed of phase propagation at the sea surface. At the other limit of very long waves for which $B \to 0$ and $kH \ll 1$, c has a maximum at $k = 0$, at which $c_{max} = \sqrt{gH}$. Although (11.18) is unbounded at large values of k, this is of little practical relevance, since very short waves are quite rapidly damped by molecular friction. The extremely short waves which lead the capillary wave trains seen in Fig. 11.2 are also of very small amplitude, and vanish at the front of the wave train, where friction destroys them as quickly as they are generated. The influence of frictional dissipation will be discussed in Section 51.

The group velocity c_g is derived from (11.16) by differentiation with respect to k; we find

$$c_g = \frac{c}{2}\left[\frac{2kH}{\sinh 2kH} + \frac{1 + 3B}{1 + B}\right], \tag{11.20}$$

which is also plotted in Fig. 11.3. For $kH \gg 1$, the minimum value of c_g (equal to 0.18 m s^{-1}) is found at $B = 0.155$, corresponding to a wavelength of 43.8 mm on a clean air–sea interface at 10°C. The minimum speed of energy propagation is thus even slower than that of phase propagation (0.23 m s^{-1}). Because of the existence of a minimum in c_g, there are two waves with the same group velocity when $c_g > c_{g\,min}$, one which is shorter than 43.8 mm and dominated by surface tension, the other longer than 43.8 mm and dominated by gravity. The same duality holds for the phase velocity, although a pair of waves with identical phase velocity will possess different group velocities, and vice-versa.

Limiting wavetypes

A number of limiting cases of (11.16) have already been alluded to above; for reference and comparison these will now be presented more systematically. We distinguish three limiting cases of special relevance.

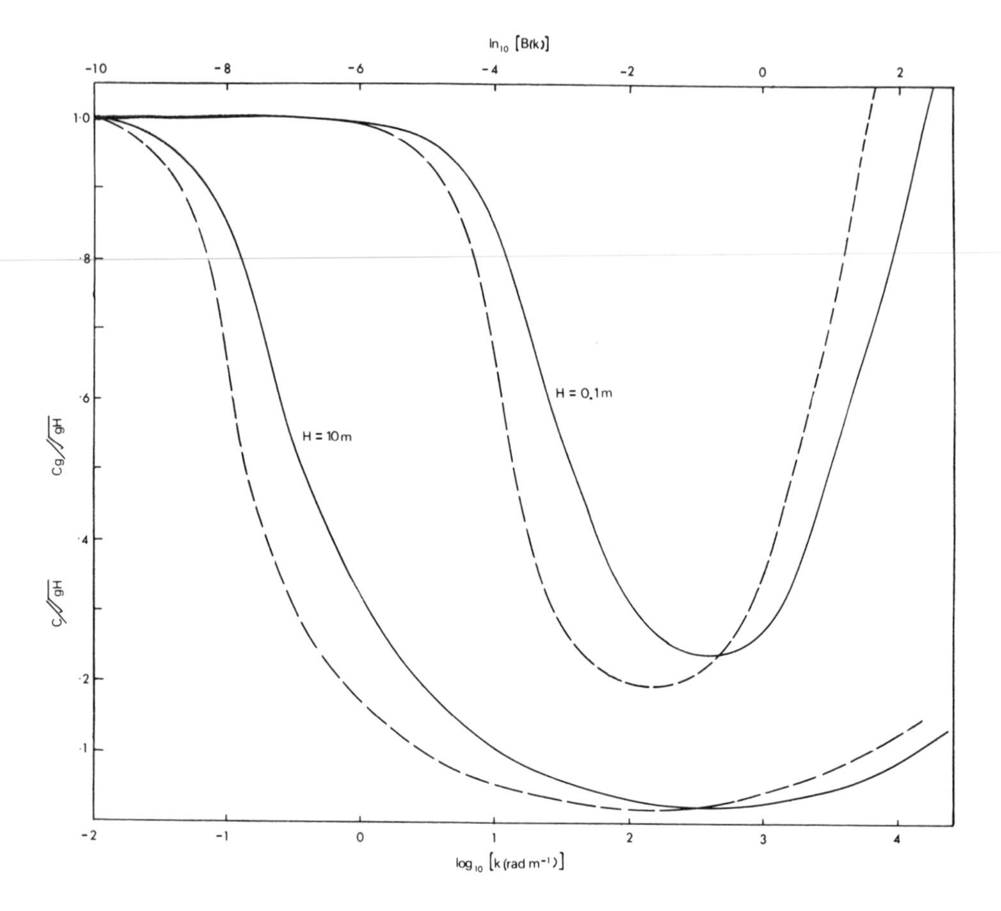

Fig. 11.3. Phase speed c(———), and group velocity c_g (– – – –) for capillary–gravity waves in a non-rotating fluid of constant depth, as a function of wavenumber k.

For $B(k) \gg 1$ and $kH \gg 1$, (11.16), (11.18) and (11.20) reduce to

$$\omega^2 = \sigma k^3/\rho_*, \tag{11.21a}$$

$$c^2 = \sigma k/\rho_*, \tag{11.21b}$$

$$c_g = 3c/2, \tag{11.21c}$$

respectively. The waves obeying these relations are "pure" capillary waves, examples of which are seen in Fig. 11.2. Since the wavelength must satisfy $\lambda \ll 17.26$ mm for $B(k) \gg 1$, the condition $kH \gg 1$ is applicable to pure capillary waves in all relevant ocean depths. As noted earlier, these waves exhibit anomalous dispersion: $c_g > c$. An observer travelling at the group velocity with a wave train of finite length would thus notice wave crests appearing at the leading edge of the wave train, then being overtaken by the wave train as a whole and eventually disappearing at its trailing edge.

Capillary waves are produced at regions of sharp curvature of the sea-surface. They are thus commonly seen around the crests of nearly breaking gravity waves (Fig. 11.2). A travelling source of capillary waves (such as a steep gravity wave) which generates a broad range of high wavenumbers always produces waves for which c_g exceeds the speed of the source U, even if the source travels faster than the phase speed of most of the capillary waves generated. A wave train of extremely short waves is always noticeable ahead of the source, from which it moves away at the relatively slow speed $c_g - U$; behind the source, on the other hand, the wave group spreads out much more rapidly, at a speed $c_g + U$, and, being made up of longer capillary waves, is much less noticeable. This asymmetry is evident in Fig. 11.2. The wave pattern around a stationary obstacle in a moving fluid (such as a fishing line) shows similar characteristics and may be obtained by a simple translation of coordinates (Lamb, 1945, Chapter 9).

The range $B(k) \ll 1$ includes "pure" gravity waves; for $kH \gg 1$ again, these waves are said to travel in deep water and (11.16), (11.18) and (11.20) become

$$\omega^2 = gk, \tag{11.22a}$$

$$c^2 = g/k, \tag{11.22b}$$

$$c_g = c/2. \tag{11.22c}$$

These waves are normally dispersive; $c > c_g$. In a travelling group of finite length, crests appear at the back of the group, progress through it and disappear at the leading edge. The wave pattern generated by a moving source (such as a ship) now lies entirely behind the source. This wave pattern has been described by various authors (Lamb, 1945, Chapter 9; Stoker, 1957, Chapter 8). The wind-generated waves commonly observed at the sea surface fall into this limiting range. A striking verification of the dispersion characteristics of these waves, as given by (11.22), and of the relevance of linear theory, has been provided by the observations of Snodgrass et al. (1966) of swell propagation across the Pacific Ocean. Since the wave energy travels at the group velocity, the time of arrival of wind waves at a distance x from a storm occurring at a time t_0 is given by

$$x/(t - t_0) = c_g. \tag{11.23}$$

In view of (11.22), we may arrange (11.23) as

$$\omega = g(t - t_0)/2x. \tag{11.24}$$

The frequency of waves reaching a distant point is then a linear function of time of arrival. A graph of ω against t yields the time of generation t_0 [the intercept of the $\omega(t)$ curve with $\omega = 0$] and the distance of the source to the point of observation [from the slope $g/2x$ of the $\omega(t)$ line]. An example of the time variation of energy density as a function of frequency is shown in Fig. 11.4. The method of estimating the source distance for waves of known dispersion characteristics by the amount of dispersion observed at the point of observation may be applied to any type of dispersive wave. The technique has been used for example to estimate the distance of pulsars (Lyne and Rickett, 1968).

The third limiting case is that of pure gravity waves in shallow water [$B(k) \ll 1$, $kH \ll 1$], for which

$$\omega^2 = gk^2H, \tag{11.25a}$$

$$c^2 = c_g^2 = gH. \tag{11.25b}$$

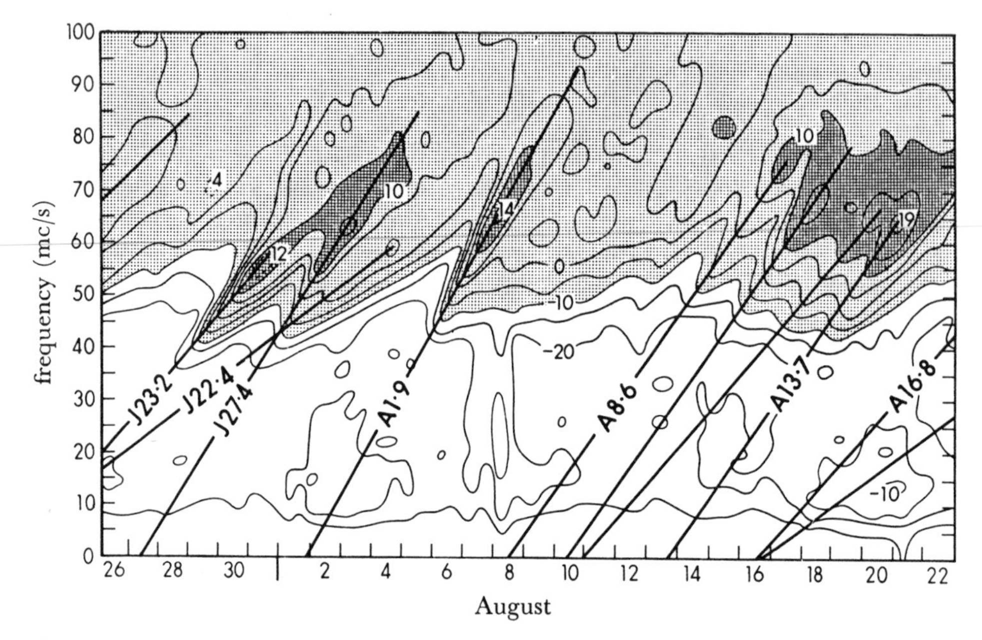

Fig. 11.4. Contours of wave energy density on a frequency–time plot for one month in Honolulu. The ridges represent arrivals of dispersed wave groups associated with storms (cf. 11.24) and are labelled according to the storm time (i.e., J27.4 means July 27, 9.6 hr G.M.T.). The ticks on the time axis denote midnight G.M.T. (From Snodgrass et al., 1966.)

These waves are nondispersive. This limit applies to the propagation of long waves for which $\omega \gg f$, such as tsunamis (earthquake-generated waves) or swell near a beach.

Particle motion

Given an oscillating pressure of the form (11.14) with depth dependence (11.15), the other wave variables may be found from the boundary condition (11.5) and (9.18b) and the momentum equations (8.67)–(8.69) (with $f = N = 0$). Let the free surface displacement be represented by

$$\eta = a \cos (kx - \omega t). \tag{11.26}$$

Then, from (9.18b) and (11.5), the constant A in (11.15) is related to the surface amplitude a by

$$A = \frac{\rho_* g a(1 + B)}{\cosh kH},$$

and the real part of (11.15) is

$$p(x, z, t) = \frac{\rho_* g a(1 + B)}{\cosh kH} \cosh [k(z + H)] \cos (kx - \omega t). \tag{11.27}$$

The velocity components are then found from (8.67)–(8.69) in which (11.27) has been substituted for p:

$$u(x,z,t) = \frac{gak}{\omega}\frac{(1+B)}{\cosh kH}\cosh\,[k(z+H)]\,\cos\,(kx-\omega t), \tag{11.28}$$

$$v(x,z,t) = 0, \tag{11.29}$$

$$w(x,z,t) = \frac{gak}{\omega}\frac{(1+B)}{\cosh kH}\sinh\,[k(z+H)]\,\sin\,(kx-\omega t). \tag{11.30}$$

From (11.27), the pressure perturbation is hydrostatic (except for the surface-tension contribution) near the free surface:

$$p(x,0,t) = \rho_* g(1+B)\eta. \tag{11.31}$$

The pressure decreases with depth, however, so that at the bottom,

$$p(x,-H,t)/p(x,0,t) = 1/\cosh\,(kH). \tag{11.32}$$

This ratio decreases rapidly with increasing kH; in deep water, the decay with depth of the pressure fluctuation is exponential. The ocean acts as a low-pass filter for pressure, and pressure gauges placed in deep water will be sensitive only to long waves, such as tsunamis and tides and will be completely blind to wind waves and swell.

The tip of the velocity vector describes an ellipse in the (x, z)-plane. The ratio of the vertical to the horizontal axis of the ellipse is

$$|w|/|u| = \tanh\,[k(z+H)], \tag{11.33}$$

which vanishes at $z=-H$, in order to satisfy the bottom boundary condition $w=0$ there, and attains its maximum value at the free surface, $z=0$. In deep water ($kH \gg 1$), $|w|/|u| \simeq 1$ and the ellipse reduces to a circle. By integrating (11.28) and (11.30) with respect to time, it is simple to establish (Exercise 11.4) that in the linear approximation individual particles are displaced in closed elliptical orbits which shrink with depth and flatten to become purely horizontal at the bottom. The particle orbits are said to be described in a prograde fashion since u and η are in phase. Currents are forwards (in the direction of wave propagation) on the crests and backwards in the troughs, as shown in Fig. 11.1.

With reference to the limiting cases of short ($kH \gg 1$) and long ($kH \ll 1$) waves, we notice that short waves do not "feel" the existence of the bottom, whereas the velocity field of long waves is dominated by the bottom constraint. For short waves, the depth dependence of the wave variables is essentially exponential: both $\cosh\,[k(z+H)]/\cosh kH$ and $\sinh\,[k(z+H)]/\cosh kH$ behave like e^{kz}. At a depth equal to half a wavelength ($z=-\pi/k$), all variables are already attenuated to about 4% of their surface value. The particle orbits are circular and become vanishingly small well above the bottom: the bottom boundary condition $w=0$ is automatically satisfied to a high degree of accuracy by the exponential approximation. For long waves on the other hand, the depth variation falls out: $\cosh\,[k(z+H)]/\cosh kH = 1$ and $\sinh\,[k(z+H)]/\cosh kH = 0$ to $0(kH)$. The bottom boundary condition makes its influence felt all the way to the surface, keeping the motion on very nearly horizontal planes. This type of motion will also be found to prevail in long barotropic waves in a rotating fluid (Section 17). The orbital motions in deep water and in water of intermediate depth [$kH \sim 0(1)$] are shown in Fig. 11.1.

As seen in the previous section, the presence of rotation modifies the dispersion relation to the more complicated form (10.22), of which (11.16) is a special case when $\sigma = 0$. The most obvious effect of rotation on the current field is the appearance of a nonzero v component along wave crests. In the two-dimensional case, the y-momentum equation reduces to

$$v = \frac{-if}{\omega} u, \tag{11.34}$$

and the current ellipses are tilted from the vertical plane by the presence of the Coriolis force (Exercise 11.5).

Short interfacial waves

As pointed out in Section 10, interfacial waves are very similar in nature to surface waves. In the absence of rotation ($f = 0$), (10.45) becomes

$$h_n = \omega^2/gk^2,$$

which, when substituted into (10.66) yields the dispersion relation for high-frequency interfacial waves between an upper layer of thickness d and density ρ_1 and a lower layer of thickness $H - d$ and density ρ_2:

$$\omega^2 = gk(\rho_2 - \rho_1)\{\rho_2 \coth\,[k(H-d)] + \rho_1 \coth\,(kd)\}^{-1}. \tag{11.35}$$

For short waves [kd and $k(H-d)$ both $\gg 1$], (11.35) gives

$$\omega^2 = gk(\rho_2 - \rho_1)/(\rho_1 + \rho_2), \tag{11.36}$$

which is analogous to (11.22a) and reduces to it when $\rho_1 = 0$. For a given wavenumber, the frequency is reduced in proportion to the decreased restoring force [the $g(\rho_2 - \rho_1)$ factor] and to the increased mass of fluid participating in the oscillation ($\rho_2 + \rho_1$). For long wavelengths [kd and $k(H-d)$ both $\ll 1$], interfacial waves are nondispersive; from (10.67) with $f = 0$,

$$c^2 = g(\rho_2 - \rho_1)d(H-d)/[\rho_2 d + \rho_1(H-d)], \tag{11.37}$$

which is to be compared to (11.25b) for long surface waves.

For a thin upper layer ($d \ll H$), the inequality $k(H-d) \gg 1$ may hold, but not $kd \gg 1$, and intermediate cases exist for which the waves are long (or short) with respect to one layer but not with respect to the other. The influence of surface tension between the two layers may also be taken into account by modifying the interfacial condition (10.54), or by working directly from the pressure continuity condition (Exercise 11.7). For an extensive discussion of interfacial waves, the reader is referred to Krauss (1966).

Exercises Section 11

1. Use the horizontal momentum equation (8.67) together with the incompressibility condition $\nabla \cdot \boldsymbol{u} = 0$ to formulate the free surface boundary condition (11.5) in terms of the vertical velocity w.

$$(\text{Ans. } w_{ztt} - gw_{xx} + \frac{\sigma}{\rho_*} w_{xxxx} = 0.) \tag{11.38}$$

2. Write the linearized surface boundary conditions at $z = \eta$ in terms of the potential function ϕ introduced in (11.11).

$$(\text{Ans. } \eta_t = \phi_z, \tag{11.39a}$$

$$\phi_{tt} + g\eta_t = 0.) \tag{11.39b}$$

3. Given a pressure gauge which is sensitive to pressure variations equivalent to ± 1 cm of water, determine from (11.32) the minimum wavelength at which a wave of 10 cm amplitude will be detectable in a 2000 m deep ocean.

(Ans. $\lambda \simeq 4200$ m.)

4. Show by integrating (11.28) and (11.30) that individual water particles describe closed elliptical orbits to a first approximation (cf. Section 13 for the next approximation).

5. Show that the presence of rotation $[2\boldsymbol{\Omega} = (0, 0, f)]$, introducing a transverse horizontal velocity component v as given by (11.34), tilts the plane of orbital motion to the left of the direction of wave propagation by an angle $\theta = \tan^{-1}(f/\omega)$.

6. From (11.16) find the period of the short gravity–capillary waves with the slowest phase and group velocities (assume $kH \gg 1$).

(Ans. $T = 0.074$ s; 0.156 s.)

7. Derive an interfacial condition for the continuity of pressure, taking into account the presence of surface tension. Use this condition to derive the dispersion relation

$$\omega^2 = \frac{gk(\rho_2 - \rho_1)}{(\rho_2 + \rho_1)} + \frac{k^3\sigma}{(\rho_1 + \rho_2)}, \tag{11.40}$$

valid for short waves.

12. NONLINEAR WAVES OF HIGH FREQUENCY

Let us now make a pause in the development of our survey of ocean waves to examine the validity of the linearizing assumption under which all the wave solutions obtained so far have been derived. In this section we give a brief discussion of how the inclusion of hitherto neglected nonlinear terms modify these small-amplitude waves.

Because of their transverse nature (cf. 8.7), the plane gyroscopic–internal gravity waves described in Section 8 are *exact* solutions of the nonlinear equations (3.9) and (3.11), of which (8.2) and (8.1) respectively are linearized versions. The terms neglected in passing from the full equations to the linearized set are of the form $\boldsymbol{u} \cdot \nabla \boldsymbol{u}$ and $\boldsymbol{u} \cdot \nabla \rho$, i.e., $u_j k_j u_i$ and $u_j k_j \rho$ for plane waves: such terms vanish since $u_j k_j = 0$ [from (8.7)]. The solutions of Section 8 are then valid for waves of arbitrary amplitude. However, the sum of two or more such solutions is *not* an exact solution of the nonlinear equations (Exercise 12.1). The normal mode solutions of Section 10, constructed by the addition of a pair of interfering plane waves, are thus limited in their validity to waves of sufficiently small amplitudes.

The validity of linearization

We have already mentioned in Section 5, as a requirement for the validity of linearization, the condition that the ratio of particle velocity to phase velocity, $|\boldsymbol{u}|/|\boldsymbol{c}|$ be much smaller than unity. Let us examine the restriction placed by this inequality on the amplitude of one of the normal mode solutions of (10.41) and (10.42). We shall limit ourselves to frequencies high enough that rotation may be neglected and to a two-dimensional situation ($\partial_y \equiv 0$). Under these restrictions, (10.39) reduces to

$$\omega U^{(h)} = k_1 P, \tag{12.1a}$$

$$V^{(h)} = 0. \tag{12.1b}$$

The solutions to the vertical eigenvalue problem (10.41) and (10.42) for a rigid-topped, uniformly stratified ocean ($N =$ constant) have already been presented in Exercise 10.8:

$$Z = \sin [n\pi(z + H)/H], \quad n = 1, 2, \ldots \tag{12.2}$$

to within an amplitude constant, with

$$n^2\pi^2/H^2 = (N^2 - \omega^2)/gh_n . \tag{12.3}$$

For $\omega \gg f$ and $k_2 = 0$, (10.45) reduces to

$$gh_n = \omega^2/k_1^2 = c^2. \tag{12.4}$$

Let us write the horizontal dependence of $U^{(h)}$ as

$$U^{(h)} = u_0 \, e^{ikx}, \tag{12.5}$$

where, for convenience, we have dropped the subscript 1 on k.

Then, from (12.1a),

$$P = cu_0 \, e^{ikx}, \tag{12.6}$$

and from (10.38)

$$\mathfrak{D} = -\rho_0 h_n \frac{n\pi}{H} \cos\left[\frac{n\pi}{H}(z+H)\right]. \tag{12.7}$$

The complete spatial dependence of the velocity components u and w are then given by

$$u(x,z) = -u_0 \frac{n\pi}{H} h_n \cos\left[\frac{n\pi}{H}(z+H)\right] e^{ikx}, \tag{12.8a}$$

$$w(x,z) = \frac{i\omega c u_0}{g} \sin\left[\frac{n\pi}{H}(z+H)\right] e^{ikx}. \tag{12.8b}$$

The vertical displacement ξ is found by integrating (12.8b) times $e^{-i\omega t}$ with respect to time:

$$\xi = -\frac{cu_0}{g} \sin\left[\frac{n\pi}{H}(z+H)\right] e^{ikx}. \tag{12.9}$$

From (12.8a), (12.9) and (12.4), the ratio of the maximum horizontal velocity $|u|$ to the wave phase speed may be written as

$$|u|/c = n\pi|\xi|/H, \tag{12.10}$$

where $|\xi|$ is the maximum amplitude of ξ as given by (12.9). The criterion for the validity of linearization is then

$$|\xi|/H \ll 1/n\pi, \quad n = 1, 2, \ldots \tag{12.11}$$

which may be interpreted as a condition on the amplitude of vertical displacement as compared to the vertical scale of the wave, H/n.

The linearization criterion for surface waves is simpler to obtain. From (11.28) and the dispersion relation (11.16) we find that at the free surface ($z = 0$)

$$|u|/c = ak \coth kH, \tag{12.12}$$

where a is the amplitude of the free surface displacement. For short waves (deep water: $kH \gg 1$), $|u|/c \ll 1$ if

$$ak \ll 1, \tag{12.13}$$

i.e., the wave slope is small. On the other hand, for long waves (shallow water: $kH \ll 1$), the relevant criterion is

$$a/H \ll 1, \tag{12.14}$$

which is similar to (12.11) and relates the surface displacement to the water depth. Note that (12.13) together with $kH \gg 1$ also implies (12.14), and that (12.14) with $kH \ll 1$ implies (12.13). Both inequalities (12.13) and (12.14) must then be satisfied for the linear surface wave solution to be valid.

Large-amplitude internal waves

As an example of the distortion of internal gravity modes by nonlinearity, let us examine wave motion in a uniformly stratified ($N =$ constant) Boussinesq fluid with a flat rigid top and a flat bottom. The solutions to the linear formulation of this problem have already been discussed above: (12.2), (12.8). We consider a two-dimensional situation ($\partial_y \equiv 0$) in the absence of rotation; the mass continuity equation is then satisfied by a stream function ψ defined by

$$u = -\psi_z, \quad w = \psi_x. \tag{12.15}$$

The only component of vorticity must then be in the y-direction, and the vorticity equation (4.2) becomes, with the help of (12.15)

$$\rho_* \left[\frac{\partial}{\partial t} \nabla^2 \psi - J(\nabla^2 \psi, \psi) \right] = -g\rho_x, \tag{12.16}$$

where $J(\nabla^2\psi, \psi)$ is the Jacobian operator: $J(\nabla^2\psi, \psi) = \psi_z \nabla^2 \psi_x - \psi_x \nabla^2 \psi_z$. The density equation (3.9) may be written as

$$\rho_t - J(\rho, \psi) = 0. \tag{12.17}$$

A standard method of exploring finite-amplitude solutions of a system such as (12.16)–(12.17) is to expand the dependent variables ρ and ψ in terms of a power series of an appropriate small parameter. Thus, here a suitable representation would be

$$\rho = \rho_0(z) + \sum_{m=1}^{\infty} \epsilon^m \rho^{(m)}(x, z, t), \tag{12.18a}$$

$$\psi = \sum_{m=1}^{\infty} \epsilon^m \psi^{(m)}(x, z, t), \tag{12.18b}$$

where the small parameter ϵ could, from (12.11), be chosen as the ratio of the vertical displacement to the depth of the fluid. Substitution of (12.18) into (12.16) and (12.17) yields a series of problems of increasing order in ϵ from which the $\rho^{(m)}$ and $\psi^{(m)}$ may be determined (Thorpe, 1968). Provided the series (12.18) converge, a solution for internal waves of finite (but still small) amplitude is obtained.

To first order in ϵ, (12.16) and (12.17) become upon substitution of (12.18)

$$\rho_* \nabla^2 \psi_t^{(1)} + g\rho_x^{(1)} = 0, \tag{12.19}$$

$$\rho_t^{(1)} + \rho_{0z} \psi_x^{(1)} = 0. \tag{12.20}$$

For a constant N and rigid boundaries the solution corresponds to that found above in (12.8):

$$\psi^{(1)} = A \sin(n\pi z/H) \sin(kx - \omega t), \tag{12.21}$$

$$\rho^{(1)} = \frac{\rho_{0z} A}{c} \sin(n\pi z/H) \sin(kx - \omega t), \tag{12.22}$$

with the dispersion relation

$$N^2/c^2 = k^2 + (n^2\pi^2/H^2), \tag{12.23}$$

which is also found from (12.3) and (12.4). To the next order in ϵ,

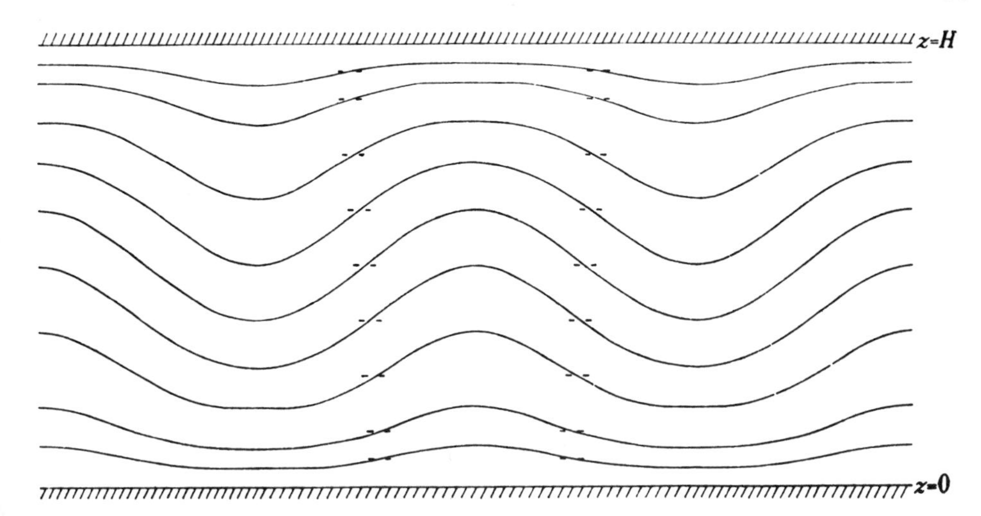

Fig. 12.1. Vertical displacement profiles for internal gravity waves of the first mode between rigid planes. The density profile is linear with depth; the amplitude is $|\xi^{(1)}(z = H/2)| = H/8$ [in the notation of (12.29)]. (From Thorpe, 1968.)

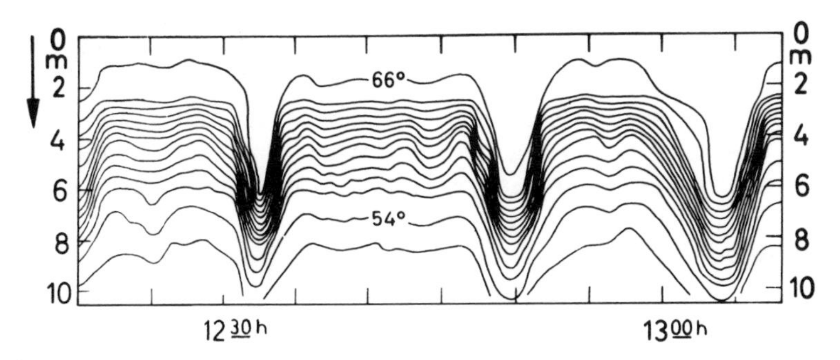

Fig. 12.2 Isotherm displacement in near-surface internal waves, showing the strong asymmetry in the wave profile induced by nonlinearity. (From Lafond, 1962.)

$$\rho_* \nabla^2 \psi_t^{(2)} + g\rho_x^{(2)} = \rho_* J(\nabla^2 \psi^{(1)}, \psi^{(1)}), \tag{12.24}$$

$$\rho_t^{(2)} + \rho_{0z}\psi_x^{(2)} = J(\rho^{(1)}, \psi^{(1)}). \tag{12.25}$$

Since both $\nabla^2\psi^{(1)}$ and $\rho^{(1)}$ are proportional to $\psi^{(1)}$, the Jacobians vanish and there are no second-order corrections. The linear solutions are thus correct to order ϵ^2 and are more accurate than would have been expected from (12.11). As the method of solution to higher orders in ϵ is straightforward, the calculation of the third-order corrections will be left as an exercise to the reader. There is, however, one nonvanishing second-order effect: the wave profile, as characterized by the position of an isopycnal, is distorted from its sinusoidal shape. Let $z = z_0 + \xi$ be the position of a line of fluid density $\rho_0(z_0)$. Since the density is invariant on a path line,

$$\rho_0(z_0) = \rho(z_0 + \xi). \tag{12.26}$$

Expanding the right side in a Taylor series about z_0 and then substituting (12.18a) and the series

$$\xi = \sum_{m=1}^{\infty} \epsilon^m \xi^{(m)}, \tag{12.27}$$

we obtain

$$\xi^{(1)} = -\rho^{(1)}/\rho_{0z}, \tag{12.28a}$$

$$\xi^{(2)} = -\{\rho^{(2)} + \xi^{(1)}\rho_z^{(1)} + (\xi^{(1)})^2\rho_{0zz}/2\}/\rho_{0z}. \tag{12.28b}$$

To first order, the waveshape is purely sinusoidal and of the form (12.22); to second order, however [with $\rho_{0zz} = 0$ and $\rho^{(2)} = 0$ (from 12.24 and 12.25)]

$$\xi = \xi^{(1)} + \epsilon\xi^{(2)} = -\frac{A}{c}\sin(n\pi z/H)\sin(kx-\omega t)$$
$$+\frac{\epsilon A^2}{2c^2}\frac{n\pi}{H}\sin(2n\pi z/H)\sin^2(kx-\omega t). \tag{12.29}$$

Let us concentrate on the first mode ($n = 1$). The correction $\xi^{(2)}$ to the first-order sinusoidal displacement $\xi^{(1)}$ is then negative in $-H/2 < z < 0$, i.e., in the upper half of the fluid, and positive in the lower half $-H < z < -H/2$. The crests are flattened and the troughs deepened in the upper half of the fluid, with the situation reversed in the lower half (Fig. 12.1). This type of distortion is evident and more pronounced in the observations shown in Fig. 12.2. It has also been shown by Griscom (1967) that interfacial waves are distorted in a similar manner. For a thin upper layer ($d \ll H$, in the notation of Sections 10 and 11), the wave crests are flattened and the wave troughs deepened by nonlinear effects. For a thin lower layer ($H - d \ll H$), the distortion is reversed.

By changing variables to

$$s = x - ct, \tag{12.30a}$$

$$\Psi = \psi + cz, \tag{12.30b}$$

the density equation (12.17) may be recast as

$$J(\rho, \Psi) = 0, \tag{12.31}$$

where the Jacobian is now defined with respect to s and z. It then follows that

$$\rho = \rho(\Psi). \tag{12.32}$$

Substitution of (12.30) and (12.32) into the vorticity equation (12.16) yields [using $(\partial/\partial t + c\partial/\partial x)\psi = 0$]

$$J(\nabla^2\Psi, \Psi) = \frac{g}{\rho_*}\frac{d\rho}{d\Psi}\Psi_s, \tag{12.33}$$

which may also be written as

$$J\left(\nabla^2\Psi + \frac{g}{\rho_*}\frac{d\rho}{d\Psi}z, \Psi\right) = 0. \tag{12.34}$$

This equation has a first integral

$$\nabla^2\Psi + \frac{g}{\rho_*}\frac{d\rho}{d\Psi}z = F(\Psi), \tag{12.35}$$

a form first obtained by Long (1953) and also discussed by Yih (1960). The function $F(\Psi)$ is in general arbitrary; it may be found for any particular case by specifying Ψ, Ψ_s and Ψ_{ss} at $s = 0$ as initial conditions. As (12.35) is simpler than the pair (12.16) and (12.17), of which it is an integral, it affords hope that exact nonlinear solutions of internal motions might be obtained by a further integration. By letting $F(\Psi)$ be a linear function of Ψ and taking $(g/\rho_*)\,d\rho/d\Psi = \text{constant}$, Magaard (1965) has derived a special solution of (12.35). However, choosing a linear form for $F(\Psi)$ is equivalent to a linearization of the vorticity equation, and it is not surprising that his solution should be identical (within the Boussinesq approximation) to the linear solution (12.21), (12.22). It is indeed readily verified that the linear solution satisfies (3.9) and (3.10) identically, so that the momentum equations (3.11) (with $\boldsymbol{\Omega} = \boldsymbol{0}$) will also be satisfied provided the pressure (which does not enter into the vorticity equation) is modified from the form it takes in the linear case (Exercise 12.3).

We have consistently restricted our attention to a Boussinesq fluid. More generally (see Exercises 12.3 and 12.4), a first integral of the full nonlinear non-Boussinesq problem can also be derived in the form (12.10), as was shown by Long (1953). The significance of the Boussinesq approximation has been critically examined by Long (1965). The Boussinesq approximation consists of neglecting terms of the first and higher orders in the small parameter N^2H/g. When the influence of some other small parameter ϵ (such as the amplitude parameter $|\xi|/H$) is examined, the neglect of terms of $0(N^2H/g)$ becomes justifiable only if $N^2H/g \ll \epsilon$. For $N^2H/g \simeq \epsilon$, corrections due to the variation of density in the inertia of the fluid and finite-amplitude effects (due to ϵ) must be considered jointly. We shall encounter a similar two-parameter dependence in the analysis of long surface gravity waves and comment further on this particular point there.

Nonlinear surface waves: the Stokes expansion

As we have noted earlier in (12.13) and (12.14), the parameters ak and a/H, where a is the amplitude of the surface displacement, must both be small for a linearized solution of the surface gravity wave problem to be valid. To determine the effect of small nonlinearities on the linear solution we therefore must use a perturbation expansion in the two small parameters. It will be convenient to formulate the problem in terms of the velocity potential ϕ introduced in (11.11). From (11.12), (3.12)–(3.14) and (11.13), the surface gravity wave problem (we ignore capillarity) is stated as

$$\nabla^2\phi = 0, \quad -H \leqslant z \leqslant \eta, \tag{12.36}$$

$$\phi_z = 0, \quad z = -H, \tag{12.37}$$

$$D\eta/Dt = \phi_z, \quad z = \eta, \tag{12.38a}$$

$$\frac{D}{Dt}(\phi_t + g\eta + \tfrac{1}{2}\boldsymbol{u}\cdot\boldsymbol{u}) = 0, \quad z = \eta. \tag{12.38b}$$

The boundary condition (12.38b) follows from (11.13) provided $Dp/Dt = 0$ at the free surface, i.e., for free waves. In order to bring out the two-parameter dependence intimated above, let us introduce nondimensional variables (indicated by primes) as follows:

$$\eta = a\eta', \tag{12.39a}$$

$$(x,y) = \lambda(x',y'), \tag{12.39b}$$

$$z = Hz', \tag{12.39c}$$

$$t = \lambda t'/\sqrt{gH}, \tag{12.39d}$$

$$\phi = a\lambda\sqrt{gH}\phi'/H. \tag{12.39e}$$

The obvious choices have been made for the geometrical scales; the time has been scaled by the time interval over which a long wave (of speed $\sqrt{gH}$) travels one wavelength; the choice of scale for the potential ensures that the linear terms of (12.36)–(12.38) are of similar magnitudes and free of parameters. Defining

$$\epsilon = a/H, \quad \mu = (H/\lambda)^2 \tag{12.40}$$

and focusing our attention on two-dimensional waves ($\partial_y \equiv 0$), we obtain, upon substituting (12.39) into (12.36)–(12.38), the nondimensional problem

$$\phi'_{z'z'} + \mu\phi'_{x'x'} = 0, \quad -1 \leqslant z' \leqslant \epsilon\eta', \tag{12.41}$$

$$\phi'_{z'} = 0, \quad z' = -1, \tag{12.42}$$

$$\mu[\eta'_{t'} + \epsilon\eta'_{x'}\phi'_{x'}] = \phi'_{z'}, \quad z' = \epsilon\eta', \tag{12.43a}$$

$$[\partial_{t'} + \epsilon\phi'_{x'}\partial_{x'}]\, \phi'_{t'} + \eta' + \frac{\epsilon}{2\mu}(\mu\phi'_{x'}\phi'_{x'} + \phi'_{z'}\phi'_{z'}) = 0, \quad z' = \epsilon\eta'. \tag{12.43b}$$

The upper boundary conditions are evaluated using a Taylor series expansion about $z' = 0$:

$$\phi'(\epsilon\eta') = \phi'(0) + \epsilon\eta'\phi'_{z'}(0) + \ldots. \tag{12.44}$$

The small parameter ak does not enter explicitly in (12.41)–(12.43). The equivalent parameter $\delta = a/\lambda$ is, however, related to ϵ and μ through

$$\delta = \epsilon\sqrt{\mu}. \tag{12.45}$$

Combinations of relative magnitudes of ϵ and μ in (12.41)–(12.43) lead to various types of wave solutions. Let us first consider propagation in deep water, such that $H/\lambda = 0(1)$. Then, $\epsilon \ll \mu = 0(1)$, and the only small parameter in the problem is ϵ. The variables η' and ϕ' will now be expanded in a series in powers of ϵ; dropping the primes,

$$(\eta, \phi) = \sum_{n=0}^{\infty} \epsilon^n(\eta^{(n)}, \phi^{(n)}). \tag{12.46}$$

The series begins with terms of order unity since both η and ϕ have been normalized with respect to the wave amplitude. To zeroth order in ϵ, we recover the linear problem; the upper boundary conditions are

$$\mu\eta_t^{(0)} = \phi_z^{(0)} \quad \text{on } z = 0, \tag{12.47a}$$

$$\phi_t^{(0)} = -\eta^{(0)} \quad \text{on } z = 0, \tag{12.47b}$$

and the solution in nondimensional variables is

$$\eta^{(0)} = \cos(kx - \omega t), \tag{12.48a}$$

$$\phi^{(0)} = \frac{\cosh[k\sqrt{\mu}(z+1)]}{\omega\cosh(k\sqrt{\mu})}\sin(kx - \omega t). \tag{12.48b}$$

The dispersion relation is obtained from (12.47a):

$$\omega^2 = \left(\frac{k}{\sqrt{\mu}}\right)\tanh(k\sqrt{\mu}), \tag{12.49}$$

which is equivalent to (11.16) with $B = 0$. To first order in ϵ, the upper boundary conditions become

$$\mu\eta_t^{(1)} - \phi_z^{(1)} = \eta^{(0)}\phi_{zz}^{(0)} - \mu\eta_x^{(0)}\phi_x^{(0)} \quad \text{at } z = 0, \tag{12.50a}$$

$$\begin{aligned}\phi_{tt}^{(1)} + \eta_t^{(1)} = &-\{[\eta^{(0)}\phi_{tz}^{(0)} + \phi_x^{(0)}\phi_x^{(0)}/2 + \phi_z^{(0)}\phi_z^{(0)}/2\mu]_t \\ &+ \phi_x^{(0)}\phi_{xt}^{(0)} + \phi_x^{(0)}\eta_x^{(0)}\} \quad \text{at } z = 0.\end{aligned} \tag{12.50b}$$

Substituting for $\phi^{(0)}$ and $\eta^{(0)}$ from (12.48) and using (12.49), (12.50) becomes

$$\mu\eta_t^{(1)} - \phi_z^{(1)} = \frac{k^2\mu}{\omega}\sin[2(kx - \omega t)], \quad z = 0 \tag{12.51a}$$

$$\phi_{tt}^{(1)} + \eta_t^{(1)} = -\left[\frac{k\omega\sqrt{\mu}}{\sinh(2k\sqrt{\mu})} - \mu\omega^3\right]\sin[2(kx - \omega t)], \text{ at } z = 0. \tag{12.51b}$$

These two conditions may be combined into one for $\phi^{(1)}$ alone:

$$\phi_{tt}^{(1)} + \frac{1}{\mu}\phi_z^{(1)} = \frac{-3\omega k\sqrt{\mu}}{\sinh(2k\sqrt{\mu})}\sin[2(kx - \omega t)], \quad z = 0. \tag{12.52}$$

In order to satisfy Laplace's equation (12.41) and the bottom boundary condition (12.42), as well as (12.52), $\phi^{(1)}$ must be given by

$$\phi^{(1)} = \frac{3}{8}\frac{\omega\mu}{\sinh^4(k\sqrt{\mu})}\cosh[2k\sqrt{\mu}(z+1)]\sin[2(kx - \omega t)] + Kt, \tag{12.53}$$

where K is a constant of integration. The velocity corrections $u^{(1)}$, $w^{(1)}$ may be written from (12.53) by inspection; the pressure $p^{(1)}$ (nondimensionalized with respect to $\rho_* ga$) is found from Bernoulli's equation (11.13) as

$$p^{(1)} = -[\phi_t^{(1)} + \tfrac{1}{2}(\phi_x^{(0)}\phi_x^{(0)} + \phi_z^{(0)}\phi_z^{(0)}/\mu)]. \tag{12.54}$$

In view of (12.53) and (12.48b), the first pressure correction is then calculated as

$$\begin{aligned}p^{(1)} = -K + \frac{k^2}{4\omega^2\cosh^2(k\sqrt{\mu})}&\left\{\left(3\frac{\cosh[2k\sqrt{\mu}(z+1)]}{\sinh^2(k\sqrt{\mu})} - 1\right)\cos[2(kx - \omega t)]\right. \\ &\left.- \cosh[2k\sqrt{\mu}(z+1)]\right\}.\end{aligned} \tag{12.55}$$

Note that $p^{(1)}$ includes time-independent terms; the average pressure $\langle p^{(1)} \rangle$ does not vanish and is slightly decreased by the presence of surface waves. The surface displacement $\eta^{(1)}$ is evaluated from (12.51) and (12.53) as

$$\eta^{(1)} = \frac{k^2[2\sinh^2(k\sqrt{\mu}) - 3]}{4\omega^2 \sinh^2(k\sqrt{\mu})} \cos 2(kx - \omega t). \tag{12.56}$$

The shape of the wave profile, as given by $\eta^{(0)} + \eta^{(1)}$ is modified from a pure sinusoid $(\eta^{(0)})$ towards a form with sharper crests and flatter troughs, similar to that seen in the internal wave distortion in (12.29). The integration constant K which appears in the potential (12.53) may be determined from a consideration of the vertical momentum flux balance inside the fluid. At any level within the fluid, the vertical flux of vertical momentum must, on the average, be sufficient to hold the weight of the fluid above that level, i.e.,

$$\langle p + \rho_* w^2 \rangle = -\rho_* gz. \tag{12.57}$$

Our development of the pressure consists of the sum of the hydrostatic pressure for a free surface with mean level at $z = 0$, plus a series of pressure corrections $p^{(n)}$ due to the wave motion. To first order in ϵ in the nondimensionalized variables, (12.57) then expresses the balance

$$\langle p^{(1)} \rangle + \langle \phi_z^{(0)} \phi_z^{(0)} \rangle / \mu = 0, \tag{12.58}$$

from which $\quad K = -k^2/4\omega^2 \cosh^2(k\sqrt{\mu}).$ $\qquad$ (12.59)

The calculation of higher order corrections to the linear theory is of considerable algebraic complexity. The steepening of wave crests and flattening of wave troughs continues to be enhanced to the next order, the surface displacement approximating a *trochoidal* curve (the curve traced by a point on a circle which rolls on a plane) (Kinsman, 1965, Chapter 5). In order to avoid secular terms the frequency must also be corrected for the wave amplitude:

$$\omega^2 = \frac{k}{\sqrt{\mu}} \tanh(k\sqrt{\mu}) \left[1 + \left(\frac{9\tanh^4 k\sqrt{\mu} - 10\tanh^2 k\sqrt{\mu} + 9}{8\tanh^2 k\sqrt{\mu}} \right) 4\pi^2 \epsilon^2 \mu + \ldots \right]. \tag{12.60}$$

Surface gravity waves thus acquire amplitude dispersion as a second-order correction: large-amplitude waves travel slightly faster than smaller waves (Whitham, 1974, Chapter 13). Expansions to fifth order in the parameter $\epsilon\sqrt{\mu}$ have been presented by Skjelbreia and Hendrickson (1961).

Surface wave instability

The success of the perturbation expansion as a means of representing waves of finite amplitude depends on the convergence of the series (12.46). Convergence proofs of Stokes' expansion, a series in powers of $\epsilon\sqrt{\mu}$, were presented by Levi–Civita (1925) and Struik (1926). Convergence does not imply stability, however! Benjamin and Feir (1967) discovered by experiment and theory that deep water surface gravity waves are unstable (the instability occurring for $kH > 1.363$). This instability will be examined further in Section 38, in which we shall return to finite-amplitude effects under the topic "wave–

wave interactions". The appearance of the instability is related to the phenomenon of amplitude dispersion, implicit in (12.60) and may be quite simply established as follows.

Let us express (12.60) in the abbreviated form (and in dimensional notation)

$$\omega = \omega^{(0)} + \omega^{(2)}a^2, \tag{12.61}$$

where $\omega^{(2)}$ is not a frequency but a function of dimension frequency/length squared. The law of conservation of wave crests (6.4) becomes

$$\frac{\partial k}{\partial t} + \frac{\partial}{\partial k}[\omega^{(0)} + \omega^{(2)}a^2]\frac{\partial k}{\partial x} + \omega^{(2)}\frac{\partial a^2}{\partial x} = 0. \tag{12.62}$$

Excluding powers of the amplitude higher than the second, and letting the energy density be proportional to the square of the wave amplitude (see next section), the energy equation (6.42) takes the form

$$\frac{\partial a^2}{\partial t} + \frac{\partial}{\partial x}(c_g^{(0)}a^2) = 0, \tag{12.63}$$

$$\text{where} \quad c_g^{(0)} = \partial\omega^{(0)}/\partial k. \tag{12.64a}$$

Equations (12.62) and (12.63) are a coupled set of equations for the wavenumber and the square of the amplitude. Expressing the second-order correction in the group velocity as

$$c_g^{(2)} = a^2\partial\omega^{(2)}/\partial k, \tag{12.64b}$$

the set of coupled equations (12.62) and (12.63) may be written in the form

$$\mathbf{AY}_x + \mathbf{IY}_t = \mathbf{0}, \tag{12.65}$$

$$\text{where} \quad \mathbf{Y} = \begin{bmatrix} a^2 \\ k \end{bmatrix}, \tag{12.66a}$$

$$\mathbf{A} = \begin{bmatrix} c_g^{(0)}, & a^2\partial c_g^{(0)}/\partial k \\ \omega^{(2)}, & c_g^{(0)} + c_g^{(2)} \end{bmatrix}, \tag{12.66b}$$

$$\mathbf{I} = \begin{bmatrix} 1 & 0 \\ 0 & 1 \end{bmatrix}. \tag{12.66c}$$

The characteristic speeds V of (12.65) are found from the determinant

$$|\mathbf{A} - V\mathbf{I}| = 0, \tag{12.67}$$

$$\text{i.e.,} \; V = (c_g^{(0)} + c_g^{(2)}/2) \pm [(c_g^{(0)} + c_g^{(2)}/2)^2 - c_g^{(0)}(c_g^{(0)} + c_g^{(2)}) + a^2\omega^{(2)}\partial c_g^{(0)}/\partial k]^{1/2}. \tag{12.68}$$

To leading order in the amplitude a,

$$V \simeq c_g^{(0)} \pm a[\omega^{(2)}\partial c_g^{(0)}/\partial k]^{1/2}. \tag{12.69}$$

For deep water waves ($kH \gg 1$), (12.60) reduces (in dimensional form) to

$$\omega^2 = gk(1 + a^2k^2), \tag{12.70}$$

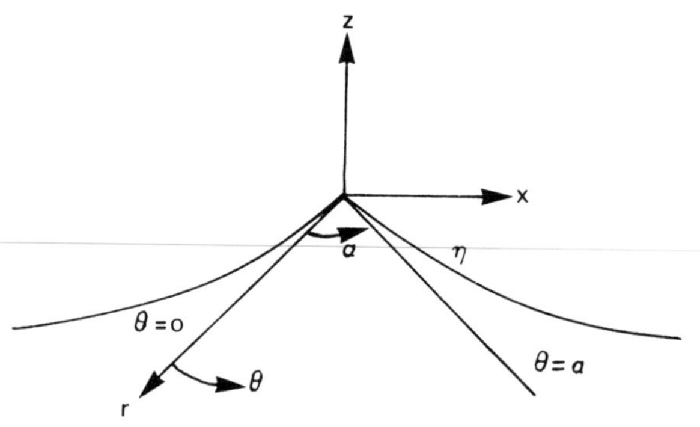

Fig. 12.3. The surface gravity wave of maximum steepness: definition sketch.

so that $\omega^{(2)} = g^{1/2}k^{5/2}/2$. As $\partial c_g^{(0)}/\partial k < 0$ for all gravity waves, which are normally dispersive, V is complex. Plane waves with complex speed of propagation grow in time and may be said to be unstable (Exercise 12.10). The case of arbitrary finite depth and the detailed analysis of this instability are discussed by Whitham (1967).

The surface wave of maximum steepness

The perturbation solution shows that the surface wave profile tends to acquire sharp crests. An argument due to Stokes (1880) permits evaluation of the maximum steepness of the crests without recourse to perturbation methods. In a coordinate system moving with the wave, the wave profile appears stationary. Let us assume that the crest is sharply peaked, forming an apex of angle α (Fig. 12.3) where η_x is discontinuous.

Steady potential flow around a corner of angle α can be expressed in terms of the velocity potential (Batchelor, 1967, Chapter 6)

$$\phi = Cr^{\pi/\alpha} \cos(\pi\theta/\alpha), \tag{12.71}$$

where (r, θ) are the plane polar coordinates defined in Fig. 12.3 and C is an amplitude constant. Near the crest, the free surface is given by the lines $\theta = 0, \alpha$, on which

$$z = -r\cos(\alpha/2). \tag{12.72}$$

On the free surface, the pressure is a constant (which we take equal to zero); in the state, Bernoulli's equation (11.13) reduces to

$$(\nabla\phi)^2/2 + gz = 0 \tag{12.73}$$

on the free surface. Substituting (12.71) and (12.72) into (12.73), we find that the latter equation can hold only if

$$2(\pi/\alpha - 1) = 1,$$

i.e., $\alpha = 120°$. This sharp angle is found only near the wavecrest; the troughs are flat and have zero slope. By taking the average of the slope of the sea surface at the crest (tan 30°) and at the trough, we obtain a gross estimate of the maximum slope of surface gravity waves:

$$\frac{2a}{\lambda/2} \simeq \frac{\tan 30^\circ + 0}{2}; \quad \text{therefore } 2a/\lambda \simeq 1/7. \tag{12.74}$$

For a more precise method of estimation, see Michell (1893).

To second order in the wave slope, the phase velocity may be written as

$$c = c^{(0)} + a^2k^2c^{(2)}. \tag{12.75}$$

From the dispersion relation (12.70) the speed of waves with maximum slope, as given by (12.74), is $c \simeq 1.1c^{(0)}$: finite-amplitude corrections to the phase speed are always small.

From (12.71) it follows that at the wave crest, $u = 0$ in a coordinate system moving at the phase speed. Hence $u = c$ at the crest for a stationary observer. A steeper wave would presumably have $u > c$ on the crest; the particles at the crest would leave the crest behind and fall on the forward face of the wave. Waves in which $u > c$ at the crest break and lose energy until $u \leqslant c$ everywhere in the wave profile. The vertical acceleration at the crest of the wave is equal to $-g/2$ (Longuet-Higgins, 1963a). A more precise criterion of the initiation of wave breaking has been presented by Banner and Phillips (1974). When the presence of a thin wind drift layer is taken into account, wave breaking occurs well before the maximum wave steepness is reached. Banner and Phillips give for the elevation of the crests in a single wave train at the point of incipient breaking

$$\eta_{max} = (c^2/2g)(1 - u_d/c)^2,$$

where u_d is the drift velocity, about 3% of the wind speed.

For standing waves, the maximum amplitude occurs when a crest makes an angle of very nearly 90°, as shown experimentally by Taylor (1953) and explored theoretically by Penney and Price (1952) and Longuet-Higgins (1973). For that steepness, the downward acceleration at the crest is just equal to that of gravity; steeper waves just fly apart at the crest.

For travelling as well as for standing waves of maximum steepness, Longuet-Higgins (1973a) has found that the wave profile may be quite accurately approximated by the curve

$$z = \ln \sec x', \tag{12.76}$$

where the origin of the nondimensional horizontal coordinate x' is at the wave trough. For the travelling wave, $|x'| \leqslant \pi/6$ (a full wavelength corresponds to the range $-\pi/6 \leqslant x' \leqslant \pi/6$); the standing wave, on the other hand, lies in the range $|x'| \leqslant \pi/4$. The amplitudes of the waves of maximum steepness are readily evaluated from (12.76) (Exercise 12.11).

Finite-amplitude waves in shallow water

Returning to the nondimensional equations (12.41)–(12.43), let us now consider the shallow water case where $\epsilon \simeq \mu \ll 1$. In shallow water, the particle orbits are nearly horizontal and the vertical velocity component $w = \phi_z$ is small. It is then appropriate to expand $\phi(x, z, t)$ about its value at the bottom ($z = -1$). An expansion in powers of $\mu(z + 1)^2$ which satisfies both Laplace's equation (12.41) and the bottom boundary condition (12.42) is

$$\phi(x,z,t) = \chi(x,t) - \mu\frac{(z+1)^2}{2!}\chi_{xx} + \mu^2\frac{(z+1)^4}{4!}\chi_{xxxx} - \dots . \tag{12.77}$$

Substitution of (12.77) into the surface boundary conditions (12.43), together with the use of the expansion (12.44), gives

$$\eta_t + \chi_{xx} = -\epsilon(\eta_x\chi_x + \eta\chi_{xx}) + \frac{\mu}{6}\chi_{xxxx} + 0(\epsilon^2, \epsilon\mu, \mu^2), \tag{12.78}$$

$$\chi_{tt} + \eta_t = -\epsilon(\chi_x\eta_x + 2\chi_x\chi_{xt}) + \frac{\mu}{2}\chi_{xxtt} + 0(\epsilon^2, \epsilon\mu, \mu^2). \tag{12.79}$$

To zeroth order in ϵ and μ, (12.78) and (12.79) give:

$$\eta_t^{(0)} + \chi_{xx}^{(0)} = 0, \tag{12.80a}$$

$$\chi_{tt}^{(0)} + \eta_t^{(0)} = 0, \tag{12.80b}$$

where the superscripts denote the zeroth-order solution in ϵ and μ. Eliminating $\eta^{(0)}$ from (12.80), we obtain the wave equation

$$\chi_{xx}^{(0)} - \chi_{tt}^{(0)} = 0, \tag{12.81}$$

with solutions

$$\chi^{(0)}(x,t) = F(x \pm t), \tag{12.82}$$

where F is an arbitrary function. The solutions (12.82) consist of long waves ($\mu \ll 1$) of infinitesimal amplitude ($\epsilon \ll 1$), subject to neither phase nor amplitude dispersion. The wave speed is $c = 1$ in nondimensional terms.

Focusing our attention on waves which travel in the positive x-direction only, for which

$$\chi^{(0)} = F(x-t), \tag{12.83}$$

$$\eta^{(0)} = \chi_x = F', \tag{12.84}$$

we look for solutions of (12.78) and (12.79) which include corrections to $\eta^{(0)}$ and $\chi^{(0)}$ to first order in ϵ and μ. From (12.83) and (12.84) the following relations will hold:

$$\chi_x = \eta + 0(\epsilon, \mu), \tag{12.85a}$$

$$\partial_x = -\partial_t + 0(\epsilon, \mu). \tag{12.85b}$$

Using these approximations to express the right-hand sides of (12.78) and (12.79) solely in terms of η, we find that

$$\eta_t + \chi_{xx} = -\epsilon(\eta^2)_x + \frac{\mu}{6}\eta_{xxx} + 0(\epsilon^2, \epsilon\mu, \mu^2), \tag{12.86a}$$

$$\chi_{tt} + \eta_t = -\frac{\epsilon}{2}(\eta^2)_t - \frac{\mu}{2}\eta_{xxt} + 0(\epsilon^2, \epsilon\mu, \mu^2). \tag{12.86b}$$

To the order quoted in ϵ and μ, we may replace ∂_x by $-\partial_t$ in (12.86b). Adding (12.86b), after this substitution, to (12.86a), we find

$$\chi_{xx} - \chi_{tt} = -\frac{3}{2}\epsilon(\eta^2)_x - \frac{\mu}{3}\eta_{xxx} + 0(\epsilon^2, \epsilon\mu, \mu^2). \tag{12.87}$$

The left side of this equation may be factored as

$$\chi_{xx} - \chi_{tt} = \left(\frac{\partial}{\partial x} + \frac{\partial}{\partial t}\right)\left(\frac{\partial \chi}{\partial x} - \frac{\partial \chi}{\partial t}\right).$$

Again to the order required, one may replace χ_t by $-\chi_x$ and then, using (12.84), χ_x by η, always to $0(\epsilon^2, \epsilon\mu, \mu^2)$. One then finally obtains, to the order quoted above,

$$\eta_t + \eta_x + \frac{3\epsilon}{4}(\eta^2)_x + \frac{\mu}{6}\eta_{xxx} = 0. \tag{12.88}$$

This equation was first obtained by Korteweg and DeVries (1895) and has since been known under their name (KdV for short). Recent derivations for water waves are found in Broer (1964) and in a paper by Peregrine (1972) where a number of shallow-water wave approximations are presented. The KdV equation has been studied with renewed interest in recent years, having been found to describe wave propagation in a number of nonlinear dispersive systems [cf. Lick (1970) and Miura (1976) for recent reviews].

Waves of permanent form travel at a constant phase velocity c. For such solutions, we may express η as

$$\eta = G(x - ct) \equiv G(s). \tag{12.89}$$

Upon substitution of (12.89) into (12.88), an ordinary differential equation for G results:

$$(1 - c)G' + \frac{3\epsilon}{4}(G^2)' + \frac{\mu}{6}G''' = 0, \tag{12.90}$$

where primes denote differentiation with respect to $s = x - ct$. This last equation may be integrated once:

$$(1 - c)G + \frac{3\epsilon}{4}G^2 + \frac{\mu}{6}G'' = \text{constant}. \tag{12.91}$$

From (12.80), $c = 1$ to zeroth order in ϵ and μ: the constant of integration vanishes. Equation (12.91) clearly shows that for waves of permanent form, a balance between amplitude (ϵ) and phase (μ) dispersion must hold at every point of the wave profile. Rearranging (12.88), we have

$$(c - 1) = \frac{3\epsilon}{4}G + \frac{\mu}{6}\frac{G''}{G}. \tag{12.92}$$

The wave speed is modified from its normalized value of unity, appropriate to infinitely long ($\mu = 0$) and small-amplitude ($\epsilon = 0$) waves. The finite-amplitude correction is proportional to the wave amplitude itself; the finite wavelength modification is proportional to the relative curvature G_{xx}/G. For waves of permanent form, the wave profile $G(x)$ must be such that the two effects add to a constant, independent of position. The solutions of (12.91) may be obtained by further integration (Karpman, 1973, Section 10; Miura, 1976). One possible solution consists of the solitary wave

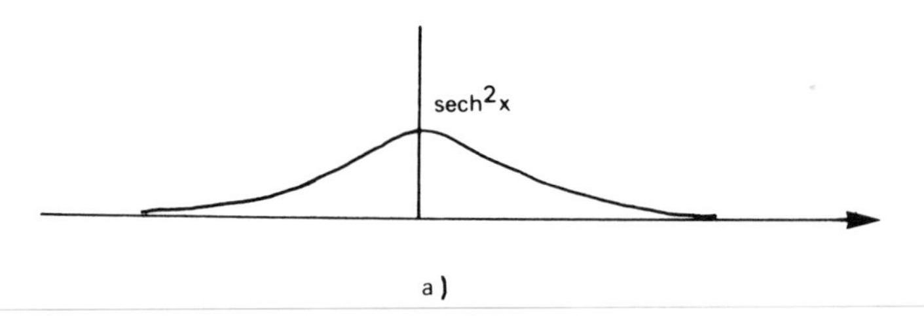

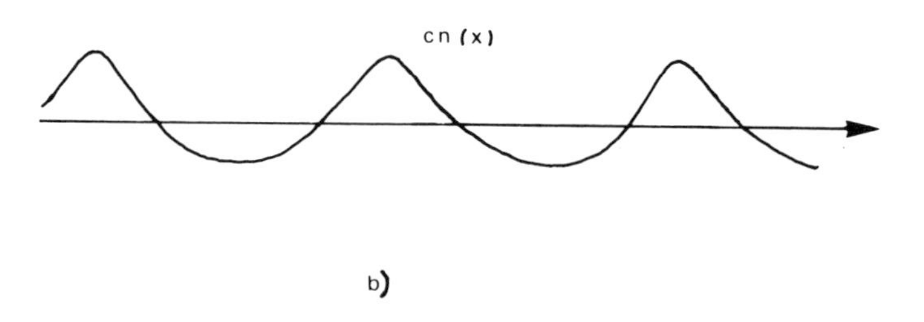

Fig. 12.4. Profiles of surface gravity waves of finite amplitude and permanent form: (a) the solitary wave; (b) a cnoidal wave.

$$\eta = \eta_0 \operatorname{sech}^2\left[\left(\frac{3\epsilon\eta_0}{4\mu}\right)^{1/2}(x-ct)\right], \tag{12.93}$$

where η_0 is the wave amplitude (Fig. 12.4a). The speed of this wave is computed from (12.92) as

$$c = 1+\epsilon\eta_0/2. \tag{12.94}$$

This wave was discovered experimentally by Russel (1844) and explained theoretically by Korteweg and DeVries (1895). Its significance has been more fully appreciated following the work of Lax (1968) who, inspired by the numerical results of Zabusky and Kruskal (1965), showed that the asymptotic solutions of (12.88) corresponding to any initial values consist of a train of "solitons" of the form (12.93).

A set of periodic solutions, expressed in terms of elliptic functions, may also be found for (12.88). These waves have sharpened crests and flattened troughs (Fig. 12.4b) as already seen from the perturbation solution (12.56). These so-called *cnoidal* waves may be found in the original paper of Korteweg and DeVries (1895) or in more recent references (Benjamin and Lighthill, 1954; Karpman, 1973; Miura, 1976).

Finite-amplitude waves of permanent form (i.e., with constant c) exist in media where amplitude and phase dispersion are both present. As such, they also occur as baroclinic modes of a stratified fluid (Long, 1953; Davis and Acrivos, 1967) and as interfacial waves (Long, 1956; Benjamin, 1966, 1967a; Gargett, 1976). We will also encounter them in Section 19 as finite-amplitude solutions for long waves in a rotating fluid.

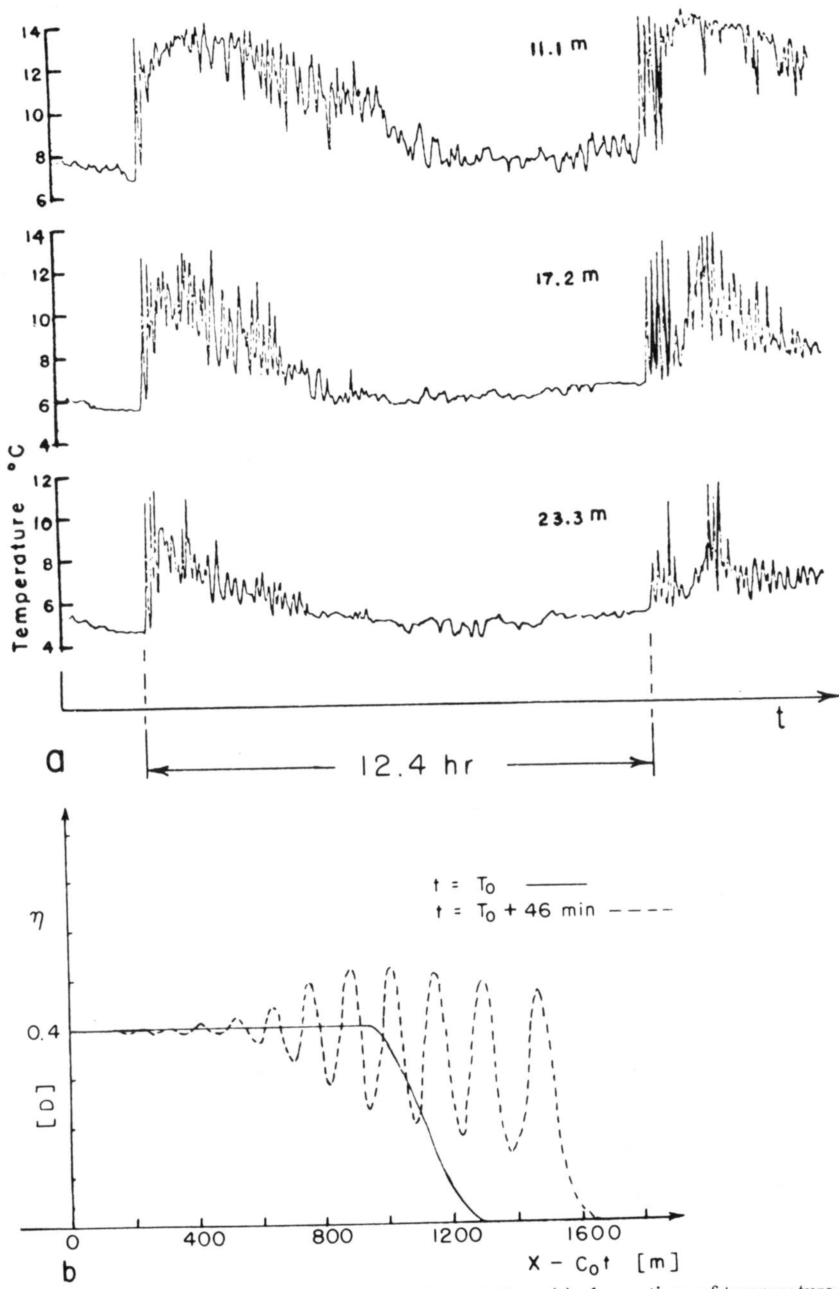

Fig. 12.5. Nonlinear internal waves in Massachussets Bay: (a) observations of temperature fluctuations as a function of time at various depths. The semi-diurnal tidal periodicity is evident (Halpern, 1971); (b) numerical results of the integration of a nonlinear wave equation similar to (12.88), from an initial condition of the form shown at $t = T_0$ (Lee and Beardsley, 1974).

An interesting application of an equation of the form (12.88) has been made by Lee and Beardsley (1974) to explain the internal wave groups observed by Halpern (1971) in Massachusetts Bay. The blocking effect of a ridge on the rising tide leads to the formation of a steep warm front (at $t = T_0$ in Fig. 12.5) which progresses landward from the obstacle, and is affected by nonlinear and phase dispersion. Taking a front of isotherm displacement at $t = T_0$ as an initial condition, a numerical integration of the wave equation shows that a sequence of waves (approximately solitons) appears as the initial front disperses. The oscillatory behaviour of temperature fluctuations is qualitatively similar to that observed by Halpern (1971).

The hydraulic limit

Another limiting case of the nondimensional equations (12.41)–(12.43) corresponds to $\mu \ll \epsilon \ll 1$. To first order in ϵ, and using (12.84) $[\eta = \chi_x + 0(\epsilon)]$, (12.78) and (12.79) become

$$\eta_t + \chi_{xx} + \epsilon(\eta\chi_x)_x = 0, \tag{12.95}$$

$$\frac{\mathrm{d}}{\mathrm{d}t}(\eta + \chi_t + \epsilon\chi_x^2/2) = 0, \tag{12.96}$$

where $\mathrm{d}/\mathrm{d}t = \partial_t + \epsilon\eta\partial_x$. Since $\chi_x = u$, the first equation (12.95) is recognized as the integrated mass continuity equation:

$$\eta_t + [(1 + \epsilon\eta)u]_x = 0. \tag{12.97}$$

The second equation (12.96) is a restatement of Bernoulli's equation (4.13) in a hydrostatic flow (where $w \ll u$) with a free surface. Thus

$$\eta + \chi_t + \epsilon\chi_x^2/2 = \text{constant} \tag{12.98}$$

on a path line. The level surface $z = 0$ is a path line. Any other level surface is also a path line since w is negligible for $\mu \ll \epsilon$.

Thus for all x along z = constant, (12.98) is satisfied. Hence, using $\chi_x = u$, (12.98) may be written, after differentiation with respect to x, as

$$u_t + \epsilon u u_x + \eta_x = 0 \tag{12.99}$$

which is the horizontal momentum equation for a hydrostatic fluid in which $w = 0$. Equations (12.97) and (12.99) are often called the hydraulic equations; they are analogous in form to the one-dimensional equations describing the flow of a compressible fluid. Nonlinear water waves in very shallow water (tides entering a river, for example) are exactly analogous to finite-amplitude sound waves and exhibit shock waves. Acoustic shock waves are discussed in Courant and Friedrichs (1948); their hydraulic analogy is analyzed in Stoker (1957) and in Whitham (1974), and illustrated in Tricker (1965).

The pair of equations (12.97), (12.99) may be written as

$$\mathbf{A}\mathbf{Y}_x + \mathbf{I}\mathbf{Y}_t = \mathbf{0}, \tag{12.100}$$

$$\text{where} \quad \mathbf{Y} = \begin{pmatrix} u \\ \eta \end{pmatrix}, \tag{12.101a}$$

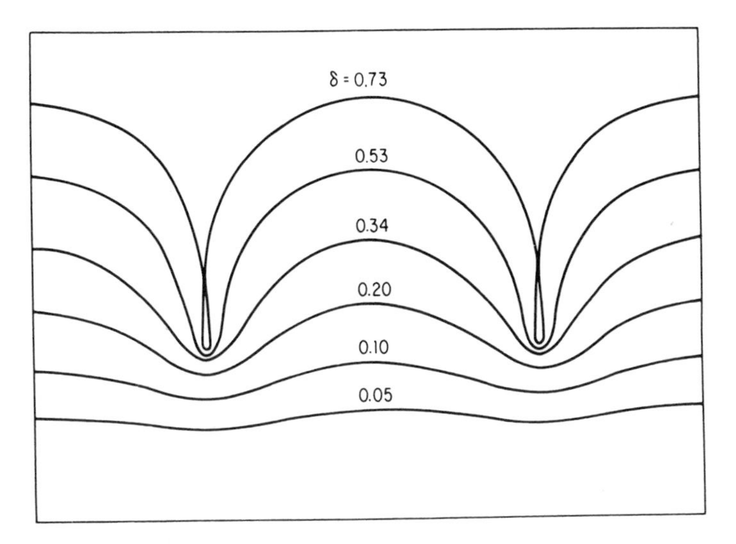

Fig. 12.6. Capillary wave profiles for a range of slope parameters $\delta = 2a/\lambda$. (From Crapper, 1957.)

$$\mathbf{A} = \begin{pmatrix} \epsilon u, & 1 \\ 1+\epsilon\eta, & \epsilon u \end{pmatrix}, \tag{12.101b}$$

$$\mathbf{I} = \begin{pmatrix} 1 & 0 \\ 0 & 1 \end{pmatrix}. \tag{12.101c}$$

The characteristic speeds V are found as the roots of a determinant of the form (12.67):

$$V = \epsilon u \pm (1+\epsilon\eta)^{1/2}. \tag{12.102}$$

The relation between the hydraulic equations in the form (12.97) and (12.99) and the limit $\mu = 0$ in the KdV equation is left to the reader as a subject of contemplation (Exercise 12.7).

From (12.102) (with the + sign), we infer that large waves invariably catch up with smaller waves. In particular, a wave crest travels faster than a trough and the forward face of a wave steepens with propagation distance. The conditions under which shock waves appear and the shape which they take under different friction laws are discussed by Whitham (1974).

Large capillary waves

Capillary waves have been excluded from the discussion of nonlinear gravity waves. Finite-amplitude effects on their wave profile and dispersion characteristics may also be examined by perturbation methods similar to those employed in the study of nonlinear gravity waves (Exercises 12.8 and 12.9). An exact nonlinear solution to the capillary wave problem has been found by Crapper (1957). The free surface displacement for large-amplitude capillary waves is shown in Fig. 12.6 for a range of values of the slope parameter $\delta = 2a/\lambda$. The crests are flattened and the troughs deepened by nonlinearity; the amplitude dispersion is also anomalous, large waves travelling more slowly than small ones, with the phase speed in deep water ($kH \gg 1$) given by

$$c = (\sigma k/\rho)^{1/2}(1 + \tfrac{1}{4}a^2k^2)^{-1/4}. \tag{12.103}$$

The above comments on nonlinear waves may seem disappointingly brief to the reader whose interest has been aroused by the fascinating properties of large-amplitude waves. Although we shall again discuss nonlinear effects in Sections 19 and 38, nonlinear waves remain peripheral to our central theme, and the curious reader is referred to the works quoted (especially Whitham, 1974) for more explicit discussions and additional information.

Exercises Section 12

1. Calculate the third-order terms $\rho^{(3)}$, $\psi^{(3)}$ and $\xi^{(3)}$ in the expansions (12.18) and (12.27) for a uniformly stratified Boussinesq fluid between rigid boundaries.

2. Show that the linear solution (12.21) and (12.22) satisfies the equations (3.9)–(3.11) *exactly* for a nonrotating Boussinesq fluid provided the perturbation pressure is equal to

$$\frac{p}{\rho_*} = Ac\frac{n\pi}{H}\cos(n\pi z/H)\sin(kx - \omega t) + \frac{A^2}{4}\left\{k^2\cos(2n\pi z/H) + \frac{n^2\pi^2}{H^2}\cos[2(kx - \omega t)]\right\}. \tag{12.104}$$

3. By substituting into (4.2) for the pressure gradients from the momentum equation (3.11) with $\boldsymbol{\Omega} \equiv \boldsymbol{0}$, derive the complete vorticity equation for two-dimensional motions in a stratified fluid, in the form

$$\rho[\nabla^2\psi_t + J(\nabla^2\psi, \psi)] + \rho_z[\psi_{zz} + J(\psi_z, \psi)] + \rho_x[\psi_{xt} + J(\psi_x, \psi)] = g\rho_x. \tag{12.105}$$

Do not make the Boussinesq approximation.

4. By making use of the transformations (12.30) and of (12.32), show that (12.105) may be written as

$$J\left\{\nabla^2\Psi + \frac{1}{\rho}\frac{d\rho}{d\Psi}\left[\frac{1}{2}(\Psi_s^2 + \Psi_z^2) + gz\right], \Psi\right\} = 0. \tag{12.106}$$

Show that the first integral of this equation may be transformed, using $\Psi' = \int\sqrt{\rho}\,d\Psi$, into

$$\nabla^2\Psi' + gz\frac{d\rho}{d\Psi'} = G(\Psi'), \tag{12.107}$$

which is identical in form to (12.35).

5. Verify that for $\epsilon \ll \mu = 0(1)$, (12.50) follows from (12.43). Carry out the substitutions indicated in the text to obtain (12.51).

6. Verify that (12.93) and (12.94) satisfy (12.90).

7. The hydraulic limit corresponds to $\mu = 0$ in (12.88). Using the transformation $\eta^* = (1 + 3\epsilon\eta/2)$ transform (12.88) into a simpler equation for which the characteristic speed

($V = \eta^*$) can be determined directly. Show that for waves propagating in the positive x-direction for which $u = \eta + 0(\epsilon)$, this result is consistent with (12.102).

8. Consider capillary waves in an infinitely deep, unstratified and nonrotating fluid in the absence of gravity. Using the exact form for p_s given in (11.2) and the kinematic boundary condition (12.38d), together with a modified form for (12.38b) which includes the influence of surface tension, but not that of gravity, formulate the problem of capillary wave motion in a form analogous to (12.36)–(12.38).

9. Choosing the wavelength λ for a vertical as well as for a horizontal scale, a period $T = (\rho/\sigma k_0^3)$ for a time scale and an appropriate scale for the potential ϕ (to be discovered), reduce the capillary wave problem to a form similar to (12.41)–(12.43) in terms of the single small parameter $\epsilon = a/\lambda$. Solve to the first nonlinear correction $\phi^{(1)}$.

10. Show that plane wave solutions of $\phi_t + V\phi_x = 0$, where V is a complex quantity, contain a part which grows exponentially in time.

11. Using (12.76), calculate the waveheight h, measured from trough ($x' = 0$) to crest ($x' = \pi/6$) of the progressive wave of maximum amplitude; show that

$$h/\lambda = 0.1374.$$

Also calculate, again from (12.76), the mean level $\bar{z}$ and show that the height of the wave crest above the mean level satisfies

$$(h - \bar{z})/\lambda = 0.0926.$$

Waves consist of oscillations about some equilibrium state and are described by the time and space dependence of these variations from equilibrium. They are also characterized by average properties, such as frequency, wavenumber, speed, energy density and flux, which generally vary over space and time scales much larger than those describing the fluctuations about equilibrium. Some of these average properties have already been discussed in general terms in Section 6. We shall now describe the more significant average properties of gravity–capillary waves discussed in the previous sections. Such properties may be considered as a hierarchy of integrals of the wave motion, starting with quantities which do not vanish in the limit of infinitesimal amplitude (such as the wavenumber and the phase and group velocities) and continuing with second-order quantities in the wave amplitude (momentum and energy densities and fluxes) and higher order, less physically familiar, quantities. Any number of such integrals may be constructed, in the form

$$I = \left\langle \int_{-H}^{\eta} F(\boldsymbol{u}, p, \rho, \ldots)\,\mathrm{d}z \right\rangle, \tag{13.1}$$

where F is a functional of the velocity, pressure and density and their derivatives. Only quadratic functions of the wave amplitude will be considered in this section; they can be constructed from the linearized solution of Section 11. The average used in (13.1) is over a whole cycle of the phase of a wave, as defined in (8.60); such an averaging process commutes with the z-integration only between constant limits of z.

Momentum density

The momentum density of surface waves in a homogeneous fluid, integrated over the depth and averaged over the phase, is the vector

$$\langle \boldsymbol{M} \rangle = \left\langle \int_{-H}^{\eta} \rho_* \boldsymbol{u}\,\mathrm{d}z \right\rangle. \tag{13.2}$$

Splitting the integral in two parts, we write

$$\langle \boldsymbol{M} \rangle = \rho_* \int_{-H}^{0} \langle \boldsymbol{u} \rangle\,\mathrm{d}z + \rho_* \left\langle \int_{0}^{\eta} \boldsymbol{u}\,\mathrm{d}z \right\rangle. \tag{13.3}$$

On the basis of linear fluctuating quantities only, the first integral vanishes, since $\langle \boldsymbol{u} \rangle = 0$. Upon expanding $\boldsymbol{u}$ in a Taylor series about $z = 0$ in the second integral, we find that to second order in fluctuating quantities,

$$\langle \boldsymbol{M} \rangle = \rho_* \langle \eta \boldsymbol{u}(x, 0, t) \rangle. \tag{13.4}$$

From the linear solution (11.26)–(11.30), (13.4) reduces to

$$\langle \boldsymbol{M} \rangle = \frac{1}{2c} \rho_* g a^2 (1 + B)(1, 0, 0). \tag{13.5}$$

The mean momentum is all forward, in the direction of wave propagation; it arises entirely

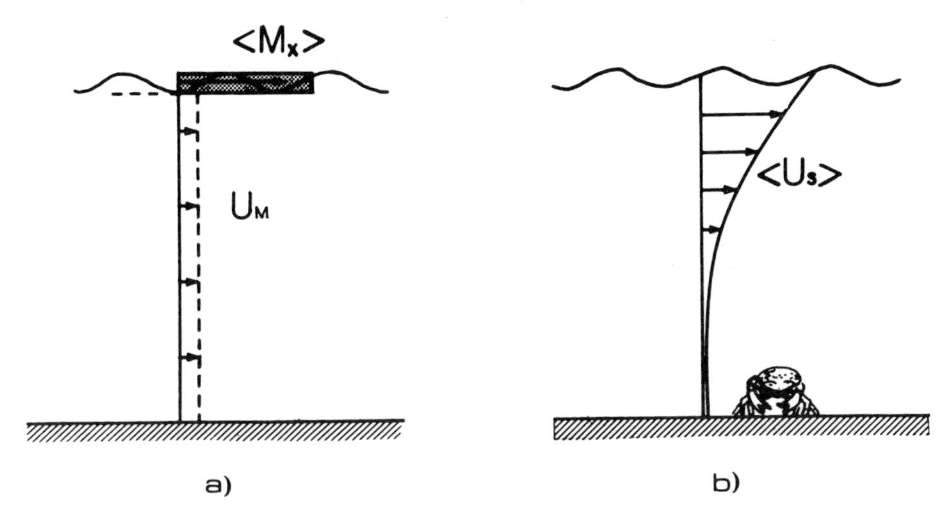

Fig. 13.1. The momentum density of small-amplitude surface waves. (a) The Eulerian point of view, at a fixed point. (b) The Lagrangian point of view, following particles.

from the second integral in (13.3), which is the contribution in the interval between the troughs and the crests of the waves, i.e., in $-a \leqslant z \leqslant a$. This concentration of mean momentum at the surface is not surprising given the oscillatory nature of the flow, which causes self-cancellation of the integral (13.2) at any level below $z = -a$, but not above $z = -a$ since, as may be deduced from Fig. 11.1, there is more water moving forwards (in the direction of $\boldsymbol{k}_h = k\hat{\boldsymbol{x}}$) than backwards at any level above $z = -a$ (indeed, for $0 \leqslant z \leqslant a$, all particle motion is forwards). After integration, all information about the vertical dependence of the mean momentum disappears, so that it is justifiable and convenient to imagine $\langle M_1 \hat{\boldsymbol{x}} \rangle$ as being carried by a mean drift velocity U_M defined through (see also Fig. 13.1)

$$\langle M_1 \rangle = \rho_* H U_M, \tag{13.6}$$

$$U_M = ga^2(1+B)/2cH. \tag{13.7}$$

The introduction of U_M presents us with a problem of interpretation: how can we speak of a drift velocity when the horizontal velocity component $u(x, z, t)$ averages out to zero at any level below $z = -a$? Waves can certainly carry momentum without the need of mean displacements, so why speak of a mean drift? The idea of a drift velocity arises more naturally from a Lagrangian interpretation of the wave motion.

The Lagrangian drift

Instead of looking at the velocity with which a series of different fluid particles flow past a point (the Eulerian point of view), suppose we concentrate our attention on the trajectories of "marked" particles. Then we discover a nonzero mean second-order drift arising from the linear solution. Let us label the velocity of a particle of fluid by $\boldsymbol{u}_{\mathrm{L}}(\boldsymbol{x}_0, t)$, where $\boldsymbol{x}_0 = (x_0, y_0, z_0)$ identifies the particle by giving its position at $t = 0$. After a time t, the new position of this particle will be

$$x(x_0, t) = x_0 + \int_0^t u_L(x_0, t')\,dt'. \tag{13.8}$$

The instantaneous Lagrangian velocity of the particle at x is of course the same as the Eulerian velocity at that point and at that instant; using (13.8),

$$u_L(x_0, t) = u(x, t)$$

$$= u\left(x_0 + \int_0^t u_L(x_0, t')\,dt', t\right). \tag{13.9}$$

For small displacements from the original position, we may expand the right side of (13.9) in a Taylor series about $x = x_0$:

$$u_L(x_0, t) = u(x_0, t) + \left[\int_0^t u_L(x_0, t')\,dt'\right] \cdot \nabla_0 u(x_0, t) + \ldots, \tag{13.10}$$

where $\nabla_0 = \partial/\partial x_0$.

The term $u_L(x_0, t')$ inside the integral must also be expanded by repeated substitutions from (13.10). To the order of the square of the fluctuating quantities, it is sufficient to replace u_L by u in the integral. Thus,

$$u_L = u + u_s, \tag{13.11}$$

where $u_s = (u_s, v_s, w_s)$ is called the *Stokes drift velocity* and represents the difference between the Lagrangian and Eulerian velocities. To the order of the square of the wave amplitude,

$$u_s = \left\{\left[\int_0^t u(x, t')\,dt'\right] \cdot \nabla\right\} u(x, t), \tag{13.12}$$

where we have chosen $x_0 = x$ in (13.10). The integral in brackets is the displacement vector ξ of particles, as estimated from the linear solution. The Stokes drift may then be expressed in the compact form

$$u_s = \xi \cdot \nabla u. \tag{13.13}$$

Substituting from the linear expressions in (11.28)–(11.30), we find

$$\langle u_s \rangle = \frac{a^2 \omega k}{2 \sinh^2 kH} \cosh\,[2k(z + H)]\;(1, 0, 0). \tag{13.14}$$

The Lagrangian momentum density is defined in the same fashion as its Eulerian counterpart in (11.2) as

$$\langle M_L \rangle = \left\langle \int_{-H}^{0} \rho_* u_L\,dz_0 \right\rangle, \tag{13.15}$$

with the vertical integration now running over particle positions at some arbitrary initial time at which the free surface position is at the equilibrium level. Since particles at the free surface always remain on that surface, the upper limit of integration does not vary with time and position. Substituting for $\boldsymbol{u}_L$ from (13.11) and noting from (12.53) that the Eulerian velocity is periodic and averages to zero, we obtain to second order in the amplitude

$$\langle \boldsymbol{M}_L \rangle = \rho_* \left\langle \int_{-H}^{0} \boldsymbol{u}_s \mathrm{d}z \right\rangle = \tfrac{1}{2}\rho_* a^2 \omega \coth kH\,(1,0,0). \tag{13.16}$$

This result is quickly reduced to that obtained in (13.5) by using the dispersion relation (11.16). In spite of being distributed quite differently through the water column (see Fig. 13.1), the Eulerian and Lagrangian momenta integrate out to the same value. To second order in wave amplitude, there is a net Lagrangian drift. Because of the spatial variation of the velocity components, the orbital paths of the particles are not quite closed. The mechanisms by which this net drift is brought about is clearly seen by imagining the orbit described by a given particle: since the horizontal velocity decays with depth, the distance covered in the upper part of the orbit ($u > 0$) is greater than the return path covered in the lower part ($u < 0$).

The net displacement over a cycle of period T is, from (13.8) and (13.11),

$$\boldsymbol{x} - \boldsymbol{x}_0 = T[\langle \boldsymbol{u}_s \rangle + \langle \boldsymbol{u} \rangle]. \tag{13.17}$$

If $\boldsymbol{u}$ is periodic so that $\langle \boldsymbol{u} \rangle = \boldsymbol{0}$, only the Stokes drift contributes to the net displacement. Since $\langle w_s \rangle = 0$, the mean vertical position of a fluid particle remains unchanged; its horizontal position varies according to (13.17). If particles are identified by a line of dye laid along a vertical line at $t = 0$, the mean slope of this line will be given by

$$\left\langle \frac{\mathrm{d}x}{\mathrm{d}z} \right\rangle = \frac{t\omega a^2 k^2 \sinh\,[2k(z+H)]}{\sinh^2\,(kH)}, \tag{13.18}$$

where t is equal to an integral number of wave periods.

Energy density

Another integral of obvious physical relevance is the energy density. It can be divided into three parts. The mean kinetic energy density per unit area of the sea surface is given by

$$\langle KE \rangle = \left\langle \int_{-H}^{\eta} \frac{1}{2} \rho_* (\boldsymbol{u} \cdot \boldsymbol{u})\, \mathrm{d}z \right\rangle. \tag{13.19}$$

From (11.28) and (11.30), $\langle KE \rangle$ is, to second order in amplitude,

$$\langle KE \rangle = \rho_* g a^2 (1 + B)/4. \tag{13.20}$$

The mean gravitational potential energy due to the presence of the waves is

$$\langle PE_g \rangle = \left\langle \int_{-H}^{\eta} \rho_* g z \mathrm{d}z \right\rangle - \int_{-H}^{0} \rho_* g z \mathrm{d}z. \tag{13.21}$$

The potential energy of the fluid at rest does not arise from wave motion and has thus been subtracted in (13.21). To second order in amplitude,

$$\langle PE_g \rangle = \rho_* g a^2/4 \tag{13.22}$$

upon using (11.26). The surface tension acts as a massless elastic membrane attached to the fluid surface. The reader may refer to Morse and Ingard (1968, Section 5.2) for a description of the forces and of the momentum and energy densities in such a membrane. The mean energy stored in surface deformation is, again to second order in amplitude,

$$\langle PE_\sigma \rangle = \left\langle \frac{\sigma}{2} \left(\frac{\partial \eta}{\partial x} \right)^2 \right\rangle = \tfrac{1}{4} \sigma a^2 k^2. \tag{13.23}$$

The total mean energy density is the sum of (13.20), (13.22) and (13.23):

$$\langle E \rangle = \tfrac{1}{2} \rho_* g a^2 (1 + B). \tag{13.24}$$

Comparing the horizontal component of $\langle \boldsymbol{M} \rangle$ in (13.5) to $\langle E \rangle$, we find the remarkable relation

$$\langle E \rangle = c \langle M_1 \rangle, \tag{13.25}$$

where c is the phase speed. This result was first discovered by Levi–Civita (1924) and further elucidated by Starr (1947a, b). This type of relation holds for many types of waves (for photons, see French, 1968; for acoustic waves, see Morse and Ingaard, 1968, Chapter 6), and is related to the manner in which work must be done on a vibratory system to generate wave motion. The existence of a mean momentum $\langle \boldsymbol{M} \rangle$ in the direction of wave propagation must be associated with the prior application of a force $F(t)$ in the same direction over some time τ:

$$\int_0^\tau F \mathrm{d}t = \langle M_1 \rangle. \tag{13.26}$$

The applied force could be exerted, for example, by the translation at a speed c of a rigid wavy boundary of wavenumber k over part of the surface of the fluid (Bretherton, 1969). Provided that this forcing can be accomplished by a force which is strictly in the direction of the translation of the rigid surface, the energy acquired by the fluid after moving a distance X would be

$$\langle E \rangle = \int_0^X F \mathrm{d}x. \tag{13.27}$$

However, $X = c\tau,$ (13.28)

so that (13.27) is also given by

$$\langle E \rangle = \int_0^{c\tau} F \frac{\mathrm{d}x}{\mathrm{d}t} \mathrm{d}t = c \int_0^\tau F \mathrm{d}t$$

$$= c \langle M_1 \rangle. \tag{13.29}$$

We thus recover (13.25). A more general derivation for waves of arbitrary amplitude, due to Levi–Civita (1925) shows that $2\langle KE \rangle = cM_1$, which reduces to (13.25) for small-amplitude waves for which $\langle KE \rangle = \langle PE \rangle$.

Momentum flux and radiation stress

Travelling waves carry momentum with them and hence produce a net momentum flux in the medium which they traverse. In an ideal nonrotating fluid, the local momentum balance is given by a rearrangement of (3.11) (with $\rho = \rho_*$, $\boldsymbol{\Omega} \equiv \boldsymbol{0}$):

$$\frac{\partial(\rho_* u_i)}{\partial t} + \frac{\partial}{\partial x_j}(p\delta_{ij} + \rho_* u_i u_j + \rho_* gz\delta_{ij}) = 0, \tag{13.30}$$

expressing a balance between the local time variations of the momentum density vector $\rho_* u_i$ and the divergence of its flux tensor $\rho_*(u_i u_j + gz\delta_{ij}) + p\delta_{ij}$. The integrated horizontal momentum flux in the presence of surface waves is the integral over the depth of the horizontal component of the flux tensor. Part of this flux is, however, not due to wave motion, but arises from the presence of the hydrostatic pressure field p_0: that part of the flux must be subtracted from the total to isolate the flux due to the waves themselves. In addition, we must not forget to include a contribution from capillarity. The mean vertically integrated horizontal momentum flux due to wave motion is then

$$T_{ij} = \left\langle \int_{-H}^{\eta} (p\delta_{ij} + \rho_* u_i u_j)\,\mathrm{d}z \right\rangle - \int_{-H}^{0} p_0 \delta_{ij}\,\mathrm{d}z + \Sigma_{ij}. \tag{13.31}$$

where i, j take the values 1, 2 corresponding to the x- and y-directions, respectively. The choice of notation (T_{ij}) for the mean momentum flux is not accidental. We shall show that this flux is identical to the interaction stress tensor introduced at the end of Section 6 and also denoted by T_{ij}. The surface tension momentum flux Σ_{ij} is given by Morse and Ingard (1968, p. 200) (with a change of sign, since the quantities treated by these authors are the negative of those discussed here), in terms of $\boldsymbol{x}$ and y components, as

$$\Sigma_{ij} = \left\langle\!\!\left\langle \begin{matrix} \frac{1}{2}\sigma(\eta_x^2 - \eta_y^2), & -\sigma\eta_x\eta_y \\ -\sigma\eta_x\eta_y, & -\frac{1}{2}\sigma(\eta_x^2 - \eta_y^2) \end{matrix} \right\rangle\!\!\right\rangle, \tag{13.32}$$

where σ is again the surface-tension coefficient.

The calculation of the components of T_{ij} is based on the work of Longuet-Higgins and Stewart (1964). For two-dimensional plane waves solutions as given in Section 10 with $\partial_y = 0$ and $v = 0$, the nondiagonal components of T_{ij} vanish. To second order in wave amplitude, (13.31) may be written as the sum of four parts:

$$T_{ij} = \left\langle \int_{-H}^{0} \rho_* u_i u_j\,\mathrm{d}z \right\rangle + \left\langle \int_{-H}^{0} (p - p_0)\delta_{ij}\,\mathrm{d}z \right\rangle + \left\langle \int_{0}^{\eta} p\,\mathrm{d}z \right\rangle \delta_{ij} + \Sigma_{ij}. \tag{13.33}$$

The first term of (13.33) is evaluated by substitution of the velocity components from (11.28) and (11.30):

$$\left\langle \int_{-H}^{0} \rho_* u_i u_j\,\mathrm{d}z \right\rangle = \langle E \rangle \left(\frac{1}{2} + \frac{kH}{\sinh 2kH} \right) \delta_{i1}\delta_{j1}. \tag{13.34}$$

From (12.57),

$$\langle p \rangle - p_0 = -\langle \rho_* w^2 \rangle, \tag{13.35}$$

from which the second integral in (13.33) may be evaluated, with the help of (11.30), as

$$\left\langle \int_{-H}^{0} (p-p_0)\,\delta_{ij}\,dz \right\rangle = \langle E\rangle \left[-\frac{1}{2}+\frac{kH}{\sinh 2kH}\right]\delta_{ij}. \tag{13.36}$$

The pressure very near the free surface is needed in the third term in (13.33), and cannot be found from (11.27), which is valid only in the domain $-H \leqslant z \leqslant 0$. Right under the free surface, the pressure may be taken as hydrostatic, with the addition of the surface tension term $p_s = -\sigma\eta_{xx}$. Thus, near $z = 0$,

$$p \simeq \rho_* g(\eta - z) + \sigma k^2 \eta + p_a, \tag{13.37}$$

where p_a is the constant atmospheric pressure. To second order in wave amplitude, the third integral in (13.33) then becomes

$$\left\langle \int_{0}^{\eta} p\,dz \right\rangle \delta_{ij} = \langle E\rangle \frac{(B+\frac{1}{2})}{(B+1)}\,\delta_{ij}. \tag{13.38}$$

Finally, from (13.32) the surface tension contribution for a two-dimensional wave ($\partial_y = 0$) becomes

$$\Sigma_{ij} = \tfrac{1}{2}\sigma\langle \eta_x^2\rangle \begin{pmatrix} 1, & 0 \\ 0, & -1 \end{pmatrix} = \frac{B\langle E\rangle}{2(1+B)} \begin{pmatrix} 1, & 0 \\ 0, & -1 \end{pmatrix}. \tag{13.39}$$

Adding all the contributions [(13.34), (13.36), (13.38) and (13.39)] to the wave momentum flux tensor, we find

$$T_{ij} = \langle E\rangle \begin{pmatrix} \dfrac{2kH}{\sinh 2kH} + \dfrac{1+3B}{2(1+B)}, & 0 \\ 0 & , \dfrac{kH}{\sinh 2kH} \end{pmatrix}. \tag{13.40}$$

A flux of momentum per unit area is also a force per unit distance, which we may call a stress (although a stress is usually understood as a force per unit area). The T_{11} term is then the rate of forward transport (per unit width of wavefront) of forward directed momentum: a normal stress in the direction of wave propagation. The T_{22} term is the sideways transport of y-directed momentum per unit distance in the direction of propagation: again a normal stress. From (13.40), $T_{22} < T_{11}$. In the short wavelength limit ($kH \gg 1$), T_{22} vanishes and we are left with

$$T_{11} = \langle E\rangle \frac{(1+3B)}{2(1+B)}. \tag{13.41}$$

From (11.20) with $kH \gg 1$, and (13.25), this is also

$$T_{11} = c_g\langle M_1\rangle. \tag{13.42}$$

That is, the flux of momentum is equal to the group velocity times the momentum density. Short waves thus exert only a forward pressure on the fluid. This type of radiation pressure is found in a photon stream, for example, where the momentum flux is equal to the momentum density times the speed of light, and in many other types of waves (see Elmore and Heald, 1969). What is most interesting about the momentum flux tensor of water waves is that it is generally anisotropic, so that it is more appropriate to speak of it as a *radiation stress* than a radiation pressure.

In Section 6, we introduced an interaction stress tensor in (6.49) to characterize the energy exchange between waves and a spatially inhomogeneous current. Omitting the zero subscripts, which are superfluous here, and under the conditions of Exercise 6.7, ($\Lambda_{ij} = \delta_{ij}; \lambda' = H$), the interaction stress tensor is expressed as

$$T_{ij} = \frac{\langle E \rangle}{\omega}\left[H\frac{\partial\omega}{\partial H}\delta_{ij} + c_{gj}k_i\right]. \tag{13.43}$$

We shall now show that this tensor is identical to that given by (13.40). From the dispersion relation (11.16),

$$\frac{H}{\omega}\frac{\partial\omega}{\partial H} = \frac{kH}{\sinh 2kH}. \tag{13.44}$$

Substituting this result in (13.43) and the components of the group velocity, as given by (11.20), it is readily verified that (13.43) is identical to (13.40). The momentum flux or radiation stress tensor thus acquires added significance as the interaction stress tensor responsible for energy exchanges with a nonhomogeneous flow.

The momentum flux tensor has also been calculated for short crested waves by Battjes (1972) and for monochromatic but not necessarily plane waves by Mei (1973).

The energy flux

The general energy equation (4.10), rewritten in Eulerian form for an incompressible fluid of uniform density ρ_*, takes the form

$$\frac{\partial}{\partial t}(\tfrac{1}{2}\rho_* u^2 + \rho_* gz) + \frac{\partial}{\partial x_j}[(p + \tfrac{1}{2}\rho_* u^2 + \rho_* gz)u_j] = 0. \tag{13.45}$$

The mean horizontal energy flux $\langle F \rangle$ is obtained from integrating the horizontal component of the flux vector in (13.45), and includes a surface tension contribution picked up at the upper boundary (Morse and Ingard, 1968, p. 199). To second order in wave amplitude,

$$\langle \boldsymbol{F} \rangle = \left\langle \int_{-H}^{0} p\boldsymbol{u}\,dz \right\rangle - \sigma\langle \eta_t \nabla\eta \rangle. \tag{13.46}$$

For plane waves, (13.46) becomes, upon using (11.26)–(11.30) and the general expression (11.20) for the group velocity,

$$\langle \boldsymbol{F} \rangle = \langle E \rangle \boldsymbol{c}_g, \tag{13.47}$$

in agreement with (6.42).

The mean properties of internal wave modes may be obtained by the same application of vertical integration and averaging. A number of subtleties enter, due to the modal structure and the distribution of buoyancy effects throughout the depth of the fluid (Garrett, 1968; Bretherton, 1969).

To second order in wave amplitude, for example, the average vertically integrated momentum of an internal wave may be written as

$$\langle M_1 \rangle = \int_{-H}^{0} \langle \rho^{(1)} u^{(1)} \rangle \, dz + \int_{-H}^{0} \rho_0 \langle u^{(2)} \rangle \, dz, \tag{13.48}$$

where $\rho^{(1)}$, $u^{(1)}$ and $u^{(2)}$ may be evaluated from a perturbation solution, as in (12.21). From the definition of the stream function for two-dimensional motion (cf. 12.15) and the linearized density equation (12.20), the first term ($\langle M_1^a \rangle$, say) is of the form [dropping superscripts (1)]

$$\langle M_1^a \rangle \propto \int_{-H}^{0} \rho_0 N^2 \langle \psi_z^2 \rangle \, dz. \tag{13.49}$$

Multiplying the linearized Boussinesq equation [derived from (12.19), (12.20)]

$$\psi_{zz} + \frac{N^2 - \omega^2}{\omega^2} k^2 \psi = 0 \tag{13.50}$$

by ψ_z and integrating, we find that

$$(\psi_z)^2 |_{-H}^{0} - k^2 \psi^2 |_{-H}^{0} + \frac{k^2}{\omega^2} \int_{-H}^{0} N^2 (\psi^2)_z \, dz = 0. \tag{13.51}$$

Hence, with $\rho_0 = \rho_*$ (the Boussinesq approximation) and for a rigid lid and flat bottom $[\psi(0) = \psi(-H) = 0]$,

$$\langle M_1^a \rangle \propto \langle (\psi^2)_z \rangle |_{-H}^{0}. \tag{13.52}$$

The integrated momentum density depends on the form of the vertical eigenfunctions. For the uniformly stratified case (N = constant), it is readily verified from (12.21) that $\langle M_1^a \rangle$ vanishes, and the mean momentum is entirely due to the contribution from $\langle u^{(2)} \rangle$. For an unbounded train of plane waves, (12.24) and (12.25) implies that $\langle u^{(2)} \rangle = 0$. However, for a finite wave group in which the amplitude is subjected to a slow modulation in space, a mean second-order flow will generally be found. This situation will be encountered in Section 37.

For internal gravity waves in an unbounded fluid, the interaction stress tensor T_{ij} may be obtained from (6.49). We first note that for a Boussinesq fluid, (8.29) implies

$$\partial \omega / \partial N = \omega / N. \tag{13.53}$$

Furthermore, if W is the vertical component of a slowly varying velocity, the density equation becomes, on the average,

$$\rho_{0t} + w \rho_{0z} = 0, \tag{13.54}$$

where the time variation of ρ_0 is over periods well in excess of a wave cycle. Differentiating (13.54) with respect to z, we find

$$\frac{\partial}{\partial t}\left(\frac{1}{\rho_0}\frac{\partial \rho_0}{\partial z}\right) + W\frac{\partial}{\partial z}\left(\frac{1}{\rho_0}\frac{\partial \rho_0}{\partial z}\right) = -\frac{1}{\rho_0}\frac{\partial \rho_0}{\partial z}\frac{\partial W}{\partial z} \tag{13.55}$$

or, since N is a function of z only,

$$\frac{1}{N}\frac{\mathrm{D}N}{\mathrm{D}t} = -\frac{1}{2}\frac{\partial W}{\partial z}. \tag{13.56}$$

From (6.47), it follows that in this case $\Lambda_{ij} = \frac{1}{2}\delta_{i3}\delta_{j3}$, and in view of (13.53), (6.49) yields T_{ij} as

$$T_{ij} = \frac{\langle E \rangle}{\omega}\left(\frac{\omega}{2}\delta_{i3}\delta_{j3} + c_{g_j}k_i\right). \tag{13.57}$$

It was shown by Garrett (1968) that T_{ij} can be identified with the radiation stress tensor as given by the average momentum flux.

A number of average relations between the momentum and the energy of surface gravity waves of arbitrary amplitude have also been derived by various authors (Starr, 1947b; Longuet-Higgins, 1974; 1975a).

Exercises Section 13

1. Show that the mean vertical component of the Stokes drift, $\langle w_s \rangle$, as evaluated from (13.12), vanishes identically.

2. A dye line is initially set vertically in a layer of fluid of uniform depth H. Show that the subsequent length $l(t)$ (averaged over a cycle) of this dye line, as it is stretched by the Stokes drift (13.14) due to shallow water waves ($kH \ll 1$), is given by

$$l(t) = \frac{H}{2}\left\{(1+\tau^2)^{1/2} + \frac{1}{\tau}\sinh^{-1}\tau\right\},$$

where $\tau = 2a^2\omega kt/H$.

3. Calculate using (13.40) the force exerted per unit length of an absorbing breakwater by surface waves of wavelength 20 m and amplitude 0.5 m in water 5 m deep.

(Ans. $0.994 \cdot 10^3$ N.)

4. Verify the result (13.47) for the energy flux.

5. Use (11.20) to rewrite the components of (13.40) in terms of the ratio c_g/c.

Ans. $$T_{11} = \frac{2c_g}{c} - \frac{1}{2}\frac{(1+3B)}{(1+B)}, \tag{13.58a}$$

$$T_{22} = \frac{c_g}{c} - \frac{1}{2}\frac{(1+3B)}{(1+B)}. \tag{13.58b}$$

6. The stress tensor T_{ij} is given in (13.40) in diagonal form, i.e., with the x-axis in the direction of wave propagation. Using the tensor transformation rule

$$T'_{ij} = T_{mn} \frac{\partial x_m}{\partial x'_i} \frac{\partial x_n}{\partial x'_j} \tag{13.59}$$

for obtaining the components T'_{ij} in the $\boldsymbol{x}'$-coordinate system in terms of those components T_{mn} in the $\boldsymbol{x}$-coordinate system, show that a rotation by an angle θ transforms (13.40) into

$$T'_{ij} = \langle E \rangle \begin{pmatrix} T_{11} \cos^2\theta + T_{22} \sin^2\theta, & \frac{1}{2}(T_{22} - T_{11}) \sin 2\theta \\ \frac{1}{2}(T_{22} - T_{11}) \sin 2\theta, & T_{11} \sin^2\theta + T_{22} \cos^2\theta \end{pmatrix}, \tag{13.60}$$

where T_{11} and T_{22} are given by (13.59).

CHAPTER 3

FREE WAVES: LONG WAVELENGTHS

14. INTRODUCTION

In this chapter, we discuss the various types of waves which are governed by the hydrostatic equations (5.36). The analysis is extended to the β-plane, where f is a linear function of latitude, as in (5.33): $f = f_0 + \beta y$. Most of the discussion will be limited to waves of small amplitude in an ocean of uniform depth without lateral boundaries.

As will be seen in Sections 15 and 16, there are two types of long waves which may be found in such a model ocean; these two types are often referred to as waves of the first and of the second class. Waves of the first class are long gravity waves, for which $\omega > f$ but for which rotation plays only a modifying role; these waves are discussed in Section 17 and represent the long wavelength limit of the normal mode solutions presented in Section 10. The second-class waves are planetary or Rossby waves, for which $\omega \ll f$; these waves do not exist in a nonrotating ocean (See Section 18). The effects of nonlinearity on both classes of waves are briefly discussed in Section 19. Next, we show in Section 20 how the β-effect can be modelled by a variable depth and hence introduce the subject of topographic planetary waves. Finally, we conclude this chapter with the topic of equatorially trapped waves (Section 21).

15. LONG WAVE EQUATIONS FOR A CONTINUOUSLY STRATIFIED FLUID

Here we discuss the linearized form of the hydrostatic β-plane equations (5.31), (5.32) and (5.36) which describe small-amplitude motions about the hydrostatic equilibrium state $p_0(z)$, $\rho_0(z)$ for which $p_{0z} = -\rho_0 g$. These equations take the form

$$u_t - fv + \frac{1}{\rho_0} p_x = 0, \tag{15.1}$$

$$v_t + fu + \frac{1}{\rho_0} p_y = 0, \tag{15.2}$$

$$p_z = -\rho g, \tag{15.3}$$

$$u_x + v_y + w_z = 0, \tag{15.4}$$

$$\rho_t + \rho_{0z} w = 0, \tag{15.5}$$

where $f = f_0 + \beta y$ $[= 2\Omega \sin \phi_0 + (2\Omega \cos \phi_0 / R) y]$ and, as in Chapter 2, p and ρ denote the perturbation pressure and density fields. We recall that in neglecting the term $\tilde{f} w$ in the momentum equation for u we have assumed that our model ocean is centered at a mid-latitude [and hence $f_0 = 0(10^{-4}\ \mathrm{rad\,s^{-1}})$ and $\beta = 0(10^{-11}\ \mathrm{m^{-1}\,rad\,s^{-1}})$] and that the horizontal length scale L is large compared to the depth H.

In making the hydrostatic approximation we have neglected, in the linear approximation, the vertical acceleration terms $\rho_0 w_t$ and $\rho_0 \tilde{f} u$ on the left side of (15.3). It is easy to show that for sufficiently strong stratification, the term proportional to $\tilde{f}$ can be neglected provided the frequency is small compared to N (Exercise 15.1). If $\tilde{f}$ is neglected a priori, then the nonhydrostatic form of (15.3) contains the additional term $\rho_0 w_t$ and this new equation can be combined with (15.5) to give the following relation between w and p, which we earlier called (8.69):

$$(\partial_{tt} + N^2)w = -\rho_0^{-1} p_{zt}. \tag{8.69}$$

On the other hand, combining (15.3) and (15.5) gives

$$N^2 w = -\rho_0^{-1} p_{zt}. \tag{15.6}$$

Comparison of (15.6) and (8.69) shows that for the hydrostatic approximation to be valid, we require

$$\omega^2 \ll N^2, \tag{15.7}$$

where ω is the frequency. Thus waves governed by (15.1)–(15.5) have "long" periods as well as long wavelengths.

Vorticity gradient equation

In order to acquire an understanding of the types of waves that can exist in the system (15.1)–(15.5) we first combine these equations into one for v alone. Eliminating p from (15.1) and (15.2) by cross-differentiation and then using (15.4) to eliminate the horizontal velocity divergence, we obtain the vorticity equation

$$(v_x - u_y)_t + \beta v - f w_z = 0. \tag{15.8}$$

Differentiating (15.8) with respect to x and then using (15.4) again to eliminate u_{tyx}, we find

$$\nabla_h^2 v_t + \beta v_x + (w_{yt} - f w_x)_z = 0. \tag{15.9}$$

From (15.3) and (15.5) it follows that

$$w = -p_{zt}/\rho_0 N^2, \tag{15.10}$$

where $N^2 = -g\rho_{0z}/\rho_0$ is the usual stability frequency. Thus (15.9) becomes

$$\nabla_h^2 v_t + \beta v_x + [N^{-2}(f p_{zxt} - p_{zytt})\rho_0^{-1}]_z = 0. \tag{15.11}$$

To eliminate the pressure we notice that (15.1) and (15.2) can be combined into the equation

$$\mathcal{L} v_t = (f p_{xt} - p_{ytt})\rho_0^{-1}, \tag{15.12}$$

where $\quad \mathcal{L} = \partial_{tt} + f^2. \tag{15.13}$

Equation (15.12) gives, upon differentiating with respect to z,

$$(f p_{zxt} - p_{zytt})\rho_0^{-1} = \mathcal{L}(v_z - g^{-1}N^2 v)_t. \tag{15.14}$$

Finally, using (15.14) in (15.11) we obtain a single equation for v:

$$\frac{\partial}{\partial t}\nabla_h^2 v + \beta\frac{\partial v}{\partial x} + \left(\frac{\partial^3}{\partial t^3} + f^2\frac{\partial}{\partial t}\right)\frac{\partial}{\partial z}\left[N^{-2}\left(\frac{\partial v}{\partial z} - \frac{N^2}{g}v\right)\right] = 0. \tag{15.15}$$

We note that (15.15) is simply the x-derivative of the vorticity equation (15.8). Therefore it represents a vorticity balance equation, describing how temporal changes in the relative vorticity of a parcel of fluid may be compensated by its latitudinal repositioning within the planetary vorticity field (the second term), and by vertical stretching of interior vortex lines (the last term). Clearly these latter two vorticity change terms are due respectively to the variability of f and to the presence of the mean density field.

It is of interest here to mention how (15.15) changes under the Boussinesq approximation, in which case ρ_0 in (15.1) and (15.2) is replaced by ρ_* [as in (5.9), for example]. Instead of (15.15), we then find

$$\frac{\partial}{\partial t}\nabla_h^2 v + \beta\frac{\partial v}{\partial x} + \left(\frac{\partial^3}{\partial t^3} + f^2\frac{\partial}{\partial t}\right)\frac{\partial}{\partial z}\left(N^{-2}\frac{\partial v}{\partial z}\right) = 0, \tag{15.15$'$}$$

where $N^2 = -g\rho_{0z}/\rho_*$ is the usual definition of the stability frequency in a Boussinesq fluid. In short, (15.15′) follows from (15.15) provided we have

$H \ll g/N^2,$

i.e. the depth scale is much less than the scale height of the stratification. This is also the condition found in Chapter 2 for nonhydrostatic motions.

Separation of vorticity equation

To study horizontally propagating wave solutions of (15.15) in an ocean of uniform depth it is customary at this point to use the method of separation of variables to obtain one equation for the *vertical structure* and one for the *x, y and t dependence*. This method for handling problems in rotating fluids was introduced by Taylor (1936) for the case of continuous stratification and by Charney (1955) and Veronis and Stommel (1956) for the case of a two-layer fluid (see Section 16). The above separation method is analogous to that used in Section 10, where the primitive, nonhydrostatic equations (10.27)–(10.30) were separated at the outset into two systems, one for the vertical and one for the horizontal structure.

If we substitute

$$v(x,y,z,t) = Z_n(z)\,v_n(x,y,t)$$

into (15.15) we obtain

$$\frac{\mathrm{d}}{\mathrm{d}z}\left[N^{-2}\left(\frac{\mathrm{d}Z_n}{\mathrm{d}z} - \frac{N^2}{g}Z_n\right)\right] + \frac{1}{gh_n}Z_n = 0 \tag{15.16}$$

and $$\nabla_h^2 v_{nt} + \beta v_{nx} - \frac{1}{gh_n}\mathcal{L}v_{nt} = 0, \tag{15.17}$$

where $1/gh_n$ is the separation constant. The factor g has been inserted so that h_n has the dimensions of length. For reasons which will be made clear later in this section, h_n is known as the equivalent depth.

The domain of (15.16) is given by $-H \leqslant z \leqslant 0$. Therefore appropriate boundary conditions must be prescribed at $z = -H$ and $z = 0$. The vertical eigenfunctions (or modes) $Z_n(z)$ and the corresponding eigenvalues h_n can then be determined for any given equilibrium stratification $N(z)$. The boundary value problem is of Liouville type, and therefore the eigenfunctions are real and orthogonal, and the eigenvalues real. As in Chapter 2 the gravest mode ($n = 0$) is identified as the barotropic mode, and the modes $n = 1, 2, 3, \ldots$ are the sequence of baroclinic modes. The substitution of a plane-wave form into (15.17) (in which for the moment f is now treated as constant) then yields the dispersion relation corresponding to the eigenfunction $Z_n(z)$, or more simply, the vertical mode number n. Since (15.17) is third order in t, it follows that such a substitution yields a cubic in the frequency ω (Exercise 15.2):

$$\frac{1}{gh_n}\,\omega^3 - \left(k_h^2 + \frac{f^2}{gh_n}\right)\omega - \beta k_1 = 0, \tag{15.18}$$

where $k_h = (k_1^2 + k_2^2)^{1/2}$. Therefore, corresponding to each mode n, there are three dispersion relations, corresponding to the three roots of this cubic. Two questions thus immediately arise. For real k_h such that $k_h H \ll 1$ (long wavelengths), is each root real? If so, what is a typical frequency scale for each root?

The answers to these questions are not immediately apparent because the cubic (15.18) does not have simple factors. However, by a simple graphical analysis (Exercise 15.3), it is fairly easy to show that all the roots are indeed real, with two of them being superinertial ($|\omega| > |f|$) and the other being subinertial ($|\omega| < |f|$). Further, in the ocean there is usually a large gap in the energy spectrum between these two frequency ranges, as may be seen from Fig. 15.1. Thus it is meaningful to analyze (15.17) by focusing our attention on time scales which are characterized by $\omega > f$ or $\omega \ll f$. This approach gives rise to two classes of wave motions: (1) long gravity waves which are the low-frequency limit ($f < \omega \ll N$) of the normal modes (with $\tilde{f} = 0$) discussed in Section 10; and (2) very low frequency planetary waves which exist because of the presence of β. Hough (1898), in his pioneering study of small-amplitude oscillations of a thin layer of fluid on a rotating sphere, classified these two types of motions as waves of the first class and waves of the second class, respectively. The first-class waves are dominated by the gravitational force and are only modified by rotation; the second-class waves, on the other hand, lose their oscillatory nature and reduce to steady currents when there is no rotation. Longuet-Higgins (1964a, 1965a) has demonstrated that a similar classification can also be used to describe long wave motions on the β-plane. For the above reasons we shall deal with each wave class separately. We do this by taking $\beta = 0$ for first-class (long gravity) waves and by imposing the condition $\omega \ll f$ for second-class (planetary) waves.

The pressure equation

In spite of these simplifications, there are fundamental problems associated with using the vorticity equation (15.15) for v as our starting point. To begin with, the boundary conditions at $z = 0$ and $z = -H$ do not take on a simple form. For example, the condition that $w = 0$ at $z = -H$ becomes

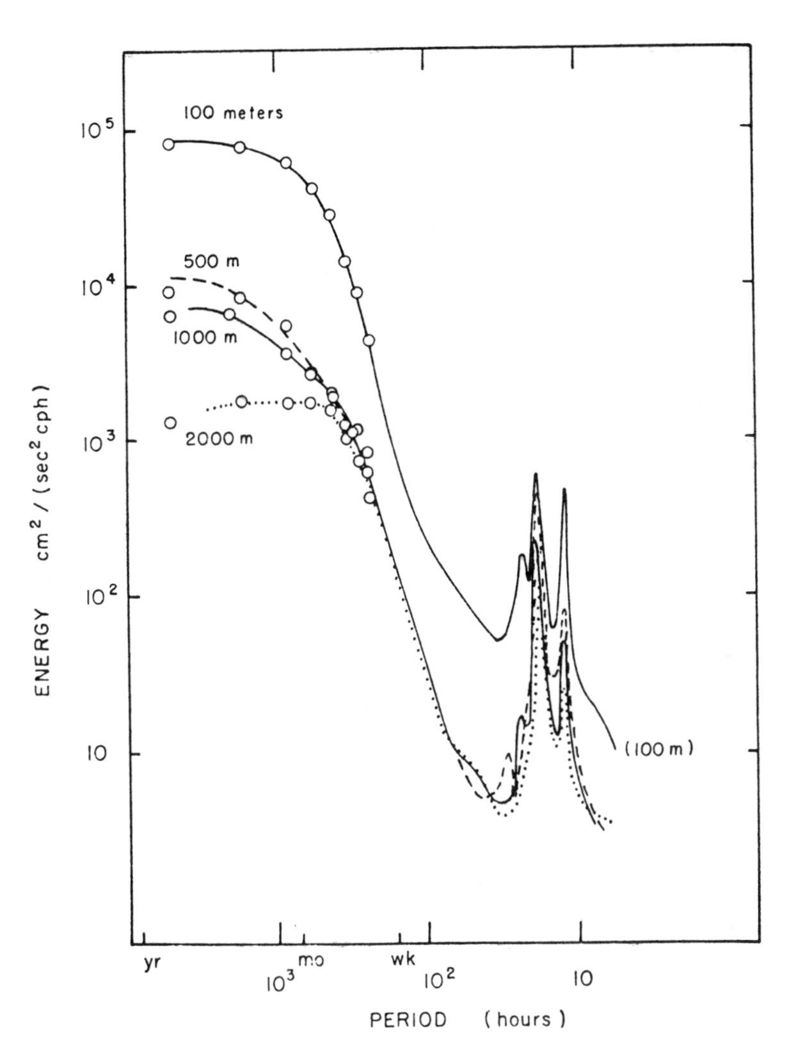

Fig. 15.1. Frequency spectra of horizontal kinetic energy density at site D (39°20′N, 70°W) as computed by Thompson (1971). (From Rhines, 1971b.) At the latitude of site D the inertial period $(2\pi/f)$ is 19 hours. Notice the large drop in the energy at all depths between the high $(\omega > f)$ and very low frequencies $(\omega \ll f)$.

$$v_z - (N^2/g)v = 0 \quad \text{at } z = -H.$$

The free surface boundary condition is even more complicated (Exercise 15.4). A second major obstacle is that once having found v, it is awkward to find all the remaining dependent variables (u, w, p and ρ). One way out of these difficulties is to use the governing equation for the pressure as our starting point. In terms of p the boundary conditions are simple, and it is easy to find u, v, w and ρ once p is known. While the pressure equation is rather complicated for *arbitrary* time scales, it reduces to a very simple form in the case of long gravity waves ($\beta = 0$ and $\omega > f$) or planetary waves ($\omega \ll f$).

To derive the pressure equation we begin by finding expressions for $\mathcal{L}u$ and $\mathcal{L}v$ from

(15.1) and (15.2):

$$\mathcal{L}u = -\frac{1}{\rho_0}(p_{xt} + fp_y), \tag{15.19}$$

$$\mathcal{L}v = -\frac{1}{\rho_0}(p_{yt} - fp_x), \tag{15.20}$$

where $\mathcal{L}$ is given by (15.13). From (15.19) and (15.20) we can find $\mathcal{L}u_x$ and $\mathcal{L}v_y$, which in turn can be used in the continuity equation (15.4) operated on by $\mathcal{L}$. After elimination of w_z, we finally obtain (Exercise 15.3):

$$\mathcal{L}[\nabla_h^2 p + \mathcal{L}\rho_0(p_z/\rho_0 N^2)_z]_t + \beta[f^2 p_x - 2fp_{yt} - p_{xtt}] = 0. \tag{15.21}$$

By repeating arguments similar to those used to derive (15.15′), we find that the Boussinesq version of (15.21) is

$$\mathcal{L}[\nabla_h^2 p + \mathcal{L}(p_z/N^2)_z]_t + \beta[f^2 p_x - 2fp_{yt} - p_{xtt}] = 0, \tag{15.21'}$$

where $N^2 = -g\rho_{0z}/\rho_*$.

The boundary condition $w = 0$ at $z = -H$ becomes, by virtue of (15.10),

$$p_{zt} = 0 \text{ at } z = -H. \tag{15.22}$$

In Section 9 it was shown that the linearized form of the surface boundary conditions (3.13) and (3.14) can be combined into one equation (cf. 9.22):

$$p_t - \rho_0 g w = 0 \text{ at } z = 0.$$

On using (15.10), this becomes

$$(N^2/g)p_t + p_{zt} = 0 \text{ at } z = 0. \tag{15.23}$$

This of course could also have been obtained directly from (9.23) by invoking the hydrostatic approximation ($\omega^2 \ll N^2$).

In analogy to what was done with v in (15.15), we now write p in the form

$$p = \Pi_n(z)p_n(x, y, t), \tag{15.24}$$

where for convenience we take Π_n to be dimensionless. Substitution of (15.24) into (15.21) gives

$$\rho_0 \frac{\mathrm{d}}{\mathrm{d}z}\left(\frac{1}{\rho_0 N^2}\frac{\mathrm{d}\Pi_n}{\mathrm{d}z}\right) + \frac{1}{gh_n}\Pi_n(z) = 0, \quad -H \leqslant z \leqslant 0 \tag{15.25}$$

and $$\mathcal{L}\left[\nabla_h^2 p_n - \frac{1}{gh_n}\mathcal{L}p_n\right]_t + \beta[f^2 p_{nx} - 2fp_{nyt} - p_{nxtt}] = 0, \tag{15.26}$$

where as in (15.16) and (15.17), $1/gh_n$ is the separation constant. The boundary conditions (15.22) and (15.23) become

$$\frac{\mathrm{d}\Pi_n}{\mathrm{d}z} = 0 \text{ at } z = -H, \tag{15.27}$$

$$\frac{\mathrm{d}\Pi_n}{\mathrm{d}z} + \frac{N^2}{g}\Pi_n = 0 \text{ at } z = 0. \tag{15.28}$$

Notice that the vertical boundary-value problem (15.25), (15.27) and (15.28) depends only on the equilibrium stratification. Therefore the same vertical eigenfunctions and eigenvalues apply for both classes of waves. By way of contrast it is interesting to note that the boundary value problem for the vertical dependence of w as discussed in Section 10 (the nonhydrostatic case) depends on the frequency as well as on the stratification [see (10.41)]. It is only for $\omega \ll N$ that the vertical eigenfunctions are independent of frequency.

The vertical modes for constant stratification

For any realistic profile of $N^2(z)$ (e.g., see Figs. 16.1 and 16.2), the boundary value problem (15.25), (15.27), (15.28) can be integrated by standard numerical techniques. Alternatively, for certain analytical forms of ρ_0 and hence N^2, (15.25) can be cast into a familiar special function equation in which case the vertical modes can be expressed in terms of known functions. Here we shall take the simplest approach possible and solve the boundary value problem for the case $N^2 = $ constant. Then (15.25) yields a constant coefficient equation, enabling us to find a solution in terms of elementary functions. The simplicity of the solution makes it easy to see the significance of the small terms which would otherwise be neglected if, for example, the Boussinesq approximation were made.

With $N^2 = $ constant, (15.25) becomes

$$\frac{\mathrm{d}^2\Pi_n}{\mathrm{d}z^2} + \frac{N^2}{g}\frac{\mathrm{d}\Pi_n}{\mathrm{d}z} + \frac{N^2}{gh_n}\Pi_n(z) = 0, \quad -H \leqslant z \leqslant 0.$$

It is now convenient to introduce the normalized depth coordinate $z' = z/H$; then $\Pi_n(z) = \Pi_n(z'H)$ and

$$\frac{\mathrm{d}^2\Pi_n}{\mathrm{d}z'^2} + 2\epsilon\frac{\mathrm{d}\Pi_n}{\mathrm{d}z'} + \lambda_n\Pi_n = 0, \quad -1 \leqslant z' \leqslant 0, \tag{15.29}$$

$$\text{where} \quad \epsilon = N^2H/2g, \quad \lambda_n = N^2H^2/gh_n. \tag{15.30}$$

The boundary conditions take the form

$$\frac{\mathrm{d}\Pi_n}{\mathrm{d}z'} = 0 \text{ at } z' = -1, \tag{15.31}$$

$$\frac{\mathrm{d}\Pi_n}{\mathrm{d}z'} + 2\epsilon\Pi_n = 0 \text{ at } z' = 0. \tag{15.32}$$

The general solution of (15.29) is given by

$$\Pi_n = \mathrm{e}^{-\epsilon z'}(A_n \cos \Lambda_n z' + B_n \sin \Lambda_n z'), \tag{15.33}$$

where $\Lambda_n = (\lambda_n - \epsilon^2)^{1/2}$. Application of the boundary conditions (15.31) and (15.32) gives

$$\epsilon A + \Lambda B = 0, \tag{15.34a}$$

$$(\Lambda \sin \Lambda - \epsilon \cos \Lambda) A + (\Lambda \cos \Lambda + \epsilon \sin \Lambda) B = 0, \tag{15.34b}$$

where for convenience the subscript n has been dropped. For a nontrivial solution the determinant of coefficients in (15.34) must be zero, which yields the following transcendental eigenvalue equation for λ:

$$\Lambda^2 \sin \Lambda - 2\epsilon\Lambda \cos \Lambda - \epsilon^2 \sin \Lambda = 0. \tag{15.35}$$

Since ϵ is typically small (for example in the deep ocean, $\epsilon = 0(10^{-3})$ for $N^2 = 10^{-5}$ rad s^{-1}, and $H = 5 \cdot 10^3$ m), we neglect terms of $0(\epsilon^2)$ in (15.35), giving the much simpler equation

$$\tan \sqrt{\lambda} = 2\epsilon/\sqrt{\lambda}, \quad 0 < \epsilon \ll 1. \tag{15.36}$$

The roots of (15.36) are most easily seen graphically and occur at the intersection of the curves $f_1(\sqrt{\lambda}) = \tan \sqrt{\lambda}$ and $f_2(\sqrt{\lambda}) = 2\epsilon/\sqrt{\lambda}$ (see Fig. 15.2). For small $\sqrt{\lambda}$, $\tan \sqrt{\lambda} \simeq \sqrt{\lambda}$ and therefore the lowest (barotropic) eigenvalue λ_0 is given by

$$\lambda_0 = 2\epsilon[1 + 0(\epsilon)]. \tag{15.37}$$

The higher (baroclinic) eigenvalues are given by

$$\lambda_n = (n\pi)^2[1 + 0(\epsilon)], \quad n = 1, 2, 3, \ldots \tag{15.38}$$

It now follows from (15.33) and (15.34) that

$$\begin{aligned} \Pi_0 &= A_0 e^{-\epsilon z'}[1 + 0(\epsilon z')] \\ &= A_0[1 + 0(\epsilon z')] \end{aligned} \tag{15.39}$$

for the barotropic vertical eigenfunction and

$$\Pi_n = A_n e^{-\epsilon z'}[\cos n\pi z' + 0(\epsilon)], \quad n = 1, 2, 3, \ldots \tag{15.40}$$

for the baroclinic vertical eigenfunctions. We note that the slowly varying decay factor is identical with that found in Chapter 2 for the nonhydrostatic case, i.e.

$$\exp(-\epsilon z') = \exp(-N^2 z/2g).$$

For small ϵ it is clear that the baroclinic eigenfunctions can be simply approximated by $A_n \cos(n\pi z/H)$. Then from (15.19) and (15.20) it follows that the baroclinic horizontal velocity components u and v have a dependence also like $\cos(n\pi z/H)$. Equations (15.10) and (15.3) on the other hand imply that w and ρ for the baroclinic modes are proportional to $\sin(n\pi z/H)$. For the barotropic mode it follows that u and v are proportional to $1 + 0(\epsilon z')$ and that w and ρ are both $0(\epsilon)$ quantities.

It is now instructive to write down the expressions for the equivalent depths. Substituting (15.37) and (15.38) into λ_n as given by (15.30) we find

$$h_0 = H[1 + 0(\epsilon)], \tag{15.41}$$

$$h_n = H(N^2H/gn^2\pi^2)[1 + 0(\epsilon)], \quad n = 1, 2, \ldots \tag{15.42}$$

The choice of the phrase "equivalent depths" for h_n is now obvious. For the barotropic mode h_0 is approximately equal to the true depth H, the small correction being proportional to $\Delta\rho_0/\rho_0$, the relative density difference throughout the depth of the fluid. Since

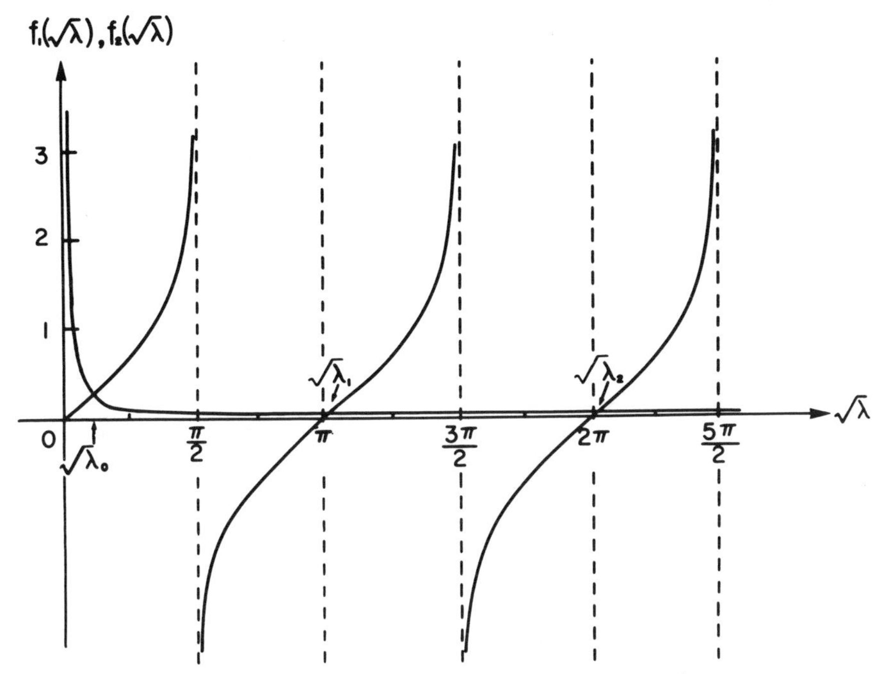

Fig. 15.2. Graphical solution of the eigenvalue equation $\tan\sqrt{\lambda} = 2\epsilon/\sqrt{\lambda}$, $0 < \epsilon \ll 1$. The eigenvalues are located at the intersection of the curves $f_1(\sqrt{\lambda}) = \tan\sqrt{\lambda}$ and $f_2(\sqrt{\lambda}) = 2\epsilon/\sqrt{\lambda}$ and are marked $\sqrt{\lambda_0}, \sqrt{\lambda_1}, \sqrt{\lambda_2}, \ldots$. As $n \to \infty$, $\sqrt{\lambda_n} \to n\pi$.

$N^2H/g = 2\epsilon \ll 1$, (15.42) shows that h_n is always much less than H; therefore we have the sequence

$$h_0(\simeq H) \gg h_1 > h_2 > \ldots . \tag{15.43}$$

Thus we speak of baroclinic motions as occurring in an ocean of "equivalent depth" h_n which is in practice much less than the true depth. The consequences of (15.43) on the phase speeds of barotropic and baroclinic waves will be discussed later.

Long wave equations for a homogeneous fluid

In view of the above results concerning the depth dependence of the barotropic mode it has become customary to consider barotropic long wave motions independently of their baroclinic brethren by means of the so-called "long wave" equations for a homogeneous, rotating fluid. The nonlinear momentum equations of this set are given by

$$u_t + uu_x + vu_y - fv + \frac{1}{\rho_*}p_x = 0, \tag{15.44a}$$

$$v_t + uv_x + vv_y + fu + \frac{1}{\rho_*}p_y = 0, \tag{15.44b}$$

$$p_z = -\rho_* g. \tag{15.45}$$

where u and v are assumed to be independent of z and p denotes the total pressure. The nonlinear terms have been included in (15.44) for the sake of future reference and also to emphasize the neglect of the terms wu_z and wv_z. Integration of (15.45) gives

$$p = \rho_* g(\eta - z) + p_a, \tag{15.46}$$

which can be used to eliminate the pressure in (15.44):

$$u_t + uu_x + vu_y - fv + g\eta_x = -\frac{1}{\rho_*} p_{ax}, \tag{15.47a}$$

$$v_t + uv_x + vv_y + fu + g\eta_y = -\frac{1}{\rho_*} p_{ay}. \tag{15.47b}$$

To close the system we obtain a conservation of mass equation by integrating the continuity equation (15.4) from $z = -H$ to $z = \eta$. For the sake of generality and for future considerations we allow H to vary, i.e. $H = H(x, y)$. We thus have

$$\int_{-H}^{\eta} u_x \, dz + \int_{-H}^{\eta} v_y \, dz + w(x, y, \eta, t) - w(x, y, -H, t) = 0. \tag{15.48}$$

Since η and H are functions of x and y, we must use the following identity to evaluate the above integrals: For any continuously differentiable functions A, B, F, we have

$$\frac{\partial}{\partial a} \int_{A(a)}^{B(a)} F(a, b) \, db = \int_{A(a)}^{B(a)} \frac{\partial F}{\partial a} db + F[a, B(a)] \frac{\partial B}{\partial a} - F[a, A(a)] \frac{\partial A}{\partial a}.$$

Thus the first integral in (15.48) gives

$$\begin{aligned} \int_{-H}^{\eta} u_x \, dz &= \frac{\partial}{\partial x} \int_{-H}^{\eta} u \, dz - u(x, y, \eta, t) \frac{\partial \eta}{\partial x} + u(x, y, -H, t) \frac{\partial}{\partial x}(-H) \\ &= \frac{\partial}{\partial x} [u(\eta + H)] - u\Big|_{z=\eta} \frac{\partial \eta}{\partial x} - u\Big|_{z=-H} \frac{\partial H}{\partial x}, \end{aligned}$$

upon recalling that u is independent of z. Similarly

$$\int_{-H}^{\eta} v_y \, dz = \frac{\partial}{\partial y} [v(\eta + H)] - v\Big|_{z=\eta} \frac{\partial \eta}{\partial y} - v\Big|_{z=-H} \frac{\partial H}{\partial y}.$$

Upon substitution into (15.48) we immediately see that the terms evaluated at $z = -H$ add up to zero because $w = -\mathbf{u} \cdot \nabla H$ at $z = -H$, and that the terms evaluated at $z = \eta$ add up to η_t because of the kinematic condition at $z = \eta$. Hence we are left with the nonlinear equation

$$[u(\eta + H)]_x + [v(\eta + H)]_y + \eta_t = 0. \tag{15.49}$$

For future reference, we note here that for $p_a =$ constant the linearized long wave equations take the form

$$u_t - fv + g\eta_x = 0, \tag{15.50a}$$

$$v_t + fu + g\eta_y = 0, \tag{15.50b}$$

$$(Hu)_x + (Hv)_y + \eta_t = 0. \tag{15.51}$$

These equations will be used in our discussion of topographic planetary waves in Section 20.

Exercises Section 15

1. To neglect $\rho_0 \tilde{f} u$ in (15.3), we require $\rho_0 \tilde{f} u \ll \rho g$. By differentiating this inequality with respect to t and using (15.3), show that

$$\frac{\omega}{N} \ll \frac{N}{\tilde{f}} \frac{H}{L}.$$

For $H/L \ll 1$ and $N \gg \tilde{f}$ (strong stratification), the right side is 0(1) and hence ω/N is small.

2. Show that (15.17) has plane-wave solutions

$$v_n = A_n \exp\,[i(k_1 x + k_2 y - \omega t)]$$

provided the cubic (15.18) holds.

3. Consider (15.18) in the form

$$k_h^2 + \beta k_1/\omega = (\omega^2 - f^2)/gh_n, \tag{15.52}$$

where k_1 and h_n are assumed to be positive; then k_h^2, βk_1, f^2 and gh_n are all positive constants. By sketching separately the graphs of the left and right sides of (15.52) as a function of real ω (allowed to take on either sign), show that there are always three real roots of (15.18) of which two are superinertial ($|\omega| > |f|$) and one is subinertial ($|\omega| < |f|$).

4. For solutions of the form

$$v = Z_n(z) \exp\,[i(k_1 x + k_2 y - \omega t)],$$

express the free surface condition

$$\rho_0 g w - p_t = 0 \text{ at } z = 0$$

in terms of $Z_n(z)$ and its derivatives.

5. Derive (15.51) by considering the horizontal mass flux through a vertical column of fluid of height $\eta + H$ and of cross-sectional area $\mathrm{d}x\mathrm{d}y$.

6. Show that (15.50) and (15.51) can be combined into the following equation for v:

$$(\nabla_h^2 - \mathcal{L}/gH) v_t + \beta v_x = 0. \tag{15.53}$$

As expected we notice that (15.53) is identical with (15.17) provided h_n in the latter equation is replaced by H.

7. Consider the barotropic form of the conservation of potential vorticity equation:

$$\frac{\mathrm{D}}{\mathrm{D}t}\left(\frac{\zeta_3 + f}{H + \eta}\right) = 0, \tag{15.54}$$

where $\zeta_3 = v_y - u_x$. Show that for an ocean with a rigid lid and flat bottom, the linearized form of this equation is

$$\nabla_h^2 \psi_t + \beta \psi_x = 0, \tag{15.55}$$

where ψ is the stream function. Notice that this is the low-frequency limit of (15.53).

16. LONG WAVE EQUATIONS FOR A TWO-LAYER FLUID

A typical vertical profile of the mean density $\rho_0(z)$ in the open ocean is shown in Fig. 16.1. Here, below a well-mixed layer of depth $\simeq 100$ m, $\sigma_t\,[\sigma_t = 10^3(\rho_0 - 1)]$ increases rapidly from 25.5 to 27.0, corresponding to a relative density change of $1.5 \cdot 10^{-3}$. Below this pycnocline region, the density slowly increases with depth, the variation being approximately exponential. The σ_t profile changes with season, but usually only in the upper hundred meters or so.

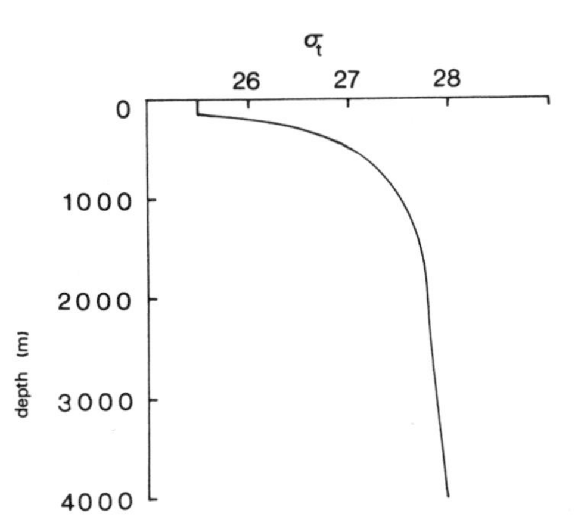

Fig. 16.1. Late spring σ_t profile from Ocean Station P (50°N, 145°W). (Adapted from Fofonoff and Tabata, 1966.) In terms of σ_t the equilibrium density in c.g.s. units is given by

$$\rho_0(z) = 1 + 10^{-3}\sigma_t(z). \tag{16.63}$$

It is thus clear that in this area of the ocean, a constant N describes only the lower part of the ocean. More generally, N is a slowly varying exponential function below the mixed layer, as can be seen in the best-fit model of Garrett and Munk (see Fig. 8.6). In shallower waters, on the other hand, either uniform stratifications or sharp near-surface pycnoclines may be found (see Fig. 16.2). To describe realistically the vertical modal structure in oceans containing a pycnoline, we must use a realistic model for $\rho_0(z)$ based on the observed density profile. This generally leads to a numerical problem of the type described in Section 15. However, a simple model for $\rho_0(z)$ which characterizes the essential features of a pycnocline and which has been frequently used in the literature is the single-step profile

$$\rho_0(z) = \begin{cases} \rho_1, & -H_1 < z \leqslant 0, \\ \rho_2, & -H \leqslant z < -H_1. \end{cases} \tag{16.1}$$

This model has already been introduced in Section 10 (cf. 10.50) in connection with interfacial gravity waves at a density discontinuity. In this so-called two-layer model (which can obviously be generalized to an n-layer model), ρ_1 and ρ_2 are constants, being representative of the density values in the mixed layer and the deep ocean, respectively. The

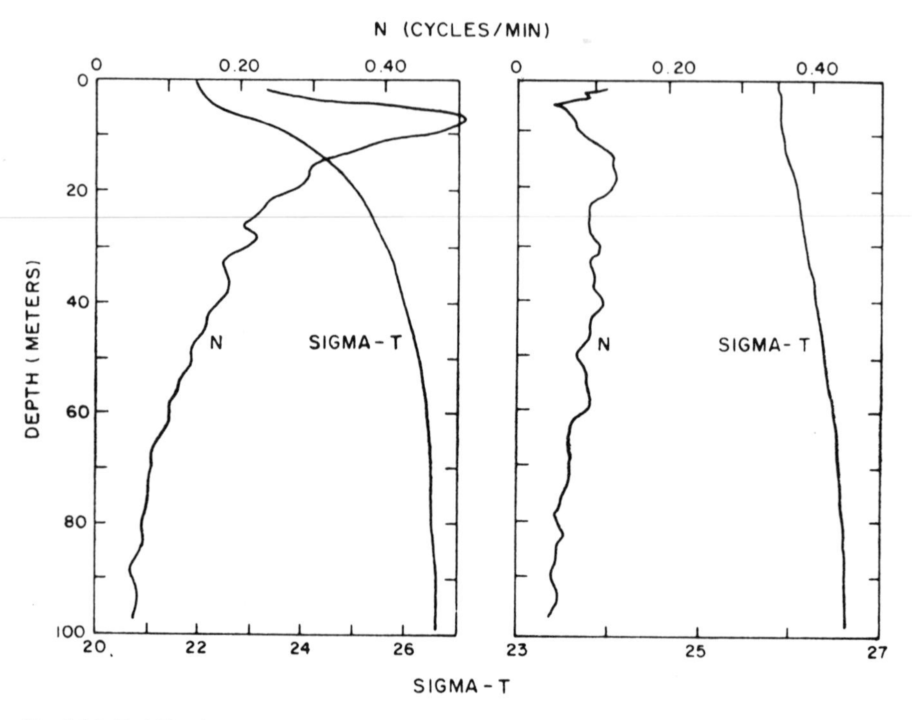

Fig. 16.2. Stability frequency, N, and σ_t profiles from 25-hour anchor stations off the Oregon coast on July 10, 1973, (a) and August 1, 1973, (b). (From Hayes and Halpern, 1976.)

quantities H_1 and $H_2 = H - H_1$ are the mean depths of these respective regions. In both Figs. 16.1 and 16.2a we note that $(\rho_2 - \rho_1)/\rho_2 \simeq 2 \cdot 10^{-3}$ and $H_1 \ll H_2$, although H_1 takes on quite a different value in each case (250 m and 10 m).

The discontinuous density model (16.1) has one obvious shortcoming, which we will now briefly discuss. In this system there are now only two modes of oscillation, a barotropic mode and the first baroclinic mode. In order to visualize the vertical velocity structure in each mode in the hydrostatic case discussed in this chapter, it is helpful to first introduce an analogous situation concerning a vibrating string. For a string of uniform density that is fixed at both ends, the normal modes of oscillation are $\sin(n\pi x/L)$, $n = 1, 2, \ldots$, where x is the distance along the string and L is the length. Consider now a "model" string consisting of two heavy, identical beads on a light string. It is a simple exercise in mechanics (and linear algebra) to show that in this system there are only two possible modes of oscillation; they are approximations to $\sin(\pi x/L)$ and $\sin(2\pi x/L)$, the gravest and second modes in the continuous system (see Exercise 16.1 and Fig. 16.3). In the gravest (or fundamental) mode (a), the beads move in the same direction, completely in phase. In the second mode (b), the beads move in opposite directions, 180° out of phase. With this illustration it now takes little imagination to visualize the possible vertical structures associated with the horizontal currents in a two-layer fluid. Corresponding to (a) and (b) in Fig. 16.3, a two-layer fluid can oscillate in the barotropic mode or the first baroclinic mode. These modes are illustrated in Fig. 16.4, for the typical situation of unequal layer depths.

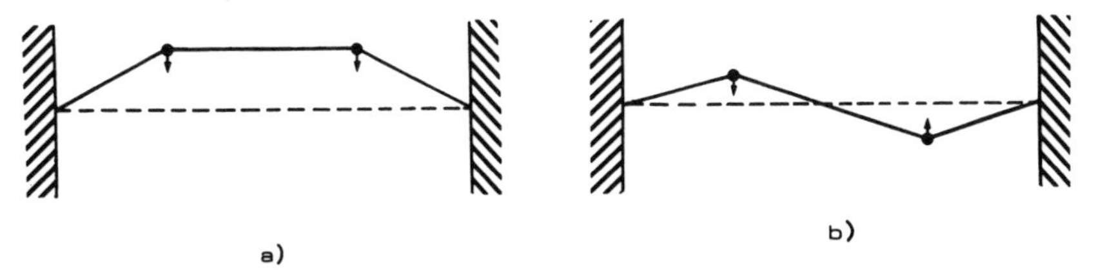

Fig. 16.3. Normal modes of oscillation of two beads fixed to an elastic string: (a) gravest mode and (b) second mode.

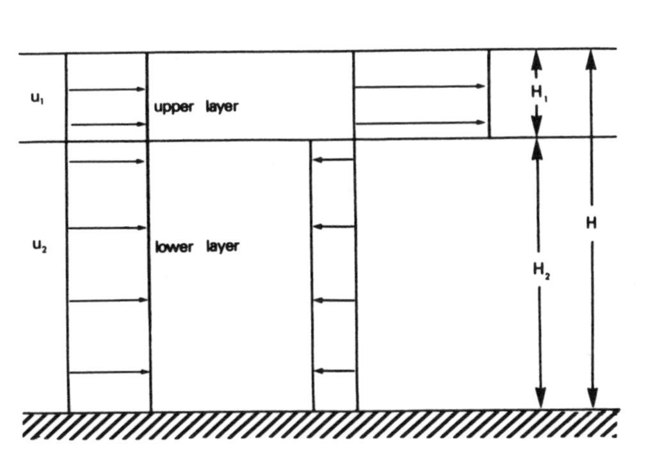

Fig. 16.4. Profiles of the horizontal currents u_1 and u_2 for the barotropic mode, (a), and baroclinic mode, (b), in a two-layer fluid of uniform depth.

Although a two-layer fluid contains only two vertical modes, such a limitation is not as serious as might first be imagined. In many situations involving long-period hydrostatic motions, observations indicate that most of the energy is usually contained in the first two modes (e.g., see Section 19). Thus, aside from circumventing mathematical difficulties, a two-layer system does contain a sufficient number of degrees of freedom to provide an adequate model of many observed vertical energy distributions.

Governing equations

In Section 15 we showed that for the barotropic mode, u_h (the horizontal velocity) and p (the perturbation pressure) depend weakly on z and that w (the vertical velocity) and ρ (the perturbation density) are small. These results motivated the introduction of the long wave equations [(15.50 and 15.51) in their linear form]; these equations describe barotropic motions whose horizontal velocities are independent of z. In the long wave equations for a two-layer fluid, it is assumed that the horizontal velocities in *each* layer are independent of z. Thus in the notation of Fig. 16.1, the (linearized) momentum equations for the upper and lower layers are upper layer: $$u_{1t} - fv_1 + \frac{1}{\rho_1} p_{1x} = 0, \tag{16.2a}$$

$$v_{1t} + fu_1 + \frac{1}{\rho_1} p_{1y} = 0, \tag{16.2b}$$

$$p_{1z} = -\rho_1 g; \tag{16.3}$$

lower layer: $$u_{2t} - fv_2 + \frac{1}{\rho_2} p_{2x} = 0, \tag{16.4a}$$

$$v_{2t} - fu_2 + \frac{1}{\rho_2} p_{2y} = 0, \tag{16.4b}$$

$$p_{2z} = -\rho_2 g, \tag{16.5}$$

where p_n $(n = 1, 2)$ is the total pressure in the nth layer. Integration of (16.3) yields

$$p_1 = \rho_1 g(\eta_1 - z) + p_a. \tag{16.6}$$

Integrating (16.5) gives

$$p_2 = \rho_2 g(\eta_2 - z - H_1) - \rho_1 g(\eta_2 - \eta_1 - H_1) + p_a, \tag{16.7}$$

where the constant of integration has been chosen to make $p_2 = p_1$ at $z = -H_1 + \eta_2$. These equations enable us to eliminate p_1 and p_2 in (16.2) and (16.4) respectively; for $p_a =$ constant, we thus find

$$u_{1t} - fv_1 + g\eta_{1x} = 0, \tag{16.8a}$$

$$v_{1t} + fu_1 + g\eta_{1y} = 0, \tag{16.8b}$$

$$u_{2t} - fv_2 + g\eta_{1x} + g'(\eta_2 - \eta_1)_x = 0, \tag{16.9a}$$

$$v_{2t} + fu_2 + g\eta_{1y} + g'(\eta_2 - \eta_1)_y = 0, \tag{16.9b}$$

where $g' = (\rho_2 - \rho_1)g/\rho_2$ is the reduced gravity for a two-layer fluid.

The equations expressing conservation of mass for each layer can be easily derived by the method used in Section 15, or as in Exercise 15.5. For a variable lower layer depth $H_2 = H_2(x, y)$, the resulting linearized equations are

$$H_1(u_{1x} + v_{1y}) + (\eta_1 - \eta_2)_t = 0, \tag{16.10}$$

$$(H_2 u_2)_x + (H_2 v_2)_y + \eta_{2t} = 0. \tag{16.11}$$

Equations (16.8)–(16.11) form a coupled set for the six unknowns u_1, v_1, η_1 and u_2, v_2, η_2; the coupling between the upper and lower layer motions comes through the terms involving η_1 and η_2, the surface and interfacial displacements. Since $g \gg g'$ for the ocean, it follows that the coupling between the layers will only be significant if $\eta_2 \gg \eta_1$. When the depth is variable the motions described by (16.8)–(16.11) are very complex since the vertical and horizontal dependences cannot, in general, be separated.

Also, as we shall see later (Section 20), simple barotropic or baroclinic motions of the sort discussed earlier do not exist in the case of variable depth. However, when $H_2 =$ constant, the analysis of (16.8)–(16.11) becomes quite straightforward, and it is then possible to express the solutions in terms of the familiar barotropic and (first) baroclinic modes.

Normal mode equations

We now show how the set (16.8)–(16.11) for the case H_2 = constant can be transformed into a normalized form in which each dependent variable represents either a barotropic or baroclinic motion. The procedure is equivalent to the separation of variables technique used in Section 15 to obtain the (vertical) barotropic and baroclinic eigenfunctions in an ocean of uniform depth. From another point of view, the normalization of the system (16.8)–(16.11) is the generalization to partial differential equations of what is commonly done to simplify systems of ordinary differential equations. Given the linear, constant coefficient system

$$A_{ij}x_j(t) = \mathbf{0}, \quad i = 1, \ldots, n, \tag{16.12}$$

where A_{ij} has n linearly independent eigenvectors, it is always possible to find a similarity transformation which takes (16.12) into a (normalized) diagonal form (Boyce and DiPrima, 1969):

$$B_{ij}\dot{y}_j(t) = \mathbf{0}, \quad i = 1, \ldots, n, \tag{16.13}$$

where $B_{ij} = 0$ for $i \neq j$ and the dot denotes the time derivative.

To illustrate the method we shall treat the simplest possible case, namely that of one-dimensional motion in a nonrotating system. In this case the basic equations are

$$u_{1t} + g\eta_{1x} = 0, \tag{16.14}$$

$$H_1 u_{1x} + (\eta_1 - \eta_2)_t = 0, \tag{16.15}$$

$$u_{2t} + g\eta_{1x} + g'(\eta_2 - \eta_1)_x = 0, \tag{16.16}$$

$$H_2 u_{2x} + \eta_{2t} = 0. \tag{16.17}$$

First we combine (16.14)–(16.17) into two equations for u_1 and u_2 alone:

$$u_{1tt} - gH_1 u_{1xx} - gH_2 u_{2xx} = 0, \tag{16.18}$$

$$u_{2tt} - gH_1(1-\delta) u_{1xx} - gH_2 u_{2xx} = 0, \tag{16.19}$$

where $\delta = (\rho_2 - \rho_1)/\rho_2 \ll 1$ is the relative density difference between the two layers. Next we add (16.18) to s times (16.19):

$$(u_1 + su_2)_{tt} - g\{H_1[1 + s(1-\delta)]u_1 + H_2(1+s)u_2\}_{xx} = 0; \tag{16.20}$$

the quantity s will be determined below.

We now define a new velocity u by the linear combination

$$u \equiv u_1 + su_2. \tag{16.21}$$

Then we notice that (16.20) will represent a single wave equation for u with propagation speed gh, provided the quantity inside the curly brackets is equal to hu, i.e.,

$$hu \equiv h(u_1 + su_2) = H_1[1 + s(1-\delta)]u_1 + H_2(1+s)u_2. \tag{16.22}$$

Since u_1 and u_2 are linearly independent, their respective coefficients on each side of (16.22) must be equal; thus

$$h = H_1[1+s(1-\delta)], \tag{16.23}$$

$$hs = H_2(1+s). \tag{16.24}$$

Dividing (16.23) by (16.24) yields a quadratic equation for s:

$$(1-\delta)H_1 s^2 + (H_1 - H_2)s - H_2 = 0. \tag{16.25}$$

The roots of (16.25) are given by

$$s^{(0)} = \frac{H_2}{H_1}[1+0(\delta)] \tag{16.26}$$

$$\text{and} \quad s^{(1)} = \frac{1}{1-\delta}\left[-1+\delta\frac{H_2}{H}+0(\delta^2)\right], \tag{16.27}$$

where $H = H_1 + H_2$. Hence (16.23) gives

$$h^{(0)} = H[1+0(\delta)] \tag{16.28}$$

$$\text{and} \quad h^{(1)} = \frac{\delta H_1 H_2}{H}[1+0(\delta)]. \tag{16.29}$$

The quantities $h^{(0)}$ and $h^{(1)}$ correspond to the equivalent depths h_0 and h_1 introduced in Section 15 for the continuously stratified ocean [cf. (15.41) and (15.42)]. As in the case of continuous stratification, the correction terms in (16.28) and (16.29) are proportional to the relative density difference δ, and hence can be neglected. Returning to (16.20), we can now write down the so-called *normal mode equations* for $u^{(n)}$:

$$u_{tt}^{(n)} - gh^{(n)}u_{xx}^{(n)} = 0, \quad n = 1, 2, \tag{16.30}$$

where $n = 1$ and $n = 2$ correspond to the barotropic and baroclinic modes, respectively. In the simple case considered above, the normal mode equations (16.30) take on a diagonal form, analogous to (16.13). To see this, we notice that (16.30) can be written in the form

$$\begin{bmatrix} \mathcal{L}_2^{(n)} & 0 \\ 0 & \mathcal{L}_2^{(n)} \end{bmatrix} \begin{bmatrix} u^{(1)} \\ u^{(2)} \end{bmatrix} = \begin{bmatrix} 0 \\ 0 \end{bmatrix},$$

$$\text{where} \quad \mathcal{L}_2^{(n)} = \partial_{tt} - gh^{(n)}\partial_{xx}.$$

It is now instructive to find u_1 and u_2 in terms of $u^{(0)}$ and $u^{(1)}$. From (16.21), (16.26) and (16.27), we find that correct to $0(\delta)$,

$$u^{(0)} = u_1 + (H_2/H_1)u_2, \quad \text{and} \quad u^{(1)} = u_1 - u_2. \tag{16.31}$$

Solving (16.31) for u_1 and u_2, we obtain

$$u_1 = \frac{1}{H}(H_1 u^{(0)} + H_2 u^{(1)}), \tag{16.32a}$$

$$u_2 = \frac{H_1}{H}(u^{(0)} - u^{(1)}). \tag{16.32b}$$

Suppose now $u^{(1)} = 0$ (no baroclinic motion); then (16.32) implies that

$$u_1 = u_2 = (H_1/H)u^{(0)} \tag{16.33}$$

for the barotropic mode, as shown in Fig. 16.4a. If $u^{(0)} = 0$, (16.32) gives

$$u_1 = -\frac{H_2}{H_1}u_2 = \frac{H_2}{H}u^{(1)}, \tag{16.34}$$

for the baroclinic mode (Fig. 16.4b). Notice that in this mode, the total mass flux is zero, i.e.,

$$u_1 H_1 + u_2 H_2 = 0.$$

We also note from (16.34) that for a deep lower layer ($H_2 \gg H_1$), the lower layer velocity will be very small. Since $H_2 \gg H_1$ in many instances, this result has led many investigators to use two-layer models in which the lower layer is motionless (see Exercise 16.2).

The velocity distributions in the barotropic and baroclinic modes as given by (16.31) and (16.34) are similar to those associated with "first" and "second" sound wave propagation in a superfluid (Landau and Lifshitz, 1959, Section 131). In a sound wave of the first type, the normal and superfluid parts of the fluid move together, as in the barotropic mode in a two-layer fluid. This sound wave corresponds to an ordinary sound wave in an ordinary fluid. In a sound wave of the second type, the normal and superfluid parts move in opposition, with the center of mass of any given volume element remaining at rest and the total mass flux being zero. This is clearly the analogue of the baroclinic mode in a two-layer fluid. However, in a superfluid, made up of the two densities ρ_s and ρ_n, the fluid is not stratified.

The analysis presented above for one-dimensional waves also carries over to the two-dimensional, rotational case. The normal mode equations for

$$u^{(n)} = u_1 + s^{(n)}u_2 \quad \text{and} \quad v^{(n)} = v_1 + s^{(n)}v_2 \; (n = 1, 2), \tag{16.35}$$

where $s^{(n)}$ are again given by (16.26) and (16.28), take the form (Rattray, 1964)

$$u_{tt}^{(n)} - fv_t^{(n)} = gh^{(n)}(u_{xx}^{(n)} + v_{xy}^{(n)}), \tag{16.36a}$$

$$v_{tt}^{(n)} + fu_t^{(n)} = gh^{(n)}(u_{xy}^{(n)} + v_{yy}^{(n)}), \tag{16.36b}$$

where the $h^{(n)}$ are given by (16.28) and (16.29). It is important to note that (16.36) holds for the β-plane as well as the f-plane. The above discussion pertaining to the behaviour of u_1 and u_2 for each mode obviously carries over to v_1 and v_2 as well.

Finally, we mention that normal mode equations can also be derived for the quantity $\eta^{(n)}$, where

$$\eta^{(n)} = \eta_1 + q^{(n)}\eta_2; \tag{16.37}$$

$q^{(0)}$ and $q^{(1)}$ are the two roots of the equation

$$(1-\delta)H_2 q^2 + [H_1 + (1-2\delta)H_2]q - \delta H_2 = 0. \tag{16.38}$$

To $0(\delta)$ the two roots of (16.38) are

$$q^{(0)} = \delta H_2/H, \tag{16.39a}$$

$$q^{(1)} = -H/H_2, \tag{16.39b}$$

corresponding to the barotropic and baroclinic modes, respectively. The equation for $\eta^{(n)}$ is complicated when $\beta \neq 0$; however, when $\beta = 0$, we have

$$\mathcal{L}\eta^{(n)} - gh^{(n)}\nabla_h^2\eta^{(n)} = 0, \tag{16.40}$$

where $h^{(0)}$ and $h^{(1)}$ are given by (16.28) and (16.29), respectively. For barotropic motions alone ($\eta^{(1)} \equiv 0$), (16.36) and (16.39b) give [for $H/H_2 = 0(1)$]

$$\eta_1 = (H/H_2)\eta_2 = 0(\eta_2). \tag{16.41a}$$

Similarly, for purely baroclinic motions ($\eta^{(0)} \equiv 0$), we find [for $H_2/H_1 = 0(1)$]

$$\eta_1 = (-\delta H_2/H_1)\eta_2 = 0(\delta\eta_2). \tag{16.41b}$$

Normal mode equations for a rigid upper surface

Since for baroclinic motions $\eta_2 \gg \eta_1$ for typical oceanic values of $\delta = (\rho_2 - \rho_1)/\rho_2$, it is natural to consider the case when free-surface fluctuations are neglected altogether and the ocean surface is regarded as being rigid. For this idealized situation, the integration of (16.3) gives

$$\begin{aligned} p_1(x,y,z,t) &= -\rho_1 gz + p_1(x,y,0,t) \\ &\equiv -\rho_1 gz + p_1^0, \end{aligned} \tag{16.42}$$

where p_1^0 is unknown, replacing the combination $-\rho_1 g\eta_1 + p_a$ which occurs in the free-surface case.

The integration of (16.6) gives

$$p_2 = \rho_2 g(\eta_2 - z - H_1) - \rho_1 g(\eta_2 - H_1) + p_1^0, \tag{16.43}$$

which is easily seen to equal p_1 at $z = -H_1 + \eta_2$. Thus in place of (16.8) and (16.9) we obtain the momentum equations

$$u_{1t} - fv_1 + \frac{1}{\rho_1}p_{1x}^0 = 0, \tag{16.44a}$$

$$v_{1t} + fu_1 + \frac{1}{\rho_1}p_{1y}^0 = 0, \tag{16.44b}$$

$$u_{2t} - fv_2 + \frac{1}{\rho_2}p_{1x}^0 + g'\eta_{2x} = 0, \tag{16.45a}$$

$$v_{2t} + fu_2 + \frac{1}{\rho_2}p_{1y}^0 + g'\eta_{2y} = 0. \tag{16.45b}$$

The conservation of mass equations are (16.11) for the lower layer (as before), but

$$H_1(u_{1x} + v_{1y}) - \eta_{2t} = 0 \tag{16.46}$$

for the upper layer.

Adding (16.11) and (16.46) we obtain

$$(H_1u_1 + H_2u_2)_x + (H_1v_1 + H_2v_2)_y = 0. \tag{16.47}$$

We now define ψ as the stream function for the vertically averaged velocity by the equations

$$\psi_x = H_1v_1 + H_2v_2 \quad \text{and} \quad \psi_y = -(H_1u_1 + H_2u). \tag{16.48}$$

The standard procedure at this stage is to reduce the system (16.44), (16.45) into two coupled equations for ψ and h, where

$$h = \eta_2 - \frac{1}{\rho_1 g} p_1^0. \tag{16.49a}$$

For the terms in (16.45) to be of the same order, $\eta_2 \simeq p_1^0/\rho_2 g'$ and therefore $\eta_2 \gg p_1^0/\rho_1 g$ for $\delta \ll 1$. Thus (16.49a) gives

$$h \simeq \eta_2. \tag{16.49a}$$

When $H_2 =$ constant, the equations for ψ and h are decoupled and represent the normal mode equations for the model.

Solving for $\mathcal{L}u_n$ and $\mathcal{L}v_n$ from (16.44) and (16.45), we find

$$\mathcal{L}u_1 = -\frac{1}{\rho_1}(p_{1xt}^0 + fp_{1y}^0), \tag{16.50a}$$

$$\mathcal{L}v_1 = -\frac{1}{\rho_1}(p_{1yt}^0 - fp_{1x}^0), \tag{16.50b}$$

$$\mathcal{L}u_2 = -\left[\left(g'\eta_{2x} + \frac{1}{\rho_2}p_{1x}^0\right)_t + f\left(g'\eta_{2y} + \frac{1}{\rho_2}p_{1y}^0\right)\right], \tag{16.51a}$$

$$\mathcal{L}v_2 = -\left[\left(g'\eta_{2y} + \frac{1}{\rho_2}p_{1y}^0\right)_t - f\left(g'\eta_{2y} + \frac{1}{\rho_2}p_{1y}^0\right)\right]. \tag{16.51b}$$

Subtracting alternate pairs in (16.50) and (16.51) and using (16.49a), we obtain

$$\mathcal{L}(u_2 - u_1) = -g'(h_{xt} + fh_y), \tag{16.52a}$$

$$\mathcal{L}(v_2 - v_1) = -g'(h_{yt} - fh_x). \tag{16.52b}$$

Thus h represents the baroclinic part of the motion, in contrast to ψ, which represents the vertically averaged or barotropic part of the motion. From (16.48) and (16.52) we can now find expressions for $\mathcal{L}u_n$ and $\mathcal{L}v_n$ in terms of ψ and h. The result is

$$\mathcal{L}u_1 = \frac{1}{H}[\mathcal{L}\psi_y + H_2g'(h_{xt} + fh_y)], \tag{16.53a}$$

$$\mathcal{L}u_2 = \frac{1}{H}[\mathcal{L}\psi_y - H_1g'(h_{xt} + fh_y)], \tag{16.53b}$$

$$\mathcal{L}v_1 = \frac{1}{H}[-\mathcal{L}\psi_x + H_2 g'(h_{yt} - fh_x)], \tag{16.54a}$$

$$\mathcal{L}v_2 = \frac{1}{H}[-\mathcal{L}\psi_x - H_1 g'(h_{yt} - fh_x)], \tag{16.54b}$$

where $H = H_1 + H_2$.

To obtain the equations for ψ and h, we first form the vorticity equations from (16.44) and (16.45) and then use (16.46) and (16.11); since H_2 = constant now, we obtain

$$(u_{1y} - v_{1x})_t - \beta v_1 - \frac{f}{H_1}\eta_{2t} = 0, \tag{16.55}$$

$$(u_{2y} - v_{2x})_t - \beta v_2 + \frac{f}{H_2}\eta_{2t} = 0. \tag{16.56}$$

Multiplying (16.55) and (16.56) by H_1 and H_2 respectively and adding, we finally obtain

$$\nabla_h^2 \psi_t + \beta\psi_x = 0, \tag{16.57}$$

upon using (16.48). Notice that (16.57) is the same as (15.55), derived directly from the barotropic potential vorticity equation (15.54). On the other hand, subtraction of (16.55) from (16.56) gives

$$(u_2 - u_1)_{yt} - (v_2 - v_1)_{xt} - \beta(v_2 - v_1) + f\left(\frac{1}{H_2} + \frac{1}{H_1}\right)h_t = 0, \tag{16.58}$$

where we have used the approximation $\eta_2 \simeq h$ [see (16.49b)]. To eliminate the velocity differences in (16.58), we use (16.52); however, since $\mathcal{L}$ and ∂_{yt} do not commute when the variations of f are taken into account, the resulting equation is somewhat complicated, rather like the lengthy pressure equation (15.26), derived for continuous stratification. When $\beta = 0$, however, (16.58) can be integrated once with respect to t, operated on by $\mathcal{L}$ and then transformed into an equation for h by means of (16.52). Thus for the f-plane, we obtain

$$\nabla_h^2 h - \frac{1}{r_{i2}^2}\left(\frac{1}{f^2}\partial_{tt} + 1\right)h = 0, \tag{16.59}$$

where $r_{i2} = (g'H_1H_2/Hf^2)^{1/2}$ (16.60)

is the *internal Rossby radius of deformation*. Along with (16.59), we note from (16.57) that ψ becomes a harmonic function when $\beta = 0$:

$$\nabla_h^2 \psi = 0. \tag{16.61}$$

Thus on the f-plane with a rigid upper surface there is no wave motion associated with ψ, the vertically averaged or barotropic part of the motion. We also recognize the internal radius of deformation r_{i2} as the dominant horizontal length scale of wave motion in a two-layer rotating fluid.

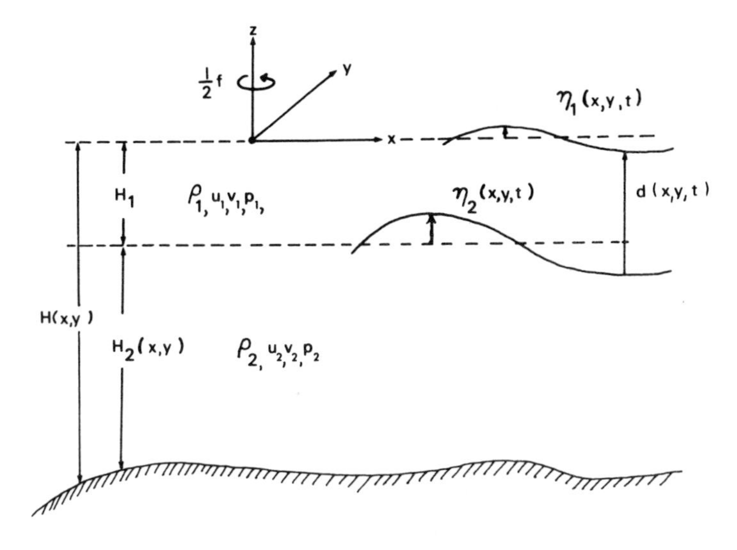

Fig. 16.5. Sketch of two-layer system. The lower layer is slightly heavier than the upper layer so that the relative density difference between the layers is small, i.e.

$\delta = 1 - \rho_1/\rho_2 \ll 1.$

Exercises Section 16

1. Find the linearized solutions which describe the modes in Fig. 16.3. Assume that: (1) the beads are of equal mass; (2) the tension in the string is constant; and (3) the distance between the beads is twice that between either bead and its nearest boundary.

2. Show that the long wave equations for the motion in a layer of fluid of density ρ_1 lying above a deep, motionless layer of density ρ_2 are

$$u_{1t} - fv_1 + g'd_y = 0, \tag{16.62a}$$

$$v_{1t} + fu_1 + g'd_y = 0, \tag{16.62b}$$

$$h_1(u_{1x} + v_{1y}) + d_t = 0, \tag{16.62c}$$

where $d(x, y, t)$ is the thickness of the upper layer (see Fig. 16.5). [Assume that $(\rho_2 - \rho_1)/\rho_2 \ll 1$.]

3. Derive the normal mode equations (16.36) for $u^{(n)}$ and $v^{(n)}$.

4. Derive (16.38) and (16.39). Show that the velocity components for barotropic and baroclinic motions are related as in (16.33) and (16.34), respectively.

We now focus our attention on long waves of the first class, whose frequencies are characterized by $\omega/f = 0(1)$. As will be seen later these waves are of fundamental importance in the theory of storm surges and tidal propagation.

Continuous stratification

For continuous stratification the governing equation for the horizontal dependence of the pressure corresponding to each vertical mode n is obtained from (15.26) by setting $\beta = 0$. This gives

$$\partial_t \mathcal{L} \left(\nabla_h^2 - \frac{1}{gh_n} \mathcal{L} \right) p_n = 0. \tag{17.1}$$

It is easy to show that for frequencies $\omega = 0(f)$ and horizontal scales of $0(L)$, the relative size of the neglected terms in (15.26) is $0(\beta L/f)$. For mid-latitude values of β and f and for $L = 10^6$ m, $\beta L/f = 0(10^{-1})$ and hence the effects of β on the first-class waves will be small.

Horizontally propagating waves of the form

$$p_n = P_n \exp\left[i(k_1 x + k_2 y - \omega t)\right], \tag{17.2}$$

where P_n is a constant, will satisfy (17.1) provided the following dispersion relation holds:

$$\omega^2 \equiv \omega_n^2 = gh_n k_h^2 + f^2, \quad n = 0, 1, 2, \ldots . \tag{17.3}$$

For $k_h^2 > 0$, propagation is possible only for $\omega_n^2 > f^2$. We notice that (17.3) is identical in form to the compatibility condition (10.45). However, in contrast to the nonhydrostatic case discussed in Section 10, here h_n is independent of frequency; h_n is determined by solving the vertical boundary value problem (15.25), (15.27) and (15.28). We also note that (17.3) follows from the dispersion relation (15.18) upon setting $\beta = 0$ in the latter.

Since the right-hand side of (17.3) depends only on the magnitude of $\boldsymbol{k}_h$ and not on the wavenumber components individually, first-class waves are isotropic, i.e., their description is invariant under a rotation of the plane about the z-axis. The phase velocity of the nth mode is given by

$$\boldsymbol{c}_n = \frac{\omega_n}{k_h^2} \boldsymbol{k}_h = \frac{(gh_n k_h^2 + f^2)^{1/2}}{k_h^2} \boldsymbol{k}_h, \tag{17.4}$$

where for definiteness we have taken the positive square root in (17.3). The group velocity takes the form

$$\boldsymbol{c}_{ng} = \frac{gh_n}{\omega_n} \boldsymbol{k}_h, \tag{17.5}$$

which is parallel to $\boldsymbol{c}_n$. Notice that

$$\boldsymbol{c}_n \cdot \boldsymbol{c}_{ng} = gh_n. \tag{17.6}$$

Further, since

$$c_n = (gh_n + f^2/k_h^2)^{1/2} \tag{17.7}$$

and $$c_{ng} = gh_n/(gh_n + f^2/k_h^2)^{1/2}, \tag{17.8}$$

it follows that

$$c_{ng} < \sqrt{gh_n} < c_n. \tag{17.9}$$

That is to say, the waves are normally dispersive ($c_g < c$), in agreement with the general property (4) proved at the end of Section 10.

When $f = 0$, these waves are nondispersive and are entirely governed by the gravitational restoring force. They are the familiar long gravity waves associated for example with tsunamis in the open ocean (the barotropic mode) or internal seiches in shallow lakes (the baroclinic modes). We note from (17.7) that in addition to introducing dispersion, rotation increases the phase speed of these waves. For our purposes here we shall call the set of all waves satisfying (17.1) "long gravity waves". However, at times we will also add the words "surface" or "internal" to indicate the barotropic mode or baroclinic modes, respectively. Long gravity waves will also be referred to as Poincaré waves (Poincaré, 1910).

For the case of constant stratification we showed that $h_0 \gg h_1 > h_2 > \ldots$ (see 15.43). Hence according to (17.7) the barotropic phase speed is much greater than each of the baroclinic speeds, the ratio of the former to the latter being $0[(\Delta\rho_0/\rho_0)^{-1/2}]$ [see (15.41), (15.42) and (17.7)]. Thus for $H = 5 \cdot 10^3$ m and $\Delta\rho_0/\rho_0 = 2 \cdot 10^{-3}$, we find from (17.7) that

$$\begin{aligned} c_0 &\geqslant 0(2 \cdot 10^2 \text{ m/s}), \\ c_n &\leqslant 0(10 \text{ m/s}). \end{aligned} \tag{17.10}$$

Similar order of magnitude results also hold for variable stratification ($N \neq$ constant).

As in Section 8, we can easily write down all the velocity components in terms of the pressure. From (15.19), (15.20) and (15.10) we obtain, upon incorporating the vertical dependence (cf. 15.24),

$$\begin{pmatrix} u_n \\ v_n \end{pmatrix} = \frac{1}{\rho_0(\omega_n^2 - f^2)} P_n \exp(iS_n)\,\Pi_n \begin{pmatrix} \omega_n k_1 + ifk_2 \\ \omega_n k_2 + ifk_1 \end{pmatrix}, \tag{17.11}$$

$$w_n = \frac{i\omega_n}{\rho_0 N^2} P_n \exp(iS_n)\,\Pi_n', \tag{17.12}$$

where $S_n = k_1 x + k_2 y - \omega_n t$

and the prime denotes differentiation with respect to z. From (15.3), the density perturbation is

$$\rho_n = -(P_n/g)\exp(iS_n)\,\Pi_n'. \tag{17.13}$$

From (17.11) and (17.12) it follows that the fluid particles trace out an elliptical path in a plane which is tilted along a line parallel to $\boldsymbol{k}_h$. For the barotropic mode ($n = 0$), w is small compared to u and v; in this case the orbital plane is nearly horizontal. Also, we note that at high frequencies ($\omega \gg f$), the current ellipses are elongated, whereas at near-inertial frequencies, they are almost circular. To help visualize the orbital motions just

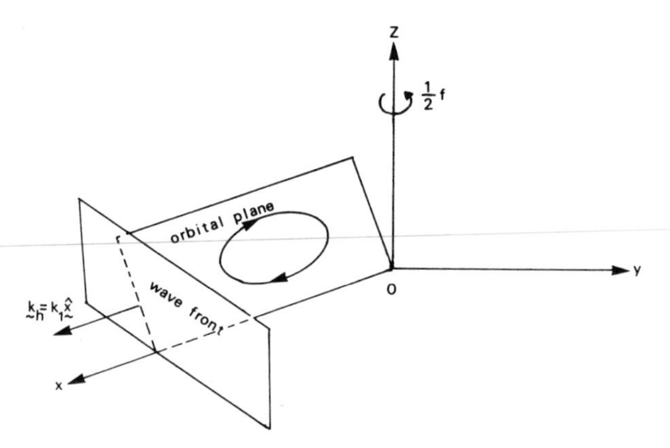

Fig. 17.1. Particle motion of long gravity waves in the Northern Hemisphere.

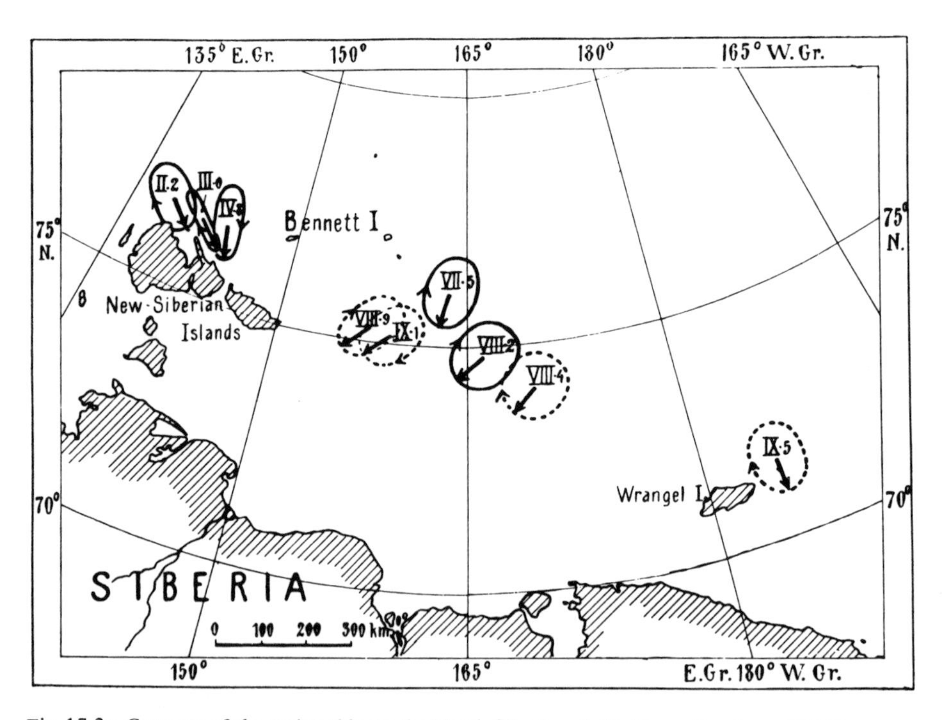

Fig. 17.2a. Currents of the spring tide on the North Siberian shelf. (From Sverdrup, 1926.)

described, it is now convenient to exploit the isotropy of long gravity waves and choose horizontal axes so that the x-axis lies along $\boldsymbol{k}_h$. Then in (17.11)–(17.13) we can set $k_2 = 0$ and take $k_1 > 0$; the velocity components in real form are

$$u_n = \frac{\omega_n k_1}{\rho_0(\omega_n^2 - f^2)} P_n \cos(k_1 x - \omega_n t)\,\Pi_n, \tag{17.14a}$$

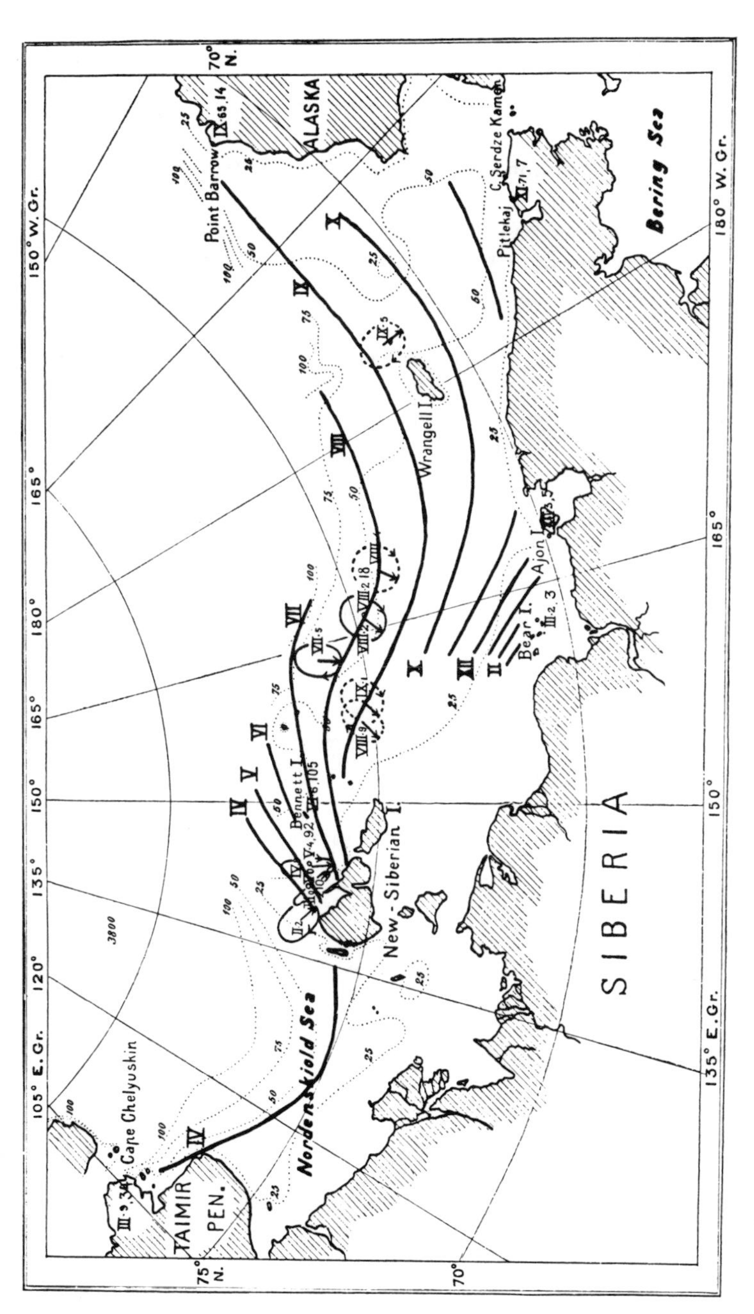

Fig. 17.2b. Co-tidal lines of the spring tide on the North Siberian shelf. (From Sverdrup, 1926.)

$$v_n = \frac{-fk_1}{\rho_0(\omega_n^2 - f^2)} P_n \sin(k_1 x - \omega_n t)\,\Pi_n, \tag{17.14b}$$

$$w_n = -\frac{\omega_n}{\rho_0 N^2} P_n \sin(k_1 x - \omega_n t)\,\Pi_n', \tag{17.14c}$$

where $\omega_n = (gh_n k_1^2 + f^2)^{1/2}$.

A typical hodograph of the orbital motion near the surface (where $\Pi > 0$ and $\Pi' > 0$) is shown in Fig. 17.1 for the case $f > 0$ (Northern Hemisphere). When $f < 0$, the particle motion is anticlockwise and the hodograph plane is tilted to the right of $\boldsymbol{k}_h$.

In his study of tides on the Northern Siberian shelf, Sverdrup (1926) constructed hodographs of the semi-diurnal barotropic tidal currents and co-tidal lines (lines of constant phase) for the "spring" tide (see Section 52). His original drawings are shown in Fig. 17.2; note that the current motions are indeed clockwise and in a horizontal plane, in accordance with the theory, as expected for a Poincaré (barotropic) wave. Also, in accordance with the theory the ellipses are nearly circular, because of the proximity of the Siberian shelf to the critical latitudes [ϕ_c as given by (10.25)] for the semi-diurnal tidal frequencies considered by Sverdrup (the M_2 and S_2 components). The co-tidal lines in Fig. 17.2b show a progressive wave advancing southeastward across the shelf, in the manner of a Poincaré wave.

We can derive a conservation of energy equation from (15.1)–(15.5) in the same manner as at the end of Section 8. In the inferior of the fluid we then have:

$$\partial_t[\tfrac{1}{2}\rho_0(u^2 + v^2 + N^2\xi^2)] + \nabla \cdot (p\boldsymbol{u}) = 0, \tag{17.15}$$

where $\xi = \int w \, dt$ is the usual vertical displacement of a particle in the fluid. We note that because of the hydrostatic approximation, the kinetic energy density does not contain a term proportional to w^2. To calculate the mean energy densities, we use the real expressions for u, v and w given in (17.14). But the definition of the mean used in Section 8 has to be slightly modified because of the modal structure of the waves. Here we take an average over both the phase and the depth:

$$\langle\langle(\cdot)\rangle\rangle = \int_{-H}^{0} \left\{\frac{1}{2\pi}\int_0^{2\pi} (\cdot)\, dS_n\right\} dz. \tag{17.16}$$

Thus now the mean energy densities represent energy per unit area in the (x, y)-plane.

$$\langle KE \rangle = \frac{P_n^2}{4}\frac{(\omega_n^2 + f^2)k_1^2}{(\omega_n^2 - f^2)^2}\int_{-H}^{0}\frac{\Pi_n^2}{\rho_0}\, dz \tag{17.17}$$

for the mean kinetic energy density. For the mean potential energy density we obtain, with inclusion of the free surface contribution $\rho(0)g\langle\xi^2\rangle/2$,

$$\langle PE \rangle = \frac{P_n^2}{4}\left[\int_{-H}^{0} \frac{\left(\dfrac{d\Pi_n}{dz}\right)^2}{\rho_0 N^2}\, dz + \frac{\Pi^2(0)}{g\rho_0(0)}\right]. \tag{17.18}$$

Using (15.25), (15.27), (15.28) and the dispersion relation (17.3), we can rewrite (17.18) as

$$\langle PE \rangle = \frac{P_n^2}{4} \frac{k_1^2}{\omega_n^2 - f^2} \int_{-H}^{0} \frac{\Pi_n^2}{\rho_0} \, dz. \tag{17.19}$$

Thus, as in the internal wave case discussed in Section 8, we observe that there is no equipartition of energy when $f \neq 0$. In fact for near-inertial frequencies, the kinetic energy is much larger than the gravitational potential energy. Adding (17.17) and (17.19), the total mean energy density $\langle E \rangle$ is given by

$$\langle E \rangle = \frac{P_n^2}{2} \frac{\omega_n^2 k_1^2}{(\omega^2 - f^2)^2} \int_{-H}^{0} \frac{\Pi_n^2}{\rho_0} \, dz. \tag{17.20}$$

Thus the energy flux is

$$\langle E \rangle \boldsymbol{c}_g = \langle E \rangle \frac{gh_n}{\omega_n} k_1 \hat{\boldsymbol{x}} = \frac{P_n^2 \omega_n k_1}{2(\omega_n^2 - f^2)} \int_{-H}^{0} \frac{\Pi_n^2}{\rho_0} \, dz \hat{\boldsymbol{x}}, \tag{17.21}$$

where $\hat{\boldsymbol{x}}$ is a unit vector along $\boldsymbol{c}_g$. It is a simple exercise to show that the energy flux as given by the pressure velocity correlation $\langle p\boldsymbol{u} \rangle$ is identical with (17.21) (Exercise 17.2).

Long gravity waves in a two-layer fluid

To describe long gravity waves in a two-layer fluid, we use the normal mode equations (16.36a, b) for the $u^{(n)}$ and $v^{(n)}$, where $n = 0$ and $n = 1$ refer to the barotropic and baroclinic modes, respectively. Since $f =$ constant in these equations we can easily eliminate $v^{(n)}$ to obtain an equation for $u^{(n)}$; after integrating the latter twice with respect to time we find that $u^{(n)}$ satisfies a Klein–Gordon equation:

$$\left[\nabla_h^2 - \frac{1}{gh^{(n)}} \mathcal{L} \right] u^{(n)} = 0. \tag{17.22}$$

The same equation also holds for $v^{(n)}$. The operator in the square brackets in (17.22) is of the same form as that inside the brackets in (17.1), the governing equation for the case of continuous stratification. However, as there are now only two modes, there are precisely two dispersion relations for horizontally propagating waves:

$$\omega_0^2 = gHk_h^2 + f^2 \qquad \text{(barotropic)}, \tag{17.23a}$$

$$\omega_1^2 = \frac{g' H_1 H_2}{H} k_h^2 + f^2 \quad \text{(baroclinic)}, \tag{17.23b}$$

where (16.28) and (16.29) have been used. The previously derived dispersion relation (10.67) reduces to (17.23b) for $\delta \ll 1$. As in the case of continuous stratification, the baroclinic wave travels much more slowly than the barotropic wave.

Figure 17.3 shows temperature and current data describing a long internal wave propagating northward along the thermocline in Babine Lake, British Columbia. The average thermocline depth throughout the length of this long, narrow and deep lake is about 15 m. However, in the left of the picture, before the wave front arrived, the thermoclinic depth is about 10 m; after the wave passed it deepened to about 20 m. The wave was generated

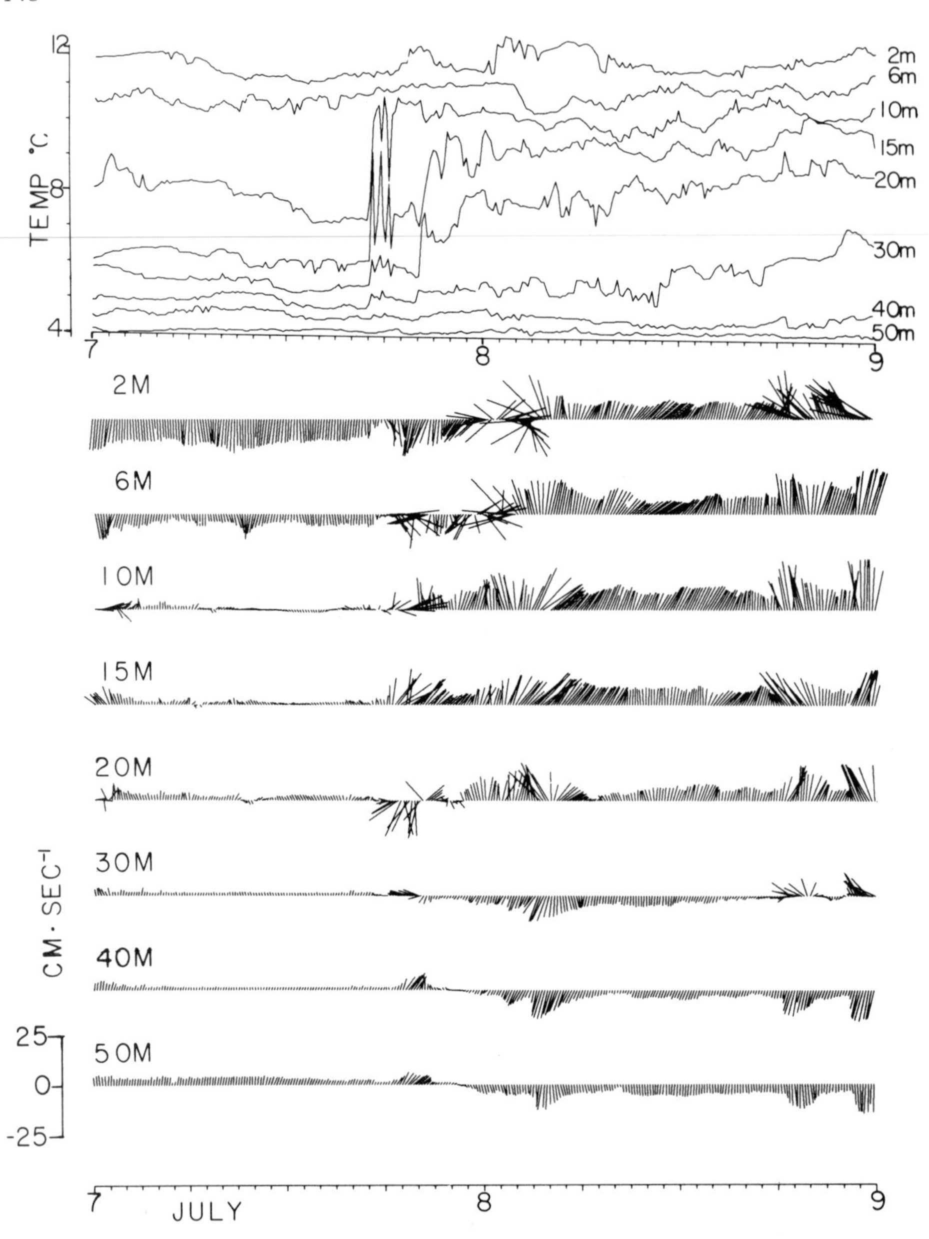

Fig. 17.3. Temperature and longshore current measurements in Babine Lake, British Columbia, showing the northward passage, during July, 1973, of an internal surge in an approximately two-layer stratification. The upper panel shows the time variations of the temperature at selected depths. The current structure at the same levels is illustrated by stick diagrams in the lower panel. Before the arrival of the surge, the surface layer (data from the 2 m and 6 m levels) moves southward towards the incoming wave. The lower layer currents are in the opposite direction, showing the phase reversal with depth expected of an interfacial oscillation. After a period of rapid fluctuations the upper-layer depth increases to about 20 m and the currents reverse their directions, again changing phase with depth at the density interface. (From Farmer, 1978.)

at the south end of the lake by winds, and in its initial stages was of small amplitude. However, nonlinear effects (mainly amplitude dispersion) eventually produced an asymmetric profile with a sharp front, almost like a turbulent bore (Farmer, 1978). Nevertheless, the baroclinicity of the currents, as predicted by a two-layer linear theory, is very much in evidence before and after the passage of the wave (see also Fig. 16.4). Although a two-layer model for the temperature stratification in Babine Lake is rather crude, as is the assumption of linearity, it is nevertheless of interest to apply (17.23b) to estimate the linear phase speed of long internal waves. For $H_1 = 15$ m and $g' = 0.35 \cdot 10^{-2}\,\mathrm{ms}^{-2}$ (corresponding to a temperature difference of 6°C between the two layers), (17.23b) gives, for $H_2 \gg H_1$ and $\omega \gg f$,

$$c_1 \simeq \sqrt{g'H_1} = 0.23\,\mathrm{m/s}.$$

Using the observed density profiles ahead of and behind the front, Farmer (1978) computed the speed of the smaller scale waves in each of these regions using a linear theory. He obtained the values 0.14 m/s and 0.21 m/s, respectively, for the wave speeds ahead of and behind the front. That is to say, the region behind the wave front is gradually catching up with the region ahead, a situation known as a supercritical flow. The observed speed of the front itself was estimated to be 0.18 m/s.

We leave the reader to examine for himself the orbital motions associated with long waves in a two-layer model. We now make a few comments on the energy equation for a two-layer fluid. Taking the scalar products of (16.8) and (16.9) with $\boldsymbol{u}_{1h}$ and $\boldsymbol{u}_{2h}$, respectively, adding the resulting equations and then using (16.10) and (16.11), we obtain

$$\frac{\partial}{\partial t}\left[\tfrac{1}{2}\rho_1(u_1^2+v_1^2)+\tfrac{1}{2}\rho_2(u_2^2+v_2^2)\right]+\frac{\partial}{\partial t}\left[\frac{1}{2H_1}\rho_1 g\eta_1^2+\frac{1}{2H_2}\rho_2 g'\eta_2^2\right]$$
$$+\left[\rho_1 g\,\frac{H_1-H_2}{H_1H_2}\,\eta_1\,\frac{\partial\eta_2}{\partial t}\right]+g\nabla\cdot\left[(\rho_1\boldsymbol{u}_{1h}+\rho_2\boldsymbol{u}_{2h})\eta_1\right]+\rho_2 g'\nabla\cdot\left[\boldsymbol{u}_{2h}(\eta_2-\eta_1)\right]=0. \tag{17.24}$$

The first square-bracketed term in (17.24) represents the (time) rate of change of the kinetic energy in the two layers; the second, the rate of change of potential energy associated with the surface and interfacial displacements; the third, an "interaction term" between the surface and interface, which vanishes when $H_1 = H_2$ or when averaged over the wave phase; and the last two terms, the divergence of the energy fluxes in the two layers. To evaluate the average energy densities for the case of barotropic motion, it is easier to deal directly with the energy equation implied by the barotropic long wave equations (15.50), (15.51); this will be done at the end of this section. We only mention here that for barotropic motions, the second term in the potential energy expression is negligible since $\eta_1 = 0(\eta_2)$ [see (16.40)]. For the baroclinic mode, we note from (16.34) that the ratio of the kinetic energy in the upper layer to that in the lower layer is very nearly $(H_2/H_1)^2$, which will be large for most mid-ocean situations. Finally, we notice that the dominant contribution to the potential energy comes from the interfacial displacement since $\eta_1 = 0(\delta\eta_2)$ for the baroclinic mode (see 16.41). We now let the reader work out detailed expressions for the various energy densities.

Before going on to the barotropic long wave equations we mention here that if the free surface is treated as being rigid and flat in a two-layer fluid, the barotropic wave disappears. This is because in terms of the vertically averaged transport ψ and interfacial displacement h, the normal mode equations are (16.61) and (16.59) respectively:

$$\nabla_h^2 \psi = 0, \tag{16.61}$$

$$\nabla_h^2 h - \frac{1}{r_{i2}^2}\left(\frac{1}{f^2}\partial_{tt} + 1\right) h = 0. \tag{16.59}$$

Therefore, ψ is harmonic rather than wavelike, and (16.59) is identical in form to (17.22) and hence also implies the dispersion relation (17.23b). We note that formally, (16.61) corresponds to waves having infinite propagation speed [cf. (17.22) and let $gh^{(0)} \to \infty$].

Long gravity waves in a homogeneous fluid

We close this section with a quick look at the "classical" unforced long wave equations for a homogeneous fluid of constant depth, i.e., (15.50) and (15.51) with $H =$ constant:

$$u_t - fv + g\eta_x = 0, \tag{17.25a}$$

$$v_t + fu + g\eta_y = 0, \tag{17.25b}$$

$$H(u_x + v_y) + \eta_t = 0. \tag{17.25c}$$

From (17.25a, b) we obtain

$$\mathcal{L}u = -g(\eta_{xt} + f\eta_y), \tag{17.26a}$$

$$\mathcal{L}v = -g(\eta_{yt} - f\eta_x). \tag{17.26b}$$

We now use (17.26) to eliminate u_x and v_y in (17.25c); the resulting equation for η is again the familiar Klein–Gordon equation

$$\mathcal{L}\eta - gH\nabla_h^2\eta = 0. \tag{17.27}$$

Thus for plane waves of the form

$$\eta = a \exp\left[i(k_1 x + k_2 y - \omega t)\right], \tag{17.28}$$

the dispersion relation is

$$\omega^2 = gHk_h^2 + f^2, \tag{17.29}$$

in agreement with (17.3) for the case $n = 0$ ($h_0 = H$), or with (17.23a). It is easy to show that u and v also satisfy (17.27). From (17.26) we observe that the orbital plane is horizontal since there is no vertical velocity associated with (17.25).

At this stage we wish to emphasize the importance of the free surface motion for these waves. If the surface were rigid and flat, (17.25) would be replaced by

$$u_t - fv + \frac{1}{\rho_*}p_x = 0, \tag{17.30a}$$

$$v_t + fu + \frac{1}{\rho_*}p_y = 0, \tag{17.30b}$$

$$u_x + v_y = 0. \tag{17.30c}$$

Equation (17.30c) now implies the existence of a stream function ψ such that

$$u = -\psi_y, \quad v = \psi_x. \tag{17.31}$$

Substituting into (17.30a, b) and eliminating the pressure, we find

$$\nabla_h^2 \psi_t = 0! \tag{17.32}$$

That is, in this case there is no wave motion on the f-plane, a result also found at the end of Section 16 (see 16.61).

From (17.25) we readily obtain the energy equation

$$\frac{\partial}{\partial t}\left[\frac{1}{2}\rho_*(u^2+v^2)+\frac{1}{2H}\rho_* g\eta^2\right]+\rho_* g\nabla\cdot(\boldsymbol{u}_h\eta) = 0. \tag{17.33}$$

Using the real part of the right-hand side of (17.28) for η, we find the mean potential energy density to be

$$\langle PE\rangle = \tfrac{1}{4}\rho_* g a^2. \tag{17.34}$$

The mean kinetic energy density is

$$\langle KE\rangle = \tfrac{1}{4}\rho_* g a^2\left(\frac{\omega^2+f^2}{\omega^2-f^2}\right). \tag{17.35}$$

Thus there is exact equipartition when $f=0$ [cf. (17.17) and (17.19)], and when $f\neq 0$ the kinetic energy is greater than the potential in the ratio $(\omega^2+f^2)/(\omega^2-f^2)$, which is very large when ω is near f. Also, we note that when $f=0$,

$$\langle E\rangle = \langle KE\rangle+\langle PE\rangle = \tfrac{1}{2}\rho_* g a^2,$$

in agreement with the result for high-frequency surface gravity waves [cf. (13.24) with $B=0$].

Exercises Section 17

1. Derive (17.19) from (17.18).
2. Show that $\langle pu\rangle$ for long gravity waves propagating in the x-direction is also given by (17.21). How do the results (17.17)–(17.21) change when a long gravity wave propagates in an arbitrary direction $\boldsymbol{k}_h$?
3. Discuss the particle paths for barotropic and baroclinic long gravity waves in a two-layer fluid.
4. Find the average kinetic and potential energy densities for baroclinic gravity waves in a two-layer fluid.
5. Derive and interpret the energy equation for long internal gravity waves in a fluid with a rigid, flat top.
6. Show that for a nonrotating fluid, the "interaction" term in (17.24) can be written as

$$\rho_1 g\,\frac{H_1-H_2}{H_1H_2}\left\{\frac{\partial}{\partial t}\left[\frac{1}{2}\eta_1^2+\frac{\rho_1H_1}{2g}(u_1^2+v_1^2)\right]+H_1\nabla_h\cdot(\boldsymbol{u}_{1h}\eta_1)\right\},$$

i.e., as the sum of an interaction energy term and a flux term.

7. For an $e^{-i\omega t}$ time dependence, show that (17.1) reduces to Helmholtz's equation. In cylindrical polar coordinates (r, θ), show that this equation has solutions of the form

$$p_n = H_s^{\pm}(k_{hn}r)\,e^{is\theta}, \quad s = 0, 1, \ldots, \tag{17.36}$$

where $k_{hn} = [(\omega_n^2 - f^2)/gh_n]^{1/2}$

and where $H_s^{\pm}$ are the Hankel functions $H_s^{(1)}$ and $H_s^{(2)}$ defined as $J_s \pm iY_s$ in terms of Bessel functions of the first kind.

8. In cylindrical polar coordinates, equations (15.1) and (15.2) for the case $f =$ constant can be written as

$$\frac{\partial u_r}{\partial t} - fv_\theta + \frac{1}{\rho_0}\frac{\partial p}{\partial r} = 0, \tag{17.37a}$$

$$\frac{\partial v_\theta}{\partial t} + fu_r + \frac{1}{\rho_0 r}\frac{\partial p}{\partial \theta} = 0, \tag{17.37b}$$

where u_r and u_θ are the velocity components in the directions of increasing r and θ, respectively. By substituting $p = p_n \Pi_n(z)$ [where p_n is given by (17.36)] into (17.37), find expressions for u_r and u_θ. Hence find real expressions for p, u_r and u_θ. By examining the phase of the expression for real p, show that the wave fronts are radial near the origin ($k_{hn}r \ll 1$), spiral at intermediate values of k_{hn} and circular far from the origin ($k_{hn}r \gg 1$).

18. PLANETARY WAVES

Recent large-scale experiments in the North Atlantic Ocean by the USSR (the "Polygon" experiment) and jointly by the USA and UK (the "MODE" experiment) have established the existence of slowly varying currents (eddies) with time scales of a few months and wavelengths of a few hundred kilometers. To understand the dynamics of these motions, it is first necessary to have a knowledge of the basic linear theory of *planetary* or *Rossby* waves, i.e., long wave motions of the second class in a flat-bottomed, stratified fluid. The theoretical complications due to nonlinearities, variable topography and mean currents will be explored in Sections 19, 20 and 44–46, respectively.

It is now well known that the low-frequency variability of the zonal circulation in the atmosphere can be explained in terms of planetary waves (Rossby, 1939; Charney, 1947). Also, eddies formed from unstable planetary waves appear to be one of the main driving mechanisms of the mean atmospheric circulation (Starr, 1968). Whether a similar situation also applies to the mean oceanic circulation at mid-latitudes is still not known. Consequently, the theoretical and observational study of planetary waves and their interactions with other waves, mean flows and topography is very much at the forefront of present-day oceanographic research (see Welander, 1973, and Thomson and Stewart, 1977, for discussions of the role played by eddies in the vorticity balance of the ocean).

Continuous stratification

We again return to the pressure equation (15.26) which holds for both subinertial ($\omega < f$) and superinertial ($\omega > f$) frequencies. To study planetary waves alone we restrict our attention to frequencies small compared to f, i.e., $\omega \ll f$. Then, provided the x- and y-scales are comparable, the dominant term multiplying β in (15.26) is $f^2 p_{nx}$. Also, $\mathcal{L} \simeq f^2$ for $\omega \ll f$. Therefore, the governing equation for the horizontal dependence of the pressure field for planetary waves in a continuously stratified ocean is

$$\left(\nabla_h^2 p_n - \frac{f^2}{gh_n} p_n\right)_t + \beta p_{nx} = 0, \tag{18.1}$$

where for any given mean stratification $\rho_0(z)$, h_n is determined by the solution of the vertical problem (15.25), (15.27), (15.28). Since the variability of f is explicitly manifested in (18.1) by the presence of the β-term, it is traditional and convenient to treat f as constant in (18.1) when considering planetary waves in mid-latitude ocean basins. In the vicinity of the equator such an approximation is not reasonable, however, since $f = \beta y$ for an equatorial β-plane. And indeed, the vanishing of f at the equator provides a mechanism for the trapping of planetary (and long gravity) wave energy in this region (see Section 21).

At very low frequencies the horizontal momentum equations (15.1) and (15.2) can be approximated by the geostrophic relations

$$v = \frac{1}{f\rho_0} p_x, \tag{18.2a}$$

$$u = -\frac{1}{f\rho_0} p_y. \tag{18.2b}$$

Because of these approximate relations planetary waves are also sometimes called quasi-geostrophic waves. Since f is variable in (18.2), we can use these equations to get an estimate of the scale of the vertical velocity associated with planetary waves. From (15.4) and (18.2) we find

$$w_z = -u_x - v_y = \frac{\beta}{f^2\rho_0} p_x . \tag{18.3}$$

Therefore, $$w = 0\left(\frac{\beta H}{f^2\rho_0} p_x\right), \tag{18.4}$$

where H is the vertical scale. Compared to u or v, w is very small, e.g.,

$$w/v = 0(\beta H/f) = 0(10^{-3}) \tag{18.5}$$

for mid-latitudes and $H = 5 \cdot 10^3$ m. This scaling is to be contrasted with that used in (5.21). Although the small vertical velocities associated with planetary waves may, over the long periods of such waves, produce appreciable vertical displacements, it remains that planetary waves basically consist of variable horizontal currents.

Substitution of the usual horizontal plane-wave form (17.2) into (18.1) gives the dispersion relation

$$\omega_n = -\frac{\beta k_1}{k_h^2 + f^2/gh_n}. \tag{18.6}$$

This relation also follows from (15.18) if in the latter the term proportional to ω^3 is neglected. The phase velocity is given by

$$c_n = \frac{1}{k_h^2(k_h^2 + f^2/gh_n)}(-\beta k_1^2, -\beta k_1 k_2), \tag{18.7}$$

which always has a *negative* x-component. That is, the phase of planetary waves always propagates in a generally westward direction. For definiteness we choose $\omega_n > 0$; then (18.6) implies that $k_1 < 0$ and hence (18.7) indicates that the phase propagates northward and westward or southward and westward according to $k_2 > 0$ or $k_2 < 0$. To vizualize the situation, it is helpful to plot the *slowness curve* (a commonly used expression for a curve of constant ω in $\boldsymbol{k}$ space) for each mode. Treating $\omega_n = \omega > 0$ as a given fixed parameter in (18.6), we rewrite this relation as

$$\begin{aligned}(k_1 + \gamma)^2 + k_2^2 &= \gamma^2 - f^2/gh_n \\ &\equiv R_n^2,\end{aligned} \tag{18.8}$$

where $\gamma = \beta/2\omega$. Equation (18.8) corresponds to a family of concentric circles centered at $(-\gamma, 0)$, with the radius of each circle being denoted by R_n (see Fig. 18.1). For the right side of (18.8) to be positive, we must have $\omega < \omega_c$, where

$$\omega_c = \beta(gh_n/4f^2)^{1/2}. \tag{18.9}$$

That is, for each mode n there is an upper cut-off frequency ω_c for the existence of real wavenumbers. For the barotropic mode ($h_n \simeq H$), (18.9) gives $\omega_c = 0(10^{-5}$ rad s$^{-1})$ at mid-latitudes; for the baroclinic modes, ω_c is much smaller.

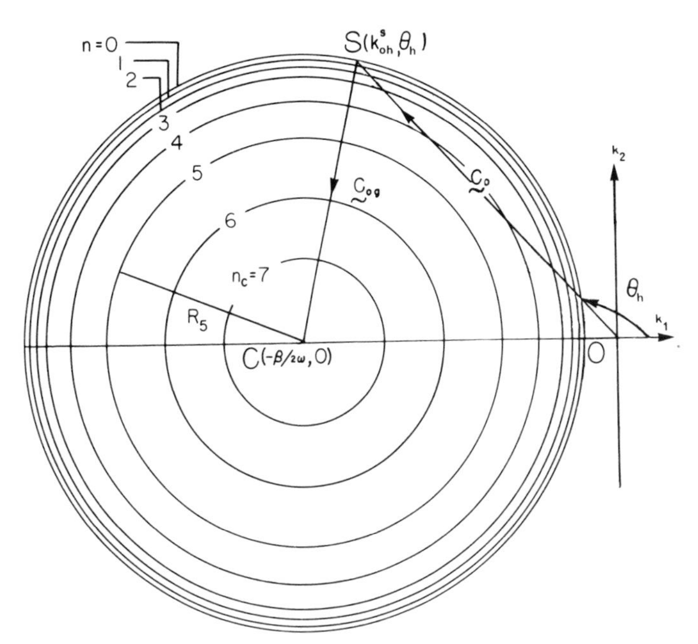

Fig. 18.1. The slowness curves (18.8) for planetary waves in a continuously stratified ocean with a flat bottom. $n = 0$ corresponds to the barotropic mode and $n = 1, 2, \ldots, n_c$, the baroclinic modes, where for illustrative purposes we have chosen $n_c = 7$. For a short barotropic wave (k_{oh}), the wave phase propagates along the ray *OS* whereas the group velocity is directed along *SC*, towards the center of the circle.

Since R_n decreases with $n[h_0 \gg h_1 > h_2 > \ldots$ according to (15.43)], it follows that for a sufficiently small value of ω there exists precisely $n_c + 1$ circles only, where $n_c \geqslant 0$ is the largest integer for which $R_n^2 > 0$. One interesting consequence of this result is that if the ocean responds linearly to a large-scale wind field with a characteristic frequency ω, then only a *finite* number of planetary wave eigenmodes can be excited: the barotropic and the first n_c baroclinic modes. However, if $\omega > \omega_c$, where ω_c is given by (18.9) for $n = 0$, then $R_n^2 < 0$ for all n and no waves would be excited. Since in practice the low-frequency end of the wind spectrum in the open ocean is broadly peaked at around three days (Fissel et al., 1976), there is somewhat of a mismatch of time scales in the atmosphere and the ocean at low frequencies. This result provides further evidence that the energy gap just below f in observed current spectra is indeed not spurious (see Fig. 15.1).

The group velocity, being equal to the wavenumber gradient of $\omega(k_1, k_2)$, is perpendicular to the slowness curve ω = constant in the direction of increasing ω and therefore is directed along a straight line passing through the center C (Fig. 18.1). From (18.6) we find

$$\boldsymbol{c}_{ng} = \frac{\beta}{(k_h^2 + f^2/gh_n)^2}(k_1^2 - k_2^2 - f^2/gh_n, 2k_1k_2). \tag{18.10}$$

In general, $\boldsymbol{c}_{ng}$ is neither parallel nor perpendicular to $\boldsymbol{c}_n$, which is indicative of the highly anisotropic nature of planetary waves. In fact considerable care must be exercised in determining the direction of $\boldsymbol{c}_{ng}$. This is because for a given frequency and direction of

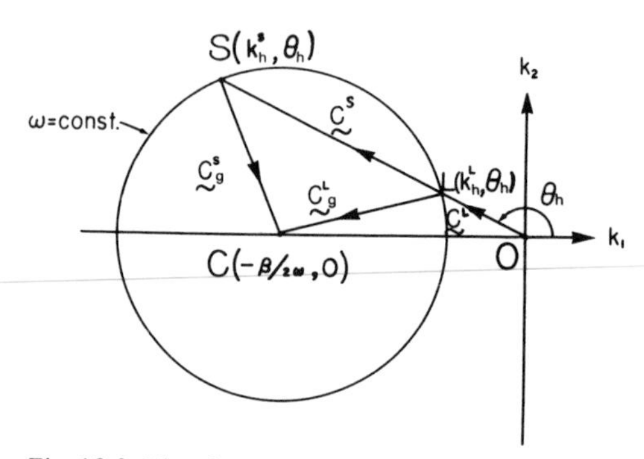

Fig. 18.2. The slowness curve for the nth (the subscript n is omitted) baroclinic mode showing the directions of the two group velocities corresponding to the short and long waves whose phase propagates along the direction defined by θ_h. The same picture also applies for the barotropic mode, except that in this case the distance CO is just slightly larger than the circle radius, as in Fig. 18.1. For a rigid flat top the circle passes through O and the long wave root $\boldsymbol{c}_0^{\mathrm{L}}$ disappears.

phase propagation, there is both a short and a long wave for each mode (see Fig. 18.2). To see this analytically, we substitute into (18.8) the polar coordinate representation for $\boldsymbol{k}_h$, viz.,

$$\boldsymbol{k}_h = k_h(\cos\theta_h, \sin\theta_h), \quad \cos\theta_h < 0.$$

Thus (18.8) gives

$$k_{hn}^{\mathrm{S}} = -\gamma\cos\theta_h + (\gamma^2\cos^2\theta_h - f^2/gh_n)^{1/2}, \tag{18.11a}$$

$$\text{or} \quad k_{hn}^{\mathrm{L}} = -\gamma\cos\theta_h - (\gamma^2\cos^2\theta_h - f^2/gh_n)^{1/2}. \tag{18.11b}$$

For the short wave, $\boldsymbol{c}_{ng} = \boldsymbol{c}_{ng}^{\mathrm{S}}(k_{hn}^{\mathrm{S}})$ is directed toward C along SC (Fig. 18.2); for the long wave, on the other hand, $\boldsymbol{c}_{ng}^{\mathrm{L}}(k_{hn}^{\mathrm{L}})$ is directed along LC. When the two roots (18.11) coalesce, $k_{hn}^{\mathrm{S}} = k_{hn}^{\mathrm{L}}$ and then the group velocities are identical. In this case the phase velocity vector is just tangent to the slowness curve and the group velocity is at right angles to it. Another interesting limiting case occurs when $\cos\theta_h = -1$. Then the phase velocity for both waves is westward; $\boldsymbol{c}_{ng}^{\mathrm{L}}$ is directed westward but $\boldsymbol{c}_{ng}^{\mathrm{S}}$ is directed eastward. The implications of this interesting result will be explored further in Section 23 where the reflection of planetary waves is considered.

For the case of constant stratification, we found that the equivalent depths are approximately given by [see (15.41) and (15.42)]

$$h_0 = H, \tag{18.12a}$$

$$h_n = N^2H^2/gn^2\pi^2. \tag{18.12b}$$

Thus (18.6) can be written as

$$\omega_0 = -\frac{\beta k_1}{k_h^2 + 1/r_e^2} \quad \text{(barotropic mode)}, \tag{18.13a}$$

$$\omega_n = -\frac{\beta k_1}{k_h^2 + (n\pi/r_i)^2} \quad \text{(baroclinic modes)}, \tag{18.13b}$$

where $\quad r_e = (gH/f^2)^{1/2} \quad$ and $\quad r_i = NH/f \qquad (18.14)$

are respectively the *external* and *internal* (*Rossby*) *radii of deformation*. The internal radius of deformation for a two-layer fluid was defined in (16.60) as $r_{i2} = (g'H_1H_2/Hf^2)^{1/2}$. Since $N^2 \sim g\Delta\rho_0/\rho_0 H_s = g'/H_s$, where H_s is the scale height of the stratification, r_i as given by (18.15) is clearly very similar to the two-layer expression.

For $H = 5 \cdot 10^3$ m and $N = 2 \cdot 10^{-3}$ rad s^{-1} (typical deep-ocean values), $r_e = 0(2000\,\text{km})$ whereas $r_i = 0(100\,\text{km})$. Thus except for very long barotropic waves, the term $1/r_e^2$ in the denominator of (18.13a) can be neglected. We shall see below that such an approximation is equivalent to assuming that the motion is horizontally nondivergent, a situation which holds exactly when the upper surface is rigid and flat. While the term $1/r_e^2$ can be neglected for barotropic planetary waves of lengths of a few hundred kilometers, the term $(n\pi/r_i)^2$ in (18.13b) is not small compared to k_h^2 for baroclinic planetary waves of the same length scales. The Rossby radii also play the following role in forced low-frequency motions in an ocean initially at rest. The quantities r_e and r_i are the horizontal scales over which the surface and internal motions, respectively, adjust to geostrophy (Mihaljan, 1963; Kraus, 1974). Also, in the context of other stratified flow problems (such as coastal upwelling), the ratio of L to r_i (where L is the horizontal scale determined by the geometry or the forcing mechanism in the problem) gives a measure of the importance of rotation. When $L/r_i \ll 1$, rotation can be safely neglected. Finally, in Section 24 we shall see how r_e and r_i serve as the trapping scales for wave energy travelling along a straight coast.

The dispersion relations (18.13a) (with no $1/r_e^2$ term) and (18.13b) (for the case $n = 1$) are plotted in Fig. 18.3 for the case $k_2 = k_1$. This figure also shows the modifications due to the presence of a sloping bottom that shoals to the north. A discussion of these dashed curves will be deferred until Section 20. Here we only wish to point out how slowly planetary waves propagate, especially as compared to long gravity waves. For $k_2 = k_1$ the phase speed of the barotropic mode is approximately (cf. 18.7)

$$c_0 = \beta/2\sqrt{2}k_1^2 \simeq 5\ \text{cm/s}$$

at mid-latitudes for $k_1 = 10^{-5}\,\text{m}^{-1}$. We leave it as an exercise for the reader to show that the baroclinic modes travel slightly more slowly. Thus both the barotropic and baroclinic modes travel a few orders of magnitude more slowly than their long gravity wave counterparts (cf. 17.10). But now the question of nonlinearity must be raised. Since the particle velocities are also observed to be a few centimeters per second, $U/c = 0(1)$ and the basic linearization assumption ($U/c \ll 1$) breaks down. A brief discussion of the importance of nonlinearities will be given in the next section.

Using (18.2), (15.10) and (15.3) we find the velocity components and pressure perturbation are given by

$$u = -\frac{ik_2}{f\rho_0} P_n \exp(iS_n)\,\Pi_n(z), \qquad (18.15a)$$

$$v = \frac{ik_1}{f\rho_0} P_n \exp(iS_n)\,\Pi_n(z), \qquad (18.15b)$$

$$w = \frac{i\omega_n}{\rho_0 N^2} P_n \exp(iS_n) \frac{d\Pi_n(z)}{dz}, \qquad (18.15c)$$

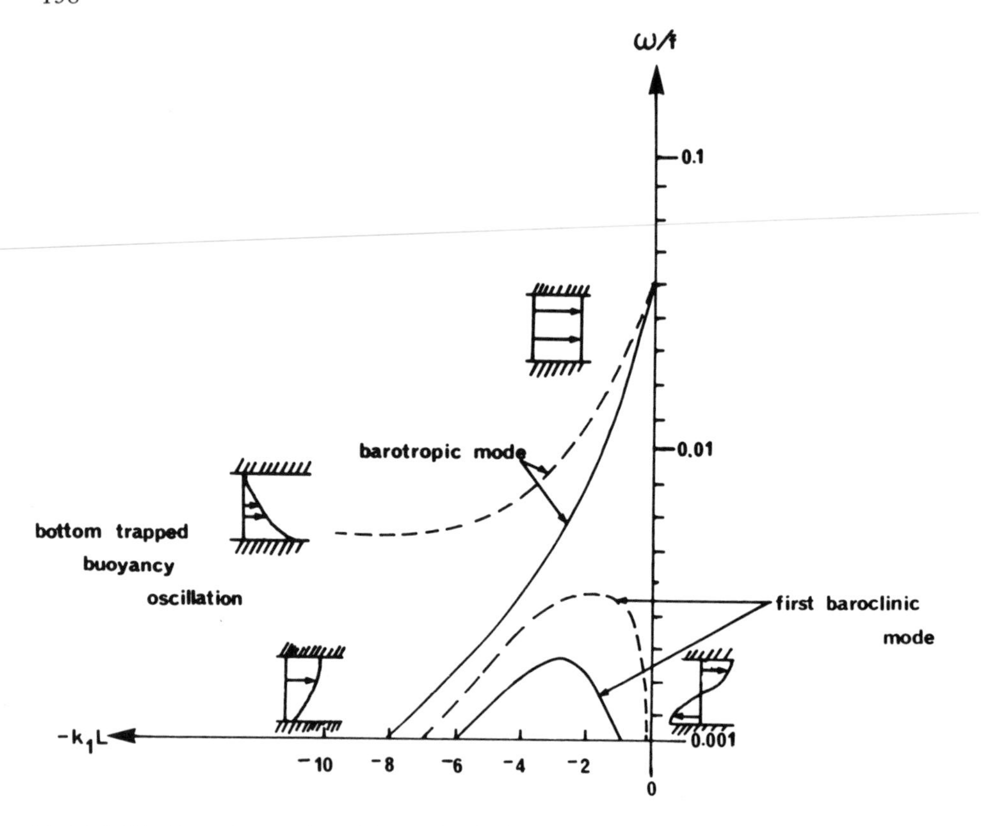

Fig. 18.3. The dispersion relations (18.13a) (with $1/r_e^2$ neglected) and (18.13b) for a latitude of $\phi_0 = 43°$, an internal Rossby radius of $r_i = 120$ km and $L = 120$ km. In both cases we have set $k_2 = k_1$ in the dispersion relations. The dashed lines are the corresponding curves for a uniformly sloping bottom that shoals to the north. For these curves the fractional depth change is $\delta_g = 0.67 \cdot 10^{-2}$, whereas, the actual bottom slope is $\alpha = 0.2 \cdot 10^{-3}$ (see Section 20). The small inserts show the nature of the vertical structure of the horizontal velocity field. (Adapted from Rhines, 1970a.)

$$\rho = -\frac{1}{g} P_n \exp(iS_n) \frac{\mathrm{d}\Pi_n(z)}{\mathrm{d}z}, \tag{18.15d}$$

where $S_n = k_1 x + k_2 y - \omega_n t$. We showed earlier that the vertical velocity is very small compared to $\boldsymbol{u}_h$ (cf. 18.5). This result also follows easily from (18.15). For $z \sim H$, we find

$$\frac{|w|}{|\boldsymbol{u}_h|} \sim \frac{\omega}{f} \cdot \left(\frac{f}{N}\right)^2 \cdot \frac{1}{k_h H}. \tag{18.16}$$

For long waves and strong stratification, $k_h H \sim f/N$ and each is much less than unity; thus the ratio (18.16) is the product of two small nondimensional parameters, ω/f and f/N, and hence is very small. From (18.15a, b) we see that planetary waves are approximately transverse:

$$\boldsymbol{u}_h \cdot \boldsymbol{k}_h = 0 \text{ to } 0(\omega/f). \tag{18.17}$$

Thus the particle motions are approximately along the wavefronts and in horizontal planes. The energy flux, as given by $\langle E\rangle \boldsymbol{c}_g$, is generally neither along $\boldsymbol{k}_h$ nor along the horizontal particle velocity $\boldsymbol{u}_h$ (see Fig. 18.2). Since $\langle p\boldsymbol{u}\rangle$ is always along the direction $\boldsymbol{u}_h$ (see Exercise 18.2), the two expressions for the energy flux differ. This "paradox" will be discussed below for the barotropic mode.

Barotropic planetary waves in a homogeneous fluid

We now seek planetary wave solutions of the long wave equations (15.50) and (15.51) with $H=$ constant. It will be shown that the dispersion relation (18.6) (with $h_0=H$) follows directly from these equations. We will also discuss the relative importance of the free surface term in the conservation of mass equation.

The basic equations are (17.25), which were used in Section 17 to study barotropic long gravity waves. In that study these equations were combined into a Klein–Gordon equation for η. However, with a variable f, such a procedure leads to a very complicated equation for η (see Section 20). A simpler approach is to combine (17.25) into one equation for v, as was first done by Longuet-Higgins (1965a). If we differentiate (17.25b) twice with respect to t and then use (17.25a and b) to eliminate u_{tt} and η_{ytt} respectively, we find

$$[(\partial_{tt}+f^2)/gH-\partial_{yy}]v_t \;=\; (f/H)\eta_{xt}+u_{xyt}. \tag{18.18}$$

Next, we eliminate η_x and η_y in (17.25a and b) by cross-differentiation and use (17.25c) in the form $u_x+v_y=-\eta_t/H$ to get

$$(f/H)\eta_{tx}+u_{tyx} \;=\; \beta v_x+v_{xxt}.$$

But the left side of this is precisely the right side of (18.18); thus we finally obtain

$$\left(\nabla_h^2-\frac{1}{gH}\mathcal{L}\right)v_t+\beta v_x \;=\; 0. \tag{18.19}$$

This is precisely the same as (15.17), which was obtained by separating the horizontal and vertical dependences in the governing equation for $v(x,y,z,t)$ in a continuously stratified fluid. For low frequencies, $\mathcal{L}\simeq f^2$ and (18.19) becomes

$$(\nabla_h^2-1/r_e^2)v_t+\beta v_x \;=\; 0, \tag{18.20}$$

where the external deformation radius r_e is given by (18.14a). For plane waves of the form

$$v \;=\; v_0\exp\left[i(k_1x+k_2y-\omega_nt)\right],$$

(18.20) yields the dispersion relation (18.13a), as expected.

If the surface fluctuations are small so that the free surface can be treated as rigid and flat, we can neglect the term η_t in (17.25c). Thus the horizontal velocity becomes non-divergent:

$$u_x+v_y \;=\; 0. \tag{18.21}$$

We can now deal directly with the linearized momentum equations (15.44a, b) which involve the pressure:

$$u_t - fv + \frac{1}{\rho_*} p_x = 0, \tag{18.22a}$$

$$v_t + fu + \frac{1}{\rho_*} p_y = 0. \tag{18.22b}$$

Eliminating p in (18.22) yields the vorticity balance equation

$$(v_x - u_y)_t + \beta v = 0. \tag{18.23}$$

This is precisely the linearized form of the vertically-averaged conservation of potential vorticity equation (cf. 4.8):

$$\frac{\mathrm{D}}{\mathrm{D}t}(\boldsymbol{\zeta}\cdot\hat{z} + f) = 0, \tag{18.24}$$

where $\boldsymbol{\zeta} = \nabla \times \boldsymbol{u}$. In this simple vorticity balance, changes in relative vorticity $\boldsymbol{\zeta}\cdot\hat{z}$ are compensated by motion of the fluid through the non-uniform planetary vorticity field $f = 2\Omega \sin\phi$. Wave motion in these circumstances may be imagined as follows. A parcel of water displaced northward from some reference latitude acquires, from (18.24), negative relative vorticity; by mass continuity, an equal amount of water must be displaced southward, where it gains positive relative vorticity. The two vortices thus created by this perturbation are influenced by each other's flow field and tend to bring each other back towards the reference latitude (Exercise 18.7).

We will show in Section 20 that if surface displacements are allowed, the equation for the conservation of potential vorticity takes the form (cf. 20.7)

$$\frac{\mathrm{D}}{\mathrm{D}t}\left(\frac{\boldsymbol{\zeta}\cdot\hat{z} + f}{H + \eta}\right) = 0,$$

which clearly reduces to (18.24) when $H = \text{constant}$ and $\eta = 0$. Thus, more generally, changes in $\boldsymbol{\zeta}$ are compensated not only by changes in f, but also by vortex stretching due to changes in the total water depth $H + \eta$.

In view of (18.21) there exists a stream function ψ such that

$$u = -\psi_y, \quad v = \psi_x. \tag{18.25}$$

Hence (18.23) becomes

$$\nabla_h^2 \psi_t + \beta \psi_x = 0, \tag{18.26}$$

which is identical to (16.53), where in the latter ψ represented the stream function for the vertically integrated velocity. Equation (18.26) has plane-wave solutions

$$\psi = \psi_0 \exp\left[i(k_1 x + k_2 y - \omega t)\right] \tag{18.27}$$

provided $\quad \omega = -\beta k_1 / k_h^2. \qquad (18.28)$

Thus for $k_h^2 \gg 1/r_e^2$, or equivalently,

$$L^2 \ll r_e^2, \tag{18.29}$$

where L is the horizontal wavelength scale, the rigid-lid approximation is valid. One important consequence of this approximation is that the slowness circle is now tangent to the k_2-axis and, therefore, the "long" wave (k_0^L) disappears. This long wave is thus associated with the presence of the free surface and, as mentioned above, allows changes in ζ to be compensated in part by vortex stretching through changes in the total water depth.

For "short" planetary waves, the energy equation is simply

$$\frac{\partial}{\partial t}\left[\tfrac{1}{2}\rho_*(u^2+v^2)\right]+\nabla_h\cdot(p\boldsymbol{u}_h) = 0. \tag{18.30}$$

There is no potential energy term since the free-surface motions have been suppressed; even if η were included, the resulting potential energy term would be very small. For ψ in the real form

$$\psi = \psi_0 \cos S_n, \tag{18.31}$$

we find, using (18.31) in (18.25) to find $\boldsymbol{u}_h$,

$$\langle E\rangle = \langle KE\rangle = \frac{1}{4H}\rho_*\psi_0^2 k_h^2. \tag{18.32}$$

From (18.28) we obtain

$$\boldsymbol{c}_g = \frac{\beta}{k_h^4}(k_1^2-k_2^2,\, 2k_1k_2). \tag{18.33}$$

Thus the energy flux as given by (18.32) and (18.33) is

$$\langle E\rangle\boldsymbol{c}_g = \frac{\beta\rho_*\psi_0^2}{4Hk_h^2}(k_1^2-k_2^2,\, 2k_1k_2). \tag{18.34}$$

From (18.22a), (18.25) and (18.31) we find

$$p = \rho_*\psi_0\left(f\cos S_n+\frac{\omega k_2}{k_1}\sin S_n\right). \tag{18.35}$$

But $\quad \boldsymbol{u}_h = \psi_0 \sin S_n(k_2,-k_1). \qquad (18.36)$

Therefore the flux as given by the pressure velocity correlation is

$$\langle p\boldsymbol{u}_h\rangle = -\frac{\omega\rho_*\psi_0^2}{2H}\left(-\frac{k_2^2}{k_1},\, k_2\right), \tag{18.37}$$

which is at right angles to $\boldsymbol{k}_h$ and, therefore, collinear with $\boldsymbol{u}_h$. Longuet-Higgins (1964b) has shown that the two flux vectors are equivalent in the sense that they differ only by the curl of a function whose average is proportional to ψ_0^2. The situation discussed above is already well known in electromagnetic theory where the Poynting vector is given as only one possible interpretation of the energy flux. A way out of this quandary has been proposed by Buchwald (1972). He shows that an alternative definition of the total energy density E is given by

$$E = T+V,$$

where $T=\frac{1}{2}\rho_0(u^2+v^2)$ is the usual kinetic energy density and V is a "spin energy" density defined by

$V = \frac{1}{2}\rho_0\beta\psi\tilde{\psi}_x,$

where $\tilde{\psi}_t = \psi$. Physically, $\tilde{\psi}_x$ represents the southward displacement of a fluid particle. Further, Buchwald derives two energy conservation equations, both of which are consistent with the energy flux propagating in the direction of $\mathbf{c}_g$. The energy flux is given by $\langle E\rangle \mathbf{c}_g$ and the equipartition principle is satisfied: $\langle T\rangle = \langle V\rangle = \frac{1}{2}\langle E\rangle$. The generalization of these results to the case of barotropic planetary waves over variable topography, for a multiply connected region, and in the presence of mean currents has been carried out by Thomson (1973). By way of contrast, Luyten (1974) has shown that for topographic planetary waves on the f-plane (Section 20), the pressure-correlation flux $\langle p\mathbf{u}\rangle$ is identical to $\langle E\rangle \mathbf{c}_g$ provided account is taken of the deformation of a wave packet along a ray. When that is done, there is no need to introduce a spin energy, as defined above.

Planetary waves in a two-layer fluid

As our starting point we use the normal mode equation (16.36) for $u^{(n)}$ and $v^{(n)}$. To facilitate the combination of these into one equation for $v^{(n)}$, we rewrite (17.36) in operator form:

$$L_1 v^{(n)} = L_2 u^{(n)}, \tag{18.38a}$$

$$L_3 v^{(n)} = L_4 u^{(n)}, \tag{18.38b}$$

where $L_1^{(n)} = gh^{(n)}\partial_{xy} + f\partial_t, \quad L_3^{(n)} = \partial_{tt} - gh^{(n)}\partial_{yy},$

$L_2^{(n)} = \partial_{tt} - gh^{(n)}\partial_{xx}, \quad L_4^{(n)} = gh^{(n)}\partial_{xy} - f\partial_t,$

in which the $h^{(n)}$ are approximately given by [cf. (16.28) and (16.29)]

$h^{(0)} = H, \quad h^{(1)} = \delta H_1 H_2/H.$

Taking L_4 of (18.38a) and $-L_2$ of (18.38b) and then adding, we obtain

$$\left(\nabla_h^2 - \frac{1}{gh^{(n)}}\mathcal{L}\right) v_t^{(n)} + \beta v_x^{(n)} = 0 \tag{18.39}$$

since L_2 and L_4 commute. This is identical in form to (15.17), the equation for $v_n(x,y,t)$ in a continuously stratified fluid. For low frequencies, $\mathcal{L} \simeq f^2$ and (18.39) reduces to

$$\left(\nabla_h^2 - \frac{f^2}{gh^{(n)}}\right) v_t^{(n)} + \beta v_x^{(n)} = 0, \tag{18.40}$$

which is the same form as the pressure equation (18.1). For horizontally propagating waves the dispersion relations are

$$\omega^{(0)} = -\beta k_1/(k_h^2 + 1/r_e^2), \tag{18.41a}$$

$$\omega^{(1)} = -\beta k_1/(k_h^2 + 1/r_{i2}^2), \tag{18.41b}$$

for the barotropic and baroclinic modes, respectively. The barotropic dispersion relation is identical to that derived for the case N = constant (see 18.13a). The baroclinic dispersion relation is the analogue of ω_1 for the case N = constant (see 18.13b). The two differ only in the numerical values of $1/r_{i2}$ and π/r_i. Numerical examples using the two different formulae will be given below in connection with planetary wave observations.

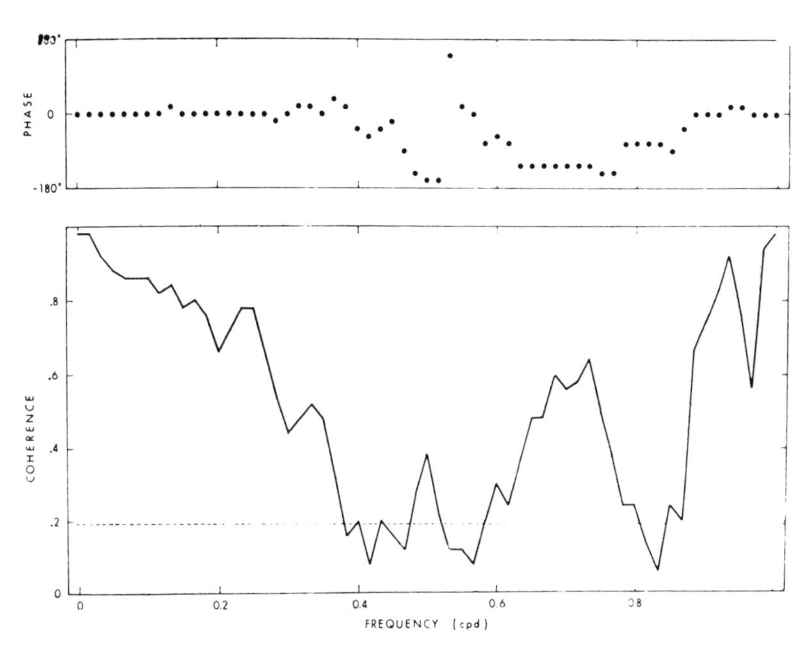

Fig. 18.4. Phase angle and coherence (the correlation as a function of frequency – see Section 31) between the tide-gauge records at Honolulu and Mokuoloe, April 1959–December 1964. A positive phase means that Honolulu leads Mokuoloe. The dashed line in the lower diagram represents the 95% confidence limit. (From Miyata and Groves, 1968; copyrighted by American Geophysical Union.)

Observational evidence of planetary waves

Among the first measurements of slowly varying currents in the open ocean were those of Swallow and his co-workers (Swallow and Hamon, 1960; Crease, 1962; Swallow, 1971), who measured the drift of neutrally buoyant floats down to depths of 4 km in the eastern and western North Atlantic. These authors found that a considerable amount of energy was contained in transient motions, with characteristic periods of 50–100 days and length scales of several hundred kilometers. N. Phillips (1966) analyzed the vertical structure of the currents measured by Crease near Bermuda in terms of a two-layer model; he found that 78% of the kinetic energy was contained in the barotropic mode. In retrospect, it appears that the high proportion of barotropic energy could be attributed to the fact that the observations were made below the main thermocline, where the baroclinic motions are weak and difficult to detect, and to the short period of observations from which the long-period baroclinic motions could not be resolved. Phillips' computations of the response of a barotropic ocean to a time-dependent wind stress produced motions which agreed well in periods and wavelengths with the observed motions, but with energy levels only about one-third of the observed values. Inclusion of baroclinicity in the form of a two-layer model did not improve the theoretical results significantly. Phillips concluded that in this region of the North Atlantic, the nonlinear interactions of these waves with the Gulf Stream must be important, at least as a possible generation mechanism.

Striking evidence of low-frequency energy in the North Pacific Ocean was presented by Miyata and Groves (1968). The co-spectrum of sea-level at Honolulu and Mokuoloe shows

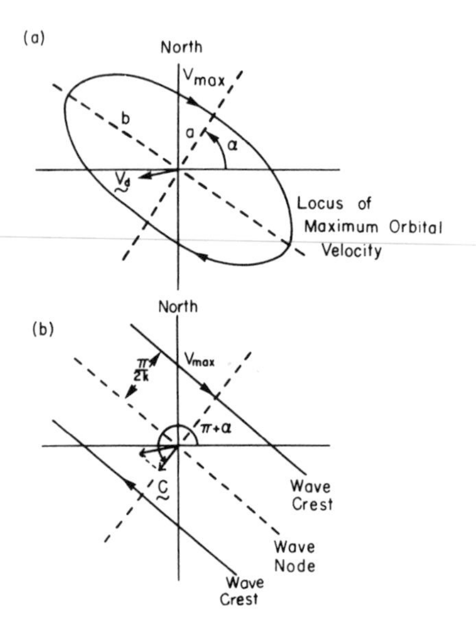

Fig. 18.5. Schematic cross-sections of (a) a horizontal current ellipse and (b) a single plane-wave representation of the horizontal currents observed at the Polygon array during 1970. (Adapted from Koshlyakov and Grachev, 1973.) Typical values of the parameters shown are $a = 90$ km, $b = 200$ km, $\alpha = 55°$, $V_d = 0.05\ \mathrm{m\,s^{-1}}$ towards 190° and $V_{\mathbf{max}} = 0.2\ \mathrm{m\,s^{-1}}$ (near the surface). To make the two figures compatible, a is identified as one-quarter of the wavelength $(\pi/2k)$, $\pi + \alpha$ as the direction of the wavenumber $\boldsymbol{k}$, and the projection of $\boldsymbol{V}_d$ (the eddy drift velocity) onto a line perpendicular to the wave crests as the phase velocity $\boldsymbol{c} = \omega \boldsymbol{k}_h/k_h^2$.

significant peaks at frequencies of 0.73, 0.50, 0.35 and 0.25 cpd (see Fig. 18.4). Longuet-Higgins (1971) has suggested that whereas the coherence peak at 0.73 cpd (= local value of f) is most likely due to inertial or near inertial currents generated by local winds, the peaks at 0.35 and 0.25 cpd may be associated with the lowest planetary wave eigenmodes of the Pacific basin. More recently though, Wunsch and Gill (1976) have shown that the peak at 0.25 cpd is most likely due to an equatorial trapped long gravity wave (Section 21). The coherence peak at 0.5 cpd, on the other hand, because of its phase, may be a manifestation of a planetary wave that is trapped by the local topography and hence propagates around the island (Rhines, 1969b; see also Section 27).

The low-frequency energy shown in Fig. 15.1 cannot be readily identified with the earlier-discussed planetary wave modes in an ocean of uniform depth, since at Site D the slope of the continental slope region is rather large $[0(10^{-2})]$. We shall show in Section 20 that slopes as small as (10^{-3}) are as important as β in the vorticity balance of barotropic planetary waves. The problem is further complicated when stratification is included because then an exact separation into normal modes is not possible and an approximate method has to be used. This will be discussed in Section 20, along with a presentation of more recent evidence of *topographic* planetary waves at Site D.

Probably the most striking evidence to date of the existence of oceanic planetary waves is contained in the horizontal current data obtained from the large-scale multiple buoy experiments conducted in the North Atlantic by the USSR in 1970 and by the USA–UK in 1973. Some of the observations of the USSR "Polygon array" experiment have been

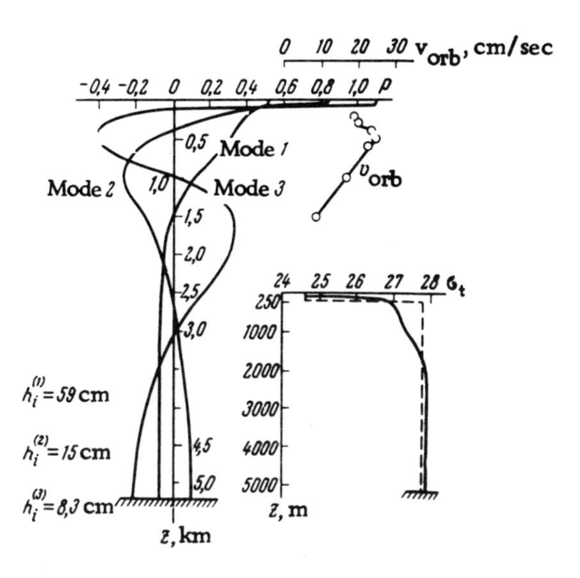

Fig. 18.6. First three baroclinic modes of oscillation of horizontal component of geostrophic flow in the Polygon test site and corresponding values of equivalent depth. Also shown are the depth dependence of the time average of the orbital velocity v_{orb} of the test site disturbance, the vertical distribution of the density σ_t in the test site, and its two-layer approximation. (From Koshlyakov, 1973.)

reported by Brekhovskikh et al. (1971) and by Koshlyakov and Grachev (1973, hereafter referred to as KG). KG described the mesoscale data (with time scales of months and length scales of hundreds of kilometers), and inferred that during the experiment a single, anticyclonic eddy (a large current vortex) a few hundred kilometers in diameter passed westward through the Polygon array (centered at 16°30′N, 33°30′W). They suggested that this may be identified as a single horizontally propagating planetary wave. A typical example of a horizontal current ellipse at a fixed depth is shown in Fig. 18.5, along with KG's simple plane wave representation. As seen in Fig. 18.6, the observed orbital velocity decreases with depth, which is what motivated KG to suggest that a single baroclinic wave may be the best fit. At the Polygon site array, $f = 4 \cdot 10^{-5}$ rad s^{-1}, $\beta = 2.2 \cdot 10^{-11}$ m^{-1} rad s^{-1}, $H = 5400$ m (with $H_1 = 250$ m and $H_2 = 5150$ m for a two-layer model – see Fig. 18.6) and $\delta = (\rho_2 - \rho_1)/\rho_2 = 3.2 \cdot 10^{-3}$. From the observed density field, the first three baroclinic modes for the vertical dependence of the pressure field were computed numerically and are shown in Fig. 18.6, along with the computed equivalent depths. For the above parameters we have computed the theoretical frequencies from (18.6) and (18.41b) for a wavelength of 360 km ($k_h = 1.75 \cdot 10^{-5}$ m^{-1}). The results are shown in Table 18.I, along with the corresponding periods, phase speeds and the observed values of these quantities.

The general agreement between theory and observation is quite encouraging; however, the results in Table 18.I clearly do not favour a description in terms of a single baroclinic wave, as suggested by KG. In fact, there is a significant mean bottom slope of $0(10^{-3})$ at the Polygon site; also there exist mean flows near the surface and in the deep ocean. McWilliams and Robinson (1974) have shown that if these features are taken into account in the theory, the best fit is a baroclinic planetary wave (in a two-layer ocean) with a period

TABLE 18.I

Single-wave description of Polygon eddy

	Observed values	Theoretical values: barotropic mode	baroclinic mode: $n = 1$, continuous stratification	baroclinic mode: two-layer model
ω (10^{-6} rad s^{-1})	0.61†	0.73	0.38	0.43
$T = 2\pi/\omega$ (days)	119†	100	190	172
c (cm s^{-1} @ 235°)	3.5†	4.2	2.2	2.5

† The phase speed is calculated by projecting V_d (the eddy drift velocity) on to the direction perpendicular to the wave crests. The period is then $\lambda (= 2\pi/k_h)$ divided by this value of c.

110 days and a speed of 3.8 cm s^{-1}. The theoretical characteristics of the barotropic wave as modified by mean flow and topography do not fit the data at all, having a relatively large phase speed (9 cm/s) and a short period (46 days).

A major deficiency of a single wave description of the Polygon eddy is that its observed elliptical polarization cannot be reproduced by a plane wave since the current oscillations associated with the latter are rectilinear, being aligned along the wave crests. This shortcoming can be overcome by taking a linear combination of a pair of waves; thus for a two-layer model the pressure field of such a combination is

$$\begin{pmatrix} p_1 \\ p_2 \end{pmatrix} = A_1 \begin{pmatrix} 1 \\ \alpha_1 \end{pmatrix} \sin(\mathbf{k}_{h1} \cdot \mathbf{x} - \omega_1 t) + A_2 \begin{pmatrix} 1 \\ \alpha_2 \end{pmatrix} \sin(\mathbf{k}_{h2} \cdot \mathbf{x} - \omega_2 t). \tag{18.42}$$

When $A_1/A_2 = 1$ the upper-layer pressure in (18.42) takes the form

$$p_1 = 2A_1 \sin(\mathbf{k}_s \cdot \mathbf{x} - \omega_s t) \cos(\mathbf{k}_d \cdot \mathbf{x} - \omega_d t), \tag{18.43}$$

where $\mathbf{k}_s = \frac{1}{2}(\mathbf{k}_{h1} + \mathbf{k}_{h2})$, $\mathbf{k}_d = \frac{1}{2}(\mathbf{k}_{h1} - \mathbf{k}_{h2})$,

$\omega_s = \frac{1}{2}(\omega_1 + \omega_2)$, $\omega_d = \frac{1}{2}(\omega_1 - \omega_2)$.

The resulting pattern is a collection of cells of alternating high and low pressure (Fig. 18.7). Lines of constant pressure are streamlines. The cell boundaries are zero-pressure lines, and interior to the cells the streamlines are closed and approximately elliptical. When $A_1/A_2 \neq 1$ the cells are not isolated, there being fluid leakage between them which tends towards parallel flow (as in a single plane wave) when $|A_1/A_2|$ becomes large or small. However, for the KG data, the case $A_1 = A_2$ seems quite reasonable. McWilliams and Robinson (1974) showed that a model of the form (18.42) with $A_1 = A_2$ and which consists of two first-mode baroclinic waves produces the best fit to the data. However, they again found that the fit improved considerably when the mean topographic slope and horizontal flows were incorporated into the theory.

Following the 1970 Polygon experiment in the eastern tropical North Atlantic, the Mid-Ocean Dynamics Experiment (MODE) was planned to further investigate oceanic

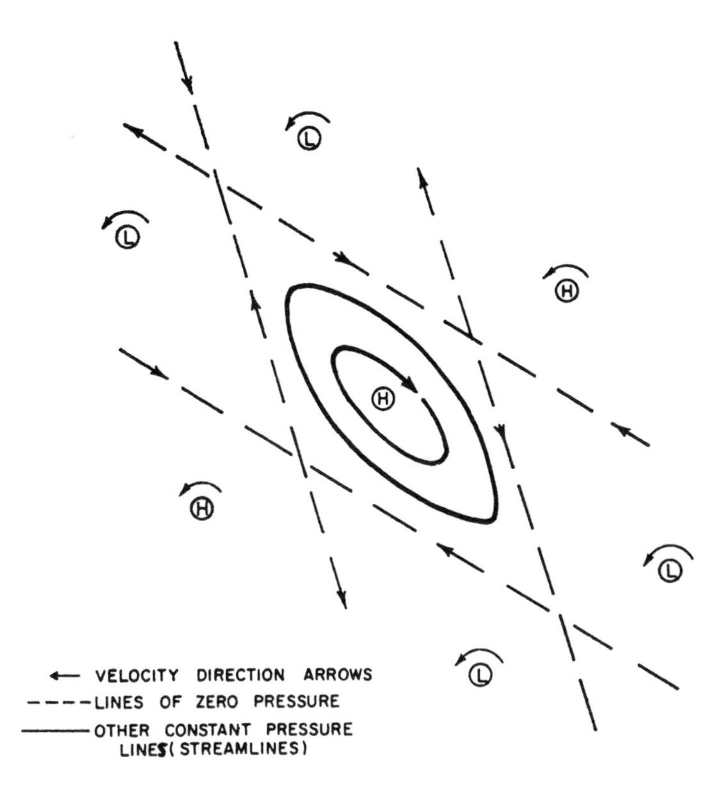

Fig. 18.7. A two-wave, horizontal pressure pattern for waves of equal amplitude in the upper layer of a two-layer fluid. The alternate high- and low-pressure cells are separated by the zero isobars, and typical isobars (streamlines) are approximately elliptical. (From McWilliams and Robinson, 1974.)

mesoscale motions and their role in the general ocean circulation in a square region (about 600 km by 600 km) centered at 28°N, 69°40′W. Several pilot studies (collectively called MODE-O) were carried out in 1971-72 to identify the energy, space and time scales; this was followed by MODE-I which started in the spring of 1973 and lasted for 180 days. In MODE-O the velocity and temperature records were dominated by 50–100 day fluctuations, with horizontal length scales of 0(100 km) (Gould et al., 1974). The vertical structure was dominated by the barotropic and the first few baroclinic modes. As part of MODE-I a large number of neutrally buoyant floats were launched at a depth of 1500 m, near the axis of the SOFAR channel. In their statistical analysis of the behaviour of these floats, Freeland et al. (1975) found that the streamline patterns unambiguously propagated westward with a mean speed of 5 cm/s (Fig. 18.8). This exceeded the rms particle speed (4 cm/s) and far exceeded the mean westward flow (0.9 cm/s). Thus although the ratio of rms speed to drift speed is of order unity, indicating strong nonlinearity, the motions exhibited the familiar westward phase propagation associated with linear planetary waves in the β-plane. McWilliams and Flierl (1976) further established the applicability of planetary waves in explaining the low-frequency behaviour of MODE-I data from current meters, moored temperature sensors, hydrographic stations and float tracks. They found that an excellent fit to the data was obtained by taking a pair of barotropic and a pair of baroclinic planetary waves, each propagating in a different direction.

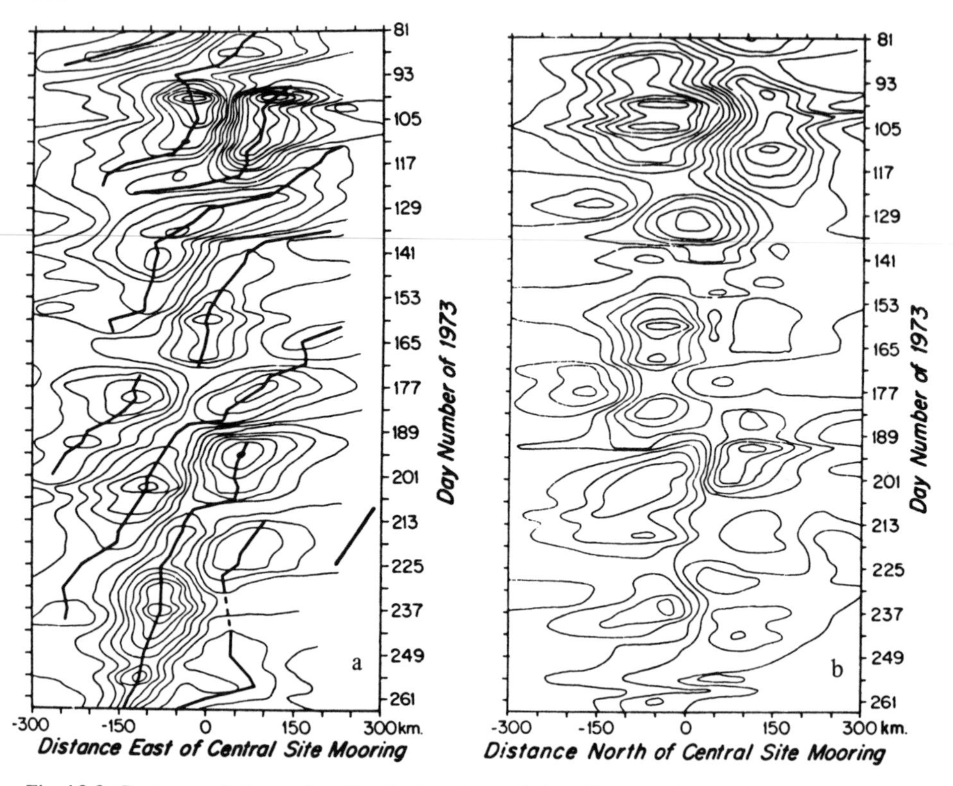

Fig. 18.8. Contours of stream function in the plane of time (downwards) and distance along a section through the MODE region (horizontal axis): (a) is a time–longitude plot and (b) a time–latitude plot. A slope of 5 cm/s is indicated by a bold line on (a). (From Freeland et al., 1975.)

Baroclinic planetary wave motions have also been recently observed in the North Pacific Ocean. From an examination of observed temperature spectra in the Northwest Pacific (near the Kuroshio), Kramareva (1973) suggested that westward propagating planetary waves were present, with periods of 7–30 days and wavelengths of a few thousand kilometers. Bernstein and White (1974) proposed that baroclinic eddies observed in temperature records from the Central North Pacific are wavelike in nature, with wavelengths of 0(500 km) and westward phase speeds of 0(5 cm/s). Emery and Magaard (1976) computed temperature spectra from monthly mean values from the same general region. They found that there was a significant amount of potential energy in the frequency range below ω_c, the cut-off frequency for baroclinic planetary waves (a period of about five months). A cross-spectral fit of a field of random set of baroclinic waves to the data indicated that the first baroclinic mode alone accounts for 65–75% of the observed spectral energy contained in periods of 1–2 years. The first mode waves with periods in this range have wavelengths of 1200–1700 km and propagate between 20° and 80° west of north.

Exercises Section 18

1. Show that the substitution

$$p_n = e^{-i\gamma x - i\omega_n t}\phi_n(x,y), \quad (\gamma = \beta/2\omega_n)$$

into (18.1) results in a Helmholtz equation for $\phi_n(x,y)$. Discuss plane wave solutions for ϕ_n of the form

$$\phi_n = A_n \exp\,[i(k_1^* x + k_2 y)],$$

where $k_1^* = k_1 + \gamma$. In particular, determine the direction of the phase and group velocities in the (k_1^*, k_2)-plane.

2. The energy equation for planetary waves in a continuously stratified fluid is also given by (17.15):

$$\frac{\partial}{\partial t}\,[\tfrac{1}{2}\rho_0(u^2+v^2) + \tfrac{1}{2}\rho_0 N^2 \xi^2] + \nabla\cdot(p\boldsymbol{u}) = 0,$$

where $\xi = \int w\,dt$. Using (18.15) and the dispersion relation (18.6), find expressions for $\langle KE\rangle$ and $\langle PE\rangle$ that are analogous to (17.17) and (17.19), respectively. Also find $\langle p\boldsymbol{u}\rangle$ for each mode and show that this is not equal to $\langle E\rangle \boldsymbol{c}_{ng}$, where $\langle E\rangle = \langle KE\rangle + \langle PE\rangle$.

3. Consider a homogeneous fluid of density ρ_* in which there is a zonal shear flow $U(y)$ that is maintained geostrophically:

$$fU + \frac{1}{\rho_*}p_{0y} = 0.$$

Write down the linearized equations for the perturbation velocities u, v and perturbation pressure p. Derive the vorticity equation for the perturbation stream function. Determine the dispersion relation for propagating waves when U is constant. Discuss your results.

4. From the nonlinear barotropic long wave equations for nondivergent flow, viz.,

$$u_t + uu_x + vu_y - fv + \frac{1}{\rho_*}p_y = 0,$$

$$v_t + uv_x + vv_y + fu + \frac{1}{\rho_*}p_y = 0,$$

$$u_x + v_y = 0,$$

derive the following equation for the stream function ψ (cf. 18.25b):

$$\nabla_h^2\psi_t + \beta\psi_x - \psi_y\nabla_h^2\psi_x + \psi_x\nabla_h^2\psi_y = 0. \qquad (18.44)$$

Show that the plane wave (18.31) with dispersion relation (18.28) is also an exact solution of (18.44).

5. For linear nondivergent barotropic planetary waves, we have shown that the stream function ψ satisfies the vorticity equation

$$\nabla_h^2\psi_t + \beta\psi_x = 0. \qquad (18.26)$$

If (18.26) is multiplied by ψ, show that, after some rearrangement, we obtain the energy equation

$$T_t + \rho_*\nabla_h\cdot \boldsymbol{F} = 0,$$

where $\quad \boldsymbol{F} = -\psi\nabla_h\psi_t - \beta(\tfrac{1}{2}\psi^2, 0), \quad T = \tfrac{1}{2}\rho_*(\psi_x^2 + \psi_y^2).$

Also show that $\langle \boldsymbol{F} \rangle$ is parallel to $\boldsymbol{c}_g$, the group velocity for barotropic planetary waves. Finally, show that the vector $\boldsymbol{F} - \langle p\boldsymbol{u}_h \rangle$ is nondivergent.

6. Show that for "slightly divergent" barotropic planetary waves characterized by $f^2/gH \ll k_h^2$ and, of course, $\omega \ll f$, the velocity components and free surface displacement may be expressed in terms of a stream function ψ:

$$u \simeq -\psi_y, \quad v \simeq \psi_x, \quad \eta \simeq f\psi/g.$$

7. Consider a homogeneous ocean with a flat bottom and a rigid top, in which the vorticity balance is given by (18.24). Imagine a perturbation about some reference latitude consisting of displacements of equal masses of fluid to equal distances north and south, respectively, of the reference latitude. Assume that the displaced fluid parcels behave like line vortices of a strength proportional to the distance of the fluid parcel from the reference latitude and of a sign appropriate to (18.24). Sketch the life history of the pair of vortices as they move in each other's flow field and show that the pattern oscillates about the reference latitude and progresses westward.

19. NONLINEAR WAVES OF LOW FREQUENCY

Our discussion of nonlinear effects in long period waves will be brief, and rather of the nature of a cursory review patterned somewhat after Section 12. Observed data on the amplitude of low-frequency waves will guide our choice of topics in which nonlinearity is relevant. Long-period waves fall into two categories: long gravity waves, specifically the tides, for which $\omega = 0(f)$, and planetary waves, for which $\omega \ll f$.

Nonlinearity in tides

Tidal motions, whose origin will be discussed in Section 52, contribute most of the energy found near the inertial period $(2\pi/f)$. The dominant tidal motion is barotropic and arises in direct response to astronomical forcing. Baroclinic tides also exist and arise from the interaction of the barotropic tide with the bottom topography.

From (17.26) and (17.29) the ratio of particle velocity in the direction of wave propagation to the phase speed is readily evaluated. The linearity criterion $|\boldsymbol{u}|/c \ll 1$ is identical to that derived in the absence of rotation (cf. 12.14):

$$a/H \ll 1. \tag{19.1}$$

Mid-ocean barotropic tides rarely have amplitudes in excess of 2 m. The criterion (19.1) therefore certainly holds except in shallow areas, where H is small and the amplitude is increased by resonance or geometrical effects. Tidal wavelengths may be estimated from the dispersion relation (17.29); they are of the order of a few thousand kilometers. Hence $kH \ll 1$ and the corollary inequality $ak \ll 1$ is also well satisfied. Mid-ocean barotropic tides are adequately treated by a linear theory.

The relevant criterion for the validity of linearization for internal tides may again be written as an inequality of the form (12.11). For uniform stratification ($N^2 = \text{constant}$), (15.40) and (15.42) approximately give $\Pi_n(z) = A_n \cos(n\pi z/H)$ and $h_n = N^2H^2/gn^2\pi^2$. With the help of the dispersion relation (17.3), it is readily established from (17.14) that (see Exercise 19.1)

$$|\boldsymbol{u}|/c \;=\; |\xi|n\pi/H, \tag{19.2}$$

where $|\xi|$ is the magnitude of the vertical displacement. The condition $|\boldsymbol{u}|/c \ll 1$ is again satisfied for waves in which the vertical displacement is only a small fraction of the scale depth $H/n\pi$. In both types of gravity modes (barotropic and baroclinic), rotation is irrelevant to the validity of linearization. Baroclinic tides commonly have much larger vertical displacements than barotropic tides (up to 100 m); they are also considerably shorter in wavelength than barotropic tides of the same frequency. Hence, internal tides will generally be influenced by amplitude and phase dispersion to a much greater degree than surface tides. We have already seen in Fig. 12.4 an example of both dispersive effects. Fig. 19.1 illustrates a more strongly nonlinear case, in which amplitude dispersion dominates completely and the internal tide looks like a shock wave.

In shallow water, the ratio a/H often becomes of $0(1)$ for barotropic tides: the wavelength remains long, however (Exercise 19.2), and the inequality $kH \ll 1$ retains its validity. Amplitude dispersion dominates over phase dispersion, and, as in the hydraulic case of Section 12, shock waves may form. In addition, nonlinearity may be introduced

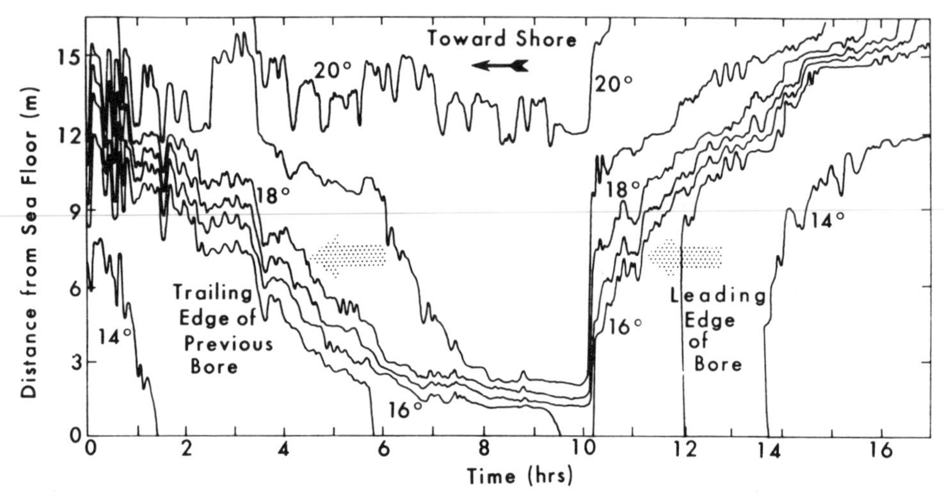

Fig. 19.1. Internal tidal bore observed in shallow water, as evidenced by isotherm positions (temperature in degrees Celsius). (From Cairns, 1967; copyrighted by American Geophysical Union.)

Fig. 19.2. The bore on the Petitcodiac River, near Moncton, New Brunswick. (From Dohler, 1964.)

through bottom friction, for which a quadratic velocity function is the favourite representation (Section 52). When nonlinearity is still weak, its effects on a tidal constituent (Section 52) may be handled by perturbation methods. Such an analysis results in the appearance of higher harmonics of the fundamental tidal frequency. Under these conditions, nonlinear interactions between tides of different frequencies may be treated by the weak wave–wave interaction methods of Section 38. In very shallow water ($a/H = 0(1)$), shock waves, known as tidal bores, often make their appearance. As sufficient tidal amplitudes for bore formation are commonly found in restricted channels of width much less than the barotropic Rossby radius r_e defined in (18.14), the velocity component transverse to the channel axis is suppressed and the effects of rotation are minimized. Long-period waves may then be described by the hydraulic equations (12.97) and (12.99), where ϵ may become of $0(1)$. Examples of tidal bores are illustrated in Tricker (1965). The bore on the Petitcodiac River, near Moncton, New Brunswick, is shown in Fig. 19.2.

Nonlinear planetary waves

From the short review of oceanic planetary wave observations presented in Section 18, it is clear that $|\boldsymbol{u}|/c = 0(1)$ and that nonlinearity should be of consequence in planetary waves. In Exercise 18.4 it was shown that a plane barotropic planetary wave is an exact solution of the nonlinear vorticity equation (18.44). Large values of velocity thus do not affect the form of a single plane barotropic planetary wave. Baroclinic waves are not exactly transverse, however, and this result does not apply to them. Furthermore, the nonlinear terms will not cancel out for a superposition of two or more plane barotropic planetary waves; even though a single wave may be an exact solution, the sum of two waves is not. As long as the current amplitudes are small, $|\boldsymbol{u}|/c = 0(10^{-1})$ say, the interactions between plane planetary waves may be treated through the perturbation methods of weak wave–wave interaction theory (Section 38). Finite-amplitude planetary waves of permanent form also exist (such as cnoidal and solitary waves), and have been discussed by Larsen (1965), Clarke (1971), Maxworthy and Redekopp (1976) and Redekopp (1977). The theory of large-amplitude planetary waves on large-scale currents has also been examined by Moore (1963) and by Larichev and Reznik (1976a, b).

The nonlinear theories of nondivergent planetary waves are based on an analysis of the equation (18.44), which can be written as

$$\nabla_h^2\psi_t - J(\nabla_h^2\psi, \psi) = -\beta\psi_x, \tag{19.3}$$

where the stream function ψ is defined as in (18.25) and $J(\nabla_h^2\psi, \psi) = \psi_y\nabla_h^2\psi_x - \psi_x\nabla_h^2\psi_y$. This equation is similar in form to (12.16), the nonlinear vorticity equation in a Boussinesq fluid. Introducing new variables s and Ψ similar to those defined in (12.30), i.e.,

$$s = x - ct, \tag{19.4a}$$

$$\Psi = \psi + cy, \tag{19.4b}$$

equation (19.3) may be reduced to a form similar to (12.34):

$$J(\nabla_h^2\Psi + \beta y\Psi, \Psi) = 0. \tag{19.5}$$

This latter equation may be integrated to give

$$\nabla_h^2\Psi + \beta y\Psi = F(\Psi), \tag{19.6}$$

where $F(\Psi)$ is a function of integration determined by initial conditions. Note also that (19.6) could be written down directly from (18.24), the equation of conservation of potential vorticity $(\boldsymbol{\zeta}\cdot\hat{z} + f)$ along a path line.

For sufficiently high velocities, the nonlinear terms in (19.3) exceed the β-term in amplitude: such a flow is more turbulent than wave-like. Because of the constraints imposed on the motion by the rigid top and bottom, this turbulence is restricted to two-dimensional flow. This two-dimensional turbulence has peculiar properties which are very relevant to atmospheric and oceanic motions (Rhines, 1975; 1977).

Letting horizontal velocity and length scales be denoted by U and k_h^{-1}, we find from (19.3) that nonlinearity dominates over the β-effect for scales such that

$$k_h \gg k_\beta = (\beta/U)^{1/2}. \tag{19.7}$$

Small eddies with wavenumbers $\boldsymbol{k}_h$ satisfying (19.7) do not propagate as linear planetary waves, but evolve by nonlinear interactions which do not include the β-effect. In the absence of dissipation, the total energy E and the total squared vorticity (the enstrophy) V of the flow field are conserved, where

$$E = \int_A \tfrac{1}{2}|\nabla_h \psi|^2 \mathrm{d}A, \tag{19.8a}$$

$$V = \int_A \tfrac{1}{2}|\nabla_h^2 \psi|^2 \mathrm{d}A, \tag{19.8b}$$

in which the integral is over the (singly connected) area A occupied by the fluid. The boundary A is rigid and, therefore, $\psi = 0$ on A. That $\partial E/\partial t = 0$ holds readily follows by integrating (18.30), for example; the conservation of V follows by integrating (19.3). Note that for convenience we have set $\rho_* = 1$. The distributions of E and V among eddies of various sizes (satisfying 19.7) may be characterized by one-dimensional spectral energy and enstrophy densities $E(k_h)$ and $V(k_h)$, where $E(k_h)\,\mathrm{d}k_h$ is the contribution to $\int \frac{1}{2}|\nabla_h\psi|^2\mathrm{d}A$ from Fourier components k_h in the range $(k_h, k_h + \mathrm{d}k_h)$. $V(k_h)$ is similarly defined for the enstrophy distribution. Thus, from (19.8) we find

$$E = \frac{1}{2}\int_0^\infty k_h^2 \int \psi\psi^* \mathrm{d}A\,\mathrm{d}k_h = \int_0^\infty E(k_h)\,\mathrm{d}k_h, \tag{19.9a}$$

$$V = \frac{1}{2}\int_0^\infty k_h^4 \int \psi\psi^* \mathrm{d}A\,\mathrm{d}k_h = \int_0^\infty V(k_h)\,\mathrm{d}k_h. \tag{19.9b}$$

From (19.9) it follows that $V(k_h) = k_h^2 E(k_h)$, and the conservation of E and V implies that

$$\frac{\partial}{\partial t}\int_0^\infty E(k_h)\,\mathrm{d}k_h = 0, \tag{19.10a}$$

$$\frac{\partial}{\partial t}\int_0^\infty k_h^2 E(k_h)\,\mathrm{d}k_h = 0. \tag{19.10b}$$

Let the energy distribution be given at some initial time, with a peak at the mean wavenumber $\bar{k}_h$, defined as

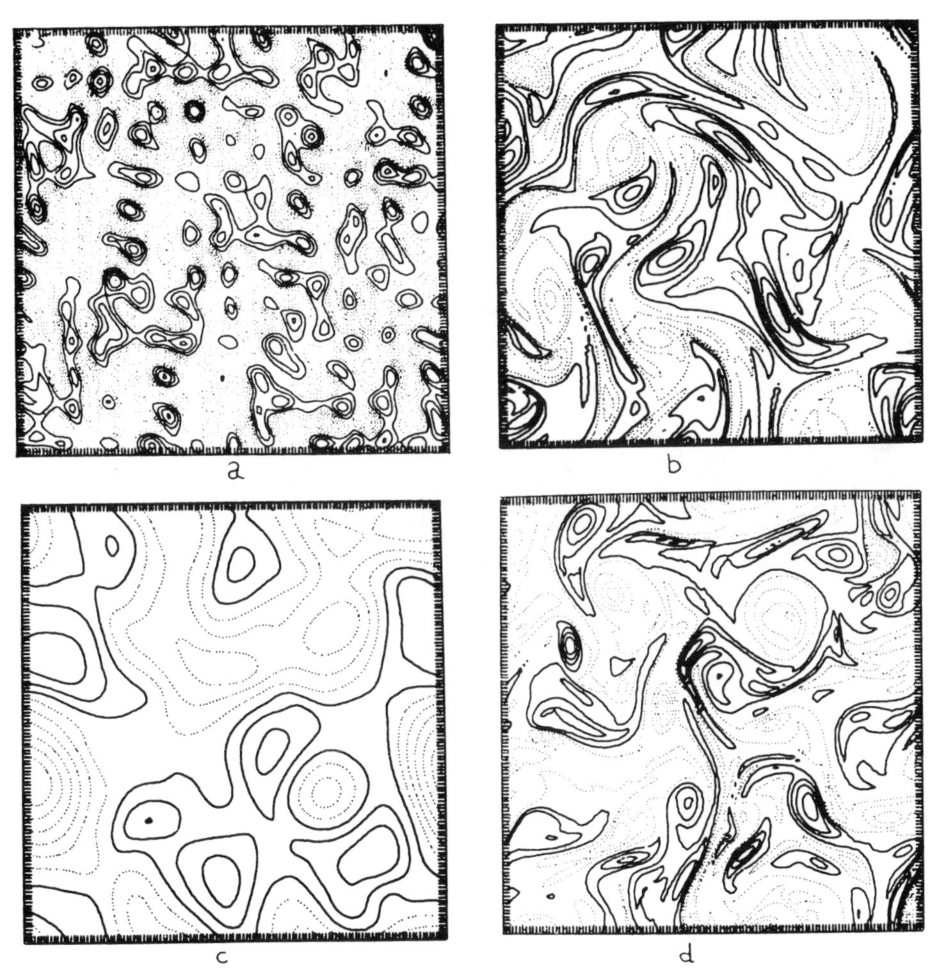

Fig. 19.3. Numerical experiments on the evolution of two-dimensional turbulence ($\beta = 0$): (a) the initial streamline field; (b) the initial vorticity contours; (c) the streamline pattern after the low-wavenumber energy cascade has been allowed to proceed for some time; (d) vorticity contours corresponding to (c). (From Rhines, 1975.)

$$\bar{k}_h = \int_0^\infty k_h E(k_h)\,\mathrm{d}k_h \Big/ \int_0^\infty E(k_h)\,\mathrm{d}k_h. \tag{19.11}$$

Nonlinear interactions between various Fourier components may be assumed to spread the energy over an increasingly larger band of wavenumbers: the spectrum "diffuses" in $\boldsymbol{k}_h$-space. This situation may be described by the inequality

$$\frac{\partial}{\partial t}\int_0^\infty (k_h - \bar{k}_h)^2 E(k_h)\,\mathrm{d}k_h > 0. \tag{19.12}$$

The second moment of the spectral distribution increases due to the widening of $E(k_h)$. In view of (19.10) and (19.11), the inequality (19.12) implies that

Fig. 19.4. The evolution of two-dimensional turbulence in the presence of a weak β-effect, presented in time–longitude plots of contours of the stream function. Time increases downwards for initial conditions at a given latitude; the tilt in the isopleths indicates westward phase propagation. The units may be taken as arbitrary in the present context. Note the initial energy cascade to larger eddies, followed by westward phase propagation and the gradual tilting of wavefronts towards the zonal direction. (From Rhines, 1975.)

$$\frac{\partial \bar{k}_h}{\partial t} < 0. \tag{19.13}$$

The mean wavenumber of the spectral distribution (or the wavenumber of the spectral peak, if we started with a sharply peaked spectrum) decreases with time: energy cascades to lower k_h, i.e., to larger scales, in complete contrast with the properties of three-dimensional turbulence. The peak of the enstrophy distribution $V(k_h)$, on the other hand, may be shown to migrate toward higher wavenumbers. The flow field evolves toward larger but more sharply defined structures, i.e., from a distribution of small-scale, low-elevation "hills" of $\psi\psi^*$ to a field of larger scales and higher "mountains". Examples are shown in Rhines (1975) and the result of one of his numerical experiments is reproduced in Fig. 19.3.

As the energy of two-dimensional oceanic turbulence cascades down to lower wavenumbers, the inequality (19.7) eventually cannot be satisfied and the β-effect, in the form of westward phase propagation, enters the picture. In the vicinity of $k_h = k_\beta$, the cascade to lower k_h continues. The dispersion relation (18.28) can be written as:

$$\omega = -\beta \cos \theta_h / k_h, \tag{19.14}$$

where θ_h is the angle which the wavenumber vector makes with the x-axis. The direction of phase propagation of waves of a given frequency ω thus shifts towards the north ($\cos\theta_h$ decreases) as k_h continues to decrease (Fig. 19.4). Ultimately, the original eddy field is transformed into an alternation of nearly zonal currents.

The relevance of two-dimensional turbulence to atmospheric dynamics is well established and its implications are still being studied intensively. Its application to the ocean is relatively recent and has been in great part stimulated by the observations of Polygon and MODE (see references in Section 18). The reader should consult Rhines' review article (1977) for a more extensive discussion.

Exercises Section 19

1. From (15.42) and (17.3), derive the relation (19.2).
2. Calculate the wavelength of a semi-diurnal tide ($\omega = 2\pi/12$ hours) at latitude 45°N in an ocean of depth: (a) $H = 5$ km; (b) $H = 10$ m.
3. Derive the inequality (19.13) from (19.10)–(19.12).

20. TOPOGRAPHIC PLANETARY WAVES

Our study of long waves in the previous sections was based on the assumption of a flat-bottomed ocean, for which an exact separation of the horizontal and vertical dependences is possible. In reality the ocean floor is not flat; apart from small-scale irregularities due to seamounts, canyons and the like, there are large regions in the open ocean where mean topographic slopes of $0(10^{-3})$ and larger exist. Although numerically small and generally of little importance for the shorter gravity waves discussed in Chapter 2, such small slopes can significantly affect the propagation of long waves, especially those of very low frequencies. Moreover, in the presence of mean slopes, an exact modal separation is not possible because the boundary condition $\boldsymbol{u} \cdot \boldsymbol{n} = 0$ at $z = -H(x, y)$ involves both the vertical and horizontal velocity components. Nevertheless, it is relatively easy to study the effects of topography on the barotropic mode alone. Therefore, the emphasis in this section will be on barotropic planetary waves in the presence of slowly varying topography. At the end of this section, however, we will briefly discuss the subject of topographic planetary waves in a stratified fluid. The study of long gravity waves in the presence of topography will be deferred to Chapter 4.

The governing equations for barotropic motions over variable depth

We shall start the analysis in this section with the nonlinear long wave equations (15.47) (in which $p_a =$ constant) and (15.49):

$$u_t + uu_x + vu_y - fv + g\eta_x = 0, \tag{20.1}$$

$$v_t + uv_x + vv_y + fu + g\eta_y = 0, \tag{20.2}$$

$$[(H+\eta)u]_x + [(H+\eta)v]_y + \eta_t = 0. \tag{20.3}$$

We have included the nonlinear terms here in order to derive an exact nonlinear equation for the conservation of integrated potential vorticity, which can be compared with (4.8), the analogous local conservation equation for a rotating stratified fluid. If η is eliminated from (20.1) and (20.2) by cross-differentiation, we obtain

$$\frac{\partial}{\partial t}(v_x - u_y) + (uv_x + vv_y)_x - (uu_x + vu_y)_y + f(u_x + v_y) + \frac{\mathrm{D}_h f}{\mathrm{D}t} = 0, \tag{20.4}$$

where $$\frac{\mathrm{D}_h}{\mathrm{D}t} \equiv \frac{\partial}{\partial t} + u\frac{\partial}{\partial x} + v\frac{\partial}{\partial y}. \tag{20.5}$$

From (20.3) it follows that

$$u_x + v_y = (H+\eta)\frac{\mathrm{D}_h}{\mathrm{D}t}\left(\frac{1}{H+\eta}\right), \tag{20.6}$$

where (20.5) has been used. Substituting (20.6) into (20.4) and rearranging the terms, we obtain an equation for the vertical component of the planetary vorticity:

$$\frac{\mathrm{D}_h}{\mathrm{D}t}\left(\frac{\zeta_3 + f}{H+\eta}\right) = 0, \tag{20.7}$$

where $\quad \zeta_3 = v_x - u_y = \boldsymbol{\zeta} \cdot \hat{z}.$

The quantity $q = (\zeta_3 + f)/(H + \eta)$ is called the barotropic or integrated potential vorticity for a nonstratified rotating fluid of variable depth, and (20.7) expresses the fact that q is conserved along particle paths. For a stratified fluid rotating with angular velocity $\boldsymbol{\Omega}$, the analogous conserved quantity is the "baroclinic" potential vorticity, $q_s = (\boldsymbol{\zeta} + 2\boldsymbol{\Omega}) \cdot \nabla\rho$ [see (4.8)]. Hence, qualitatively speaking, (20.7) represents a depth-averaged version of (4.8). If $\nabla\rho$ is approximated by $\nabla\rho_0 = \rho_{0z} z$, then

$$q_s \simeq (\zeta_3 + f)\rho_{0z},$$

since $f = 2\Omega_3$. The similarity between the two conservation equations is now clear: at a fixed latitude, the relative vorticity of an element of fluid is increased and the vortex lines stretched when the total depth $H + \eta$ increases or where the stratification decreases (i.e., the vertical separation between isopycnals increases).

From (20.7) we see that vortex stretching and compression are due to two aspects of the same process: changes in the total depth through variations in the mean depth H and in the free surface displacement η. Since changes in H over large horizontal distances are much larger than free surface fluctuations, the dynamics of topographic planetary waves are not seriously affected by neglecting η with respect to H and thereby treating the ocean surface as rigid. Then, (10.7) reduces to

$$\frac{D_h}{Dt}\left(\frac{\zeta_3 + f}{H}\right) = 0, \tag{20.8}$$

and the current motion is horizontally nondivergent (cf. 20.3):

$$(Hu)_x + (Hv)_y = 0. \tag{20.9}$$

Thus for a rigid-top ocean, we can introduce a mass transport stream function ψ such that

$$Hu = -\psi_y, \quad Hv = \psi_x. \tag{20.10}$$

Using (20.10) we can then write the linearized form of (20.8) as

$$\left(\frac{\psi_{xt}}{H}\right)_x + \left(\frac{\psi_{yt}}{H}\right)_y - \left(\frac{f}{H}\right)_x \psi_y + \left(\frac{f}{H}\right)_y \psi_x = 0,$$

$$\text{or} \quad \nabla_h \cdot \left(\frac{1}{H}\nabla_h \psi_t\right) + \left[\nabla_h \psi \times \nabla\left(\frac{f}{H}\right)\right] \cdot \hat{z} = 0. \tag{20.11}$$

We saw in Section 18 that the neglect of the free surface was valid provided (cf. 18.29)

$$f^2/gHk_h^2 \ll 1, \tag{20.12}$$

where k_h^{-1} is the wavelength scale. For $H = 5 \cdot 10^3$ m and $k_h^{-1} = 10^5$ m, we find

$$f^2/gHk_h^2 = 0(10^{-3}),$$

which indeed concurs with (20.12). In fact it is only for extremely long waves (wavelengths of order 10,000 km) that the divergence parameter $f^2/gHk_h^2 = 0(1)$ and the rigid-lid approximation breaks down.

Rather than using the potential vorticity equation (20.11) to find topographic planetary wave solutions, an equation for η alone could be used. If (20.1)–(20.3) are linearized, the resulting momentum equations solved for u and v, and these expressions are then substituted into $(Hu)_x + (Hv)_y + \eta_t = 0$, we obtain

$$(H\eta_{xt})_x + (H\eta_{yt})_y + f(\eta_y H_x - \eta_x H_y) - \frac{1}{g}\mathcal{L}\eta_t - \beta H[\eta_x + 2f\mathcal{L}^{-1}(\eta_{yt} - f\eta_x)] = 0, \tag{20.13a}$$

where $\mathcal{L} = \partial_{tt} + f^2$. It is important to notice that when $\beta = 0$, this equation is third order in time because of the topographic variations. Thus in addition to the gravity wave modes corresponding to ∂_{tt} in $\mathcal{L}$, there will be another mode, the topographic planetary wave, whenever the coefficient of f is not zero. This term is the analogue of the second term in (20.11). To filter out the gravity wave modes in the above equation for η, we make the usual low-frequency approximation $\mathcal{L} \simeq f^2$ and hence $\mathcal{L}^{-1} \simeq f^{-2}$. Then for an $e^{-i\omega t}$ time dependence, the terms in the square bracket combine into

$$-\eta_x + (2/f)\eta_{yt} = -\eta_x[1 + 0(\omega/f)].$$

Hence for $\omega \ll f$, we obtain

$$\nabla_h \cdot (H\nabla_h \eta_t) + H^2\left[\nabla_h \eta \times \nabla_h\left(\frac{f}{H}\right)\right]\cdot \hat{z} - \frac{f^2}{g}\eta_t = 0. \tag{20.13b}$$

The similarity between (20.11) and (20.13b) is remarkable, with (20.13b) containing the extra term $-f^2\eta_t/g$, which represents the stretching of vortex lines by the motion of the free surface. Upon comparing this term with the first term in (20.13b), we see that it can be neglected provided

$$f^2/gHk_h^2 \ll 1,$$

where k_h^{-1} is the horizontal length scale. Notice that this is identical to (20.12). Even with the term $-f^2\eta_t/g$ neglected, however, the equations for ψ and η are not identical; but one can be obtained from the other, correct to $0(\omega/f)$, by a simple transformation (see Exercise 20.2).

Equation (20.8) implies that a depth which monotonically decreases with y in a uniformly rotating homogeneous fluid is in some sense equivalent to a variable Coriolis parameter of the form $f = f_0 + \beta y$ in a fluid of uniform depth. This fact has been exploited by experimentalists to study planetary waves in the laboratory (e.g., see N. Phillips, 1965; Ibbetson and Phillips, 1967; Beardsley, 1969; Holton, 1971). In mid-latitude oceans, however, it is the interplay between topographic variations and the β-effect which is of fundamental importance for the propagation of planetary waves. That is to say, as fluid elements move across contours of f/H, their relative vorticity changes. Hence, wherever the contours of f/H are crowded together (for example, near continental margins or mid-ocean mountain ranges – see Fig. 20.1) we could expect to find substantial evidence of planetary wave motions. The importance of topography on the propagation of barotropic planetary waves was first investigated by Veronis (1966) and then more systematically by Rhines (1967, 1969a, b). The effects of stratification have also been studied (e.g., Rhines, 1970a; Needler and LeBlond, 1973) and will be discussed briefly at the end of this section. Evidence of the existence of topographically dominated planetary waves in the ocean has been reported by Thompson (1971), Thompson and Luyten (1976) and Kroll and Niiler (1976).

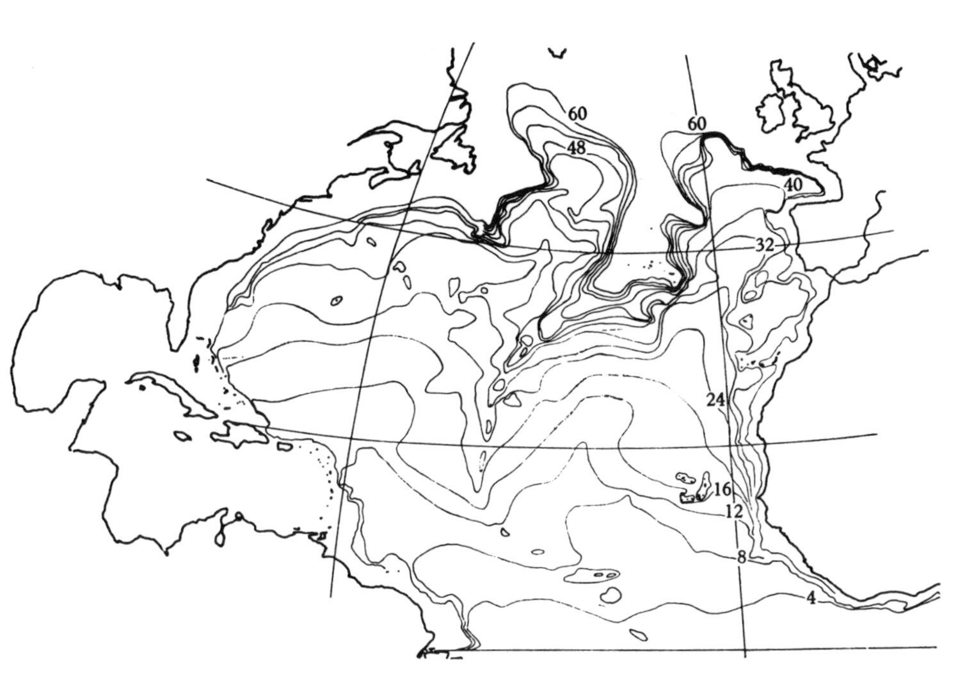

Fig. 20.1. Contours of f/H ("isotrophes") for the North Atlantic Ocean. The contours are separated by 4° of latitude at $H = 4000$ m (the numbers refer to this base). Many closed contours, and all contours north of 60°N, $H = 4000$ m have been omitted. (From Rhines, 1969a.)

Relative importance of topographic and Coriolis parameter variations

Since $f = f(y)$, we first examine (20.11) when $H = H(y)$ only. With the usual simple harmonic time dependence $e^{-i\omega t}(\omega > 0)$, (20.11) for this topography reduces to

$$\nabla_h^2\psi - \frac{H_y}{H}\psi_y + \frac{i}{\omega}\left(\beta - f\frac{H_y}{H}\right)\psi_x = 0. \tag{20.14}$$

We assume that the topographic variations are characterized by one horizontal length scale l, so that $-H_y/H = 0(1/l)$; also we assume $l \geqslant L$, the length scale of the waves (usually k_h^{-1}). Setting $(x', y') = (x, y)/L$, $\omega' = \omega/f_0$ and $R_0 = f_0/\beta = R \tan\phi_0$, we find (20.14) can be written as

$$\left\{\nabla_h'^2 + 0\left(\frac{L}{l}\right)\partial_{y'} + \frac{i}{\omega'}\left[\frac{L}{R_0} + \left(1 + \frac{L}{R_0}y'\right)0\left(\frac{L}{l}\right)\right]\partial_{x'}\right\}\psi(x', y') = 0. \tag{20.15}$$

Since $\omega' \ll 1$, it is clear that the dominant topographic term in (20.15) is the one multiplied by $1 + Ly'/R_0$ (the nondimensional Coriolis parameter). Also, since $Ly'/R_0 \leqslant 0(10^{-1})$ it follows upon examining the square-bracketed term that the transition from "β" planetary waves to "topographic" planetary waves occurs when $L/l \sim L/R_0$, or $R_0/l \sim 1$. Even more important, we can say that topography dominates β when $R_0/l > 1$. Since $1/l \sim |H_y|/H$, this inequality gives

$$|H_y| > H/R_0 \tag{20.16}$$

as the criterion on the bottom slope for topography to be dominant. For $H \sim 5 \cdot 10^3$ m and $R_0 \sim 6 \cdot 10^6$ m, (20.16) gives $|H_y| > 10^{-3}$. The above argument can be easily extended to the more general case $H = H(x, y)$, and therefore, we conclude that (slowly varying) topography in the ocean dominates β when the slope $|\nabla_h H| > 10^{-3}$. Since there are many regions in the ocean where the bottom slope is greater than 10^{-3} (for example, the continental slope regions, the Mendocino escarpment, the mid-Atlantic ridge), it is clear that the inclusion of topographic variations is crucial in the study of barotropic planetary waves. As we shall see later, topography is also very important in the stratified case.

The exponential depth profile

Let us now consider the specific depth profile defined by $H_y/H = -1/l$; this gives $H = H_0 \exp(-y/l)$ (shoaling water to the north for $l > 0$). Equation (20.14) now reduces to

$$\nabla_h^2 \psi + (1/l)\psi_y + (i/\omega)(\beta + f/l)\psi_x = 0, \tag{20.17}$$

in which we treat f as constant (the traditional approximation). For zonally propagating planetary waves of the form

$$\psi(x,y) = \psi_0 \exp(ik_1 x),$$

(20.17) yields the dispersion relation

$$\omega = -(\beta + f/l)/k_1. \tag{20.18}$$

corresponding to westward propagation at a higher frequency than the usual planetary wave. Equation (20.18) shows that in the Northern Hemisphere ($f > 0$), an equivalent β is given by f/l, with $l > 0$. Thus in a uniformly rotating ocean ($\beta = 0$), the phase of a pure topographic waves moves with the shallow water on the right. Equation (20.18) also indicates that if $l < 0$ (deeper water to the north), the β-effect is opposed by topography. And as $l \to -f/\beta$, $\omega \to 0$ and the motion reduces to a steady current pattern. Finally, when $l < -f/\beta$, the direction of phase propagation is eastward.

For two-dimensional planetary waves

$$\psi(x,y) = \psi_0 \exp[i(k_1 x + k_2 y)],$$

(20.17) gives the complex dispersion relation

$$-k_1^2 - k_2^2 + ik_2/l - (k_1/\omega)(\beta + f/l) = 0.$$

However, if we put $k_2 = k_{2r} + i/2l$, this reduces to the real dispersion relation

$$\omega = -(\beta + f/l)k_1/(k_1^2 + k_{2r}^2 + 1/4l^2). \tag{20.19}$$

Thus the phase has a westward (eastward) component according to $\tilde{\beta} \equiv \beta + f/l > 0 \; (< 0)$. However, even more important, we note that the presence of the term $1/4l^2$ in the denominator of (20.19) makes it similar in form to the dispersion relation for divergent barotropic planetary waves in an ocean of constant depth [see (18.13a)]. Hence for a given direction of phase propagation, topography gives rise to two group velocities, both of which are directed toward the center of the slowness circle. It also follows that there exists an upper cut-off frequency for topographic planetary waves.

With k_2 complex, ψ is slowly modulated by a factor proportional to $H^{1/2}$:

$$\psi(x,y,t) = \psi_0 \exp\left[-y/2l + i(k_1x + k_{2r}y - \omega t)\right] \propto H^{1/2} \exp\left[i(k_1x + k_{2r}y - \omega t)\right]. \tag{20.20}$$

This is analogous to the two-dimensional solution for internal waves in a non-Boussinesq fluid in which $\rho_{0z}/\rho_0 =$ constant [cf. (8.23) and (8.26)]:

$$w = W \exp\left[N^2z/2g + i(k_1x + k_{3r}z - \omega t)\right] \propto \rho^{-1/2} \exp\left[i(k_1x + k_{3r}z - \omega t)\right].$$

That is to say, in each case the wave amplitude varies slowly in the direction of the variation of the medium. The modulation factor arises in each case because of the existence in the respective governing equations of a term proportional to the first derivative of ψ or w in the direction of inhomogeneity. For the case of planetary waves, this term is identified as $(1/l)\psi_y$ in (20.17) [or as $(-H_y/H)\psi_y$ in (20.14)]. However, compared to the other topographic term, $(if/\omega l)\psi_x$, it is $0(\omega/f)$ and therefore can be neglected for low-frequency waves. The neglect of this term is what Rhines (1969a) calls the "first approximation". What we refer to as the traditional approximation [putting $f =$ constant in (20.14) or in (20.17)], Rhines calls the "second approximation". The order in which these approximations are made is relevant to giving a proper account of the energetics of topographic planetary waves. Veronis (1966) has shown that if the "first approximation" is not made, then the "second approximation" must not be made either if the total average energy of a closed system is to be conserved. This situation is analogous to that which prevails in a non-Boussinesq fluid, where the slow variation of the density must be taken into account to preserve energy. Thus letting $f =$ constant after ignoring the term $(1/l)\psi_y$ in (20.17) is similar to putting $\rho = \rho_*$ after neglecting $-N^2Z'/g$ in (10.41).

From (18.22) and (20.9) we now obtain the energy equation

$$\frac{\partial}{\partial t}\left[\tfrac{1}{2}\rho_* H(u^2 + v^2)\right] + \nabla_h \cdot (Hp\mathbf{u}_h) = 0,$$

which clearly reduces to (18.30) when $H =$ constant. To find u and v we use (20.10), in which (20.20) is employed. Upon using the real parts of u and v, we find that the average kinetic energy per unit area is constant:

$$\langle E \rangle = \tfrac{1}{2}\rho_* H\langle u^2 + v^2\rangle = \frac{1}{4H_0}\rho_*|\psi_0|^2(k_1^2 + k_{2r}^2 + 1/4l^2),$$

where here the average is only over the phase and not over the depth. Finally we mention that for $H = H_0 e^{-y/l}$ and $\beta = 0$, Luyten (1974) has shown that the energy flux as given by $\langle E\rangle \mathbf{c}_g$ is identical to $\langle Hp\mathbf{u}_h\rangle$ to order L/l, provided that one does not make the "first approximation". In contrast to the approach used here, his analysis starts with a (non-dimensional) equation for the pressure whose solution is proportional to $H^{-1/2}$.

More general profiles

Solutions of the form (20.20) can also arise for other depth profiles $H = H(y)$. If we make the substitution

$$\psi(x,y) = H^{1/2} \exp(ik_1x)\Psi(y)$$

in (20.14), we obtain the following equation for $\Psi(y)$:

$$\Psi'' + [E - V(y)]\,\Psi = 0, \tag{20.21}$$

where $E = -k_1^2 - k_1\beta/\omega$,

$$V = \frac{-k_1 f}{\omega}\left(\frac{H_y}{H}\right) + \frac{1}{4}\left(\frac{H_y}{H}\right)^2 - \frac{1}{2}\left(\frac{H_y}{H}\right)_y,$$

where E is taken as >0. Equation (20.21) is analogous to (10.48) for the depth dependence of the internal wave eigenfunctions in a nonuniformly stratified ocean. By comparison with the one-dimensional Schrödinger equation in quantum mechanics, it is clear that the waves are propagating or decaying (evanescent) according to whether the "potential" V is less than or exceeds the basic "energy" E of a planetary wave in a region of uniform depth. Situations for which $V > E$ thus lead to the trapping of planetary wave energy by topography, similar to internal wave trapping by pycnoclines; one such topographic trapping situation will be explored in Section 24.

When straight contours of H are no longer oriented in the east–west direction, the contours of f/H are no longer straight. However, the analysis of (20.11) in such a case is still fairly straightforward providing we employ a coordinate system aligned with the topography. That is, suppose $H = H(\eta)$, where ξ, η are new plane coordinates rotated at an angle θ to the positive x-direction:

$$\begin{pmatrix}\xi\\ \eta\end{pmatrix} = \begin{pmatrix}\cos\theta, & \sin\theta\\ -\sin\theta, & \cos\theta\end{pmatrix}\begin{pmatrix}x\\ y\end{pmatrix}. \tag{20.22}$$

Thus when $\theta = \pi/2$ say, η is directed westward. Under the transformation (20.22), equation (20.11), with an $e^{-i\omega t}$ time dependence, reduces to

$$\psi_{\xi\xi} + \psi_{\eta\eta} + \frac{i}{\omega}\left[\beta(\cos\theta\,\psi_\xi - \sin\theta\,\psi_\eta) - f\frac{H_\eta}{H}\psi_\xi\right] = 0. \tag{20.23}$$

In deriving (20.23) we have used the "first approximation". The analysis of plane-wave solutions of (20.33) is left as an exercise for the reader. We only wish to remark here that it is now possible for the center of the slowness circle to be displaced from the k_1-axis.

For more general depth profiles which are neither straight nor possess some symmetric properties, equation (20.9) can be analyzed by the method of ray theory in the asymptotic limit as $\omega \to 0$, corresponding to short waves ($k_h \to \infty$). A number of examples based on this technique have been worked out by Smith (1970).

Topographic planetary waves in a stratified ocean

We now consider the propagation of planetary waves in a stratified fluid with a rigid top and a uniformly sloping bottom. The analysis presented here essentially follows that of Rhines (1970a). The sea floor is described by (see Fig. 20.2)

$$z = -H(y) = -H_0(1 - \delta_s y/L). \tag{20.24}$$

In terms of δ_s, the fractional change in depth over the horizontal scale L, the actual bottom slope is $\mathrm{d}z/\mathrm{d}y = \alpha = H_0\delta_s/L$. Although $L/H \gg 1$ for long waves, we assume that α is sufficiently small so that $\delta_s \ll 1$ also; this is generally the case in practice. For $H_0 = 5 \cdot 10^3$ m,

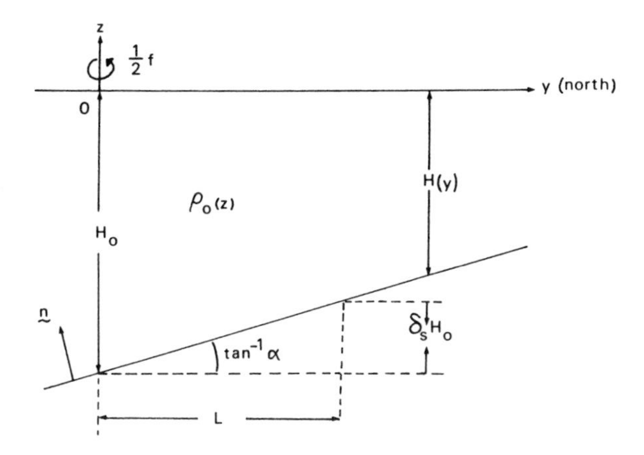

Fig. 20.2. The geometry for planetary waves in a stratified fluid with a uniformly sloping bottom and a rigid top. The parameter δ_s is the fractional change in depth over a horizontal distance L. The actual bottom slope is $\alpha = \delta_s H_0/L$. $\boldsymbol{n} = (0, -\alpha, 1)$ is a vector normal to the bottom.

$L = 10^5$ m and $\alpha = 0(10^{-3})$, $\delta_s = L\alpha/H_0 = 0(2 \cdot 10^{-2})$. The governing equations used here are (15.1)–(15.5) with the Boussinesq approximation:

$$u_t - fv + \frac{1}{\rho_*} p_x = 0,$$

$$v_t + fu + \frac{1}{\rho_*} p_y = 0,$$

$$p_z = -\rho g,$$

$$\rho_t + \rho_{0z} w = 0,$$

$$\nabla \cdot \boldsymbol{u} = 0.$$

Requiring that $\boldsymbol{u} \cdot \boldsymbol{n} = 0$ on the top and bottom gives the boundary conditions

$$w = 0 \quad \text{at} \quad z = 0, \tag{20.25a}$$

$$w = (\delta_s H_0/L)v \quad \text{at} \quad z = -H(y). \tag{20.25b}$$

According to (20.24), $H(y) = 0$ at $y = L/\delta_s$, thereby creating a singular corner in the domain. One standard way of overcoming this difficulty is to introduce vertical walls at $y = 0$, L say, and hence consider zonally propagating waves. The boundary conditions $v = 0$ at $y = 0, L$ then quantify the north–south wavenumber k_2 (Rhines, 1970a). Alternatively, we can simply restrict y to span an interval of $0(L)$ centered about the origin, thereby obtaining a solution that is locally valid in the vicinity of scale depth H_0. We shall adopt the latter approach.

The major obstacle preventing us from finding exact, horizontally propagating wave solutions of the above system is (20.25b), the bottom boundary condition. Because of the latter, the separation of the equations into the usual barotropic and baroclinic modes is not possible. However, since δ_s is small, an approximately separable solution may be found by means of an asymptotic expansion in powers of δ_s.

Let us define a set of nondimensional (primed) quantities by the equations

$$(x, y) = L(x', y'), \quad z = H_0 z', \quad t = f_0^{-1} t',$$
$$(u, v) = U(u', v'), \quad w = (H_0 U/L) w',$$
$$p = \rho_* U f_0 L p', \quad \rho = (\rho_* f_0 UL/gH_0)\rho'. \tag{20.26}$$

Substituting (20.26) into the above governing equations and boundary conditions, we obtain, after dropping the primes,

$$u_t - (1 + \delta_s \hat{\beta} y)v + p_x = 0, \tag{20.27a}$$
$$v_t + (1 + \delta_s \hat{\beta} y)u + p_y = 0, \tag{20.27b}$$
$$p_z = -\rho, \tag{20.28}$$
$$\rho_t - B(z)w = 0, \tag{20.29}$$
$$u_x + v_y + w_z = 0, \tag{20.30}$$
$$w = 0 \quad \text{at} \quad z = 0, \tag{20.31a}$$
$$w = \delta_s v \quad \text{at} \quad z = -1 + \delta_s y. \tag{20.31b}$$

Two nondimensional combinations of the chosen scales appear in (20.27)–(20.31); they are

$$\hat{\beta} = \beta H_0 / f_0 \alpha, \text{ a planetary vorticity factor} \tag{20.32}$$

$$\text{and} \quad B(z) = (N^2 H_0^2 / f_0^2)/L^2 = r_i^2/L^2, \text{ the Burger number.} \tag{20.33}$$

The quantity $\hat{\beta}$ is clearly a measure of the importance of β as compared to topography; for $\alpha = 0(10^{-3})$, $\hat{\beta} = 0(1)$, in which case topography and β are of equal importance. For large slopes, $\hat{\beta} \ll 1$ and β will play a minor role. The Burger number $B(z)$ is the square of the ratio of the internal (Rossby) radius of deformation to the horizontal length scale. For $N^2 = 0(10^{-3}\ \text{rad s}^{-1})$, $H_0 = 0(5 \cdot 10^3\ \text{m})$ and $L = 0(10^5\ \text{m})$, we find $B(z) = 0(1)$. For very large horizontal scales, $B(z) \ll 1$ and stratification will play a minor role. However, at this stage we assume that both $\hat{\beta}$ and B are of order unity. The only small parameter in (20.27)–(20.31) is then δ_s. We therefore expand u in the form

$$u = u^{(0)} + \delta_s u^{(1)} + \ldots \tag{20.34}$$

and similarly for v, w, p and ρ. Each of the dependent variables is assumed to have an $e^{-i\omega t}$ dependence, where ω is as yet unknown. Therefore, we also write

$$\omega = 0 + \delta_s \omega^{(1)} + \ldots \tag{20.35}$$

where we a priori set $\omega^{(0)} = 0$ in order to focus attention on low-frequency waves ($\omega \ll f$). Finally, we expand each term in the bottom boundary condition (20.45) in a Taylor series about $z = -1$.

Substituting expansions of the form (20.34) and (20.35) into (20.27)–(20.32) and collecting like powers of δ_s, we find that to $0(\delta_s^0)$,

$$-v^{(0)} + p_x^{(0)} = 0, \tag{20.36a}$$
$$u^{(0)} + p_y^{(0)} = 0, \tag{20.36b}$$

$$p_z^{(0)} = -\rho^{(0)}, \tag{20.37}$$

$$w^{(0)} = 0, \tag{20.38}$$

$$u_x^{(0)} + v_y^{(0)} = 0, \tag{20.39}$$

and $$w^{(0)} = 0, \quad \text{at} \quad z = 0, \tag{20.40a}$$

$$w^{(0)} = 0, \quad \text{at} \quad z = -1. \tag{20.40b}$$

Since $w^{(0)} \equiv 0$ by (20.38), which is a consequence of taking $\omega^{(0)} = 0$, the boundary conditions (20.40) are identically satisfied. Thus to this order the motion is geostrophic, with $p^{(0)}(x, y, z)$ serving as the geostrophic stream function. In view of (20.36), equation (20.39) is automatically satisfied, and we are left with three equations [(20.36a, b) and (20.37)] for the four unknowns $p^{(0)}$, $u^{(0)}$, $v^{(0)}$ and $\rho^{(0)}$. To solve for these quantities we must proceed to the next order.

To $0(\delta_s)$ we find the "quasi-geostrophic" equations

$$-i\omega^{(1)}u^{(0)} - v^{(1)} - \hat{\beta}yv^{(0)} + p_x^{(1)} = 0, \tag{20.41a}$$

$$-i\omega^{(1)}v^{(0)} + u^{(1)} + \hat{\beta}yu^{(0)} + p_y^{(1)} = 0, \tag{20.41b}$$

$$p_z^{(1)} = -\rho^{(1)}, \tag{20.42}$$

$$-i\omega^{(1)}\rho^{(0)} - B(z)w^{(1)} = 0, \tag{20.43}$$

$$u_x^{(1)} + v_y^{(1)} + w_z^{(1)} = 0, \tag{20.44}$$

with boundary conditions $$w^{(1)} = 0 \quad \text{at} \quad z = 0, \tag{20.45a}$$

$$w^{(1)} = v^{(0)} \quad \text{at} \quad z = -1. \tag{20.45b}$$

It is now a fairly simple matter to combine (20.41)–(20.44) into one equation for $p^{(0)}$. It turns out to be a vorticity balance equation since $p^{(0)}$ acts as a stream function for the zeroth-order flow. The boundary conditions (20.45) can also be expressed in terms of $p^{(0)}$. Then, after solving this system for $p^{(0)}$, the geostrophic velocities and zeroth-order density can be found using (20.36) and (20.37), respectively. We now eliminate $p^{(1)}$ from (20.41) by cross-differentiation, then use (20.44) to eliminate $u_x^{(1)} + v_y^{(1)}$ in favour of $w_z^{(1)}$, which in turn we write in terms of $\rho^{(0)}$ through (20.43). The result is, after using (20.36) and (20.37),

$$-i\omega^{(1)}[\nabla_h^2 p^{(0)} + (B^{-1}p_z^{(0)})_z] + \hat{\beta}p_x^{(0)} = 0, \quad -1 \leqslant z \leqslant 0. \tag{20.46}$$

The boundary conditions for $p^{(0)}$ are

$$p_z^{(0)} = 0 \quad \text{at} \quad z = 0, \tag{20.47a}$$

$$i\omega^{(1)}p_z^{(0)} = Bp_x^{(0)} \quad \text{at} \quad z = -1. \tag{20.47b}$$

The vorticity balance equation (20.46) is the nondimensional analogue of the low-frequency limit of the pressure equation (15.21) derived at the beginning of this chapter. We notice that through the use of a perturbation approach, we have arrived at a zeroth-order problem which can be solved by the separation of variables. The difference between the situation here and the earlier analyses for a flat-bottomed ocean is in the boundary condition at $z = -1$. The frequency $\omega^{(1)}$, which is to be determined, now occurs both in the differential equation and in this boundary condition. This, we shall see, gives rise to some rather interesting results.

It is instructive at this point to first examine the case when topography plays a dominant role. Setting $\hat{\beta} = 0$ and taking $N =$ constant for simplicity, (20.46) and (20.47a) then have the wavelike solution

$$p^{(0)} = \exp\left[i(k_1x + k_2y)\right] \cosh\left(\sqrt{B}k_hz\right). \tag{20.48}$$

Substitution of (20.48) into (20.47b) gives the dispersion relation

$$\omega^{(1)} = -k_1B\frac{\coth\left(\sqrt{B}k_h\right)}{\sqrt{B}k_h}. \tag{20.49}$$

Since $B > 0$, (20.49) implies that $\omega^{(1)}/k_1 < 0$, i.e., the phase velocity again has a westward component (the phase travels with shallow water to the right). We now consider two limiting cases of (20.48) and (20.49).

For waves long compared to the internal Rossby radius, $\sqrt{B}k_h \ll 1$; in this limit,

$$p^{(0)} = \exp\left[i(k_1x + k_2y)\right] \tag{20.50}$$

$$\text{and} \quad \omega^{(1)} = -k_1/k_h^2. \tag{20.51}$$

Thus the current components $u^{(0)}$ and $v^{(0)}$ are uniform with depth (cf. 20.36), implying barotropic motion, and the dispersion relation is independent of stratification, as for barotropic planetary waves. Indeed, in terms of the dimensional quantities given in (20.26), equation (20.51) can be written as

$$\omega = -\frac{\alpha f_0}{H_0}\frac{k_1}{k_1^2 + k_2^2}, \tag{20.52}$$

which is of the form (18.28) for nondivergent barotropic planetary waves. The "equivalent" β in (20.52) is $\alpha f_0/H$, which is analogous to the factor f/l which appears in the dispersion relation for an exponential depth profile (cf. 20.18). We conclude that long topographic planetary waves are virtually unaffected by stratification.

For short waves ($\sqrt{B}k_h \gg 1$),

$$p^{(0)} = \tfrac{1}{2}\exp\left[i(k_1x + k_2y) - \sqrt{B}k_hz\right] \tag{20.53}$$

$$\text{and} \quad \omega^{(1)} = -\sqrt{B}k_1/k_h. \tag{20.54}$$

The currents derived from (20.53) have a strong vertical shear, with a dimensional vertical scale given NH/f_0, i.e., the internal radius of deformation. The frequency of these "bottom-trapped" buoyancy oscillations is proportional to the stability frequency, N. In dimensional form, (20.54) can be written as

$$\omega = -\alpha N \sin\phi, \tag{20.55}$$

where $\sin\phi = k_1/k_h$. For $\alpha = 0(10^{-2})$ and $N = 0(10^{-3}\ \text{rad s}^{-1})$, (20.55) gives $\omega = 0(10^{-5}\ \text{rad s}^{-1})$, corresponding to a period of about a week. From this limiting case we conclude that stratification strongly affects short topographic planetary waves, giving rise to a buoyancy oscillation trapped near the bottom. Rhines (1970a) also showed that there exists a boundary trapped buoyancy wave in a uniformly rotating and uniformly stratified fluid

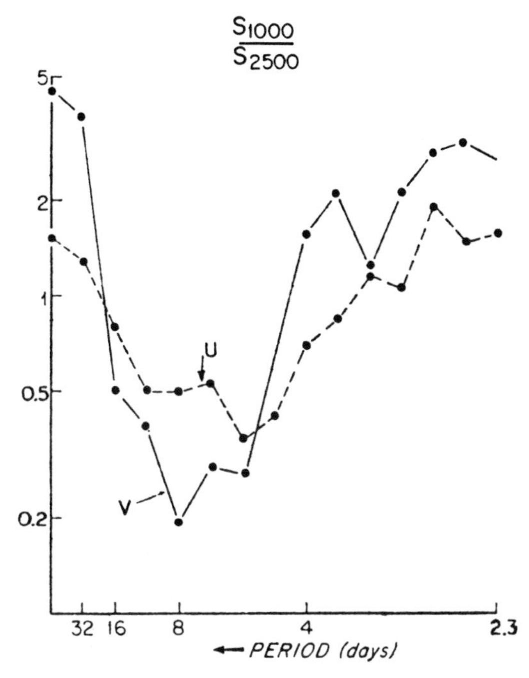

Fig. 20.3. Ratios of energy spectra at 1000 m to those at 2500 m from Site S (39°N, 70°W). The spectra were estimated from 10 months of 1972 current data. (From Thompson and Luyten, 1976.)

lying over a bottom of *arbitrary slope* $\alpha = \sin \beta_b$, where β_b is the inclination of the bottom to the horizontal. Unlike the asymptotic solution presented here, this solution is exact; but remarkably enough, the frequency is also given by (20.55). For large slopes, corresponding to $\beta_b = 0(\pi/4)$, this wave has a frequency of order N, i.e., much higher than the planetary wave frequency.

Thompson and Luyten (1976) have recently presented convincing evidence that bottom-trapped motions probably occur at Site S, on the continental rise south of Cape Cod. Figure 20.3 shows the ratios of the spectra at 1000 m depth at Site S to those at 2500 m, for u and v separately. We notice that while there is more energy at the upper level for low and high frequencies, the lower level has more energy at intermediate frequencies, which together with the high observed coherences at these frequencies are indicative of the presence of bottom-trapped motions with periods of around eight days. Around Site $S, \alpha \simeq 10^{-2}$ and $N \simeq 10^{-3}$ rad s^{-1}; using these values in (20.55), approximately the observed periods are obtained. Using the exact dispersion relation (20.49) together with (20.48), Thompson and Luyten estimated the wavelength of a bottom-trapped wave of 8 days period to be 90 km.

When $\beta \neq 0$, (20.46) and (20.47a) have wavelike solutions of the form

$$p^{(0)} = \exp\left[i(k_1 x + k_2 y)\right] \cos mz, \tag{20.56}$$

along with the usual planetary wave dispersion relation

$$\omega^{(1)} = -\hat{\beta} k_1/(k_h^2 + m^2/B). \tag{20.57}$$

To determine m and hence the vertical modal structure, one must solve the transcendental equation

$$m \tan m = -(Bk_h^2 + m^2)/\hat{\beta}, \tag{20.58}$$

which is obtained by substituting (20.56) into (20.47b). The solution of (20.58) is discussed at length by Rhines (1970a), both for $\hat{\beta} > 0$ as well as for $\hat{\beta} < 0$ (which occurs when $\alpha < 0$ – shoaling water to the south). A few of the main results are mentioned here. First, for $\hat{\beta} > 0$ (as in Fig. 20.2), the topography reinforces the β-effect, increasing the frequency of all modes. The first-mode solution of (20.58) corresponds to a barotropic-like motion; the corresponding dispersion relation for long wavelengths reduces to that given by (20.19). At short wavelengths, however, this mode reduces to a bottom-trapped oscillation. The dispersion curve corresponding to this mode is shown as a dashed curve in the upper left part of Fig. 18.3. The higher modes have a sinusoidal vertical profile and therefore represent the baroclinic modes. At short wavelengths, however, they have the property that the fluid at the bottom can no longer negotiate the slope, and the waves adjust by moving a node in $\boldsymbol{u}_h$ to the bottom (see lower dashed curve in Fig. 18.3, which corresponds to the first baroclinic mode).

When $\hat{\beta} < 0$ ($\alpha < 0$, corresponding to deeper water to the north), the bottom slope opposes β, and if α is sufficiently large some of the waves will propagate eastward. Nevertheless, the lowest mode remains barotropic at long wavelengths and bottom-trapped at short wavelengths.

For bottom slopes of other orientations the picture becomes quite complex and we refer the reader to Robinson and McWilliams (1974) for details. For two-dimensional topography or even one-dimensional nonplanar topography, no simple analysis is possible. Some work has been done on the problem of propagation over sinusoidally varying bottoms (Rhines and Bretherton, 1973, for the barotropic case; McWilliams, 1974 for the baroclinic case). McWilliams (1976) has also discussed planetary wave propagation in a two-layer ocean in the presence of general topographic variations and non-uniform currents under the assumption that the inhomogeneities of the medium occurred on spatial scales much larger than the wavelength of the waves under consideration.

Exercises Section 20

1. Complete the derivation of (20.7).
2. Show that the substitution

$$\hat{z} \times \nabla_h \psi = \frac{gH}{f}\left(\hat{z} \times \nabla_h \eta - \frac{i\omega}{f} \nabla_h \eta\right)$$

into (20.11) gives (20.13b) (in which both have ∂_t replaced by $-i\omega$), provided $\omega \ll f$ and the term $-f^2\eta_t/g$ is neglected in (20.13b).

3. Find the slowness curve implied by (20.19) and hence determine the upper cut-off frequency for topographic planetary waves. Estimate this frequency for bottom slopes of $0(10^{-3})$, $0(10^{-2})$.
4. Derive equation (20.21).
5. Discuss the plane-wave solutions of (20.23) which take the form

$$\psi(\xi, \eta) = \psi_0 e^{i(k\xi + l\eta)}.$$

Assume that $H_\eta/H = $ constant.

21. EQUATORIALLY TRAPPED WAVES

Thus far our discussion of long waves in a rotating fluid has been concerned with their propagation in mid-latitude ocean basins. The propagation of planetary waves in a polar basin, where the β-plane approximation breaks down (see Section 5), has been studied by LeBlond (1964) and will be discussed later in Section 29. However, the f-plane equations are valid at the poles, and therefore our discussion of long gravity waves certainly is relevant to these regions. Another important geographical area as far as the oceans are concerned is the equatorial region, where at the equator itself, the Coriolis parameter vanishes. We now examine the effect of this on the propagation of long waves along the equator.

In this section we shall show that a narrow band of latitudes centered on the equator can act as a waveguide for westward propagating planetary waves and zonally propagating long gravity waves. Stern (1963) and Bretherton (1964b) were among the first to show that low-frequency oscillations can be trapped in such an equatorial region; however, in both studies it was assumed that the motion is independent of the longitude. Here, on the other hand, we do not make this restriction and instead present a summarized version of some of the more general results obtained, for example, by Blandford (1966), Matsuno (1966), Munk and Moore (1968) and Lighthill (1969). In each of these studies, the authors considered zonally propagating long waves of the first and second classes that are equatorially trapped in an ocean of constant depth. Applications of the theory to recent observations will also be given.

The analysis will be based on the long wave equations (15.1)–(15.5) in which $f = \beta y$, corresponding to a β-plane centered at the equator. Further, we shall work with the governing equation for $v(x, y, z, t)$ (cf. 15.15). After separating out the vertical dependence of this equation, we obtained (15.17) for $v_n(x, y, t)$, where n is the vertical mode number, $n = 0$ corresponding to the barotropic mode and $n = 1, 2, \ldots$ to the sequence of baroclinic modes. We rewrite (15.17) as

$$-i\omega\left\{\partial_{xx} + \partial_{yy} - [(\beta y)^2 - \omega^2]\,\frac{1}{gh_n}\right\}v_n + \beta v_{nx} = 0, \tag{21.1}$$

where an $e^{-i\omega t}$ time dependence, with $\omega > 0$, has been assumed. For any given stratification $N(z)$, the equivalent depths h_n are obtained by integrating (15.16) together with appropriate boundary conditions.

Since $f = 0$ at the equator, we should not neglect the Coriolis terms associated with the horizontal component of the Earth's rotation to be consistent with our discussion in Section 5. The problem of wave propagation is, however, not separable in terms of vertical and horizontal dependences when $\tilde{f}$ is retained, and it is then very difficult to proceed with the analysis. The influence of $\tilde{f}$ is important within $\pm\,1°$ of latitude from the equator; we shall find some justification in neglecting $\tilde{f}$ in the fact that the trapping scales of equatorial planetary waves are somewhat in excess of this distance.

For zonally propagating waves, we substitute

$$v_n(x,y) = V_n(y)\exp(ik_1 x) \tag{21.2}$$

into (21.1) and seek solutions V_n which are analytic at $y = 0$ and which tend to zero as $y \to \pm\,\infty$. The equation for V_n is

$$V_n'' + \left(\frac{\omega^2}{gh_n} - k_1^2 - \frac{\beta k_1}{\omega} - \frac{\beta^2 y^2}{gh_n}\right) V_n = 0, \quad |y| < \infty. \tag{21.3}$$

For each vertical mode n, (21.3) has turning points at the critical latitudes $y = \pm y_c$ where the coefficient of V_n vanishes. For $|y| < y_c$ ($> y_c$) the behaviour of V_n is oscillatory (damped). Thus the zonally propagating waves are trapped between two parallels of latitude symmetric about the equator. The form of (21.3) is analogous to (10.48), describing internal wave trapping near a sharp pycnocline, and also to (24.6), which, as we shall see, describes wave trapping over an escarpment.

In solving (21.3) it is first convenient to introduce a non-dimensional frequency, wavenumber and length for each vertical mode n:

$$\hat{\omega} = T\omega, \quad \hat{k} = Lk_1, \quad \hat{y} = y/L, \tag{21.4}$$

$$\text{where} \quad T = (gh_n)^{-1/4}(2\beta)^{-1/2}, \quad L = (gh_n)^{1/4}(2\beta)^{-1/2}. \tag{21.5}$$

Using (21.4) and (21.5), equation (21.3) takes the form

$$\phi_n''(\hat{y}) + (\nu + 1/2 - \hat{y}^2/4)\phi_n(\hat{y}) = 0, \quad |\hat{y}| < \infty, \tag{21.6}$$

$$\text{where} \quad \phi_n(\hat{y}) = V_n(\hat{y}L), \tag{21.7}$$

$$\nu + 1/2 = \hat{\omega}^2 - \hat{k}^2 - \hat{k}/2\hat{\omega}. \tag{21.8}$$

The solution of (21.6) which is analytic at $y = 0$ is given by the parabolic cylinder function (cf. Magnus and Oberhettinger, 1949)

$$\phi_n(\hat{y}) = \mathrm{D}_\nu(\hat{y}) = \exp(-\hat{y}^2/4) \sum_{j=0}^{\infty} a_j \hat{y}^j, \tag{21.9}$$

where the infinite series represents a linear combination of a pair of confluent hypergeometric functions. For arbitrary ν, D_ν is not bounded as $\hat{y} \to \pm\infty$; however, for $\nu = m = 0, 1, 2, \ldots$, the series in (21.9) terminates and D_m becomes proportional to a Hermite polynomial of degree m:

$$\mathrm{D}_m(\hat{y}) = \exp(-\hat{y}^2/4)\,\mathrm{He}_m(\hat{y}), \tag{21.10}$$

$$\text{where} \quad \mathrm{He}_m(\hat{y}) = (-1)^m \exp(\hat{y}^2/2) \frac{\mathrm{d}^m}{\mathrm{d}y^m} \exp(-\hat{y}^2/2);$$

$$\mathrm{He}_0 = 1, \quad \mathrm{He}_1 = \hat{y}, \quad \mathrm{He}_2 = \hat{y}^2 - 1, \ldots. \tag{21.11}$$

For these discrete values of ν, $\phi_n \to 0$ as $\hat{y} \to \pm\infty$, as required.

From (21.10) and (21.11) we see that the north–south mode number m corresponds to the number of modal lines, all parallel to the equator, that each solution possesses. For a given $\nu = m$, (21.8) is a cubic in $\hat{\omega}$ which can be solved for the three frequency functions $\hat{\omega} = \hat{\omega}_j(\hat{k})$, $j = 1, 2, 3$, subject to the restriction $\hat{\omega}_j > 0$. For the case $m = 0$, (21.8) can be easily factored:

$$(\hat{\omega} + \hat{k})(\hat{\omega}^2 - \hat{k}\hat{\omega} - 1/2) = 0. \tag{21.12}$$

The root $\hat{\omega} = -\hat{k}$, shown as a dashed line in Fig. 21.1, is inadmissible in that it gives rise to a velocity component u_n which is unbounded as $|\hat{y}| \to \infty$ (Matsuno, 1966). The other

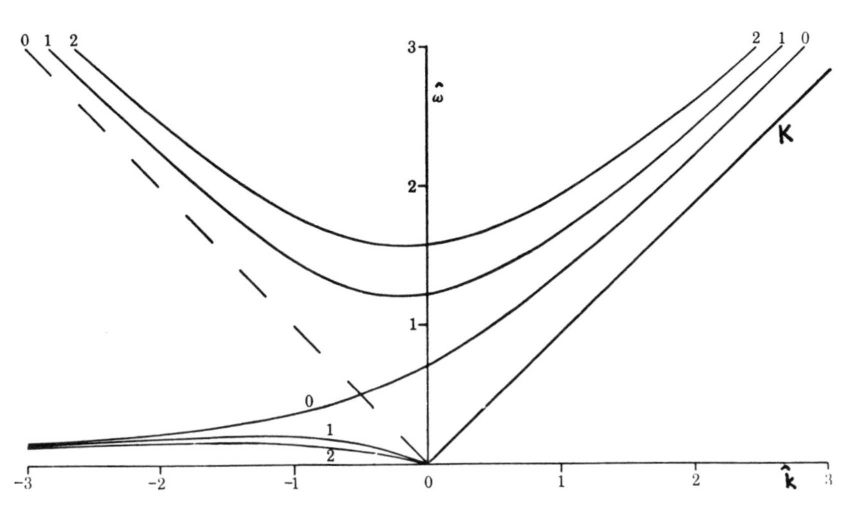

Fig. 21.1. Equatorial wave dispersion relations corresponding to modes $\nu = m = 0, 1, 2$, as computed from (21.8). The length and time scales are given by (21.5). The curved lines are, from top to bottom, the gravity waves, the mixed gravity–planetary wave (Yanai wave), and the planetary waves. The dashed line is an inadmissible solution for $m = 0$, and the line labelled K is the Kelvin wave, for which $v_n \equiv 0$. Notice that in each of the gravity wave and the planetary wave curves, there is a $\hat{k}$ at which the tangent has zero slope, corresponding to zero group velocity. (From Lighthill, 1969.)

root of (21.12) with $m = 0$ and $\hat{\omega} > 0$ is admissible, however, and corresponds to a mixed gravity–planetary wave. For large positive $\hat{k}$ it is a high-frequency wave dominated by gravity; for $\hat{k}$ large and negative it behaves like a low-frequency planetary wave. In the literature this mixed wave is often referred to as a Yanai wave. For each positive integer m, the cubic (21.8) gives two gravity waves and one planetary wave (see Fig. 21.1). Of the pair of gravity waves, one travels westward ($k < 0$) and the other eastward, ($k > 0$) whereas the phase of the planetary wave travels only westward, as usual.

There is, however, one further trapped equatorial wave not contained in the above solutions. Moore (1968) showed that there also exists an eastward-propagating Kelvin wave whose solution is given by

$$\left.\begin{aligned} v_n &\equiv 0, \\ u_n &= \mathrm{D}_0(\hat{y})\, \mathrm{e}^{i(\hat{k}\hat{x} - \hat{\omega}\hat{t})}, \end{aligned}\right\} \tag{21.13}$$

where $\hat{\omega} = \hat{k}$. It is interesting to note that $\hat{\omega} = \hat{k}$ is also a root of (21.8) when $\nu = m = -1$, even though the latter equation is derived on the a-priori implicit assumption that $v_n \neq 0$. Kelvin waves in their more familiar form as waves trapped against a vertical wall will be discussed in Section 24. The role played by equatorial Kelvin waves in generating coastal Kelvin waves has recently been explored by Anderson and Rowlands (1976a).

For $\hat{\omega} \ll 1$ and $\hat{k} \ll 1$, equation (21.8) with $\nu = m$ reduces to

$$\hat{\omega}/\hat{k} = -1/(2m + 1), \tag{21.14}$$

implying that low-frequency equatorial planetary waves are nondispersive, in contrast to their mid-latitude relatives. Waves of these time and length scales can be excited by winds associated with the Southwest Monsoon in the Indian Ocean. When such waves encounter the west coast of Africa, they "deposit" their energy, mostly in a baroclinic boundary current which flows northward along the coast. Lighthill (1969) has suggested that this is a possible mechanism for the generation of the Somali Current that flows northward along the Somali coast after the onset of the monsoon season. Lighthill's theory has been explored in greater depth, both analytically and numerically, by Cox (1976) and by Anderson and Rowlands (1976b).

Equatorial planetary waves are also thought to be relevant to the dynamics of the equatorial undercurrent, which is a prominent feature of the Pacific and Atlantic equatorial current systems (Knauss, 1966; Philander, 1973). The eastward-flowing equatorial undercurrent is concentrated just below the surface, having a thickness of about 200 m and a maximum speed of about 1 m/s at a depth of about 100 m. It is laterally confined between 2°N and 2°S. For an ocean with constant stratification, the dimensional phase speed for low-frequency baroclinic planetary waves trapped on the equator is given approximately by

$$c_{mn} = HN/n\pi(2m+1). \tag{21.15}$$

For the equatorial Pacific region, $H = 4 \cdot 10^3$ m and a depth-averaged value for N is $(2\pi/3.4) \cdot 10^{-3}$ rad s^{-1} (Munk and Moore, 1968). Hence (21.15) gives a maximum speed of

$$c_{01} = 2.3\ \mathrm{m\,s^{-1}}.$$

For a wavelength of 2800 km, the period is 14 days; the critical latitude is given by $y_c =$ 225 km or 2° of latitude. These theoretical values led Munk and Moore to suggest that a rectified flow generated by weakly interacting low-frequency planetary waves may help to maintain the equatorial undercurrent. A meandering motion has also recently been observed in the Atlantic undercurrent. It has been suggested that these meanders are likely a manifestation of unstable planetary waves superimposed on a laterally and vertically sheared equatorial flow (Düing et al., 1975; Philander, 1976).

El Niño is a well known oceanographic phenomenon characterized by the appearance of abnormally warm water off the coast of Peru. It usually coincides with the Southern Hemispheric summer, when the trade winds are weak, and with reduced upwelling off the Peruvian coast (Wyrtki et al., 1976b). It has been proposed that such a large-scale, transient phenomenon may also excite westward-propagating planetary waves (e.g., see Harvey and Patzert, 1976). During the 1975 El Niño expedition (Wyrtki et al., 1976b) current meters were deployed at the equator about 300 km west of the Galápagos Islands with the expectation that such waves, if generated, may be detected. From the current meter records, Harvey and Patzert indeed observed a 25-day period oscillation which moved westward at 0.5 m/s with a wavelength of about 1000 km. These characteristics agree very well with a (vertical) first-mode baroclinic planetary wave whose dispersion relation is given by (21.14) with $m = 1$.

In a recent theoretical paper, McCreary (1976) has shown that El Niño, far from being a local event, is really an ocean–atmosphere interaction phenomenon occurring on an oceanic scale. Using a two-layer frictionless model, McCreary examined the response of a laterally bounded ocean to variations of the large-scale wind stress. His solutions may be interpreted in terms of equatorial Kelvin waves [solution (21.13)] carrying a disturbance

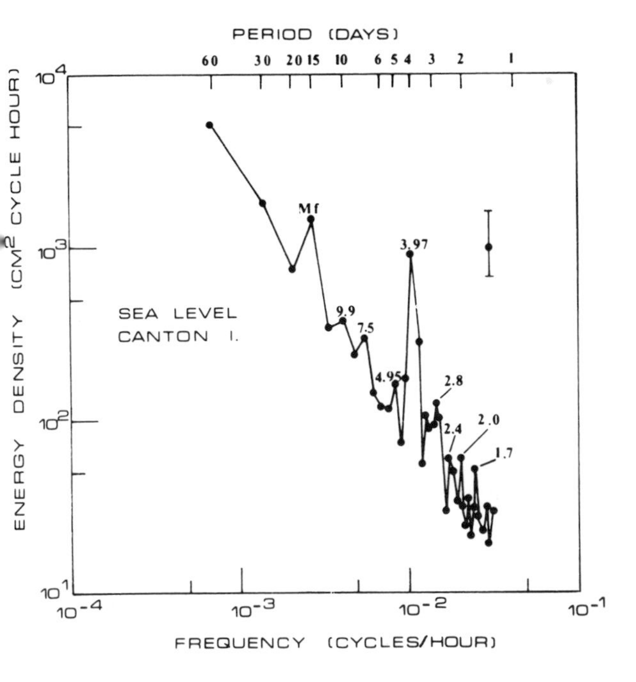

Fig. 21.2. The sea-level spectrum from four years of data at Canton Island (2°46′S, 171°43′W). The bar denotes the 95%-confidence level. Note the prominent peak at four days; *Mf* is the fortnightly tide. (From Wunsch and Gill, 1976.)

towards the eastern boundary and exiciting poleward-propagating coastal Kelvin waves (see Section 24) there. These Kelvin waves combine to produce the anomalous situations in current and water properties which are associated with the name El Niño. A numerical simulation of El Niño by Hurlburt et al. (1976) leads to a similar interpretation. These works illustrate the importance of the equatorial waveguide in providing a transoceanic connection between events which occur on opposite sides of an ocean basin.

In a series of papers by G.W. Groves and his collaborators (e.g. Groves and Miyata, 1967; Miyata and Groves, 1968), a prominent four-day peak in the spectra of sea-level records from several tropical Pacific islands was noted (see Fig. 18.4). Wunsch and Gill (1976) re-examined the sea-level records at seventeen Pacific islands and a few corresponding meteorological records. The four-day oscillation was indeed pronounced at many of these stations (Fig. 21.2) and was found to be equatorially confined (Fig. 21.3). On the basis of the theory outlined above, Wunsch and Gill found that this observed peak agrees both in period and latitudinal structure with the $n = 1, m = 2$ equatorially trapped internal gravity wave mode. From coherence studies it was further found that this mode is probably generated by the meridional wind and that the peak occurs because of the existence of a zero group velocity for a small range of negative wavenumbers (see Fig. 21.1). Wunsch and Gill also speculated that a number of other peaks in the sea-level spectra could be identified as other trapped first-mode baroclinic internal gravity waves ($n = 1, m = 0, 1, 3, 4, \ldots$).

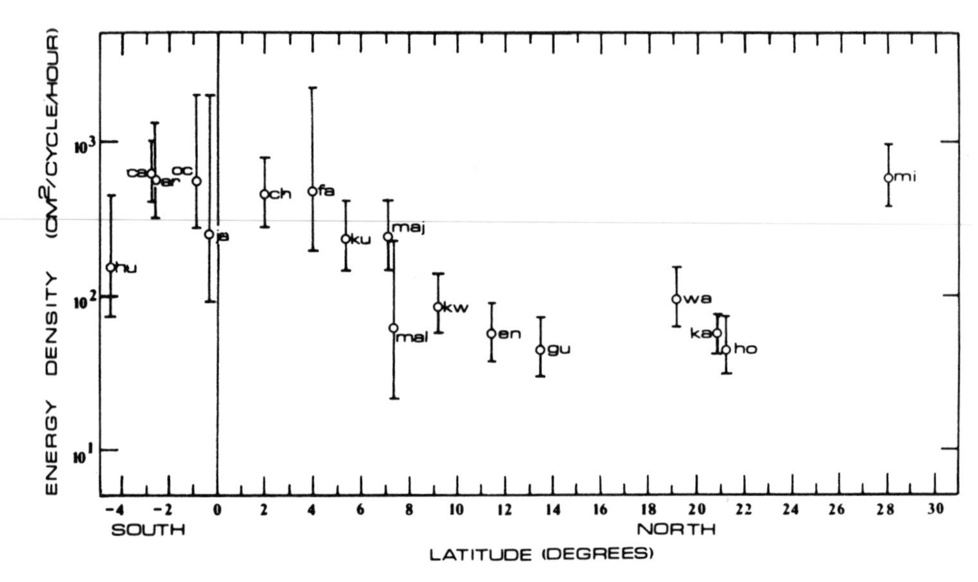

Fig. 21.3. Four-day energy as a function of latitude. For locations north of Majuro (i.e. mal, kw. . .) there was no discernible peak at this period. Energies there represent the background continuum. Remaining points show energy in the peak plus background continuum. Error bars denote 95%-confidence limits. (From Wunsch and Gill, 1976.)

Exercises Section 21

1. Derive the expression for c_{mn} in (21.15).

2. In a rigid-topped Boussinesq ocean show that at low frequencies and low wave-numbers there is no barotropic equatorial trapped wave. Explain this result physically.

3. Derive the equatorial Kelvin wave solution (21.13). From the equatorial Pacific Ocean, $H = 4 \cdot 10^3$ m and a depth-averaged value for N is $(2\pi/3.4) \cdot 10^{-3}$ rad s^{-1}. Use these to estimate the phase speed of the barotropic ($n = 0$) and gravest-mode baroclinic ($n = 1$) Kelvin wave.

CHAPTER 4

FREE WAVES: LATERAL BOUNDARY EFFECTS

22. THE ADDITION OF LATERAL BOUNDARIES

Following Section 10, our presentation divided into two paths, taking us through the theory of high-frequency, short gravity waves in Sections 11–13 and through a parallel exposition of the properties of low-frequency, long gravity waves and planetary waves in Chapter 3. The two paths come together in this chapter, in which the geometrical complexity of our model ocean is gradually augmented by adding an increasing number of lateral boundaries.

The horizontally unbounded ocean of the previous section is first delimited by a single rectilinear vertical boundary, and the reflection from this wall of the normal mode solutions obtained above is examined in Section 23. Wave trapping at a vertical wall in a rotating fluid gives rise to Kelvin waves (Section 24). The partial reflection and trapping of waves by escarpments is also discussed in Sections 23 and 24, respectively. Modelling an oceanic boundary more realistically by a sloping bottom, we then examine in Section 25 coastal trapped waves of high and low frequencies (edge waves and continental shelf waves, respectively). Finally, passing to nonrectilinear boundaries, waves around seamounts and islands are considered in Section 26. Wave diffraction at boundaries with points of very high curvature, such as a semi-infinite vertical barrier, is briefly reviewed in Section 27.

Wave propagation in channels is studied by adding a second lateral boundary (Section 28); the addition of a third boundary closes one end of the channel and provides a model of a semi-enclosed sea. Lastly, the wave oscillations of completely enclosed basins are examined in Section 29; both analytical solutions in basins of simple geometry and numerical results for more realistic and complicated model oceans are considered.

This chapter completes the description of the basically linear modal structure of free waves in the ocean. Later chapters will use the results obtained in the first part of the book to study the interactions between these waves and their oceanic environment, the interactions between the waves themselves, and the energy sources and sinks for the various types of waves.

23. WAVE REFLECTION FROM VERTICAL WALLS AND ESCARPMENTS

In this section we examine how the normal modes of a flat-bottomed rotating ocean reflect off a vertical wall. The reflection problem will be stated generally, but explicit solutions will be written down only for long gravity waves and for planetary waves, which are of special interest. Reflection by a wall which does not reach the surface (i.e., by an abrupt escarpment) is also examined.

Consider a stratified fluid of uniform depth H rotating about the vertical axis and bounded at $x = 0$ by a rigid vertical wall. The fluid occupies the half-space $x \geqslant 0$. The normal mode solutions satisfying top and bottom boundary conditions obey (10.41) and (10.42). The horizontal dependence of the wave field is given by (10.39) and (10.40), to which we now add the lateral boundary condition $u = 0$ at $x = 0$, i.e.,

$$U^{(h)}(0,y)\mathfrak{D}(z) = 0. \tag{23.1}$$

This condition may be satisfied by the superposition of an incident and a reflected wave, both of which must be of the same vertical mode if (23.1) is to hold at all values of z. As the phenomena of most interest (such as the reflection of tides and planetary waves) occur when rotation is important, we concentrate our attention on long-period waves, leaving the reflection of high-frequency waves as an exercise for the reader.

The reflection of long gravity waves

In our treatment of long gravity waves in Section 15, we separated the pressure into vertically and horizontally dependent factors:

$$p(x,y,z,t) = p_n(x,y,t)\Pi_n(z), \tag{23.2}$$

where n refers to the vertical mode number. The vertical dependence satisfies (15.25) together with the boundary conditions (15.27) and (15.28). The horizontal dependence, on the other hand, obeys

$$\partial_t \mathcal{L}\left[\nabla_h^2 - \frac{1}{gh_n}\mathcal{L}\right]p_n = 0, \tag{23.3}$$

where $\mathcal{L} = \partial_{tt} + f^2$ and h_n is the equivalent depth. For plane waves of the usual form (cf. 17.2), the dispersion relation is given by (17.3):

$$\omega_n^2 = gh_n(k_1^2 + k_2^2) + f^2. \tag{23.4}$$

Let us consider a plane Poincaré wave propagating towards the wall at $x = 0$ (Fig. 23.1). The pressure field of this wave may be written (dropping the mode number subscript)

$$p_i = A \exp\left[i(-k_1 x + k_2 y - \omega t)\right]\Pi(z), \tag{23.5}$$

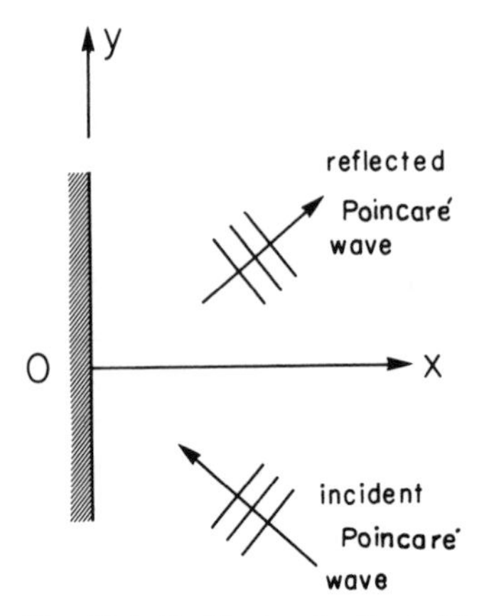

Fig. 23.1. Incident and reflected Poincaré waves.

where ω and k_1 are positive, and the subscript i denotes incident wave properties. The horizontal velocity component normal to the wall associated with (23.5) is obtained from (17.11):

$$u_i = \frac{A(-\omega k_1 + ifk_2)}{\rho_0(\omega^2 - f^2)} \exp\,[i(-k_1x + k_2y - \omega t)]\Pi(z). \tag{23.6}$$

The reflected wave of the same vertical mode is written in the form

$$p_r = RA \exp\,[i(l_1x + l_2y - \sigma t)]\Pi(z), \tag{23.7a}$$

where R is an amplitude reflection coefficient. The reflected wave velocity component normal to the wall is

$$u_r = \frac{RA(\sigma l_1 + ifl_2)}{\rho_0(\sigma^2 - f^2)} \exp\,[i(l_1x + l_2y - \sigma t)]\Pi(z). \tag{23.7b}$$

The sum $u_i + u_r$ vanishes at $x = 0$ for all y, z and t provided $l_2 = k_2$ and $\sigma = \omega$; the dispersion relation (23.4) then implies $l_1 = k_1$. In addition, we must have R of the form

$$R = \frac{\omega k_1 - ifk_2}{\omega k_1 + ifk_2}. \tag{23.8}$$

Note that $|R| = 1$, so that perfect reflection occurs; however, because R is complex the incident Poincaré wave undergoes a phase change upon reflection. The linear combination of $p_i + p_r$, due to the superposition of two plane Poincaré waves (which we shall also call a Poincaré wave), may be written as

$$\begin{aligned} p &= A \exp\,[i(k_2y - \omega t)]\,[\exp\,(-ik_1x) + \exp\,\{i(k_1x + 2\arg B)\}]\,\Pi(z) \\ &= 2A \exp\,[i(k_2y - \omega t + \arg B)]\,\cos\,(k_1x + \arg B)\Pi(z), \end{aligned} \tag{23.9}$$

where $B = \omega k_1 - ifk_2$. The real part of p obtained from (23.9) is sketched in Fig. 23.2 for the case $k_2 < 0$. We note that the wave fronts are perpendicular to the wall and that there are a set of nodal lines parallel to the wall. However, because $\arg B$ changes sign with k_2 there is a significant asymmetry between a "left-bounded" Poincaré wave (one moving with the wall on the left, with $k_2 > 0$) and a "right-bounded" Poincaré wave ($k_2 < 0$). For $f > 0$, the distance from the wall to the first node of the right-bounded wave is less than $\pi/2k_1$, whereas for the left-bounded wave it is greater than $\pi/2k_1$ (see Fig. 23.3). When $f = 0$, $\arg B$ is always zero and this asymmetry disappears. Later, we shall see other instances of this "splitting" of the right- and left-bounded waves by the rotation and recognize it as a common feature of all long gravity waves that travel along a coast. One important consequence of this asymmetry is that the superposition of a left-bounded and right-bounded Poincaré wave cannot produce a standing wave. Thus one cannot solve the problem of reflection of a Poincaré wave by a barrier normal to the wall without the addition of the whole spectrum of Poincaré waves and a Kelvin wave (Section 24). Hence the mathematical analysis associated with long surface waves in a rotating basin with sharp corners is far from trivial! (For example, see Section 28 for the solution of the seemingly innocuous problem of long waves in a semi-infinite channel.)

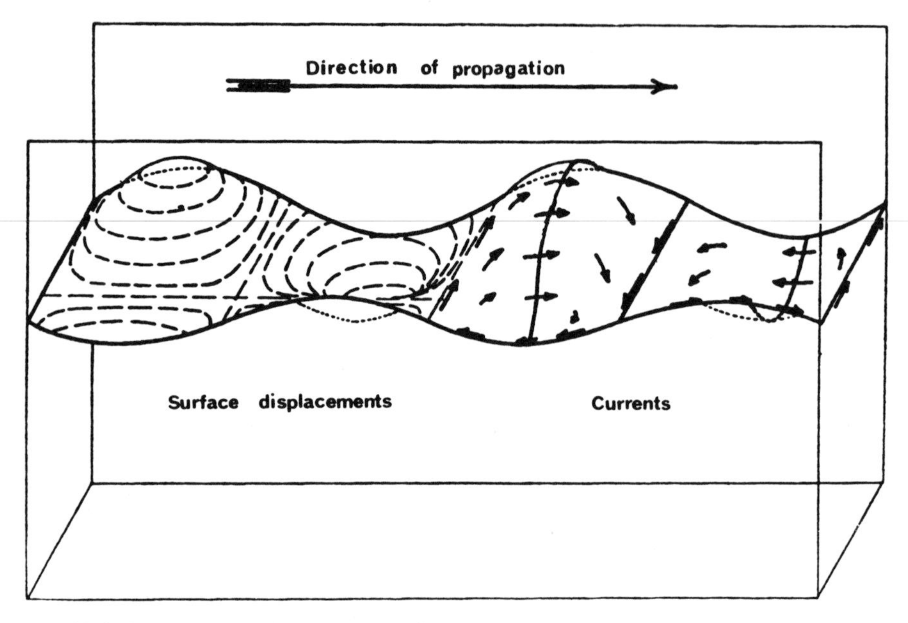

Fig. 23.2. Sketch of a barotropic Poincaré wave in a channel.

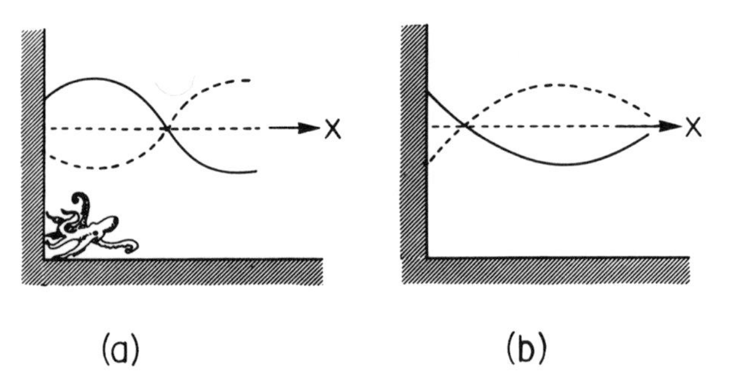

Fig. 23.3. Transverse profiles of surface elevation for barotropic Poincaré waves in the Northern Hemisphere ($f > 0$): (a) a left-bounded wave; (b) a right-bounded wave. The surface position is shown at two times, half a period apart.

The reflection of planetary waves

The reflection of planetary waves from a vertical wall may be treated as above provided a few elementary precautions are taken. First, in view of the highly anisotropic nature of these waves, we cannot expect that reflection about a north–south boundary ($x = 0$) will tell us all there is to know about the reflection properties of planetary waves. Let us then choose a general orientation for the wall, along the line

$$y - \alpha x = 0, \tag{23.10}$$

as shown in Fig. 23.4. Secondly, since the direction of the group velocity is in general

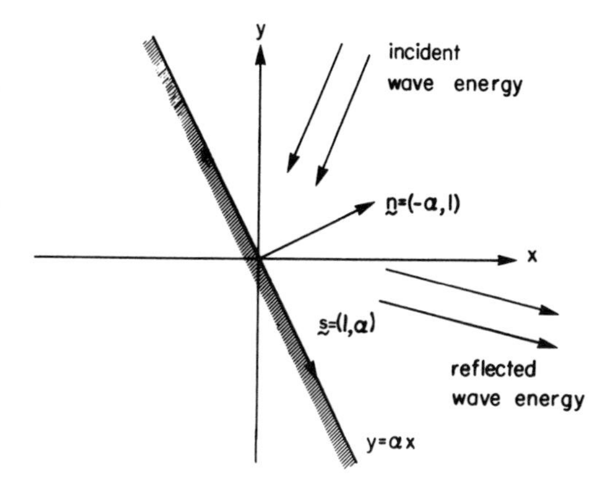

Fig. 23.4. Planetary wave reflection by a vertical wall. The vectors ***n*** and ***s*** are respectively normal and parallel to the wall.

different from that of the phase velocity, an incident planetary wave is defined in terms of the direction of $\boldsymbol{c}_g$ and not that of $\boldsymbol{c}$.

With the pressure field of the wave again separated into its horizontal and vertical dependences as in (23.2), the horizontal dependence satisifies (18.1):

$$\partial_t \left(\nabla_h^2 - \frac{f^2}{gh_n}\right) p_n + \beta p_{nx} = 0. \tag{23.11}$$

The dispersion relation for plane waves is given by

$$\omega_n = -\beta k_1/(k_1^2 + k_2^2 + f^2/gh_n), \tag{23.12}$$

where again, n refers to the vertical mode number and h_n is the equivalent depth. For sufficiently low-frequency waves, the horizontal velocity components are approximately given by the geostrophic relations (18.2):

$$u = -\frac{1}{f\rho_0} p_y, \tag{23.13a}$$

$$v = \frac{1}{f\rho_0} p_x, \tag{23.13b}$$

from which we have now dropped the modal subscript. The pressure is then a stream function and the boundary condition of vanishing normal velocity on the wall may be simply stated as

$$p(x, y, z, t) = 0 \quad \text{on} \quad y = \alpha x. \tag{23.14}$$

The incident wave is written as

$$p_i = A \exp\,[i(k_1 x + k_2 y - \omega t)]\,\Pi(z); \tag{23.15}$$

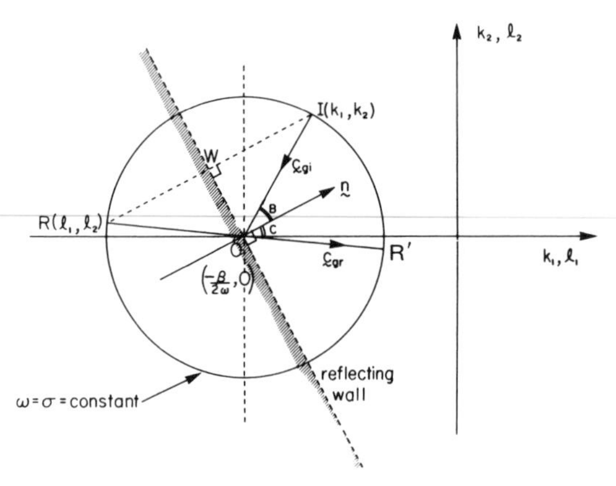

Fig. 23.5. The slowness curve ω = constant showing the incident and reflected waves. The reflecting wall depicted in Fig. 23.4 is superimposed here to help visualize the reflection process. The group velocity vector undergoes an optical reflection (i.e. Snell's law holds) at the wall so that $\sphericalangle B = \sphericalangle C$.

the locus of the wavenumber vector $\boldsymbol{k} = (k_1, k_2)^\dagger$ and of the corresponding incident group velocity $\boldsymbol{c}_{gi}$ are shown in Fig. 23.5. The reflected wave is of a similar form, with identical vertical dependence:

$$p_r = RA \exp [i(l_1 x + l_2 y - \sigma t)] \Pi(z), \tag{23.16}$$

where R is an amplitude reflection coefficient. The dispersion relation (23.12) is satisfied by both incident and reflected waves. The boundary condition (23.14) requires that

$$\exp [i(k_1 + \alpha k_2)x - \omega t] + R \exp [i(l_1 + \alpha l_2)x - \sigma t] = 0, \tag{23.17}$$

which holds for all x and t if

$$\sigma = \omega, \tag{23.18a}$$

$$l_1 + \alpha l_2 = k_1 + \alpha k_2, \tag{23.18b}$$

$$R = -1. \tag{23.18c}$$

From (23.18c), the reflected wave amplitude is equal to that of the incident wave, but a change of phase of 180° takes place upon reflection. From (23.18a), $\boldsymbol{l}$ lies on the same slowness circle as $\boldsymbol{k}$ and (23.18b) implies that (l_1, l_2) in fact correspond to the point R such that RI intersects the wall orthogonally at W (see Fig. 23.5). Thus from Fig. 23.5 it follows that the group velocity of the reflected wave, $\boldsymbol{c}_{gr}$, is directed along ROR' away from the wall such that $\sphericalangle C = \sphericalangle B$. That is to say, the group velocity undergoes Snell's law of reflection at the wall. It is also fairly easy to show (Exercise 23.5) that the energy flux normal to the wall is conserved in magnitude (but reversed in direction) upon reflection.

From (18.10), the group velocity of the incident wave is

---

† In this chapter, all wave propagation is in the horizontal plane. The subscript h on the wavenumber vector will not be used.

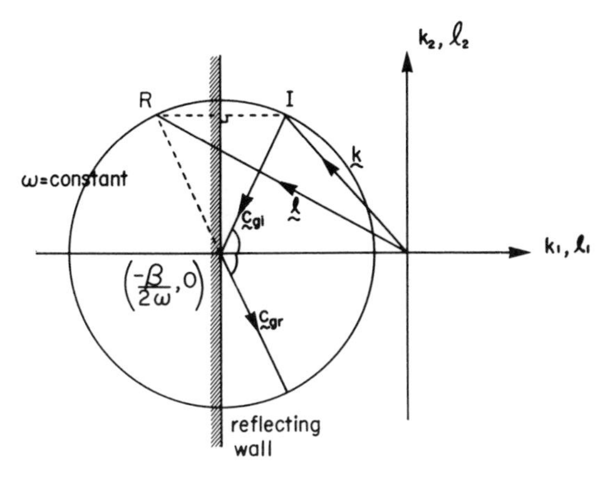

Fig. 23.6. Reflection of planetary waves from a north–south wall.

$$c_{gi} = \beta(k_1^2 - k_2^2 - f^2/gh_n, 2k_1k_2)/(k_1^2 + k_2^2 + f^2/gh_n)^2. \tag{23.19}$$

For nearly nondivergent waves ($k^2 = k_1^2 + k_2^2 \gg f^2/gh_n$), $|c_{g_i}| \propto k^{-2}$ and, similarly, $|c_{g_r}| \propto l^{-2}$. To the same degree of approximation, (18.32) implies that the energy density satisfies

$$\langle E_i(k)\rangle \propto k^2, \quad \langle E_r(k)\rangle \propto l^2, \tag{23.20}$$

so that the product $|c_g|\langle E\rangle$ is independent of the magnitude of the wavenumber. These qualitative results provide a simple physical explanation for the generally observed concentration of small-scale energy on the western side of ocean basins. A relatively long wave whose energy is "quickly" propagating westward will be transformed into a relatively short and slowly propagating wave upon reflection from a north–south boundary (Fig. 23.6). However, since $l^2 > k^2$, (23.20) implies that the reflected wave will be more energetic, i.e. have more energy per unit volume of fluid. Thus the western boundary acts as a *source* for small-scale motions that have a high concentration of energy. In an analogous manner, it is clear that the opposite situation arises for barotropic planetary wave energy incident upon an eastern boundary of an ocean basin. Hence the eastern boundary acts as a *sink* for small-scale motions. This observation was first made by Pedlosky (1965b) and has been developed further by Platzman (1968) and Gates (1970) in their studies of the role of transient motions in the theory of large-scale oceanic circulation. In laboratory demonstrations of planetary waves, this westward intensification of short-wave energy has also been clearly observed (Ibbetson and Phillips, 1967).

In the above discussion no specific reference has been made as to whether the incident wave corresponding to a given frequency and direction of phase propagation is the so-called "short" or "long" wave (cf. 18.11). This is because the reflection properties hold for both cases. However, we do wish to remark that the presence of a free surface enhances the phenomenon of westward intensification of small-scale energy. This is because a very "long" wave, $\boldsymbol{k}^{\mathrm{L}}$, which is absent in a rigid-topped ocean model, can be converted into a very "short" wave $\boldsymbol{l}^{\mathrm{S}}$, upon reflection from a western wall (see Fig. 23.7).

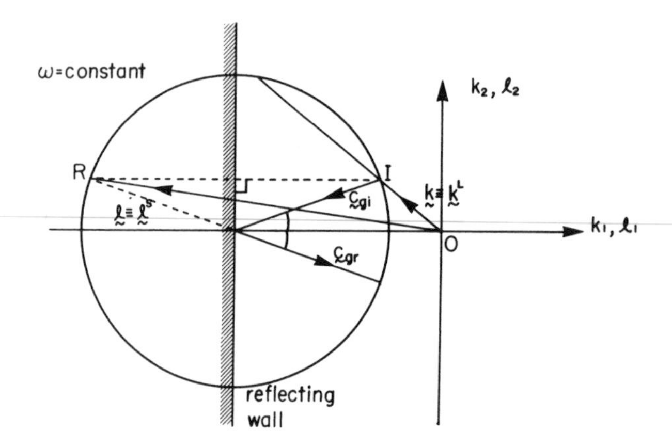

Fig. 23.7. The reflection of a very long planetary wave (wavenumber $\boldsymbol{k}^{\mathrm{L}}$) by a north–south wall to form a short reflected planetary wave.

An alternative formulation which is useful for the study of the reflection of planetary waves involves the introduction of the "carrier wave" transformation (Longuet-Higgins, 1965a)

$$p(x, y, t) = \Psi(x, y)\, e^{-i(\gamma x + \omega t)}, \tag{23.21}$$

where $\gamma = \beta/2\omega$. As shown in Exercise 18.1, Ψ satisfies

$$[\nabla_h^2 + (\gamma^2 - f^2/gh_n)]\Psi = 0, \tag{23.22}$$

which is Helmholtz's equation and hence indicates that wave propagation for Ψ is isotropic. In terms of the (k_1^*, k_2)-plane, where $k_1^* = k_1 + \gamma$, the slowness curve is given by

$$K^2 \equiv k_1^{*2} + k_2^2 = \gamma^2 - f^2/gh_n, \tag{23.23}$$

which, for a given ω, is a circle centered on the origin. Hence, corresponding to a given direction of phase propagation $\boldsymbol{K} = (k_1^*, k_2)$, there is only one wave, and it can be shown that the group velocity for this wave is in a direction opposite to that of the phase velocity, i.e., directed toward the origin of the (k_1^*, k_2)-plane.

Direct observational verification of the reflection properties of planetary waves is still at a rudimentary stage. On the basis of a pattern of depressions of the 20°C isothermal surface measured in the Bermuda–Bahamas area, Beckerle and Delnore (1973) noted horizontal scales of 200–500 km and suggested that this spatial pattern may be the result of the interference between incident and reflected barotropic or first-mode baroclinic planetary waves. Similar eddy patterns have been observed by Kinder et al. (1975) in the Bering Sea.

Reflection by an escarpment

An examination of topographic charts of the world oceans reveals that the large-scale mid-ocean features in any one basin consist of either submarine mountain ranges or escarpments which separate the basin into regions of approximately uniform depth. Thus it is of interest to determine whether these features are effective barriers for the propagation of long-wave energy from one part of the ocean to another. A submarine escarpment is a wall which does not reach the sea surface, and the vertical wall considered earlier may be thought of as a special case of such an escarpment.

The propagation of short gravity waves over a step presents difficulties: the boundary condition of vanishing horizontal velocity is to be applied over only part of the depth. For a treatment of the barotropic case, the reader may refer to Newman (1965a, b). The baroclinic case may be handled by normal modes (Larsen, 1969a) or by the method of characteristics (Sandstrom, 1969). For barotropic long waves ($kH \ll 1$), the details of the vertical structure are ignored and only mass and momentum fluxes remain to be matched across the depth discontinuity: the problem becomes much easier. As an example, the reflection of barotropic planetary waves by a depth discontinuity is analyzed here; the application of the same method to long gravity waves is straightforward and left as an exercise.

Consider the depth distribution

$$H(y) = \begin{cases} H_1, y < 0 \\ H_2, y > 0 \end{cases} \tag{23.24}$$

where H_1 and H_2 are constants. A nondivergent barotropic planetary wave approaches the depth discontinuity from the side $y<0$ and is in part reflected back and in part transmitted beyond $y=0$. The discontinuity has been placed on $y=0$, since this is the simpler case, where the contours of H and f are parallel.

Rewriting the stream function $\psi(x,y)$ as

$$\psi(x,y) = \Psi(y) \exp(ik_1 x), \tag{23.25}$$

the wave equation (20.21) for $H=$ constant becomes the one-dimensional Schrödinger equation:

$$\Psi_{yy} - (V-E)\Psi = 0, \tag{23.26}$$

where $\quad E = -k_1^2 - \beta k_1/\omega > 0 \qquad (23.27a)$

and $\quad V \equiv 0. \qquad (23.27b)$

The solution which represents the sum of an incident, a reflected and a transmitted wave is

$$\Psi(y) = \Psi_0 \begin{cases} \exp(ik_2 y) + R\exp(-ik_2 y), & y<0 \qquad (23.28a) \\ T\exp(ik_2 y) \quad , & y>0, \qquad (23.28b) \end{cases}$$

where R and T are amplitude reflection and transmission coefficients, respectively, and $k_2 = -\sqrt{E} < 0$, so that the group velocity has a northward component (Fig. 23.8) and the wave therefore impinges on the escarpment from $y<0$. At $y=0$, the normal mass flux Hv must be continuous; hence, according to (20.10), ψ_x is continuous at $y=0$, and thus Ψ as well. Hence

$$\mathbf{[}\Psi\mathbf{]} = 0 \quad \text{at} \quad y = 0, \tag{23.29}$$

where the bold square brackets denote the difference in the value of Ψ across $y=0$. To determine the second matching condition, we integrate (20.11) from $y=-\epsilon$ to $y=\epsilon$ and then take the limit as $\epsilon \to 0$. Invoking (23.29), we find the following condition, which ensures pressure continuity across $y=0$:

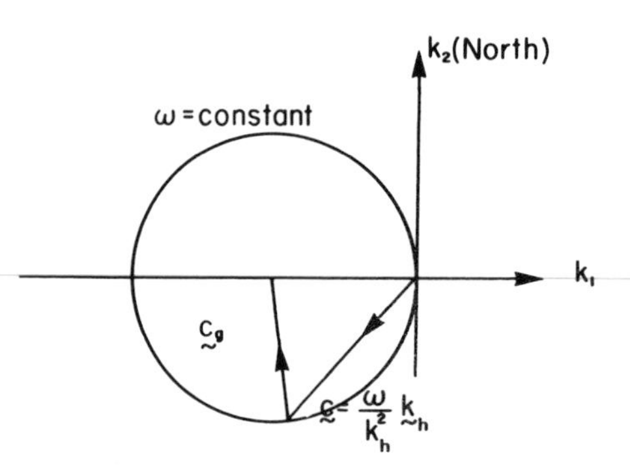

Fig. 23.8. The slowness curve for nondivergent barotropic planetary waves, showing that k_2 must be negative for northward energy propagation towards a zonal escarpment.

$$\left[\frac{1}{H}\frac{d\Psi}{dy}\right] = \frac{k_1 f}{\omega}\Psi(0)\left[\frac{1}{H}\right] \quad \text{at} \quad y = 0. \tag{23.30}$$

Since the left-hand side is proportional to $[u]$, (23.30) implies that there is a discontinuity in the velocity component along the escarpment, i.e., a vortex sheet. Application of (23.29) and (23.30) to (23.28) gives

$$1 + R = T, \tag{23.31a}$$

$$\frac{T}{H_2} - \frac{(1-R)}{H_1} = -\frac{ik_1 f}{k_2\omega} T \frac{H_1 - H_2}{H_1 H_2}, \tag{23.31b}$$

from which we obtain

$$T = 2[\Delta + 1 + i(\Delta - 1)k_1 f/k_2\omega]^{-1}, \tag{23.32}$$

$$R = \frac{[1 - \Delta - i(\Delta - 1)k_1 f/k_2\omega]}{[1 + \Delta + i(\Delta - 1)k_1 f/k_2\omega]}, \tag{23.33}$$

where $\Delta = H_1/H_2$. The transmission and reflection coefficients are complex, indicating that phase changes occur at the discontinuity in depth. Since the group velocity is independent of the depth for nondivergent waves, the fraction of energy transmitted beyond the step is $T_E^0 = TT^*$ and that reflected is $R_E^0 = RR^*$. For a strong discontinuity, with shallow water to the north, $\Delta \gg 1$, and (23.32) implies that $TT^* \ll 1$. In that case, very little energy is transmitted, as one would expect. The energy transmission coefficient T_E^0 is plotted in Fig. 23.9 for small depth changes $[\Delta = 0(1)]$ and will be discussed later.

The problem of wave propagation from a region of constant depth, through a gently sloping transition region, to an area of a different depth may be examined by the methods applied above to a discontinuous escarpment (Rhines, 1969a). Let us consider a nondivergent planetary wave incident from the south ($y < 0$) onto a weak zonal exponential slope which bridges the depths H_1 and $H_2 = H_1 \exp(\delta_s)$ (Fig. 23.10).

For small δ_s, the slope in the variable-depth region $0 < y < a$ is approximately linear and the *fractional change in depth* is

$$[H_1 \exp(\delta_s) - H_1]/H_1 \simeq \delta_s. \tag{23.34}$$

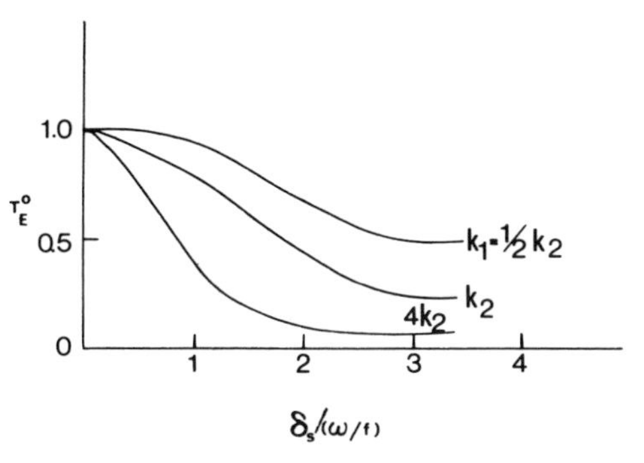

Fig. 23.9. The energy transmission coefficient $T_E^0 = TT^*$, as given by (23.40), againt $\delta_s/(\omega/f)$, with the ratio k_1/k_2 as a parameter. Here, δ_s is the fractional change in depth (cf. 23.34); for small depth changes, $\delta_s = 1 - \Delta$. (From Rhines, 1969a.)

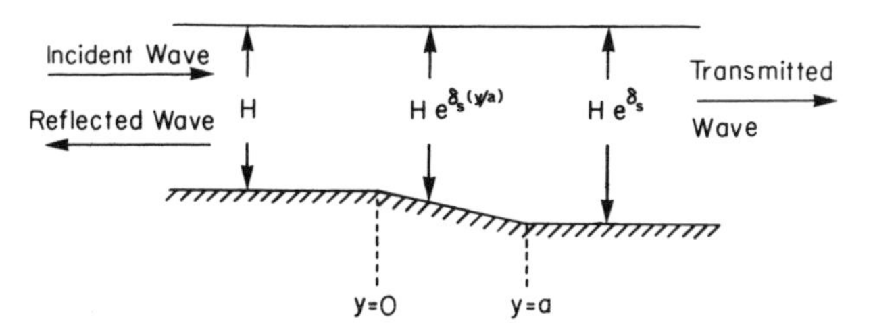

Fig. 23.10. The geometry of a gradual zonal depth transition.

Since $\delta_s > 0$ the topographic variation in $0 < y < a$ opposes the β-effect, and provided the slope $H_1\delta_s/a$ is larger than $0(10^{-3})$ the region $0 < y < a$ will act as a potential barrier [$V > 0$ in (23.26)]. For $H_1 = 5 \cdot 10^3$ m and $a = 0(10^5$ m) say, this will be true provided $\delta_s > 0(2 \cdot 10^{-1})$. Thus it is conceivable that even fairly small depth changes [of 0(1 km)] will be effective in reflecting planetary wave energy.

For a bottom slope which is weak enough to apply Rhines' "first approximation" (see Section 20), we can neglect $(-H_y/H)\psi_y$ in (10.14) and, again using the wave form (23.25), the governing equation for the y-dependence becomes (23.26), where E is still given by (23.27a) but V is now

$$V(y) = \begin{cases} 0, & y < 0 \\ -f\delta_s k_1/\omega a, & 0 < y < a \\ 0, & y > a. \end{cases} \tag{23.35}$$

The solution of (23.26) which represents an incident and reflected wave in $y < 0$ and a transmitted wave in $y > a$ is written as

$$\Psi = \Psi_0 \begin{cases} \exp(ik_2 y) + R\exp(-ik_2 y), & y < 0 \\ A\exp(\mu y) + B\exp(-\mu y), & 0 < y < a \\ T\exp(ik_2 y), & y > a, \end{cases} \tag{23.36}$$

where $\mu = (-f\delta_s k_1/\omega a - E)^{1/2} > 0$ (23.37)

and k_2, R, and T have the same meaning as in the previous problem; A and B are constants.

The mass flux continuity condition (23.29) applies at $y = 0, a$; in the absence of depth discontinuities the pressure continuity condition (23.30) reduces to

$$[\Psi_y] = 0 \quad \text{at} \quad y = 0, a. \tag{23.38}$$

Applying the conditions (23.29) and (23.38) to the solution (23.36) readily gives the following set of four equations for R, A, B and T:

$$1 + R = A + B,$$

$$ik_2(1 - R) = \mu(A - B),$$

$$A\,e^{\mu a} + B\,e^{-\mu a} = T \exp(ik_2 a),$$

$$\mu(A\,e^{\mu a} - B\,e^{-\mu a}) = Tik_2 \exp(ik_2 a).$$

Upon solving the first equation for R, substituting this into the second and then using the second and third equations to find $A = A(T)$ and $B = B(T)$, the fourth equation can be used to find T:

$$T = 2i \exp(-ik_2 a)/[2i \cosh \mu a + (-\mu/k_2 + k_2/\mu) \sinh \mu a].$$

Hence the portion of energy transmitted is

$$T_E = |T|^2 = 4/[4 + (\mu/k_2 + k_2/\mu)^2 \sinh^2 \mu a], \tag{23.39}$$

which is plotted in Fig. 23.11 for the case $k_1 = k_2$. Note that for a fixed frequency and fractional depth change, T_E varies little until a is relatively large. This means that a narrow but smooth rise in the bottom may be approximated by a discontinuous escarpment ($a = 0$) in reflection problems. We also note that for a given δ_s, T_E is small for low

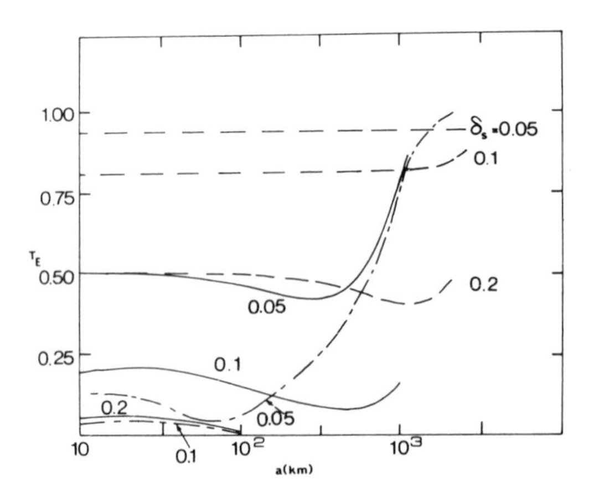

Fig. 23.11. The energy transmission coefficient T_E defined by (23.39), as a function of step width a, for $k_1 = k_2$ and latitude $\phi_0 = 45°$. – – – –, $\omega/f = 0.1$ (wavelength $= 5.7 \cdot 10^6$ m); ——, $\omega/f = 0.025$ (wavelength $= 1.4 \cdot 10^6$ m); –·–·–·–, $\omega/f = 0.01$ (wavelength $= 5.7 \cdot 10^5$ m). (From Rhines, 1969a.)

frequencies, as expected, since the latter corresponds to short wavelengths which are most susceptible to internal reflection by the escarpment. Of particular interest is the limit $\mu a \to 0$, corresponding to a sharp drop from H_1 to $H_2 = H_1 \exp(\delta_s)$ at $y = 0$:

$$\lim_{\mu a \to 0} T_{\rm E} = T_{\rm E}^0 = 1/[1 + \tfrac{1}{4}\delta_s^2(k_1/k_2)^2/(\omega/f)^2]. \qquad (23.40)$$

The reader may verify (Exercise 23.7) that with $H_1/H_2 = \Delta = \exp(-\delta_s) = 1 - \delta_s + 0(\delta_s^2)$, the transmission coefficient for a discontinuous step given by (23.32) gives the same expression for the fraction of energy transmitted as in (23.40).

A plot of $T_{\rm E}^0$ as given by (23.40) versus the ratio $\delta_s/(\omega/f)$ is shown in Fig. 23.9 for different values of k_1/k_2 to illustrate the effect of direction of incidence on transmission over the escarpment. As expected, the transmission is lowest for near-glancing angles of incidence ($\mathbf{c}_g$ nearly east–west). Note that $T_{\rm E}^0$ is even in δ_s and hence is also less than unity for an abrupt (but small) rise to the north ($\delta_s < 0$). However, when $\delta_s < 0$ and $a \neq 0$, i.e., for a gradual rise to the north, the topography enhances the β-effect. In this case the escarpment becomes a potential well ($V < 0$) and $T_{\rm E}$ oscillates between unity and a value less than unity. The value unity corresponds to perfect transmission, which occurs when an integral number of half waves fit over the slope and no reflected wave is needed to satisfy the matching conditions. Such a phenomenon is called "centered" scattering and is known as the Ramschauer-Townsend effect in quantum physics (Schiff, 1955, Chapter 3).

The reflection of planetary waves by ridges may be handled by an extension of the above methods (Exercise 23.8). Rhines (1969a) finds that in general $T_{\rm E}$ is quite large for a narrow ridge. Because of the anisotropy of planetary waves (waves propagating their energy westward are longer than waves of the same frequency propagating their energy eastward) a ridge will be a more effective obstacle to waves generated to its east than to its west.

Exercises Section 23

1. Discuss the reflection of surface and internal gravity waves of arbitrary wavelength by a vertical wall. Use the separation technique introduced in Section 10 and assume that the vertical eigenfunctions and eigenvalues have already been determined.

2. Find real expressions for u and v for a barotropic Poincaré wave satisfying a vanishing normal velocity condition at a vertical wall $x = 0$. Show that the surface elevation η is in phase with v (the longshore velocity component) but in quadrature with u.

3. Determine ω as a function of $k[= (k_1^2 + k_2^2)^{1/2}]$ and the angle of incidence for a Poincaré wave satisfying a wall boundary condition, and hence show that the spectrum is continuous in its dependence on these quantities. For a given k, find the low-frequency cut-off for the existence of such Poincaré waves.

4. Show that for almost nondivergent barotropic planetary waves the energy flux normal to the wall is conserved upon reflection.

5. Discuss the changes (in both direction and magnitude) of the particle velocities of incident and reflected almost nondivergent barotropic planetary waves.

6. Determine the reflection and transmission coefficients for a long gravity wave incident upon an abrupt step in a nonrotating fluid.

7. Calculate from (23.32), with $\Delta = 1 - \delta_s$, the fraction of energy transmitted over a discontinuous step; for small δ_s, the result should agree with (13.40).

8. Derive a transmission coefficient for nondivergent barotropic planetary waves over a ridge of rectangular cross-section.

24. WAVE TRAPPING BY VERTICAL WALLS AND ESCARPMENTS

In the presence of rotation, propagating waves can be trapped in the vicinity of a vertical wall, or of an escarpment, with their amplitude decaying away from the supporting bathymetric feature. The classical Kelvin wave is found along a vertical wall; double Kelvin waves occur at escarpments.

The Kelvin wave

A distorted gravity wave of a special type was discovered by Lord Kelvin (Kelvin, 1879). The Kelvin wave is a solution of the wave propagation problem in the presence of a vertical side boundary, complementary to and simpler than the Poincaré waves discussed in the previous section.

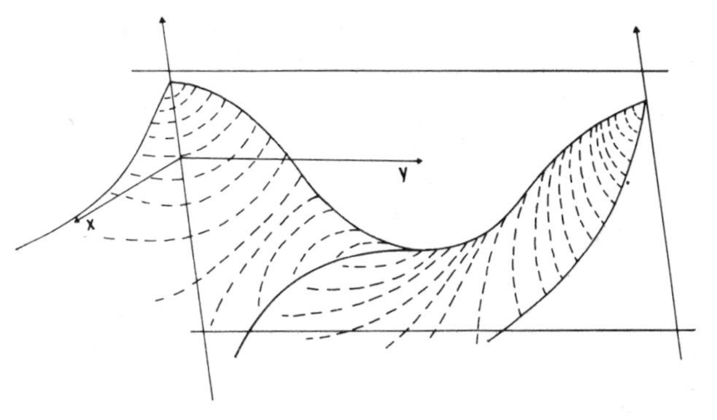

Fig. 24.1. The sea surface topography of the barotropic Kelvin wave (which propagates in the negative y-direction in the Northern Hemisphere).

Consider a stably stratified fluid in an ocean of uniform depth on the f-plane, bounded on one side by a vertical wall at $x = 0$ (Fig. 24.1). The vertical dependence of the wave motion is determined by (10.41) and (10.42), and its horizontal variation by (10.39) and (10.41), together with the lateral boundary condition (23.1). Solutions to this problem (Poincaré waves) have already been found above; the Kelvin wave distinguishes itself from these solutions in that it satisfies (23.1) identically: in a fluid of uniform depth the velocity component normal to the boundary vanishes *everywhere* for a Kelvin wave, and not just at a countable infinity of values of x, as for Poincaré waves. With $U^{(h)} \equiv 0$ in (10.39) and (10.40), the horizontal dependence equations reduce to

$$fV^{(h)} = P_x, \tag{24.1a}$$

$$i\omega V^{(h)} = P_y, \tag{24.1b}$$

$$gh_n V_y^{(h)} = i\omega P, \tag{24.1c}$$

where $\omega > 0$. For waves which travel along the boundary and have an $\exp(ik_2 y)$ dependence, the solution of (24.1) is

$$V^{(h)}(x, y, t) = V_0 \exp[fk_2 x/\omega + i(k_2 y - \omega t)], \tag{24.2a}$$

$$P(x,y,t) = \frac{\omega V_0}{k_2} \exp\,[fk_2x/\omega + i(k_2y - \omega t)], \tag{24.2b}$$

$$\omega^2 = gh_nk_2^2. \tag{24.3}$$

The dispersion relation (24.3) is independent of the Coriolis parameter. Substituting for h_n as given by (24.3) into (10.41), it is also clear that the vertical dependence is unaffected by rotation. Thus, for any vertical mode and at any frequency, the depth dependence and propagation velocity of a Kelvin wave are exactly the same as those of a wave travelling in a nonrotating fluid. In particular, there is no singularity at $\omega = f$ and Kelvin waves can exist for frequencies on both sides of the local inertial frequency as well as at the inertial frequency itself. The Coriolis force is exactly balanced, as seen in (24.1a), by a pressure gradient normal to the boundary. If the solution (24.2) is to remain finite at large values of x, where $x > 0$, fk_2 must be negative. In the Northern Hemisphere, with $f > 0$, this means that $k_2 < 0$ and that a Kelvin wave travels with the boundary on its right when looking in the direction of propagation (Fig. 24.1). The energy flux is entirely along the wall for a Kelvin wave, and this solution may be considered as the limit of grazing angle of incidence of the reflection problem considered in the previous section.

For any one mode, the decay scale may also be written, with the help of (24.3), as $r = \sqrt{gh_n}/f$, which will be recognized as effectively the Rossby radius of deformation, introduced in (18.14) and (18.15). For a deep ocean ($H = 5$ km) at latitude 30°N, $\sqrt{gH}/f \simeq 3000$ km: the barotropic Kelvin wave reaches far from the coast to occupy a substantial fraction of the width of a typical ocean. Continental shelf regions usually extend only about a hundred kilometers from the coast, so that a steeply sloping continental rise is practically indistinguishable from a vertical wall on the scale of the Rossby radius. It is then not surprising to find that much of the energy of long waves travelling along continents is propagated in the form of barotropic Kelvin waves. Munk et al. (1970) estimated that along the coast of California more than two-thirds of the semi-diurnal and about half of the diurnal tidal amplitudes can be accounted for in terms of propagating barotropic Kelvin waves. The remainder consists of contributions from free Poincaré waves and from a forced wave (Section 52).

In shallower waters, the amplitude decay from the coast is more noticeable. Figure 24.2 shows successive positions of a travelling wave crest (co-tidal lines) and amplitudes (co-range lines) for the M_2 (lunar semi-diurnal) tide in the English Channel and the Irish Sea. The mean depth of the western portions of these areas is only about 100 m, giving a barotropic Rossby radius of about 280 km. The tidal amplitudes are clearly larger on the right of the direction of propagation in Fig. 24.2, a consequence of the nature of the Kelvin wave as well as of geometrical amplification by shoaling coastal embayments.

The Rossby radius of deformation for baroclinic modes is much smaller than its barotropic counterpart (cf. 18.14). In a two-layer system, typical of mid-latitude regions in the summer, the interfacial Rossby radius is found from (16.60) as

$$r_{i2} = \left[\frac{g(\rho_2 - \rho_1)}{f^2\rho_2} \cdot \frac{H_1H_2}{H_1 + H_2}\right]^{1/2}, \tag{24.4}$$

where H_1, ρ_1 are the thickness and density of the upper layer and H_2, ρ_2 the same quantities in the lower layer. Values of r_{i2} of the order of 5 km are commonly observed in large lakes (Csanady, 1971b, 1973a; Lee, 1975). Internal Kelvin waves have been

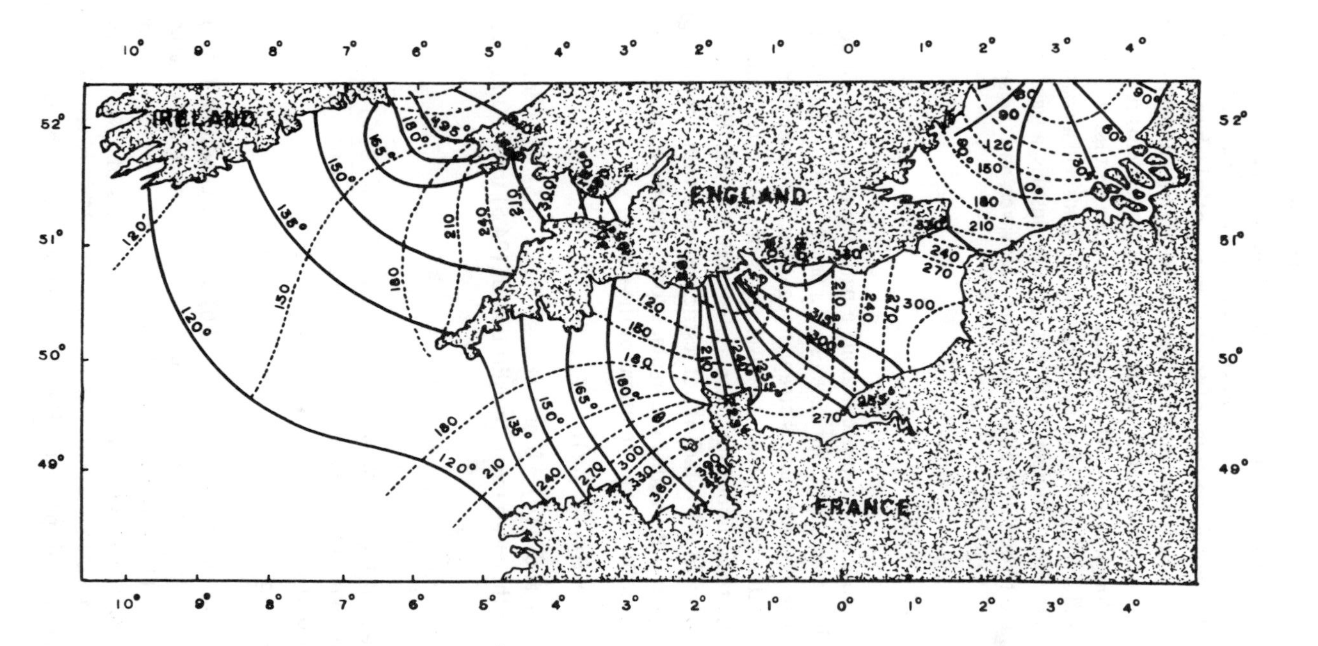

Fig. 24.2. Co-tidal (——, phase in degrees referred to the upper culmination of the moon in Greenwich) and co-range (– – – –, amplitude in cm) lines in the English Channel. (From Defant, 1961, p. 382.)

observed to be trapped in a narrow coastal band in Lake Ontario by Csanady and Scott (1974).

Internal Kelvin waves have also been observed by Forrester (1974) in the St. Lawrence estuary, where they arise upon reflection of a surface tide from an abrupt change in depth.

The role of internal Kelvin waves propagating along a coast on time-dependent upwelling phenomena has been discussed by Gill and Clarke (1974).

The shorter the Rossby radius, the more sensitive Kelvin waves will be to small-scale coastal bathymetric variations in the direction normal to the coast. We shall see in the next section how such depth gradients modify Kelvin waves, and also how they allow the existence of a whole spectrum of trapped waves of which the Kelvin wave, at long wavelengths, is only the lowest mode. For a discussion of secondary effects such as that of the Earth's curvature or of bends in the coastline, the reader is referred to Miles (1972). The influence of longshore coastal irregularities about a rectilinear form has been investigated by Mysak and Tang (1974) using stochastic methods. On a coast such as that of northern Siberia, where there are deep gulfs and large peninsulas, a significant fraction of a Kelvin wave is found to be scattered away as Poincaré waves, and the speed of the Kelvin wave is appreciably decreased in the process.

Trapping by an escarpment

Wave trapping may also occur along a partial vertical wall, or equivalently, an abrupt transition in ocean depth. We called such a depth variation an escarpment in the previous section. Let us consider slopes steep enough that β may be neglected in (20.11). We also restrict our attention, at first, to wavelengths much shorter than the barotropic Rossby radius, so that the upper surface may be taken as rigid. The escarpment is taken along the x-axis $[H = H(y)]$, which is not restrictive, since with $\beta = 0$ the problem has horizontal symmetry and the actual geographical orientation of the escarpment is arbitrary. Let us look for barotropic wave solutions of the form

$$\psi = \Psi(y) \exp [i(k_1 x - \omega t)], \tag{24.5}$$

where $\omega > 0$ and $\Psi(y) \to 0$ as $y \to \pm \infty$. Thus (24.5) represents a trapped wave which travels along the escarpment. Substitution of (24.5) into (20.11) with $f =$ constant gives

$$\left(\frac{1}{H}\Psi_y\right)_y + \left[\frac{H_y}{H^2}\frac{k_1 f}{\omega} - \frac{k_1^2}{H}\right]\Psi = 0. \tag{24.6}$$

Examination of the nature of the coefficient of Ψ in (24.6) shows that the solution will be of a decaying nature on both sides of the escarpment, since $H_y \to 0$ as $y \to \pm \infty$, and oscillatory over the escarpment provided

$$H_y f k_1 / \omega H > k_1^2. \tag{24.7}$$

The behaviour of the solutions of (24.6) and the condition (24.7) under which wave propagation is possible are similar to those of internal gravity waves in a pycnocline, as deduced from (10.48). Provided (24.7) is satisfied, trapped waves may propagate along the escarpment. The solution of (24.6) then consists of a set of even and odd eigenfunctions (Fig. 24.3). In the Northern Hemisphere, (24.7) is satisfied provided $H_y k_1$ is sufficiently large and positive. The phase associated with each of the eigenfunctions

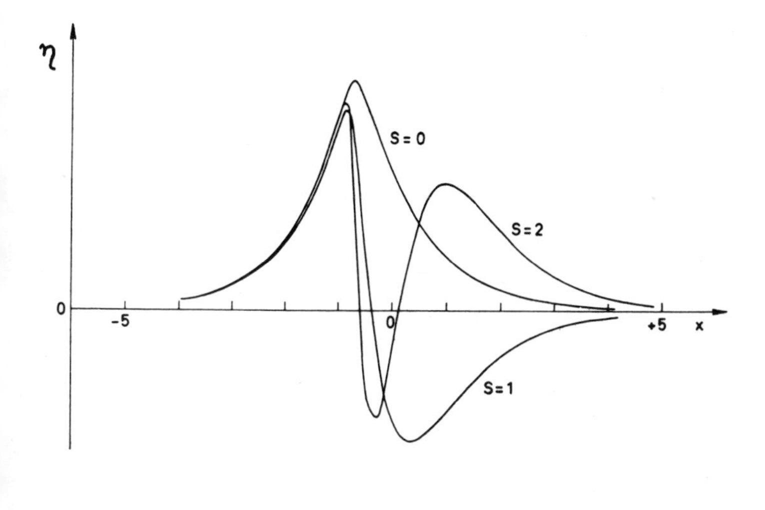

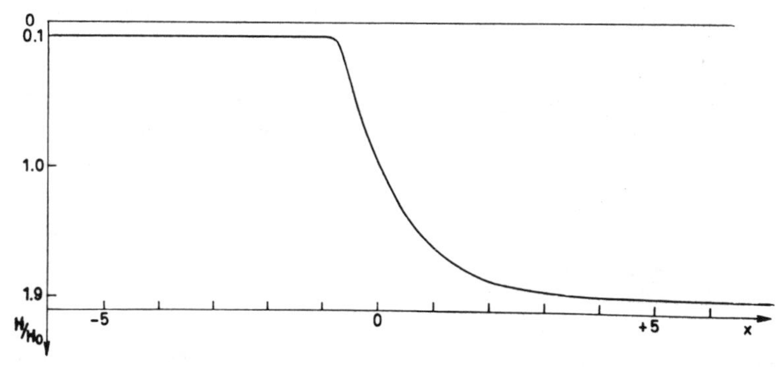

Fig. 24.3. Surface displacements for double Kelvin waves. The depth profile considered is shown in the lower part. The upper panel shows $\eta(x)$ for the first three eigenfunctions (labelled $S = 0, 1, 2$). (From Saint-Guily, 1976.)

propagates with the shallow water on its right (Rhines, 1967, 1969a; Buchwald and Adams, 1968; Saint-Guily, 1976). It may also be shown that these waves are dispersive. For the lowest mode, the group velocity is opposite in direction to the phase velocity. For the higher modes, the phase and group velocities are approximately equal in magnitude and in the same direction for long wavelengths ($|k_1|a \ll 1$, where a is a measure of the width of the escarpment) but are in opposite directions for short wavelengths. As the width of the escarpment tends to zero, all the higher modes reduce to steady currents, leaving the oscillation described below.

For the abrupt single-step escarpment where the depth is given by (23.24), with $H_2 > H_1$, (24.6) reduces to

$$\Psi_{yy} - k_1^2 \Psi = 0 \tag{24.8}$$

on either side of the escarpment. The solution for which $\Psi \to 0$ as $y \to \pm \infty$ is

$$\Psi = \begin{cases} A \exp(|k_1| y) \;, & y < 0 \\ B \exp(-|k_1| y), & y > 0. \end{cases} \tag{24.9}$$

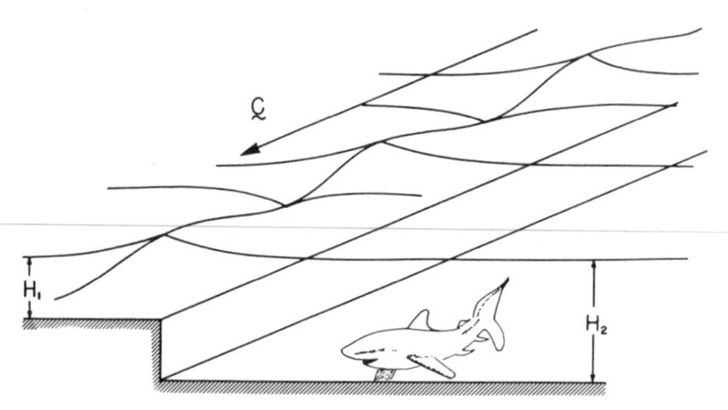

Fig. 24.4. The sea surface topography of a double Kelvin wave travelling along an abrupt escarpment in the Northern Hemisphere.

Application of the continuity conditions (23.29) and (23.30) gives

$$\frac{\omega}{f} = \frac{H_2 - H_1}{H_2 + H_1} \operatorname{sgn} k_1 . \tag{24.10}$$

From (24.10), $|\omega/f| < 1$: the frequency is always subinertial. Note that for both signs of k_1, the group velocity is identically zero, and no energy propagates along the escarpment in this nondivergent model. As remarked above, the phase propagates with the shallow water to the right in the Northern Hemisphere (Fig. 24.4). Because of its double exponential behaviour (cf. 24.9), this wave and, by extension, similar waves trapped on a continuous escarpment, is sometimes called a "double Kelvin wave". We note that the major defining characteristic of ordinary Kelvin waves ($v = 0$) is not found in double Kelvin waves.

When the β-effect is included in the single-step depth profile (23.24), the dispersion relation for trapped waves depends on the orientation of the escarpment. The waves then have nonzero group velocity and can, therefore, transport energy. Although the phase propagation is still in the same direction as before, $\boldsymbol{c}_g$ can be in either direction, depending on the wavelength and the orientation of the escarpment. However, for long wavelengths, $\boldsymbol{c}_g$ is parallel to $\boldsymbol{c}$ for any orientation. According to Rhines (1969a), the β-effect can be neglected in practice since, for typical escarpments, it becomes relevant only for wavelengths in excess of about 6000 km.

The effects of a free surface on trapped waves over the single-step escarpment have been discussed by Longuet-Higgins (1968a). The phase and group velocities are unidirectional at long wavelengths and of opposite directions at short wavelengths. Nondivergent double Kelvin waves over continuous depth profiles have been studied by Longuet-Higgins (1968b), Odulo (1975a, b) and Saint-Guily (1976). Except for the special solution found by Saint-Guily (1976) for a wide continental shelf, these studies have been largely concerned with a qualitative analysis or a numerical integration of the governing equation for the surface amplitude.

Double Kelvin waves have not yet been detected at sea. The expected currents associated with such waves for moderate wind stress forcing near the Mendocino escarpment, for example, are quite small, i.e., of 0(1 cm/s) (Mysak, 1969). However, double Kelvin waves have been successfully generated and studied in the laboratory (Caldwell and

Fig. 24.5. Plan view of current patterns associated with double Kelvin waves in a laboratory set-up. The wavemaker is the structure visible in the lower left corner; the escarpment follows a circle, roughly through the eddies, with deep water at the top of the picture. The white straight marks are radial lines above the tank. The instantaneous streamline pattern which manifests itself as a pair of prominent eddies is due to a double Kelvin wave travelling towards the right (with shallow water on its right). (From Caldwell and Longuet-Higgins, 1971).

Longuet-Higgins, 1971). Figure 24.5 shows the current pattern associated with a lowest-mode long double Kelvin wave in a rotating circular tank of radius 61 cm with a uniformly sloping escarpment. The two large vortices or "eddies", making up one wavelength, are centered over the model escarpment bridging two uniform depths. By forcing the waves at different frequencies, Caldwell and Longuet-Higgins were able to obtain an experimental dispersion curve which agreed very well with the theoretical curve for the case of divergent waves [for the laboratory situation, the divergence parameter $f^2/gHk_1^2 = 0(1)$].

The trapping of waves by ridges, considered as a pair of escarpments of opposite slopes, may be treated by an extension of the methods applied to a single escarpment. We find it convenient to delay the discussion of this topic to Section 26.

Nonlinear Kelvin waves

Kelvin waves which propagate along a sloping oceanic boundary are weakly dispersive (Smith, 1972). The normal dispersion correction to the wave speed $c = \sqrt{gh_n}$, as given by (24.3), occurs as a correction of the order of the divergence parameter f^2l^2/gH, where l is the shelf width. By balancing this phase dispersion against amplitude dispersion, Smith

(1972) obtained a modified KdV equation (cf. 12.88) for nonlinear barotropic Kelvin waves. The influence of nonlinearity is, however, detectable only over extremely long length scales and would manifest itself only in closed basins. A similar analysis has also been presented by Grimshaw (1976a). Both of these authors also extend their analysis to continental shelf waves, to be discussed in the next section. A study of nonlinear internal Kelvin waves by the method of characteristics has been presented by Bennett (1973).

Exercise Section 24

1. Suppose that over the abrupt escarpment model (23.24) there is a two-layer fluid, with the interface occurring at a mean depth $z = -d$, where $d < H_1 < H_2$. For $\beta = 0$ and the rigid-lid approximation, find the dispersion relation for a *baroclinic* double Kelvin wave. Show that for long waves (wavelengths much larger than the internal Rossby radius r_{i2}), the relation reduces to (24.10) and that the currents are barotropic. For short waves, show that the currents are bottom-trapped, like the bottom-trapped buoyancy oscillations discussed in Section 20 (Rhines, 1977).

25. COASTAL TRAPPED WAVES

In the previous section we encountered our first example of coastal trapped waves: surface (or barotropic) and internal (or baroclinic) Kelvin waves. In this section we introduce topographic variations normal to the coast in order to model the continental shelf/slope region and ask again, what kind of trapped waves can exist? However, a fundamental difficulty now arises. For a depth-varying, continuously stratified fluid that is rotating about the vertical axis, an exact separation of the horizontal and vertical dependences is not possible. This is because the bottom boundary condition $\boldsymbol{u} \cdot \boldsymbol{n} = 0$ involves both the vertical and horizontal velocity components. Therefore, in this section we shall deal mostly with (barotropic) coastal trapped waves in a homogeneous fluid. At the end of this section we shall touch on the question of coastal trapped waves in a stratified fluid.

It is apparent that the difficulties concerning vertical modal separation that were encountered in Section 20 in connection with topographic planetary waves also occur here. Therefore, our emphasis on barotropic coastal trapped waves parallels the earlier emphasis on barotropic topographic planetary waves. However, in contrast to the discussion in Section 20, we shall be concerned here with both first-class (gravity) and second-class (planetary) wave motions.

Even in a homogeneous fluid, there are only a few known solutions for waves of arbitrary length. Stokes (1846) discovered the classical solution for the fundamental-mode edge wave on a uniformly sloping beach. It took more than a century before Stokes' work was extended by Eckart (1951) and Ursell (1952) to include the whole spectrum of possible modes. Ursell's solution was generalized to include rotation by Saint-Guily (1968). The influence of both the vertical and horizontal components of the Coriolis force on the fundamental-mode edge wave was investigated by Johns (1965a). Nevertheless, all these solutions are for one very restricted type of topography: a linearly varying depth profile where the depth tends to infinity far from the coast. To handle more general profiles, the bulk of the theory of trapped coastal waves has centered on the analysis of the so-called shallow-water equations (the barotropic long wave equations in our previous terminology). The use of such equations is quite reasonable since the observed wavelengths of trapped waves are long compared to the depth of the water, at least in the shallow areas where most of the amplitude is concentrated. We shall thus follow suit in this section and base our discussion on the unforced barotropic long wave equations derived in Section 15.

The shallow-water equations for a rotating fluid

From (15.50) and (15.51), the linearized shallow-water equations are:

$$u_t - fv + g\eta_x = 0, \tag{25.1a}$$

$$v_t + fu + g\eta_y = 0, \tag{25.1b}$$

$$(Hu)_x + (Hv)_y + \eta_t = 0. \tag{25.2}$$

In (25.1) we treat f as a constant on the assumption that the slope of the bottom topography is sufficiently large so as to dominate the β-effect, i.e., $|\nabla H| > H/R_0$ [see (20.16)]. For shelf depths $H = 0(2 \cdot 10^2\,\text{m})$ and $R_0 = R \tan \phi_0 = 0(6 \cdot 10^6\,\text{m})$ (R = radius of Earth), $H/R_0 = 0(10^{-4})$. Since typically $|\nabla H| = 0(5 \cdot 10^{-3})$ for the shelf region, the assumption of constant f is very reasonable. At the same time, we assume that the bottom slope is sufficiently small [e.g., less than $0(10^{-1})$] in order that the vertical motions may be neglected.

From (25.1) we obtain

$$\mathcal{L}u = -g(\eta_{xt} + f\eta_y), \tag{25.3a}$$

$$\mathcal{L}v = -g(\eta_{yt} - f\eta_x), \tag{25.3b}$$

where $\mathcal{L} = \partial_{tt} + f^2$.

Operating on (25.2) by $\mathcal{L}$ and then using (25.3), we arrive at an equation for η alone:

$$\nabla_h \cdot (H\nabla_h \eta_t) + f(H_x\eta_y - H_y\eta_x) - g^{-1}\mathcal{L}\eta_t = 0, \tag{25.4}$$

which also follows from (20.13a) when $\beta = 0$.

We assume that the coastal topography varies only in the direction normal to the coast, so that the equation of the bottom is of the form

$$z = -H(x), \quad 0 \leqslant x < \infty, \tag{25.5}$$

where $x = 0$ corresponds to the mean coastline. In practice there are also longshore variations in the coastal topography, but introducing such complications at this stage would hinder rather than help an understanding of coastal trapped waves. The study of the effects of such variations, however, is a very active area of research today. We shall return to this topic later in the book (e.g., Section 34). With this assumption we can now seek travelling wave solutions of the form

$$\eta = F(x)\, e^{i(ky - \omega t)}, \tag{25.6}$$

where k, the longshore wavenumber, is taken to be positive; ω is allowed to have either sign. Since we are dealing with one-dimensional wave propagation throughout this section, we have, for notational convenience, simply used k (rather than k_2) for the wavenumber. This is in keeping with the convention used in Sections 11–13, where only propagation of gravity waves parallel to the x-axis was considered. Substitution of (25.5) and (25.6) into (25.4) gives the amplitude equation

$$(HF')' + \left(\frac{\omega^2 - f^2}{g} - k^2H - \frac{fk}{\omega}H'\right)F(x) = 0, \quad 0 \leqslant x < \infty. \tag{25.7}$$

The boundary conditions we apply to F are

$$F \to 0 \quad \text{as} \quad x \to \infty, \tag{25.8}$$

$$\text{and} \quad H(-\omega F' + fkF) = 0 \quad \text{at} \quad x = 0. \tag{25.9}$$

The first condition ensures that for all depth profiles $H(x)$ which increase less rapidly than exponentially as $x \to \infty$, the energy is trapped against the coast; the second ensures that there is no mass flux through the coastal boundary [see (25.3a)]. If $H(0) \neq 0$, then (25.9) reduces to

$$fkF(0) - \omega F'(0) = 0, \tag{25.10}$$

whereas if $H(0) = 0$ (as in the case of a uniformly sloping beach), (25.9) will hold provided F is differentiable at $x = 0$:

$$|F'(0)| < M \text{ (a constant).} \tag{25.11}$$

Since differentiability implies continuity, (25.11) guarantees that the sea surface as well as its slope (and thus also the velocity components u, v) are well behaved at the coastline.

We note that since (25.4) is third order in the time derivative (through $\mathcal{L}$), or equivalently, the coefficient of F in (25.7) is cubic in ω, it is conceivable that for a given $k > 0$ there exist three distinct roots for ω, corresponding to three different waves. What kinds of solutions can we expect to find for various classes of topographies $H(x)$?

Historical review

A partial answer to the above question was first given by Stokes (1846) who considered the case of a sloping beach of constant inclination β_b:

$$H(x) = (\tan\beta_b)x = \alpha x, \quad 0 \leqslant x < \infty. \tag{25.12}$$

Using the usual potential theory for surface gravity waves described briefly in Section 11, Stokes found the solution

$$F(x) = F_0 \exp[(-k\cos\beta_b)x] \tag{25.13}$$

with the corresponding dispersion relation

$$\omega^2 = gk\sin\beta_b. \tag{25.14}$$

This wave, the classical Stokes edge wave, can travel in either direction along the coast with speed $\omega/k = (gk^{-1}\sin\beta_b)^{1/2}$, which for a small angle of inclination ($\beta_b \ll 1$) is much slower than the deep-water gravity wave speed, $(g/k)^{1/2}$. Further, in view of (25.13) the energy of the Stokes edge wave is effectively confined to within one wavelength of the shoreline. It also follows that away from the coast, this wave is circularly polarized.

It was not until over a hundred years later that Eckart (1951) established, on the basis of long wave theory, that the Stokes edge wave is the gravest of an infinity of modes whose energy is trapped against the coast. Eckart obtained the following dispersion relation for the nth mode:

$$\omega_n^2 = gk(2n+1)\tan\beta_b, \quad n = 0, 1, 2, \ldots, \tag{25.15}$$

which for $n = 0$, agrees with (25.14) for very gentle bottom slopes ($\beta_b \ll 1$). For the nth mode wave, there are n nodal lines parallel to the coast; this oscillatory behaviour is superimposed on the same exponential decay (25.13) (with $\cos\beta_b$ replaced by unity) as for the fundamental mode ($n = 0$) found by Stokes.

Ursell (1952) derived and experimentally verified the exact theory for these waves without recourse to the long wave approximations; in particular he found that

$$\omega_n^2 = gk\sin(2n+1)\beta_b, \quad n = 0, 1, 2, \ldots, \tag{25.16}$$

$$\text{provided} \quad (2n+1)\beta_b \leqslant \pi/2. \tag{25.17}$$

Thus for a given beach inclination β_b, (25.17), which ensures a finite velocity everywhere, implies that only a finite number of edge wave modes are physically permissible, a restriction that is absent in the shallow-water theory. However, we note that for small slopes and low mode numbers such that $(2n+1)\beta_b \ll 1$, (25.16) reduces approximately to the shallow-water result (25.15). The restriction in the number of possible modes is of little practical relevance: for $\beta_b = 10^{-2}$, solutions are possible up to $n = 78$. Nobody has ever observed that many modes!

Reid (1958) studied the effect of the Coriolis force on shallow-water edge waves and showed that rotation gives rise to slightly different phase speeds for right- and left-bounded edge waves. (In the direction of phase propagation, a right-bounded wave has the coast on its right.) For a given mode n he established the existence of a "quasi-geostrophic" wave which for a given hemisphere can propagate in one direction only. This wave corresponds to the third ω-root mentioned in the paragraph following (25.11) and represents a low-frequency wave of the "second class". The edge waves, on the other hand, represent waves of the "first class" (see Chapter 3). It is Reid's unified solution for a sloping beach that we present below.

Wave trapping and ray theory

The question naturally arises as to whether first- and second-class trapped waves also exist on a more realistic shelf/slope topography, where $H(x) \to H_0$, a constant, as $x \to \infty$. With the Coriolis force present we can always expect rotational trapping and hence second-class trapped waves. A detailed discussion of the ray theory of rotational trapping in the presence of topography is given by Smith (1971). The physical mechanism for the trapping of first-class waves, on the other hand, is based on refraction and is akin to the phenomenon of total internal reflection in optics. The reader will recall that internal reflection also occurred for internal waves propagating through a region of variable $N(z)$ (see Section 9). According to ray theory (Section 6), far from the coast in a region of gentle slopes a long gravity wave propagates with a speed proportional to the square root of the local depth. Thus for a topography $H = H(x)$ which slowly increases with x, the rays of such a wave are always gradually refracted towards the coast (Fig. 25.1). At the coast, the wave is reflected and the refraction process starts all over again. For a given coastal angle θ_0 (see Fig. 25.1), there is a *caustic line* parallel to the coast which consists of the envelope of rays which at that distance from the coast are parallel to the coastline. The x-dependence of the oscillations is in the form of a standing wave between the shore and the caustic and exponentially decaying outwards from the caustic. The over-all pattern of refracted and reflected long waves on a *uniformly sloping* beach is what constitutes a classical edge wave. Far from the coast, where the depth is eventually constant in practice, this trapping mechanism fails; we shall discuss this situation in more detail later. It is apparent that this type of wave trapping can also occur on sloping topographies adjacent to nonrectilinear boundaries such as islands and seamounts. This topic will be explored in depth in the next section. For a careful mathematical account of ray theory as applied to topographic trapping, we refer the reader to Hoogstraten (1972).

Trapped waves on a sloping beach in the presence of rotation

The linear beach profile was the first one investigated and the wave solutions trapped against it illustrate most of the fundamental properties of coastal trapped waves. For the semi-finite sloping beach (25.12), equation (25.7) for the amplitude $F(x)$ takes the form

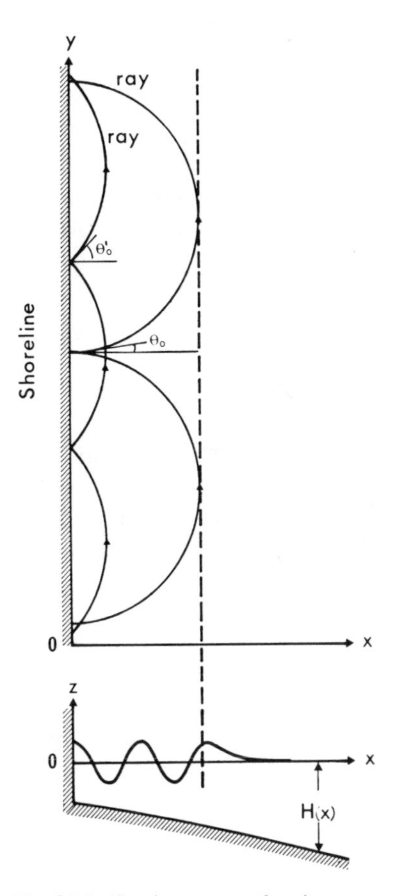

Fig. 25.1. Gravity wave refraction on a sloping bottom. Top: plan view showing two rays, making angles of incidence θ_0 and θ_0' with the coast. The position of the caustic line for one of the rays is shown by a dashed line. Bottom: side view, showing the oscillatory behaviour inshore of the caustic and the exponential decay offshore.

$$(xF')' + (\mu k - k^2 x)F(x) = 0, \quad 0 \leqslant x < \infty, \tag{25.18}$$

where $$\mu = \frac{\omega^2 - f^2}{\alpha g k} - \frac{f}{\omega}. \tag{25.19}$$

The appropriate boundary conditions for this case are (25.8) and (25.11). For a given $k > 0$, (25.18) and the boundary conditions represent an eigenvalue problem for the eigenvalue μ and corresponding eigenfunction $F \equiv F_\mu(x)$. Once these are determined, (25.19) is used to obtain the dispersion relation $\omega = \omega_\mu(k)$ corresponding to each μ.

One standard way of solving this problem is by the method of power series; while relatively straightforward, it is a tedious calculation because of the presence of the regular singular point at $x = 0$. The solution of (25.18), together with the coastal boundary condition (25.11), is proportional to a Laguerre function. However, in order to satisfy the boundary condition (25.8), μ must be a positive odd integer, in which case the solution reduces to an exponential function multiplied by a Laguerre polynomial. Thus one of the remarkable mathematical features of edge wave theory is that even though the domain is semi-infinite, the spectrum (i.e., μ) is discrete, a feature which was enlarged upon by Ursell (1952).

An alternative and elegant method of solving (25.18) and the boundary conditions (25.8) and (25.11) is by means of the Laplace transform. Since the coefficients of (25.18) are linear in x, this equation transforms into a first-order differential equation for the Laplace transform of $F(x)$ which can be readily integrated. The inverse Laplace transform is then evaluated by elementary contour integration techniques to give the eigenfunctions $F_\mu(x)$ and the eigenvalues μ. The details of this calculation are left as an exercise (25.2). The solution for $F(x)$ is given by the integral

$$F(x) = \frac{1}{2\pi i} \int_{-i\infty+\gamma}^{i\infty+\gamma} e^{skx} \frac{(s-1)^{(\mu-1)/2}}{(s+1)^{(\mu+1)/2}} \, ds, \tag{25.20}$$

where $\gamma > 1$ so that the path of integration passes to the right of the singularities at $s = \pm 1$. For arbitrary μ, these singularities are branch points and it can be shown that as $x \to \infty$, the singularity at $s = 1$ gives contributions to $F(x)$ which are proportional to e^{kx}. This clearly violates the boundary condition (25.8). However, we can circumvent this difficulty by choosing

$$(\mu - 1)/2 = n = 0, 1, 2, \ldots,$$

$$\text{or equivalently,} \quad \mu = 2n + 1 = 1, 3, 5, \ldots, \tag{25.21}$$

which eliminates the singularity at $s = 1$. Then for the choice (25.21), $(\mu + 1)/2 = n + 1$ and the singularity at $s = -1$ in the integrand of (25.20) is a pole of order $n + 1$. Thus without loss of generality we can take $\gamma = 0$ and

$$F(x) \equiv F_n(x) = \frac{1}{2\pi i} \int_{-i\infty}^{i\infty} e^{skx} \frac{(s-1)^n}{(s+1)^{n+1}} \, ds. \tag{25.22}$$

To evaluate the integral in (25.22) for $x > 0$, consider the closed contour $C = C_i + C_R$ shown in Fig. 25.2. Upon application of Cauchy's integral formula, we obtain

$$\frac{1}{2\pi i} \int_C \frac{e^{skx}(s-1)^n}{(s+1)^{n+1}} \, ds = \frac{1}{n!} \left\{ \frac{d^n}{ds^n} e^{skx}(s-1)^n \right\}_{s=-1}. \tag{25.23}$$

However, as $R \to \infty$ it is easy to show that $\int_{C_R} \to 0$ and hence $\int_C \to \int_{-i\infty}^{i\infty}$ as $R \to \infty$. Thus combining (25.22) and (25.23) in the limit $R \to \infty$ we finally have

$$F_n(x) = \frac{1}{n!} \left\{ \frac{d^n}{ds^n} e^{skx}(s-1)^n \right\}_{s=-1}, \quad n = 0, 1, 2, \ldots. \tag{25.24}$$

Simplifying (25.24) gives:

$$F_0(x) = e^{-kx},$$

$$F_1(x) = e^{-kx}(1 - 2kx),$$

$$F_2(x) = e^{-kx}(1 - 4kx + 2k^2x_2^2),$$

$$\vdots$$

$$F_n(x) = e^{-kx} L_n(2kx), \tag{25.25}$$

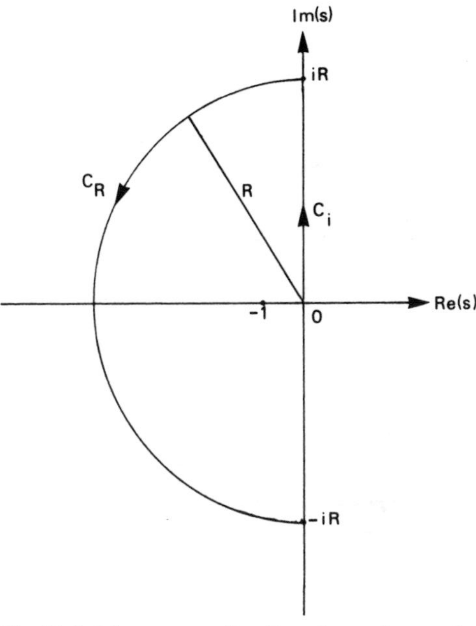

Fig. 25.2. The contour $C = C_i + C_R$ used to evaluate the inverse Laplace transform integral (25.20).

where $L_n(z)$ is the nth-degree Laguerre polynomial in powers of z. Hence the surface elevation $\eta_n(x, y, t)$ corresponding to the nth mode has n nodal lines parallel to the coast and a simultaneous decaying behaviour with the e-folding distance k^{-1}.

Combining (25.19) and (25.21) we obtain the following implicit dispersion relation for each mode:

$$\omega_n^3 - [f^2 + (2n+1)g\alpha k]\omega_n - fg\alpha k = 0, \quad n = 0, 1, 2, \ldots. \tag{25.26}$$

Thus for any $k(> 0)$ and mode n there are three roots for ω_n implied by (25.26); we shall denote these frequency functions or dispersion relations by $\omega_{jn}(k), j = 1, 2, 3$. The reader may verify (Exercise 25.3) that the roots $\omega_{jn}(k)$ are real and have the following properties:

$$\sum_{j=1}^{3} \omega_{jn} = 0, \tag{25.27a}$$

$$\prod_{j=1}^{3} \omega_{jn} = fg\alpha k. \tag{25.27b}$$

The properties (25.27) imply that when $f > 0$, two of the roots are negative and the third is positive. Thus in the Northern Hemisphere the positive root, which we shall denote by $j = 1$, is a left-bounded wave, moving in the positive y-direction, and the two negative roots, which we shall denote by $j = 2, 3$ are right-bounded waves [see (25.6)]. Further, when $f = 0$, (25.26) reduces to

$$[\omega^2 - (2n+1)g\alpha k]\omega_n = 0,$$

which we rewrite as

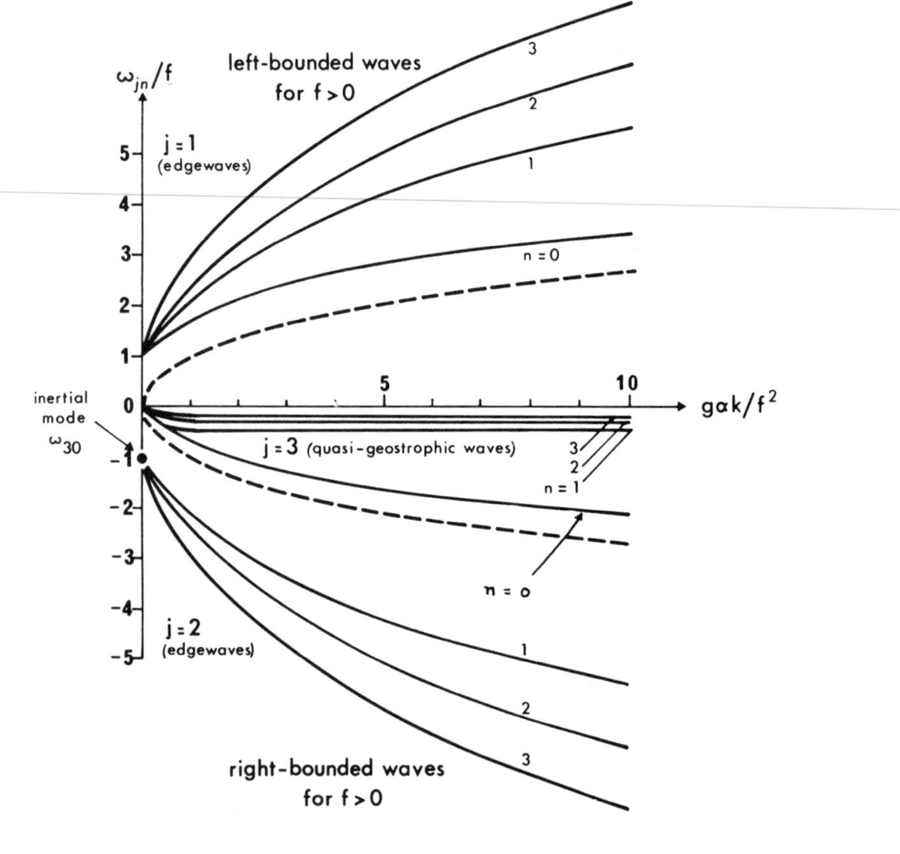

Fig. 25.3. The dispersion relations for edge and quasi-geostrophic waves on a semi-infinite sloping beach of slope α. The dashed curve (a parabola) corresponds to the lowest mode ($n = 0$) Stokes' edge wave solution in the absence of rotation ($\omega_{j0}^2 = g\alpha k, j = 1, 2$), and is shown for comparison. The dispersion curves are labelled by mode number n. Note the existence of the ω_{30} inertial wave solution at $k = 0, \omega = -f$. (Adapted from Reid, 1958.)

$$\omega_{1n} = [(2n+1)g\alpha k]^{1/2},$$

$$\omega_{2n} = -[(2n+1)g\alpha k]^{1/2},$$

in agreement with Eckart's solution, (25.15), and $\omega_{3n} = 0$, corresponding to steady currents. Thus the roots ω_{jn}, $j = 1, 2$, are identified with the first-class edge wave modes, as modified by rotation, whereas the root ω_{3n} is recognized as a second-class wave mode trapped by rotation. The three frequency functions implied by (25.26) for $n = 0, 1, 2, 3$ are plotted in Fig. 25.3. For the gravest mode ($n = 0$), (25.26) can be factored to give

$$[\omega_{10}/f - (1/2 + a)][\omega_{20}/f - (1/2 - a)][\omega_{30}/f + 1] = 0, \tag{25.28}$$

where $a = (1/4 + g\alpha k/f^2)^{1/2}$.

The edge wave roots exhibit the phenomenon of rotational "splitting" of the frequencies: the left-bounded wave ($j = 1$) moves faster than the right-bounded wave ($j = 2$) by an amount f/k; however, the group velocity, though modified by rotation, has the same magnitude for each direction. As $k \to 0$, $\omega_{10} \to f$, whereas $\omega_{20} \to 0$. The third root of (25.28), ω_{30}, corresponds to an inertial oscillation of infinite wavelength ($k = 0$), which may be obtained from an analysis of the primitive equations (25.1) and (25.2) for u, v, and η. For the higher mode numbers ($n \geqslant 1$), the edge wave roots are still slightly asymmetric about $\omega = 0$; however, they all have the property that $\omega_{1n} \to f$ and $\omega_{2n} \to -f$ as $k \to 0$. The second-class waves, on the other hand, have the property that $\omega_{3n} \to 0$ as $k \to 0$ and that $|\omega_{3n}/f| \ll 1$, for all $k > 0$; hence Reid (1958) introduced the term "quasi-geostrophic" to describe these low-frequency waves. They are effectively topographic planetary waves of the type discussed in Section 20 that are trapped against the coast. Consequently, they have a relatively large vorticity compared to that of the rotationally modified edge waves. Also, in contrast with edge waves, for a fixed k the speed of phase propagation of the quasi-geostrophic waves decreases with increasing mode number n (see Fig. 25.3).

Observations of edge waves

For a long time edge waves were regarded as a mere curiosity of hydrodynamics. Accordingly, in his discussion of the gravest-mode solution of Stokes, Lamb (1945, p. 447) states that "it does not appear that the type of motion here referred to is very important." However, there is now considerable evidence which indicates that edge waves are common in occurrence and of practical importance. We shall defer a discussion on the observations of quasi-geostrophic waves until later and consider first the evidence for the existence of first-class edge waves.

Munk et al. (1956) examined the sea-level records taken from the eastern U.S. coast during the passage of several hurricanes and storms during 1954 and found that these storms excited gravest-mode ($n = 0$) edge waves, with typical amplitudes of one meter, periods of around six hours and wavelengths of several hundred kilometers. Detailed initial value calculations for this type of edge wave generation have been done by Greenspan (1956) for the case $f = 0$, and by Kajiura (1958) for $f \neq 0$. Observations from the California coast show a different situation: a continuum of edge wave noise, in the period range 10–30 minutes, always seems to be present. Munk et al. (1956) suggested that these waves may be generated by atmospheric internal gravity waves travelling just above the sea surface. Very high-frequency edge waves have also been observed by Huntley and Bowen (1973) off the South Devon coast (see Fig. 25.4). It is also now well established that earthquake-generated long waves (tsunamis) incident from the open ocean upon the continental shelf/slope region can also excite edge waves (see, for example, Munk et al., 1956; Aida, 1967; Fuller and Mysak, 1977). Their presence has also been indirectly detected through their role in exciting seiche action in coastal bays (Lemon, 1975). Laboratory experiments on edge waves have been reported by Ursell (1952) and Galvin (1965).

Finally, edge waves have been shown to be of fundamental importance in the dynamics and the sedimentology of the near-shore zone through their interaction with ocean swell and surf to produce rip current patterns. Rip currents are known and feared by many ocean-beach bathers; they appear in seaside folklore as the dreaded "undertow". While it has long been obvious that the rip currents must form the return flow for water thrown ashore by breaking waves, a satistactory explanation of their dynamics and

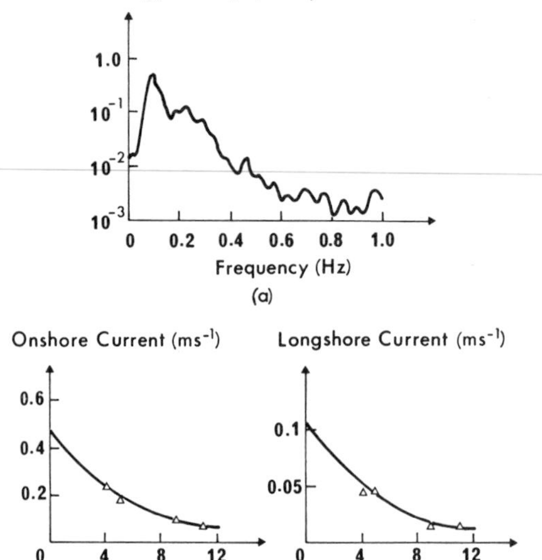

Fig. 25.4. Edge wave measurements taken from Slapton Beach, South Devon, England, at an offshore distance of 9.0 m, in August, 1972. (a) A typical onshore current spectrum. The broad peak at 0.2 Hz corresponds to the observed breaker frequency in the surf zone; the subsequent decay at higher frequencies is typical of wind wave spectra. The additional sharp peak at half the breaker frequency (0.1 Hz) is suggestive of a subharmonic edge wave generated by the breakers. (b) Measurements (triangles) of onshore and longshore current amplitudes at 0.1 Hz, which exhibit the pure decaying behaviour of a gravest-mode edge wave. The solid lines show the theoretical decay for the gravest-mode edge wave with a frequency of 0.1 Hz and a corresponding wavelength of 32 m, as determined by the dispersion relation for edge waves on the profile (25.29) with $H_0 = 7.05$ m and $a = 3.4 \cdot 10^{-2}\,\mathrm{m}^{-1}$. (From Huntley and Bowen, 1973.)

of their spacing is quite recent and is to be found in an interaction process between edge waves and the incoming swell, as described by Bowen (1969b) and Bowen and Inman (1969).

An explanation of rip current formation has also led to a better understanding of the formation of the striking cusp patterns seen along many ocean beaches (Fig. 25.5) and of their relation to edge waves. Equally impressive crescentic bars (Fig. 25.6) occurring offshore between headlands have also been related to edge waves (Bowen and Inman, 1971).

However, the theory which describes the basic features of the interaction between the incoming swell and edge waves which lead to the development of rip currents, cusps and crescentic bars is based on wave–wave interaction theory, which will be discussed in Section 37.

Trapped waves on other topographies

From the above mention of tsunami-generated edge waves it is clear that the semi-infinite sloping beach model is not adequate to describe phenomena that involve trapped

Fig. 25.5. Beach cusps from Musquodoboit Harbour, Nova Scotia. (Photograph courtesy of E. M. Owens.)

waves that are generated by or connected to deep-sea motions for which a finite depth is found far from the coast. Also, in practice a typical shelf width can be comparable to or even less than the wavelength, in which case the refraction mechanism for edge waves is no longer effective. Finally there also arises the question of the validity of the shallow-water theory when the depth tends to infinity, as for a semi-infinite sloping beach. For these and other reasons, recent studies of trapped waves have included more realistic shelf/slope topographies.

Generally speaking, all the shelf/slope models studied fall into two categories. In the interval $0 \leqslant x < L < \infty$ the depth profiles are either (a) concave upward or (b) concave downward. However, in each case the depth far from the coast is constant (see Fig. 25.7). Since in practice most continental shelf/slope topographies fall into type (b), the bulk of the literature deals with topographies of this type. Accordingly, we shall derive explicit solutions for three type (b) topographies below. However, an interesting solution for a type (a) topography was found by Ball (1967). He investigated the behaviour of edge waves and quasi-geostrophic waves over an exponential depth profile of the form

$$H(x) = H_0(1 - e^{-ax}), \quad 0 \leqslant x < \infty. \tag{25.29}$$

Huntley and Bowen (1973) found that (25.29) is a reasonably accurate model of Slapton Beach, South Devon, where they detected lowest-mode edge waves (Fig. 25.4). The profile (25.29) leads to a hypergeometric equation for $F(x)$ whose appropriate solutions for

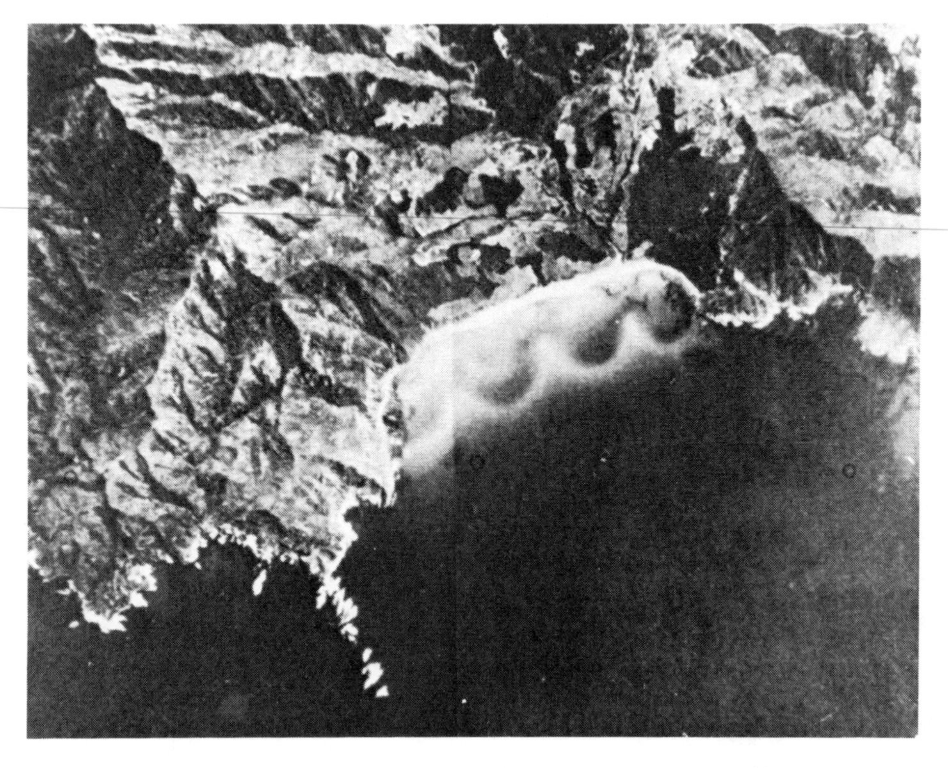

Fig. 25.6. Crescentic bars from the coast of Algiers. (From Bowen and Inman, 1971.)

trapped modes are hypergeometric (Jacobi) polynomials. There are two attractive limiting cases of this profile:

(1) For large a, (25.29) models a profile of constant depth with an abrupt edge arbitrarily close to the vertical. In this case the solutions degenerate into the barotropic Kelvin wave mode (Section 24).

(2) For large H_0 and small a such that the slope $\alpha = aH_0$ is kept constant, (25.29) models a uniform slope for arbitrarily large distances from the coast; in this case the solutions degenerate into Reid's solutions for a semi-infinite beach.

It is interesting to note that in the short wave limit, the frequency of the second-class waves is approximately given by $\omega_n = f/(2n + 1)$, which is the same as that obtained by Reid (1958) for a uniformly sloping beach. Such short topographic planetary waves are so strongly trapped near the shore that they are not influenced by the shape of the deep-ocean bathymetry. Finally, Ball also noted that for long wavelengths such that $ka \ll 1$, *trapped* edge waves cannot exist, which of course is due to the breakdown of the refraction trapping mechanisms discussed above. This feature of long edge waves has also been explored further by Clarke (1974) for a variety of both type (a) and type (b) shelves. Grimshaw (1974), on the other hand, derived upper and lower bounds for the gravest-mode edge wave on an arbitrary topography of type (a). However, rotation is ignored in his work so that the results are only valid for high frequencies.

Huthnance (1975) has recently studied the qualitative theory of the amplitude equation (25.7) as applied to edge waves and quasi-geostrophic waves over any depth

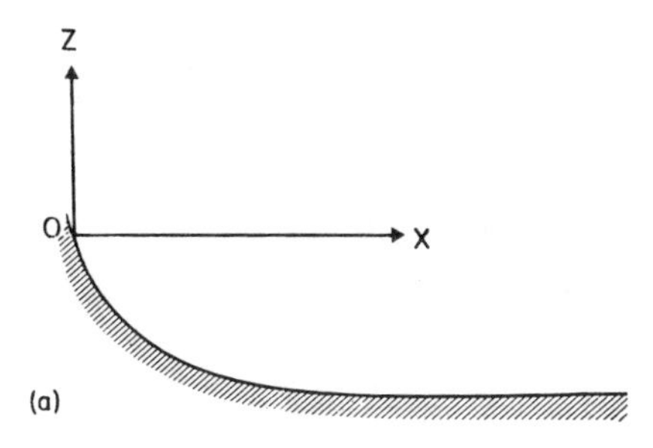

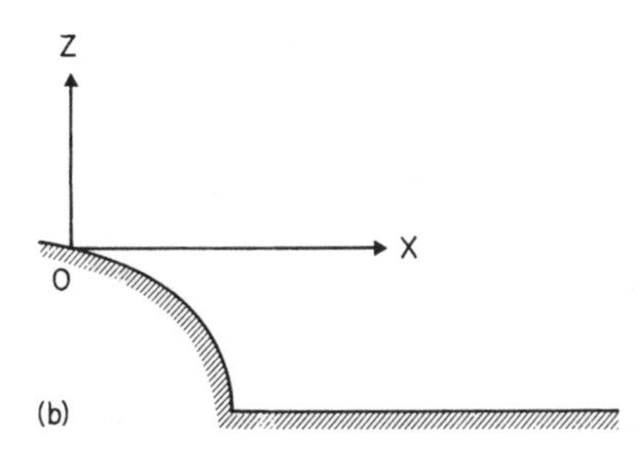

Fig. 25.7. The two basic shelf/slope topographies treated in the literature: (a) profiles with upward concavity; (b) profiles concave downward.

profile $H(x)$ that monotonically increases to H_0 as $x \to \infty$. His theory, which covers both (a) and (b) type profiles, will be discussed later in this section.

Trapped waves on a sloping shelf of finite width

A type (b) topography which readily illustrates the effect of a sloping shelf of finite width is given by

$$H(x) = \begin{cases} H_1 x/l, & 0 \leqslant x < l \\ H_2 \text{ (constant)}, & x > l. \end{cases} \tag{25.30}$$

This model was introduced by Robinson (1964) to study very low frequency and very long trapped quasi-geostrophic waves. Equation (25.30) represents a shelf region ($0 \leqslant x < l$) of uniform slope $\alpha = H_1/l$ which sharply drops off to a deep-sea region ($x > l$) of constant depth. For the topography (25.30), the appropriate solution of (25.7) which satisfies (25.8) and (25.11) is

$$F(x) = \begin{cases} A\, \mathrm{e}^{-kx} L_\nu(2kx), & 0 \leqslant x < l \\ B\, \mathrm{e}^{-Kx}, & x > l, \end{cases} \tag{25.31}$$

where $\nu = (-1 + \mu)/2$, with μ as given by (25.19), and

$$K = [k^2 + (f^2 - \omega^2)/gH_2]^{1/2}, \tag{25.32}$$

which must be real and positive for trapped waves. $L_\nu(z)$ is the Laguerre function, which has the series representation

$$L_\nu(z) = 1 - \nu z - \nu(-\nu + 1)z^2/(2!)^2 - \ldots$$

and which is related to the confluent hypergeometric function by the identity $L_\nu(z) = {}_1F_1(-\nu; 1; z)$. For $\nu = n = 0, 1, 2, \ldots, L_\nu$ reduces to a Laguerre polynomial of degree n. An extensive discussion of the properties of the Laguerre function is contained in Pinney (1946).

Examination of (25.31) and (25.32) clearly shows that for the high-frequency edge waves ($\omega^2 \gg f^2$), trapping is impossible when k is small. When $K^2 < 0$ the spectrum is continuous. These very long waves are the so-called leaky edge wave modes first discussed by Snodgrass et al. (1962) for a flat shelf. They are also known as topographically modified Poincaré waves (Munk et al., 1970). For quasi-geostrophic waves ($\omega^2 < f^2$), $K > 0$ for all $k > 0$ and trapping occurs at all wavelengths.

At the edge of the shelf ($x = l$), we require that η and Hu be continuous. This leads to two homogeneous equations for A and B. For a nontrivial solution the determinant of coefficients must vanish, which results in the following implicit dispersion relation:

$$L_\nu(2\kappa)\{[1 + \delta\Delta(1 - \sigma^2)/\kappa^2]^{1/2} + 1/\sigma - \Delta(1 + 1/\sigma)\} + 2\Delta L'_\nu(2\kappa) = 0, \qquad (25.33)$$

where $\sigma = \omega/f$, $\kappa = kl$, $\delta = f^2l^2/gH_1$ and $\Delta = H_1/H_2$. The square-root term inside the curly brackets is proportional to K [see (25.32)] and the quantity ν is related to σ, κ and δ by

$$\sigma^3 - [1 + (2\nu + 1)\kappa/\delta]\sigma - \kappa/\delta = 0, \qquad (25.34)$$

which is equivalent to the cubic for ω given by (25.26).

The relationship (25.33) was derived and analyzed by Mysak (1968a). For small Δ, corresponding to a large drop from the shelf to the deep-sea region, the waves on the shelf are weakly coupled to those in the deep-sea region. Hence a first approximation for the implicit dispersion relation is obtained by setting

$$L_\nu(2\kappa) = 0. \qquad (25.35)$$

This relation implies that the shelf waves have a node at the edge of the shelf, so that they are only weakly coupled to any offshore motions. Since, typically, $\Delta = 4 \cdot 10^{-2}$ ($H_1 = 200$ m, $H_2 = 5000$ m), this approximation introduces an error of only a few percent. For a given κ (the nondimensional wavenumber), (25.35) is satisfied for a countably infinite number of discrete values of $\nu = \nu_0, \nu_1, \nu_2, \ldots$. The first three of these have been tabulated as a function of κ by Mysak (1968a). Given these values of ν_n, the cubic (25.34) can then be solved for the frequency functions $\sigma = \sigma(j, \nu_n, \kappa)$. For each mode ν_n, the roots $j = 1$ and 2 correspond to the edge wave dispersion relations, whereas the root $j = 3$ corresponds to the quasi-geostrophic dispersion relation. In the limit of a wide shelf, or equivalently, as $\kappa \to \infty$, the numbers $\nu_n \to n = 0, 1, 2, \ldots$: the values for a semi-infinite shelf. In computing the values of ν_n, however, the only values of κ which are allowed are those for which $K > 0$ (trapped wave condition). It was pointed out by Munk et al. (1970) that in using the approximation (25.35), Mysak omitted (except for the case $\omega = -f$) the very long ($K \ll 1$) Kelvin-like wave mode implicit in (25.33). Like the barotropic Kelvin wave discussed in Section 24, this wave is trapped against the wall at $x = l$

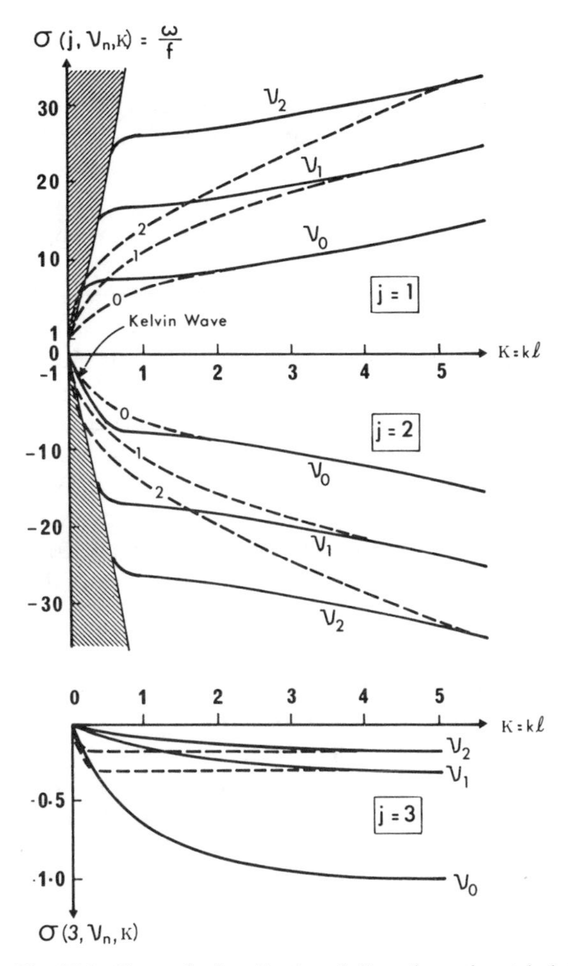

Fig. 25.8. Plots of edge ($j = 1$ and 2) and continental shelf or quasi-geostrophic ($j = 3$) wave dispersion relations for a shelf of finite width (solid lines) and semi-infinite width (dashed lines) with $\delta = f^2 l^2/gH_1 = 0.027$ [$f = 0.73 \cdot 10^{-4}$ rad s^{-1}, $H_1 = 200$ m, $l = 10^5$ m (or $\alpha = 2 \cdot 10^{-3}$)]. The different modes ν_n or n are indicated on the curves. There is no gravest-mode quasi-geostrophic dispersion relation shown since it is an inertial oscillation with $k = 0$. The shaded region corresponds to the continuous spectrum of topographically modified long gravity waves; it is bounded by the hyperbolae $\omega/f = \pm\,(1 + gH_2k^2)^{1/2}$ (with $H_2 = 5 \cdot 10^3$ m), corresponding to $K = 0$. (Adapted from Mysak, 1968a.)

but its speed is slightly modified by the presence of the shelf. A careful analysis of (25.33) reveals that the gravest-mode edge wave for $j = 2$, which is a right-bounded wave for $f > 0$ and $K \geqslant 1$, continuously merges into this barotropic Kelvin wave mode for $K \ll 1$ (see Fig. 25.8). It is this curve which separates the right-bounded edge waves from the low-frequency quasi-geostrophic waves, for which Robinson (1964) coined the term "continental shelf waves". However, Robinson only considered the very low frequency end of the spectrum ($\sigma \ll 1$, $\kappa \ll 1$) under the additional approximation $\delta \ll 1$, in which case the waves are nondispersive and nondivergent. Mysak (1967a) also analyzed this end of the frequency spectrum for continental shelf waves travelling around a large circular continent with the same shelf/slope topography.

In Fig. 25.8 the frequency functions $\sigma(j, \nu_n, \kappa)$ are plotted for the first three modes ν_0, ν_1, ν_2 and are compared with the semi-infinite shelf solutions of Reid (Fig. 25.3). Note that for $\kappa \leqslant 1$ the curves for both edge waves and shelf waves are numerically quite distinct from those due to Reid, thus revealing the importance of a finite shelf width and of a finite-depth deep-sea region. The shaded region in the upper diagram corresponds to the frequency–wavenumber region defined by $K^2 \leqslant 0$ and contains the continuous spectrum of topographically modified long gravity waves. Clearly the energy of these waves is not exponentially trapped against the coast.

Huthnance (1975) has proved that for any depth profile $H(x)$ which monotonically increases to a constant value, the spectrum of barotropic trapped modes is of the form shown in Fig. 25.8. There is always an infinite discrete set of edge waves, a single "Kelvin wave" (at large wavelengths) and an infinite discrete set of low-frequency ($|\omega| < |f|$) continental shelf waves. He showed that if H'/H is also bounded for all $x \geqslant 0$, the group velocity of each shelf wave mode is in the direction opposite to that of the phase velocity for some range of wavenumbers. Since $H(0) = 0$ for the topography (25.30), the shelf wave curves shown in Fig. 25.8 do not have this property. However, later in this section we shall see that for an exponential shelf for which H'/H is then bounded [see (25.55)], c and c_g for the shelf waves are of opposite sign at short wavelengths (see Fig. 25.11).

Trapped waves on a flat shelf

Snodgrass et al. (1962) introduced the flat shelf model [defined below by (25.36)] for a continental shelf/slope region in order to investigate irrotational ($f = 0$) edge waves on the California shelf. They were among the first to make the distinction between the continuous spectrum of leaky modes and the classical discrete spectrum of trapped wave motions. The existence of these two types of modes is of course well known in the theory of wave propagation in layered media (Brekhovskikh, 1960; Wait, 1962; Tolstoy, 1973, Chapter 3). An analysis of data from bottom pressure gauges at La Jolla and at San Clemente Island, about 100 km seaward, showed that in the period range of 5 minutes to 5 hours there was a fairly even partition between the leaky and trapped modes, with amplitudes of about half a centimeter. However, in a later experiment (Munk et al., 1964), it was found that most of the energy in this period range was contained in the first few trapped modes and that it was fairly evenly split between northward- and southward-propagating waves. Buchwald and de Szoeke (1973) also used the flat shelf model to study the generation of edge waves off the East Australian coast. The flat shelf model and more complicated multi-step models have been used extensively in the Japanese literature on edge waves (see, for example, Aida, 1967; Aida et al., 1968). J. Larsen (1969) studied the propagation of low-frequency quasi-geostrophic waves and the Kelvin wave for this topography. Although Larsen derived the dispersion relation valid for all frequencies, he did not discuss the rotationally modified edge wave modes. In their lengthy study of tidal propagation along the coast of California, Munk et al. (1970) performed a thorough analysis of all types of waves which travel along the flat shelf profile. The dispersion curves they obtained are very similar to those which were found for a sloping shelf of finite width (Fig. 25.8). The only qualitative difference between the results for the two profiles is that only the lowest-mode shelf wave appears for the flat shelf; all the higher modes disappear and the shelf wave spectrum is reduced to a single wave associated with the single-step change in topography.

In the interest of simplicity, we analyze the two classes of waves separately, first considering edge waves on a flat shelf, with $f=0$ (thereby removing the Kelvin and shelf wave modes) and then shelf waves in a rotating ocean with a rigid lid (thereby removing the edge and Kelvin wave modes).

(1) Edge waves. For the flat shelf topography given by

$$H(x) = \begin{cases} H_1, & 0 \leqslant x < l \\ H_2, & x > l. \end{cases} \tag{25.36}$$

with $H_2 > H_1$, and for $f=0$, equation (25.7) for the wave amplitude becomes

$$F_i'' + \left(\frac{\omega^2}{gH_i} - k^2\right) F_i = 0, \quad i = 1, 2, \tag{25.37}$$

where the subscripts $i = 1, 2$ refer to the shelf and to the deep-sea regions, respectively. Following Buchwald and de Szoeke (1973) we now introduce the nondimensional quantities

$$\chi = x/l, \quad \kappa = kl, \quad \Omega = \omega l/(gH_1)^{1/2} \tag{25.38}$$

and $\quad \gamma = (H_2/H_1)^{1/2} > 1.$ $\qquad$ (25.39)

In terms of these quantities, the amplitude variations $\hat{F}_i(\chi) \equiv F_i(\chi l)$ for the two regions are governed by

$$\hat{F}_1''(\chi) + (\Omega^2 - \kappa^2)\hat{F}_1(\chi) = 0, \quad 0 \leqslant \chi < 1, \tag{25.40a}$$

$$\hat{F}_2''(\chi) + (\Omega^2/\gamma^2 - \kappa^2)\hat{F}_2(\chi) = 0, \quad \chi > 1. \tag{25.40b}$$

The long wave propagation speeds in the deep and shallow parts of the basin are $(gH_2)^{1/2}$ and $(gH_1)^{1/2}$ respectively, or in nondimensional form, γ and 1. The nondimensional wave speed, Ω/κ, for the coupled system will be denoted by c. Depending on whether (a) $c > \gamma$, (b) $1 < c < \gamma$ or (c) $1 > c$, we shall find (a) leaky modes, for which the energy may propagate away from the coast, (b) trapped modes, for which wave propagation occurs parallel to the coast and the energy decays away exponentially in the deep-sea region, and (c) virtual modes, for which no wave propagation is possible.

The appropriate boundary conditions on the boundaries of the domain $0 \leqslant \chi < \infty$ are

$$\hat{F}_1'(0) = 0 \tag{25.41}$$

$$\text{and} \quad \hat{F}_2 \begin{Bmatrix} \text{bounded for leaky modes} \\ \to 0 \quad \text{for trapped modes} \end{Bmatrix} \text{as } \chi \to \infty. \tag{25.42}$$

At the depth discontinuity ($\chi = 1$), the continuity of surface elevation and of normal flux impose the matching conditions

$$\hat{F}_1(1) = \hat{F}_2(1), \tag{25.43a}$$

$$\hat{F}_1'(1) = \gamma^2 \hat{F}_2'(1). \tag{25.43b}$$

(a) Leaky modes: $c > \gamma$.

The solutions of (25.40), subject to the conditions (25.41)–(25.43) may be written as

$$\hat{F}_1(\chi) = A \cos \epsilon \cos \mu_1 \chi,$$

$$\hat{F}_2(\chi) = A \cos \mu_1 \cos [\mu_2(\chi - 1) + \epsilon], \tag{25.44}$$

where $\mu_1 = \Omega \cos \theta_1 = \kappa \cot \theta_1$, $\mu_2 = \Omega\gamma^{-1} \cos \theta_2 = \kappa \cot \theta_2$, $\tan \epsilon = (\mu_1/\gamma^2\mu_2) \tan \mu_1$ and A is an arbitrary amplitude constant.

Combining (25.44) with the wave form (25.6), viz., $\exp [i(\kappa Y - \Omega\tau)]$, where $Y = y/l$ and $\tau = t(gH_1)^{1/2}/l$, we see that these solutions represent long waves approaching the shelf at an angle θ_2 to the normal, being reflected and refracted at $\chi = 1$ and reflected from the coast at $\chi = 0$, with $\theta_1(<\theta_2)$ the angle between the wavenumber vector and the normal to the coast on the shelf. The conservation of phase at the edge of the shelf implies that Snell's law holds:

$$\sin \theta_2 = \gamma \sin \theta_1. \tag{25.45}$$

(b) Trapped modes: $1 < c < \gamma$.

Total internal reflection now occurs at $\chi = 1$, and $\theta_2 > \sin^{-1}(\gamma)$. The solutions in this case are expressed as

$$\hat{F}_1(\chi) = A \cos \mu_3 \chi, \tag{25.46a}$$

$$\hat{F}_2(\chi) = A \cos \mu_3 \exp [-\nu(\chi - 1)], \tag{25.46b}$$

where $\mu_3^2 = \Omega^2 - \kappa^2, \quad \nu^2 = \kappa^2 - \Omega^2/\gamma^2, \quad \nu > 0,$ (25.47)

and $\mu_3 \tan \mu_3 = \nu\gamma^2.$ (25.48)

It is theoretically possible to eliminate μ_3 between (25.47) and (25.48) to arrive at an implicit form for the dispersion relation $\Omega = \Omega(\kappa)$.

A detailed discussion of the relation $\Omega(\kappa)$ is given by Snodgrass et al. (1962) and by Buchwald and de Szoeke (1973). For each mode n, the dispersion curves are similar to the edge wave curves shown in Fig. 25.8. However, the asymmetry in Fig. 25.8 due to the presence of rotation does not arise here since we have taken $f = 0$ (see Fig. 25.9). We also note that all the modes are contained in the wedge bounded by the lines $\Omega = \kappa$ and $\Omega = \gamma\kappa$; the trapped waves travel at a velocity c intermediate between that of waves in depth H_1 and that of waves in depth H_2 $(1 < c < \gamma)$.

The trapping by total internal reflection may also be viewed in terms of travelling waves. Combining (25.46a) with the travelling wave form (25.6), the oscillations on the shelf may be written as the sum of an incident and reflected wave:

$$\eta_1(x, y, t) = \frac{A}{2} \{\exp [i(\mu_3 x/l + ky - \omega t)] + \exp [i(-\mu_3 x/l + ky - \omega t)]\}. \tag{25.49}$$

At $x = l$, the phase of the reflected wave differs from that of the incident wave by $2\mu_3$. This is the phase shift, ϕ_1, which occurs upon total internal reflection. Using (25.48) and the definition of γ [see (25.39)], we have

$$\phi_1 = 2\mu_3 = 2 \tan^{-1}(H_2\nu/H_1\mu_3), \tag{25.50}$$

an expression which will come in useful later.

(c) Virtual modes: $c < 1$.

In this case, the χ-dependence is exponential everywhere; there are no solutions corresponding to real wavenumbers and hence no propagating waves (see Exercise 25.7).

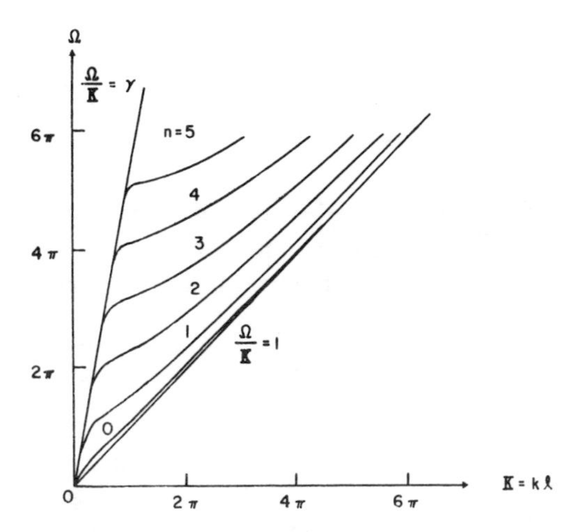

Fig. 25.9. The dispersion relations of the first six trapped edge wave modes ($n = 0, 1, \ldots, 5$) on a flat shelf with $\gamma^2 = H_2/H_1 = 16$ and no rotation ($f = 0$). Because $f = 0$, the curves for $\Omega < 0$ are a mirror image of the curves shown here. (From Buchwald and de Szoeke, 1973.)

(2) Shelf waves. The amplitudes of observed shelf waves usually do not exceed a few centimeters (Fig. 25.12) and hence their analysis may be performed under the restrictions of the rigid-lid hypothesis discussed in Section 20. We recall that provided (cf. 20.12)

$$f^2L^2/gH \ll 1, \tag{25.51}$$

where L is a typical horizontal length scale of the motion, variations in the free surface of a topographic planetary wave play a negligible role in the vorticity balance. For a typical mid-latitude shelf, $f = 10^{-4}\,\mathrm{rad\,s^{-1}}$, $L = 5 \cdot 10^4\,\mathrm{m}$ (a characteristic shelf width) and $H = 2 \cdot 10^2\,\mathrm{m}$; therefore, $f^2L^2/gH = 0(10^{-2})$, in agreement with (25.51). In the deep-sea region where $H = 5 \cdot 10^3\,\mathrm{m}$, (25.51) remains satisfied for $L < 10^6\,\mathrm{m}$.

For a rigid-topped ocean the motions are of course horizontally nondivergent; therefore, we introduce the transport stream function $\psi(x, y, t)$ such that (cf. 20.10)

$$Hu = -\psi_y, \quad Hv = \psi_x. \tag{25.52}$$

For topographic planetary waves, ψ satisfies (20.11). In this equation we take $f =$ constant, $H = H(x)$ and ψ of the form

$$\psi = \Psi(x)\, \mathrm{e}^{i(ky - \omega t)}, \quad k > 0. \tag{25.53}$$

Then equation (20.11) reduces to

$$\left(\frac{1}{H}\Psi'\right)' + \left[-\frac{k^2}{H} - \frac{fk}{\omega}\frac{H'}{H^2}\right]\Psi(x) = 0, \quad 0 \leqslant x < \infty. \tag{25.54}$$

For the step topography (25.36), (25.54) gives

$$\Psi_i'' - k^2\Psi_i(x) = 0, \quad i = 1, 2 \tag{25.55}$$

in each region. At the coast $Hu = 0$; hence (25.52) and (25.53) imply that

$$\Psi_1(0) = 0. \tag{25.56}$$

Far from the coast

$$\Psi_2 \to 0 \quad \text{as} \quad x \to \infty. \tag{25.57}$$

For Hu to be continuous at $x = l$, (25.52) and (25.53) imply that

$$\Psi_1(l) = \Psi_2(l). \tag{25.58}$$

To obtain the jump condition (i.e., the magnitude of the discontinuity in v) at $x = l$, we integrate (25.54) over a small neighbourhood about $x = l$, as in the manner of Section 23. From (23.30),

$$\left[\frac{1}{H}\Psi'\right] = -\frac{kf}{\omega}\Psi(l)\left[\frac{1}{H}\right] \quad \text{at} \quad x = l, \tag{25.59}$$

where the bold square brackets denote the difference in the value of the quantity inside across $x = l$ (cf. 23.30). It can readily be shown that (25.59) also implies that the pressure is continuous at $x = l$. The solution of (25.55) which satisfies (25.55)–(25.58) is given by

$$\Psi_1 = A_1 \sinh kx, \quad 0 \leqslant x < l \tag{25.60a}$$

$$\Psi_2 = A_1 \sinh kl\, \mathrm{e}^{-k(x-l)}, \quad x > l. \tag{25.60b}$$

The application of (25.59) yields the dispersion relation

$$\omega/f = -(\gamma^2 - 1)/(1 + \gamma^2 \coth kl), \tag{25.61}$$

where $\gamma = (H_2/H_1)^{1/2} > 1$, as defined in (25.39). Graphically, this dispersion relation is like the ν_0 curve shown in the lower part of Fig. 25.8. For $f > 0$, (25.61) clearly shows that $\omega < 0$ and hence the wave is right-bounded. The topographic nature of the wave is made evident by changing the relative magnitudes of H_1 and H_2: for propagation along a trench ($\gamma < 1$), the wave reverses its direction of phase propagation. It also should be noted that the group and phase velocities are always in the same direction (as is the case for edge waves on a flat shelf).

The relation (25.61) was first obtained by J. Larsen (1969) as a limiting case of his divergent solutions for trapped quasi-geostrophic waves; it also falls out as a special case of the dispersion relation obtained by Niiler and Mysak (1971) for shelf waves in the presence of a mean laterally sheared coastal current (see Section 45).

Shelf waves on an exponential shelf

We noted in Section 20 that for the case of nondivergent waves, an exponentially varying depth profile leads to a constant coefficient differential equation for the stream function Ψ. This fact was exploited by Buchwald and Adams (1968) in their study of nondivergent shelf waves off the East Australian coast. They considered a depth profile of the form (see Fig. 25.10)

$$H(x) = \begin{cases} H_1 \mathrm{e}^{2bx}, & 0 \leqslant x < l \\ H_2, & x > l, \end{cases} \tag{25.62}$$

where $H_2 = H_1 \mathrm{e}^{2bl}$, so that H is continuous at $x = l$. Appropriate values for the shelf near Sydney are $H_1 = 67\,\mathrm{m}$, $H_2 = 5 \cdot 10^3\,\mathrm{m}$, $l = 8 \cdot 10^4\,\mathrm{m}$ and $b = 3.4 \cdot 10^{-5}\,\mathrm{m}^{-1}$. The solution

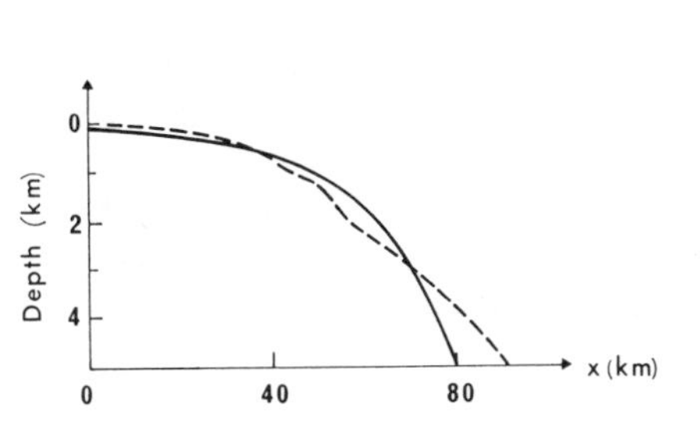

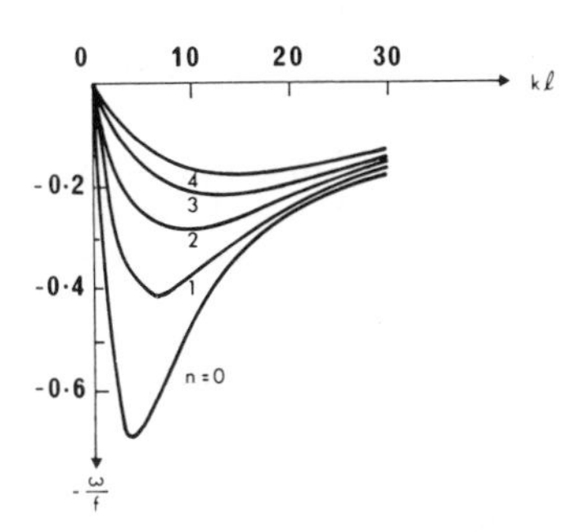

Fig. 25.10. Comparison of the shelf/slope topography near Sydney (dashed line) with the exponential model given by (25.35). (From Buchwald and Adams, 1968.)

Fig. 25.11. The dispersion relations of the first five modes of non-divergent shelf waves on an exponential shelf with decay parameter $bl = 2.7$ and $f > 0$. (From Buchwald and Adams, 1968.)

of (25.54) for $\Psi(x)$ and the conditions (25.56)–(25.59) is left as an exercise. The resulting dispersion relation $\omega = \omega(k)$ is given implicitly by two equations for ω, k and m:

$$m^2 + k^2 + b^2 + 2bfk/\omega = 0 \tag{25.63}$$

$$\text{and} \quad \tan ml = -m/(b+k). \tag{25.64}$$

As in the case of edge waves on a flat shelf, it is possible to eliminate m from (25.63) and (25.64) to obtain a single relation between ω and k in the form (see Buchwald and Adams, 1968 for details):

$$\omega/f = G(kl; bl). \tag{25.65}$$

The dispersion curves given by (25.65) corresponding to the first five modes for the case $bl = 2.7$ are shown in Fig. 25.11. We note that for $f > 0$, each mode is right-bounded [$\omega < 0$, as may be inferred directly from (25.63)]. However, in contrast to all the other dispersion relations for shelf waves seen so far, we notice that for short waves ($kl \gg 1$), the group velocity of each mode is in the opposite direction to the phase velocity. The group velocity changes sign by passing through zero at an intermediate wavenumber. This result is simply an example of Huthnance's theorem stated earlier: when H'/H is bounded, c and c_g will be of opposite sign for some k. The occurrence of an extremum in group velocity is usually associated with an Airy phase (see Section 50), for which the decay rate is slower than at other wavenumbers. In this case, the Airy phase does not propagate ($c_g = 0$), and one would expect a resonant type of behaviour at those wavenumbers where the group velocity is equal to zero. Observations by Cutchin and Smith (1973) of low-frequency motions on the Oregon shelf, where a depth profile very similar to (25.62) is found, have shown peaks of the coherence spectrum which nearly coincide with the frequencies at which the group velocity vanishes. The situation is far from clear cut, but it is tempting to think, as do Cutchin and Smith, that this coincidence is not accidental.

Because of the change of sign of c_g at short wavelengths, a long shelf wave travelling along an exponential shelf can back-scatter energy, in the form of shorter shelf waves, upon encountering topographical or coastal irregularities. For the case of a small coastal bump this problem has been considered by Buchwald (1977). Thus shelf wave scattering by topography provides a mechanism through which energy can be transferred between different scales of motion and in different directions.

The effect of a free surface on shelf waves over an exponential shelf has been examined by Buchwald (1973b). While not important in oceanic situations, horizontal divergence is relevant to the interpretation of laboratory experiments, for which the parameter $f^2L^2/gH = 0(1)$. Caldwell et al. (1972) have successfully generated and tested the theory of divergent shelf waves in a rotating tank, both at long and short wavelengths. The vortex motions associated with the waves are similar to those shown in Fig. 24.5 for double Kelvin waves.

Observations of shelf waves

Continental shelf waves were first observed along the Australian coast by Hamon (1962, 1963, 1966). Hamon (1962) presented the spectra of daily mean sea level and atmospheric pressure fluctuations at Sydney and Coff's Harbour (situated 500 km to the north of Sydney) on the East Australian coast, and at Lord Howe Island, situated about 800 km eastward of Sydney. Surprisingly, he found that even at very low frequencies (periods greater than a few days) the daily mean† sea level on the shelf did not respond as an inverse barometer. That is, corresponding to an increase of 1 mbar (10^2 Pa) in the atmospheric pressure, the sea surface did not decrease by 1 cm (0.01 m), as would be expected for static deformations. In particular, Hamon found that at Sydney and Coff's Harbour the sea level was depressed only by about half the expected amount. He also found that the spectra of the adjusted sea level (defined as the sea level minus the negative of the atmospheric pressure measured in centimeters of water) were peaked at 9 and 5 days, corresponding to the winter and summer peaks in the atmospheric pressure spectrum (see Fig. 25.12). At Lord Howe Island, on the other hand, the sea surface did respond as an inverse barometer. Hamon (1962, 1963) also performed a coherence and a lag analysis between the two coastal stations and found that for periods longer than 3 days the adjusted sea level at Sydney led that at Coff's Harbour by about 1 day (see Figs. 25.13 and 25.14). This result suggested the presence of a low-frequency, nondispersive left-bounded wave travelling northward along the continental shelf. To test this hypothesis further, Hamon (1966) performed a more extensive lag analysis of the adjusted sea level between Eden (37°S) and five stations to the north. These data suggested that a northward-travelling wave propagates along the entire East Australian coast (Fig. 25.14). The slope of the dashed line in Fig. 25.15 gives an average speed of 4 m s^{-1} for this wave.

Robinson (1964) showed that the observed nonbarometric sea-level behaviour on the East Australian continental shelf could be due to a resonant response of the adjusted sea level over a shelf topography of the form (25.30) to pressure fluctuations of large-scale moving weather systems. Mysak (1967b) further substantiated this explanation on the basis of spectral methods (Chapter 5). Mysak showed that provided linear bottom friction (see Chapter 8) is included with a friction coefficient of $0(10^{-8}\,s^{-1})$, the theoretical response spectrum of the adjusted sea level that is forced by a numerical model for the observed atmospheric pressure spectrum compares quite favourably with the dominant

† Henceforth in this discussion on shelf observations we shall omit the phrase "daily mean" for convenience.

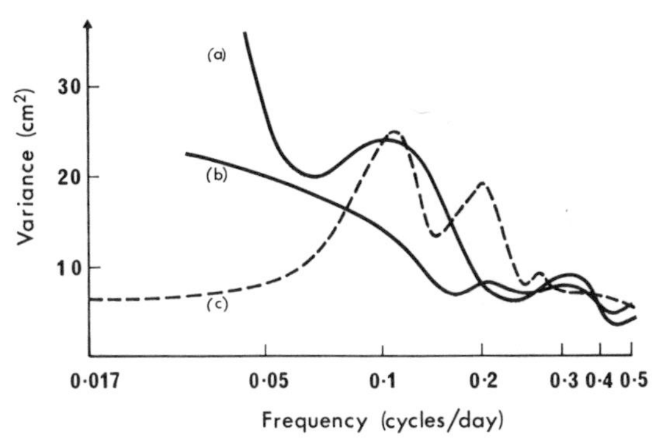

Fig. 25.12. The spectra of adjusted sea levels at Sydney during (a) "winter" (April–September 1958) and (b) "summer" (October 1957–March 1958), and (c) the atmospheric pressure for the period July 1957–December 1958. (Adapted from Hamon, 1962.)

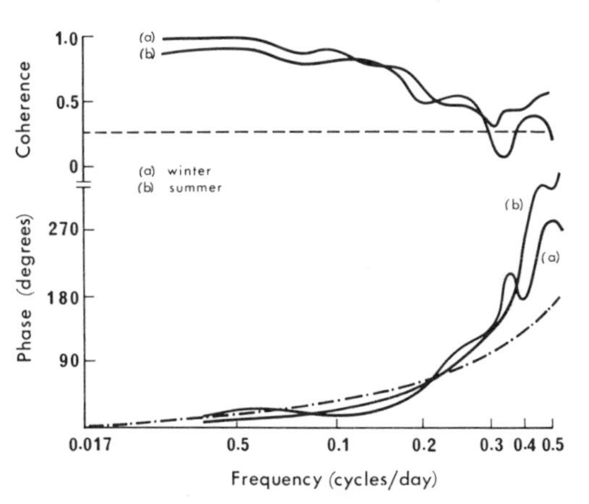

Fig. 25.13. Coherence and phase between adjusted sea levels at Sydney and Coff's Harbour. The dashed line indicates the 95%-confidence limit for coherence. The dot–dash line stands for a phase difference corresponding to Sydney leading Coff's Harbour by 1 day at all frequencies (see also Fig. 4.23). (From Hamon, 1962; 1963.)

features of the observed adjusted sea-level spectrum. Robinson (1964) attributed the observed lead to a low-frequency, nondispersive continental shelf wave. In the limiting case $\kappa^2 = k^2l^2 \ll 1$, $\sigma^2 \ll 1(\omega^2 \ll f^2)$ and $\delta = f^2l^2/gH_1 \ll 1$, equation (25.35) reduces to (see Exercise 25.5).

$$J_0(2\sqrt{q}) = 0, \quad (q = -fkl/\omega). \tag{25.66}$$

From the first zero (lowest-mode solution) of this equation, Robinson calculated a theoretical phase speed of $2.5\,\mathrm{m\,s^{-1}}$, a value considerably lower than the observed speed. Upon addition of a continental slope region of finite constant slope to produce a topography closer to type (b) (see Fig. 25.10), Mysak (1967a) found that the gravest-mode speed was increased by about 30%. A comparable speed was also obtained by Buchwald and Adams (1968) using the exponential shelf model (25.62): they obtained a speed of

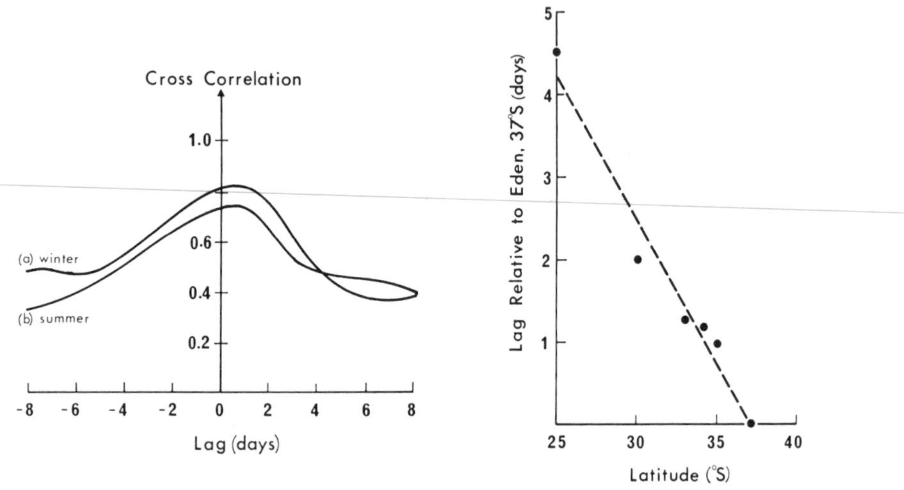

Fig. 25.14. The cross-correlation between adjusted sea levels at Sydney and Coff's Harbour; positive lag means Sydney leads Coff's Harbour. (From Hamon, 1962; 1963.)

Fig. 25.15. Cumulative lags in adjusted sea levels as a function of latitude for the East Australian coast. (From Hamon, 1966.)

$2.8\,\mathrm{m\,s^{-1}}$ for the lowest-mode phase speed in the (nondispersive) limit $kl \ll 1$. In another paper on shelf wave generation, Adams and Buchwald (1968) proposed that the observed nonbarometric behaviour might be due to the low-frequency response of the sea surface to localized moving wind stresses, rather than to atmospheric pressure variations. The question of forced low-frequency motions on the shelf will be examined further in Section 55.

However, the topographic modifications discussed above do not totally eliminate the discrepancy between the theory and the observations. This is perhaps not too surprising since no account was taken of the main oceanic features off the East Australian coast: the East Australian current and its associated stratification (Hamon, 1965). These offshore features were crudely modelled by Mysak (1967a), who introduced into the topographic model (25.30) a two-layer deep-sea stratification and a basic horizontal deep-sea current of constant magnitude flowing southward in the upper layer alongside the edge of the shelf (see Fig. 7 in Mysak, 1967a). Mysak found that for a relative density difference of $2.5 \cdot 10^{-3}$ between the two layers and a current speed of $1\,\mathrm{m\,s^{-1}}$, the northward-propagating, lowest-mode shelf wave at low frequencies had a speed of $4\,\mathrm{m\,s^{-1}}$, in excellent agreement with Hamon's observations.

Shelf waves have subsequently been observed in many other parts of the world. They have been detected on the West Australian coast (Hamon, 1966), the Oregon coast (Mooers and Smith, 1968; Cutchin and Smith, 1973), the North Carolina coast (Mysak and Hamon, 1969), the west coast of Scotland (Cartwright, 1969) the North Mediterranean coast (Saint-Guily and Rouault, 1971), Lake Ontario (Csanady, 1976), and the Florida Strait (Schott and Düing, 1976; Brooks and Mooers, 1977). Also, in the more recent studies, current meter records rather than sea-level fluctuations have been analyzed to confirm the existence of shelf waves and to investigate their properties more closely (see Section 55).

Over the past few years it has become increasingly evident that shelf waves are intimately connected with other dynamic features of coastal and deep-ocean regions. For example, they appear to interact (either passively or actively) with western boundary currents such as the East Australian current (as mentioned briefly above) and the Gulf Stream (Niiler and Mysak, 1971). Also they appear to play an important role in the phenomena of coastal upwelling (Gill and Clarke, 1974; Suginohara, 1974; Kishi and Suginohara, 1975; Peffley and O'Brien, 1976). Unfortunately, a discussion of these topics is beyond the scope of this book. We instead refer the reader to a recent review on coastal trapped waves by LeBlond and Mysak (1977) which includes a lengthy discussion on the role of shelf waves in coastal and deep-ocean dynamics.

Coastal trapped waves in a stratified fluid

As was mentioned at the beginning of this section, in a stratified fluid of variable depth it is not generally possible to separate the vertical and horizontal dependences of the dependent variables in order to obtain simple barotropic and baroclinic modes. Some special cases are, however, more accessible to analysis, and a number of problems where the stratification and the bottom slope occur in different regions have been considered.

Edge waves in a stratified fluid have only been briefly touched upon in the literature. This is perhaps not surprising, since edge waves are generally closely confined to the shoreline and are thus unlikely to be affected by the deep-sea stratification. Mysak (1968b) showed that edge waves travelling along an abrupt topography of the form (25.30), with a two-layer fluid in the deep-sea region, are only weakly coupled to long internal waves propagating in the deep ocean.

Greenspan (1970), on the other hand, found solutions for coastal trapped surface and internal waves in a continuously and uniformly stratified fluid lying over the semi-infinite sloping beach profile given by (25.12). In his paper, the rotation of the Earth is not included ($f = 0$), but no hydrostatic approximation is made. Greenspan found a lowest-mode solution which is almost identical to the fundamental mode of Stokes' solution (see Exercise 25.11). The dispersion relation and the horizontal currents are unaffected by the stratification for that mode, but perturbations of the pressure and density fields arise from the presence of the mean density gradient. The higher-mode solutions are much more complicated and may be examined in the original paper. More recently, Odulo (1974) has studied this lowest-mode edge wave in a rotating stratified fluid.

The solutions of Mysak (1967a, 1968b) and of Gill and Clarke (1974) for coastal trapped wave propagation in a nonstratified fluid on a sloping shelf adjacent to a stratified deep-sea region of uniform depth do not describe trapped wave propagation in a stratified fluid, but merely the coupling of trapped wave solutions in a homogeneous fluid to offshore baroclinic wave motions. Two important papers which elegantly deal with the topic of low-frequency ($\omega^2 \ll f^2$) coastal trapped waves over topography with *on-shelf* stratification are those of Kajiura (1974) and Wang and Mooers (1976). Kajiura used a two-layer model for the stratification with the interface lying over both the shelf [modelled by the single-step topography (25.36)] and the deep-sea region. Wang and Mooers, on the other hand, solved the relevant two-dimensional boundary value problem numerically for realistic continuous stratification over an exponential shelf (cf. 25.62). Both studies illustrate the nature of the coupling between barotropic and baroclinic motions due to the presence of topography. For very long wavelengths or weak stratification the low-frequency waves are essentially like the barotropic shelf waves. For short wavelengths or strong stratification, the waves are a mixture of internal Kelvin wave modes trapped against the vertical coast (at $x = 0$) and topographic planetary waves, bottom-trapped over the shelf/slope region. In the continuously stratified case there is no simple characterization of the current structure in terms of horizontal and vertical modes. Special contributions to this problem have also been made by Allen (1975), who examined the case of weak coupling (internal Rossby radius much less than the shelf width), and by Wang (1975), who studied a two-layer fluid model with both a flat and an exponential shelf.

Nonlinear edge waves and shelf waves

Nonlinear corrections to the linear edge wave solutions for a sloping beach (see 25.12) have recently been presented by Guza and Bowen (1976) and independently by Whitham (1976). Guza and Bowen showed, on the basis of an expansion in the small-amplitude parameter

$$\epsilon = a\omega^2/g\alpha^2, \tag{25.67}$$

where a is the amplitude and α the bottom slope, that shallow-water edge waves are modified by nonlinearities in a way similar to surface gravity waves in deep water (see Section 12). The wave crests tend to be sharpened and the frequency increases with amplitude. The latter result causes amplitude dispersion, with the waves propagating faster at large amplitudes. However, in the more thorough analysis by Whitham, it is shown that at third order, the shallow-water solution is not uniformly valid as $x \to \infty$. That is, the third-order correction does not decay far from the coast. However, Minzoni (1976) has shown that this secularity can be eliminated if a more realistic depth profile is used, i.e., one that tends to a constant far from the coast.

The effects of nonlinearities on long shelf waves ($kl \ll 1$, where l = shelf width) have been examined in two complementary papers by Smith (1972) and Grimshaw (1977a). In both cases, weak phase dispersion is balanced by weak amplitude dispersion and a Korteweg-de Vries equation (see Section 12) for the amplitude function is derived. Smith carried out an experiment in the small divergence parameter $\epsilon_1^2 = f^2l^2/gH_0$ (H_0 is some reference depth), whereas Grimshaw held this parameter (albeit small) fixed and expanded in the small parameter $\epsilon_2^2 = l^2/L^2$ where L is the wavelength. Grimshaw then examined the limit $\epsilon_1 \to 0$ a posteriori. The two approaches give the same amplitude equation in the common parameter regime ϵ_1 and ϵ_2 both small but $\epsilon_2^2 \ll \epsilon_1^2$.

Grimshaw (1977b) also examined shelf waves for side-band instability (Section 12). The modulated shelf wave is described by a nonlinear Schrödinger equation, from which it follows that long shelf waves are stable to side-band modulations. However, as the wavenumber is increased one moves into regions (in wavenumber space) of instability.

Exercises Section 25

1. Derive the Stokes edge wave solution (25.13) and (25.14) using potential theory. Show that far from the coast, the associated currents are circularly polarized.

2. Derive the solution (25.20). (Hint: It is first convenient to put $\chi = kx$ (>0) in (25.18).)

3. Show that the roots of (25.26) are real and establish the properties (25.27).

4. Discuss the dispersion relation for quasi-geostrophic waves on a semi-infinite sloping beach that is obtained by invoking the filtering approximation $\omega^2 \ll f^2$ in (25.26).

5. Show that for $k^2l^2 \ll 1$, $\omega^2 \ll f^2$ and $f^2l^2/gH_1 \ll 1$, equation (25.18) for $0 \leqslant x \leqslant l$ and $\alpha = H_2/l$ reduces to a variant of Bessel's equation, with solutions $J_0(2\sqrt{q\xi})$, where $q = -fkl/\omega$ and $\xi = x/l$. Hence the dispersion relation in this limiting case for the topography (25.30), with $\Delta = H_1/H_2 \ll 1$, is approximately given by

$$J_0(2\sqrt{q}) = 0. \tag{25.68}$$

For those familiar with special functions, it is fairly easy to show that (25.35) also reduces to (25.68) in this limit.

6. Taking the phase change of an edge wave on a flat shelf undergoing total internal reflection at $x = l$ to be ϕ_1, with $\phi_1/2 = \tan^{-1}(H_2\nu/H_1\mu_3)$, show that by matching the phase of a standing wave of offshore wave number μ_3 in the interval $0 \leqslant x \leqslant l$ we obtain (25.48). Assume that the reflection at $x = 0$ is without change of phase.

7. Show that for a flat shelf and $f = 0$, there cannot exist any propagating modes with phase speed less than $\sqrt{gH_1}$.

8. Derive (25.61) and show that ω/f (for $f > 0$) is a monotone decreasing function of kl. Also, discuss the limits $kl \to 0$ (narrow shelf) and $kl \to \infty$ (wide shelf). In the second limit you should recover a form equivalent to (24.10), the escarpment solution.

9. Derive the relations (25.63) and (25.64).

10. Show that for $\gamma^2 \gg 1$, $kl \ll 1$ and $\gamma^2 \ll kl$ the speed of a shelf wave on a flat shelf is given by $\omega/k = -fl$. How does this compare with the lowest-mode solution of (25.68)?

11. Show that in a non-Boussinesq, uniformly stratified fluid with a free surface, the perturbation pressure of the lowest-mode edge wave is given by

$$p = A\rho_0(z) \exp\left[-k(x \cos\beta_b - z \sin\beta_b) + i(ky - \omega t)\right],$$

where A = constant, β_b is defined as in (25.12) and $\omega^2 = gk \sin\beta_b$.

12. From (25.67) determine the frequency, amplitude and bottom slope scales for which nonlinear effects will be important for edge waves.

26. WAVE TRAPPING BY SEAMOUNTS AND ISLANDS

The study of wave propagation in the presence of nonrectilinear obstacles presents additional complications. This topic may be conveniently divided into two parts: scattering problems, in which the far-field properties of waves scattered and/or reflected from an obstacle are of special interest, and trapping problems which focus on the behaviour of wave motion trapped in the immediate vicinity of the obstacle.

In a typical scattering problem, an incident wave impinges upon one or more obstacles of given shapes and dimensions. The scattered wave field is then specified by the geometry of these obstacles and the need to satisfy certain conditions on the boundaries of the obstacles. Many of the scattering problems encountered in the theory of ocean waves may be reduced to similar problems already treated for other kinds of waves, abundant examples of which are given in Jones (1964) for electromagnetic waves. Some scattering problems for surface gravity waves are discussed in Stoker (1957). Rhines (1969b) has discussed the scattering of planetary waves by simple obstacles. We shall return to this topic in Section 27.

Wave propagation around seamounts and islands may be considered as an extension to nonrectilinear coasts of the situation described in the previous section. Only a relatively simple case, that of trapping of long gravity waves by axisymmetric obstacles, is discussed in detail below. Qualitative extensions to more complicated situations will be mentioned briefly.

Considerations from ray theory

The simplest way to account for topographic trapping of gravity waves on a sloping bottom is in terms of wave refraction. As already seen in Fig. 25.1, wave fronts and rays issuing from a coast bordered by a sloping bottom are refracted back towards the shore because of the increase in propagation velocity with distance from the coast. The methods of ray theory will be liberally exploited in this section to help our understanding of the peculiarities of wave trapping by seamounts and islands. Exploring first in more detail the geometrical ray trapping along a sloping rectilinear coast, let us consider gravity waves in a nonrotating fluid. The local value of the frequency ω is given by (11.16), with $B = 0$:

$$\omega^2 = gk \tanh kH, \tag{26.1}$$

where $k = (k_1^2 + k_2^2)^{1/2}$.

From the ray equations (6.15), it is clear that if H is not a function of y or t, k_2 and ω are invariant along a ray. Thus,

$$k_2 = k \sin\theta = \text{constant}, \tag{26.2}$$

$$\omega = kc = \text{constant}, \tag{26.3}$$

and Snell's law holds in the form

$$\sin\theta/c = \sin\theta_o/c_o,$$

where the subscript o refers to values at the shore, $x = 0$. Rays leaving from the coast are turned back where $\theta = \pi/2$, at which position there is a *caustic*, or envelope of rays, located at that value of x_o where

$$c(x_o) = c_o/\sin\theta_o. \tag{26.4}$$

No ray leaving the coast at an angle θ_o can propagate beyond a distance x_o at which (26.4) holds: such waves are trapped near the coast. The position of the caustic (or of the turning point as it is also called), as given by (26.4), depends on the longshore wave-number k_2 through the angle θ_o.

It is obviously not necessary for the coast to be rectilinear or for the depth to vary only normally to the shore for this ray reversal phenomenon to occur: coastal trapping will occur at any frequency for which a continuous caustic is found offshore. Similarly, in the absence of a coast, waves may still be trapped by completely submerged bottom topography, such as a ridge, if continuous caustics are present on both sides of such a bathymetric feature. Should the coastline or the ridge terminate abruptly, or the bottom slope flatten out, the caustics may diverge to infinity: trapping no longer occurs and wave energy is scattered away from the guiding bathymetry. Shen et al. (1968) have exploited the methods of ray theory to obtain the spectra of waves trapped over a variety of bottom profiles, and provided general rules for constructing rays over arbitrary topographies. The two turning point problem of wave trapping by a pair of parallel caustics, such as would occur over a ridge, has been investigated by McKee (1975) who constructed solutions which are uniformly valid between and beyond the caustics. A discussion of a variety of turning point problems has been presented by Smith (1975). A recent review of the application of ray theory to gravity wave propagation in a fluid of variable depth has been presented by Shen (1975).

Gravity wave trapping around axisymmetric obstacles

Let us consider shallow-water waves in a nonrotating homogeneous fluid, for which the analysis is particularly simple. The results are qualitatively similar for short waves, and the effects of rotation will be discussed later. The governing equation for the free surface elevation in a nonrotating fluid of variable depth is (20.13) with $f = 0$ and $\beta = 0$; after one time-integration, (20.13) in this case becomes

$$(\partial_{tt} - gH\nabla_h^2)\eta - g\nabla_h H \cdot \nabla_h \eta = 0. \tag{26.5}$$

In a nonrotating fluid, the horizontal velocity vector is related to η through a simplified form of (15.50):

$$\boldsymbol{u}_t = -g\nabla_h \eta. \tag{26.6}$$

The simplest type of seamount is the axisymmetric one, for which, in plane polar coordinates (r, θ), $H = H(r)$ only. Considering motions which are harmonic in time, with $\eta \propto e^{-i\omega t}$, where $\omega > 0$, (26.5) takes the form

$$\left(\partial_{rr} + \frac{1}{r}\partial_r + \frac{1}{r^2}\partial_{\theta\theta} + \frac{H_r}{H}\partial_r + \frac{\omega^2}{gH}\right)\eta = 0. \tag{26.7}$$

Assuming separable solutions of the form

$$\eta(r,\theta) = M(\theta)S(r), \tag{26.8}$$

(26.7) splits into a pair of ordinary differential equations:

$$M''(\theta) = -m^2 M(\theta), \tag{26.9}$$

$$\left[\frac{\mathrm{d}^2}{\mathrm{d}r^2} + \left(\frac{1}{r} + \frac{1}{H}\frac{\mathrm{d}H}{\mathrm{d}r}\right)\frac{\mathrm{d}}{\mathrm{d}r} + \left(\frac{\omega^2}{gH} - \frac{m^2}{r^2}\right)\right] S(r) = 0, \tag{26.10}$$

where m is a separation constant, and primes indicate θ-derivatives. The solution of (26.9) is

$$M(\theta) = a\,\mathrm{e}^{im\theta} + b\,\mathrm{e}^{-im\theta}, \tag{26.11}$$

where a and b are constants of integration. For $m > 0$ and an $\mathrm{e}^{-i\omega t}$ time-dependence with $\omega > 0$, (26.11) describes a sum of two waves, one propagating around the origin in a counterclockwise direction (the a-term) and the other in a clockwise direction (the b-term). The separation constant m is an angular wavenumber and must be an integer if $M(\theta)$ is to remain single-valued: $m = 0, 1, 2, \ldots$. The form of radial dependence $S(r)$ depends on the depth profile $H(r)$; $S(r)$ may be either of oscillatory or of exponential character, the transition between the two types of behaviour occurring at critical radii which coincide with the caustic circles of ray theory.

It is readily established that in an ocean of finite depth there must always exist at least one caustic (or critical) circle, on which the radial dependence $S(r)$ passes from an exponential to an oscillatory form as r increases. Let us define $\Sigma(r)$ through

$$S'(r) = \Sigma(r)/rH. \tag{26.12}$$

Equation (26.10) for the radial dependence then takes the form

$$\frac{1}{rH}\Sigma'(r) = -\left[\frac{\omega^2}{gH} - \frac{m^2}{r^2}\right] S(r). \tag{26.13}$$

Plotting the directions of the trajectories of the system (26.12) and (26.13) in the phase plane (Σ, S) shows (Exercise 26.1) that oscillatory solutions, for which Σ and S are bounded and the trajectories enclose the origin of the phase plane, exist only for

$$\omega^2/gH > m^2/r^2. \tag{26.14}$$

Therefore, provided that H increases less rapidly than r^2 as $r \to \infty$, i.e.,

$$H = 0(r^{2-\epsilon}), \quad \epsilon > 0 \quad \text{as} \quad r \to \infty, \tag{26.15}$$

which is certainly the case when H is bounded, the solutions of (26.10) will be oscillatory at large r. Gravity waves can thus never be fully trapped by cylindrical topography in a realistic ocean, since radial phase (and also energy) propagation is always possible at large enough values of r. The form of $H(r)$ near the origin will determine the position and the number of caustic circles surrounding the centre of the axisymmetric seamount. In the simplest case where $H = H_o$, a constant, there exists only one caustic circle, located at $r = m\sqrt{gH_o}/\omega$.

Waves whose amplitude decays exponentially in a central region and then oscillates beyond a turning point are called "leaky" modes. The amount of energy leakage from the central core area is proportional to the radial energy flux beyond the outermost caustic. If the wave amplitude has decayed sufficiently at the distance of that caustic, the outward energy flux may be quite small, and hence, near the origin, the attenuation rate due to energy leakage may not be noticeable over a cycle and the waves may seem genuinely trapped over a time scale which is much longer than a wave period but much shorter than the time scale of energy leakage.

Spiral waves

In order to understand the wave solutions in the presence of bathymetric variations, it is useful at first to look at spiral waves, satisfying (26.9) and (26.10), in an ocean of uniform depth (Exercise 17.8). Defining

$$k = \omega/\sqrt{gH}, \tag{26.16}$$

the general solution of (26.10), with $\mathrm{d}H/\mathrm{d}r = 0$, is

$$S(r) = AJ_m(kr) + BY_m(kr), \tag{26.17}$$

where J_m and Y_m are Bessel functions of the first and second kind, respectively, and A and B are constants. To facilitate the interpretation of the solutions, consider the relatively short wave case, $m \gg 1$ (but $kH \ll 1$ still holds), for which we rely partly on the results of geometrical optics. From Abramowitz and Stegun (1965, Chapter 9) asymptotic forms for J_m and Y_m at large m are given as

$$J_m(kr) \sim \left(\frac{2}{m\pi \tan u}\right)^{1/2} \cos\left[m(\tan u - u) - \pi/4\right], \tag{26.18a}$$

$$Y_m(kr) \sim \left(\frac{2}{m\pi \tan u}\right)^{1/2} \sin\left[m(\tan u - u) - \pi/4\right], \tag{26.18b}$$

$$u = \cos^{-1}(m/kr)$$

for $kr > m$, i.e. in the oscillatory region, and as

$$J_m(kr) \sim \left(\frac{1}{2m\pi \tanh v}\right)^{1/2} \exp\left[-m(v - \tanh v)\right], \tag{26.19a}$$

$$Y_m(kr) \sim \left(\frac{2}{m\pi \tanh v}\right)^{1/2} \exp\left[m(v - \tanh v)\right], \tag{26.19b}$$

$$v = \cosh^{-1}(m/kr), \tag{26.19c}$$

for $kr < m$, the exponential region. The transition from one type of solution to the other occurs at a caustic circle, occurring at $kr = m$ as indicated earlier. The solutions of (26.9) for $kr > m$ may also be written in terms of Hankel functions, a representation which brings out their travelling wave nature more clearly. Using

$$H_m^{(1)} = J_m + iY_m; \quad H_m^{(2)} = J_m - iY_m, \tag{26.20}$$

one finds from (26.18) that for $m \gg 1, kr > m$,

$$H_m^{(1,2)}(kr) \sim \left(\frac{2}{m\pi \tan u}\right)^{1/2} \exp\{\pm i[m(\tan u - u) - \pi/4]\}. \tag{20.21}$$

The plus sign belongs to $H_m^{(1)}$, the minus sign to $H_m^{(2)}$. As we shall see, the $H_m^{(1)}$ solutions represent outward wave propagation, those with $H_m^{(2)}$, inward propagation. Following Longuet-Higgins (1967), the wavelike solution outside the critical circle $r = m/k$ may be interpreted as follows. The wavenumber in any direction is the rate of change of phase in that direction [see (6.3)]; thus, the radial wavenumber k_r is found from (26.21) as

$$k_r = m \frac{\mathrm{d}}{\mathrm{d}r}[\tan u - u].$$

Carrying out the differentiation and substituting for u from (26.18c), we obtain

$$k_r = k \sin u = (k^2 - m^2/r^2)^{1/2}. \tag{26.22}$$

With a tangential dependence of the form $\mathrm{e}^{im\theta}$, the tangential wavenumber is $k_\theta = m/r$. The total wavenumber is thus $k_r^2 + k_\theta^2 = k^2$, which from (26.16) is a constant. This result is of course not unexpected, since no refraction takes place in a medium of uniform depth. By this token, the rays must be rectilinear in the wavelike region ($kr > m$). For $k_r < m$, and on the caustic circle ($k_r = m$), the wave phase travels only tangentially: a smooth matching of phase propagation across the caustic circle implies that the rays must be tangential to this circle (a stricter geometrical proof is left as Exercise 26.2). The curves of equal phase (wave fronts) take on the spiral pattern shown in Fig. 26.1. The wave fronts are radial right on the critical (caustic) circle $r = m/k$, where phase propagation is tangential (Fig. 26.2a), but since the tangential wavenumber decreases as $1/r$, propagation tends to be more and more outwards as r increases and the wave fronts become nearly cylindrical at large r. These waves have been called spiral waves by Platzman (1971). Just how spiral wave fronts can arise in spite of the rectilinearity of the rays is shown by a simple geometrical construction in Fig. 26.2b.

The cylindrical seamount

The trapping of gravity waves around a centrally symmetric feature can now be described with the help of the uniform depth solutions. Let us consider wave motion around an isolated circular seamount of radius $r = a$ and uniform depth H_1, in an ocean of uniform depth H_2 (Longuet-Higgins, 1967). Since the solution must remain finite at the origin, only the J_m part of (26.17) is retained over the seamount:

$$\eta = AJ_m(k_1 r) \exp[i(m\theta - \omega t)], \quad 0 \leqslant r < a, \tag{26.23}$$

$$\text{with} \quad k_1 = \omega/\sqrt{gH_1}. \tag{26.24}$$

Since, from (26.20),

$$J_m = \tfrac{1}{2}(H_m^{(1)} + H_m^{(2)}), \tag{26.25}$$

two travelling waves interfere over the seamount, one ($H_m^{(1)}$) with outward phase propagation, the other ($H_m^{(2)}$) with inward propagation. Both waves rotate around the island in the same direction (i.e., with the same value of m). The radial dependence is oscillatory for $m/k_1 < r < a$ and exponential within the inner critical circle $0 \leqslant r \leqslant m/k_1$, as shown in Fig. 26.3.

Off the seamount, only the wave with outward phase propagation satisfies the far-field radiation condition (that energy should leak away from the seamount). Hence,

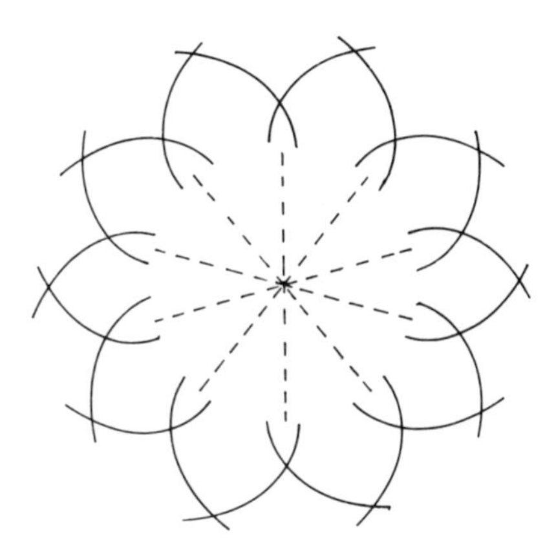

Fig. 26.1. The spiral wave pattern corresponding to $J_m(kr)\,e^{im\theta}$ for $m = 10$. The radial dependence is exponential inside the circle $r = m/k$ (inside which the crests are shown as dotted lines). The pattern consists of an incoming and an outgoing wave, each of which may be represented by an appropriate Hankel function, which add up to produce J_m as per (26.20).

$$\eta = BH_m^{(1)}(k_2 r)\exp[i(m\theta - \omega t)], \quad \text{for } r > a, \tag{26.26}$$

$$\text{where} \quad k_2 = \omega/\sqrt{gH_2}. \tag{26.27}$$

The constants A and B are related by the matching conditions at $r = a$, where the pressure must be continuous, which implies in the hydrostatic case that η is continuous. The normal transport $H\boldsymbol{u} \cdot \boldsymbol{n}$ must be continuous, where $\boldsymbol{n}$ is the normal to the depth discontinuity. In the absence of rotation, it is clear from (26.6) that the radial velocity is proportional to η_r, so that both η and $H\eta_r$ must be continuous at $r = a$. Hence the two equations

$$AJ_m(k_1 a) = BH_m(k_2 a), \tag{26.28}$$

$$AH_1 k_1 J_m'(k_1 a) = BH_2 k_2 H_m'(k_2 a), \tag{26.29}$$

must be satisfied, where $H_m = H_m^{(1)}$ for short, and the primes indicate differentiation with respect to the argument. Nontrivial solutions of (26.28) and (26.29) exist only if

$$H_2 k_2 J_m(k_1 a) H_m'(k_2 a) - H_1 k_1 J_m'(k_1 a) H_m(k_2 a) = 0. \tag{26.30}$$

Since k_1 and k_2 are related to ω through (26.24) and (26.27), the solution of (26.30) gives the eigenfrequencies of the trapped waves. These frequencies have been computed by Longuet-Higgins (1967) and by Summerfield (1969) for a range of values of the depth ratio H_1/H_2. The number of roots of (26.30) never exceeds m, and decreases as the depth contrast between the seamount and the surrounding ocean becomes smaller, i.e., as $H_2/H_1 \to 1$. The roots are complex and fall into two groups. In the first group, the imaginary part of each eigenfrequency is small; these waves have a long decay time and are strongly trapped over the seamount. The imaginary part of the frequency is large in the second group; these waves are rapidly radiated away from the seamount. Upon inspection of Fig. 26.3, it is evident that the waves are strongly trapped around the seamount if $H_m^{(1)}(kr)$ is of exponential character and decays between the edge of the seamount ($r = a$) and the

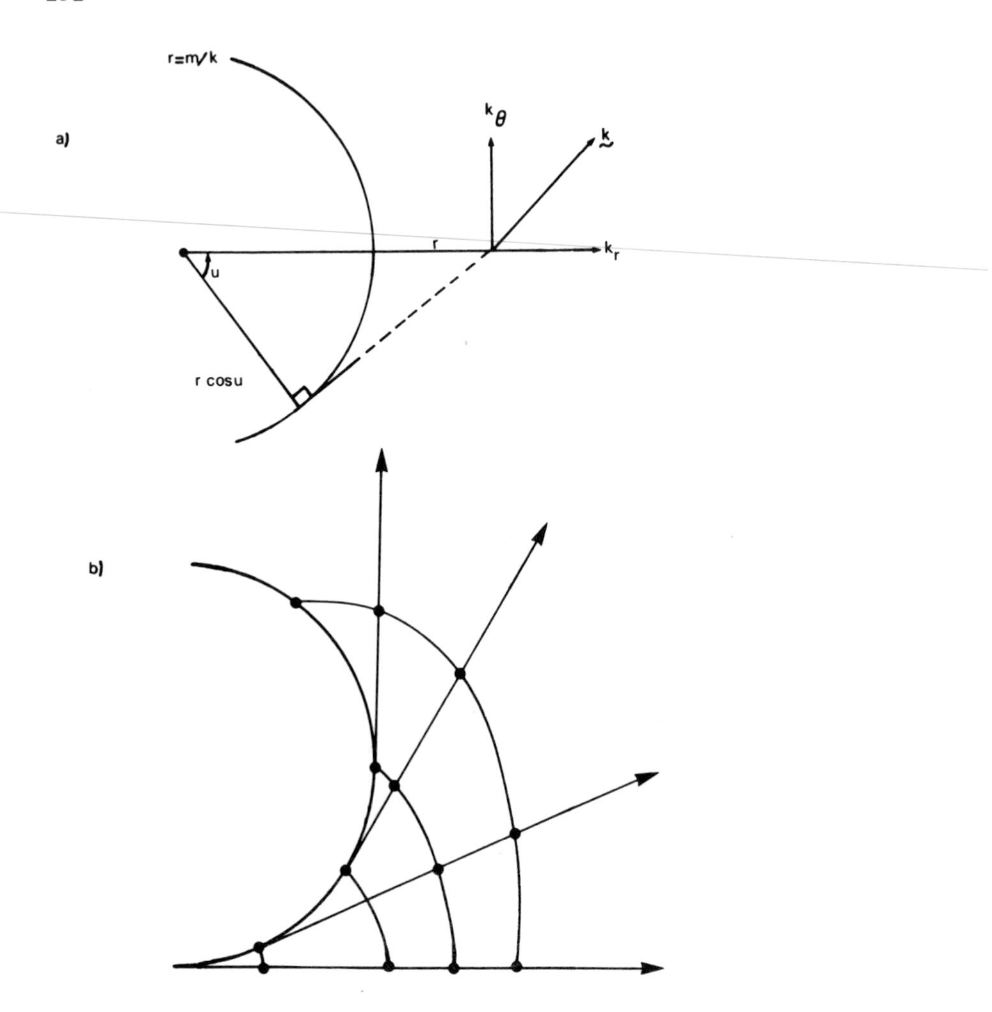

Fig. 26.2. (a) The relation between the angle u and the wavenumber vector; the circle $r = m/k$ is a caustic curve. (b) The construction of spiral wavefronts using rectilinear rays tangent to a circle. Equal distances are marked out on the rays, and joined by curves of equal phase. The spiral pattern emerges from the necessity of tying up the points of a certain phase which have travelled directly outwards with those which have travelled along the circumference of the circle.

outer critical radius ($r = m/k_2 > a$). The greater the difference between these two radii, the greater the outward amplitude decay and the stronger the trapping. The frequencies and decay times of these trapped waves are given by the first group of roots of (26.30). When the outer critical radius lies over the seamount ($m/k_2 < a$), there is no region of exponential decay and the wave energy rapidly leaks off to infinity: the complex frequencies of these very leaky modes fall in the second group of roots of (26.30).

Computation of the eigenfrequencies from (26.30) is a rather involved numerical procedure, and a simpler and more direct approximate method is available, based on geometrical optics. As noted earlier, the waves are strongly trapped if their amplitude decays sufficiently in a region of exponential dependence interior to the outer critical circle

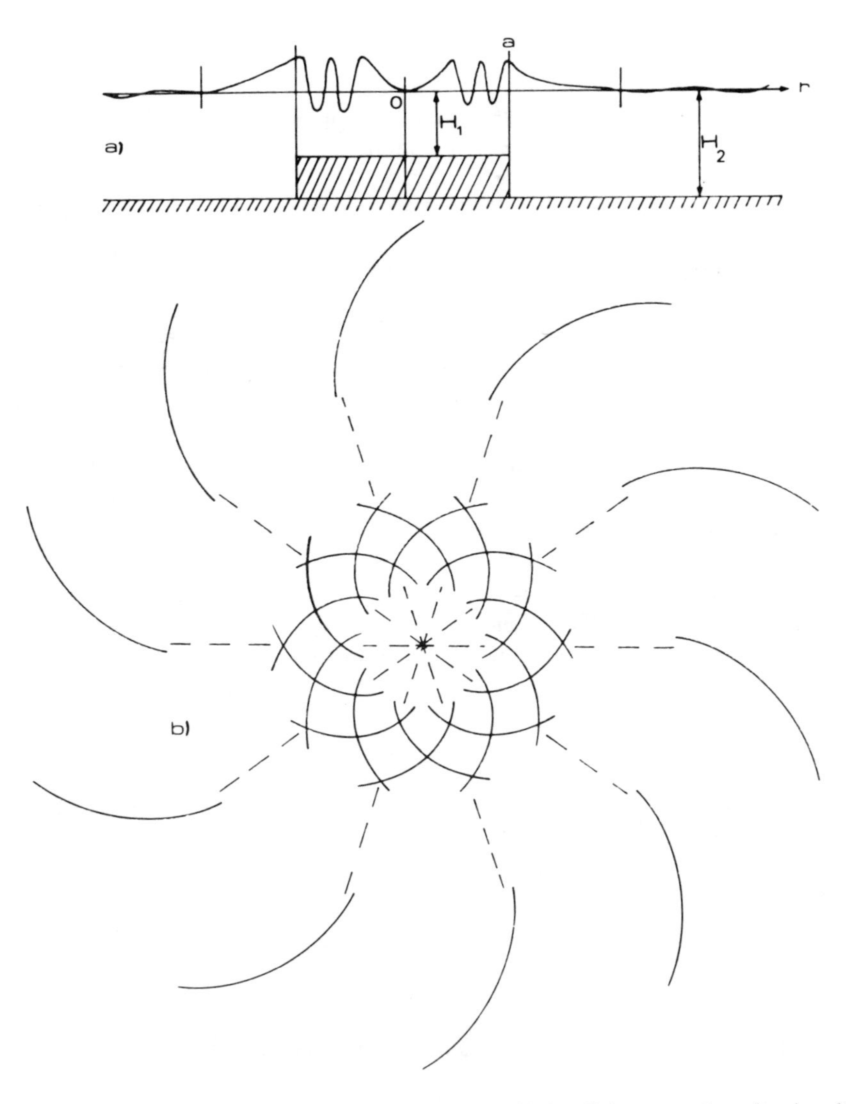

Fig. 26.3. Free waves trapped by a circular seamount. (a) A radial cross-section, showing alternation of regions with exponential and oscillatory radial dependence. (b) Plan view of the wave crests; the radial dependence is exponential on the dotted lines.

$r = m/k_2$. Glancing back at Fig. 26.3, it is clearly necessary that for efficient trapping, the radial dependence of η changes from sinusoidal to exponential across the depth discontinuity: the ray paths interior to $r = a$ must approach the edge of the seamount at an angle which exceeds the critical angle for internal reflection. For large enough m, the circular discontinuity may be considered locally straight on the scale of a wavelength and the results of Section 25 for wave reflection from a rectilinear discontinuity may be used. Snell's law (25.45) implies that for critical reflection of the interior rays, $\gamma \sin\theta_1 \geqslant 1$, i.e., $m/ak_1 \geqslant 1/\gamma$. With $\gamma = (H_2/H_1)^{1/2}$, this condition becomes

$$m/k_1a \geqslant (H_1/H_2)^{1/2}. \tag{26.31}$$

In addition, the distance between the outer critical radius and the edge of the seamount must be large enough (i.e., equal to a large number of deep-water wavelengths) for a sufficient amount of attenuation to take place. This condition may be written as

$$k_2\left(\frac{m}{k_2}-a\right) \gg 1, \quad \text{or} \quad m \gg 1 + k_2 a. \tag{26.32}$$

Since $k_1/k_2 = (H_2/H_1)^{1/2}$, the condition (26.31) may be expressed as

$$m > k_2 a, \tag{26.33}$$

which is certainly satisfied if (26.32) is obeyed. The criterion (26.32) is essentially a short wave condition, and we can thus expect that our approximate method, based on ray theory, will yield only the first group of eigenfrequencies. The method of finding these frequencies is based on phase matching of waves reflected back and forth between the inner critical circle and the depth discontinuity, and is similar to that used by Shen et al. (1968) to match rays trapped between caustics and other reflecting surfaces.

Consider a ray which travels inward from the edge of the seamount. On its inward path the wave is described by the Hankel function $H_m^{(2)}$; it is transformed upon reflection into the other Hankel function $H_m^{(1)}$, changing its phase by $-\pi/2$ in the process. The change of phase experienced by the time the ray has returned to $r = a$ is then $-\pi/2$ plus a phase shift proportional to the total path length covered; from (26.18) the total phase shift is

$$\phi_o = 2[m(\tan u_a - u_a) - \pi/4] \tag{26.34}$$

$$\text{with} \quad u_a = \cos^{-1}(m/k_1 a). \tag{26.35}$$

Notice that the angle u_a is the complement of the angle of incidence of interior rays upon the discontinuity (Fig. 26.2a). Our ray is then reflected back at $r = a$ with a change of phase ϕ_1, say. Since the radial dependence in $m/k_1 < r < a$ is of a standing wave nature, the phase must match again as the ray leaves $r = a$ to repeat its journey. Hence,

$$\phi_o + \phi_1 = 2n\pi, \quad n = 1, 2, 3, \ldots. \tag{26.36}$$

For ϕ_1, we use the result (25.50) obtained for a rectilinear discontinuity:

$$\phi_1 = 2\tan^{-1}(\nu H_2/\mu_3 H_1), \tag{26.37}$$

where we recall that μ_3 is the wavenumber on the shallow shelf ($H = H_1$) normal to the discontinuity, and ν the exponential decay rate [as given by (25.47) with $\Omega = 0$] in deep water ($H = H_2$). Both μ_3 and ν were nondimensionalized by the same length, the width of the shelf, so that dimensional wavenumbers may equally well be used in (26.37). In the present context, with the tangential wavenumber at the discontinuity given by m/a, the radial part of the shallow-water wavenumber (on the seamount) is (cf. 26.22)

$$\mu_3 = (k_1^2 - m^2/a^2)^{1/2}.$$

In view of the definition of u_a in (26.35), this may also be written

$$\mu_3 = \frac{m}{a}\tan u_a. \tag{26.38}$$

The decay rate ν is obtained from the second of the relations in (25.47), with $\Omega = 0$, as

$$\nu = \left[\frac{m^2}{a^2} - \frac{H_1}{H_2} k_1^2\right]^{1/2}$$

or using (26.35) again,

$$\nu = \frac{m}{a}\left[1 - \frac{H_1}{H_2}\sec^2 u_a\right]^{1/2}. \tag{26.39}$$

Thus, taking tangents of both sides of (26.36), using the expressions (26.37)–(26.39) for ϕ_1, μ_3 and ν, we finally obtain for the phase consistency relation,

$$\tan\left[m(\tan u_a - u_a) - \pi/4\right] = \frac{H_2}{H_1}\left[\left(\frac{H_2}{H_1} - 1\right)\cot^2 u_a - 1\right]^{1/2}. \tag{26.40}$$

This transcendental relation may be solved graphically, as shown in Fig. 26.4, to obtain approximate frequencies which are compared in Table 26.I with the values obtained from a numerical integration of (26.30). The agreement between the two methods of computation is satisfactory.

Against the background provided by the elementary example of trapping by a symmetric cylindrical seamount, we can now examine further complications, such as may be brought about by more complex bathymetry, asymmetry or rotation.

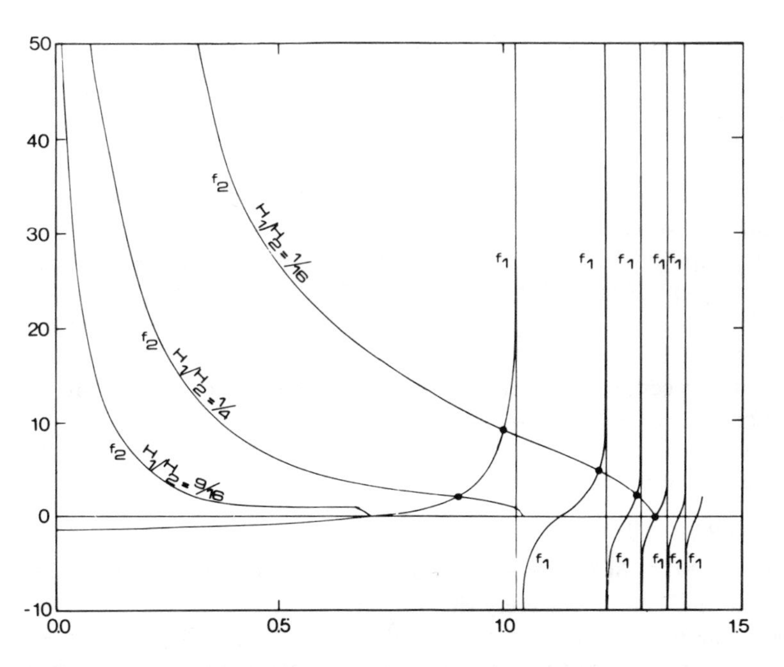

Fig. 26.4. The graphical solution of (26.40) for three depth ratios and $m = 4$. f_1 denotes the right-hand side of (26.40), f_2 its left-hand side. The eigenvalues of u_a are found at the intersection of f_1 and f_2; given a value of u_a, ω can then be found from (26.18c) and (26.24).

TABLE 26.I

The eigenfrequencies of free trapped waves around a circular seamount of radius a. Values of $ak = a\omega/\sqrt{gH_1}$ are given for two depth ratios and a few values of m. On the right, appear the results of the approximate calculation based on (26.40). The exact values, obtained from (26.30) are shown on the left

	$m = 1$	$m = 2$	$m = 3$	$m = 4$
$\frac{H_1}{H_2} = \frac{1}{16}$	3.408 3.656	4.909 4.921	6.196 6.209	7.423 7.447
		7.683 8.219	9.436 9.456	10.823 10.798
			11.989 12.818	13.942 14.027
				16.000 17.426
$\frac{H_1}{H_2} = \frac{1}{4}$		4.897 4.796	5.474 5.632	6.843 6.799

Other axially symmetric seamounts and islands

As indicated earlier (cf. 26.15), exponential-like solutions are found whenever the depth increases faster than r^2. It is then not only around cylindrical seamounts, such as examined above, that ray trapping by total internal reflection is possible, but also around any symmetric bathymetry with sides sloping down more rapidly than r^2. From the view-point of ray theory, we recover the necessity of the existence of caustic circles surrounding a bathymetric feature in order for trapping to occur.

A central island of radius $b < a$ is comfortably accommodated on a cylindrical sill of radius a, and its presence is of little importance to wave motion around the sill if $b < m/k_1$, i.e., if the island lies completely within the inner exponential region. Around larger islands, free trapped waves are still possible, but the structure and frequency of such waves is modified by the presence of the island. As the island occupies more of the upper surface of the seamount and the width of the "standing wave" zone between the shore of the island and the drop-off to deep water decreases, wave trapping becomes possible only for increasingly short waves, i.e., for larger values of m. Why? Consider ϕ_1, which is the inverse tangent of the right-hand side of (26.40); for the same depth contrast and internal angle of incidence $(\pi/2 - u_a)$, ϕ_1 is unchanged by the intrusion of a circular island on a circular seamount. In order to keep ϕ_o also unchanged and maintain the phase condition (26.36), it is necessary to increase the wavenumber, since the path length over the shelf is decreased by the presence of the island.

For a purely cylindrical island, rising like a column from the sea floor, it is also possible to find solutions of the form

$$\eta = Y_m(k_2 r) \exp[i(m\theta - \omega t)] \tag{26.41}$$

which satisfies the condition of zero normal velocity on $r = b$ (Chambers, 1965). The radial dependence of the solutions is everywhere in the nature of a standing wave and there is no energy leakage from the island. On the other hand, this standing wave pattern fills the whole sea, and it is difficult to conceive how it could be set up in a finite time! Such waves are not trapped in the sense that we have discussed so far, not even in the sense of the slowly leaking modes described earlier, since there is no sharp decrease of energy away from the island. Chambers nevertheless applied his work to explain the occurrence of peaks in the gravity wave spectrum observed in the vicinity of Guadelupe Island (see Exercise 26.4).

Asymmetric seamounts

Symmetric seamounts or islands are seldom met outside the laboratory. The most striking effect to which asymmetry gives rise is a phenomenon of mode splitting: depending on the type of asymmetry, each eigenfrequency of the symmetric seamount is either unchanged by asymmetry, or split into a "fine structure" of two or more adjacent frequencies. Longuet-Higgins (1967) and especially Summerfield (1969) have considered oscillations around elliptical seamounts. Their analysis is too lengthy to reproduce, and besides, the origin of the mode splitting may be accounted for geometrically by ray theory.

Any free mode may be represented by a ray which returns with the same phase (modulo 2π) to its starting point after completing one or more circuits around the seamount. For circular symmetry, the resulting ray pattern may be rotated arbitrarily around the centre of symmetry without changing its shape. Such a rotation may obviously not be performed in the absence of circular symmetry, and it will be seen that the asymmetry introduces some preferred directions, to which the ray patterns become bound: the arbitrariness in orientation of the ray pattern for any particular mode of the circular geometry is removed and a finite number of ray patterns, all slightly different, and all of an orientation imposed by the asymmetry, arise in place of the original symmetric pattern.

Let us illustrate the situation by contrasting elliptic with circular topography, and let us start with an enclosed basin of uniform depth, for purposes of presentation of the argument. The simplest mode of the circular basin has a modal line along any diameter (Fig. 26.5a), and a linear ray pattern corresponding to reflection of a ray back on itself. In an ellipse, however, a ray will be reflected upon itself only along the axes; since the

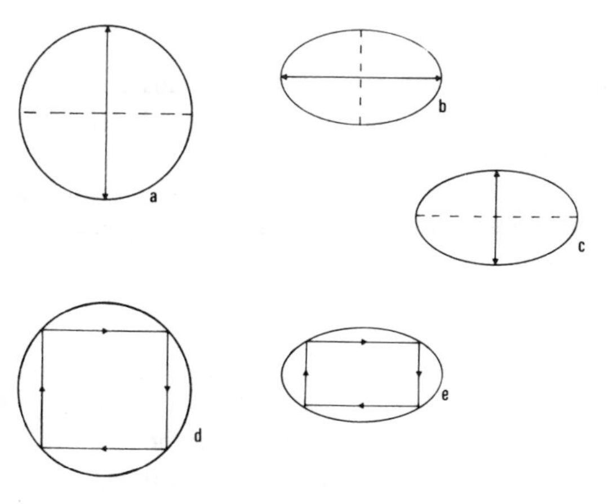

Fig. 26.5. The influence of asymmetry, and geometrical mode splitting. (a) Transverse mode in a circular basin. (b and c) The corresponding pair of modes in an ellipse. (d) A mode with $m = 2$ in a circle; (e) The same mode (modified, but not split) in an ellipse.

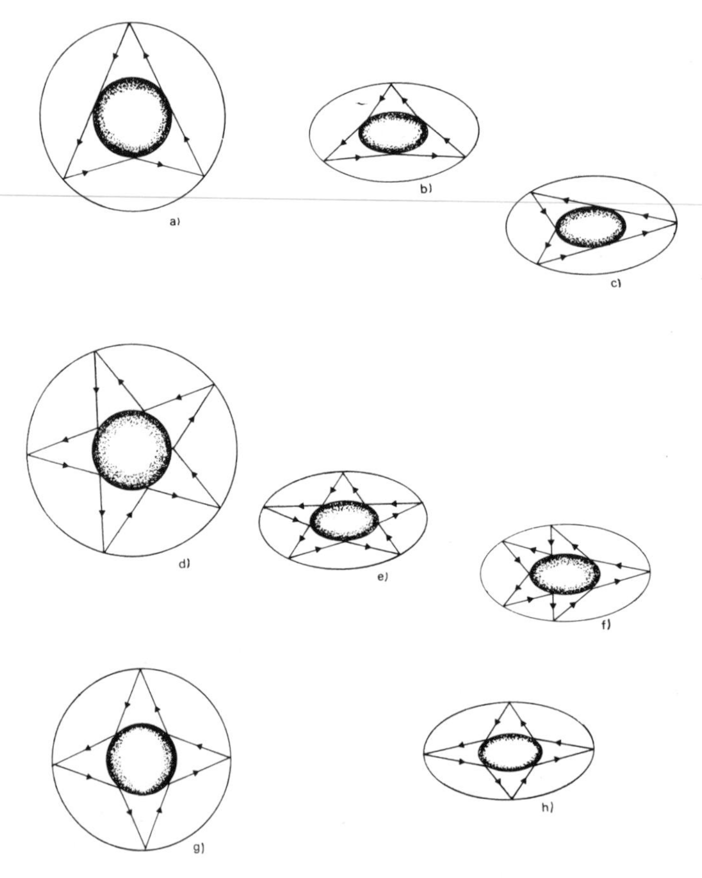

Fig. 26.6(a) The $m = 3$ mode in a circular annulus. (b and c) The corresponding pair of modes in an elliptical annulus: note the asymmetry in the modal patterns imposed by the geometry. (d, e and f) A similar comparison for the mode $m = 5$. In contrast, the mode $m = 4$ is not split (g, h).

axes are of unequal length, the longitudinal (along the major axis) and transverse oscillations have different frequencies (Figs. 26.5b and c). The asymmetry has imposed preferred directions of oscillation and splits the basic cross-basin mode of the circle into two different modes. Seen in this light, the origin of the mode-splitting phenomenon is evident. It should not be concluded, however, that all the modes are split. Suppose that the boundary of the basin, in plane polar coordinates (r, θ) is given by

$$r = r_0(1 + q \cos p\theta), \qquad (26.42)$$

and thus represents a shape which departs from a circle of radius r_0 through a modulation of the radial distance of the boundary with amplitude $q(< 1)$ and angular frequency p, where p is an integer $\geqslant 2$. Modes which have the same basic symmetry as the seamount, i.e., for which $m = NP$, where $N = 1, 2, \ldots$, are not split. Thus, for an ellipse, $p = 2$; the cross-basin mode $m = 1$ is split, but the next mode ($m = 2$), as shown in Figs. 26.6d, e, is not split. The reader may refer to Jeffreys (1924) for a more complete discussion of oscillations in an elliptical nonrotating basin.

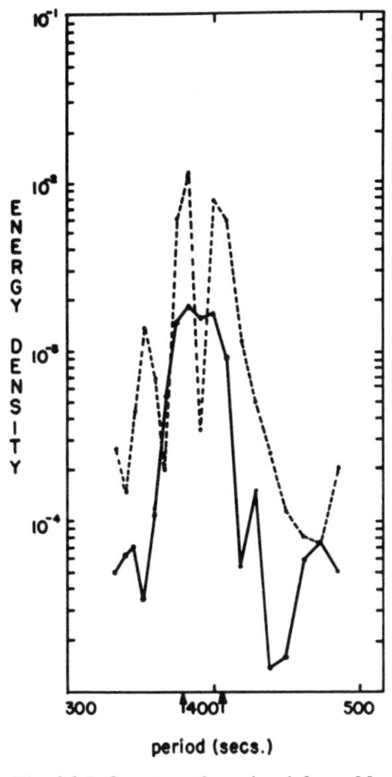

Fig. 26.7. Spectra of sea level from Macquarie Island. The solid line and the dotted line correspond to two different 5-hour time series. The arrows indicate periods of 6.3 and 6.8 minutes. (From Summerfield, 1969.)

On a cylindrical seamount, the rays are reflected back and forth between the outer edge and the inner critical circle; the rays must now be traced between concentric circles or, in a less symmetrical situation, concentric ellipses. The ray patterns for these two geometries are contrasted in Fig. 26.6 for a few of the lower modes. The influence of asymmetry in establishing preferential axes, and introducing differences between oscillations which were indistinguishable in the circular geometry, is again evident.

Observations of waves trapped around islands

The recent interest in wave trapping around islands has been in great part stimulated by observations of short period oscillations around Macquarie Island. The early records (Radok, 1964) revealed short period ($\simeq 6$ min) fluctuations, lasting for several days and fluctuating in amplitude with a beat of about three hours (Fig. 26.7). Macquarie Island is shaped like a thin rectangle, about 37 km in length and 5 km wide, with a circumference of about 80 km. The locally observed period of a wave of mode number m travelling in water of depth H_1, around an island of circumference D may be written as

$$T_m = D/m\sqrt{gH_1}, \qquad (26.43)$$

an estimate based on ray theory, and which completely ignores any possible mode splitting due to the marked asymmetry of the island. Nevertheless, with $D = 80$ km and $H_1 = 100$ m, Longuet-Higgins (1967) found

$$T_m = 42/m \text{ minutes,}$$

which would indicate that the observed waves correspond to the mode $m = 7$. Further measurements (Summerfield, 1967) have revealed the existence of two distinct oscillations, with periods of 6.3 and 6.8 minutes (Fig. 26.7). This frequency split has been accounted for by Summerfield (1969) in terms of wave trapping around an elliptical island resembling Macquarie. The three-hourly modulation would then follow from a beat between the individual waves, at a frequency $\Delta\omega = (\omega_1 - \omega_2)/2 \simeq 0.35$ cycles/ hour.

Trapping by ridges

We have already seen in Section 24 how waves could be trapped along an escarpment, and inferred that trapping over a ridge could be studied by an extension of the analytical method used for an escarpment. A ridge may also be considered as an extremely elongated asymmetrical seamount. The methods of ray theory used advantageously in the above paragraphs are also useful in understanding wave trapping over a ridge.

On a steep-sided ridge, the wave trapping process is readily understood in terms of multiple reflections, from one side back to the other, of rays incident upon the edges of the ridge at an angle exceeding the critical angle. The ridge becomes a waveguide. For smooth-sided ridges the trapped rays are turned back at caustic lines running parallel to the ridge. The dispersion relation for waves trapped over a flat-topped ridge may be derived by the same type of phase consistency requirement used to obtain (26.40) (see Exercise 26.5). As is common in waveguide theory (Tolstoy, 1973, Chapter 3), a cut-off frequency is found, below which no propagation is possible. A more interesting and less obvious property of gravity waves trapped over a long ridge is the occurrence of a minimum in the group velocity at a finite, nonzero wavenumber k_0. An example of the dispersion curves, drawn from Buchwald (1968a), is shown in Fig. 26.8. Waves travelling at the minimum group velocity will form the trailing edge of any wave group propagating on the ridge. Such waves will not disperse away from each other since $\partial c_g/\partial k$ vanishes at $k = k_0$, and, because of the absence of dispersion at that wavenumber, the wave amplitude at k_0 should be expected to decrease less rapidly with distance from the source than at other wavenumbers. At a sufficiently long distance from the source, a pronounced spectral peak could be expected to appear at k_0. This type of phenomenon is familiar in seismology, where it is called the Airy phase (Ewing et al., 1957, Chapter 4); it arises in that context from wave trapping in low-velocity layers of the Earth's crust, in a fashion entirely analogous to the trapping of gravity waves by a ridge. The Airy phase will be encountered again below (Section 50) in the theory of impulsively generated surface gravity waves. No definite indications of the existence of an Airy phase in waves travelling over oceanic ridges have yet been found. De Szoeke (1971) looked for this effect in his study of gravity waves travelling on the Norfolk Island ridge, but could not find it.

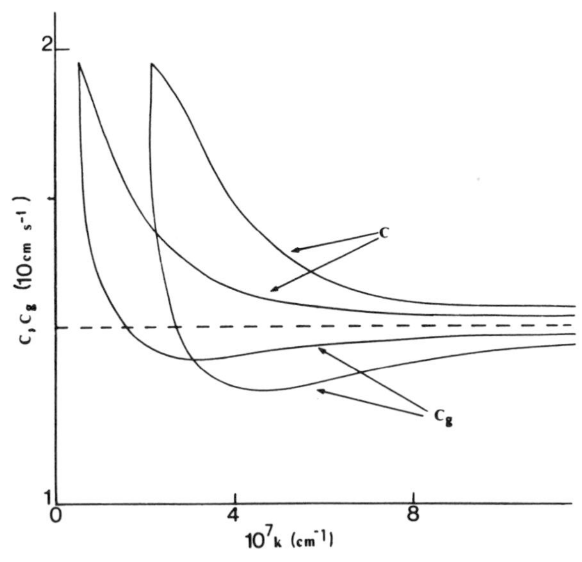

Fig. 26.8. Dispersion curves for gravity waves trapped on a flat-topped ridge: graphs of c and c_g for the first two modes for a ridge of width 100 km, depth 2000 m, between deep basins of depths 5000 m and 4000 m, respectively.

The influence of rotation

The Coriolis force has been left out of the entire discussion so far. It must, however, be included if the results are to be extended to oscillations of lower frequencies. It was noted in the previous section that the Coriolis force introduces an asymmetry in the speed of propagation of edge waves: in the Northern Hemisphere, a left-bounded edge wave has its speed of propagation increased by the Earth's rotation, while a right-bounded edge wave has its speed diminished (cf. 25.28). Rotation will also split the frequency of a pair of trapped modes travelling in different directions around a seamount or an island. There are thus two sources of frequency fine-structure for topographically trapped waves: geometrical asymmetry and rotation. The relative effect of the two sources will depend on the degree of asymmetry and on the frequency ratio ω/f. For $\omega \gg f$ and $m \gg 1$, Longuet-Higgins (1967) has shown, using the approximate phase-matching arguments used above in (26.36), that the frequency split around a circular seamount is $\Delta\omega \sim f/m$. At the latitude of Macquarie Island (54° 30′S), $|f| = 1.18 \cdot 10^{-4}$ rad s^{-1} and $\Delta\omega$ for the seventh mode is 0.06 cycle/hour, about an order of magnitude too small to account for the observed modulation of the short waves, which must then be attributed to the ellipticity of the island.

In addition to modifying the gravity edge wave modes, the Earth's rotation allows for trapping of low-frequency ($\omega < f$) second-class waves. The situation is similar to that which prevails along rectilinear coasts, as described in the previous section. Wave trapping by a cylindrical island is possible only if free surface effects are included, and if the radius of the island exceeds $\sqrt{gH}/f$, the barotropic Rossby radius (Longuet-Higgins, 1969c). If the island has sloping sides, however, trapping is possible even in a rigid-topped ocean providing the frequency is sufficiently small (Rhines, 1969b; Longuet-Higgins, 1970c). For axisymmetric islands, there is a double infinity of discrete modes whose wavelengths and frequencies are determined by the azimuthal wavenumber and the topography. The frequency is always less than f and, in the Northern Hemisphere, the waves progress clockwise around the island in accordance with the general rule that the phase should travel with the shallow water on its right. The 0.5 cycle/day (cpd) oscillation observed in the sea-level records of Oahu (see Fig. 18.4) has been identified by Longuet-Higgins (1971) as a trapped Rossby wave because its phase does propagate clockwise around the island. Note the negative phase difference at the 0.5 cpd coherence peak in Fig. 18.4, corresponding to clockwise propagation. The low-frequency oscillations visible at 0.25 and 0.35 cpd have already been discussed in Sections 18 and 21. The broad peak in coherence centered on $\omega \simeq f = 0.73$ cpd (the local inertial frequency) deserves special consideration.

Recent laboratory experiments on resonant long-period waves around islands performed by Caldwell and Eide (1976), give results which are in agreement with calculated resonant frequencies at frequencies higher than inertial, and for Kelvin-type waves at all frequencies. The results obtained for subinertial shelf wave resonant frequencies, however, disagree with calculated values; this discrepancy remains unexplained.

Trapped inertial oscillations

Let us set $\omega = f$ and choose cylindrical coordinates (r, θ), with radial and tangential velocities denoted by u and v, respectively. The long wave equations over a flat bottom take the form

$$u_t - fv = -g\eta_r, \tag{26.44a}$$

$$v_t + fu = -g\eta_\theta/r, \tag{26.44b}$$

$$(ru)_r + v_\theta = -r\eta_t/H. \tag{26.44c}$$

At the local inertial frequency, all variables have a harmonic time-dependence of the form $\exp(-ift)$, and the above set of equations becomes

$$f(iu + v) = g\eta_r, \tag{26.45a}$$

$$rf(iv - u) = g\eta_\theta, \tag{26.45b}$$

$$H[(ru)_r + v_\theta] = irf\eta. \tag{26.45c}$$

Combining (26.45a) and (26.45b), we find

$$r\eta_r + i\eta_\theta = 0. \tag{26.46}$$

This has the general solution

$$\eta(r, \theta) = F(i\theta + \ln r), \tag{26.47}$$

where F is an arbitrary function. In the absence of islands or boundaries, the only analytic function F which is bounded everywhere is F = constant. In that case, the momentum equations give

$$u = iv: \tag{26.48}$$

the motion is entirely horizontal and circularly polarized. The velocity vector rotates in a clockwise sense. These inertial waves are recognizable as a limiting case of the gyroscopic waves discussed in Section 9.

Let us now put an island (of radius b) at the origin and seek solutions of (26.45) which satisfy $u(b, \theta) = 0$ and which remain bounded in $b \leqslant r < \infty$. Solutions of the form

$$F(s) \propto e^{-ms}, \quad m = 0, 1, 2, \ldots \tag{26.49}$$

for the arbitrary function F of (26.47) are bounded as $r \to \infty$ and are periodic in θ. From (26.47) and (26.49), the surface elevation for the mth mode can be written in the form

$$\eta(r, \theta) = Ab^m r^{-m} \exp(-im\theta), \tag{26.50}$$

where A is the amplitude of η at the island ($r = b$). Eliminating v between (26.45a) and (26.45c), we find that

$$(ru)_r - iu_\theta = \frac{irf}{H}\eta - \frac{g}{f}\eta_{r\theta}. \tag{26.51}$$

Substituting (26.50) into (26.51) we see that for $u(r, \theta) = u(r) \exp(-im\theta)$ a simple ordinary differential equation for $u(r)$ is obtained. Integration of the latter together with the condition $u(b) = 0$ gives

$$u(r) = \frac{ifA}{2rH}(r^2 - b^2), \quad m = 0, \tag{26.52a}$$

$$u(r) = \frac{ifAb}{H}\left[\ln\left(\frac{r}{b}\right) - \frac{gH}{2f^2}\left(\frac{1}{r^2} - \frac{1}{b^2}\right)\right], \quad m = 1, \tag{26.52b}$$

$$u(r) = \frac{iAb^m f}{2H}\left[\frac{b^{2-2m} r^{m-1} - r^{1-m}}{m-1} + \frac{gHm}{f^2}\left(r^{-m-1} - b^{-2m} r^{m-1}\right)\right], m \geqslant 2. \tag{25.52c}$$

There are no solutions for $u(r)$ which remain bounded as $r \to \infty$ and hence no trapped wave solutions. However, if one assumes a priori that the non-divergent approximation may be made for islands small enough that $f^2b^2/gH \ll 1$, only the terms preceded by the factor gH/f^2 remain in the solution (26.52) for $u(r)$. In that case, the $m = 1$ mode remains bounded. This approximation always breaks down at sufficiently large values of r and the logarithmic singularity prevails as $r \to \infty$. Longuet-Higgins examined the non-divergent case and argued that since the phase always travels clockwise around the island [see (26.50) combined with $\exp(-ift)$], the phase difference between Honolulu and Mokuole in Fig. 18.4 could be accounted for by locally excited inertial, or near-inertial oscillations.

Our discussion has been limited to barotropic waves. For the baroclinic modes, an equation of the form (26.5) may be written, in which the actual depth is replaced by the equivalent depth h_n. Unfortunately, as we have remarked earlier, the separation of the vertical and horizontal dependencies is not generally possible in a stratified fluid with a sloping bottom and one cannot simply extend the results to internal modes by replacing

H by h_n. Since h_n depends on the stratification as well as on the actual depth, local maxima of the stability frequency N are analogous to seamounts, and internal wave trapping by raised isopycnals could arise and might be treated according to the above methods. However, tilted isopycnals are always associated with geostrophic currents, and the analysis would not be adequate without the addition of a mean geostrophic flow.

Exercises Section 26

1. Show geometrically that the system (26.12), (26.13) has oscillatory solutions, corresponding to trajectories encircling the origin of the (Σ, S) plane, whenever $\omega^2/gh > m^2/r^2$.

2. Show that deep-water rays around a cylindrically symmetric seamount are tangent to the radial boundary of the seamount.

3. Consider a two-stage seamount, consisting of a right cylinder sitting on top of a similar, broader cylinder. Sketch the form of the solution for the surface elevation as a function of radial distance for various relative diameters of the two cylinders.

4. Show that there is a symmetric nonrotating mode ($m = 0$) for standing wave oscillations of the form (26.41). Write down the dispersion relation for that mode. Guadelupe Island is located at 29°N, and may be idealized for the purpose of this problem by a cylinder of radius 50 km in water of depth 1000 m. Calculate the lowest eigenfrequency in the $m = 0$ mode (Chambers, 1965).

5. Use the phase matching technique that yielded equation (26.40) to find a dispersion relation for trapped gravity waves propagating on a rectangular ridge (Buchwald, 1968a).

27. WAVE DIFFRACTION AND SCATTERING

Waves are diffracted and/or scattered when they encounter obstacles whose boundaries have radii of curvature comparable to or less than the wavelength of the incident field. Although these two words are often used interchangeably, strictly speaking, diffraction occurs when waves encounter a sharp edge or corner (with zero radius of curvature) and scattering, when they encounter bluntly shaped objects or irregular surfaces. For numerous examples of the diffraction and scattering of electromagnetic and acoustic waves, we refer the reader to Jones (1964) and to Morse and Ingard (1968), respectively. We mention these important references since the mathematical methods used to handle the diffraction and scattering problems considered can readily be adapted to related oceanic situations. It is important to emphasize that the simple ray theory of Section 6 is inapplicable to the study of diffraction and scattering, although an extension of ray optics to a geometrical theory of diffraction, for example, is available (Keller, 1962b; Christiansen, 1975).

In this section we present a survey of selected ocean wave diffraction problems, without going into the details of their solutions. Our discussion of scattering will be even more restricted, covering only that which occurs near irregular coastlines. Scattering by islands, seamounts and the like has been thoroughly discussed by Jonsson et al. (1976) and Rhines (1969b) for the cases of gravity waves and planetary waves, respectively.

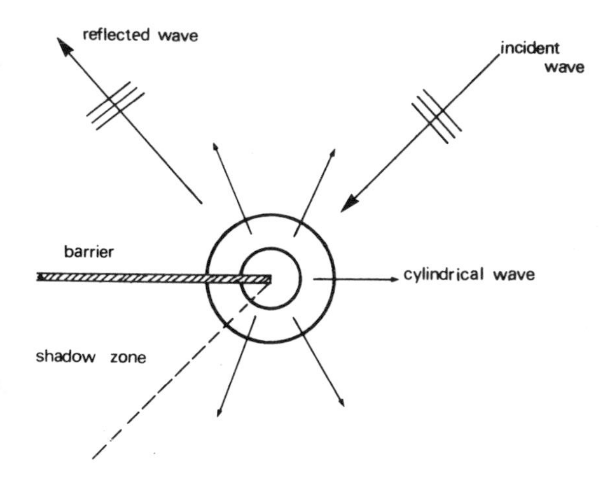

Fig. 27.1. The diffraction of surface gravity waves by a semi-infinite vertical barrier in the absence of rotation.

The classic example of a diffractive situation is the Sommerfeld diffraction problem: a plane wave incident upon a semi-infinite rigid barrier normal to the plane containing the incident wavenumber (Fig. 27.1). In the context of electromagnetic or acoustic waves, say, the total solution consists of the incident wave, a wave reflected from the barrier and a diffracted wave which, far from the barrier, takes the form of an outward propagating cylindrical wave. Only the diffracted wave penetrates the geometrical shadow zone behind the barrier. The Sommerfeld diffraction problem may be solved by a variety of analytical techniques.

Diffraction in a nonrotating fluid

The diffraction of surface gravity waves may be formulated in terms of a potential, in a form which is identical to the electromagnetic or acoustic problem. Diffraction by a vertical barrier is as described above (Fig. 27.1) and the theory has been verified experimentally by Putnam and Arthur (1948). The results are directly applicable to an evaluation of surface gravity wave agitation in the lee of breakwaters. Since the horizontal dependence equations (10.29) and (10.30) reduce, in the absence of rotation, to Helmholtz's equation

$$\nabla_h^2 P + \frac{\omega^2}{gh_n} P(x, y) = 0 \tag{27.1}$$

for any vertical mode, the diffraction properties of the baroclinic modes are identical to those of the surface mode.

Diffraction of gravity waves from a variety of types of vertical barrier has been considered by many authors; references may be found in Porter (1972).

Diffraction of long gravity waves in a rotating fluid

The introduction of rotation complicates the boundary condition on a vertical reflecting wall (Section 23); it also allows the existence of a Kelvin wave travelling (in the Northern Hemisphere) with the wall on its right. The diffracted wave field for long-period waves incident on a semi-infinite vertical barrier then includes, in addition to a reflected wave and a cylindrical scattered wave, a right-bounded Kelvin wave (Fig. 27.2).

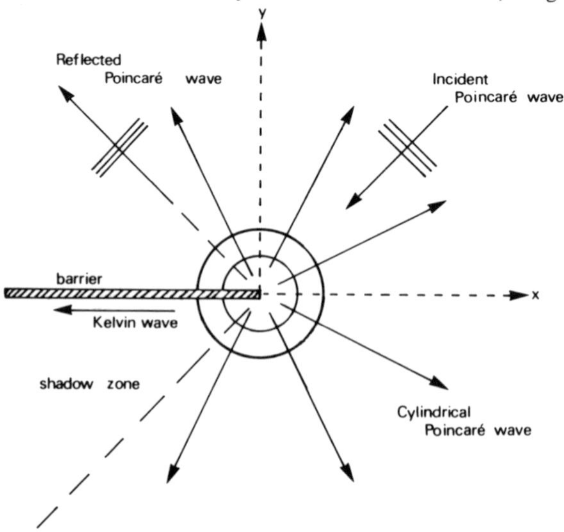

Fig. 27.2. The diffraction pattern due to a Poincaré wave incident on a semi-infinite barrier.

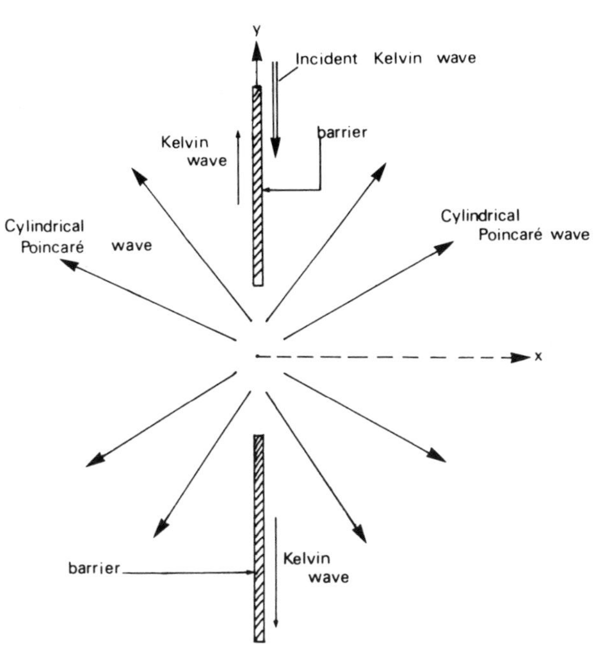

Fig. 27.3. The diffracted Kelvin and Poincaré waves due to a Kelvin wave incident upon a gap.

The Sommerfeld diffraction problem in a rotating fluid was first considered by Crease (1956a) for the case of normal incidence. Modelling the British Isles by a semi-infinite vertical barrier, he showed that the amplitude of the Kelvin wave generated in the North Sea by the diffraction of a Poincaré wave incident from the North Atlantic could exceed the amplitude of the incident wave. This coastal amplification was proposed by Crease as a possible explanation for the origin of some of the large-amplitude storm surges (Section 52) causing flooding on the eastern British coast. The Sommerfeld diffraction problem for Poincaré waves of arbitrary incidence angle was solved by Chambers (1964) and the extension to diffraction by a wedge was carried out by Roseau (1967). Crease (1958) also considered the diffraction of long waves by two parallel barriers, either both semi-infinite or one infinite and one semi-infinite. The diffraction of Kelvin waves by a gap has been examined by Buchwald and Miles (1974) (Fig. 27.3). The diffraction of Kelvin waves by a semi-infinite submarine escarpment between regions of different depths has been considered by Pinsent (1971). The diffracted field then consists of a Kelvin wave and of a double Kelvin wave travelling along the depth discontinuity (Fig. 27.4).

In an ocean of constant depth, the basic equations are separable and the baroclinic modes satisfy the same equations (10.29) and (10.30) as the barotropic mode, with a different equivalent depth h_n. The results obtained for surface waves are thus also applicable to internal waves. The Sommerfeld diffraction problem for internal waves in a medium with uniform stratification ($N =$ constant) has been solved by Manton et al. (1970).

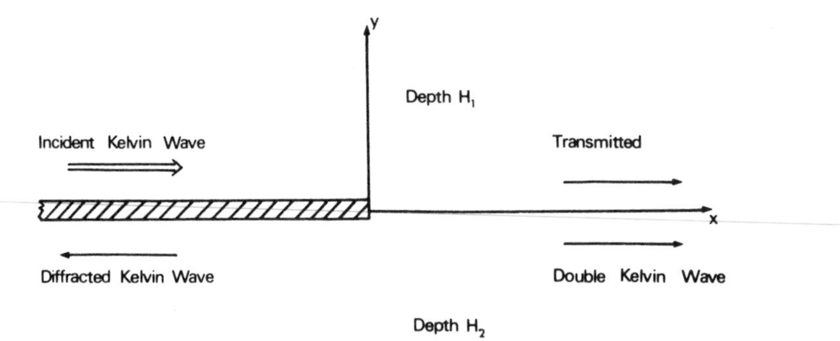

Fig. 27.4. The diffraction of a Kelvin wave by a semi-infinite escarpment with unequal depths of water across the axis of the barrier. In the Northern Hemisphere a double Kelvin wave is transmitted only if $H_2 > H_1$ (shallow water on right).

Diffraction of planetary waves

Under the transformation introduced in Exercise 18.1, planetary wave propagation becomes isotropic and is described by a Helmholtz equation. However, since the group velocity is opposite in direction to the phase velocity in the transformed wavenumber plane, proper care must be given to the correct radiation condition. Diffraction by a semi-infinite vertical wall has been examined by Mysak and LeBlond (1972) and by Siew and Hurley (1972). McKee (1972) also solved this problem and, in addition, studied the diffraction through a gap in a north–south wall considered as an idealized model of the Drake Passage. Diffraction of planetary waves by a wedge has been examined by Hall (1974).

Diffraction and scattering by coastal variations

In recent years considerable attention has been devoted to the effects of coastal variations on the propagation of the barotropic tide, as modelled by Kelvin and Poincaré waves. The types of coastal variations considered in these studies generally fall into two categories: (1) sharp bends; (2) small irregularities on an otherwise rectilinear coast.

One of the first studies which dealt with coastlines of the first category is that of Buchwald (1968b) in which the Wiener-Hopf method was used to obtain the diffracted wave field due to a Kelvin wave incident on a right-angle corner. He showed that for subinertial frequencies ($\omega < f$) a Kelvin wave propagates around the corner without change in amplitude. However, at higher frequencies ($\omega > f$) the amplitude of the incident Kelvin wave is reduced as it passes around the corner and cylindrical Poincaré waves are generated at the corner. The same results also hold for bends of all other angles except $\pi/(2n + 1)$, $n = 1, 2, \ldots$, in which case the superinertial Kelvin wave propagates around the corner without a reduction in amplitude (Packham and Williams, 1968). The analysis of Packham and Williams was further extended by Miles (1972) to obtain explicit results for the phase of the transmission coefficient. In particular, Miles found that for the bend at Cape Mendocino, California, the phase shift for a Kelvin wave is 1.3 hours.

Pinsent (1972) was the first to present the theory of the scattering of Poincaré waves by coastlines of the second category. He also discussed the attenuation of a Kelvin wave as it propagates along an irregular coastline, with the energy being radiated into a diffusely scattered Poincaré wave field. Pinsent's analysis is based on an application of the Born approximation. That is, the total wave field is expanded in powers of the small parameter $\epsilon = \xi/r$ (ξ = amplitude of coastal irregularities and r = deformation radius), with the zeroth-order solution representing the incident field, and the higher-order corrections, the scattered fields. However, as will be seen in Section 34, such an approach gives a uniformly valid solution only for coastal irregularities that occur in a finite segment of the coast (i.e., that have compact support). To deal with extensive irregular coastlines, other techniques must be used. One such other approach involves treating an infinitely long coast as straight except for small deviations which are represented by a stationary, random zero-mean function. Then a modified form of the theory of wave propagation in random media can be used. The two problems considered by Pinsent [Kelvin wave generation and Kelvin wave attenuation] have been treated in this manner by Howe and Mysak (1973), Mysak and Tang (1974) and Mysak and Howe (1978).

28. WAVES IN CHANNELS AND BAYS

Continuing our ascent on the ladder of increasing geometrical complexity, we now examine the essentials of wave propagation between parallel, or nearly parallel, boundaries; that is, a channel. We shall consider in succession the propagation of waves travelling in one direction only, then the interference of waves travelling in opposite directions, and finally the problem of wave reflections from a closed or open boundary. As a channel is basically a waveguide, we shall find that many of the results obtained for ocean waves have equivalents in other physical systems, such as in acoustic or electromagnetic waveguides. The analysis will focus on long waves in a rotating fluid of uniform depth, a topic adequate to illustrate most phenomena of interest. The influence of topographical variation in the cross-channel direction on both barotropic and baroclinic modes will be discussed briefly. The problem of interference of high-frequency gravity waves (for which we may take $f = 0$) in a channel of uniform depth is straightforward and left as an exercise; the same problem in a channel of varying depth is more complex and will be examined below.

In this section we shall not explicitly discuss the phenomena of transverse seiches in channels, although the mathematical description of such motions can be formally obtained from the solutions given below by setting the long-channel wavenumber equal to zero. For an interesting discussion of internal transverse seiches in long lakes (such as Lake Ontario and Lake Michigan) we refer the reader to Csanady (1973a). The problem of internal seiches in a channel of circular cross-section has been investigated by Yang and Yih (1976).

The channel of uniform depth

Wave propagation in a channel of uniform depth, i.e., of rectangular cross-section, is governed by the separated equations (10.41) and (10.42) for the vertical dependence and (10.39) and (10.40) for the horizontal dependence. We shall concentrate our attention on long wave propagation, in which case the equivalent depths h_n, as determined by the solution of the vertical problem (15.25), (15.27), (15.28), are independent of frequency. With the pressure in the form $p(x, y, z, t) = p_n(x, y, t)\Pi_n(z)$, the equation for p_n is (23.3):

$$\partial_t \mathcal{L}\left[\nabla_h^2 - \frac{1}{gh_n}\mathcal{L}\right]p_n = 0, \tag{28.1}$$

where $\quad \mathcal{L} = \partial_{tt} + f^2.$

Let us position the walls of the channel at $x = 0, l$. The boundary conditions to be satisfied there are

$$u(0, y, z, t) = u(l, y, z, t) = 0. \tag{28.2}$$

The velocity component u is related to the pressure through (15.19):

$$\begin{aligned} \mathcal{L}u &= -\rho_0^{-1}(p_{xt} + fp_y), \\ &= -(p_{nxt} + fp_{ny})\rho_0^{-1}\Pi_n(z). \end{aligned} \tag{28.3}$$

The conditions (28.3) apply to each vertical mode n.

We first note that the Kelvin wave solution (24.2) will automatically satisfy (28.2) since $u \equiv 0$ for a Kelvin wave. The Poincaré wave solution (23.9) satisfies half of (28.2): $u(0, y, z, t) = 0$. Thus we merely have to impose an additional constraint on this solution to satisfy $u(l, y, z, t) = 0$. Substituting (23.9) into (28.3) and then putting $u = 0$ we find that for each vertical mode n there is a cross-channel velocity node at those values of x given by

$$\tan(k_1 x + \arg B) = -fk_2/\omega k_1. \tag{28.4}$$

Since $B = \omega k_1 - ifk_2$, the right-hand side of (28.4) is equal to tan (arg B). The condition (28.4) is therefore certainly satisfied at $x = 0$, as we have established in Section 23; it is also satisfied at $x = l$ if

$$k_1 l = m\pi, \quad m = 1, 2, 3, \ldots \tag{28.5}$$

Introducing into (23.9) a modified amplitude constant C defined by

$$C = 2A\cos(\arg B)\exp(i\arg B),$$

the pressure field of the Poincaré wave for the mth cross-channel mode may be written as

$$p_m = C\left[\cos k_1 x + \frac{fk_2}{\omega k_1}\sin k_1 x\right]\exp\left[i(k_2 y - \omega t)\right]\Pi(z), \tag{28.6}$$

where $k_1 = m\pi/l$. For convenience we have dropped the vertical-mode subscript n. The horizontal velocity components, as derived from (15.19) and (15.20) take the form

$$\mu_m = \frac{iC\omega k_1}{\rho_0(\omega^2 - f^2)}\left(1 + \frac{k_2^2 f^2}{\omega^2 k_1^2}\right)\sin k_1 x \exp\left[i(k_2 y - \omega t)\right]\Pi(z), \tag{28.6}$$

$$v_m = \frac{C}{\rho_0}\left[\frac{k_2}{\omega}\cos k_1 x + \frac{f(k_1^2 + k_2^2)}{(\omega^2 - f^2)k_1}\sin k_1 x\right]\exp\left[i(k_2 y - \omega t)\right]\Pi(z). \tag{28.7b}$$

In view of (28.5), the dispersion relation (23.4), now takes the form

$$k_2^2 = \frac{(\omega^2 - f^2)}{gh_n} - \frac{m^2\pi^2}{l^2}. \tag{28.8}$$

Clearly, for given values of the quantities ω, f, h_n and l, $k_2^2 > 0$ only for values of m which are smaller than a maximum value which we shall call m_0. Modes for which $m > m_0$ are evanescent: they decay down-channel from their point of origin. It is quite conceivable that under some conditions, such as for very narrow channels, $m_0 = 0$ and all Poincaré modes are evanescent. Conditions for propagation may also be expressed in terms of the frequency ω. For $k_2^2 > 0$, it may be seen, by rearranging terms in (28.8), that the condition

$$\omega^2 \geqslant f^2 + \frac{m^2\pi^2 gh_n}{l^2} \equiv \omega_c^2 \tag{28.9}$$

must be satisfied. The second term on the right-hand side of this inequality is recognized as the frequency of the mth transverse seiche mode ($k_2 = 0$) in a nonrotating channel ($f = 0$). Thus ω must not only exceed both f and the frequency of that cross-channel seiche mode, but ω^2 must be greater than the sum of the squares of these two frequencies for Poincaré waves to propagate.

The dispersive nature of Poincaré waves is evident from the form of the phase velocity (c) and group velocity (c_g), derived from (28.8):

$$c^2 = gh_n + \frac{f^2}{k_2^2} + \frac{m^2\pi^2 gh_n}{l^2 k_2^2}, \tag{28.10}$$

$$c_g = gh_n/c. \tag{28.11}$$

The phase speed always exceeds the long wave speed $\sqrt{gh_n}$; the energy, however, always travels more slowly than $\sqrt{gh_n}$. At the cut-off frequency given by (28.9), (28.8) implies that $k_2^2 = 0$. Therefore, (28.10) and (28.11) give $c = \infty$ and $c_g = 0$ when $\omega = \omega_c$. This type of geometric dispersion is a common property of waveguides (Tolstoy, 1973, Chapter 3).

The special case of wave propagation in a nonrotating channel is obtained by setting $f = 0$ in all the above results. The Kelvin wave degenerates to a surface gravity wave and the Poincaré waves are simply a Fourier series of higher cross-channel gravity modes.

Let us now consider the interference of waves travelling in opposite directions. In a linear theory, the sum of any number of waves which are solutions of the problem is of course also a solution. Consider first the interference pattern produced by two Kelvin waves, one with current amplitude V_0 at $x = 0$, the other with the same amplitude at $x = l$, travelling through each other. The sum of two pressure fields of the form (24.2b) has the horizontal and time dependence

$$P(x, y, t) = \frac{\omega V_0}{k_2}\left[\exp(ik_2 y + fk_2 x/\omega) + \exp(-ik_2 y - fk_2(x - l)/\omega)\right] e^{-i\omega t}, \tag{28.12}$$

where $\omega^2 = gh_n k_2^2$ (cf. 24.3). Curves of equal pressure in the x, y-plane lie on

$$|P|^2 = \frac{2\omega^2 V_0^2}{k_2^2} \exp(fk_2 l/\omega)\{\cos 2k_2 y + \cosh[2f(x - l/2)/\sqrt{gh_n}]\} = \text{constant}. \quad (28.13)$$

The dispersion relation (24.3) has been used in the derivation of (28.13). For the barotropic mode, the curves $|P|^2 = \text{constant}$ are also curves of constant amplitude of the surface displacement η. In tidal terminology, these curves are called co-range or co-amplitude lines. Nodal points (where $|P| = 0$) are found only on the axis of the channel (at $x = l/2$) and are spaced half a wavelength apart. Maximum amplitudes of

$$|P| = \frac{\omega V_0}{k_2}\{2(\exp(fk_2 l/\omega)[1 + \cosh(fl/\sqrt{gh_n})]\}^{1/2} \quad (28.14)$$

are found on the edges of the channel, also half a wavelength apart, but between the nodal points. The curves of equal phase (co-phase lines) are more complicated to calculate; the phase may be shown to rotate about nodal points in the beautiful *amphidromic* pattern shown in Fig. 28.1a. The nodal points are often called *amphidromic* points, or *amphidromes.*

Setting $f = 0$ in (28.12), we can draw the amplitude interference pattern appropriate for the nonrotating case (Fig. 28.1b). The overwhelming influence of rotation on the interference pattern is obvious from a comparison of Figs. 28.1a and b. As ω increases, corresponding to a decreasing rotational influence, the wavelength increases in proportion and the amphidromic pattern of Fig. 28.1a stretches out. If, instead of keeping the width of the channel constant and letting the amphidromic points diverge, we keep the wavelength constant and let the width of the channel decrease, the interference system at higher frequencies may be deduced from Fig. 28.1a by squeezing in the walls towards the centre. As the width of the channel becomes very small, the pattern of Fig. 28.1a approaches that of Fig. 28.1b.

The interference pattern of Poincaré waves is more complicated, because of the more detailed transverse structure of these waves. Addition of the pressure fields of two Poincaré waves travelling in opposite directions can obviously not yield any long-channel nodal lines since these two waves have different amplitude distributions across the channel (Fig. 23.3). Only nodal *points* exist and their position is found by the same method as that used for Kelvin waves. Adding two waves of the form (28.6) of opposite signs of k_2 and of the same amplitude, we find for the sum of the horizontal pressure distributions

$$p = 2C\left[\cos k_1 x \cos k_2 y + \frac{ifk_2}{\omega k_1} \sin k_1 x \sin k_2 y\right] e^{-i\omega t}. \quad (28.15)$$

The amplitude of this pressure field is given by

$$|p|^2 = 4C^2\left[\cos^2 k_1 x \cos^2 k_2 y + \frac{f^2 k_2^2}{\omega^2 k_1^2} \sin^2 k_1 x \sin^2 k_2 y\right]. \quad (28.16)$$

The sum of two positive terms does not vanish unless both terms are zero. The nodal points thus lie at $2k_2 y = \pi, 3\pi, 5\pi, \ldots$ on the lines $\sin k_1 x = 0$ (nodal lines of cross-channel velocity) and at the intermediate positions $2k_2 y = 0, 2\pi, 4\pi, \ldots$ on the lines $\cos k_1 x = 0$ (antinodes of cross-channel velocity).

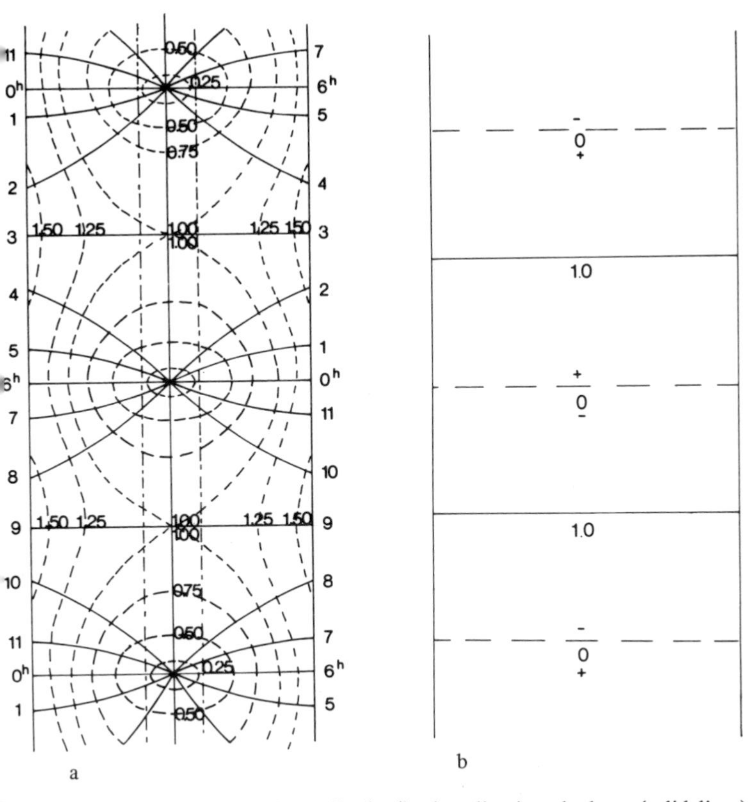

Fig. 28.1.(a). Curves of equal amplitude (broken lines) and phase (solid lines) of a pair of Kelvin waves propagating in opposite directions. The phase rotates clockwise around amphidromic points located half a wavelength apart along the center line of the channel.

(b) The interference pattern of two gravity waves in the absence of rotation. Regions of equal phase occupy rectangles a half wavelength long; the phase changes by 180° across the nodal lines labelled by 0.

The thin strip shown in part (a) illustrates how the behaviour of the phase variation in a very narrow rotating channel approximates that in a nonrotating channel. For a narrow channel, the change of phase takes place in a distance which is much smaller than a half wavelength; in the limit, as the width becomes very small, the change of phase is all concentrated near the node, as in the absence of rotation.

Reflection from the end of a channel

Upon examining the nodal structure of the long-channel velocity component v associated with a pair of Kelvin or Poincaré waves travelling in opposite directions, we find that there are no nodal lines of v, just nodal points. It is therefore not possible for a pair of such waves to satisfy the boundary condition $v = 0$ at the closed end of a channel. This difficulty was discovered by Taylor (1920), and it is only resolved by involving a whole spectrum of Poincaré waves into the reflection process. A complete set of orthogonal functions in $(0, l)$ consists of the Poincaré waves, as given by (28.6), (28.7) and (28.8) with $m = 1, 2, \ldots$, and the Kelvin wave described by (24.2) and (24.3). Any long-channel velocity distribution may then be synthesized from the contributions of a Kelvin wave and of a complete spectrum of Poincaré waves.

The reflection problem for a Kelvin wave incident onto the end of a channel is readily formulated as follows. Let the incident Kelvin wave (of amplitude equal to C_0 at $x = l$) arrive from $y > 0$ onto a wall at $y = 0$. The total pressure field in $y > 0$, will be given by the sum of the incident Kelvin wave, a reflected Kelvin wave, and a sum of Poincaré waves:

$$p(x, y, t) = C_0 \exp\,[fk_2(l-x)/\omega - i(k_2 y + \omega t)] + RC_0 \exp\,[fk_2 x/\omega + i(k_2 y - \omega t)]$$

$$+ \sum_{m=1}^{\infty} C_m \left[\cos\left(\frac{m\pi x}{l}\right) + \frac{fk_{2m} l}{\omega m l} \sin\left(\frac{m\pi x}{l}\right)\right] \exp\,[i(k_{2m} y - \omega t)], \qquad (28.17)$$

where R is a reflection coefficient for the Kelvin wave and C_m is the amplitude of the mth Poincaré mode. The wavenumber (k_{2m}) of the mth Poincaré mode is determined from (28.8); that of the Kelvin wave (k_2) is given by

$$k_2 = \omega/\sqrt{gh_n}\,.$$

The long-channel velocity component for the superposition of all these waves is found by substituting (28.17) multiplied by $\Pi(z)$ into (15.20):

$$\rho_0 v/\Pi(z) = \frac{k_2}{\omega} C_0 \{R \exp\,[fk_2 x/\omega + ik_2 y] - \exp\,[fk_2(l-x)/\omega - ik_2 y]\} e^{-i\omega t}$$

$$+ \sum_{m=1}^{\infty} C_m \left\{\frac{k_{2m}}{\omega} \cos\left(\frac{m\pi x}{l}\right) + \frac{fl}{m\pi} \frac{1}{gh_n} \sin\left(\frac{m\pi x}{l}\right)\right\} \exp\,[i(k_{2m} y - \omega t)], \qquad (28.18)$$

where the dispersion relation (28.8) has been used. The boundary condition $v(x, 0, z, t) = 0$ is then satisfied for all z and t provided

$$k_2 C_0 [R \exp\,(fk_2 x/\omega) - \exp\,\{fk_2(l-x)/\omega\}]$$

$$+ \sum_{m=1}^{\infty} C_m \left[k_{2m} \cos\left(\frac{m\pi x}{l}\right) + \frac{fl\omega}{m\pi g h_n} \sin\left(\frac{m\pi x}{l}\right)\right] = 0 \qquad (28.19)$$

for all x in the interval $[0, l]$.

The simplest way to solve for the unknown coefficients R, C_m ($m \geqslant 1$), is through the method of "collocation" or "point matching", used by Brown (1973), and explained by Laura (1970). Let us truncate the sum in (28.19) at the Nth term; the number of unknown coefficients is then $N + 1$(R and $C_1 \ldots C_N$). If equation (28.19) is made to hold at $N + 1$ points on $[0, l]$, there then result $N + 1$ inhomogeneous linear equations for the same number of unknowns. Solutions are readily obtained by inverting the matrix of coefficients. As N is increased, a (normally) converging sequence of values is found for R and for each of the C_m's. The method is of course valid only if these sequences converge. Secondary tests may be applied: (1) the values of $v(x, 0, z, t)$ should also approach zero at values of x intermediate between those at which v has already been forced to vanish; (2) the wave amplitudes, as calculated by the collocation process, should be consistent with energy conservation.

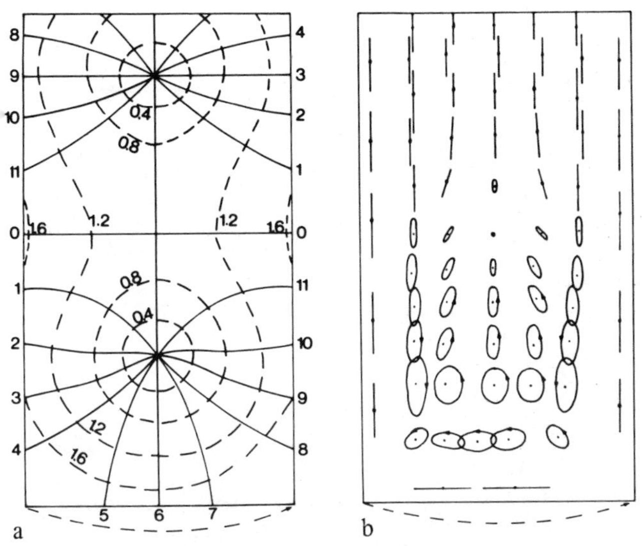

Fig. 28.2(a). The amphidromic pattern and (b) particle trajectories near the closed end of a channel for a reflected Kelvin wave. Curves of equal amplitude are given by the broken lines; curves of constant phase by the solid lines. (From Taylor, 1920.)

The simplest type of reflection occurs when the incident Kelvin wave has a frequency below the cut-off frequency ω_c for Poincaré wave propagation (28.9). All Poincaré modes are then evanescent and, although their existence is essential in order to satisfy the boundary condition (28.19), their presence is felt only near their plane of generation, at $y = 0$. An example of the amphidrome modification and, of more interest, of the particle trajectories near the closed end is shown for this case in Fig. 28.2. For higher frequencies, the amphidromic pattern diverges from that due to a pair of Kelvin waves, acquiring more and more lateral structure as more Poincaré modes are reflected from the closed end of the channel. Figure 28.3, taken from Brown (1973), shows the distortion of the amphidromes with increasing frequency.

Wave interference in channels of finite length has been applied with success to explain tidal amplitude and phase patterns in rectilinear gulfs. Some recent applications are the works of Godin (1965) in the Labrador Sea and of Hendershott and Speranza (1971) in the Adriatic Sea and the Gulf of California. Many examples may be found in texts providing regional descriptions of the tides (Defant, 1961; Dietrich, 1963). (See also Section 52.) In realistic cases, tidal forcing at the mouth of a gulf (at $y = D$, say), through pressure or current variations there, will excite Poincaré waves as well as a Kelvin wave, which will all (except the evanescent modes) propagate into the channel and be reflected at its head. The total long-channel velocity distribution may then be written as above (cf. 28.18), with the addition of an extra set of Poincaré waves, generated at the mouth of the channel. Let the amplitude of the mth Poincaré mode generated at the mouth be denoted by D_m. Then, assuming that the imposed long-channel velocity at $y = D$ excites only the nth vertical mode and therefore has a vertical dependence of the form $\Pi_n(z)/\rho_0 \equiv \Pi(z)/\rho_0$, the coefficients C_0, R, C_m and D_m are determined from the conditions

Fig. 28.3. The influence of frequency on the amphidromic pattern for Kelvin waves reflected from the closed end of a channel. The closed end of the channel is at the bottom of the panels and lines of constant phase and amplitude are labelled as in Figs. 28.1 and 28.2. The width of the channel is $l = 500\,\text{km}$, the latitude 54.46°N and the cut-off period $T_c = 2\pi/\omega_c$ (cf. 28.9) is equal to 8.46 hr for $m = 1$. The Kelvin wave frequency decreases from panels A to H as follows: A: 12.0 hr; B: 10.0 hr; C: 9.0 hr; D: 8.6 hr; E: 8.1 hr; F: 8.0 hr; G: 7.0 hr; H: 6.0 hr. In panels A–D, all Poincaré waves are evanescent, and at a sufficiently large distance from the reflecting end the interference pattern becomes that of a pair of Kelvin waves travelling in opposite directions. In the last four panels (E–H) the $m = 1$ Poincaré mode propagates and the field far from the reflecting end consists of two Kelvin waves and a Poincaré wave. The asymmetry introduced by the reflected Poincaré wave is evident in these four panels. (From Brown, 1973.)

$$v(x, 0, z, t) = 0 \quad \text{and} \quad v(x, D, z, t) = F(x)e^{-i\omega t}\Pi(z)/\rho_0, \tag{28.20}$$

where $F(x)$ is the cross-channel dependence of the forcing velocity. The collocation procedure described earlier is just as easily applied to this more complicated problem.

In sufficiently narrow canals, no Poincaré modes will propagate for a given wave frequency ω if $\omega < \omega_c$. For a long enough channel, only one set of Poincaré waves enters each boundary condition in that case, the other set having decayed to an insignificant level while propagating from the other end of the channel, and the calculations of the amplitude coefficients are simplified accordingly.

Barotropic waves in narrow channels

A very narrow channel may be defined as being only a small fraction of a wavelength in width, so that the inequality $k_2 l \ll 1$ is satisfied. We have already seen, at least by graphical means (Fig. 28.1), how the influence of rotation is reduced by a narrowing of the channel: the proximity of the side walls inhibits the effect of the Coriolis force and the cross-channel velocity component is much smaller in amplitude than the long-channel component. Narrow channels may thus be analyzed under the assumption that the Coriolis force is negligible and that $f = 0$.

In narrow channels of uniform rectangular section, the behaviour of travelling or standing oscillations may be determined from (28.1) with the boundary condition of vanishing normal velocity on the walls. With $f = 0$, that problem is trivial and deserves no further attention. Of more interest is the phenomenon of wave propagation in narrow channels of varying width, $l(y)$, and depth, $H(y)$. Only the barotropic mode will be considered at present; as we have repeatedly remarked earlier, the vertical and horizontal dependences of the motion are not separable in the presence of bottom slopes.

Let us then consider a channel bounded laterally by rigid walls at $x = l_1(y)$ and $x = -l_2(y)$, and thus of total width $l_1 + l_2 = l(y)$. For long barotropic waves, in which the horizontal velocity components are independent of depth, the vertically integrated continuity equation (15.51) may be used as a starting point:

$$(Hu)_x + (Hv)_y = -\eta_t. \tag{28.21}$$

Integrating this across the width of the channel, we have

$$\int_{-l_2}^{l_1} (Hu)_x \mathrm{d}x + \int_{-l_2}^{l_1} (Hv)_y \mathrm{d}x = -\int_{-l_2}^{l_1} \eta_t \mathrm{d}_x. \tag{28.22}$$

On the side walls the normal velocity component vanishes:

$$u - v\frac{\mathrm{d}l_1}{\mathrm{d}y} = 0, \text{at } x = l_1, \tag{28.23a}$$

$$u + v\frac{\mathrm{d}l_2}{\mathrm{d}y} = 0, \text{at } x = -l_2. \tag{28.23b}$$

With the help of (28.23), (28.22) becomes

$$\frac{\partial}{\partial y}\left[\int_{-l_2}^{l_1} Hv \,\mathrm{d}x\right] = -\frac{\partial}{\partial t}\int_{-l_2}^{l_1} \eta \,\mathrm{d}x. \tag{28.24}$$

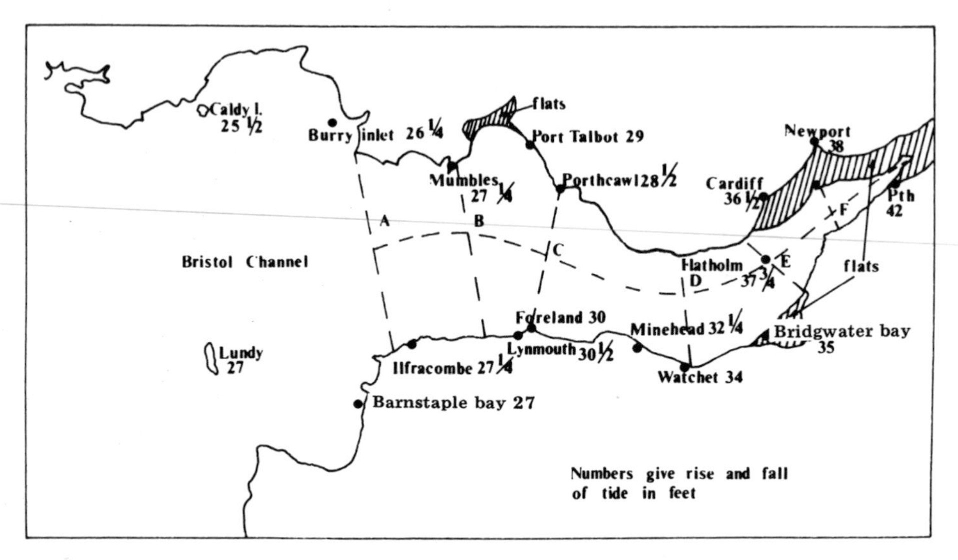

Fig. 28.4. Tidal ranges (high tide minus low tide) in the Bristol Channel. (From Taylor, 1921b.)

If the channel is narrow enough and the walls only slowly convergent (or divergent), the flow may be assumed to be confined to the down-channel axis and not to vary at all across the channel. In that case, (28.24) reduces to

$$(Hlv)_y + l\eta_t = 0. \tag{28.25}$$

The downstream momentum balance is given by a simplified form of (15.50b) with $f = 0$:

$$v_t + g\eta_y = 0. \tag{28.26}$$

These last two equations may be combined by eliminating v:

$$l\eta_{tt} - g(Hl\eta_y)_y = 0. \tag{28.27}$$

For a harmonic time dependence ($\eta \propto e^{-i\omega t}$), this equation reduces to

$$\eta'' + [(Hl)'/Hl]\eta' + (\omega^2/gH)\eta = 0, \tag{28.28}$$

where primes denote y-differentiation.

As a simple example, we may use Taylor's (1921b) calculation of tidal amplitudes in the Bristol Channel. The geography is illustrated in Fig. 28.4, and good fits to the depth and width of the channel are given by the linear functions

$$H = \alpha y, \quad l = \beta y, \tag{28.29}$$

where y is measured from the head of the channel, at which both the depth and breadth vanish, and $\alpha = 3.5 \cdot 10^{-4}$ and $\beta = 0.4$. Inserting (28.29) into (28.28) gives the equation

$$y^2\eta'' + 2y\eta' + \frac{\omega^2}{g\alpha} y\eta = 0. \tag{28.30}$$

The solution which remains finite at $y = 0$ is

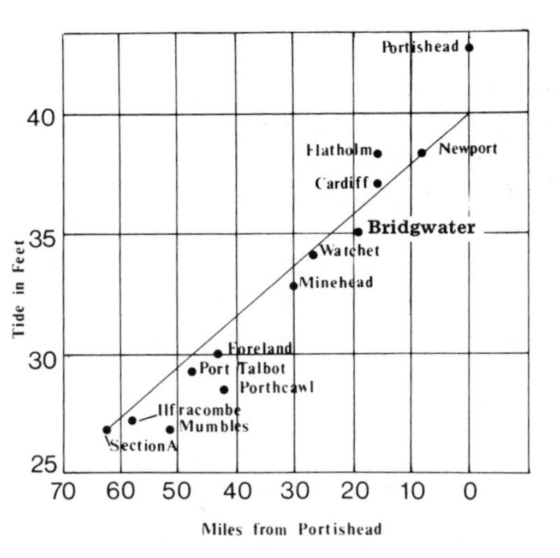

Fig. 28.5. Tidal ranges as observed (points) in the Bristol Channel compared to the range calculated from (28.31) (solid line). (From Taylor, 1921b.)

$$\eta = Ay^{-1/2} J_1 [2\omega(y/g\alpha)^{1/2}], \tag{28.31}$$

where the amplitude coefficient A may be fitted to the tidal amplitude at the entrance to the channel. The results of the computation are shown in Fig. 28.5, and are in good agreement with observations. The question of tidal propagation in the Bristol Channel has, however, been re-examined recently by Bennett (1975), who found that a considerable loss of energy occurs at the head of the channel, and that the situation is not quite as simple as described by Taylor (see Exercise 28.5).

For a narrow channel which is also very slowly varying along its axis, such that the scales of variation of H and l are much longer than a wavelength, the geometry may be taken, in a first approximation, as uniform. A solution of (28.27) is then

$$\eta = F\left(\frac{y}{\sqrt{gH}} - t\right). \tag{28.32}$$

The solution behaves everywhere like a nondispersive travelling wave. Letting $y = \sqrt{gH}Y$, we take, as the next approximation, a solution of the form

$$\eta = \xi(Y)F(Y-t). \tag{28.33}$$

where $\xi(Y)$ is a slowly varying envelope function. Substitution of (28.33) into (28.27) yields

$$\frac{\xi''}{\xi} + \frac{\xi'}{\xi}\left[\frac{H'}{2H} + \frac{l'}{l}\right] + \frac{F'}{F}\left[\frac{2\xi'}{\xi} + \frac{H'}{2H} + \frac{l'}{l}\right] = 0, \tag{28.34}$$

where primes indicate Y-derivatives. Consistent with our approximation is the fact that the length scale of the slow modulation factor ξ is much longer than that of F; this permits us to drop the first two terms of (28.34). Integrating what is left of that equation, we find for the amplitude factor ξ,

$$\xi \propto l^{-1/2}H^{-1/4}. \tag{28.35}$$

This simple result has commonly been glorified with the name of Green's law (Green, 1837) and may also be derived from simple energy considerations (Exercise 28.6).

Channels of nonrectangular cross-section

The properties of barotropic waves in an unstratified channel in which the depth is a function of the cross-channel direction may be examined by a modification of the analysis presented in Section 25 for coastal trapped waves. With a surface displacement of the form (25.6),

$$\eta = F(x)\, e^{i(ky-\omega t)}, \tag{28.36}$$

in which $k \equiv k_2$, the cross-channel amplitude dependence is given by (25.7). The boundary condition (25.9) now applies at $x = 0$ and at $x = l_2$.

Let us take the flat shelf model examined in Section 25 as a simple example. The depth profile is now given by

$$H(x) = \begin{cases} H_1, & 0 \leqslant x < l_1 \\ H_2, & l_1 < x \leqslant l_2. \end{cases} \tag{28.37}$$

The equation for the surface amplitude of barotropic edge waves in the absence of rotation is given by (25.37). With the nondimensional quantities defined in (25.38) and (25.39), the surface amplitude $\hat{F}(\chi)$, where $\chi = x/l_1$ obeys equations (25.40a) and (25.40b) in the two regions. The solutions of these equations have already been discussed. In the absence of rotation, the boundary conditions of vanishing normal velocity at $x = l_2$ ($\chi = l_2/l_1$) reduces to

$$\hat{F}_2'(l_2/l_1) = 0. \tag{28.38}$$

Of the solutions of (25.40) presented earlier, only the leaky mode as given by (25.44b) can satisfy (28.38). Substituting (25.44b) into (28.38), we find, in the notation of Section 25,

$$\sin\,[\mu_2(l_2/l_1 - 1) + \epsilon] = 0. \tag{28.39}$$

Hence $\quad \mu_2(l_2/l_1 - 1) + \epsilon = n\pi, \quad n = 1, 2, 3, \ldots. \qquad (28.40)$

The parameter ϵ is defined through $\tan\epsilon = (\mu_1\gamma^2/\mu_2)\tan\mu_1$; the dispersion relation (28.40) then involves the nondimensional parameters μ_1 and μ_2, which are both related to the wavenumber and the frequency through the definition following (25.44). The presence of the second lateral boundary allows only those leaky mode solutions which interfere in a manner such as to place a node of cross-channel velocity at $x = l_2$. For any given frequency, the wavenumber is quantized by the finite width of the channel: only those wavenumbers propagate for which (28.40) is satisfied.

For other types of depth profiles, the solution of (25.7) must be sought in terms of special functions, or obtained numerically. The influence of the second boundary is nevertheless of the same nature: it discretizes the spectrum of the oscillations.

The Kelvin wave is also modified by depth gradients. In particular, the cross-channel velocity no longer vanishes identically when $H = H(x)$ and corrections to the flow field and to the speed of propagation are found. This problem has been discussed by Smith (1972).

Combined stratification and topography

The coupling of internal waves and surface waves in a channel of nonuniform depth has been examined by Niiler (1968) and by Parrish and Niiler (1971). In the models studied, a two-layer stratification and depth profiles of the form (28.37) and (25.30) were used. These models provide idealizations of the stratification and topography of the Florida Straits. Niiler applied the theory of cross-channel oscillations to account for the observed variability of the Florida current at tidal frequencies. The experiments conducted by Parish and Niiler verified the existence of mode coupling at those frequencies where the dispersion curves $\omega(k)$ of the internal and surface wave models intersected.

Second-class waves

The reflection of plane planetary waves from a rectilinear wall has already been examined in Section 23. In the notation of that section, with x pointing eastwards and y northwards, we now add a second wall, at $y = \alpha x + y_0$, parallel to that at $y = \alpha x$. The two walls are separated by a width $D = y \cos [\tan^{-1} \alpha]$ (Fig. 28.6). As before, we take the pressure field to consist of a wave incident on $y = \alpha x$ plus a wave reflected

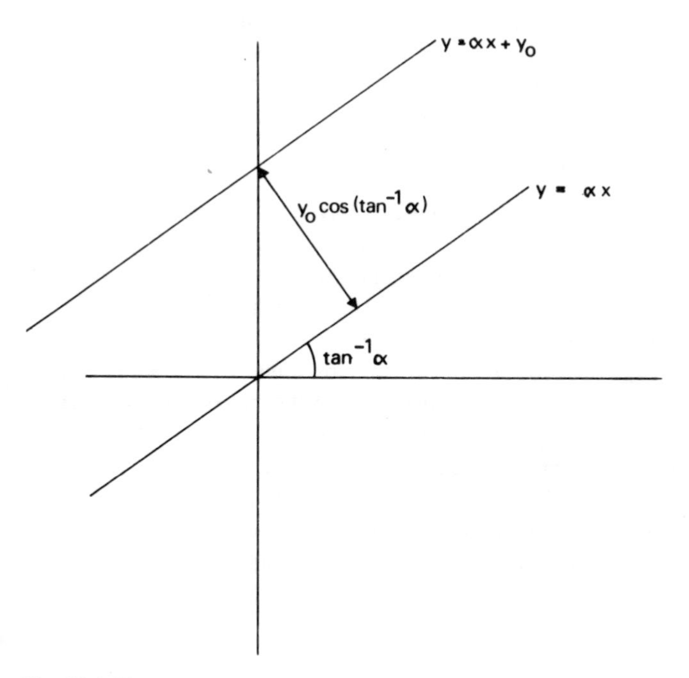

Fig. 28.6. The geometry of planetary wave propagation in a channel of width $y_0 \cos (\tan^{-1} \alpha)$.

from $y = \alpha x$ [cf. (23.15) and 23.16)]; the horizontal dependence of this superposition is

$$p = A[\exp\{i(k_1x + k_2y - \omega t)\} - \exp\{i(l_1x + l_2y - \omega t)\}], \tag{28.41}$$

where, as in (23.18b),

$$l_1 + \alpha l_2 = k_1 + \alpha k_2 \tag{28.42}$$

to satisfy conservation of crests on $y = \alpha x$. In addition, from (23.14), p must now also vanish on $y = \alpha x + y_0$, so that

$$\exp[i\{(k_1 + \alpha k_2)x + k_2 y_0\}] = \exp[i\{(l_1 + \alpha l_2)x + l_2 y_0\}] \tag{28.43}$$

must hold for all x. Using (28.42), we end up with the periodicity condition

$$l_2 - k_2 = \pm 2n\pi/y_0, \quad n = 1, 2, \ldots \tag{28.44}$$

This last relation determines the cross-channel structure of the waves as well as the number of modes which may propagate between the pair of walls at any given frequency. The influence of the second wall, as expressed through (28.44), is most simply interpreted in terms of the propagation diagram of Fig. 23.5, reproduced here in a simplified form as Fig. 28.7. The condition (28.42) which must be satisfied by the wavenumbers $\boldsymbol{k}$ and $\boldsymbol{l}$ upon reflection from one boundary ($y = \alpha x$, say) is that the chord which joins the points $\boldsymbol{k}$ and $\boldsymbol{l}$ on the circle of constant ω must be normal to the reflecting plane ($y = \alpha x$). This condition may be satisfied by an infinity of pairs of $\boldsymbol{k}$ and $\boldsymbol{l}$. The extra condition (28.44) quantizes the $\boldsymbol{k}$, $\boldsymbol{l}$ pairs allowed as solutions of (28.42) and lying on the same frequency circle, by insisting that k_2 and l_2 be separated by multiples of a wavenumber based on the channel width. As $2\pi/y_0$ is the north–south component of the channel width wavenumber, (28.44) appears as a standard interference condition, which ensures that the boundary condition $p = 0$ is repeated on a series of parallel lines.

A glance at Fig. 28.7 shows that only a finite number of modes of a given frequency may be accommodated in a given channel. For the graphical example chosen, only the first three modes can propagate in any one direction. By analogy with a similar behaviour found in Poincaré waves, higher modes will be evanescent and decay along the channel. Further, since the diameter of the circle of constant ω [given by $(\beta^2/\omega^2 - 4f^2/gh_n)^{1/2}$] determines the maximum value of the difference between k_2 and l_2, it may well turn out that for a high enough frequency (small diameter) no propagating modes are possible, a situation which is again analogous with that encountered with Poincaré waves. This possibility is examined further in Exercises 28.7 and 28.8.

Satisfying a third boundary condition, $p = 0$, at the closed end of a channel is quite straightforward, and is accomplished by superposing an additional pair of waves with wavenumber vectors $\boldsymbol{m}$, $\boldsymbol{n}$ to the two in (28.41). These two new waves satisfy the relations (28.42) and (28.44) necessary for propagation in the channel. They also lie on the opposite side of the diameter normal to the wall (Fig. 28.8). Corresponding members (A and C, B and D) of the two pairs then lie on lines parallel to the wall $y = \alpha x$, and thus perpendicular to an end wall at $x = -\alpha y$. They thus satisfy boundary conditions of the form

$$\exp[i(k_1x + k_2y)] = \exp[i(m_1x + m_2y)]$$

on $x = -\alpha y$; i.e., for both pairs,

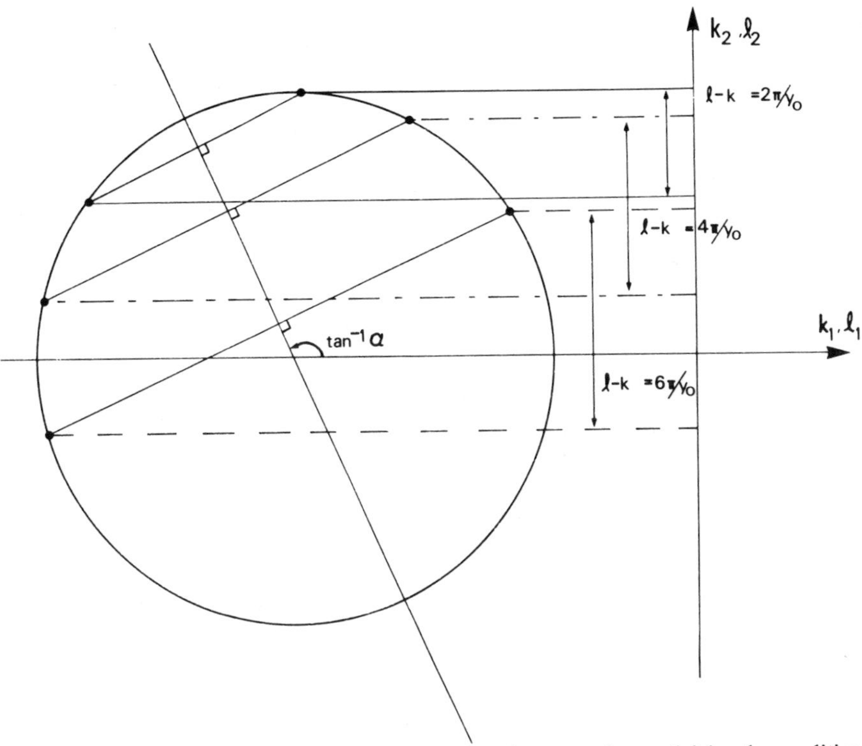

Fig. 28.7. The location on the slowness curve of pairs of wavenumbers satisfying the condition (28.44) for propagation along a channel oriented along a line making an angle $\tan^{-1}\alpha$ with the x-axis. The three pairs shown have a net southward component of energy propagation. Three pairs with a northward component of energy propagation also exist and may be found as mirror images of the pairs shown.

$$\alpha k_1 - k_2 = \alpha m_1 - m_2, \quad \text{and} \quad \alpha l_1 - l_2 = \alpha n_1 - n_2. \tag{28.45}$$

It is quite simple now to pass to the solution for the eigenfrequencies of a completely closed basin by applying a quantization condition on the pairs $(\boldsymbol{n}, \boldsymbol{l})$ and $(\boldsymbol{k}, \boldsymbol{m})$, of a form similar to (28.44). The reader is encouraged to anticipate the results of the next section by taking this step immediately.

The propagation of planetary waves in channels of non-rectangular cross-section may be analyzed by replacing the trapping boundary condition (25.8) with a constraint on the normal velocity at a second boundary. A very simple example is provided by an exponential profile similar to (25.62):

$$H(x) = H_1\, e^{2bx}, \quad 0 \leqslant x \leqslant l, \tag{28.46}$$

where $b > 0$.

Under the rigid-lid approximation, and with $H(x)$ as given by (28.46), the amplitude $\psi(x)$ of a wave of the form (25.53) obeys the equation (cf. 25.54)

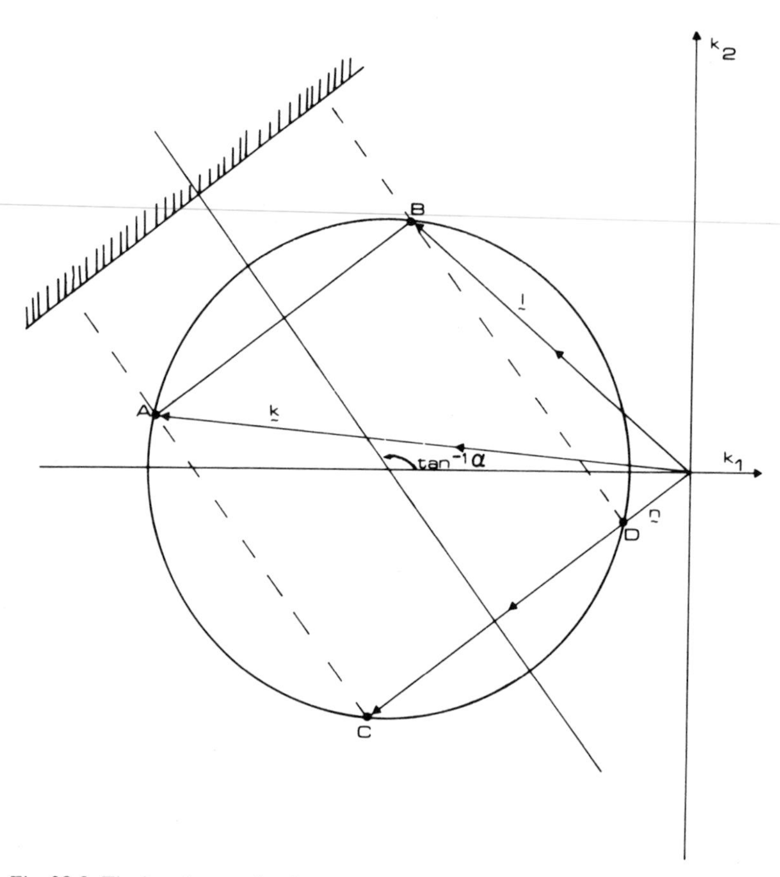

Fig. 28.8. The location on the slowness curve of two pairs (A, B) and (C, D) of wavenumbers satisfying (28.44). These two pairs propagate their energy in opposite directions and interfere to satisfy the required boundary condition on a wall normal to the channel axis. The pairs (A, C) and (B, D) satisfy the relations (28.45).

$$\psi'' - 2b\psi' - \left(k^2 + \frac{2fkb}{\omega}\right)\psi = 0, \quad 0 \leqslant x \leqslant l \tag{28.47}$$

and the boundary conditions

$$\psi = 0 \quad \text{on} \quad x = 0, l. \tag{28.48}$$

The solution of (28.47) may be written as

$$\psi(x) = A_1 \exp(\gamma_1 x) + A_2 \exp(\gamma_2 x), \tag{28.49}$$

where $$\gamma_{1,2} = b \pm (b^2 + k^2 + 2bkf/\omega)^{1/2}. \tag{28.50}$$

The boundary condition (28.48) is satisfied at $x = 0, l$ if

$$A_2 = -A_1 \quad \text{and} \quad \gamma_1 = \gamma_2 + \frac{2m\pi i}{l}, \quad m = 1, 2, \ldots \tag{28.51}$$

Thus, substituting for γ_1 and γ_2 and rearranging terms, a dispersion relation is obtained in the form

$$\left(k + \frac{fb}{\omega}\right)^2 = b^2\left(\frac{f^2}{\omega^2} - 1\right) - \frac{4m^2\pi^2}{l^2}. \tag{28.52}$$

This relation holds only for subinertial waves ($\omega < f$); for any frequency $\omega > 0$ and bottom slope parameter b there is, as for Poincaré waves (cf. 28.2), a maximum value of m, m_0 say, above which no propagation takes place. Furthermore, since

$$k = -\frac{fb}{\omega} \pm \left[b^2\left(\frac{f^2}{\omega^2} - 1\right) - \frac{4m^2\pi^2}{l^2}\right]^{1/2}, \tag{28.53}$$

in the Northern Hemisphere ($f > 0$) the sign of k is always opposite to that of b: $bk < 0$. The phase therefore propagates with the shallow water to its right. The group velocity is obtained by differentiating (28.52):

$$\frac{\partial\omega}{\partial k} = \pm\frac{[b^2(f^2/\omega^2 - 1) - 4m^2\pi^2/l^2]^{1/2}}{fbk/\omega^2}. \tag{28.54}$$

Since $bk < 0$, the long waves (the plus sign) propagate their energy towards negative values of y, in the same direction as the phase, and the short waves have their group velocity in the direction of increasing y.

The general properties of waves of the first and second class propagating along a channel of nonrectangular cross-section including the β-effect, have been discussed by Odulo (1975a). A number of references on second-class waves propagating in an ocean of non-uniform depth have already been given in Sections 20 and 25. The propagation of planetary waves in a two-layer fluid with sloping bottom in the presence of parallel boundaries has been examined by Helbig and Mysak (1976).

Harbours and resonance

Wave motion in semi-enclosed channels or harbours is usually excited by forcing through the open end. The response of the channel to the forcing, which may be defined as the ratio of the maximum amplitude within to that at the mouth, is in general dependent upon both the shape of the channel and upon the forcing frequency. Green's law (28.35) provides a simple example of geometrical dependence. The simplest instance of a resonant response is that occurring in the *quarter-wave resonator.* In a nonrotating, nonstratified fluid, the simplest oscillation which satisfies lateral boundary conditions in a channel of finite length D, open at $y = 0$ and closed at $y = D$, is that for which the velocity is solely along the channel. For this case, the surface elevation and velocity components are given by

$$\begin{aligned} \eta &= A\cos\left[\omega(y - D)/\sqrt{gH}\right] e^{-i\omega t}, \\ v &= iA\sqrt{g/H}\sin\left[\omega(y - D)/\sqrt{gH}\right] e^{-i\omega t}, \\ u &= 0. \end{aligned} \tag{28.55}$$

The amplitude constant A is determined by matching with some given forcing function $\eta(0, t) = \eta_0\, e^{-i\omega t}$ at the mouth:

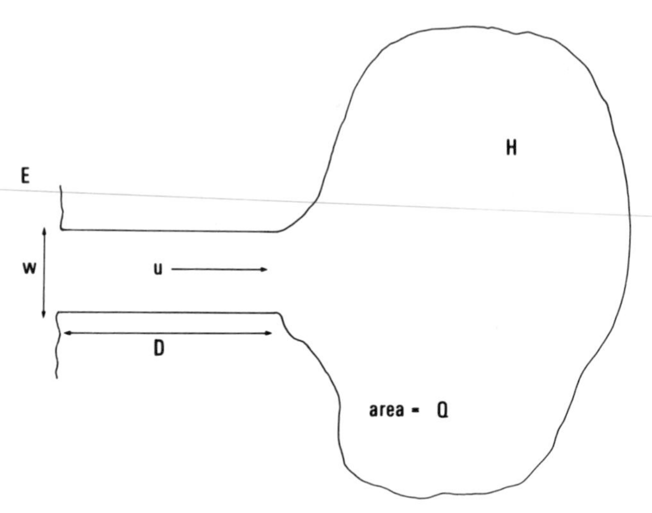

Fig. 28.9. A harbour which behaves like a Helmholtz resonator.

$$A = \eta_0/\cos(\omega D/\sqrt{gH}). \tag{28.56}$$

This solution is obviously inadequate whenever $\cos(\omega D/\sqrt{gH})$ vanishes: A is then unbounded. In that case, resonance is said to occur, and the response of the channel to exterior forcing is limited by dissipative phenomena, which are neglected here. Realizing that $k = \omega/\sqrt{gH}$ is the wavenumber of long surface waves in a nonrotating fluid, we see that resonance occurs for channel lengths $D_n = (2n-1)\pi/2k$, $n = 1, 2, \ldots$ The shortest such resonating channel is one-quarter wavelength long. In such channels, the position of the mouth corresponds to a nodal line of η for free oscillations; forcing η to be finite at the mouth makes it unbounded in the rest of the basin.

A rather different, but equally simple resonator is that commonly named after Helmholtz [Rayleigh, 1896 (see collected works, 1945, Chapter 16)]. Its geometry is sketched in Fig. 28.9; a channel of length D, width W and depth H connects the basin (of area Q) to the open sea. Given an oscillation of the form $\eta_0 = A_0 \exp(-i\omega_0 t)$ in the open sea, the water level η_1 in the basin will always lag somewhat behind η_0 because of the constriction through which the flow must pass. Let us assume that the basin is small enough that η_1 is independent of position. The rate of flow u in the channel may then, in the absence of friction, be written as proportional to the pressure difference across its ends:

$$\frac{\partial u}{\partial t} = \frac{g}{D}(\eta_0 - \eta_1). \tag{28.57}$$

On the other hand, the rate at which the basin fills up is proportional to the volume flowing through the channel:

$$\frac{\partial}{\partial t}(Q\eta_1) = WHu. \tag{28.58}$$

Combining (28.57) with (28.58), we obtain the following equation for u:

$$u_{tt} + \left(\frac{gWH}{QD}\right)u = -\frac{ig\omega_0 A_0}{D} \exp(-i\omega_0 t). \tag{28.59}$$

Resonance now occurs at $\omega_0^2 = gWH/QD$, at which the response is

$$u = \frac{gA_0}{2D} t \exp(-i\omega_0 t),$$

$$\eta_1 = \frac{A_0}{2}[\exp(-i\omega_0 t) + i\omega_0 t \exp(-i\omega_0 t)]. \tag{28.60}$$

That the behaviour of these two elementary resonators is so different is not surprising: in the quarter-wave resonator, the phase lag is entirely due to wave propagation; in the Helmholtz resonator, the lag is associated with the time taken to fill up or empty the basin. Real harbours will seldom be reducible to either of the above elementary geometries; in addition, higher order oscillations might be excited in them and affect their response to external stimulation. A complete theory of harbour seiches (a seiche being an oscillation of a closed or semi-closed body of water) requires adequate knowledge of the possible motions inside the harbour as well as of the form of the wave field scattered from the mouth of the harbour into the open sea.

One approach to the problem of harbour seiches is that employed by Buchwald and Williams (1975) which made use of the Galerkin technique to obtain a harbour response. The equivalent circuit method developed by Miles (1971a, 1974a) simplifies the understanding of the problem and will now be discussed in more detail.

Consider again the harbour shown in Fig. 28.9. The surface displacement inside the basin ($x \epsilon B$) may be attributed to the flux of mass through the mouth. Let I be the total volumetric flow through the mouth (the current, in an electrical analogy) and $If(s)$ the current per unit width, where the profile function $f(s)$ satisfies

$$\int_{\mathrm{M}} f(s)\,\mathrm{d}s = 1, \tag{28.61}$$

and the integral over M is across the mouth of the harbour. We may then write

$$\eta(\boldsymbol{x}) = I \int_{\mathrm{M}} G_{\mathrm{B}}(\boldsymbol{x}, s) f(s)\,\mathrm{d}s \quad (\boldsymbol{x}\epsilon\mathrm{B}), \tag{28.62}$$

where $G_{\mathrm{B}}(\boldsymbol{x}, s)$ is the Green's function for the harbour due to forcing by a point source of velocity at the mouth (Miles, 1971a).

In the exterior domain ($\boldsymbol{x}\epsilon\mathrm{E}$), the surface displacement is written in two parts:

$$\eta(\boldsymbol{x}) = \eta_0(\boldsymbol{x}) - I \int_{\mathrm{M}} G_{\mathrm{E}}(\boldsymbol{x}, s) f(s)\,\mathrm{d}s \quad (\boldsymbol{x}\epsilon\mathrm{E}). \tag{28.63}$$

The function $\eta_0(\boldsymbol{x})$ is the wave field which would exist in the absence of the harbour: as the width of the mouth tends to zero, $\eta \to \eta_0$ for $\boldsymbol{x}\epsilon\mathrm{E}$; the second term is the response of the open sea, in the form of scattered waves, to the oscillating mass flow at the mouth of the harbour.

Since the two expressions for the surface displacement (28.62) and (28.63) must match across the mouth, they provide, in combination, an integral equation for $f(s)$, the current distribution. The following averaging procedure is, however, simpler to use. Multiplying both (28.62) and (28.63) by $f^*(\sigma)$ (the complex conjugate of f) and integrating over the width of the mouth, we find that at the mouth of the basin itself,

$$V_B = Z_B I = V_0 - Z_E I, \tag{28.64}$$

where $$V_B = \int_M \eta f^*(s)\, ds; \quad V_0 = \int_M \eta_0 f^*(s)\, ds; \tag{28.65}$$

and $$Z_{E,B} = \int_M \int_M f^*(s) G_{E,B}(s, \sigma) f(\sigma)\, d\sigma\, ds. \tag{28.66}$$

The averaged elevations V_B and V_0 may be considered as analogous to voltages, and the flow I, to an electric current, those quantities being related through (28.64) as if they were characteristics of the circuit of Fig. 28.10.

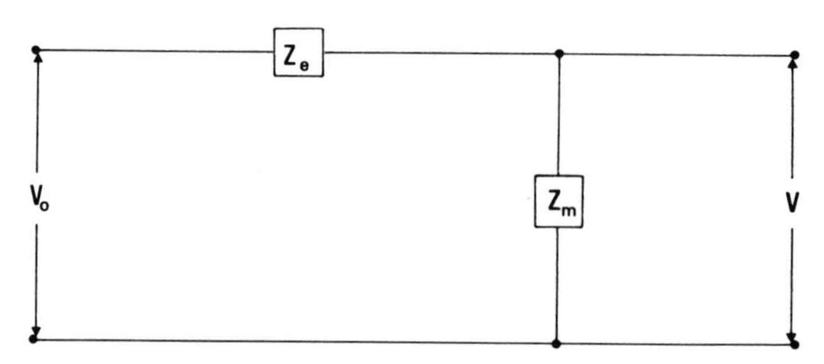

Fig. 28.10. The equivalent circuit diagram for harbour forcing.

The response of the harbour, expressed in terms of the current resulting from the application of a certain voltage V_0, is, from (28.64),

$$I = V_0/(Z_B + Z_E). \tag{28.67}$$

The maximum response occurs at the minimum of $Z_B + Z_E$, and this is where resonance phenomena occur. The form of the frequency dependence $I(\omega)$, and hence of the resonance peaks, will depend on the modal structure of the harbour oscillations (which enter Z_B as a sum of series impedances Z_n, each equivalent to an inductor and a capacitance in parallel), on the frictional forces at work (which will enter Z_B as resistive elements), and on the wave-scattering properties which make up Z_E. Examples are found in Mile's works (1971a, 1974a), and in Garrett (1975), who extended Miles' theory to include direct body forcing, such as by tidal forces. Earlier applications of circuit theory may also be found in the works of Neuman, as described by Defant (1961, p. 169). The only instance in which $Z_B + Z_E$ vanishes is when both the harbour and the open sea are finite in area: the eigenvalues of the coupled basins are found at the zeros of $Z_B + Z_E$, and their response to a given forcing increases linearly with time at resonance. For an unbounded outer ocean this situation does not occur and the resonance peaks are finite in height. For a nondissipative basin, Z_B is purely imaginary (reactive impedance) and Z_E is complex: hence their sum cannot vanish.

As a straightforward example of the above method, consider a rectangular basin of depth H with unrestricted access to a nonrotating sea of depth H_E. No Helmholtz mode is expected in the basin (because of the unrestricted access). Assuming for the sake of argument that no cross-basin modes of oscillation are present, let us take $f(s) = 1/W$, where W is the width of the basin. The current and amplitude are thus independent of the cross-basin direction.

From (28.64) and (28.65), with subscripts M for values at the mouth, we find

$$Z_B = V_B/I = \eta_M/WHv_M; \tag{28.68}$$

substituting for η_M and v_M from (28.55), we find

$$Z_B = i \cot\,[\omega D/\sqrt{gH}]/W\sqrt{gH}. \tag{28.69}$$

The basin impedance is thus purely reactive. For the exterior wave-scattering impedance Z_E, it can be shown that the Green's function for wave radiation due to a time harmonic current source with spatial form

$$v(0, s) = H_E^{-1}\delta(s-\sigma)$$

is
$$G_E = \frac{\omega}{2gH_E} H_0^{(1)}(\omega|s-\sigma|/\sqrt{gH_E}), \tag{28.70}$$

where $H_0^{(1)}$ is the zeroth-order Hankel function, describing outward phase propagation. For small values of $\omega W/\sqrt{gH_E}$., corresponding to channel openings much narrower than the length of the scattered wave, the Hankel function is approximated by (Abramowitz and Stegun, 1965, p. 360)

$$H_0^{(1)}(z) \simeq \left[1 + \frac{2i}{\pi}\left\{\ln\,(z/2) + \gamma\right\}\right] \quad \text{for small } z, \tag{28.71}$$

where $\gamma = 0.5772$ is Euler's constant. Substituting G_E into (28.66), we thus find, again with $f(s) = 1/W$,

$$Z_E = \frac{\omega}{2gH_E W^2} \int\limits_{-W/2}^{W/2} \int\limits_{-W/2}^{W/2} H_0^{(1)}(\omega|s-\sigma|/\sqrt{gH_E})\, d\sigma\, ds \tag{28.72}$$

$$\simeq \frac{\omega}{2gH_E}\left[1 + \frac{2i}{\pi}\left(\ln \frac{\omega W}{2\sqrt{gH_E}} + \gamma - \frac{3}{2}\right)\right]. \tag{28.73}$$

The response of the basin, as characterized by $I/V_0 = (Z_B + Z_E)^{-1}$, may be calculated using (28.69) and (28.73) as a function of frequency (Exercise 28.11). The amplitude of the response is given by the magnitude of $(Z_B + Z_E)^{-1}$, its phase by $\tan^{-1}[-\mathrm{Im}\,(Z_B + Z_E)/\mathrm{Re}(Z_B + Z_E)]$. Note that when $Z_B = 0$, i.e. for $\cos\,[\omega D\sqrt{gH}] = 0$, the magnitude of the response is entirely controlled by the radiative impedance. It is the presence of a non-zero value of Z_E which ensures that the resonance peaks (at $Z_B = 0$) are of finite height and width. Since Z_E is approximately proportional to ω, the magnitude of the resonant response decreases roughly as $1/\omega$ for the sequence of values of ω for which $Z_B = 0$. As an example of a response function, we show in Fig. 28.11 the results obtained by Lemon (1975) for Port San Juan, an inlet on the coast of Vancouver Island, British Columbia.

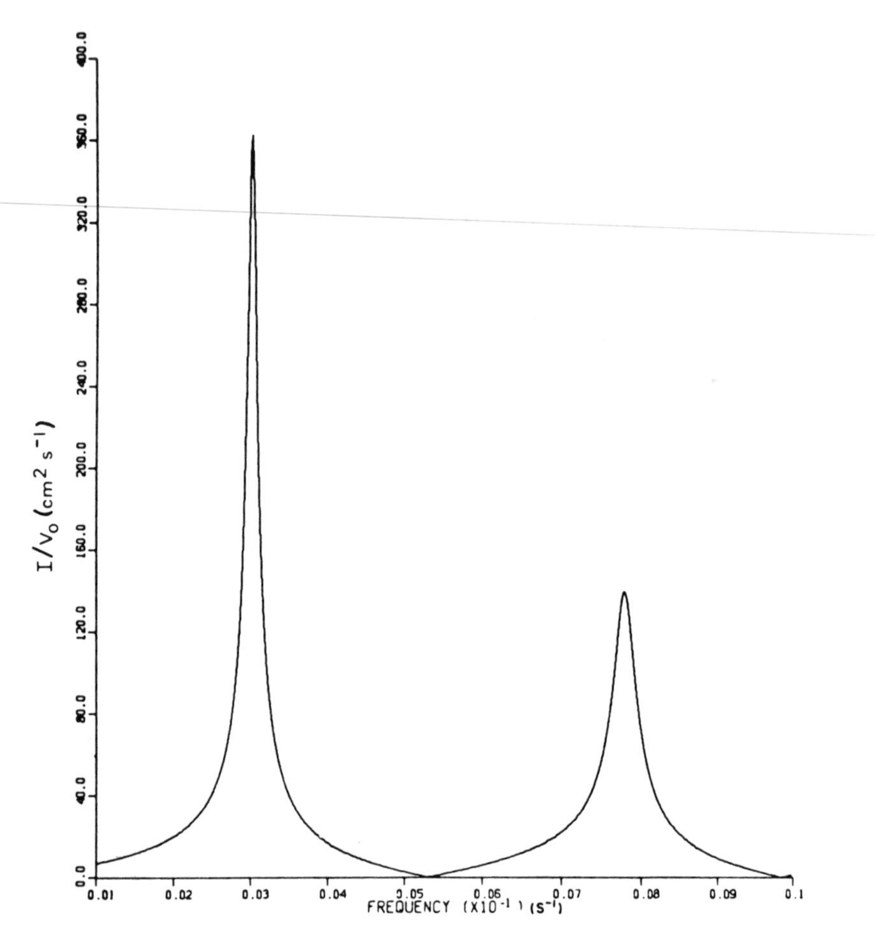

Fig. 28.11. The response of San Juan Harbour, British Columbia, as a function of frequency. The quantity plotted is the volume flux through the mouth of the inlet (I) for an incoming wave of unit amplitude (V_0). (From Lemon, 1975.)

As an intriguing phenomenon, we mention that the narrowing of the harbour mouth does not affect the mean-square response of a Helmholtz resonator to a random excitation in the neighbourhood of its resonant frequency. This curiosity was designated as "the harbour paradox" by Miles and Munk (1961), who discovered it and thought that it applied to the higher modes as well. It was shown by Garrett (1970a) that only the Helmholtz mode possesses this peculiar property. The paradox, of course, lies in the fact that in the limit of zero harbour width, the amplitude of the excitation in the basin remains the same. The solution of the paradox lies in adding frictional forces to the problem, which by acting on the flow in the channel which connects the sea and the basin, will reduce the rate of flow of fluid and hence the amplitude of the response. The Helmholtz resonance of harbours has been investigated in greater detail by Miles and Lee (1975).

Exercises Section 28

1a. Which Poincaré modes can propagate in the Adriatic Sea? (width $\simeq 150$ km, depth $\simeq 200$ m, latitude 45°N).

1b. Answer the same question for the North Sea (width $\simeq 500$ km, depth $\simeq 50$ m, latitude 55°N).

2. Under the hydrostatic approximation, the energy flux (cf. 13.46) is given by

$$F = \tfrac{1}{2}\rho g H \langle \eta \boldsymbol{u} \rangle .$$

Show that the total energy flux of a Kelvin wave in a channel of width D is given by

$$F = \tfrac{1}{2}\, \rho g^{3/2} H^{-1/2} \eta_0^2 \int_0^D \exp\,(2fx/\sqrt{gh_n})\,\mathrm{d}x .$$

Calculate the corresponding result for Poincaré waves.

3. Look at the amplitude of interfering Kelvin waves, as given by (28.12), very near an amphidromic point; show analytically that the phase rotates clockwise around that point.

4. Draw the amphidromic system of nodal points (co-range and co-phase lines) for intersecting Poincaré waves of equal amplitude and of mode $n = 1$, using (28.15).

5. Show that the mean energy flux associated with a standing wave solution of (28.27), such as used by Taylor to describe the tides in the Bristol Channel, vanishes. Find a solution of (28.27) (with H and l given by 28.29) in terms of Hankel functions that represent travelling waves, and obtain an expression for the energy flux in terms of the ratio of up-channel to down-channel wave amplitudes (Bennett, 1975).

6. Show that the conservation of energy flux in narrow channels which vary smoothly enough that no energy is reflected or dissipated leads directly to Green's law (28.35).

7. Show why the condition (28.44) is inapplicable for a north–south channel and replace it by another condition in terms of the east–west wavenumber component.

8. Consider a channel of width $D = 2000$ km and depth $H = 5$ km, oriented at an angle $\tan^{-1}\alpha = 45°$, at a latitude of 45°N. What is the maximum frequency of planetary waves which can propagate along that channel?

9. Derive a general expression for the maximum (cut-off) frequency for propagation of planetary waves up an arbitrary rectilinear channel of width D and orientation α.

10a. Consider a harbour of the general form shown in Fig. 28.9 with an area $Q = 100\ \mathrm{km}^2$ connected to the ocean by a channel of length $D = 2$ km, width $W = 2$ km and depth $H = 100$ m. What is the resonance period for the Helmholtz mode?

(Ans. 33.1 min.)

10b. Assuming that the same harbour is nearly square in plan area, and of the same depth as the channel which connects it to the sea, find the period at which quarter-wave resonance occurs.

(Ans. 25 min.)

11. Use (28.69) and (28.73) to calculate as a function of frequency the response $(Z_\mathrm{B} + Z_\mathrm{E})^{-1}$ of a narrow rectangular inlet of uniform depth $H = 100$ m, length $D = 10$ km and width $W = 1$ km, adjacent to an ocean of depth $H_\mathrm{E} = 300$ m.

29. WAVES IN CLOSED BASINS

We conclude Chapter 4 and hence our description of free modes in the ocean with a discussion of first- and second-class wave oscillations in closed basins. This subject has attracted the attention of many scientists since the time of Laplace, and accordingly there is an extensive literature on this topic. For the most part we shall deal with long-wave oscillations in a rectangular basin of uniform depth, such a basin being representative of many simple shapes that arise in practice. During the last century many analytical techniques (mostly approximate in nature) have been developed to compute the eigenfrequencies of basins of various shapes and with variable topography. However, with the arrival of the modern computer, many of these techniques have been superseded by finite-difference, finite-element or other numerical procedures. In keeping with the analytical approach used in this book, we shall focus our attention on oscillations in the rectangular or circular basins, easily described analytically.

First-class oscillations in a rectangular basin of constant depth

For a uniformly rotating fluid of constant depth which is stably stratified, we can use the separation technique discussed in Section 10 to find the natural frequencies of both barotropic and baroclinic first-class (gravity) oscillations of any closed basin. This was formally done in Section 10 for the case of a rectangular basin; we first "solved" the horizontal problem (10.39) and (10.40) and then substituted the result (10.46) for h_n into the vertical problem (10.41)–(10.43). The solution of the vertical problem for a given $\rho_0(z)$ would then yield the values of the natural frequencies (eigenfrequencies). However, this approach glosses over the difficulties inherent in solving the horizontal boundary-value problem when $f \neq 0$. Therefore, we shall here reverse the procedure and assume a priori that for a given stratification $\rho_0(z)$, the vertical problem (10.41)–(10.43) has been solved for $Z_n(z)$ and the equivalent depths h_n, $n = 0, 1, 2, \ldots$. For arbitrary wavelengths, h_n depends on the frequency; however, as most observed oscillations in lakes and enclosed seas have long periods, the hydrostatic approximation may be invoked ($\omega^2 \ll N^2$), in which case h_n becomes independent of frequency. We shall thus henceforth treat this case only and therefore go directly over to the long-wave equations introduced in Chapter 3. This was also the approach taken in Section 23 on wave reflection.

In the long-wave approximation, the horizontal dependence of the pressure, $p_n(x, y, t)$, satisfies (23.3) for gravity wave motions. Setting

$$p_n(x, y, t) = p_n(x, y)\, \mathrm{e}^{-i\omega t}, \quad \omega > 0, \tag{29.1}$$

we find that $p_n(x, y)$ obeys Helmholtz's equation:

$$\nabla_h^2 p_n + \left(\frac{\omega^2 - f^2}{g h_n}\right) p_n = 0, \tag{29.2}$$

where h_n is determined from the vertical problem (15.25), (15.27), (15.28), for $H_n(z)$, vertical dependence of the pressure $[p = p_n(x, y, t)\Pi_n(z)]$. We wish to solve (29.2) in the rectangular domain $0 \leqslant x \leqslant a, 0 \leqslant y \leqslant b$, subject to the usual boundary condition of zero normal flow at the boundary. From (15.19) and (15.20), which give u and

v in terms of p, we can write this condition as

$$-i\omega p_{nx} + f p_{ny} = 0 \quad \text{at} \quad x = 0, a, \tag{29.3a}$$

$$f p_{nx} + i\omega p_{ny} = 0 \quad \text{at} \quad y = 0, b. \tag{29.3b}$$

Equations (29.3a) and (29.3b) ensure that $u = 0$ at $x = 0$, a and $v = 0$ at $y = 0, b$, respectively.

With h_n known, the set (29.2) and (29.3) represents a boundary-value problem for the eigenfrequency ω and eigenfunction $p_n(x, y)$. We shall use the notation $\omega \equiv \omega_n^{lm}$ for the eigenfrequencies, where n refers to the vertical mode number and l and m to the (horizontal) mode numbers in the x- and y-directions, respectively. Similarly, we set $p_n \equiv p_n^{lm}$ for the eigenfunctions.

Most of the literature on long-wave (gravity) oscillations in rectangular basins deals with barotropic motions; for this case, $n = 0$ and $h_0 = H$, the true depth. Also, in (29.2) and (29.3) the pressure p_n is replaced by η, the free surface elevation. Our discussion of the solutions of (29.2) and (29.3) thus includes these free surface oscillations as a special case.

In the absence of rotation ($f = 0$), the boundary-value problem (29.2), (29.3) is equivalent to that for the standing wave oscillations of a vibrating membrane with a free boundary. The solution in this zero-rotation case is

$$p_n^{lm} = A_n^{lm} \cos(l\pi x/a) \cos(m\pi y/b), \quad l, m = 0, 1, 2, \ldots, \tag{29.4}$$

where A_n^{lm} is an arbitrary constant, together with the eigenfrequencies

$$\overset{0}{\omega}{}_n^{lm} = \pi[gh_n(l^2/a^2 + m^2/b^2)]^{1/2}, \tag{29.5}$$

where the 0 over ω indicates that $f = 0$.

For over a century, the formula (29.5) with $h_n = H$ has been used to compute the approximate natural periods of the surface oscillations in lakes, which are commonly referred to as seiches. In long and narrow lakes the observed seiches are often one-dimensional, that is the velocity component across the lake is zero (and hence $m = 0$ for a lake extending along the x-axis, say). In this case (29.5) gives Merian's formula for the period:

$$T_l = 2a/l\sqrt{gH}. \tag{29.6}$$

For a brief survey of observed seiches in various lakes we refer the reader to Defant (1961, Chapter 6). Here we shall only present the striking data of Bergsten (1926) who observed a unimodal, one-dimensional seiche in Lake Vättern, Sweden (Fig. 29.1), which is a narrow and uniformly shaped lake of length 124 km. For an interesting study of wind-generated internal seiches in Lake Windermere, England, we refer the reader to Heaps and Ramsbottom (1966).

When $f \neq 0$ in the boundary-value problem (29.2), (29.3) has no simple solution in terms of standing waves because rotation gives rise to the "splitting" of the modes, a phenomenon discussed in Sections 25 and 26 (see also Fig. 29.2 below). Now the solution consists of a complex mixture of Kelvin- and Poincaré-type waves that propagate positively (in the same direction as rotation) or negatively around the basin. Further, the modes are of two types: antisymmetric and symmetric. In the former, the surface elevation (in the

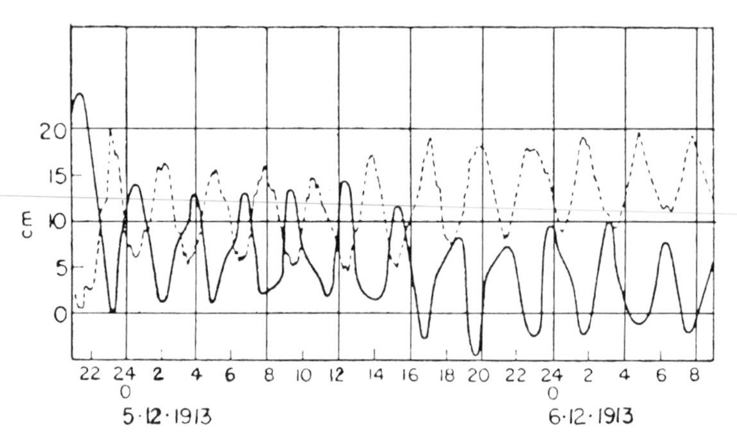

Fig. 29.1. Simultaneous recordings of the water levels at both ends of Lake Vättern, Sweden, which are indicative of a uninodal seiche. The displacements at one end of the lake are out of phase by 180° with the displacements at the other end. (From Defant, 1961.)

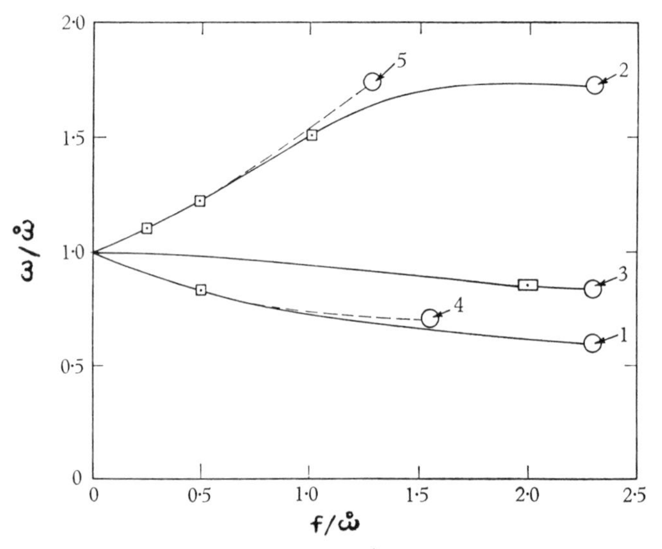

Fig. 29.2. Frequency of oscillation ($\omega/\overset{0}{\omega}$) for long waves in square and rectangular basins of constant depth, as a function of the Coriolis parameter ($f/\overset{0}{\omega}$); the normalizing frequency $\overset{0}{\omega}$ denotes the slowest zero-rotation frequency as given by (29.5). Note that when $f = 0$ (no rotation) the "split" modes as given by curves *1* and *2* degenerate into the point $\omega/\overset{0}{\omega} = 1$ on the vertical axis. The curves *1* and *2* correspond to the slowest positive and negative antisymmetric modes in a square, and curve 3 corresponds to the slowest positive antisymmetric mode in a 2 × 1 rectangle. Also shown for comparison are Corkan and Doodson's (1952) results for a square (⊡) and Taylor's (1920) result for a 2 × 1 rectangle (▭). Curves *4* and *5* show the results of Van Dantzig and Lauwerier's (1960) perturbation solution for the slowest positive and negative modes in a square, valid to $0[(f/\overset{0}{\omega})^2]$. (From Rao, 1966.)

case of the barotropic mode, say) at two points diagonally opposite with respect to the center is the same in magnitude but opposite in sign; also, for the antisymmetric mode $l+m$ is odd. For a symmetric mode, both the magnitude and sign of the two diagonally opposite elevations are the same and $l+m$ is even. A detailed discussion of the numerous attempts to solve the exact problem (29.2), (29.3) for the barotropic mode is beyond the scope of this book, and we refer the reader to Rao (1966) for a brief survey of the earlier work (see also Fig. 29.2). With the exception of Corkan and Doodson (1952), all the work prior to the 1960's is valid only for small rotation rates (as measured relative to $\overset{0}{\omega}{}_0^{10}$ or $\overset{0}{\omega}{}_0^{01}$, the slowest zero-rotation frequency for the barotropic mode) and deals only with the lowest modes. Corkan and Doodson (1952) treated the case of a square basin and numerically integrated the basic equations (29.2), (29.3) to find the eigenfrequencies of the slowest positively and negatively propagating antisymmetric modes for a wide range of rotation rates. One of the interesting results they found was that as the rotation rate increases the negative waves become unstable in the sense that these waves tend to transform into positive waves. Rao (1966), however, has carried out the most extensive calculations for the long wave barotropic eigenfrequencies and eigenfunctions of a rotating rectangular basin. His solutions for the lowest modes are shown in Figs. 29.3 and 29.4. He has also determined the higher modes (see Fig. 29.4) and confirms the results of Corkan and Doodson with regard to the stability of the negatively propagating modes. Rao's method of solution is rather complex, however, as it involves a partitioning of the velocity field into a rotational and irrotational part, and then an expansion of each part in terms of a certain set of orthogonal functions; we again refer the reader to the original paper by Rao (1966) for details. Pnueli and Pekeris (1968), on the other hand, have shown that variational methods can be readily used to find the normal modes for basins of the form of rectangles and also of sectors of circles. In particular they found that for large rotation rates such that $f^2 \gg \omega^2$, the wave motion is rotationally trapped, and similar to that of a Kelvin wave travelling around the basin; that is, the wave moves in the same direction as rotation and its amplitude falls off rapidly away from the boundary. In the next subsection we shall give an explicit solution of such a wave motion in a circular basin of constant depth.

there is a considerable body of literature on the application of the two-layer fluid equations to study first-mode internal seiches and waves in rectangular basins. For a review of this topic we refer the reader to Csanady (1975). However, we point out that Csanady recommends the use of normal mode equations, of the form discussed in Section 16, to describe independent barotropic and baroclinic motions in a rectangular but variable-depth lake. It should by now be evident to the reader that such an approach should be treated with caution as it excludes a priori, the possibility of any topographic coupling between barotropic and baroclinic motions (e.g., see Niiler, 1968). For examples of the propagation of wind-stress-generated internal Kelvin waves in two-layer lakes of constant depth, we refer the reader to Yuen (1969) and Kanari (1975).

First-class oscillations in other basins

While the horizontal problem for gravity oscillations in a rotating rectangular basin of constant depth does not have an explicit separable solution, the corresponding problem for a circular basin *is* separable in terms of plane polar coordinates (r, θ) whose origin

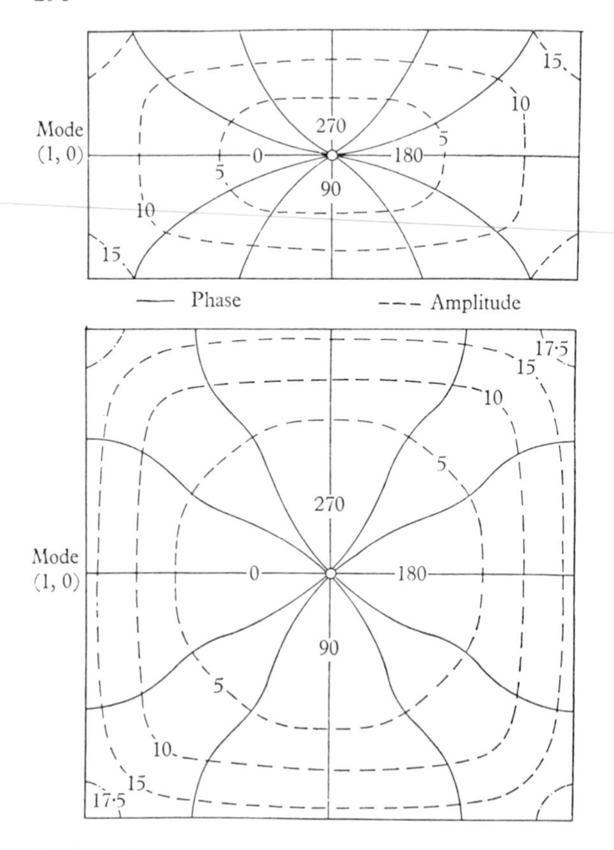

Fig. 29.3. The modal structure of the slowest positive antisymmetric mode in a 2 × 1 rectangle and in a square for the case $f/\overset{0}{\omega} = 2$. The dashed and solid lines denote contours of amplitude (co-amplitude lines) and phase (co-tidal lines), respectively. This mode, in both basins, consists of one wave travelling in the positive direction (counterclockwise) about an amphidromic point at the center of the basin. The amplitude of oscillation is zero at the amphidromic point and increases outward; it reaches a maximum value at the corner. (From Rao, 1966.)

lies at the center of the basin. This solution is discussed by Lamb (1945, Section 209) and more extensively by Howard (1960) and Csanady (1972). When $f = 0$ some of the properties of the solutions are analogous to those of the solutions for trapped oscillations around a circular island (Section 26).

In terms of the polar coordinates (r, θ), (29.2) takes the form

$$\left[\partial_{rr} + \frac{1}{r}\partial_r + \frac{1}{r^2}\partial_{\theta\theta} + \frac{\omega_n^2 - f^2}{gh_n}\right] p_n(r, \theta) = 0, \tag{29.7}$$

which is defined for the circular domain $0 \leqslant \theta < 2\pi$, $0 \leqslant r \leqslant a$. At $r = a$ we require that the radial velocity, u_r, vanishes. Using (17.37) it follows that in terms of p_n this condition takes the form

$$-i\omega_n \frac{\partial p_n}{\partial r} + \frac{f}{r}\frac{\partial p_n}{\partial \theta} = 0 \quad \text{at} \quad r = a, \tag{29.8}$$

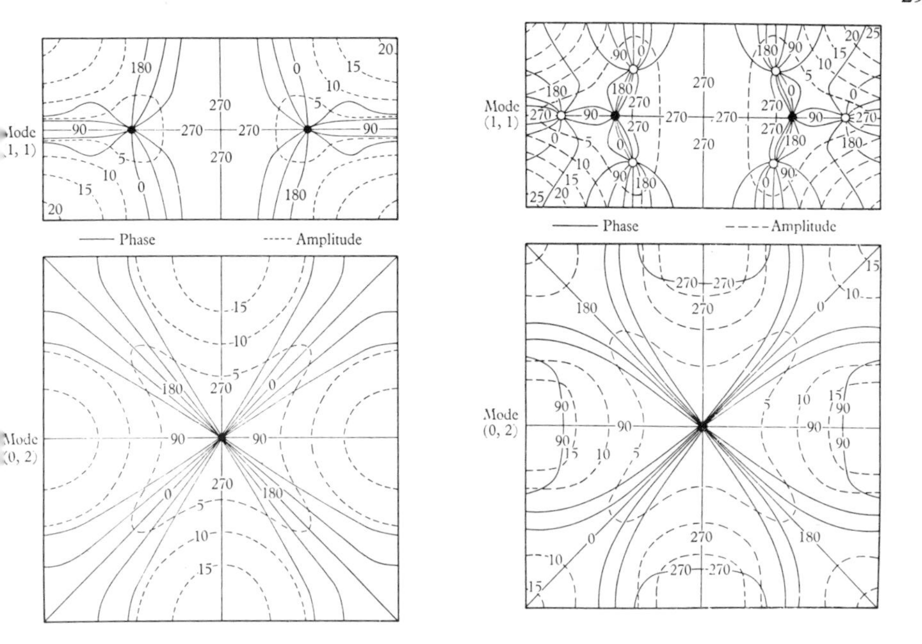

Fig. 29.4. The modal structure in rectangular and square basins of (a) the slowest negative symmetric mode for the case $f/\overset{0}{\omega} = 1$; (b) the slowest negative symmetric mode at twice the rotation rate as in (a): $f/\overset{0}{\omega} = 2$. As the rotation rate increases further, the structure becomes even more complicated and eventually becomes unstable, with the motion turning into a positive wave. (From Rao, 1966.)

At $r = 0$, we require that p_n be bounded. For periodic solutions of the form

$$p_n = F(r)\,\mathrm{e}^{im\theta}, \quad m = 0 \pm 1, \pm 2, \ldots, \tag{29.9}$$

(29.7) and (29.8) give

$$F'' + \frac{1}{r}F' + \left(K^2 - \frac{m^2}{r^2}\right)F = 0, \quad 0 \leqslant r \leqslant a, \tag{29.10}$$

$$-\omega_n F'(a) + (mf/a)F(a) = 0, \tag{29.11}$$

where $K^2 = (\omega_n^2 - f^2)/gh_n$. On combining (29.9) with the assumed time-dependence $\exp(-i\omega_n t)$, $\omega_n > 0$, we see that for $m > 0$, say, the solution represents a right-bounded or counterclockwise travelling wave which is a positive wave when $f > 0$. Further, for a fixed $|m|$, (29.9) implies that there are $|m|$ nodal diameters across the basin.

The solution of (29.10) which is bounded at $r = 0$ is given by

$$F(r) = \begin{cases} J_m(Kr), & \omega_n^2 > f^2 \\ I_m(K'r), & \omega_n^2 < f^2, \end{cases} \tag{29.12}$$

where $K'^2 = (f^2 - \omega_n^2)/gh_n$. In the application of (29.11) it is now convenient to distinguish between the cases $m = 0$ and $m \neq 0$.

Let us first consider the case of axial symmetry, $m = 0$. When $\omega_n^2 < f^2$, (29.11) implies $I_0'(K'a) = 0$, which has no solution for $a > 0$. When $\omega_n^2 = f^2$, an examination of the primitive governing equations for the horizontal dependence reveals that no solution exists in this case also. [It is interesting to recall that there was no bounded solution for the related exterior problem ($r \geqslant a$) considered in Section 27]. Finally, when $\omega_n^2 > f^2$, (29.11) gives

$$J_0'(Ka) = 0, \tag{29.13a}$$

or equivalently,

$$J_1(Ka) = 0. \tag{29.13b}$$

Thus if λ_l denotes the lth zero of $J_1(\lambda)$, then (29.13) implies that the eigenfrequencies are given by $Ka = \lambda_l$, or equivalently,

$$\omega_n = \left(f^2 + \frac{gh_n\lambda_l^2}{a^2}\right)^{1/2}. \tag{29.14}$$

Also, it follows that for a given l there are l nodal concentric but non-uniformly spaced circles with centres at $r = 0$. Finally, it is also easy to show that even with axial symmetry, the axial velocity component $u_\theta \neq 0$.

Let us now consider the case $m \neq 0$. When $\omega_n^2 < f^2$, (29.11) and (29.12) give

$$(K'a)I_m'(K'a) = (fm/\omega_n)I_m(K'a). \tag{29.15}$$

Since I_m and I_m' are both positive, we note that if (29.15) is to have a solution at all then the parameter $\mu \equiv fm/\omega_n > 0$. (In fact μ will be greater than unity since $\omega_n^2 < f^2$.) Thus in the Northern Hemisphere, $m > 0$ and such a solution represents a wave which propagates counterclockwise (i.e., is right-bounded); further, since $I_m(K'r)$ is a monotone increasing function of r, the wave amplitude falls off away from the boundary at $r = a$. Such a solution effectively represents a Kelvin type wave travelling around the inside of the basin; since $u_r \neq 0$ it is not a true Kelvin wave, however. To find the root of (29.15), we pick a positive value of the parameter μ and then determine graphically for a given m the value of $K'a$ for which the left-hand side equals the right-hand side (see Fig. 29.5). Suppose this root is $K'a = z_0$. Then squaring and solving for ω_n^2, we obtain

$$\begin{aligned}\omega_n^2 &= f^2 - gh_n z_0^2/a^2 \\ &= \mu^2\omega_n^2/m^2 - gh_n z_0^2/a^2,\end{aligned}$$

since $\mu = fm/\omega_n$. Hence we finally obtain the following expression for the eigenfrequencies:

$$\omega_n = \sqrt{gh_n}\, z_0 m/a(\mu^2 - m^2)^{1/2}.$$

For the case $\omega_n^2 > f^2$, $m \neq 0$, the implicit frequency relation akin to (29.15) now involves $J_m(Ka)$ and its derivative. This relation for $f > 0$, say has an infinity of roots $(Ka)_l$ corresponding to a given pair $\mu, m > 0$ and another infinite set of roots for μ, $m < 0$. It can then be shown that for a given $m, -m$ pair of solutions, the positive wave ($m > 0$) travels more slowly (i.e., has a lower frequency) than the negative wave ($m < 0$) (see Exercise 29.3).

We conclude this subsection with a short survey of the earlier work on first-class oscillations in various other types of basins. The problem treated immediately above has

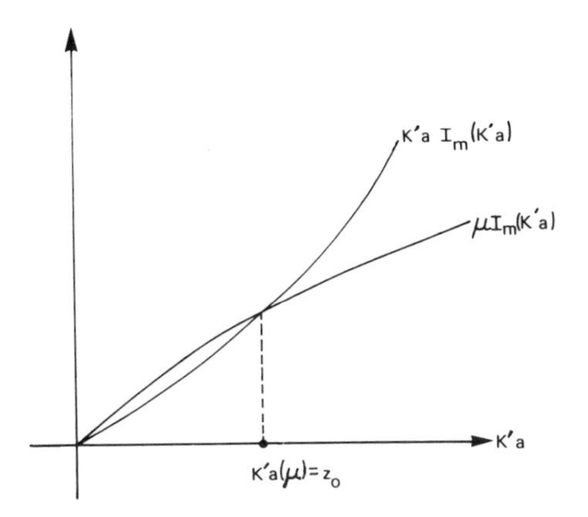

Fig. 29.5. The graphical solution of (29.15).

also been worked out by Lamb (1945) for the case of a paraboloidal-shaped bottom profile: $H = H_0(1 - r^2/a^2)$. Oscillations (with $f = 0$) in an elliptic basin of constant depth have been discussed by Jeffreys (1924). Seiches in rectangular basins of nonuniform depth have been studied by Hidaka (1932) and D. J. Clarke (1971). Seiches in a curved lake have been looked at by Johns and Hamzah (1969). In contrast to the above analytically oriented calculations, we refer the reader to Platzman (1972) and Rao et al. (1976) for examples of sophisticated numerical calculations of the free surface modes in a variety of natural basins. Finally, one-dimensional oscillations in multichannel basins have been discussed by Defant (1961) by means of electrical circuit analogues (similar to that used in the study of harbour resonance in Section 28), and by Easton (1971) using numerical methods.

Second-class oscillations in a rectangular basin

To study both barotropic and baroclinic planetary oscillations in a stratified basin of uniform depth and of limited horizontal extent, we start directly from (23.11), the planetary wave equation for $p_n(x, y, t)$. As in the above study of first-class basin oscillations, we put

$$p_n(x, y, t) = p_n(x, y)\, e^{-i\omega t}, \quad \omega > 0;$$

substituting this into (23.11), we obtain

$$\nabla^2 p_n + 2i\gamma p_{nx} - (f^2/gh_n)p_n = 0, \tag{29.16}$$

where $\gamma = \beta/2\omega > 0$ and the equivalent depths h_n are determined from (15.25), (15.27) and (15.28), the vertical boundary-value problem for $\Pi_n(z)$, where

$$p = p_n(x, y, t)\Pi_n(z).$$

We argued in Section 23 that for sufficiently low frequencies ($\omega \ll f$), the pressure behaves like a stream function and therefore must vanish on the lateral boundaries. Thus for the rectangular domain $0 \leqslant x \leqslant a, 0 \leqslant y \leqslant b$

$$p_n(x,y) = 0 \quad \text{at} \quad x = 0, a \quad \text{and} \quad y = 0, b. \tag{29.17}$$

It is now convenient to introduce the carrier wave transformation (Exercise 19.1)

$$p_n(x,y) = \psi_n(x,y)\,\mathrm{e}^{-i\gamma x}; \tag{29.18}$$

then (29.16) becomes

$$\nabla_h^2\psi_n + (\gamma^2 - f^2/gh_n)\psi_n = 0, \quad 0 \leqslant x \leqslant a, 0 \leqslant y \leqslant b \tag{29.19}$$

and (29.17) gives

$$\psi_n = 0 \quad \text{at} \quad x = 0, a \quad \text{and} \quad y = 0, b. \tag{29.20}$$

With f treated as a constant, the boundary-value problem (29.19), (29.20) is equivalent to that which describes the oscillations of a clamped membrane! On combining the solution for ψ_n and ω with $\mathrm{e}^{-i\gamma x}$ and the time-dependence, we obtain the eigenfunctions

$$p_n^{lm}(x,y,t) = A_n^{lm} \sin\frac{l\pi x}{a} \sin\frac{m\pi x}{b}\,\mathrm{e}^{-i(\gamma x - \omega t)}, \tag{29.21}$$

with eigenfrequencies

$$\omega \equiv \omega_n^{lm} = \frac{\beta}{2(l^2\pi^2/a^2 + m^2\pi^2/b^2 + f^2/gh_n)^{1/2}}. \tag{29.22}$$

Thus the solution for the pressure field has two types of nodal lines: those which correspond to the nodes of the envelope function and which remain fixed in position for all time, and those which correspond to the carrier wave $\mathrm{e}^{-i(\gamma x + \omega t)}$ and which lie north–south and progress westward at a speed ω/γ (see Fig. 29.6).

For the case of slightly divergent barotropic waves the solution (29.21), (29.22) was first derived by Longuet-Higgins (1965a). He also considered such oscillations in a triangular basin and a rectangular basin of arbitrary orientation (see Fig. 29.6). Buchwald (1973a), on the other hand, argued that for the same type of low-frequency barotropic oscillations in a closed basin, the equation for the stream function ψ should also include a small additional term proportional to the quantity

$$(1/A)\int_S \psi\,\mathrm{d}x\,\mathrm{d}y,$$

where A is the area of the basin. For a basin such as the North Pacific, this extra term changes the eigenfrequencies only by at most one percent. Thus it does not appear to be a significant correction. Slightly divergent barotropic planetary waves in closed basins have also been considered by Larichev (1974).

We mention here that Ball (1965) has studied second-class oscillations in a uniformly rotating shallow fluid by using a paraboloidal bottom to model the β-effect. In particular he exploits the rigid-top approximation and shows that for a small range of rotation rates the motion is unstable. Such unstable motions always exert a couple tending to oppose the rotation of the container. The astute reader will appreciate this as another example of Le Chatelier's principle. Saint-Guily (1972), on the other hand, found that in a circular, uniformly rotating basin with a variable depth, second-class oscillations are stable when the free surface is retained.

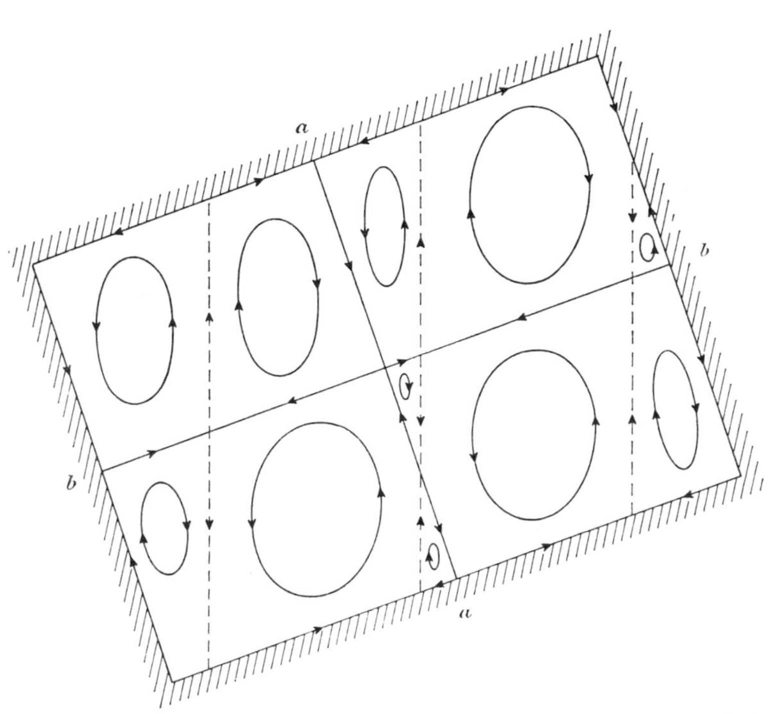

Fig. 29.6. Streamlines for barotropic planetary waves in a rectangular basin of arbitrary orientation: the mode $l = m = 2$, $n = 0$. The solid and dashed lines represent, respectively, the nodal lines for the envelope function and a westward-travelling carrier wave. (From Longuet-Higgins, 1965a.)

The significance of barotropic second-class oscillations in closed basins with regard to the wind-generated ocean circulation has been considered by a number of authors. For example, Pedlosky (1965a, 1967) has studied both the linear and nonlinear time-dependent response of a rigid-topped square basin to fluctuating winds. If bottom friction is the dominant amplitude-limiting mechanism in a resonant response, Pedlosky finds that the structures of the steady and fluctuating ocean circulations are strongly dependent on the forcing frequency. If the nonlinearities are the main amplitude-limiting mechanism at resonance, however, he finds that the normal mode oscillations can never rectify in a way to produce intense boundary currents. When the motions are driven at nonresonant frequencies steady boundary layers can be produced. The effects of western coastal orientation on divergent barotropic planetary wave reflection and the resulting large-scale wind-driven circulation have been investigated by Gates (1970) by means of a numerical integration of the primitive equations.

Low-frequency planetary oscillations in a rectangular basin which spans the equator were first studied by Rattray and Charnell (1966). Since $f = \beta y$ for an equatorial basin, the basic solutions, as we saw in Section 21, take the form of parabolic cylinder functions. However, in contrast to the case of an unbounded ocean considered in Section 21, lateral boundary conditions also have to be applied at $x = 0, L$, say. The final solution is rather complex and will not be presented here. For the Pacific Ocean, Rattray and Charnell computed the barotropic mode to have a period of 2.7 days, which is very close to the observed spectral peak of 2.9 days, at Oahu (see Fig. 18.4). Moore (1968), on the other hand, considered baroclinic oscillations of both gravity and planetary type in an equatorial basin. Finally, Mofjeld and Rattray (1971) considered barotropic oscillations of both classes in an equatorial basin. They found that the gravity oscillations are dynamically similar to either Poincaré or Kelvin waves. The Poincaré waves, however, are restricted to latitudes below the critical (inertial) latitude (see 10.25). The planetary oscillations consist primarily of Rossby waves that propagate westward under modulating envelopes that produce equatorial trapping. Finally, we note that LeBlond (1964) has studied nondivergent planetary waves in a symmetric polar basin, both with and without topography.

Laplace's tidal equations

In many ocean basins of the world the horizontal length scale L is comparable with the Earth's radius R. Hence any solution for the oscillations of such large basins derived from the β-plane (or f-plane) equations is of questionable validity in view of one of the fundamental β-plane approximations (see 5.24):

$$(L/R)^2 \ll 1.$$

To study the oscillations of such large basins, we need to return to the original conservation equations written in terms of spherical coordinates. One set of celebrated equations that comes from these was first introduced by Laplace (1775) to study barotropic long-wave oscillations in the ocean. We shall now briefly discuss the so-called Laplace's tidal equations (LTE) and their generalization to include stratification. However, even these simplified equations are very difficult to solve analytically. Hence we shall not present any explicit solutions here but rather give a qualitative discussion of their solutions. For further details we refer the reader to the recent papers by Miles (1947b, c), whose presentation we follow in part below.

Laplace (1775) made the following idealizations and approximations in order to study tidal motion (see Section 52) in the ocean:

(1) Sea water is nondiffusive and homogeneous (only undamped barotropic modes are possible).

(2) The Earth is a spherical planet with a uniform gravitational field. The ocean bottom is rigid (no Earth tides occur).

(3) Tidal flows represent only small deviations from a state of uniform rotation (the motions are linear).

(4) The ocean is shallow and the horizontal component of the Earth's rotation is neglected in the Coriolis force (the motions are hydrostatic).

On the basis of these assumptions, the resulting equations for u, v and η are

$$\partial_t(u, v) + f(-v, u) = -g\nabla(\eta - \bar{\eta}), \tag{29.23a}$$

$$\nabla \cdot [H(u, v)] + \partial_t \eta = 0. \tag{29.23b}$$

Here $\bar{\eta}$ is the equilibrium value of the surface elevation η that would be produced by external fields (such as solar and lunar gravitational fields) in the absence of inertial (other than centrifugal) forces; ∇ is the *spherical* gradient operator on the spherical surface $r = R$. The components (u, v, f) form a right-handed triad with u and v being the velocity components in the southward and eastward directions. Equations (29.23) are known as *Laplace's Tidal Equations* (LTE).

Lamb (1945) discusses some solutions of LTE for rather special boundary or topographic configurations which are not terribly relevant for tidal prediction in the ocean. On the other hand, numerical solutions for the real oceans are reviewed by Hendershott and Munk (1970) and Hendershott (1977). Here we shall consider free oscillations in a global ocean of constant depth H. For cyclic wave motions of the form

$$\eta = F(\mu)\, e^{i(s\lambda - \omega t)}, \quad \mu = \cos(\pi/2 - \phi),$$

where $s = 0, \pm 1, \pm 2, \ldots$ and λ and ϕ are the longitude and latitude, respectively, (29.23) yield

$$(\mathcal{L} + \epsilon)F(\mu) = 0, \tag{29.24a}$$

where here $\epsilon = 4\Omega^2 R^2/gH$ ($\epsilon = 0$ corresponds to a rigid-topped ocean) and

$$\mathcal{L} = \frac{\partial}{\partial \mu}\left(\frac{1-\mu^2}{\sigma^2 - \mu^2}\right)\frac{\partial}{\partial \mu} + \frac{s(\sigma^2 + \mu^2)}{\sigma(\sigma^2 - \mu^2)^2} - \frac{s^2}{(1-\mu^2)(\sigma^2 - \mu^2)}, \tag{29.24b}$$

where $\sigma = \omega/2\Omega$.

The equation (29.24) has apparent singularities at $\mu = \pm\sigma$ [the so-called inertial or critical latitudes: see (10.25)] and regular singularities at $\mu = \pm 1$ (the poles). For a given s and σ, with ϵ as the eigenvalue, the complete eigenfunctions of (29.24) are usually referred to as Hough functions; however, unlike most Sturm-Liouville equations, (29.24) admits negative eigenvalues if and only if $\sigma^2 < 1$ (subinertial frequencies). These eigenfunctions are regular in $|\mu| < 1$, $0[(1-\mu^2)^{s/2}]$ as $\mu \to \pm 1$, and are mutually orthogonal for $|\mu| < 1$.

Laplace found rather cumbersome power series solutions to (29.24) which led him to develop the theory of infinite continued fractions. Hough (1897, 1898) obtained more elegant series solutions in terms of Legendre polynomials. Hough also showed that in the limit of slow rotation ($\Omega \to 0$), the eigenfrequencies are given by

$$\omega = \pm\, [m(m+1)gH]^{1/2}/R \tag{29.25a}$$

$$\text{and} \quad \omega = -2\Omega s/m(m+1) \quad (m = s, s+1, \ldots, \infty), \tag{29.25b}$$

corresponding to first- and second-class oscillations, respectively.

The rate of convergence of Hough's expansions increases with decreasing ϵ; unfortunately, for the real ocean $\epsilon \simeq 25$, so that Hough's solutions are not of much practical use. Flattery (1967) and Longuet-Higgins (1968c), however, have recently obtained more extensive solutions for large ϵ. Longuet-Higgins (1968c) also used his LTE solutions to check the validity of earlier equatorial β-plane solutions. In the same spirit, Longuet-Higgins (1964a, 1965a) also compared mid-latitude β-plane solutions with those of the exact LTE for both $\epsilon = 0$ and $\epsilon \neq 0$.

The solutions of LTE for closed ocean basins are, needless to say, even more complicated. Apart from the many purely numerical solutions (e.g., see Doodson, 1958; Christensen, 1973a; Platzman, 1975; Hendershott, 1976), most of the analytical work appears to have been done by Longuet-Higgins (1966) and Longuet-Higgins and Pond

(1970). In the latter two papers oscillations in a hemispherical basin bounded by meridians of longitude were considered for the case of $\epsilon = 0$ and $\epsilon \neq 0$, respectively. Also, in Longuet-Higgins (1966) a comparison between LTE solutions and β-plane solutions is made, and in a later paper Moore's (1968) equatorial β-plane solutions are validated. The importance of realistic topography in the solutions of LTE has as yet hardly been touched upon (e.g., see Christensen, 1973b).

One of the main drawbacks in the application of LTE to the real ocean rests with the barotropic assumption (1), which eliminates any internal or baroclinic modes. Laplace himself studied such oscillations in the rather special case of an isothermal, perfect atmosphere for which (29.24) also apply but with H replaced by the scale height $H_0 \equiv c_T^2/g$ (c_T = isothermal speed of sound). We now consider the more general case of any radially stably stratified inviscid fluid described by $p_0(r)$, $\rho_0(r)$, $c_0(r)$, the basic pressure, density and sound speed, respectively, for which

$$p_0'(r) = -\rho_0(r)g$$

and the stability frequency N is defined as

$$N^2 = -\frac{g\rho_0'}{\rho_0} - \frac{g^2}{c_0^2} > 0. \qquad (29.26)$$

Note that (29.26) includes the compressibility term proportional to $1/c_0^2$. If we also assume that the hydrostatic approximation holds for the perturbation pressure p and density ρ and that p has the form

$$\frac{p - p_0}{\rho_0 g} = P\,e^{i(s\lambda - \omega t)},$$

then it can be shown that under the approximations (2)–(4), $p = p(r, \mu)$ satisfies the second-order partial differential equation

$$\mathfrak{L}p = \frac{\epsilon g H}{\rho_0}\frac{\partial}{\partial r}\left(\frac{\rho_0}{N^2}\frac{\partial p}{\partial r}\right), \qquad (29.27)$$

where $\mathfrak{L}$ is the operator given in (29.24b). Equation (29.27) is analogous to the pressure equation (15.21) derived for the β-plane and for an incompressible fluid.

Separable solutions of (29.27) can be obtained which satisfy finiteness conditions at $\mu = \pm 1$ and the appropriate boundary conditions at the spherical surfaces $r = R$, $r = R - H$. However, there arises the following difficulty: (29.27) is hyperbolic everywhere in $\mu^2 < 1$ if $\sigma^2 > 1$ ($\omega^2 > 4\Omega^2$) and hyperbolic or elliptic according to $\mu^2 < \sigma^2$ or $\mu^2 > \sigma^2$ if $\sigma^2 < 1$. This is to say, the equation changes character at the inertial latitudes $\mu^2 = \sigma^2$ for subinertial frequencies. Accordingly, the problem is not well posed, and even to this day the consequences of this difficulty are not fully understood and are usually ignored.

For recent contributions to this and to other mathematical difficulties associated with LTE, we refer the reader to Miles (1974b) and also to Stewartson and Rickard (1969) and Pekeris (1975).

Exercises Section 29

1. Using (29.6) and the seiche data for Lake Vättern shown in Fig. 29.1 compute the mean depth of this lake.
2. What is the period of the fundamental-mode, longitudinal seiche in your bath tub?
3. Discuss the solutions of the frequency relation for seiches in a rotating circular lake of constant depth when $\omega^2 > f^2$ and $m \neq 0$. In particular, show that a positive wave moves more slowly than a negative wave.
4. Find the eigenfrequencies and eigenfunctions for planetary wave oscillations in a circular basin of constant depth.
5. Show that the eigenfrequencies for low-frequency barotropic divergent oscillations in a basin with variable depth are complex.
6. Show that (29.22) also represents the (second-class) eigenfrequencies of a rectangular basin of arbitrary orientation. Also derive this result by using the geometric methods discussed in Section 28 in connection with planetary waves in a channel.
7. Show that a power series solution of (29.24) [in powers of μ or of $(1-\mu^2)^{1/2}$] leads to a three-term recursion formula, for the solution of which Laplace developed the method of infinite continued fractions. Can you think of any other equations of mathematical physics where such a technique can be used?
8. We note from (29.25) that $|\omega|\uparrow$ as $m\uparrow$ for first-class oscillations, whereas $|\omega|\downarrow$ as $m\uparrow$ for second-class oscillations. Is this behaviour also true for any other solutions discussed earlier in the text?

CHAPTER 5

STATISTICAL AND PROBABILISTIC METHODS

30. INTRODUCTION

Anyone who has gazed at the restless sea knows that it is rarely possible to describe it simply in terms of plane waves. The real ocean surface at any one instant resembles a rather irregular spatial pattern of troughs and crests of varying sizes, shapes and orientations. At a later moment, another very complex pattern emerges, often bearing little resemblance to the earlier pattern. To describe and to understand the dynamics of the random character of the sea surface, statistical and probabilistic methods have to be invoked. Methods based on the theory of random noise (Rice, 1945) and time series analysis (Yaglom, 1962) were first used in the study of surface gravity waves about thirty-five years ago (for a survey of the early literature, see Cartwright, 1962; Kinsman, 1965). In the last two decades or so, these methods have also been used to describe the irregular fluctuations associated with currents, temperature, salinity, and pressure measurements taken from the ocean's interior. The reader is undoubtedly fully aware by now that much of the data that have been presented to illustrate the theory have been expressed in terms of time series analysis concepts.

The analysis of tides followed at first a different course, based on Kelvin's harmonic method, a special kind of Fourier analysis in which a time series is resolved into contributions from a few chosen frequencies (the tidal forcing frequencies, see Section 52). Harmonic analysis (Section 32) survives in the more routine handling of the tides and in the preparation of tidal predictions. The modern theory of time series analysis, however, now incorporates harmonic analysis as a special case, particularly suited to handling line spectra (Godin, 1972).

In Section 31 we give the basic definitions that are frequently used in the theory and applications of time series analysis (e.g., stationary random function, power spectrum, coherence). Then in Section 32 we discuss the application of power spectral methods to computing wave spectra. The slow space–time evolution of wave spectra is then briefly discussed in Section 33. Finally, in Section 34 we give a short survey of the topic of stochastic differential equations, with application to wave propagation in random media.

31. BASIC DEFINITIONS AND CONCEPTS FROM TIME SERIES ANALYSIS

Random functions and their moments

Let us begin with Kolmogorov's definition of a random function, which is based on the idea of a probability space (see Kolmogorov, 1950; Chorin, 1975). The more conventional definition, which is based on the concept of an infinity of joint probability distribution functions (see Yaglom, 1962), will be given later for comparison.

Let α be a parameter which ranges over a set A, a space of events or probability (measure) space. Let $p(\alpha)$ be a given probability density function (or probability measure) defined on A. Thus $p(\alpha)\,\mathrm{d}\alpha$ represents the probability of event α lying in the interval α, $\alpha + \mathrm{d}\alpha$ and

$$\int_A p(\alpha)\,\mathrm{d}\alpha = 1, \tag{31.1}$$

where the integration over the set A is in the sense of Lebesgue. Then a random function f of the scalar variable t, written $f(t, \alpha)$, is defined as the totality or ensemble of all realizations $f(t, \alpha_1)$, $f(t, \alpha_2)$, . . . where $\alpha_i \in A$. Further, for fixed t, f is required to be a measurable function of α. In set theoretic notation,

$$f(t, \alpha) = \{f(t, \alpha_i) | \alpha_i \in A\}.$$

For this discussion we shall think of t as representing the time: however, we could also take t to be a spatial coordinate or a wave amplitude.

The statistics (or equivalently, the moments) of f are completely determined by the density function $p(\alpha)$. The *mean* (first moment) of f, which is denoted by $\bar{f}$, is defined by the integral

$$\bar{f} = \int_A p(\alpha) f(t, \alpha)\,\mathrm{d}\alpha. \tag{31.2}$$

We note that in general, the mean is a function of t: $\bar{f} = \bar{f}(t)$. The *auto-covariance function*, denoted by $\Gamma(t, t')$, is defined by

$$\Gamma(t, t') = \int_A p(\alpha)[f(t, \alpha) - \bar{f}(t)][f(t', \alpha) - \bar{f}(t')]\,\mathrm{d}\alpha. \tag{31.3}$$

In particular, $\Gamma(t, t)$ is called the variance of f. When $\bar{f} = 0$ the autocovariance function is then the same as the second moment, which is defined by the integral

$$\int_A p(\alpha) f(t, \alpha) f(t', \alpha)\,\mathrm{d}\alpha.$$

The integral operator

$$\int_A p(\alpha)\{\cdot\}\mathrm{d}\alpha \tag{31.4}$$

is called the ensemble or probability average (mathematical expectation), and is often denoted by script $\mathcal{E}$:

$$\mathcal{E}\{\cdot\} \equiv \int_A p(\alpha)\{\cdot\}\,\mathrm{d}\alpha. \tag{31.5}$$

We note that any random function f can always be written in the form

$$\begin{aligned} f(t, \alpha) &= \mathcal{E}\{f\} + f'(t, \alpha) \\ &= \bar{f}(t) + f'(t, \alpha), \end{aligned} \tag{31.6}$$

where $f'(t, \alpha)$ is called the random or fluctuating part of $f(t, \alpha)$. Since (31.1) and (31.4) imply that $\mathcal{E}(\mathcal{E}\{f\}) = \mathcal{E}\{f\}$, it follows from (31.6) that $\mathcal{E}\{f'\} = 0$. If f represents, say, a realization of the amplitude of a random wave field, then $\mathcal{E}\{f\}$ and f' are sometimes called the coherent and incoherent parts of the field, respectively.

Stationary random functions

Let us now restrict $f(t, \alpha)$ to be a real random function. We say that f is a *stationary random function* (or *stationary random process*) in the *wide sense* if

$$\bar{f}(t) = m_0 = \text{constant}, \tag{31.7}$$

and $$\Gamma(t, t') = \Gamma(t-t') = \Gamma(\tau), \quad \tau = t-t'. \tag{31.8}$$

On the other hand, f is said to be a *stationary random function* in the *strict sense* if (31.7) and (31.8) hold and the nth moment of f (for all $n \geqslant 3$) is a function of $n-1$ variables. That is to say, *all* the moments are translationally invariant. These two definitions of stationarity coincide when f is Gaussian, in which case all the statistics of f are completely determined by the first and second moments. In practice, whenever one speaks of a stationary random process only stationarity in the wide sense is usually meant; we shall also adopt this convention here. Wide sense stationarity is also called weak stationarity.

From the definitions, it follows that for a stationary random process,

$$\Gamma(\tau) = \Gamma(-\tau), \quad |\Gamma(\tau)| \leqslant \Gamma(0). \tag{31.9}$$

Since $\mathcal{E}\{(f-\bar{f})^2\} = \Gamma(0)$, $\Gamma(0)$ represents the variance of f. For most processes, $\Gamma(\tau) \to 0$ as $|\tau| \to \infty$; the decay time which characterizes this behaviour is often called the *correlation scale* (see also Section 32). For example, a Markov stationary random process with zero mean has the auto-covariance function

$$\Gamma(\tau) = (\Gamma_0/2\tau_0)\exp(-|\tau|/\tau_0), \tag{31.10}$$

where Γ_0 is a constant and τ_0 is the correlation scale. Taking the limit $\tau_0 \to 0$ in (31.10) gives the auto-covariance function of a zero-mean white noise random process:

$$\Gamma(\tau) = \Gamma_0 \delta(\tau), \tag{13.11}$$

where $\delta(\tau)$ is the Dirac delta function.

Finally, we define the *auto-correlation function* $R(\tau)$ as

$$R(\tau) = \Gamma(\tau)/\Gamma(0); \tag{31.12}$$

from (31.9) we see that $|R(\tau)| \leqslant 1$.

The ergodic theorem

Because of the practical difficulty in computing an ensemble average, one often invokes the *ergodic theorem*, which allows such averages to be replaced with time averages. The theorem will merely be stated here; the conditions under which it is valid are discussed in standard texts on stochastic processes (cf. for example, Yaglom, 1962). The ergodic theorem states that, "If f is an ergodic stationary random function, then the first moment and the auto-covariance function obtained by ensemble averages are identical to the corresponding time averages computed for any given realization." That is,

$$m_0 = \mathcal{E}\{f(t,\alpha)\} \equiv \lim_{T\to\infty} \frac{1}{T} \int_{-T/2}^{T/2} f(t,\alpha)\, dt, \tag{31.13}$$

$$\Gamma(\tau) = \mathcal{E}\{[f(t+\tau,\alpha)-m_0][f(t,\alpha)-m_0]\} \equiv$$

$$\lim_{T\to\infty} \frac{1}{T} \int_{-T/2}^{T/2} [f(t+\tau,\alpha)-m_0][f(t,\alpha)-m_0]\, dt. \tag{31.14}$$

As mentioned earlier, the significance of this theorem is that it enables us to estimate m_0 and $\Gamma(\tau)$ by means of time averages. Consider a finite-length time series extending over the time interval $(0, T)$ say, which is assumed to represent part of an ergodic stationary random process. Then according to the ergodic theorem, we have the following estimates for m_0 and $\Gamma(\tau)$:

$$m_0 \simeq \frac{1}{N} \sum_{n=1}^{N} f(n\Delta), \tag{31.15}$$

and $$\Gamma(\tau) \simeq \frac{1}{N} \sum_{n=1}^{N-\tau/\Delta} [f(n\Delta+\tau)-m_0][f(n\Delta)-m_0], \tag{31.16}$$

where Δ is a small time interval much less than T, τ is a multiple of Δ and N is chosen so that $N\Delta = T$. Clearly, from (31.16) it follows that the smaller τ/Δ is compared to N, the better the estimate of $\Gamma(\tau)$. The computational aspects of time series analysis will be discussed further in Section 32. As well as the references quoted there, we also refer the reader to Denman (1975) for an elementary account of the calculation of time series statistics.

Random functions as random variables

At this point we now review the random-variable definition of a random function (e.g., see Yaglom, 1962; Phillips, 1966). For a fixed realization α, the random function $f(t,\alpha)$ is defined as an infinite family of random variables: $f(t_1,\alpha), f(t_2,\alpha), \ldots, f(t_n,\alpha)$, where t_i has any value and n has any value. To specify all the statistics of f we first specify an infinity of joint probability distribution functions (pdf's). Thus we have, for example, that the mean of f is defined by the Stieljes integral

$$\bar{f}(t) = \int_{-\infty}^{\infty} u\, dF_t(u),$$

where $F_t(u) = \text{Prob}\,\{f(t) < u\}$ is the single pdf. The explicit dependence on α has been suppressed since α is fixed. The second moment is defined by

$$\overline{f(t_1)f(t_2)} = \int_{-\infty}^{\infty}\int_{-\infty}^{\infty} uv\, dF_{t_1 t_2}(u,v),$$

where $F_{t_1 t_2}(u, v) = \text{Prob}\,\{f(t_1) < u \text{ and } f(t_2) < v\}$ is the joint pdf. The auto-covariance function is defined also in terms of $F_{t_1 t_2}$ in an obvious manner. From this viewpoint f is said to be a stationary random function (in the strict sense) if all the pdf's are translationally invariant:

$$F_{t_1+\tau,\, t_2+\tau, \ldots}(u, v, \ldots) = F_{t_1 t_2 \ldots}(u, v, \ldots).$$

However, the defining properties of a stationary random function in the wide sense are identical with those introduced earlier [(31.7), (31.8)]. Hence in this case the two paths cross and we now proceed with the important concept of the "power spectrum" or simply the "spectrum".

The power spectrum

Before proceeding with the definition of the power spectrum for a (real) stationary random function f, we first assume without loss of generality that $\bar{f} = 0$. This is no serious restriction since if $\bar{f} \neq 0$, we would simply consider the stationary random function $f - \bar{f}$ instead. A zero-mean random function is also called a *centered* random function.

For any realization α we define the truncated Fourier transform of $f(t, \alpha) \equiv f(t)$ by

$$\hat{f}_T(\omega) = \int_{-T/2}^{T/2} f(t)\, e^{-i\omega t}\, dt. \tag{31.17}$$

The power spectrum, or more simply, the *spectrum* of f is then defined as

$$S(\omega) = \lim_{T \to \infty} \frac{1}{2\pi T} \,|\hat{f}_T(\omega)|^2. \tag{31.18}$$

In analogy with the definition of the auto-covariance, some authors call $S(\omega)$ the auto-spectrum. It follows immediately from (31.17) and (31.18) that $S(\omega)$ is real and positive and $S(\omega) = S(-\omega)$. Since (31.17) represents a Fourier coefficient of f at a frequency ω, $S(\omega)$, being proportional to the square of this coefficient, represents a measure of the "power" or "energy" per unit frequency interval. A powerful result on which the Blackman–Tukey method (see Section 32) for computing spectra is based, is known as the *Wiener–Khinchin theorem*: "For any ergodic stationary real random process f with zero mean, $S(\omega)$ is given by the Fourier transform of the auto-covariance function of f." That is,

$$S(\omega) = \frac{1}{2\pi} \int_{-\infty}^{\infty} \Gamma(\tau)\, e^{i\omega\tau}\, d\tau. \tag{31.19}$$

The proof of this is left as Exercise 31.1. By the inverse Fourier transform rule, we thus also have

$$\Gamma(\tau) = \int_{-\infty}^{\infty} S(\omega)\, e^{-i\omega\tau}\, d\omega. \tag{31.20}$$

Thus $S(\omega)$ and $\Gamma(\tau)$ form a Fourier transform pair; a few specific examples are given in Table 31.I. From (31.20) we immediately see that the variance of f is given by

TABLE 31.I

Auto-covariance functions and corresponding spectra for various ergodic stationary random processes with zero mean

Process	$\Gamma(\tau)$	$S(\omega)$
Markov	$(\Gamma_0/2\tau_0)\exp(-\lvert\tau\rvert/\tau_0)$	$\Gamma_0/2\pi(1+\omega^2\tau_0^2)$
White noise	$\Gamma_0\delta(\tau)$	$\Gamma_0/2\pi$
Finite-band white noise	$\Gamma_0 \sin(\tau/\tau_0)/\pi\tau$	$\frac{1}{2\pi}\begin{cases}\Gamma_0, & \lvert\omega\rvert < \tau_0^{-1}\\ 0, & \lvert\omega\rvert > \tau_0^{-1}\end{cases}$
Gaussian	$\Gamma_0 \exp(-\tau^2/\tau_0^2)/(\pi\tau_0^2)^{1/2}$	$(\Gamma_0/2\pi)\exp(-\omega^2\tau_0^2/4)$
Narrow band	$\Gamma_0 \cos\omega_0\tau$	$(\Gamma_0/2)[\delta(\omega+\omega_0)+\delta(\omega-\omega_0)]$

$$\overline{f^2} = \Gamma(0) = \int_{-\infty}^{\infty} S(\omega)\,\mathrm{d}\omega. \tag{31.21}$$

Thus, as mentioned above, (31.21) also shows that $S(\omega)$ represents the density ("energy") of contributions to $\overline{f^2}$ in the interval ω, $\omega + \mathrm{d}\omega$.

To many readers it may appear that we have been overly precise in the various definitions given above. Nevertheless, it is important that the reader should at least be made aware of the fundamentals of the theory of time series analysis, on which so much of modern oceanography depends. It is possible to introduce the concept of power spectrum without ever mentioning random functions, the ergodic theorem, etc. (e.g., see Gossard and Hooke, 1975). First, the auto-covariance function $\Gamma(\tau)$ for the function f is introduced via the time-average definition in (31.14b) in which $m_0 = 0$. Then the power spectrum $S(\omega)$ is simply introduced as the Fourier transform of $\Gamma(\tau)$, as in (31.19). However, the reader may now appreciate the subtleties and assumptions suppressed in such an elementary approach.

Generalizations

The above definitions can be immediately extended to random functions that depend on position $\boldsymbol{x}$ as well as on t, viz. $f(\boldsymbol{x}, t, \alpha) \equiv f(\boldsymbol{x}, t)$. If f is a centered stationary random function of $\boldsymbol{x}$ and t in the wide sense, then the first and second moments take the form

$$\mathcal{E}\{f(\boldsymbol{x},t)\} = m_0 = \text{constant} = 0, \tag{31.22a}$$

$$\mathcal{E}\{f(\boldsymbol{x}+\boldsymbol{r}, t+\tau)\,f(\boldsymbol{x},t)\} = \Gamma(\boldsymbol{r},\tau). \tag{31.22b}$$

In particular, we note that $\Gamma(\boldsymbol{0},\tau)$ is equivalent to $\Gamma(\tau)$ for any fixed $\boldsymbol{x}$. It is important to remark here that in the context of waves, some authors (e.g., Phillips, 1966) refer to a random function which satisfies (31.22) as one that is homogeneous in space and stationary in time. As before, if the process is ergodic with respect to both $\boldsymbol{x}$ and t, then the *wavenumber–frequency spectrum* is given by

$$S(\boldsymbol{k}, \omega) = \frac{1}{(2\pi)^4} \iiiint_{-\infty}^{\infty} \Gamma(\boldsymbol{r}, \tau)\, e^{i(\boldsymbol{k}\cdot\boldsymbol{r}+\omega\tau)} d\boldsymbol{r}\, d\tau, \tag{31.23}$$

and conversely,

$$\Gamma(\boldsymbol{r}, \tau) = \iiiint_{-\infty}^{\infty} S(\boldsymbol{k}, \omega)\, e^{-i(\boldsymbol{k}\cdot\boldsymbol{r}+\omega\tau)} d\boldsymbol{k}\, d\omega. \tag{31.24}$$

Of course is ergodicity is not invoked, then $S(\boldsymbol{k}, \omega)$ can be defined directly in terms of truncated Fourier transforms, as before. That is,

$$\text{if} \quad \hat{f}_{LT}(\boldsymbol{k}, \omega) = \iiint_{-L/2}^{L/2} \int_{-T/2}^{T/2} f(\boldsymbol{x}, t)\, e^{-i(\boldsymbol{k}\cdot\boldsymbol{x}+\omega t)} d\boldsymbol{x}\, dt, \tag{31.25}$$

$$\text{then} \quad S(\boldsymbol{k}, \omega) \equiv \lim_{\substack{L\to\infty \\ T\to\infty}} \frac{1}{(2\pi)^4 L^3 T} |\hat{f}_{LT}(\boldsymbol{k}, \omega)|^2. \tag{31.26}$$

However, we shall, unless otherwise specified, tacitly assume ergodicity and take (31.23) as the defining property of the spectrum. Nevertheless, the relation (31.26) in discretized form is commonly used in practice to estimate the power spectrum from a given data record (see Section 32). From (31.24) we see that the variance is given by

$$\overline{f^2} = \Gamma(\boldsymbol{0}, 0) = \iiiint_{-\infty}^{\infty} S(\boldsymbol{k}, \omega)\, d\boldsymbol{k}\, d\omega.$$

Thus $S(\boldsymbol{k}, \omega)$ can be interpreted as the contribution of "energy" to the variance per unit volume of wavenumber–frequency space.

The *frequency spectrum* $S(\omega)$ defined earlier can be obtained by integrating $S(\boldsymbol{k}, \omega)$ over all $\boldsymbol{k}$ and hence $S(\omega)$ is sometimes called a *reduced spectrum*. Thus

$$S(\omega) = \iiint_{-\infty}^{\infty} S(\boldsymbol{k}, \omega)\, d\boldsymbol{k}. \tag{31.27}$$

Similarly, the *wavenumber spectrum* $S(\boldsymbol{k})$ is given by

$$S(\boldsymbol{k}) = \int_{-\infty}^{\infty} S(\boldsymbol{k}, \omega)\, d\omega. \tag{31.28}$$

Suppose $f(\boldsymbol{x}, t, \alpha)$ is a centered stationary random function of t but not necessarily of $\boldsymbol{x}$. Then a useful statistic is the covariance of f between positions $\boldsymbol{x}_1$ and $\boldsymbol{x}_2$ (the cross-covariance), denoted by $\Gamma_{12}(\tau)$ and defined as

$$\Gamma_{12}(\tau) = \mathcal{E}\{f(\boldsymbol{x}_1, t+\tau, \alpha)\, f(\boldsymbol{x}_2, t, \alpha)\}. \tag{31.29}$$

The normalized version of $\Gamma_{12}(\tau)$ is known as the *cross-correlation function*, $R_{12}(\tau)$, where

$$R_{12}(\tau) = \Gamma_{12}(\tau)/\Gamma_{12}(0). \tag{31.30}$$

An example of $R_{12}(\tau)$ for the low-frequency sea-level oscillations at Sydney and Coff's Harbour is shown in Fig. 25.14. The *cross-spectrum* of f between $\boldsymbol{x}_1$ and $\boldsymbol{x}_2$ is given by

$$S_{12}(\omega) = \frac{1}{2\pi}\int_{-\infty}^{\infty} \Gamma_{12}(\tau)\, e^{i\omega\tau}\, d\tau$$

$$\equiv S_c(\omega) + iS_q(\omega), \qquad (31.31)$$

where $S_c(\omega) = \text{Re}\,[S_{12}(\omega)]$ and $S_q(\omega) = \text{Im}\,[S_{12}(\omega)]$ are respectively called the *co-spectrum* and *quadrature spectrum* of f between $\boldsymbol{x}_1$ and $\boldsymbol{x}_2$, or in other words the spectra of the in-phase and out-of-phase part of $f(\boldsymbol{x}_1, t+\tau, \alpha)$ and $f(\boldsymbol{x}_2, t, \alpha)$. Finally, the *coherence* and *phase* of f between $\boldsymbol{x}_1$ and $\boldsymbol{x}_2$ are respectively defined as

$$\left.\begin{aligned} C_{12}(\omega) &= S_c/(S_c^2 + S_q^2)^{1/2}, \\ \theta_{12}(\omega) &= \tan^{-1}(S_q/S_c). \end{aligned}\right\} \qquad (31.32)$$

The coherence is like a frequency-dependent cross-correlation coefficient and is clearly always less than unity. Examples of C_{12} and θ_{12} are shown in Figs. 18.4 and 25.13. Of course when $\boldsymbol{x}_1 = \boldsymbol{x}_2$, the above definitions all coincide with the corresponding previous ones for a random function $f(t)$, and $C_{12} = 1, \theta_{12} = 0$.

Finally, in an analogous fashion, we can define a covariance and cross-spectrum for two centered random functions, say $f(t, \alpha)$ and $g(t, \alpha)$. Thus

$$\Gamma_{fg}(\tau) = \mathcal{E}\{f(t+\tau, \alpha)\, g(t, \alpha)\} \qquad (31.33)$$

and $$S_{fg}(\omega) = \frac{1}{2\pi}\int_{-\infty}^{\infty} \Gamma_{fg}(\tau)\, e^{i\omega\tau} d\tau, \qquad (31.34)$$

with obvious definitions for the coherence and phase. The latter functions are very useful, for example, in studying, as a function of frequency, the relationship between sea level "f" and pressure "g" at a given position.

Exercises Section 31

1. Derive (31.19) using (31.14), (31.17) and (31.18). You may assume that the various limiting processes as $T \to \infty$ are valid.
2. Check the entries for the spectra in Table 31.I by evaluating the appropriate Fourier integrals.

Calculation of wave spectra

The construction of spectra and covariance functions from real data proceeds according to well-known algorithms. The reader is referred to Jenkins and Watts (1968) or Rayner (1971) for detailed instruction. Briefly stated, two methods of computing energy spectra are widely used. The Blackman and Tukey (1958) method first involves the calculation of the auto-covariance function (see Section 31), which is then smoothed and Fourier transformed to yield the spectrum. The more recently developed Fast Fourier Transform (FFT) technique allows direct calculation of Fourier coefficients of a time series without intermediate estimation of the auto-covariance function. In the FFT technique the series is first split into very small pieces (down to a single point wherever possible), for which the Fourier coefficients are very easily computed, and then relations between the Fourier coefficients for the short series are used to construct the spectrum of the whole series (Jenkins and Watts, 1968, p. 313). For a series of N points, the FFT method requires only $2N \log_2 N$ operations, compared to N^2 needed for the Blackman and Tukey method. The FFT method has understandably become very popular in the analysis of long geophysical time series; however, because knowledge of the auto-covariance function is often extremely useful, the Blackman and Tukey method, although slower, is sometimes preferred, as it yields the latter function as well as the spectrum.

In practice, geophysical time series are always finite in length and sampled at discrete points (even if the instruments measure continuously, the data are always digitized into a format suitable for numerical analysis). The integrals of Section 31 must then be replaced by sums over the sampling points: the Fourier integrals become Fourier series. Consider a centred time series $y(t)$ consisting of data given at an even number N of regularly spaced points, at times $t = 0, \Delta, 2\Delta, \ldots, (N-1)\Delta$. These data can be synthesized as a Fourier series of the harmonics of the fundamental frequency $f_1 = 1/N\Delta$ (in cycles per second). Because the data points are separated by a finite time interval, only a finite number of harmonics will contribute to the Fourier series. The highest frequency which can be detected with data sampled at intervals Δ is $f_{N/2} = 1/2\Delta$, the $\frac{1}{2}N$th harmonic. This frequency is called the *Nyquist frequency*. The discrete Fourier series of such a digitized signal is then the finite sum

$$y(n\Delta) = \sum_{m=1}^{N/2} a_m \exp(i\omega_m n\Delta), \quad n = 0, 1, \ldots, N-1, \tag{32.1}$$

where $\omega_m = 2\pi m/N\Delta$. The Fourier coefficients are given by

$$a_m = \frac{1}{N} \sum_{n=0}^{N/2} y(n\Delta) \exp(-i\omega_m n\Delta). \tag{32.2}$$

From these Fourier coefficients, one may construct a *sample spectrum* which is a measure of the distribution of the variance over the different Fourier components. From Jenkins and Watts (1968, p. 210), this sample spectrum is, for the discrete sampling case:

$$S(\omega_m) = \frac{1}{2\pi T}\left|\Delta \sum_{n=0}^{N-1} y(n\Delta) \exp(-i\omega_m n\Delta)\right|^2, \tag{32.3}$$

where $T = N\Delta$ is the length of the time series. In the limit of large T, the values of ω_m become very close together and $S(\omega_m)$ approaches a continuous function. A comparison of (32.3) with (31.18) shows that the discrete sample spectrum $S(\omega_m)$ tends to the continuous spectrum $S(\omega)$ in the limit of large T. All the spectra presented and discussed here arise from discretely sampled series; in most cases it will be assumed that the series consist of enough points that they may be interpreted as continuous spectra and the results of Section 31 applied to them. The only exception will be encountered in the discussion on tides (Section 52).

The obvious way to present a spectrum is to plot $S(\omega)$ versus ω. In some cases, however, when ω varies over a wide range of values, a logarithmic frequency scale is more convenient. It is then advisable to plot $\omega S(\omega)$ versus $\log \omega$, so that the total area under the plotted curve remains equal to the total variance of the time series. It is also possible to illustrate the distribution of variance with frequency by plotting the square of the Fourier coefficients $|a_m|^2$ at the discrete frequencies ω_m. The resulting graph is called a periodogram, or a Fourier line spectrum; if the lines are closely spaced, their tips may be joined to obtain a curve of variance against frequency; an example is shown in Fig. 25.12. These various representations of the distribution of variance with frequency are of course equivalent.

Line spectra are often encountered in connection with tidal analysis. As will be seen below (Section 52), the tides may be represented as the response of the ocean to a countable number of precisely known frequencies. It is then natural to decompose a tidal time series into the forcing frequencies rather than in regularly spaced Fourier coefficients. Let such a time series for the surface displacement be represented as

$$\eta(t_i) = \sum_n a_n \cos(\omega_n t_i + \phi_n), \tag{32.4}$$

where the ω_n are the tidal forcing frequencies and a_n and ϕ_n the amplitude and phase of the response at these frequencies, and the sampling times are $t_1, t_2, \ldots, t_n$. The constants a_n and ϕ_n may be determined by least-squares fitting; for example, see Godin (1972, p. 209). The harmonic decomposition of the tide may then be given in a table (Table 52.II) which shows a_n and ϕ_n for the dominant ω_n, or shown as a line spectrum (Fig. 52.10).

The interpretation of wave spectra

Spectral analysis is performed in order to increase the understanding of serial data. Properties of the data which are evident from a perusal of the original time series should also manifest themselves in its spectral representations. In addition, features in the data that are invisible to the naked eye are often revealed through statistical treatment, but recognition directly from the spectra, say, of such features and of their significance depends greatly on experience. Let us then examine some properties of spectra and correlation functions.

The auto-correlation function $R(\tau)$ defined in (31.11) starts at unity for zero lag, $\tau = 0$.

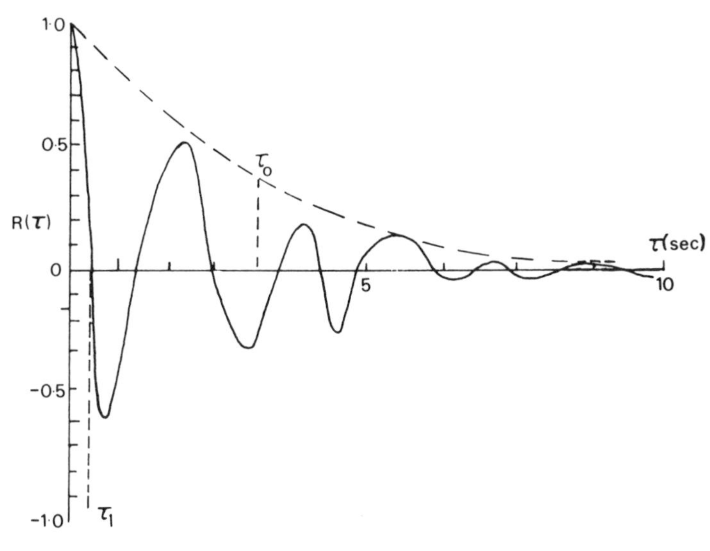

Fig. 32.1. The correlation function $R(\tau)$ for surface gravity waves. τ_0 is the e-folding time of the exponential envelope of $R(\tau)$; τ_1 is the time lag at which the first zero crossing occurs. (From Kinsman, 1965.)

The resemblance between a time series and its shifted image decreases as the lag τ increases and usually vanishes at large τ. From an actual correlation curve, such as that shown in Fig. 32.1, one can recognize a decay scale τ_0 ($\simeq 2.5$ s) characterizing the attenuation of correlation with lag time. Analytic examples of decay correlation functions have been given in Table 31.I. A different correlation time may be deduced from the position τ_1 of the first zero crossing of $R(\tau)$. The auto-covariance function $\Gamma(\tau)$ of a purely sinusoidal function vanishes for $\tau_1 = T/4$; it is then natural to associate the first zero crossing of $\Gamma(\tau)$ with the dominant period of a time series. In Fig. 32.1, $\tau_1 \simeq 0.4$ s, corresponding to an angular frequency $\omega = 2\pi/4\tau_1 = 3.9$ rad s^{-1}. A glance at the spectrum calculated from this correlation function (Fig. 32.2) shows a peak at $\omega \simeq 4$ rad s^{-1}. The presence of further zero crossings at points which are about $2\tau_1$ apart (in Fig. 32.1) reflects the continuing dominance of the peak frequency at larger time lags.

The energy spectrum $S(\omega)$ is equally rich in possibilities of interpretation. For example, spikes in a spectrum are direct evidence of the presence of a high energy content in narrow frequency bands centered on such spikes. The inertial frequency $\omega = f$ commonly shows up as a pronounced peak in horizontal current data (Fig. 15.1). Tidal frequencies at diurnal and semi-diurnal periods are also common features of current and sea-level spectra (Fig. 15.1). Spectra often show regularities, such as power-law regions, where $S(\omega) \propto \omega^{\alpha}$ for some power α. These regularities may usually be related to some peculiarities of the time series from which the spectrum was constructed and often, beyond the data, to basic properties of the physical system sampled. For example, consider a time series contaminated for some reason (perhaps instrumental) by a number of sharp spikes of much larger amplitude than the rest of the signal. Such spikes approximate Dirac δ-functions, the Fourier transform of which is flat (independent of ω) at all frequencies. If there are no higher order discontinuities in the data, the high-frequency part of the spectrum will be

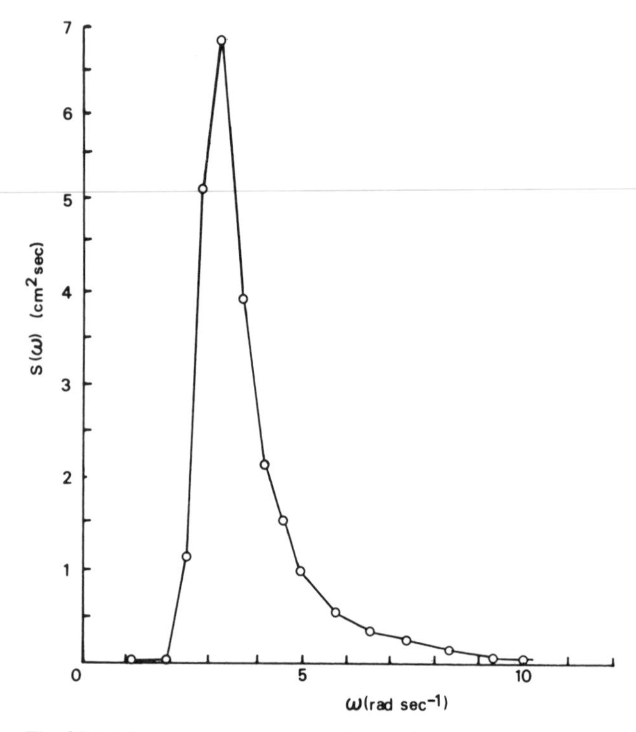

Fig. 32.2. The spectrum corresponding to the correlation function of Fig. 32.1.

dominated by the contributions from the spikes and, from (31.26), the spectrum will be flat at high frequencies. In a similar way, if the worst discontinuities in the signal are in the form of step functions, the Fourier transform of which is $1/\omega$, then the high-frequency part of the spectrum (from 31.26) will fall as ω^{-2}. Continuing in the same vein, the dependence of the high-frequency part of the spectrum of a signal with discontinuities in its nth derivative will be of the form $\omega^{-2(n+1)}$ (Lighthill, 1962a).

As an example of the kind of circumstances which would lead to the presence of discontinuities in a time series, consider the influence of fine structure on temperature measurements. The fine structure in question consists of step-like variations superimposed upon a large-scale vertical gradient (Stommel and Fedorov, 1967; Pingree, 1969). In the presence of vertical oscillations, temperature changes due to the passage of these steps past a fixed instrument will be discontinuous and lead to a spectrum proportional to ω^{-2}. This question has been examined in more detail by Garrett and Munk (1971a).

A simple example of how a spectral form may be related to the physics of the phenomenon under inspection is provided by the equilibrium range of the surface gravity wave spectrum. A large part of the spectrum of Fig. 32.3 (which is the same as that in Fig. 32.2, but replotted on a log-log scale) shows a steep ω^{-5} dependence. This power law was explained by Phillips (1958) as follows. Consider waves of length and period sufficiently large that capillary effects may be neglected. Under equilibrium conditions, such waves will lose energy, mainly through breaking, at the same rate as they acquire it from the wind. The rate of energy loss through breaking, which depends only on g (gravity) and ω, the frequency itself, will thus determine the shape of the spectrum in that range of

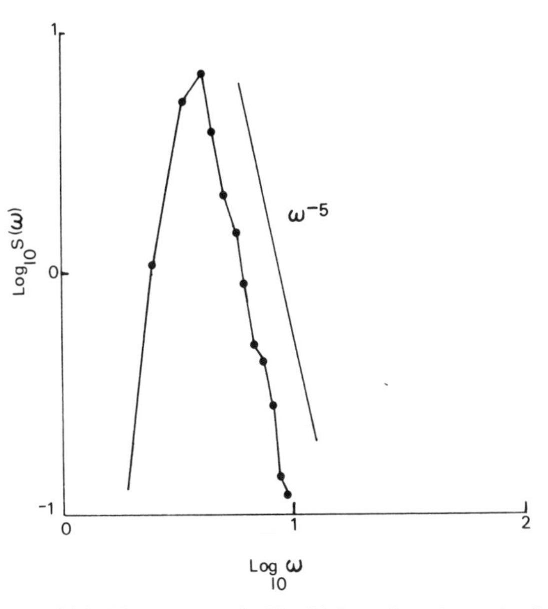

Fig. 32.3. The spectrum in Fig. 32.2, replotted on a log-log scale. A curve of slope ω^{-5} is shown for comparison.

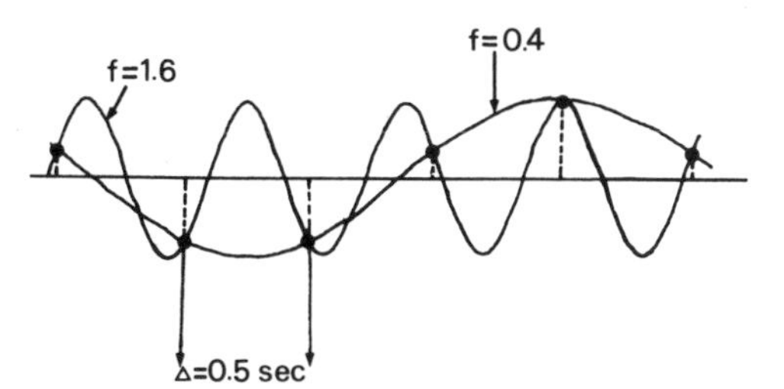

Fig. 32.4. Smapling of sinusoidal waves at intervals $\Delta = 0.5$ s. The Nyquist frequency is $f_{N/2} = 1$ cps. The two curves show how points contributing to a high frequency $f = 1.6$ cps will be interpreted as part of a lower frequency signal ($f = 0.4$ cps).

frequencies. Taking $S(\omega)$ as being the spectrum of the surface displacement, we see from (31.17) and (31.18) that $S(\omega)$ must have the dimensions $[L^2T]$, and that

$$S(\omega) = \alpha g^2\omega^{-5}/(2\pi)^4, \tag{32.5}$$

with α a dimensionless constant, is the only combination of powers of g and ω with the dimensions of $S(\omega)$ [the factor $(2\pi)^{-4}$ is introduced for later convenience].

As a final illustration, the effect of white noise on a spectrum is seen (from Table 31.I) to be similar to that of delta functions. Instrumental noise will thus tend to flatten out the high-frequency end of a spectrum, provided the sampling frequency is high enough to record random noise processes.

The matter of *aliasing* should also be mentioned. It was assumed above that the

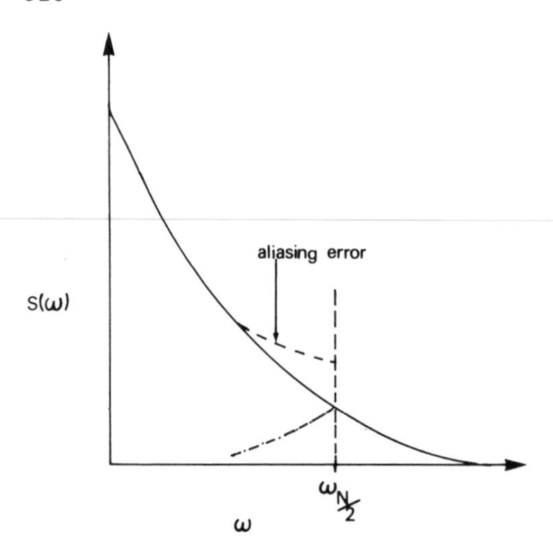

Fig. 32.5. The effect of aliasing on a spectrum. That part of the spectrum at $\omega > \omega_{N/2} = 2\pi f_{N/2}$ is reflected upon $\omega_{N/2}$ and added to $S(\omega)$ to produce an aliasing error.

sampling was dense enough that discrete spectra could be treated as continuous. This assumption breaks down at high frequencies: no frequency higher than the Nyquist frequency, $f_{N/2}$ can be resolved by discrete, evenly spaced sampling. Information concerning frequencies $f > f_{N/2}$ is lost through digitization of the time series. This is a minor loss if one is interested only in the lower frequency part of a signal. However, the high frequencies come back to haunt us and contaminate (or *alias*) the spectrum at frequencies less than $f_{N/2}$. The manner in which this *aliasing* occurs is readily seen in Fig. 32.4. With a sampling interval $\Delta = 0.5$ s, $f_{N/2} = 1$ cps; a higher frequency oscillation ($f = 1.6$ cps say) will be interpreted as contributing to a frequency below $f_{N/2}$ (in this case, $f = 0.4$ cps). The contributions to the spectrum above $f_{N/2}$ are reflected about $f_{N/2}$; contributions to $f_{N/2} + f'$ add themselves to low frequencies at $f_{N/2} - f'$, as seen in Fig. 32.5. If the shape of the high-frequency spectrum is known, it may be corrected for. A more sensible solution to the problem of aliasing is to design instruments with a sluggish response at high frequency, or which average over the signal between sampling times, so that the high-frequency response is eliminated and aliasing does not occur.

Wave statistics

Wavenumber and frequency spectra characterize the spatial and temporal dependencies of a phenomenon. A number of other statistical distributions of wave fields are found useful. For example, an important statistic of a wave record is the distribution of displacement amplitudes in the sample. This distribution may be characterized by the probability that any given wave have an amplitude which exceeds a certain value. Similar probability distributions may be constructed for wave slopes or currents. This kind of statistics has received considerable theoretical and observational attention for surface gravity waves. Theoretical treatments of the statistical geometry of the sea surface are found in a series of papers by Longuet-Higgins (references may be found in Longuet-

Higgins, 1975b) and by Cartwright and Longuet-Higgins (1956). Cox and Munk (1954) have reported on the statistics of surface wave slopes. A summary of empirical results may be found in Wiegel (1964). The statistical properties of internal wave fields have been discussed by Miropolsky (1973) and Borisenko and Miropolsky (1974).

Exercises Section 32

1. In the equilibrium range of surface gravity waves for which (32.5) holds, show that the wave number spectrum $S(k)$ is proportional to k^{-4}. This may be achieved either through dimensional analysis or by noticing that breaking waves exhibit discontinuities in the slope of the free surface.

2. Calculate the correlation function of a harmonic signal such as that given by (32.4); then use the results of Table 31.I to find the spectrum.

Oceanic wave fields may be engendered through the action of (1) surface phenomena, such as storms; (2) boundary interactions, such as that which transforms barotropic into internal tides at continental shelf margins; and (3) internal processes like baroclinic instability. These waves spread away from their point of origin and eventually perish at the hand of dissipative processes.

Since the different Fourier components which make up the wave spectrum will not be generated at the same rate, nor propagate or dissipate equally rapidly, it is to be expected that the form of the spectrum will change through the life history of a wave field. The question thus arises as to why spectra (and other statistical properties) take the shapes they do at various stages in the life of a wave field. Is it possible to derive predictive (as opposed to purely empirical or phenomenological) theories for the statistical properties of varying wave fields? More specifically, given the statistics of the forcing functions (such as the wind, the atmospheric pressure), the boundary and initial conditions and the coefficients of the governing differential equations, each of which will, in general, have a random component, can we compute the solution and its statistics in terms of the statistics of the forcing? In all generality, this question leads to very difficult mathematical problems, such as are encountered in the theory of turbulent flow (Hinze, 1975). Despite these great difficulties it is nevertheless the belief of some schools of thought (e.g., see Monin and Yaglom, 1971) that we should study only stochastic fluid mechanics problems, since, in the real world of fluids, nothing is ever purely deterministic (i.e., nonrandom). As this is not a manual of turbulence, we shall examine only two broad classes of stochastic problems which are of relevance to ocean waves: (1) the transport theory of the space–time evolution of spectra (see below), which is capable of handling weak nonlinear energy transfers between waves and their environment as well as between different wave modes; and (2) the theory of linear stochastic initial boundary-value problems (see Section 34).

At the beginning of this chapter, we introduced and interpreted the so-called two-point correlation theory of stationary random functions. The property of stationarity, taken in a wide sense, allowed us to invoke the ergodic theorem and to replace ensemble averages by averages over space or time. As soon as we begin to consider unsteady situations, in which wave spectra evolve in space and time, the stationarity hypothesis becomes untenable. Should the spectral evolution be slow enough, however, so that changes in wave properties take place over a time period which is long compared to the sampling period over which reliable statistics may be obtained, time (and space) series may still be considered stationary over the sampling period.

It is thus assumed that the time and space scales of spectral evolution greatly exceed the longest period and wavelength which may be resolved in the data. Under such conditions, the ray-theory equations derived in Section 6 are applicable. Let a wave field be represented by a continuous spectrum of plane waves with displacements of the form

$$\eta(\boldsymbol{x}, t; \boldsymbol{k}, \omega) = a(\boldsymbol{x}, t; \boldsymbol{k}, \omega)\, e^{i(\boldsymbol{k} \cdot \boldsymbol{x} - \omega t)}, \tag{33.1}$$

where a, $\boldsymbol{k}$, and ω are slowly varying functions of $\boldsymbol{x}$ and t. A frequency-wavenumber spectrum of η may be defined according to (31.23) or (31.26). We shall refer to $S(\omega, \boldsymbol{k})$ as the energy spectrum, even though the definition lacks factors such as the density, or gravity, which must eventually be included to recover a quantity with the dimensions of energy per unit volume. As we will be dealing mostly with situations in which wave propagation takes place in a time-independent medium, for which, from (6.15c), the frequency ω remains invariant along a ray, we shall restrict our attention to the wavenumber spectrum $S(\boldsymbol{k})$, obtained by integration of $S(\omega, \boldsymbol{k})$ over ω (cf. 31.28). The wavenumber spectrum is allowed to vary slowly in space and time and we can make that dependence explicit by writing it as $S(\boldsymbol{k}; \boldsymbol{x}, t)$. Corresponding to the energy spectrum, there is an action density spectrum given by

$$n(\boldsymbol{k};\boldsymbol{x},t) = S(\boldsymbol{k};\boldsymbol{x},t)/\omega_0, \tag{33.2}$$

where ω_0 is the frequency observed in a frame of reference moving with the fluid. The amount of wave action contained in a small element δk of wavenumber space is $n(\boldsymbol{k}; \boldsymbol{x}, t)\delta k$. For a continuous spectrum the equation for the conservation of wave action (6.44) becomes (replacing $\langle E_0\rangle/\omega_0$ by $n\delta k$)

$$\frac{\partial}{\partial t}(n\delta k) + \nabla \cdot (\boldsymbol{c}_g n\delta k) = 0. \tag{33.3}$$

The elemental volume of wavenumber space δk may be assumed to be centered on some wavenumber $\boldsymbol{k} = (k_1, k_2, k_3)$. If $\mathrm{d}\boldsymbol{k}_1$, $\mathrm{d}\boldsymbol{k}_2$ and $\mathrm{d}\boldsymbol{k}_3$ are directed line elements along the three components of $\boldsymbol{k}$, the volume element is then given by

$$\delta k = |(\mathrm{d}\boldsymbol{k}_1 \times \mathrm{d}\boldsymbol{k}_2) \cdot \mathrm{d}\boldsymbol{k}_3|. \tag{33.4}$$

As a wave train undergoes refraction, each component of the wavenumber vector varies according to (6.15b) and the volume element δk changes as well. The changes in δk may be found from the equation for the conservation of crests (6.4). Let the wavenumber change from $\boldsymbol{k}$ to $\boldsymbol{k} + \mathrm{d}\boldsymbol{k}$; equation (6.4) holds for the changed wavenumber as well as for $\boldsymbol{k}$. Thus

$$\frac{\partial}{\partial t}(\boldsymbol{k} + \mathrm{d}\boldsymbol{k}) + \nabla[\omega(\boldsymbol{k} + \mathrm{d}\boldsymbol{k}, \lambda)] = 0. \tag{33.5}$$

Expanding $\omega(\boldsymbol{k} + \mathrm{d}\boldsymbol{k}, \lambda)$ in a Taylor series about $\omega\boldsymbol{k}$, and subtracting (6.4) from (33.5), we find that $\mathrm{d}\boldsymbol{k}$ must satisfy

$$\frac{\partial}{\partial t}\mathrm{d}\boldsymbol{k} + \nabla(\boldsymbol{c}_g \cdot \mathrm{d}\boldsymbol{k}) = 0. \tag{33.6}$$

It is relatively straightforward to combine the three components of this equation to find how δk [as given by (33.4)] varies in time and space. The result is

$$\frac{\partial}{\partial t}\delta k + \nabla \cdot (\boldsymbol{c}_g \delta k) = 0. \tag{33.7}$$

Combining (33.3) and (33.7), we find that

$$\frac{\partial n}{\partial t} + \boldsymbol{c}_g \cdot \nabla n = 0, \tag{33.8}$$

where $n = n(\boldsymbol{k}; x, t)$. Thus, the spectral density of wave action remains conserved along a ray, a property which is not in general shared by the action density of a single wavenumber component, which obeys (6.44).

In view of the dependence of the action density spectrum on wavenumber, it is sometimes useful to bring out this dependence explicitly. If $\boldsymbol{k}$ is considered to be a function of $\boldsymbol{x}$ and t in (33.8) this equation may be written, after carrying out the differentiations, as

$$\frac{\partial n}{\partial t} + \left(\frac{\mathrm{d}\boldsymbol{k}}{\mathrm{d}t}\right) \cdot \left(\frac{\partial n}{\partial \boldsymbol{k}}\right) + \left(\frac{\mathrm{d}\boldsymbol{x}}{\mathrm{d}t}\right) \cdot \left(\frac{\partial n}{\partial \boldsymbol{x}}\right) = 0, \tag{33.9}$$

where $\mathrm{d}\boldsymbol{k}/\mathrm{d}t = \partial \boldsymbol{k}/\partial t + (\boldsymbol{c}_g \cdot \nabla)\boldsymbol{k}$. This latter equation is concisely expressed as

$$\frac{\mathfrak{D}}{\mathfrak{D}t} n(\boldsymbol{k};\boldsymbol{x},t) = 0, \tag{33.10}$$

where $\mathfrak{D}/\mathfrak{D}t$ denotes differentiation along a ray path in six-dimensional $(\boldsymbol{x}, \boldsymbol{k})$-space. In the presence of sources of wave action, (33.10) is directly generalized into a "radiation balance equation"

$$\mathfrak{D}n/\mathfrak{D}t = \Sigma, \tag{33.11}$$

where the source term Σ may include forcing, dissipation and interaction processes, and will be a function of $\boldsymbol{x}$, $\boldsymbol{k}$, t as well as of n and of integrals of n over parts of the wavenumber range. Some of the main source terms will be examined in later sections; only a qualitative overview of their nature is offered at this point.

Forcing processes may be defined as those which increase the action density of waves by transferring energy from the medium in which they propagate (or through one of its boundaries) to the waves themselves. The most important such process is that associated with wave generation by fluctuating atmospheric winds and pressure fields. A vast literature is devoted to the topic of short gravity wave production by the wind (Kinsman, 1965; Barnett and Kenyon, 1975). Internal oscillations may also be generated by atmospheric fluctuations (Thorpe, 1975). At the lower end of the meteorological variability spectrum, storms, pressure fronts and seasonal fluctuations generate low-frequency ($\omega < f$) oceanic disturbances which may propagate as planetary waves. Other direct surface forcing mechanisms such as earthquakes, volcanic eruptions or human activities may be of local and temporary relevance, but are relatively minor on a global scale. Further, within the bulk of the ocean, hydrodynamic instability mechanisms, such as the Kelvin–Helmholtz shear instability (Chandrasekhar, 1961) or baroclinic instability (Robinson and McWilliams, 1974) are capable of feeding energy to wave motions (see Chapter 7). Finally, waves may also appear through the interaction of currents with bottom topography, as lee waves for a constant current, or as propagating waves for a slowly varying current (Bell, 1975).

Dissipative processes are interpreted as those which remove action density through nonadiabatic processes along a ray path in $(\boldsymbol{x}, \boldsymbol{k})$-space: they exclude refractive effects. Energy-removing processes include friction and mixing (both molecular and turbulent), which are most effective at small scales and will consequently tend to reduce the high-wavenumber (and usually high-frequency) tail of propagating wave spectra. [Recall how quickly capillary waves decay from their source (cf. Fig. 11.2).] Other dissipative processes such as wave breaking or absorption at critical levels (Section 41), where the phase speed equals that of the current, may be effective at larger scales, or may even be tuned to some particular band of wavenumbers.

By interaction processes, we mean those energy exchanges, through nonlinearity, between the wave field and itself or other wave fields of the same or different natures. These interactions work in either direction, adding or removing energy from the waves under study, depending on circumstances; they are most efficient at transferring energy when certain resonance conditions are satisfied, and in practice, only such resonant interactions are studied (Section 38). The subject has been extensively reviewed by Hasselmann (1967a, 1968).

Equation (33.15) is recognized to be of the same form as the Boltzmann transport equation of statistical mechanics (Kittel, 1969, p. 406), with n the distribution function of particles in position–velocity space and Σ the collision terms. The transport equation for wave action may be formally integrated along a ray path in $(\boldsymbol{x}, \boldsymbol{k})$-space:

$$n(\boldsymbol{k};x,\,t) \;=\; n(\boldsymbol{k}_0, \boldsymbol{x}_0, t) + \int_{t_0}^{t} \Sigma(\boldsymbol{k}(t');\boldsymbol{x}(t'),\,t')\,\mathrm{d}t'. \tag{33.12}$$

This equation may not, however, be considered a solution of the action transfer problem, since Σ is in general a functional of the spectrum of action density. The form of this functional must be specified for each one of the transfer processes. The greatest success in this direction has been achieved in the study of surface gravity waves (Hasselmann, 1968; Willebrand, 1975), for which the methods were developed and where the best data are available for verification of theoretical ideas. Müller and Olbers (1975), among others, have attempted a similar specification of internal wave sources.

To conclude, we remark that whenever the propagation medium is time-independent, ω is constant along a ray path, and (33.11) describes the energy spectral density $S(\boldsymbol{k};\boldsymbol{x},\,t)$, with Σ representing the energy sources. Most of the literature (see Kinsman, 1965) on the evolution of surface wave spectra refers to energy rather than to action density. For this case the two points of view are equivalent.

Exercises Section 33

1. *Two-scale analysis of wave propagation*. The slow variation of wave properties assumed to prevail under the assumptions of ray theory may be brought out more explicitly by introducing additional space and time variables $T = \epsilon t$, $\boldsymbol{X} = \epsilon \boldsymbol{x}$, so that a plane wave may be written as

$$\eta(\boldsymbol{x}, t; X, T) \;=\; a \exp\left[i(\boldsymbol{k}\cdot\boldsymbol{x} - \omega t)\right],$$

where a, $\boldsymbol{k}$ and ω are functions of X and T. The small parameter ϵ is of the order of the ratio of a wavelength to the scale length of spatial variation of the medium. Show that for weak sources $[\Sigma \sim 0(\epsilon)]$, the governing equation (33.11) for wave propagation can be expressed in terms of the long scales X and T.

2. Derive (33.17) for the wavenumber volume element δk from (33.4) and (33.6).

34. LINEAR STOCHASTIC INITIAL BOUNDARY-VALUE PROBLEMS

In Section 33 we discussed qualitatively the radiation balance equation which describes the slow space–time evolution of wave action spectra. Although this equation is in principle applicable to nonlinear and almost stationary problems, its solution often presents formidable mathematical difficulty. Therefore it is now useful to look at only *linear* stochastic initial boundary-value problems (IBVP) which can arise in various wave problems of a random nature. After categorizing the various types of stochastic IBVP's, we shall briefly discuss the spectral response of the ocean to stationary random forcing functions and the propagation of coherent waves in stationary random media.

The stochastic initial boundary-value problem

In line with the above, we now consider the following scalar linear IBVP which describes some physical system (e.g., forced long wave motion in a semi-infinite ocean):

$$\left.\begin{aligned} \mathcal{L}\phi(\boldsymbol{x},t) &= f(\boldsymbol{x},t) \text{ in } V, \quad t \geqslant 0, \\ BC: B_i\phi &= g_i \text{ on } \partial V, \\ IC: \phi(\boldsymbol{x},0) &= I_1(\boldsymbol{x}), \phi_t(\boldsymbol{x},0) = I_2(\boldsymbol{x}), \ldots, \end{aligned}\right\} \tag{34.1}$$

where $\mathcal{L}$ and B_i are linear differential (or possibly integro-differential) operators, and V and ∂V denote the spatial domain under consideration and its boundary. Suppose that one or more of the coefficients in $\mathcal{L}$ and/or B_i, or one or more of the "forcing" functions f, g_i and I_i is a random function whose statistics are known. Then (34.1) is called a *random or stochastic IBVP*. A random solution of (34.1) is a stochastic process which satisfies (34.1) with probability one.

In contrast to the case of nonrandom or deterministic IBVP's, there is little known about the existence and uniqueness of random solutions of (34.1) except for very special random coefficients and forcing functions. Even for such cases, the mathematical theory is by no means elementary (Birkhoff et al., 1973; Arnold, 1974). Thus we shall assume from the outset that a random solution of (34.1) exists and satisfies the equation with probability one. The next step therefore is to relate the statistics of this solution to the known statistics of the given random functions. Even this can be a very complicated procedure if we take all the given inputs $f, g_i, \ldots$ and coefficients to be random functions. In practice one usually treats specific types of random IBVP's in which only some of these inputs and coefficients are random functions. For example, first suppose that only f is random and in fact represents a stationary random process. This leads to the following problem in statistical communication theory for which the mathematical theory is well developed (Davenport and Root, 1958; Stratonovich, 1963): Given the (input) spectrum of f, what is the (output or response) spectrum of ϕ? We shall briefly discuss this problem and its application to ocean waves below. If, on the other hand, only the initial conditions are random, we have a random IVP, whose theory has only recently been developed (Birkhoff et al., 1973) and as yet has not been applied to ocean wave problems. For example, the problem of determining the statistics of tsunamis generated by random initial motions of the ocean boundary in a localized region has not yet been solved. If only the boundary ∂V itself is random, leading to random coefficients in the operators B_i, we can consider two types of random BVP's in the wave context: the scattering of waves by random surfaces (Beckmann and Spizzichino, 1963; Howe and Mysak, 1973) and wave propagation along random boundaries (Mysak and Tang, 1974). Finally, if only the coefficients in the differential equation are random, then we have what are technically called stochastic differential equations. Such equations first arose in the study of Brownian motion in the presence of a white noise random velocity field, and the mathematical theory of these equations for such a random process is now well developed (Arnold, 1974). More generally, stochastic differential equations with random coefficients representative of a variety of stochastic processes arise in the study of random eigenvalue problems (Boyce, 1968), turbulent diffusion (Lo Dato, 1973) and wave propagation in continuous random media (Tatarski, 1961; Frisch, 1968; Dence and Spence, 1973). We shall discuss the last topic at the end of this section.

Other stochastic problems which could arise in oceanography but which are not included in the above list are what might be called geometrically random BVP's. These include the propagation of waves in a uniform medium with a random distribution of discrete scattering objects, such as islands (Frisch, 1968; Lax, 1973; Bell, 1975), the settling of particles that are suspended in a fluid, and the flow of liquids in porous media.

The response spectrum for a uniform medium

Suppose $\mathcal{L}$ in (34.1) is simply a constant-coefficient differential operator of order n. Then for $\phi = \phi_0 \mathrm{e}^{i(\boldsymbol{k}\cdot\boldsymbol{x}-\omega t)}$,

$$\mathcal{L}\phi = P_n(\boldsymbol{k}, \omega)\,\phi, \tag{34.2}$$

where P_n is an nth-degree polynomial in $\boldsymbol{k}$ and ω. In the absence of forcing, $P_n(\boldsymbol{k}, \omega) = 0$ represents an implicit form of the dispersion relation. The quantity

$$H(\boldsymbol{k}, \omega) = 1/P_n(\boldsymbol{k}, \omega) \tag{34.3}$$

is called the transfer function corresponding to $\mathcal{L}$. Now suppose that $f(\boldsymbol{x}, t)$ is a stationary random function for all t and $\boldsymbol{x}$ in $V = R^3$; then in the absence of any boundary or initial conditions, (34.1) implies that ϕ will also be a stationary random function of $\boldsymbol{x}$ and t. Under these assumptions the spectra of f and ϕ are related by the equation

$$S_\phi(\boldsymbol{k}, \omega) = |H(\boldsymbol{k}, \omega)|^2 S_f(\boldsymbol{k}, \omega). \tag{34.4}$$

To establish (34.4) we first take the truncated Fourier transform [defined in (31.25)] of $\mathcal{L}\phi = f$; this gives, in view of (34.2) and (34.3),

$$\hat{\phi}_{LT}(\boldsymbol{k}, \omega) = H(\boldsymbol{k}, \omega) f_{LT}(\boldsymbol{k}, \omega) + IP, \tag{34.5}$$

where IP represents the terms obtained from integration by parts that are evaluated at x, y and $z = \pm L/2$, $t = \pm T/2$. We assume that IP is bounded as $L, T \to \infty$. Hence on forming the product $|\hat{\phi}_{LT}|^2$ from (34.5) and invoking the definition of the spectrum (see 31.26), we immediately arrive at (34.4).

Equation (34.4) represents a very special case of the transport equation discussed in Section 33. Nevertheless, for certain simple wave systems, (34.4) represents a very useful formula for the response spectrum in terms of the forcing spectrum. It is interesting to note from (34.4) that even if S_f is finite for all $\boldsymbol{k}$ and ω, S_ϕ is infinite at the poles of

$H(\boldsymbol{k}, \omega)$, corresponding to ω and $\boldsymbol{k}$ satisfying the dispersion relation. This is called a resonant response. However, if the system is dissipative, then $P_n \neq 0$ for any real $\boldsymbol{k}$ and ω and S_ϕ is bounded. Thus an important topic to be considered in Chapter 8 will be the modelling of dissipation in ocean wave propagation.

In practice one rarely encounters random wave generation problems in which the forcing functions are stationary random functions of all the spatial variables. For example, in the generation of edge waves or shelf waves by the atmospheric pressure, one might consider the pressure, as represented by $f(x, y, t)$, to be a stationary random function in y and t and deterministic in x, where x is the distance normal to the coast. Then one is led to a deterministic boundary-value problem in x for the different modes in the offshore direction. In this case (34.4) will contain spectra that depend on x and the right-hand side of this equation will consist of a sum over all these modes. Such an approach to the study of shelf wave generation was considered by Mysak (1967b). Another example of this type of problem has been considered by Käse and Tang (1976). They studied the generation of internal waves by a surface wind stress that is stationary in x, y and t; the resulting deterministic boundary-value problem is one for the vertical modes. The problem of the wind-generation of surface gravity waves (Section 51) is likewise a mixed problem, being a deterministic initial value problem in time and random in x and y.

Stochastic differential equations and wave propagation in random media

Let us now consider linear equations of the form

$$(L+M)\phi = f, \tag{34.6}$$

where L is a deterministic differential operator, f a deterministic forcing function and M a random differential operator whose coefficients are centered (zero-mean) stationary random functions. An example of an unforced version of (34.6) is the two-dimensional internal wave equation with a randomly varying Brunt–Väisälä frequency:

$$\{\partial_{xxtt} + \partial_{zztt} + f^2\partial_{zz} + N_0^2[1+\mu(z)][\partial_{xx} - g^{-1}(\partial_{ztt} + f^2\partial_z)]\}\psi = 0, \tag{34.7}$$

where ψ is the stream function, $N_0 =$ constant and $\mu(z)$ is a stationary random function with $\mathcal{E}\{\mu\} = 0$. The random form of $N^2(z)$ given in (34.7) was introduced by Miropolsky (1972) and independently by McGorman and Mysak (1973) as a model of oceanic fine-structure in the deep ocean. The solution of (34.7), therefore, will describe the effects of microstructure on the propagation of internal waves. To cast (34.7) in the form (34.6), we write

$$L = \partial_{xxtt} + \partial_{zztt} + f^2\partial_{zz} + N_0^2[\partial_{xx} - g^{-1}(\partial_{ztt} + f^2\partial_z)], \tag{34.8a}$$

$$M = -N_0^2\mu(z)g^{-1}(\partial_{ztt} + f^2\partial_z) + N_0^2\mu(z)\,\partial_{xx}, \tag{34.8b}$$

and clearly, $\mathcal{E}\{M\} = 0$. For equations of the form (34.6) it is of primary interest to determine $\mathcal{E}\{\phi\}$, the mean or coherent field. A question immediately arises: should we first solve (34.6) for ϕ and then average, or first average and then try to find $\mathcal{E}\{\phi\}$? Further, for a given equation, do the two approaches give the same result? Equally important (and difficult) is the problem of solving (34.6) in a manner such as to yield the second and higher moments, e.g., $\mathcal{E}\{\phi^2\}$. For a discussion of this topic we refer the reader to the recent reviews by Morrison and McKenna (1973) and Chow (1975).

Assuming $(L+M)^{-1}$ exists, the first approach gives

$$\phi = (L+M)^{-1}f.$$

Hence on averaging, we obtain

$$\mathcal{E}\{\phi\} = \mathcal{E}\{(L+M)^{-1}\}f \tag{34.9}$$

since f is deterministic. Thus (34.9) gives an explicit solution for $\mathcal{E}\{\phi\}$ provided we can simplify $\mathcal{E}\{(L+M)^{-1}\}$.

On the other hand if we average (34.6) from the outset we obtain

$$L\mathcal{E}\{\phi\} = -\mathcal{E}\{M\phi\}+f \tag{34.10}$$

since both L and f are deterministic. We now encounter the familiar closure problem which nearly always occurs when the second approach is used. Averaging (34.6) has produced two unknowns, $\mathcal{E}\{\phi\}$ and $\mathcal{E}\{M\phi\}$, and yet we only have one equation, (34.10). Therefore, to close the problem, we need another equation for these two unknowns. We proceed by rewriting (34.6) in the form

$$\phi = -L^{-1}M\phi + L^{-1}f, \tag{34.11}$$

where L^{-1} is the integral operator inverse to L, whose kernel is the causal Green's function corresponding to L:

$$L^{-1}f(\boldsymbol{x},t) = \int G(\boldsymbol{x},\boldsymbol{x}';t,t')f(\boldsymbol{x}',t')\,\mathrm{d}\boldsymbol{x}'\mathrm{d}t',$$

$$LG = \delta(\boldsymbol{x}-\boldsymbol{x}')\delta(t-t'),$$

and G satisfies the appropriate radiation condition. Multiplying (34.11) by M and averaging, we obtain

$$\mathcal{E}\{M\phi\} = -\mathcal{E}\{ML^{-1}M\phi\}. \tag{34.12}$$

Thus not only have we succeeded in increasing the number of equations by one, but also the number of unknowns by one. Continuing in this manner we shall always end up with one more unknown than number of equations. Since we wish to solve for $\mathcal{E}\{\phi\}$, we note that we can close the problem by making the *closure or truncation approximation*

$$\mathcal{E}\{ML^{-1}M\phi\} \simeq \mathcal{E}\{ML^{-1}M\}\mathcal{E}\{\phi\}. \tag{34.13}$$

Then (34.10) and (34.12) can be combined into one equation for $\mathcal{E}\{\phi\}$:

$$[L-\mathcal{E}\{ML^{-1}M\}]\mathcal{E}\{\phi\} = f. \tag{34.14}$$

Since the approximation (34.13) involves making some assumption about the solution, which we do not know a priori, this second approach has been termed "dishonest" by Keller (1962a). On the other hand, Bourret (1962, 1965) argues that (34.13) (which he calls the one "ficton" approximation) is justified on the grounds that in practice the correlation scale of $\mathcal{E}\{ML^{-1}M\}$ is either much smaller or larger than the scales associated with ϕ and hence the average of the two processes is the same as the product of their averages (the local independence hypothesis). However, it turns out that (34.14) can be obtained from the first solution (34.9) provided the randomness is "small", which clearly is an assumption that does not depend on the solution we are seeking. Hence the first approach has been termed "honest" by Keller (1962a).

To simplify $\mathcal{E}\{(L+M)^{-1}\}$ for small M, we proceed as follows (Keller, 1967): Letting I be the identity operator, we have

$$\begin{aligned}(L+M)^{-1} &= [L(I+L^{-1}M)]^{-1} \\ &= (I+L^{-1}M)^{-1}L^{-1}. \end{aligned} \tag{34.15}$$

Now provided the norm (assumed defined for the appropriate function space under consideration) of the operator $L^{-1}M$ is less than unity, we can expand the operator $(I+L^{-1}M)^{-1}$ using the binomial expansion. Thus

$$(L+M)^{-1} = \sum_{0}^{\infty} (-L^{-1}M)^n L^{-1}$$

and hence $$\mathcal{E}\{(L+M)^{-1}\} = [I + L^{-1}\mathcal{E}\{ML^{-1}M\} + \ldots]L^{-1} \tag{34.16}$$

since $\mathcal{E}\{M\} = 0$. Finally, taking the inverse of both sides of (34.16) and again using the binomial expansion, we have

$$[\mathcal{E}\{(L+M)^{-1}\}]^{-1} = L - \mathcal{E}\{ML^{-1}M\} \tag{34.17}$$

correct to second order in M. Thus, upon rewriting (34.9) as

$$[\mathcal{E}\{(L+M)^{-1}\}]^{-1}\mathcal{E}\{\phi\} = f,$$

we finally obtain, on using (34.17) in the left-hand side,

$$[L - \mathcal{E}\{ML^{-1}M\}]\mathcal{E}\{\phi\} = f,$$

which is identical to (34.14).

When the random part of (34.6) is not small, we cannot generally use the binomial expansion† but must resort to using the second, "dishonest" approach to solving (34.6) for $\mathcal{E}\{\phi\}$. This approach may of course be used to obtain equations for the second and higher moments of ϕ as well, but as before, some closure assumption must eventually be made to obtain a closed set of equations. For a discussion of different types of closure approximations we refer the reader to Richardson (1964). For an account of recent attempts to use a method "half way" between the "honest" and "dishonest" approaches, see Chorin (1974).

We are finally in a position to make a few general remarks about wave propagation in continuous random media. In earlier studies of this topic (Tatarski, 1961), efforts were focused on ordinary perturbation solutions for the free wave field in the presence of small random inequalities in the medium. That is, solutions of (34.6), written in the form

$$(L+\epsilon M)\phi = 0 \quad (0 < \epsilon \ll 1), \tag{34.18}$$

were obtained by setting

$$\phi(\boldsymbol{x}, t; \epsilon) = \phi_0(\boldsymbol{x}, t) + \epsilon\phi_1(\boldsymbol{x}, t) + \ldots \tag{34.19}$$

and then solving the sequence of equations

---

† When the random coefficient is large, it is sometimes possible to introduce a change of random variable that results in a new stochastic equation with a small random part (see Howe, 1974, for an example of this approach).

$$L\phi_0 = 0,$$
$$L\phi_1 = -M\phi_0, \qquad (34.20)$$
$$L\phi_2 = -M\phi_1,$$
$$\vdots \qquad \vdots$$

for $\phi_0, \phi_1, \phi_2, \ldots$. Finally, $\mathcal{E}\{\phi\}$, $\mathcal{E}\{\phi^2\}$, etc. were obtained by averaging the solution (34.19). However, the difficulty with this approach is that even for very small ϵ, (34.19) is not a good representation of the solution if the irregularities in the medium extend over a large region of space. And if the irregularities extend over all space (i.e., do not have compact support), the solution is secular (i.e., becomes unbounded as $|\boldsymbol{x}| \to \infty$). Thus to obtain $\mathcal{E}\{\phi\}$ in an infinite random medium, one must resort to other methods.

It was difficulties of this sort that led Keller (1967) to derive (34.14), which he then used (with $f = 0$) to obtain the dispersion relation for a coherent plane wave of the form

$$\mathcal{E}\{\phi\} = \phi_0 e^{i(\boldsymbol{k}\cdot\boldsymbol{x} - \omega t)}. \qquad (34.21)$$

Provided M is a random operator whose coefficients are stationary random functions of both $\boldsymbol{x}$ and t, Keller showed that the following expression gives the dispersion relation of a coherent wave of the form (34.21):

$$e^{-iS}[L - \mathcal{E}\{ML^{-1}M\}]e^{iS} = 0, \qquad (34.22)$$

where $S = \boldsymbol{k} \cdot \boldsymbol{x} - \omega t$. When $M = 0$ (a deterministic medium), (34.22) gives the deterministic dispersion relation, which may be assumed to yield, for real $\boldsymbol{k}$, a real relation of the form $\omega = \omega(\boldsymbol{k})$. Thus a coherent wave propagates unattenuated with speed c_0, say, in the nonrandom case. However, when $M \neq 0$, there is a correction to this real relation which involves an integral over a functional of the auto-covariance function characterizing the random medium. This correction is generally complex, and for any conservative oscillatory system, Howe (1973) has shown that this term always implies that there is a net transfer of energy from the coherent field to the incoherent field. This is to say the coherent wave amplitude is attenuated in the direction of the group velocity. It should be noted that this damping has nothing to do with dissipation in the usual sense, but is rather a phase interference effect, caused by the different components of the coherent wave being scattered by the random inhomogeneities in the medium. Estimates of the attenuation of internal waves due to scattering by fine-structure [as modelled by (34.7)] and by random currents have been given by Tang and Mysak (1976) and Thomson (1976), respectively. Thomson (1975a) has also considered the energy loss in planetary waves due to scattering from random topography. The correction term due to M changes the phase speed from c_0 to c. For Helmholtz's equation and an isotropic random medium, it can be shown that $c < c_0$ (Frisch, 1968). This is to be expected on the grounds that as a wave travels from A to B through random irregularities, its path length is effectively increased and hence its net speed is decreased. For waves in anisotropic random media, however, c can be either larger or smaller than c_0 (Keller and Veronis, 1969; Thomson, 1975a).

Exercises Section 34

1. The mean field equation (34.14) (with $f = 0$) can also be derived by considering two coupled equations for $\mathcal{E}\{\phi\}$ and ϕ', where $\phi' = \phi - \mathcal{E}\{\phi\}$ is the usual fluctuating field.

From (34.6) with $f = 0$, show that $\mathscr{E}\{\phi\}$ and ϕ' satisfy

$$L\mathscr{E}\{\phi\} = -\mathscr{E}\{M\phi'\}, \tag{34.23}$$

$$L\phi' = -M\mathscr{E}\{\phi\} - (I - \mathscr{E})\{M\phi'\}. \tag{34.24}$$

By the method of successive approximations or otherwise, show that (34.24) has the formal solution

$$\phi' = -\sum_0^{\infty} (-1)^n S^n L^{-1} M\mathscr{E}\{\phi\}, \tag{34.25}$$

where $S = L^{-1}(I - \mathscr{E})M$. Hence show that $\mathscr{E}\{\phi\}$ satisfies the *Dyson equation*

$$\left[L - \sum_0^{\infty} (-1)^n \mathscr{E}\{MS^n L^{-1} M\}\right] \mathscr{E}\{\phi\} = 0. \tag{34.26}$$

Notice that upon taking only the first term in the summation in (34.26), we arrive at the left-hand side of (34.14). Since this approach involves an averaging over the fluctuating field together with M, it has been termed the "smoothing method" by Frisch (1968). The lowest-order approximation, resulting in (34.14), is called the first-order smoothing approximation. This approach was developed independently by Howe (1971) who called (34.14) the "binary collision approximation" since to this order, multiple scattering effects are neglected.

2. Derive and write down the formal solutions to (34.20) for $L = \nabla^2 + k_0^2$ and $M = \mu(\boldsymbol{x})k_0^2$. Show explicitly that (34.19) converges only if ϵ is small and the randomness $\mu(\boldsymbol{x})$ has compact support (Frisch, 1968).

CHAPTER 6

WAVE INTERACTIONS

35. INTRODUCTION

We now begin to explore some of the interaction mechanisms through which waves trade energy with their surroundings and with other waves. Such mechanisms are of intrinsic interest; their elucidation is also a prerequisite to the integration of the transport equation (33.11). The interactions considered in this chapter may be described as action-conserving. The action density of individual Fourier components and, in more complex cases, that of a spectral band of components, remains unaffected by the interaction process. The action-conserving interactions discussed below are of two types.

(1) Purely refractive modifications of a spectrum $n(\boldsymbol{k};\boldsymbol{x},t)$ in answer to inhomogeneities of the medium of propagation. The transport equation is in that case homogeneous and takes the form (33.10).

(2) Weak wave–wave interactions, in which action density is exchanged between different spectral components, but for which the action density integrated over wavenumber space remains invariant. In that case $\int \Sigma \delta k$ vanishes.

Following the initial discussion on refractive processes (Section 36), we shall look at the reverse side of the coin: the influence of waves on the mean state (Section 37). Weak interactions between individual waves will be discussed in Section 38 and the effect of such interactions on spectral evolution is examined briefly in Section 39.

36. WAVE REFRACTION

The geometrical theory of wave refraction has already been discussed in Section 6, under the heading of ray theory. Some elementary examples of refraction by inhomogeneities in the medium of propagation have already been seen in Section 9. In this section, we concentrate our attention on the refraction of waves by currents. We first consider the behaviour of long surface gravity waves in a horizontally sheared current, following with the more interesting case of the refraction of internal gravity waves by a vertically sheared flow. An example of wave solutions near a caustic (where ray theory fails) will also be given. Finally, we discuss the modifications of short waves by the currents due to much longer waves upon which they are superimposed.

Refraction by a shear flow

The simple case of wave propagation in a unidirectional steady shear flow is amenable to precise analysis and illustrates rather well the fates encountered by waves refracted by currents. Let us consider first the behaviour of long surface gravity waves travelling in a horizontal current $U = [U(y), 0, 0]$ of the form shown in Fig. 36.1a. The wavenumber

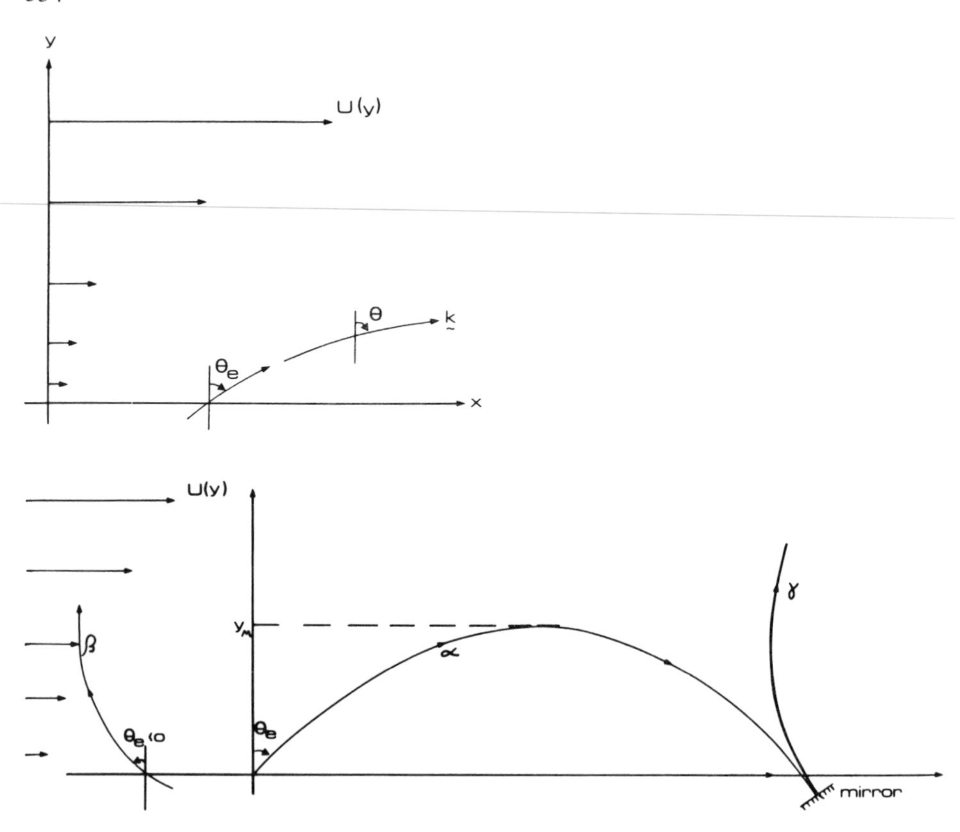

Fig. 36.1. (a) Geometry of ray propagation in a shear current. (b) The behaviour of various types of rays (explanation in the text).

$\boldsymbol{k} = (k_1, k_2, 0)$ lies in the horizontal plane. Defining the angle θ as that between the y-axis and the vector $\boldsymbol{k}$ (Fig. 36.1a), the components of the wavenumber are given by $k_1 = k \sin\theta$ and $k_2 = k \cos\theta$. Since the medium is time-independent, it follows from (6.15c) that ω is invariant along a ray. Hence, from (6.16).

$$kc_0 + kU \sin\theta \;=\; \omega \;=\; \text{constant}. \tag{36.1}$$

Further, since the homogeneity is in the y-direction only, (6.19a) gives

$$k_1 \;=\; k \sin\theta \;=\; \text{constant}. \tag{36.2}$$

Eliminating k between (36.1) and (36.2), and letting the subscript e (for entry) refer to values of wave properties at the point of entry into the shear flow, we obtain

$$\frac{c_0}{\sin\theta} + U(y) \;=\; \frac{c_e}{\sin\theta_e} + U_e, \tag{36.3}$$

which is Snell's law of refraction, modified by the presence of the mean current. With the help of a dispersion relation, such as (6.17), it is generally possible to express c_0 as a function of wavenumber and hence to eliminate k between (36.2) and (36.3) to solve for $\theta(y)$. For long surface gravity waves, which are nondispersive, $c_0 = \sqrt{gH}$, and k does not enter (36.3). Let us assume that H = constant and that $U_e = 0$. The angle $\theta(y)$ is then found from (36.3) as

$$\sin\theta(y) = \frac{\sin\theta_e}{1 - \dfrac{U(y)}{\sqrt{gH}}\sin\theta_e}. \tag{36.4}$$

The behaviour of the rays is readily deduced from (36.4). For $0 < \theta_e < \pi/2$, θ must increase as U increases: an incident ray is refracted in the direction of the current (Fig. 36.1b). If the current is sufficiently strong, the right-hand side of (36.4) approaches unity as y increases and the refraction continues until $\theta = \pi/2$, at which angle the ray is turned back by total internal reflection. This phenomenon occurs at $y = y_M$ such that

$$\frac{U(y_M)}{\sqrt{gH}} = \frac{1}{\sin\theta_e} - 1. \tag{36.5}$$

We also note that when $\theta = \pi/2$, the y-component of the wavenumber, $k_2 = k\cos\theta$, vanishes. The length scale of the waves is obviously no longer small compared to that of the shear flow at that point, and ray theory must fail. As may be seen by drawing a few adjacent rays near the point of total internal reflection, an *envelope of rays,* or *caustic,* occurs at that point. We shall return to this situation below and show how valid solutions may be found near the caustic. For negative angles of attack, $\theta_e < 0$, $|\theta|$ decreases as U increases: an incident ray is bent towards the normal to the current (Fig. 36.1b).

Finally, we note that these rays are not reversible. A mirror (in the form of a reflecting dock, for example) put across the path of a ray emerging from the current, such as ray α in Fig. 36.1b would reflect the ray back on itself with a negative value of θ_e, sending it along path γ. This irreversibility is typical of refraction by current variations and is not found in refraction by variations of scalar field variables (see Exercise 36.1).

How does the wave energy vary along a ray? Consider first a single Fourier component, with a given value of $\boldsymbol{k}$. For the simple case considered ($c_0 = \sqrt{gH}$, no x and t dependence) the conservation of wave action equation (6.44) reduces to

$$\frac{\partial}{\partial y}\left[\langle E_0\rangle\sqrt{gH}\cos\theta/\omega_0\right] = 0, \tag{36.6}$$

which integrates to

$$\langle E_0\rangle\cos\theta/k = \text{constant}. \tag{36.7}$$

Using (36.2) to eliminate k we find that

$$\langle E_0\rangle\sin 2\theta = \text{constant}. \tag{36.8}$$

For a continuous spectrum $S(\boldsymbol{k};\boldsymbol{x},t)$ of bandwidth δk, we find from (33.8) that, in a coordinate system moving with the fluid, S/ω_0 = constant. Replacing ω_0 by $k\sqrt{gH}$ and using (36.2) we find that

$$S(\boldsymbol{k};\boldsymbol{x},t)\sin\theta = \text{constant}. \tag{36.9}$$

Under the present circumstances, the bandwidth δk is found from (33.7) to obey

$$\delta k \cos\theta = \text{constant}. \tag{36.10}$$

Since $\langle E_0\rangle = S\delta k$, we recover (36.8).

It is interesting to note that, whereas the behaviour of $\langle E_0\rangle$ is consistent with that of S for rays which refract towards the normal to the current (for which $\theta \to 0$), the two functions show radically different behaviours near the point of total internal reflection ($\theta \to \pi/2$). $\langle E_0\rangle$ becomes unbounded towards $\theta = \pi/2$, in spite of the fact that S decreases continuously to that level, because the bandwidth δk is singular at $\theta = \pi/2$. This singularity is associated with the breakdown of ray theory at the point of total internal reflection.

Finally, with (36.1) and (36.2) in mind, consider what happens as a wave penetrates a shear current where $dU/dy > 0$. As U increases, $k \sin\theta$ remaining constant, it is conceivable that the term $kU \sin\theta$ may become as large as ω itself. At that "critical layer", the intrinsic frequency $\omega_0 = kc_0$ vanishes: the waves no longer propagate with respect to the fluid. For the long gravity waves discussed above, and again with $U_e = 0$, (36.3) gives

$$U_c = \sqrt{gH}/\sin\theta_e$$

for the current speed at the critical layer. Recalling the value of U at the point of total internal reflection $y = y_M$ [see (36.5)], it is clear that $U(y_M) < U_c$; hence the critical layer is never reached by the waves in this case.

The difference in behaviour of waves near a critical level to that near a level of total internal reflection is most clearly illustrated with internal gravity waves. In the absence of rotation ($f = 0$), the dispersion relation for internal gravity waves is found from (8.56):

$$\omega_0^2 = \frac{N^2(k_1^2 + k_2^2)}{(k_1^2 + k_2^2 + k_3^2)}. \tag{36.11}$$

Let assume that waves which satisfy (36.11) in a reference frame where the fluid is at rest are propagating in a vertically sheared horizontal current $\boldsymbol{U} = [U(z), 0, 0]$. Let us take $k_2 = 0$, for simplicity. The frequency ω seen by a stationary observer is equal to

$$\omega = \omega_0 + k_1 U, \tag{36.12}$$

where ω is constant along a ray. The vertical component of the group velocity is found by differentiating (36.12) with respect to k_3. Using (36.11), and with $k_2 = 0$,

$$\partial\omega/\partial k_3 = c_{g3} = -k_3\omega_0/(k_1^2 + k_3^2). \tag{36.13}$$

Thus c_{g3} vanishes either when $k_3 = 0$ or when $\omega_0 = 0$. When $k_3 = 0$, $\omega_0 = N$; we have already noted in Section 9 that internal gravity waves suffer total cuspidal reflection at a level $\omega_0 = N$. The other case occurs at a critical level, where $\omega = k_1 U$. At the critical level, k_3 becomes unbounded.

By using (36.11), with $k_2 = 0$, the vertical component of the group velocity may be rewritten

$$c_{g3} = -\left(1 - \frac{\omega_0^2}{N^2}\right)^{1/2} \frac{\omega_0^2}{Nk_1}. \tag{36.14}$$

Let us expand ω_0 in a Taylor series near its value at the critical level $z = z_c$:

$$\omega_0(z - z_c) = \omega_0(z_c) + (z - z_c)\frac{\partial\omega_0}{\partial z} + \ldots$$

$$\simeq (z - z_c)k_1 \frac{\partial U}{\partial z}. \tag{36.15}$$

Substituting into (36.14) we find that, near the critical level,

$$c_{g3} \simeq -\frac{k_1}{N}\left(\frac{\partial U}{\partial z}\right)^2 (z - z_c)^2. \tag{36.16}$$

From the ray equation (6.15a), $\mathrm{d}z/\mathrm{d}t = c_{g3}$. Integrating this relation, we find that the time taken to travel from a level z_1 to another level z_2 nearer the critical level is

$$t_2 - t_1 \simeq \frac{N}{k_1(\partial U/\partial z)^2}\left[\frac{1}{(z_2 - z_c)} - \frac{1}{(z_1 - z_c)}\right]. \tag{36.17}$$

As z_2 approaches z_c, the transit time becomes unbounded: a wave group never reaches the critical level.

In contrast, the same type of argument in the vicinity of a level (z_m) of total internal reflection gives $t_2 - t_1 \propto |z - z_m|^{1/2}$. A group of waves travels to the level z_m and is reflected in a finite time. The above argument was first presented by Bretherton (1966), and illustrates rather succinctly the difference in wave propagation in the vicinity of the two different levels where $c_{g3} = 0$. The critical level will be scrutinized more closely in Section 41. For the moment, let us take a closer look at wave solutions in the vicinity of a caustic.

Waves near a caustic

Let us go back to the simple case of long surface gravity waves explored above. The position of the caustic $(y = y_M)$ is given by (36.5). At the caustic, there is an abrupt transition between an irradiated region $(y < y_M)$ to a domain into which no wave with an angle of entry $\theta_e > \theta_M$ can penetrate (θ_M is that angle of entry into the shear flow for which total internal reflection occurs at y_M). Ray theory does not provide a description of the behaviour of the wave amplitude across the transition. Further, even though we have concluded that reflection takes place at the caustic, we remain ignorant of the amplitude of the reflected wave and of any phase change which might occur upon reflection.

In order to take a closer look at the situation in the vicinity of the caustic line, we reformulate the problem ab initio for surface gravity waves, following a simplified version of the work of McKee (1974).

We first postulate a zeroth-order state consisting of a unidirectional shear flow $U(y)$ and a hydrostatic pressure field $p_0(z)$, which identically satisfy the momentum and continuity equations. The perturbations about this steady state satisfy the linearized equations

$$u_t + Uu_x + vU_y = -g\eta_x, \tag{36.18a}$$

$$v_t + Uv_x = -g\eta_y, \tag{36.18b}$$

$$H(u_x + v_y) = -(\eta_t + U\eta_x), \tag{36.19}$$

where η is the surface displacement and u, v the perturbation velocity components in the horizontal. Assuming for all dependent variables a dependence of the form $\phi(y)$ exp $[i(kx - \omega t)]$, it is possible to express the velocity components in terms of η and η_y:

$$u = \left[gk\eta - \frac{gU_y\eta_y}{(\omega - kU)}\right] \Big/ (\omega - kU), \tag{36.20a}$$

$$v = -ig\eta_y/(\omega - kU). \tag{36.20b}$$

Substituting for both u and v from (36.20) into the continuity equation (36.19), we obtain

$$\eta_{yy} + \frac{2kU_y\eta_y}{(\omega - kU)} - \left[k^2 - \frac{(\omega - kU)^2}{gH}\right]\eta = 0. \tag{36.21}$$

Let us now introduce some scaled variables defined as follows:

$$k' = \sqrt{gH}\,k/\omega; \quad y' = y/L; \quad U' = U/\sqrt{gH}. \tag{36.22}$$

The primed variables are dimensionless; L is a characteristic horizontal scale associated with the shear flow. Introducing the parameter γ and the variable $\delta(y')$ through

$$\gamma = L\omega/\sqrt{gH}; \quad \delta(y') = 1 - k'U'(y'); \tag{36.23}$$

(36.21) takes the nondimensional form (dropping primes)

$$\eta_{yy} - \frac{2\eta_y\delta_y}{\delta} - \gamma^2(k'^2 - \delta^2)\eta = 0. \tag{36.24}$$

Let us look at waves with lengths much shorter than the scale length of the shear flow: $\gamma \gg 1$. A WKB approximation to the solution of (36.24), equivalent to the ray theory solution of (6.2) is found by writing

$$\eta(y) = A(y, \gamma)\, e^{i\gamma S(y)}, \tag{36.25}$$

where the amplitude $A(y, \gamma)$ is a slowly varying function of y and is written, for large γ, as

$$A(y, \gamma) = A_0(y) + \sum_{m=1}^{\infty} \frac{A_m(y)}{(i\gamma)^m}. \tag{36.26}$$

Substituting the assumed dependence of η, as given by (36.25), into (36.24), we obtain an equation involving the variables A and S:

$$A_{yy} - \frac{2\delta_y A_y}{\delta} + i\gamma\left[(S_yA)_y + S_yA_y - \frac{2\delta_y}{\delta}AS_y\right] - \gamma^2(S_y^2 + k'^2 - \delta^2)A = 0. \tag{36.27}$$

Let us now solve (36.27) to successive orders in γ, using (36.26) for $A(y, \gamma)$. To $0(\gamma^2)$, we find

$$S_y^2 = \delta^2 - k'^2. \tag{36.28}$$

Since S_y is essentially a wavenumber in the y-direction, (36.28) is equivalent to $(k_1^2 + k_2^2)gH = (\omega - k_1 U)^2$ in dimensional form (and with subscripts on the wavenumber components to differentiate them from each other). Equation (36.28) is called the *Eikonal equation*; it was encountered earlier in the form (6.50). The transition from sinusoidal spatial variation in y to exponential behaviour occurs at $\delta = k'$, where S_y^2 changes sign. Next, to $0(\gamma)$, (36.27) yields

$$A_0^2 S_y/\delta^2 = \text{constant}. \tag{36.29}$$

This result is readily shown to be equivalent to (36.8). The energy density is proportional to the square of the wave amplitude: $\langle E_0 \rangle \propto A_0^2$; S_y is a wavenumber in the y-direction (in our previous notation, $k \cos\theta$). Using the definition of δ (36.23), δ^2 may be written $\omega_0^2/\omega^2 = k^2 gH/\omega^2$. Performing all these substitutions, (36.29) becomes

$$\langle E_0 \rangle \omega^2 \cos\theta / kgH = \text{constant}.$$

Since ω and $k \sin\theta$ are constants, this latter expression is identical to (36.8).

The WKB solution, of the form (36.25), allows a description of the waves on either side of the caustic, but fails to match the solutions across $S_y = 0$. A uniformly valid asymptotic solution for large γ may be found by a method proposed by Ludwig (1966). This solution takes the form

$$\eta(y) = F(y, \gamma) C[-\gamma^{2/3}\xi(y)] + i\gamma^{-1/3} G(y, \gamma) C'[-\gamma^{2/3}\xi(y)], \tag{36.30}$$

where both F and G have expansions in γ similar to that shown for $A(y, \gamma)$ in (36.26). The function $C(\mu)$ is a linear combination of the Airy functions $Ai(\mu)$ and $Bi(\mu)$, and satisfies

$$\frac{d^2}{d\mu^2} C(\mu) - \mu C(\mu) = 0. \tag{36.31}$$

In (36.30), C' denotes the derivative of the function C with respect to its argument. Equation (36.24) is solved to successive decreasing orders in γ by substituting into it (36.30) for η. Both to $0(\gamma^2)$ and to $0(\gamma^{5/3})$, we obtain

$$\xi \xi_y^2 = \delta^2 - k'^2, \tag{36.32}$$

which is equivalent to the Eikonal equation (36.28). To $0(\gamma)$, after making use of (36.31), (36.24) becomes

$$2\xi\xi_y (G_0)_y + G_0 \left[\xi\xi_{yy} + \xi_y^2 - \frac{2\delta_y}{\delta} \xi\xi_y \right] = 0. \tag{36.33}$$

This equation may be integrated to yield

$$G_0 = a\delta/\sqrt{\xi\xi_y}, \tag{36.34}$$

where a is a constant. Similarly, to $0(\gamma^{2/3})$, we find

$$F_0 = b\delta/\sqrt{\xi_y}, \tag{36.35}$$

where b is another constant. The higher order terms F_m, G_m, may be found by continuing the procedure (McKee, 1974).

Let us choose the position of the caustic at $y = 0$, for convenience. Integrating (36.32), we obtain

$$\frac{2}{3}\xi^{3/2} = \int_y^0 [\delta^2(t) - k'^2]^{1/2}\,dt, \quad y < 0; \tag{36.36a}$$

$$\frac{2}{3}(-\xi)^{3/2} = \int_0^y [k'^2 - \delta^2(t)]^{1/2}\,dt, \quad y > 0. \tag{36.36b}$$

Thus, $\xi > 0$ for $y < 0$, and $\xi < 0$ for $y > 0$. The Airy function C appearing in (36.30) must now be chosen so that $\eta(y)$ remains finite everywhere. The function $Bi(\mu)$ increases exponentially for large positive argument (Abramowitz and Stegun, 1965, p. 449, eq 10.4.63) and is thus discarded. If $(\delta^2 - k'^2)$ is assumed to have a simple zero at the caustic, ξ is proportional to $-y$ for small y. To avoid a singularity at the caustic, we must then put $\boldsymbol{a} = 0$. The leading term in the uniform asymptotic expansion is therefore found by substituting for F_0 from (36.35) into (36.30):

$$\eta \simeq \frac{b\delta}{(\xi_y)^{1/2}} Ai[-\gamma^{2/3}\xi(y)]. \tag{36.37}$$

The y-dependence of η near the caustic is illustrated in Fig. 36.2. The maximum amplitude is attained somewhat before the waves reach the caustic. Far enough from the caustic, the Airy function may be replaced by its asymptotic expansion for large argument (Abramowitz and Stegun, 1965, p. 448), the leading terms being

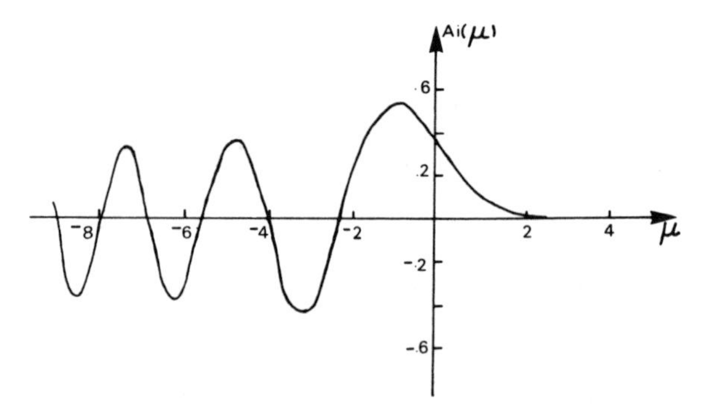

Fig. 36.2. The Airy function, $Ai(y)$, showing the variation in wave amplitude across a caustic at $y = 0$.

$$Ai(\mu) \sim \tfrac{1}{2}\pi^{-1/2}\mu^{-1/4} \exp(-\tfrac{2}{3}\mu^{3/2}), \tag{36.38a}$$

$$Ai(-\mu) \sim \pi^{-1/2}\mu^{-1/4} \sin(\tfrac{2}{3}\mu^{3/2} + \pi/4). \tag{36.38b}$$

Thus, for $y > 0$, $\xi < 0$ and the wave amplitude decays exponentially away from the caustic. For $y < 0$, on the other hand, the dependence of η is oscillatory in space. Substituting for ξ from (36.36) into the argument of the Airy function in (36.37), and using (36.38), we can express η, for $y < 0$ as

$$\eta \propto \exp\left\{i\left[\gamma\int_y^0 (\delta^2 - k'^2)^{1/2}\,\mathrm{d}t + \pi/4\right]\right\} + \exp\left\{i\left[-\gamma\int_y^0 (\delta^2 - k'^2)^{1/2}\,\mathrm{d}t + \frac{3\pi}{4}\right]\right\}, \tag{36.39}$$

i.e., the sum of an incident and reflected wave of equal amplitudes but differing in phase by $\pi/2$. Reflection at a caustic is thus equivalent, in terms of phase change, to reflection on a "soft" wall (on which $\eta = 0$), and differs from reflection by a "hard" wall (on which $v = 0$ and hence $\eta_y = 0$), at which no phase change occurs.

Beyond the caustic ($y > 0$), (36.38a) shows that the wave amplitude decays exponentially. This behaviour, as well as the propagating wave form on the $y < 0$ side, is as predicted by the WKB solution. However, the WKB solution is singular at $y = 0$, whereas the uniformly valid solution (36.37) is not, and successfully matches the two sides of the caustic.

Waves on currents due to other waves

The ray-theory method is also useful to consider interactions between two waves, provided the period and wavelength of one of these waves greatly exceed those of the other, so that the shorter, higher frequency wave may be considered as propagating in a current which is slowly varying in time and in space. Consider a field of short high-frequency plane waves superimposed upon waves of a much longer wavelength and lower frequency (the properties of which will be identified by the subscript l). The coordinate system in which the fluid is at rest, and in which the spectral density of the wave action of the short waves $n(\boldsymbol{k};\boldsymbol{x},t)$ is constant along a ray [cf. (33.8)], is one which moves with the particle velocity of the long waves, $\boldsymbol{u}_l$. This coordinate system is not inertial, being subjected to the acceleration of the long wave particle motion. The frequency ω_0 of the short waves, as observed in that coordinate system, is to be expected to be influenced by the acceleration of the frame of reference, a fact to which we will return below. Thus, in these coordinates,

$$\frac{\mathrm{d}}{\mathrm{d}t}\frac{S(\boldsymbol{k};\boldsymbol{x},t)}{\omega_0} = 0. \tag{36.40}$$

The wavenumber $\boldsymbol{k}$ and the bandwidth δk are unaffected by a translation or an acceleration of the reference frame. It is most convenient to examine them in a reference frame travelling at the long wave phase velocity $\boldsymbol{c}_l$. In these coordinates, the spatial variations of the wavenumber and of the bandwidth are found from the steady-state forms of (6.19a) and (33.7) respectively:

$$(c_{gj} + u_{lj} - c_{lj})\frac{\partial k_i}{\partial x_j} = -k_j\frac{\partial u_{li}}{\partial x_j}, \tag{36.41}$$

$$\frac{\partial}{\partial x_j}[(c_{gj} + u_{lj} - c_{lj})\delta k] = 0. \tag{36.42}$$

In order to proceed further, let us be more specific: both long and short waves will be taken as deep-water surface gravity waves. In a reference frame which is unaccelerated, both waves satisfy the dispersion relation

$$\omega^2 = gk. \tag{36.43}$$

Let the long waves propagate in the x-direction. Their surface displacement is given by $\eta_l = a_l \cos S$, where $S = k_l x - \omega_l t$. The horizontal component of the particle velocity is $u_l = a_l k_l c_l \cos S$. We shall assume that the short wave propagation vector makes an angle θ with the x-axis: $\boldsymbol{k} = (k \cos\theta, k \sin\theta)$.

From (36.42), the bandwidth obeys

$$\delta k(c_g \cos\theta + u_l - c_l) = \text{constant}. \tag{36.44}$$

On the basis of our earlier discussion of long surface gravity waves in a shear flow, where we saw that δk becomes unbounded at a caustic, we expect that caustics will occur where $c_g \cos\theta + u_l - c_l = 0$. With $\langle E_0 \rangle = S\delta k$, (36.40) and (36.44) combine into

$$\frac{\langle E_0 \rangle}{\omega_0}(c_g \cos\theta + u_l - c_l) = \text{constant}, \tag{36.45}$$

a result first obtained by Gargett and Hughes (1972) in their study of the interaction of short surface waves with long internal waves, a problem identical in its formulation to the one examined here, except for the dispersion relation of the long waves. Gargett and Hughes also examined in some detail the refraction of the short waves by the long wave currents and the situation near caustics.

The refraction of the short waves is described by the wavenumber equation (36.41). With the chosen orientation of the short waves relative to the long waves, (36.41) yields

$$k \sin\theta = \text{constant}, \tag{36.46a}$$

$$(c_g \cos\theta + u_l - c_l)\frac{\partial k \cos\theta}{\partial x} = -k \cos\theta \frac{\partial u_l}{\partial x}. \tag{36.46b}$$

Long gravity waves travel faster than short gravity waves; with $c_g \cos\theta \ll c_l$, (36.46b) becomes approximately

$$(c_l - u_l)\frac{\partial k \cos\theta}{\partial x} = -k \cos\theta \frac{\partial}{\partial x}(c_l - u_l),$$

which integrates to

$$k \cos\theta\, (c_l - u_l) = \text{constant}. \tag{36.47}$$

Using $u_l = a_l k_l c_l \cos S$, this relation is also written

$$k \cos\theta = \text{constant}/c_l(1 - a_l k_l \cos S). \tag{36.48}$$

For long waves of small amplitude, $a_l k_l \ll 1$, so that to first order in $a_l k_l$, (36.48) becomes

$$k \cos\theta \propto (1 + a_l k_l \cos S). \tag{36.49}$$

For short waves which are collinear with the long waves, $\cos\theta = 1$; from (36.49), the magnitude of the wavenumber, in that simpler case, is seen to vary with position along the long wave profile. The short waves are alternately stretched and compressed by divergences and convergences of the long wave horizontal velocity. On the crests

Fig. 36.3. Surface wave interaction with internal waves: note the band of short, peaked waves with crests running along the rough band. The length of the surface waves in the rough band near the bottom left corner is of order 30 cm. Two more rough bands are visible in the distance, with much smoother water in between. Notice also that these bands appear darker, because of decreased specular reflection, than the smooth areas between them. (From Gargett and Hughes, 1972.)

($\cos S = 1$), the short waves are shorter than on the average (over S), while they are longer on the troughs of the long waves ($\cos S = -1$). For waves which are not colinear, $\theta \neq 0$; from (36.46a) and (36.49),

$$\tan\theta \propto (1 - a_l k_l \cos S). \tag{36.50}$$

The short waves tend to line up with the direction of propagation of the long waves on the crests of the latter. An illustration of this behaviour is seen in Fig. 36.3, taken from Gargett and Hughes' observations of the interactions between short surface gravity waves and internal waves.

Let us now go back to the frequency ω_0, as observed in the noninertial reference frame moving with the long wave particle motion. The acceleration of the coordinate system adds itself to that of gravity, so that the dispersion relation (36.43) is modified to (Garrett and Smith, 1976):

$$\omega_0^2 = \left(g + \frac{\partial^2 \eta_l}{\partial t^2}\right) k, \tag{36.51a}$$

$$= g(1 - a_l k_l \cos S) k. \tag{36.51b}$$

For the collinear case, substituting for k from (36.49), we find that $\omega_0^2 = g\langle k \rangle$, where the average is over the phase of the long waves. Neglecting c_g with respect to c_l in (36.45), we then find that to first order in $a_l k_l$, $\langle E_0 \rangle \propto (1 + a_l k_l \cos S)$. The short wave amplitude is related to the energy density by

$$\langle E_0 \rangle = \frac{1}{2}\rho_* \left(g + \frac{\partial^2 \eta_l}{\partial t^2}\right) a^2, \qquad (36.52)$$

so that to first order in $a_l k_l$, $a \propto (1 + a_l k_l \cos S)$. The short wave amplitude is amplified on the crests of the long waves. The short wave slope, which is proportional to ak, becomes proportional to $(1 + 2a_l k_l \cos S)$.

In the more general case where the long and the short waves are not collinear, we find from (36.46a) and (36.50) that to first order in the long wave slope, $k \propto (1 + \frac{1}{2}a_l k_l \cos S)$, so that ω_0 now varies with position along the long wave profile. The variations in the energy density and in the short-wave amplitude are readily evaluated, as above.

The consequences of the interaction process between long and short gravity waves have been discussed extensively. Since the short-wave slopes are largest on the long-wave crests, one should also expect that the short waves would dissipate more rapidly on those crests (cf. Section 51). Phillips (1963b) argued that since the energy dissipated by the short waves had been acquired from the long waves in the first place, the result of the interaction would be to damp the long waves. However, Longuet-Higgins (1969a) reasoned that as the short waves dissipate, they also give up their momentum to the long waves and thus exert a stress which is in phase with the orbital velocity of the long waves and which should lead to their growth (the stress exerted by waves on currents will be discussed more fully in the next section). Longuet-Higgins showed that, provided the short waves were continuously regenerated by the wind, the input of energy to the long waves due to this "maser-type" mechanism would greatly exceed the long-wave damping proposed by Phillips (1963b). Hasselmann (1971) pointed out that one aspect of the interaction had been overlooked from the previous works. The momentum density $\langle M \rangle$ of the short waves is related to the energy density through (13.25): $|\langle M \rangle| = \langle E_0 \rangle / |c|$. For the collinear case, from above,

$$\langle M \rangle \propto (1 + 2a_l k_l \cos S). \qquad (36.53)$$

Since the momentum density is also equal to the averaged mass flux of the short waves, there must be a transfer of mass from the long waves to the short waves to make $\langle M \rangle$ a maximum at the long wave crests. This transfer of mass is accompanied by a transfer of the potential energy of the exchanged mass from the long waves to the short waves. The result of this aspect of the interaction is to cancel out the effect of Longuet-Higgins' (1969a) maser mechanism. It was recently pointed out by Garrett and Smith (1976) that long-wave growth can result from the interaction, provided short-wave generation (rather than dissipation) is correlated with the orbital velocity of the long waves.

The consequences of the interaction may also be viewed from the point of view of the short waves. Let us consider the short waves as lying in the saturated part of the wind–wave spectrum where the amplitude is limited by wave breaking. That portion of the spectrum is then of the form (32.5). Breaking of the short waves will occur predominantly on the crests of the long waves. Since the short-wave energy is greater on the crests than it is elsewhere, the average energy density of short waves will be diminished by the presence of long waves. Instead of breaking everywhere, as they would in the absence of long waves, the short waves break only over a fraction of the sea surface and their spectral density is lower than would be predicted by (32.5). As we shall see later (Section 51), long waves develop on the sea surface only after strong winds have blown for a long time over a long fetch. In a gradually developing wave field, generated by a steady wind, short waves appear first and quickly reach their saturation energy levels. As longer waves appear and grow, the interaction described above will lead to a decrease of the energy level of the short waves. This phenomenon, whereby short waves overshoot, at an early stage of wave generation, the energy level which they will attain later on, is commonly observed in measurements (Fig. 51.3).

The modulation of a field of short waves by the current pattern of long waves is of particular interest when the long waves are internal gravity waves. The latter waves would not normally be observable visually at the sea surface, but their influence on short gravity waves produces refractive effects which are clearly recognizable. This observation opens the door to an interesting diversion.

Surface effects of internal waves

Internal waves are commonly observed through their surface effects, especially in areas where a shallow pycnocline is to be found. These surface manifestations are often detectable from sea level (Curtin and Mooers, 1975), but are most striking and distinctive from high altitudes (Shand, 1953; Apel et al., 1975; Gargett, 1976). What is seen at the surface is certainly not the free surface displacement due to internal wave motion: this is of the order of the relative density difference across the pycnocline (see Section 16), and as such gives rise to extremely small surface slopes. Rather, the surface manifestations are to be attributed to the interactions between the internal wave flow and short surface waves. The refractive interaction mechanism just discussed will enhance and shorten surface waves in narrow bands over which the root-mean-square slope of the sea surface will be increased. Specular reflection of sunlight will be reduced over such bands and enhanced in the intermediate bands in which the surface waves are lengthened and diminished by the interaction. When viewed by an observer facing the sun, the rough bands will thus appear darker than the smoother areas between them. On the other hand, again because of the greater root-mean-square surface slope in the rough areas, light scattering will be more intense from such areas than in the smooth areas, so that to an observer facing away from the sun (or under cloudy conditions) the relative brightness of rough and smooth bands will be reversed.

Another mechanism leading to variations in sea surface roughness in the presence of internal waves was proposed much earlier (Ewing, 1950). It has long been known that the spreading of an organic or oily surface film on a water surface leads to a striking reduction of short gravity–capillary wave activity. When spread in a continuous monomolecular layer, with the molecules standing close packed like cigarettes in a pack, a film becomes very difficult to compress and strongly resists any deformation of the surface, thus inhibiting wave action (cf. Section 51).

On the other hand, a discontinuous film, broken up in numerous small patches of oily substance will not be endowed with such rigidity and will have little or no effect (depending on the size of the patches) on surface wave motion. Coastal waters are commonly contaminated with oily substances of organic origin. In the presence of internal waves, continuous films will be formed by compaction in regions of surface convergence and will be broken up again in divergent regions. Smooth bands of reduced wave activity, called slicks, will form and propagate with the internal waves. These slicks will appear bright or dark depending on the angle of viewing, as explained above.

Finally, it may also be possible to detect internal waves visually, even under flat calm conditions, when neither of the above mechanisms will operate. It is not unusual, in the vicinity of river mouths, to find a thin turbid fresh surface layer, coloured by its sediment content (or sometimes by plankton blooms). Whenever internal waves travel on the interface between this layer and the denser, cleaner and bluer water underneath, one observes from the surface an alternation of coloured and bluer bands. The coloured bands occur over the troughs of the internal waves, where a maximum thickness of turbid water is found; the blue bands lie over the crests of the internal waves, where the thickness of the upper layer is reduced to a minimum and the clean blue water is visible beneath it.

Exercises Section 36

1. Consider long ($kH \ll 1$) surface wave refraction in a stationary fluid of monotonically increasing $H(x)$. Show that rays incident from shallow water always reflected back into shallow water by gradual refraction. Note that for such scalar refraction, the rays are reversible, i.e., can be traced in either direction, which is not the case for refraction by a current.

2. Short gravity waves ($kH \gg 1$) are incident from a stationary fluid onto a jet of triangular velocity profile:

$$U(y) = y\frac{U_m}{L}, \quad 0 \leqslant y \leqslant L$$

$$= 0, \qquad y \leqslant 0 \quad \text{and} \quad y > L.$$

Find, as a function of angle of incidence $\theta_e (> 0)$, the maximum incident wavelength of waves which are reflected back at $y = L$ by refraction in the current.

3. Consider the interaction of short gravity waves with much longer internal gravity waves propagating along a shallow pycnocline. Where, in relation to the internal wave profile, will the maximum amplitudes of the short waves be found? Assume that no caustics occur.

37. THE INFLUENCE OF WAVES ON THE MEAN STATE

In the previous section, we examined the influence of inhomogeneities in the mean state, and particularly of shear currents, on wave propagation. Let us now reverse our point of view and consider how wave motion can alter the mean properties of the fluid. Our analysis will centre on the effects of surface gravity waves, for which a variety of examples and applications are available. We have already discussed some of the mean properties of gravity waves in Section 13 and found that a wave field has associated with it mean fluxes of mass, momentum and energy. Any net divergence of these fluxes, due to changes in wave properties along ray paths, will produce local accumulations of mass, momentum and energy which will manifest themselves as changes in mean sea level and as mean currents. This section may be considered as an extension of Section 13, in which the average wave fluxes are now incorporated into a set of averaged dynamical equations for gravity wave motion in a horizontally sheared flow.

The averaged integrated equations for gravity waves

Following the methods of Section 13, we now formulate the average equations for the depth-integrated, time-averaged mass, momentum and energy balance in the presence of surface gravity waves. The derivation and interpretation of the averaged equations for these waves evolved from a series of papers by Longuet-Higgins and Stewart (1960, 1961, 1962, 1964) and by Whitham (1962); they are also presented by Phillips (1966), Hasselmann (1971) and Garrett (1976).

Let us consider an unstratified, non-rotating fluid with a free surface at $z = \eta$. A horizontal flow $\boldsymbol{U} = (U, V, 0)$ driven by a pressure field P is superimposed upon the velocity $\boldsymbol{u} = (u, v, w)$ and pressure field p due to short gravity waves. We first consider the average mass balance by integrating (over the depth) and averaging (over the phase) the continuity equation

$$\nabla \cdot (\boldsymbol{U} + \boldsymbol{u}) = 0. \tag{37.1}$$

Using the rule for differentiation of integrals

$$\frac{\partial}{\partial a} \int_{A(a)}^{B(a)} F(a, b)\, \mathrm{d}b = \int_{A}^{B} \frac{\partial F}{\partial a}\, \mathrm{d}b + F[a, B(a)] \frac{\partial B}{\partial a} - F[a, A(a)] \frac{\partial A}{\partial a}, \tag{37.2}$$

and the boundary conditions

$$w = \eta_t + u\eta_x + v\eta_y \qquad \text{at } z = \eta, \tag{37.3a}$$

$$w = -uH_x - vH_y \qquad \text{at } z = -H, \tag{37.3b}$$

we obtain from (37.1), after vertical integration and averaging,

$$\rho_* \frac{\partial}{\partial t} \langle \eta \rangle + \nabla_h \cdot (\langle M \rangle + M^m) = 0, \tag{37.4}$$

$$\text{where} \quad \langle M \rangle = \left\langle \int_{-H}^{\eta} \rho_* \boldsymbol{u}_h \, \mathrm{d}z \right\rangle, \quad \text{with} \quad \boldsymbol{u}_h = (u, v, 0), \tag{37.5a}$$

$$M^m = \left\langle \int_{-H}^{\eta} \rho_* U \, dz \right\rangle. \tag{37.5b}$$

The angle brackets denote an average over the phase of the gravity waves. The horizontal divergence operator is $\nabla_h = (\partial/\partial x, \partial/\partial y)$. The average properties of the gravity waves, as well as those of the mean current are, in general, functions of time and position over scales well in excess of the period and the wavelength of the surface gravity waves. Note from (37.4) that the mean mass flux (or momentum density) of the waves enters the mean mass balance. Thus, we expect part of the mean surface displacement to be caused by spatial variations of the wave momentum density.

The momentum equation, in the form (13.30), is subjected to the same two operations: integration and averaging. The vertical component of momentum obeys

$$\frac{\partial}{\partial t}(\rho_* w) + \frac{\partial}{\partial x_j}(\rho_* u_j w) + \frac{\partial}{\partial z}(\rho_* w^2 + p) + \rho_* g = 0, \tag{37.6}$$

with $j = 1, 2$. Integrating (37.6) over the depth of the fluid and using the boundary conditions (37.3), we obtain

$$\frac{\partial}{\partial t} \int_{-H}^{\eta} \rho_* w \, dz + \frac{\partial}{\partial x_j} \int_{-H}^{\eta} \rho_* w u_j \, dz + p(\eta) - p(-H) + \rho_* g(\eta + H) = 0. \tag{37.7}$$

The first term represents the vertical acceleration of the centre of mass of a fluid column; it may not vanish locally even when averaged over a cycle, although its large-scale spatial average over the whole fluid container is of course zero. The surface pressure $p(\eta)$ will be taken as zero. Rearranging (37.7) and taking averages, we find

$$\langle p(-H) \rangle = \rho_* g(\langle \eta \rangle + H) + \frac{\partial}{\partial x_j} \left\langle \int_{-H}^{\eta} \rho_* w u_j \, dz \right\rangle + \frac{\partial}{\partial t} \left\langle \int_{-H}^{\eta} \rho_* w \, dz \right\rangle. \tag{37.8}$$

For waves travelling over a flat bottom, the two integrals in (37.8) vanish, to second order in wave amplitude, since w is out of phase with η and with u_j (cf. Section 11). In that case, (37.8) implies that the pressure at the bottom is simply equal to the hydrostatic pressure at the bottom of a fluid of depth $H + \langle \eta \rangle$. Over a sloping bottom, however, w and u_j or η are not strictly out of phase, as pointed out by Mei (1973), and there is a non-vanishing contribution from the two integrals. It is thus not generally possible to relate the mean bottom pressure to the mean surface displacement without knowledge of the flow field. Having once mentioned the existence of this contribution to the bottom pressure, we will continue under the assumption that any bottom slopes to be encountered in the following discussion will be small enough that the surface gravity waves may be well represented by the flat bottom solutions. The last two terms of the right-hand side of (37.8) are henceforth ignored.

The horizontal momentum equation

$$\frac{\partial}{\partial t}(\rho_* u_j) + \frac{\partial}{\partial x_j}(\rho_* u_i u_j + p\delta_{ij}) + \frac{\partial}{\partial z}(\rho_* u_i w) = 0, \tag{37.9}$$

where the subscripts i and j refer only to horizontal components, may be integrated to yield

$$\frac{\partial}{\partial t}\int_{-H}^{\eta}\rho_* u_i \, dz + \frac{\partial}{\partial x_j}\int_{-H}^{\eta}(\rho_* u_i u_j + p\delta_{ij})\, dz - p(-H)\frac{\partial H}{\partial x_j}\delta_{ij} = 0. \tag{37.10}$$

Using the simplified form of (37.8), from which the integral terms are neglected, the bottom pressure term occurring in (37.10) may be averaged and rearranged as

$$\langle p(-H)\rangle\frac{\partial H}{\partial x_j} = \frac{\partial}{\partial x_j}[\tfrac{1}{2}\rho_* g(\langle\eta\rangle + H)^2] - \rho_* g(\langle\eta\rangle + H)\frac{\partial\langle\eta\rangle}{\partial x_j}. \tag{37.11}$$

With the help of (37.11), averaging over the rest of the terms (37.10) gives

$$\frac{\partial}{\partial t}\left\langle \int_{-H}^{\eta}\rho_* u_i \, dz\right\rangle + \frac{\partial}{\partial x_j}\left\langle \int_{-H}^{\eta}[\rho_* u_i u_j + (p - p_0)\delta_{ij}]\, dz\right\rangle = -\rho_* g(\langle\eta\rangle + H)\frac{\partial\langle\eta\rangle}{\partial x_i}, \tag{37.12}$$

where p_0 is the hydrostatic pressure under the mean free surface $z = \langle\eta\rangle$. Now expanding the horizontal flow field into $\boldsymbol{u}_h + \boldsymbol{U}$, we recognize the first term as the time derivative of $\langle \boldsymbol{M}\rangle + \boldsymbol{M}^m$. The second integral yields four terms, one of which,

$$\left\langle \int_{-H}^{\eta}[\rho_* u_i u_j + (p - p_0)\delta_{ij}]\, dz\right\rangle,$$

is the radiation stress T_{ij} encountered earlier in (13.40) as the momentum flux of the waves. The conservation equation for the total averaged momentum is thus

$$\frac{\partial}{\partial t}(\langle M\rangle_i + M_i^m) + \frac{\partial}{\partial x_j}\left[T_{ij} + \left\langle \int_{-H}^{\eta}\rho_*(u_i U_j + u_j U_i)\, dz\right\rangle + \left\langle \int_{-H}^{\eta}\rho_* U_i U_j \, dz\right\rangle\right]$$

$$= -\rho_* g(\langle\eta\rangle + H)\frac{\partial\langle\eta\rangle}{\partial x_i}. \tag{37.13}$$

This equation, for the total mean momentum, includes contributions to both the mean wave momentum $\langle \boldsymbol{M}\rangle$ and to the current momentum $\boldsymbol{M}^m$. In order to find an equation for the momentum of the current alone, and hence find out what forces are exerted by the waves on the current, we will subtract, following Garrett (1976), a wave mean momentum equation from (37.13). In a coordinate system at rest, the wave action equation, by a simple Galilean transformation of (6.44), becomes

$$\frac{\partial}{\partial t}\frac{\langle E_0\rangle}{\omega_0} + \nabla\cdot\left[(\boldsymbol{U} + \boldsymbol{c}_{g0})\frac{\langle E_0\rangle}{\omega_0}\right] = 0. \tag{37.14}$$

From (6.19a), the variation of the wavenumber, in the same coordinate system, is

$$\frac{\partial \boldsymbol{k}}{\partial t} + (\boldsymbol{U} + \boldsymbol{c}_{g0})\cdot\nabla\boldsymbol{k} = -\boldsymbol{k}\cdot\nabla\boldsymbol{U}. \tag{37.15}$$

Since $\langle \boldsymbol{M}\rangle = \langle E\rangle\boldsymbol{k}/\omega_0$, a combination of (37.14) and (37.15) gives

$$\frac{\partial}{\partial t}\langle M\rangle_i = -\frac{\partial}{\partial x_j}[\langle M\rangle_i\{U_j + (c_{g0})_j\}] - \langle M\rangle_j\frac{\partial}{\partial x_i}U_j. \tag{37.16}$$

Subtracting (37.16) from (37.13) we obtain for the mean flow momentum alone

$$\frac{\partial M_i^m}{\partial t} + \frac{\partial}{\partial x_j}\left\langle \int_{-H}^{\eta} \rho_* U_i U_j \, dz \right\rangle + \rho_* g(H + \langle\eta\rangle)\frac{\partial\langle\eta\rangle}{\partial x_i} = \Gamma_i, \tag{37.17}$$

where the mean force Γ_i exerted by the mean flow on the waves is

$$\Gamma_i = \frac{\partial}{\partial x_j}\left\{[U_j + (c_{g0})_j]\langle M\rangle_i - T_{ij} - \left\langle \int_{-H}^{\eta} \rho_*(u_i U_j + u_j U_i)\, dz \right\rangle\right\} + \langle M\rangle_j \frac{\partial U_j}{\partial x_i}. \tag{37.18}$$

For a depth independent mean flow, the term $U_j\langle M\rangle_i$ cancels with one of the terms in the integral and Γ_i simplifies to

$$\Gamma_i = \frac{\partial}{\partial x_j}[(c_{g0})_j\langle M\rangle_i - U_i\langle M\rangle_j - T_{ij}] + \langle M\rangle_j \frac{\partial U_j}{\partial x_i}. \tag{37.19}$$

The force $\mathbf{\Gamma}$ is simply the contribution to the rate of change of the momentum of the current due to the presence of the waves; its various terms can be traced to individual contributions to the total momentum balance (37.13) and to the wave momentum balance (37.16).

A vertically integrated averaged equation for the total energy density may be derived by following the procedure used above. As the balance of mean wave energy is given by (37.14), we can obtain by subtraction, as for the case of momentum, an equation for the energy of the current alone. The derivation is left as an exercise to the reader; it may also be found in Phillips (1966) and in Hasselmann (1971).

Restriction to small-amplitude waves

Let us now focus our attention on small-amplitude waves, for which solutions have been given in Sections 11 and 13. We already know from (13.5) and (13.40) that both $\langle M\rangle$ and T_{ij} are of $0(a^2)$, where a is the wave amplitude. The mean displacement of the sea level, $\langle\eta\rangle$, is also at least of $0(a^2)$; although we have obtained $\langle\eta\rangle = 0$ to second order in Section 12 for a uniform wave train of finite amplitude, we now admit the possibility that the interaction between waves and currents (or even the mere presence of a current) may lead to the presence of mean sea-level displacements.

To second order in the wave amplitude, the mass conservation equation (37.4) is unchanged. The mean force Γ_i is also formally unchanged, but may take simplified forms in limiting cases. For example, in deep water, $c_{g0} = c_0/2$, and $T_{ij} = \langle E_0\rangle\delta_{ij}/2$; using $\langle M\rangle = \langle E_0\rangle \boldsymbol{k}/\omega_0$, Γ_i reduces to

$$\Gamma_i = \frac{\partial}{\partial x_j}[-U_i\langle M\rangle_j] + \langle M\rangle_j \frac{\partial U_j}{\partial x_i}. \tag{37.20}$$

In shallow water, on the other hand, $c_{g0} = c_0$, and

$$T_{ij} = \langle E_0\rangle \begin{pmatrix} 3/2, & 0 \\ 0, & 1/2 \end{pmatrix},$$

provided $\boldsymbol{k} = (k, 0)$. After some manipulation we find that

$$\Gamma_i = -\frac{1}{2}\frac{\partial}{\partial x_i}\langle E_0\rangle - U_i \frac{\partial}{\partial x_j}\left(\frac{k\langle E_0\rangle}{\omega_0}\delta_{1j}\right) + \frac{k\langle E_0\rangle}{\omega_0}\left(\frac{\partial U_j}{\partial x_i} - \frac{\partial U_i}{\partial x_j}\right)\delta_{i2}. \tag{37.21}$$

Applications to deep water waves [based, however, on the total momentum equation (37.13)] may be found in Longuet-Higgins and Stewart (1964) for example. An elegant application to the reinforcing of Langmuir circulations by the mean forces due to surface waves in deep water has been given by Garrett (1976). For waves in deep water, one should recall that the wave momentum $\langle \boldsymbol{M}\rangle$, being concentrated at the free surface, between $z = \langle\eta\rangle$ and $z = \langle\eta\rangle + \eta$, occupies only a small fraction of the total water depth. A precise consideration of the mass balance in that case is more appropriately based on a two-layer model: the upper layer, between wave crests and troughs, contains all the mean wave momentum; the lower layer occupies the rest of the fluid column and contains only mean flow momentum. The two layers are coupled by a mean vertical velocity W (not included above) which enters the mass balance equation as one of the terms which balance the divergence of the mean wave momentum. In very shallow water ($kH \ll 1$) this two-layer representation is no longer necessary. The examples presented below will be limited to shallow-water cases, where the mean flow may be taken as strictly horizontal.

Let us consider first a one-dimensional situation, with nearly plane waves advancing (in the x-direction) in shallow water of uniform depth. Let us also assume that there is no independent mean flow in the absence of waves, although the waves themselves may induce some mean motion. In that case, $\boldsymbol{U}$ is at most of order of the square of the wave amplitude and, to that order, the mean flow momentum equation reduces to

$$\frac{\partial M^m}{\partial t} + \rho_* gH \frac{\partial\langle\eta\rangle}{\partial x} = -\frac{1}{2}\frac{\partial}{\partial x}\langle E\rangle. \tag{37.22}$$

Note that there is no need for a subscript on $\langle E\rangle$ now: $\langle E\rangle = \langle E_0\rangle$. The mass continuity equation (37.4) takes the one-dimensional form

$$\rho_* \frac{\partial\langle\eta\rangle}{\partial t} + \frac{\partial}{\partial x}\left[\langle M\rangle + M^m\right] = 0. \tag{37.23}$$

Since everything is one-dimensional, the subscripts have been dropped. The mean wave momentum is related to the energy density by $\langle M\rangle = \langle E\rangle/c$. Thus, we may rewrite (37.23) as

$$\rho_* \frac{\partial\langle\eta\rangle}{\partial t} + \frac{\partial M^m}{\partial x} = -\frac{1}{c}\frac{\partial\langle E\rangle}{\partial x}\cdot \tag{37.24}$$

From (37.22) and (37.24), we see that the modulations in wave energy through a wave-group produce forcing terms for the mass and momentum balance of the mean state. To second order in wave amplitude, and with a constant wave frequency and group velocity, the wave energy equation (37.14) reduces to

$$\frac{\partial\langle E\rangle}{\partial t} + c_g \frac{\partial\langle E\rangle}{\partial x} = 0. \tag{37.25}$$

Thus, $\langle E\rangle$ is invariant along a ray and is then a function of $(x - c_g t)$. The dependent

Fig. 37.1 The depression of the mean sea level, $\langle\eta\rangle$, under a group of large waves in shallow water.

variables $\langle\eta\rangle$ and M^m will be assumed to have similar functional dependences. We may then replace $\partial/\partial t$ by $-c_g\partial/\partial x$ in (37.22) and (37.24). Eliminating $\partial M^m/\partial x$ between these equations we then obtain

$$\rho_* \frac{\partial\langle\eta\rangle}{\partial x} = -\frac{\left(\frac{1}{2}+\frac{c_g}{c}\right)}{(gH-c_g^2)}\frac{\partial\langle E\rangle}{\partial x}. \tag{37.26}$$

Since it is consistent to assume that $\langle\eta\rangle = 0$ when $\langle E\rangle = 0$, this equation integrates to

$$\rho_*\langle\eta\rangle = -\frac{\left(\frac{1}{2}+\frac{c_g}{c}\right)}{(gH-c_g^2)}\langle E\rangle. \tag{37.27}$$

Since $gH > c_g^2$, the mean sea level is always depressed under the waves (Fig. 37.1). The simplest dynamical interpretation of this result is that the bigger waves, having a larger value of radiation stress, tend to push fluid away from underneath themselves; from (37.24), part of the push is counterbalanced by the pressure field associated with the variation in $\langle\eta\rangle$.

In shallow water, we may replace c by c_g in the numerator of the right-hand side of (37.27). The denominator has to be treated with more care. For $kH \ll 1$, we find from (10.16) and (10.20) that, to $0(k^2H^2)$,

$$\omega = k\sqrt{gH}\left(1-\frac{k^2H^2}{6}\right),$$

$$c_g = \sqrt{gH}\left(1-\frac{k^2h^2}{6}\right).$$

Hence, (37.27) becomes

$$\rho_*\langle\eta\rangle = -\frac{3}{2}\frac{\langle E\rangle}{gk^2H^3}. \tag{37.28}$$

Along a ray, ω is constant; hence, for small kH, $k \propto H^{-1/2}$. Similarly, for very small variations in H, the energy flux $\langle E \rangle c_g$ is preserved along a ray; for small kH, $\langle E \rangle$ is then also proportional to $H^{-1/2}$. Hence, from (37.28), $\langle \eta \rangle \propto H^{-5/2}$, a result already obtained by Longuet-Higgins and Stewart (1964). The mean sea-level depression under large waves is amplified as these waves travel into shallow water. We shall return to this point shortly.

In the absence of an independent mean current, with M^m induced by the waves themselves, there is no real point in separating the mean wave momentum $\langle M \rangle$ from the mean flow momentum M^m. Under these circumstances we shall continue our analysis, using the same total mean momentum equation (37.13) rather than the pair (37.16), (37.17). The same results are obtained directly from (37.13) in a simpler way than by addition of the solutions of the two momentum equations for the individual components of the total mean momentum.

Wave "set-down" and "set-up"

Let us examine in more detail the "set-down" of mean sea level predicted by (37.27) when gravity waves travel into shallow water. For steady, one-dimensional conditions, and in the absence of an independent mean flow, (37.13) reduces to

$$\frac{\partial \langle \eta \rangle}{\partial x} = -\frac{1}{\rho_* g H} \frac{\partial}{\partial x} T_{11}. \tag{37.29}$$

For a given depth profile $H(x)$ and T_{11} given from (13.40), and with the conditions that ω remain constant and that the energy flux $\langle E \rangle c_g$ be invariant (in the absence of reflection) along a ray, (37.29) may be integrated to find the set-down near the shore arising from a known deep water incident wave field. Even for the linearized theory, the fit between calculated and observed values of $\langle \eta \rangle$ is quite good nearly up to the point where the waves break. The results of tank experiments (Bowen et al., 1968) and of field observations (Saville, 1961) are shown in Fig. 37.2 and Fig. 37.3, respectively.

Beyond the breaker line (in the surf zone), linear theory becomes untenable. We nevertheless continue, rather heuristically, to use some of the results valid for small-amplitude waves, for lack of a good description of breaking waves. Thus, in shallow water, (37.29) reduces to

$$\frac{\partial \langle \eta \rangle}{\partial x} = -\frac{3}{2\rho_* g H} \frac{\partial \langle E \rangle}{\partial x}. \tag{37.30}$$

Breakers continuously lose energy as they travel shorewards. Similarity arguments, as well as observations suggest that the wave amplitude is proportional to the local depth in the surf zone:

$$a = \alpha(H + \langle \eta \rangle) \simeq \alpha H, \tag{37.31}$$

where α is a function of the bottom slope and varies between 0.3 and 0.6 (Longuet-Higgins, 1970a). Observations by Munk (1949a) indicate that swell tends to break when $a/H \sim 0.39$, giving a value of α lying in that range. Continuing to use $\langle E \rangle = \frac{1}{2}\rho_* g a^2$, (37.29) becomes (with $T_{11} = 3\langle E \rangle/2$)

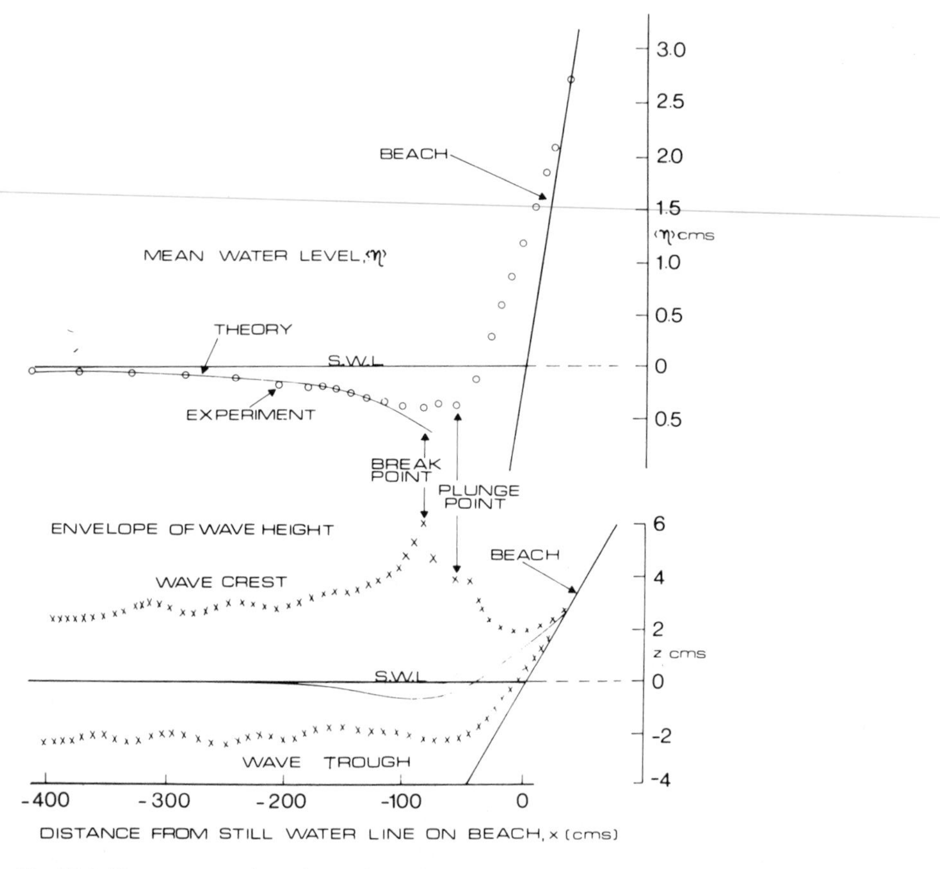

Fig. 37.2. Wave set-up and set-down, from the wave tank experiments of Bowen et al. (1968). Profiles of the mean water level and of the envelope of the wave height for a typical experiment. The theoretical wave set-down is estimated from (37.29). Wave period = 1.14 s; α_0 = 3.23 cm, a_b = 4.28 cm; $H = 0.82x$; S.W.L. stands for "stationary water level", the mean level in the absence of waves. (Copyrighted by American Geophysical Union.)

$$\frac{\partial\langle\eta\rangle}{\partial x} = -\frac{3}{2}\alpha^2\frac{\partial H}{\partial x}. \tag{37.32}$$

With x increasing shorewards, $\partial H/\partial x < 0$ near the beach and there is a mean set-up of the sea level. The seaward pressure force resulting from the surface slope just balances, on the average, the momentum flux divergence of the incoming waves. This set-up has been observed by Saville (1961) (Fig. 37.3) and Bowen et al. (1968) (Fig. 37.2). An important consequence of the existence of mean sea-level slopes induced near the shore by wave action is the impediment which they present to estimating the stress exerted by a constant wind on the surface of an enclosed body of water by the equilibrium surface slope which would be necessary to balance the wind stress. Since a wind always generates waves, corrections due to wave set-up (or set-down, depending on the location of the measurements) must be applied to this technique.

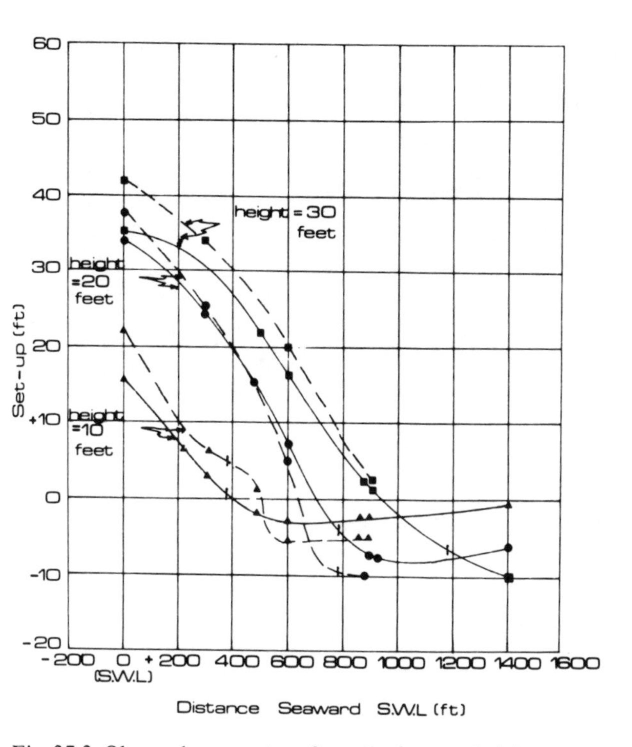

Fig. 37.3. Observed wave set-up from the large-scale laboratory experiments of Saville (1961). Values of $\langle\eta\rangle$ are plotted for two wave periods ($T = 9.25$ s ——; $T = 15.0$ s – – – –) and three different incident wave amplitudes. The beach has a linear slope $H = 0.033x$ for $x < 3000$ ft and is flat at $H_0 = 100$ ft for $x > 3000$ ft. Vertical tick marks are placed at $H = 2.6a$, the approximate breaker location (corresponding to $\alpha = 0.38$ in (37.31).

Rip currents

Consider the mean continuity equation (37.4) for plane surface gravity waves approaching a beach normally, as in the above discussion. The mean flow induced by the waves, M^m, may be represented in very shallow water by a depth-independent mean current U:

$$M^m \;=\; \rho_* HU. \qquad (37.33)$$

The compensating current U balances the shoreward mass flux of the waves. In the surf zone, continuing to use $\langle M\rangle = \langle E\rangle/c$ and $\langle E\rangle = \rho_* g a^2/2$ [with a given by (37.31)], we find by integration of (37.4),

$$U \;=\; -\tfrac{1}{2}\alpha^2\sqrt{gH} + \text{constant}. \qquad (37.34)$$

The constant of integration must be taken as zero if U is to vanish where $H = 0$. In the surf zone, the return current is comparable to the wave speed, and its influence on the momentum and energy balances may no longer be assumed to be negligible. Furthermore, if U varies along the beach, its extreme values will exceed the one-dimensional result (37.34), which may then be considered as an average value in the longshore direction. This is what happens in practice: the return flow of the fluid thrown onto the beach by the breakers is found to be concentrated in strong "rip-currents" which interfere with the progress of the incoming waves. The nonlinear coupling between the shoaling waves and the return flow which they induce becomes an essential part of the shore-zone dynamics. Thus, a correct theory of rip current generation must include the full nonlinearity in (37.13). In addition, friction and energy dissipation should be included in the problem. The full nonlinear problem remains unsolved, although some progress has been made by James (1974a, b). The coupling of waves and rip currents has also been investigated, through perturbation methods, by LeBlond and Tang (1974).

The first analytical model of rip current formation was presented by Bowen (1969b); a version of this model is discussed below. Although this model sins in the temerity of many of the assumptions on which it is based, it provides an easily understood picture of the basic dynamics of the problem. A rectilinear coast of constant slope is considered: taking the beach at $x = 0$ ($x < 0$ over the ocean), the depth profile is written as

$$H(x) = -sx. \tag{37.35}$$

Strictly speaking, the shoreline should be at the value of x where $H + \langle\eta\rangle = 0$; the influence of set-up will be neglected in estimating the position of the shoreline. It turns out to be of little relevance to the form and amplitude of the induced currents. The nearshore area is divided into two zones by the position of the breaker line $x = -x_b$. In both zones, we use the results of linear shallow-water wave theory (the linearity is really an outrageous assumption in the surf zone; for more realistic representations, see James, 1974a). Thus, everywhere, we use

$$T_{ij} = \langle E\rangle \begin{pmatrix} 3/2, & 0 \\ 0, & 1/2 \end{pmatrix}, \qquad \langle E\rangle = \tfrac{1}{2}\rho_* g a^2. \tag{37.36}$$

Currents and surface slopes are generated in the surf zone by the strong divergence of the radiation stress due to dissipation. We choose, for analytical simplicity, a simple friction term of the form $-R(\langle M\rangle_i + M_i^m)$, where R is a constant friction coefficient. In spite of our earlier comments, we also neglect the nonlinear terms in the mean momentum equation (37.13). The simplified momentum balance in the surf zone then becomes

$$\frac{1}{\rho_* g H}\frac{\partial}{\partial x_j}T_{ij} = -\frac{\partial\langle\eta\rangle}{\partial x_i} - R(\langle M\rangle_i + M_i^m). \tag{37.37}$$

Continuity implies that the currents generated in the surf zone must continue beyond the breakers. In the latter region, any current generated by the nonbreaking waves is of second order in the wave amplitude and, especially immediately seaward of the breakers, is much smaller than the currents arising from continuity with the relatively large flows set up in the surf zone. Seaward of the breaker line ($x = -x_b$) it is then assumed that the only currents and slopes of relevance are those arising by continuity with those of the surf zone. For $x < -x_b$ we thus use an unforced version of (37.37), obtained by putting the left-hand side equal to zero.

In both zones, the steady mass flux equation (37.4) reduces to

$$\frac{\partial}{\partial x_j}(\langle M\rangle_j + M_j^m) = 0. \tag{37.38}$$

The problem is two-dimensional, and we may introduce a transport stream function ψ, defined by

$$\langle M\rangle_1 + M_1^m = -\frac{\partial\psi}{\partial y}; \quad \langle M\rangle_2 + M_2^m = \frac{\partial\psi}{\partial x}. \tag{37.39}$$

Substituting for T_{ij} from (37.36) into (37.37) and cross-differentiating to eliminate $\langle\eta\rangle$, we find in the surf zone $(-x_b < x \leqslant 0)$ that

$$2\frac{\partial^2 a^2}{\partial x\partial y} + \frac{1}{H}\frac{\partial H}{\partial x}\frac{\partial a^2}{\partial y} = 4RH\nabla_h^2\psi. \tag{37.40}$$

We continue to relate the wave amplitude to the depth through (37.31), and now introduce a longshore modulation of the form

$$a = \alpha H(1 + \epsilon\cos\lambda y), \tag{37.41}$$

where ϵ is small.

Bowen and Inman (1969) considered the case where this modulation arises because of the superposition of edge waves on the incoming swell; the modulation could also arise from the interference of two incident wavefields (Dalrymple, 1975). Finally, it could be imagined as a parameterization of the nonlinear interaction process of the waves and the rip currents. In any case, substitution of (37.41) into (37.40) gives

$$\nabla_h^2\psi = -\frac{5}{2}\frac{\epsilon\alpha^2\lambda}{R}\frac{\partial H}{\partial x}\sin\lambda y + 0(\epsilon^2). \tag{37.42}$$

Taking a linear bottom slope $(\partial H/\partial x = -s)$, and assuming that ψ has the same y-variation as the forcing term, (37.42) becomes

$$\psi_{xx} - \lambda^2\psi = \frac{5}{2}\frac{\epsilon s\alpha^2\lambda}{R} \equiv P \tag{37.43}$$

for $-x_b < x \leqslant 0$. The solution of (37.43) for which $\psi = 0$ on $x = 0$ (the normal transport must vanish at the beach) is

$$\psi_a(x, y) = \left[A\sinh\lambda x + \frac{P}{\lambda^2}(\cosh\lambda x - 1)\right]\sin\lambda y, \tag{37.44}$$

where ψ_a stands for the stream function in the surf zone and A is a constant to be determined. The offshore (in $x < -x_b$) stream function ψ_b satisfies (37.43) with $P = 0$. The solution which decays as $x \to -\infty$ is

$$\psi_b = C\mathrm{e}^{\lambda x}\sin\lambda y, \tag{37.45}$$

where C is another constant. The constants A and C are found from matching ψ (the normal velocity) and $\langle\eta\rangle$ at $x = -x_b$. From the y-component of (37.37) and the definition of ψ, the second condition implies that there is a discontinuity in $\partial\psi/\partial x$ proportional to that assumed to exist for $(\partial/\partial x_j)T_{2j}$ (which is taken as zero for $x < -x_b$). Thus, at $x = -x_b$

$$R\left(\frac{\partial\psi_a}{\partial x} - \frac{\partial\psi_b}{\partial x}\right) = -\frac{1}{\rho_* gH}\frac{\partial}{\partial x_j}T_{2j}. \tag{37.46}$$

A discontinuity in the velocity component parallel to the matching line is not unphysical since no lateral friction has been included and hence lateral slip is possible. Substituting for T_{ij} in the surf zone, (37.46) becomes

$$\frac{\partial \psi_a}{\partial x} - \frac{\partial \psi_b}{\partial x} = \frac{\epsilon \alpha^2 H \lambda}{2R} \sin \lambda y + 0(\epsilon^2) \qquad \text{at} \qquad x = -x_b . \tag{37.47}$$

The constants A and C are then readily determined from the matching condition $\psi_a = \psi_b$ at $x = -x_b$ and from the condition (37.47):

$$A = PQ/\lambda^2, \; C = \frac{P e^{\Lambda}}{\lambda^2} (\cosh \Lambda - 1 - Q \sinh \Lambda),$$

$$\text{where} \quad Q = \frac{(\sinh \Lambda + \cosh \Lambda - 1 - \Lambda/5)}{\sinh \Lambda + \cosh \Lambda},$$

$$\Lambda = \lambda x_b .$$

Note that for $\Lambda < 5$, i.e., for a longshore wavelength in excess of the width of the surf zone, $C < 0$ and the seaward flow, as calculated from (37.39) and (37.45), occurs at the minima of the wave amplitude a, as per (37.41): the rip currents are found at positions of low wave intensity, as could be expected. Contours of the stream function ψ are shown in Fig. 37.4 for $\Lambda = 0.2$.

The main artifact of the above theory of rip currents, and also of some of its refined versions, is the wave amplitude modulation (37.41). Bowen and Inman (1969) have shown, mainly in laboratory studies, that longshore amplitude variations could result from an interaction of the incoming surf with edge waves of the same frequency resonating between headlands. The details of this interaction have been examined by Guza and Davis (1974) using the techniques described in the next section. Many refinements have been added to the above rip current model, by changing the form of the bottom friction coefficient R (Bowen, 1969b), by adding lateral viscosity (Bowen, 1969b, Hino, 1974), and by including some nonlinear effects (Bowen, 1969b; Tam, 1973; LeBlond and Tang, 1974).

Longshore currents

Let us now leave aside our assumption of normal wave incidence and consider the currents generated by waves incident at some angle to the beach. A rotation by an angle $-\theta$ (Fig. 37.5) is necessary to express the radiation stress tensor in beach coordinates. From (13.60) and the shallow-water diagonal form (37.36), we find that after rotation, the radiation stress tensor becomes

$$T_{ij} = \frac{\langle E \rangle}{2} \begin{pmatrix} 3 \cos^2 \theta + \sin^2 \theta, & \sin 2\theta \\ \sin 2\theta, & 3 \sin^2 \theta + \cos^2 \theta \end{pmatrix} . \tag{37.48}$$

The assumption already introduced in our discussion of rip currents will be adhered to here: significant forcing by the divergence of T_{ij} occurs in the surf zone only. Now however, T_{ij} is no longer diagonal. Under the simplest conditions, when the wave amplitude and angle of attack are independent of the longshore direction, that is

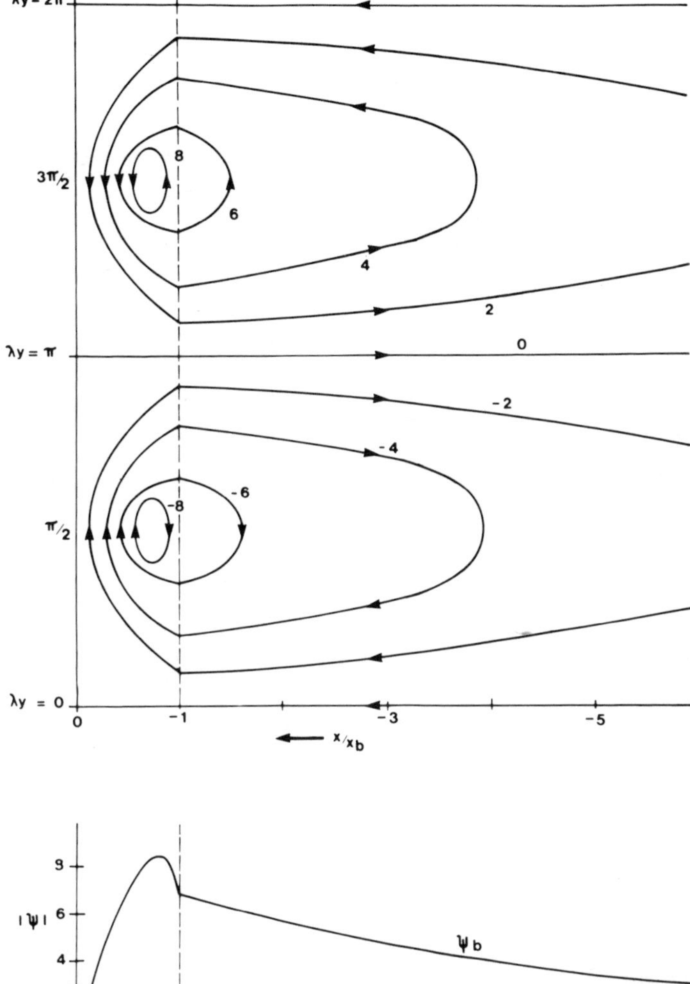

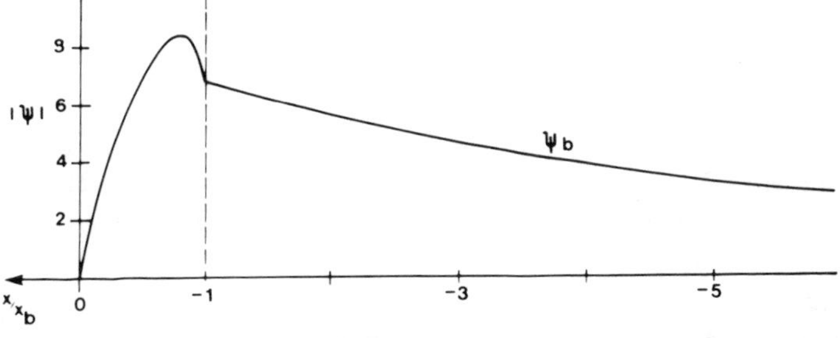

Fig. 37.4. Rip current circulation pattern: (a) contours of the streamfunction from (37.44) and (37.45) over one cycle in the longshore direction; (b) the amplitude of the streamfunction as a function of the offshore coordinate. ($\Lambda = 0.2$.)

$$\frac{\partial}{\partial y}\langle E\rangle = \frac{\partial \theta}{\partial y} = 0,$$

the y-component of (37.37) becomes

$$\frac{1}{\rho_* g H}\frac{\partial}{\partial x} T_{12} = -R(\langle M\rangle_2 + M_2^m). \tag{37.49}$$

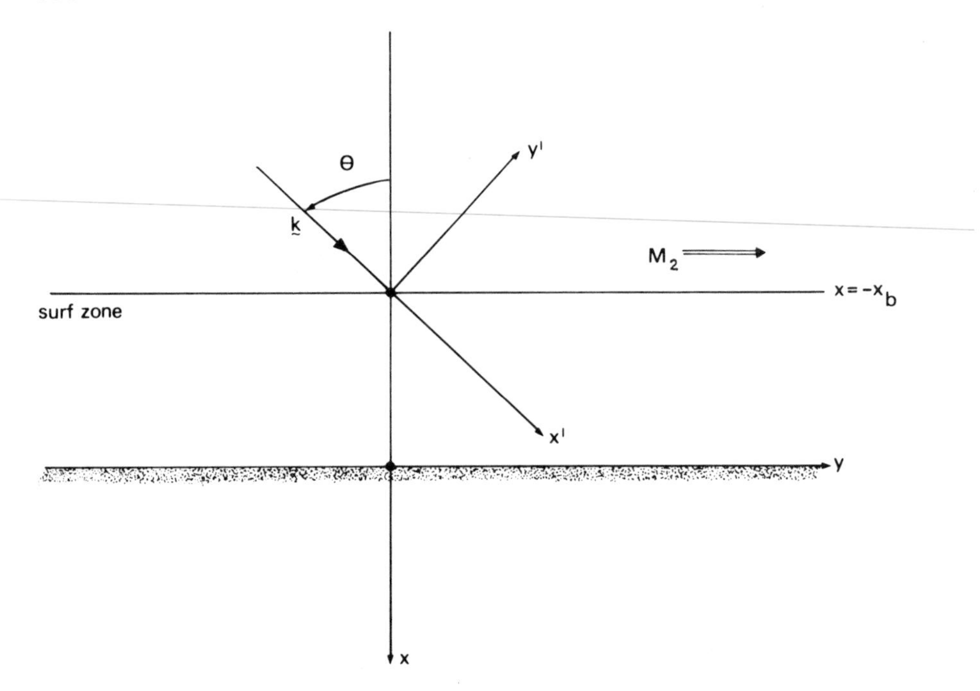

Fig. 37.5. The generation of longshore currents by obliquely incident waves: definition diagram.

Since the wave amplitude (and hence also T_{12}) vanishes at the shore line, the integral of (37.49) across the surf zone yields the total longshore momentum transport in terms of the beach profile $H(x)$ and the incident component of the radiation stress. Substituting for T_{12} from (37.48) and integrating across the surf zone, we find that

$$(\tfrac{1}{2}\langle E\rangle \sin 2\,\theta)_{x=-x_b} = g\rho_* \int_{-x_b}^{0} RH(\langle M\rangle_2 + M_2^m)\,\mathrm{d}x. \tag{37.50}$$

The longshore transport reaches a maximum value for $\theta = 45°$. The velocity profile across the surf zone may be determined, given $H(x)$ and using (37.49) (Exercise 37.2). The reader is referred to the original papers by Longuet-Higgins (1970a, b) and by Bowen (1969a) and to the review articles quoted earlier for further details.

The presence of longshore and rip currents in the highly turbulent surf zone leads to fluxes of suspended sediments. According to Longuet-Higgins (1972), the volume transport rate of sediment in the surf zone is proportional to the total longshore momentum flux T_{12} imparted to the surf zone at $x = -x_b$. Any longshore variation of the sand volume transport leads to erosion or deposition on the beach, thus modifying the shoreline and hence the geometry of the whole wave forcing problem in the coastal zone. Calculations of beach deformation by wave-induced longshore currents have been examined by Komar (1973) and Rea and Komar (1975). The formation of beach cusps and crescentic bars by rip currents and their extensions off the breaker zone have been discussed by Bowen and Inman (1971) and Hino (1974). (See Fig. 25.5 for an example of a beach cusp pattern.)

Our presentation of the mean effects of waves on the medium through which they propagate has focused on surface gravity waves, for which the theory has been well explored and where a number of interesting consequences have been discovered. The analysis leading to mean equations of the form (37.5), (37.13) and (37.20) can be carried out in an analogous fashion for other types of wave fields. The mean stress exerted by the waves on the medium is of particular interest; its form has been studied for internal gravity waves by Garrett (1968) and by Bretherton (1969). Longshore current generation by shoaling interfacial waves has been examined by Hogg (1971) and Thomson (1975b). The reader is invited to examine for himself a similar problem for long waves in a rotating fluid.

Exercises Section 37

1. Following the procedure used to derive the mean momentum equation (37.13), derive a balance equation for the total mean energy density per unit area of the sea surface due to surface gravity waves and a depth-independent horizontal mean current.

2. Using (37.49), calculate the velocity profile, as a function of distance from shore, of a longshore current on a linearly sloping rectilinear beach.

3. Show that, in general, it is not only the existence of a finite angle of incidence which contributes to the generation of longshore currents, but also the longshore variation of the angle of wave incidence and of wave energy.

General formulation

In the first part of this chapter, we have considered the interaction of travelling waves with their surroundings under the assumptions of ray theory, namely that the length and period of the waves under study be much smaller than the characteristic distance and time over which the medium itself varies. Under such restrictions it also turned out to be possible to examine some wave–wave interactions, one wave being considerably longer and of greater period than the other. These methods are patently inapplicable to the study of interactions between waves of comparable wavelengths and periods; a different formalism is needed. The appropriate theory developed here is usually referred to as that of "weak wave–wave interactions". The influence of these interactions on the shape of wave spectra will be examined in the next section.

The theory of weak wave–wave interactions may be viewed as an outgrowth of the perturbation methods used to study weakly nonlinear waves (Sections 12 and 19). Consider a wave equation of the form

$$L(u)+M(u,u) = 0, \tag{38.1}$$

where $L(u)$ is linear in u and $M(u, u)$ is quadratic in u and its derivatives (in general, M could be any nonlinear operator; the choice of a quadratic operator is merely for simplicity). Boundary conditions on u are assumed either linear and homogeneous or periodic. We assume that associated with (38.1) there is a small nondimensional parameter ϵ (e.g., the wave slope, the Rossby number, . . .), in terms of which, u can be expanded as a power series of the form

$$u = \epsilon u_1 + \epsilon^2 u_2 + \ldots . \tag{38.2}$$

Generally speaking, ϵ characterizes the relative size of the (small) nonlinear term in (38.1). Substituting (38.2) into (38.1), we find that to order ϵ, u_1 satisfies the linear equation

$$L(u_1) = 0. \tag{38.3}$$

In order to keep u_1 real, we write a plane wave solution for u_1 as

$$\begin{aligned} u_1 &= \tfrac{1}{2}[A \exp\{i(\boldsymbol{k}\cdot\boldsymbol{x}-\omega t)\} + A^* \exp\{-i(\boldsymbol{k}\cdot\boldsymbol{x}-\omega t)\}] \\ &= \mathrm{Re}\,[A \exp\{i(\boldsymbol{k}\cdot\boldsymbol{x}-\omega t)\}], \end{aligned} \tag{38.4}$$

where both $\boldsymbol{k}$ and ω are real. Substitution of (38.4) into (38.3) yields a relation dispersion between ω and $\boldsymbol{k}$:

$$\Delta(\omega, \boldsymbol{k}) = 0. \tag{38.5}$$

To the next order in ϵ, (38.1) and (38.2) give

$$L(u_2) = -M(u_1, u_1); \tag{38.6}$$

the corrections to the linear solution u_1 are then found as the particular solution of a non-homogeneous linear equation. The process is by now familiar from Section 12, and can be continued to any order [although convergence of the series (38.2) must be established] to refine the solution. (As we have noted in Section 12, however, convergence does not imply stability!)

Now suppose that u_1 is the sum of *two* plane waves: $u_1 = u_a + u_b$, of generally different wavenumber and frequency (with corresponding subscripts). Both u_a and u_b satisfy the linear wave equation (38.3) and hence their wavenumber and frequency also satisfy (38.5). To second order in ϵ, (38.6) takes the form

$$L(u_c) = -M(u_a + u_b, u_a + u_b), \tag{38.7}$$

where $u_2 = u_c$ may now be considered as the result of the interaction between u_a and u_b through the nonlinear terms in the wave equation (38.1). From the expansion (38.2) we see that in general, the resultant interaction wave part of the solution ($\epsilon^2 u_c$) is much smaller than the primary wave part $[\epsilon(u_a + u_b)]$. Another source of nonlinearity in wave propagation can arise through the boundary conditions (as for surface gravity waves: see Section 12). The solution u_c to (38.7) is expected to be of the form (38.4), with $\boldsymbol{k}_c$ and ω_c equal to the wavenumber and frequency, respectively of the forcing term. Allowing for all possible combinations of products of u_a, u_b and their derivatives, the right-hand side of (38.7) includes terms with frequencies $\omega_c = 0, \pm 2\omega_a, \pm 2\omega_b, \pm(\omega_a \pm \omega_b)$, and similarly for the wavenumber vectors. Provided both signs are allowed for all frequencies and wavenumbers, all these cases are covered and described most elegantly by

$$\omega_a + \omega_b + \omega_c = 0, \tag{38.8a}$$

$$\boldsymbol{k}_a + \boldsymbol{k}_b + \boldsymbol{k}_c = 0, \tag{38.8b}$$

provided $a = b$ is allowed. The solution of (38.7) may then be written

$$u_c = \sum \frac{C \exp[i(\boldsymbol{k}_c \cdot \boldsymbol{x} - \omega_c t)] + C^* \exp[-i(\boldsymbol{k}_c \cdot \boldsymbol{x} - \omega_c t)]}{\Delta(\omega_c, \boldsymbol{k}_c)}, \tag{38.9}$$

where the sum is taken over all the sign combinations of (38.8) which occur in the nonlinear forcing term. With each choice of signs goes a value of the amplitude factor C. The interaction product u_c remains of order C and thus in the total solution u, $\epsilon^2 u_c$ is a small correction to the sum of the interacting waves, *unless* $\Delta(\omega_c, \boldsymbol{k}_c) = 0$, when resonance occurs. At resonance, the solution to (38.7) becomes secular and its amplitude grows linearly in time and/or space. A resonant wave is thus a free wave satisfying the dispersion relation and propagating at the frequency and wavenumber of the interference pattern produced by two primary interacting waves. Since it keeps up with the interference pattern, it is continuously excited and keeps growing as it travels. Nonresonant waves are forced waves which do not obey the dispersion relation and must be continuously excited to be kept in existence. After the interaction between the two primary waves has proceeded for some time (or over some distance), one would expect that the secondary waves which arise out of the coupling will be dominated by the resonant solution. It then becomes legitimate to consider only the resonant interactions as the most significant outcome of the wave–wave coupling. The conditions for resonance are given by (38.8), where each pair of frequencies and wavenumbers satisfies the dispersion relation (38.5). Whether the conditions can be satisfied between (in general) n interacting waves depends on the form of the dispersion relation. For nondispersive waves, all interactions are resonant, since all waves travel at the same speed, but this is not the case for dispersive waves, as we shall see in the examples below.

The above outline pleasantly glosses over the algebraic tedium of deriving the perturbation equation (38.7), verifying the possibility of resonance and calculating the strength of the coupling. We shall limit ourselves to illustrative examples of maximum simplicity.

Planetary wave interactions

The nonlinear equation describing nondivergent planetary waves in a fluid of uniform depth is the conservation of potential vorticity equation (18.24):

$$\frac{\mathrm{D}}{\mathrm{D}t}(\zeta_3 + f) = 0. \tag{38.10}$$

Introducing a stream function ψ as defined in (18.25), we find, upon expanding (38.10), that ψ satisfies

$$\nabla_h^2 \psi_t + J(\psi, \nabla_h^2 \psi) + \beta \psi_x = 0, \tag{38.11}$$

where $J(a, b)$ is the Jacobian: $J(a, b) = a_x b_y - a_y b_x$. Consider two plane waves of the form (38.4):

$$\psi_a = C_a \cos S_a,$$

$$\psi_b = C_b \cos S_b, \tag{38.12}$$

where C_a, C_b are real amplitude factors and†

$$S = kx + my - \omega t. \tag{38.13}$$

Each one of these plane waves satisfies (38.11) identically, provided the following dispersion relation is satisfied:

$$\omega = -k\beta/(k^2 + m^2). \tag{38.14}$$

This is the dispersion relation (18.28) found earlier for small-amplitude, nondivergent, barotropic planetary waves. Although ψ_a and ψ_b are both solutions of the nonlinear equation (38.11) their sum is not: substitution of $\psi_a + \psi_b$ into (38.11) leaves quadratic terms in C_a and C_b. Introducing an interaction product ψ_c so that the sum $\epsilon(\psi_a + \psi_b) + \epsilon^2 \psi_c$ (ϵ is the small nondimensional ordering parameter) will satisfy (38.11), we find upon substitution that ψ_c must obey

$$\partial_t \nabla_h^2 \psi_c + \beta \partial_x \psi_c = -\frac{C_a C_b}{2}\left[(k_a^2 + m_a^2) - (k_b^2 + m_b^2)\right](k_b m_a - k_a m_b)$$
$$\cdot \left[\cos(S_a + S_b) - \cos(S_a - S_b)\right], \tag{38.15}$$

an equation of the general form (38.7). Resonance can, in principle, occur for either of $-S_c = S_a + S_b$ or $-S_c = S_a - S_b$ [the negative sign is chosen with S_c to keep the general form (38.8) for the resonance conditions]. Suppose that resonance takes place

† We use k and m here for wavenumber components rather than the usual k_1 and k_2 to avoid double subscripts.

with the plus sign; then, each pair of ω and $\boldsymbol{k}$ satisfies the planetary wave dispersion relation (38.14) as well as

$$-\boldsymbol{k}_c = \boldsymbol{k}_a + \boldsymbol{k}_b, \tag{38.16a}$$

$$-\omega_c = \omega_a + \omega_b. \tag{38.16b}$$

We shall see later that there are nontrivial solutions under these conditions and that resonant triads of planetary waves can actually exist. The companion solution, for which

$$-\boldsymbol{k}_c = \boldsymbol{k}_a - \boldsymbol{k}_b, \tag{38.17a}$$

$$-\omega_c = \omega_a - \omega_b, \tag{38.17b}$$

might also be resonant if ω_c and $\boldsymbol{k}_c$ are related through (38.14), in which case $\Delta(\omega_c, \boldsymbol{k}_c) = 0$ in the notation of (38.5). When resonance does not occur, $\Delta(\omega', \boldsymbol{k}_c') \neq 0$, where primes denote a nonresonant wave, and the dispersion relation is not satisfied. For small ϵ, the resonant wave (of order $\epsilon^2 t$) should rapidly dominate the response and overwhelm any nonresonant forced wave (which remains of order ξ^2). Let us now continue with the resonant case. We may write the solution of (38.15) as

$$\psi_c = C_c(t) \cos S_c, \tag{38.18}$$

where the amplitude C_c obeys

$$(k_c^2 + m_c^2) \frac{\partial}{\partial t} C_c = -\frac{1}{2} [(k_a m_b - k_b m_a)(k_a^2 + m_a^2 - k_b^2 - m_b^2)] C_a C_b. \tag{38.19}$$

The coefficient of $C_a C_b$ is called the coupling coefficient, which we denote by D. Multiplying (38.19) by C_c, we find that

$$\frac{1}{2}(k_c^2 + m_c^2) \frac{\partial}{\partial t} C_c^2 = D C_a C_b C_c, \tag{38.20}$$

so that the direction of energy transfer from the primary waves, ψ_a and ψ_b, to the interaction product ψ_c depends on the sign of D as well as on that of $C_a C_b C_c$. Evidently, the strength of the coupling, as characterized by the magnitude of D, will depend on just what combinations of wavenumbers and frequencies are found at resonance.

Resonance occurs when (38.16) is satisfied, with each combination of ω and $\boldsymbol{k}$ being related through (38.14). One very simple solution is found when the tips of the three wavenumber vectors form an equilateral triangle with centre at the origin. The condition (38.16a) is then automatically satisfied and, since $k^2 + m^2$ is also the same for each wave, (38.16b) subject to (38.14), is also satisfied. A special case of this type of resonant triplet is shown in Fig. 38.1a. The choice $\boldsymbol{k}_a = (-1, \sqrt{3})\kappa$, $\boldsymbol{k}_b = (-1, \sqrt{3})\kappa$, $\boldsymbol{k}_c = (2, 0)\kappa$ and $\omega_a = \omega_b = \beta/4\kappa$, $\omega_c = -\beta/2\kappa$ satisfies (38.14) and the wavenumber vectors clearly form an equilateral triangle centred on the origin. The sum of the two primary waves (a and b) of unit amplitude is $2 \cos(\kappa x - \omega_a t) \cos(\sqrt{3}\kappa y)$; the interaction product (c) has the form $C_c \cos[2(\kappa x - \omega_a t)]$, and travels at the same speed ω_a/κ as the interference pattern due to the primary waves. The two patterns are phase-locked in resonance. Since the magnitude of the wavenumber is the same for all three waves, the coupling coefficient, D, vanishes (see 38.19): the three waves co-exist, but do not exchange energy.

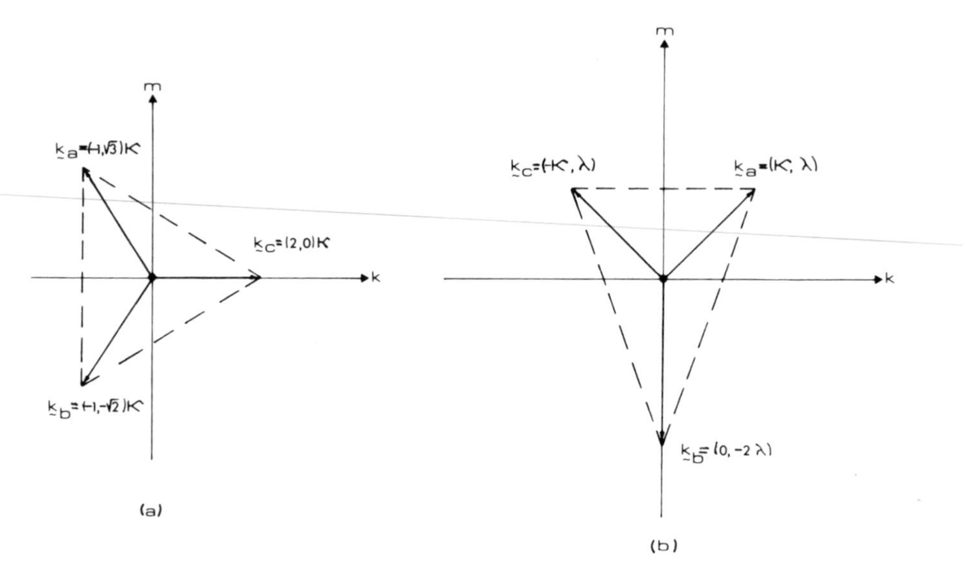

Fig. 38.1. Resonant planetary wave triplets; the resonant conditions (38.14) and (38.16) are readily verified; positive east–west wavenumbers give negative frequencies. The wave with wavenumber $\boldsymbol{k}_b$ in (b) has zero frequency and is a steady zonal current.

Another simple solution of the resonance conditions is shown in Fig. 38.1b and consists of a triplet of wavenumbers with vertices arranged in an isosceles triangle symmetric about the m-axis. The choice $\boldsymbol{k}_a = (\kappa, \lambda)$, $\omega_a = \omega = -\beta\kappa/(\kappa^2 + \lambda^2) < 0$; $\boldsymbol{k}_b = (0, -2\lambda)$, $\omega_b = 0$ describes the interaction of a Rossby wave with a steady zonal flow to produce a wave of wavenumber $\boldsymbol{k}_c = (-\kappa, \lambda)$ and frequency $\omega_c = -\omega$. In that case, the coupling coefficient D takes the value $D = \kappa\lambda(\kappa^2 - 3\lambda^2)$. Since $u = -\psi_y$, the zonal current is given by $u_b = 2\lambda C_b \sin(2\lambda y)$. With $\kappa, \lambda, C_a, C_c$ all positive, the energy transfer to the interaction product of the original wave and of an easterly ($C_b > 0$) current is positive if $\lambda < \kappa/\sqrt{3}$, i.e., whenever the angle between the wavenumber vector $\boldsymbol{k}_a$ and the direction of the current is less than 30°.

More generally, the nine quantities ($k_a, k_b, k_c; m_a \ldots; \omega_a \ldots$) are related through six relations [obtained from (38.14) and (38.16)]. Given two of these quantities, k_c and m_e, say, these six relations form a system of equations with one degree of freedom. The geometrical problem of determining all the pairs of wavenumbers $\boldsymbol{k}_a, \boldsymbol{k}_b$ which can satisfy the resonance conditions for a given $\boldsymbol{k}_c$ has been solved by Longuet-Higgins and Gill (1967). Starting from the dispersion relation between ω_c and $\boldsymbol{k}_c$, viz.,

$$\omega_c(k_c^2 + m_c^2) + k_c\beta = 0, \tag{38.21}$$

we use (38.16b) to express ω_c in terms of ω_a and ω_b, and then the dispersion relation (38.14) to express these frequencies in terms of the wavenumbers:

$$\left[\frac{k_b}{(k_b^2 + m_b^2)} + \frac{k_a}{(k_a^2 + m_a^2)}\right](k_c^2 + m_c^2) + k_c = 0. \tag{38.22}$$

Rewriting (38.16a) as

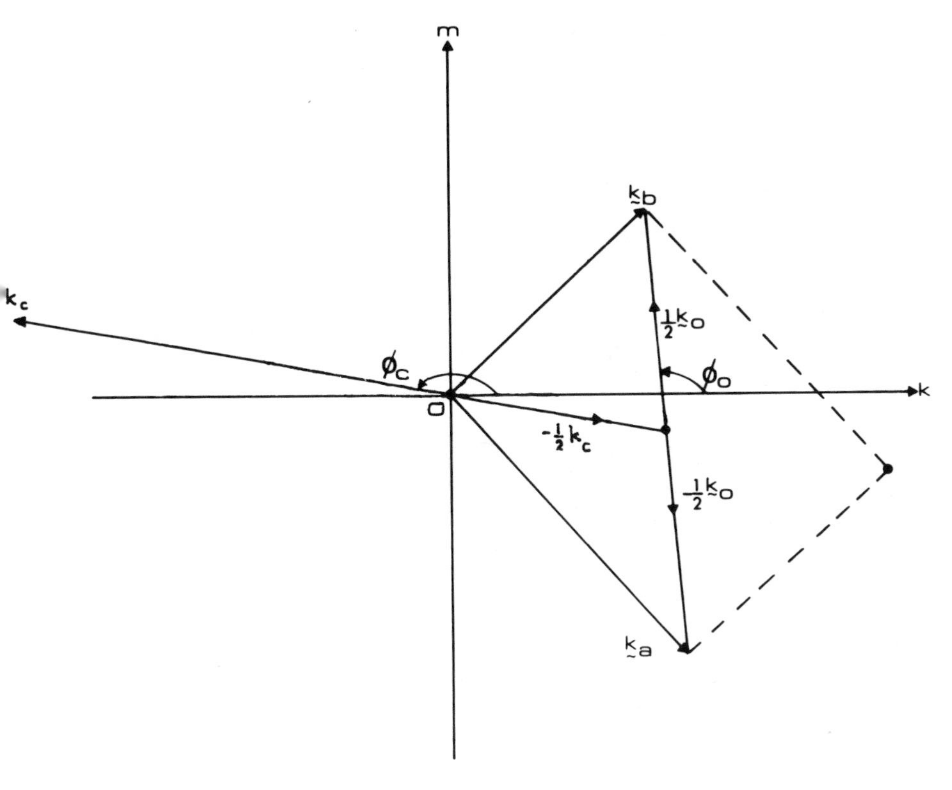

Fig. 38.2. The geometrical relationship between a triplet ($\boldsymbol{k}_a$, $\boldsymbol{k}_b$, $\boldsymbol{k}_c$) of resonant planetary waves.

$$\boldsymbol{k}_a + \boldsymbol{k}_b = -\boldsymbol{k}_c + \boldsymbol{0},$$

where $\boldsymbol{0}$ is the origin, we note that the ends of the vectors $\boldsymbol{k}_a, \boldsymbol{k}_b, -\boldsymbol{k}_c$ and $\boldsymbol{0}$ form a parallelogram (Fig. 38.2) whose diagonals intersect at the tip of the vector $-\frac{1}{2}\boldsymbol{k}_c$. Upon defining $\boldsymbol{k}_0 = \boldsymbol{k}_b - \boldsymbol{k}_a$, we may write

$$\boldsymbol{k}_a = -\tfrac{1}{2}\boldsymbol{k}_c - \tfrac{1}{2}\boldsymbol{k}_0, \tag{38.23a}$$

$$\boldsymbol{k}_b = -\tfrac{1}{2}\boldsymbol{k}_c + \tfrac{1}{2}\boldsymbol{k}_0. \tag{38.23b}$$

Substituting into (38.22), we find an equation for k_0 and m_0 in terms of k_c and m_c:

$$k_c[3(k_c^2 + m_c^2)^2 + 2(k_c^2 + m_c^2)(k_0^2 + m_0^2) - (k_0^2 + m_0^2)^2 + 4(k_c k_0 + m_c m_0)^2]$$
$$- 8k_0(k_c^2 + m_c^2)(k_c k_0 + m_c m_0) = 0. \tag{38.24}$$

Introducing polar coordinates

$$(k, m) = \kappa\,(\cos\phi, \sin\phi), \tag{38.25}$$

(38.24) becomes

$$[3\kappa_c^4 + 2\kappa_c^2\kappa_0^2 - \kappa_0^4 + 4\kappa_c^2\kappa_0^2\cos^2(\phi_0 - \phi_c)]\cos\phi_c - 8\kappa_c^2\kappa_0^2\cos(\phi_0 - \phi_c)\cos\phi_0 = 0. \tag{38.26}$$

Defining the angle ϕ' through

$$\phi' = \phi_0 - \phi_c, \tag{38.27}$$

we obtain

$$[(3\kappa_c^4 - \kappa_0^4) - 2\kappa_c^2\kappa_0^2 \cos 2\phi'] \cos \phi_c + 4\kappa_c^2\kappa_0^2 \sin 2\phi' \sin \phi_c = 0. \tag{38.28}$$

When $\cos \phi_c = 0$ [i.e., $\boldsymbol{k}_c = (0, m_c)$], the possible solutions are:

(1) $\kappa_0 = 0$, in which case $\boldsymbol{k}_0 = \boldsymbol{0}; \boldsymbol{k}_a = \boldsymbol{k}_b = -\frac{1}{2}\boldsymbol{k}_c = -\frac{1}{2}(0, m_c)$;

(2) $\phi' = \pm \pi/2$, for which $\boldsymbol{k}_0 = (k_0, 0)$. This is the case illustrated in Fig. 38.1b in which a and c are taken as the primary waves and b the product of the interaction;

(3) $\phi' = 0$, for which $\boldsymbol{k}_0 = (0, m_0)$.

None of these three cases are of great interest, since the coupling coefficient D vanishes for each one of them.

When $\cos \phi_c \neq 0$, we may rewrite (38.28) as

$$(\kappa_0/\kappa_c)^4 + 2(\kappa_0/\kappa_c)^2 \sec 2\alpha \cos [2(\phi' - \alpha)] - 3 = 0, \tag{38.29}$$

where $\tan 2\alpha = -2 \tan \phi_c$. This equation has only one real root for $(\kappa_0/\kappa_c)^2$:

$$(\kappa_0/\kappa_c)^2 = -B + (B^2 + 3)^{1/2}, \tag{38.30}$$

where $B = \sec 2\alpha \cos [2(\phi' - \alpha)]$. Given k_c, m_c, and a value of ϕ', (38.30) uniquely determines κ_0; ϕ_0 is found from (38.27). The vector $\boldsymbol{k}_0$ is then given by (38.25) and $\boldsymbol{k}_a$, $\boldsymbol{k}_b$ by (38.23). The locus on which $\boldsymbol{k}_a$ and $\boldsymbol{k}_b$ lie is shown in Fig. 38.3 for a selection of values of ϕ_c in the range $-\pi/2 \leqslant \phi_c \leqslant 0$. When ϕ_c is in the third quadrant the locus can be obtained by reflection about the m-axis; for $0 \leqslant \phi_c \leqslant \pi$, the figures are mirror images about the k-axis of their counterparts for negative ϕ_c. The quartic locus has axes of symmetry passing through the point $-\frac{1}{2}\boldsymbol{k}_c$ and making angles $\alpha, \alpha + \pi/2$ with the vector $\boldsymbol{k}_c$. The wavenumbers $\boldsymbol{k}_a$ and $\boldsymbol{k}_b$ lie at diametrically opposite points on the curve; two examples of possible pairs, beside the special PO, QR pairs discussed below, are shown in Fig. 38.3c. When $\phi' = 0$ or π, $\sin 2\phi' = 0$ and (38.28) factors to

$$(\kappa_0^2 - \kappa_c^2)(\kappa_0^2 + \kappa_c^2) = 0,$$

so that $\kappa_0 = \kappa_c$. From (38.23) and (38.27), $\boldsymbol{k}_b = \boldsymbol{0}$, $\boldsymbol{k}_a = -\boldsymbol{k}_c$ and the origin (O) and the tip of $-\boldsymbol{k}_c$ (P) both lie on the curve. Similarly, at $\phi' = \pm \pi/2$, (38.29) factors to

$$(\kappa_0^2 - 3\kappa_c^2)(\kappa_0^2 + \kappa_c^2) = 0,$$

and thus, $\kappa_0 = \sqrt{3}\kappa_c$; these points are labelled Q and R. It is readily verified from (38.23) that the points P, Q and R are intersections of the circle of radius κ_c with the quartic (38.29); the fourth intersection is the point S, obtained by reflection of $\boldsymbol{k}_c$ about the m-axis. These points are useful reference locations on the resonance locus since they determine positions where the coupling coefficient D changes sign. Using (38.23), D may be written

$$D = \tfrac{1}{4}\kappa_0\kappa_c(\kappa_a^2 - \kappa_b^2) \sin \phi', \tag{38.31}$$

and its sign may be followed around the resonance locus (Exercise 38.3).

This rather lengthy exposé of the geometry of the resonant interactions between planetary wave triads should be sufficient to familiarize the reader with the essential

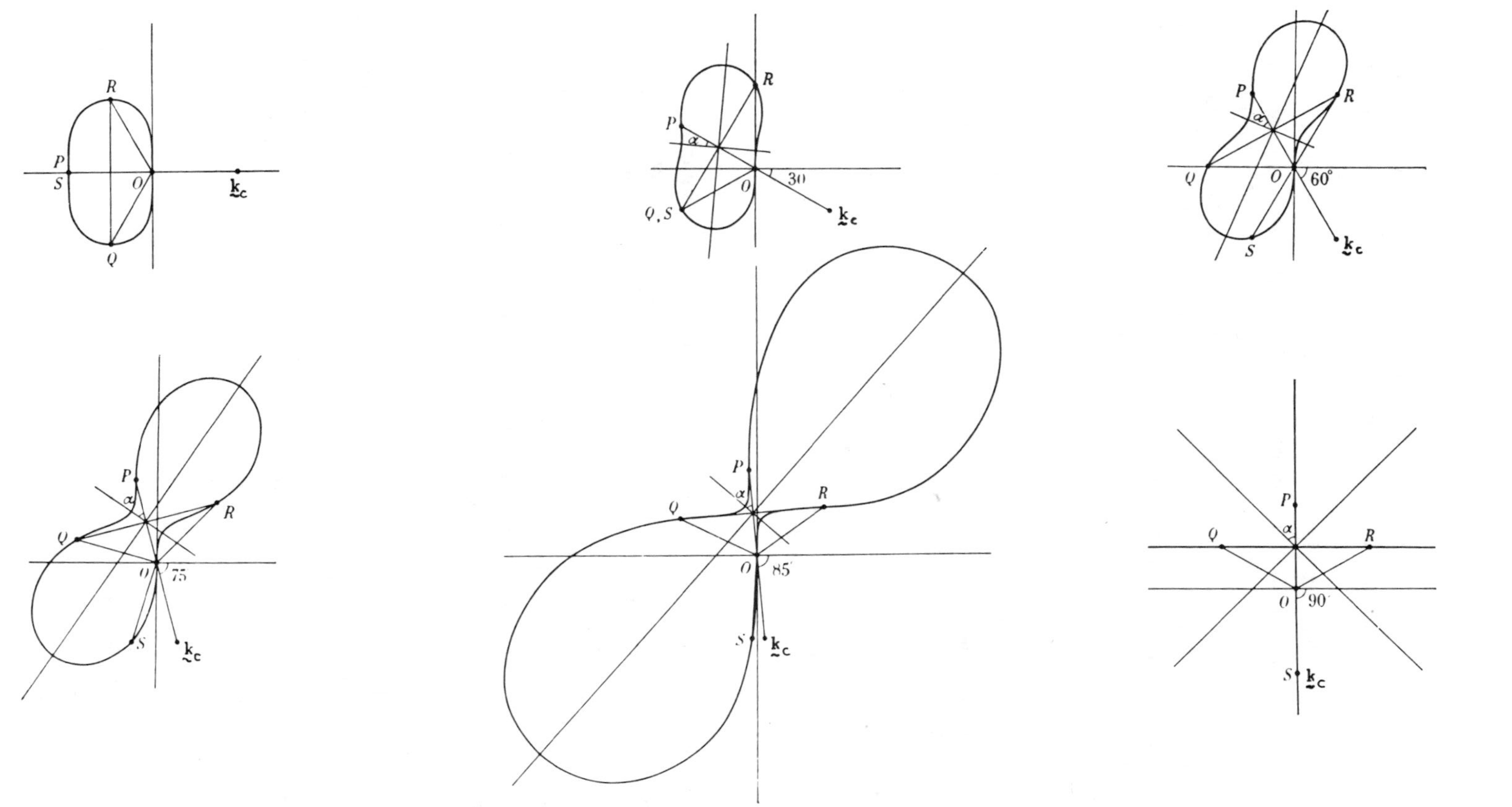

Fig. 38.3. The locus of pairs of wavenumbers $\boldsymbol{k}_a$, $\boldsymbol{k}_b$ which interact with a given wavenumber $\boldsymbol{k}_c$ (shown for various orientations in the fourth quadrant) for nondivergent planetary waves. The angle α is defined in the text; the different figures correspond to different values of the angle ϕ_c (negative in this example). The points P, Q, R, S are also defined in the text. The pairs $\boldsymbol{k}_a$, $\boldsymbol{k}_b$ lie at opposite ends of the diagonals of the quartic curve: pairs situated at the points P, O, and at R, Q, satisfy the conditions for resonance. (From Longuet-Higgins and Gill, 1967.)

points of the study of weak wave–wave interactions. This type of detailed analysis will not be repeated for other kinds of waves. It should also be obvious to the reader by now that resonant interactions are not limited to multiplets of travelling waves: one simple example (Fig. 38.1b) involved two waves and a steady, but spatially modulated, current. Thus, for triplet resonances satisfying (38.16) and appropriate dispersion relations, one of the wavenumber or frequency components may vanish, the latter case representing purely spatial variations of the medium. Wave interactions with variable topography or currents and with time-varying media can be incorporated within the same theoretical framework and, in particular, the problems of wave propagation in random media discussed in Section 34 may also be approached from the point of view of weak wave–wave interactions. Interactions between planetary waves have been shown to be capable of generating steady zonal flows (Newell, 1969). Weak interactions with a turbulent flow are also tractable, simply by considering interactions with single Fourier components of the turbulence.

Finally, it is relatively simple to show (Exercise 38.2) that the total energy and the total vorticity squared (also called enstrophy: cf. Section 19) of a resonant planetary wave triad remain constant. Energy (and wave-action) and enstrophy are interchanged between the components of the triplet, but not lost to (or gained from) other nonresonant waves or the ambient fluid: the triplet forms a closed system. The resonance conditions are thus equivalent to conservation laws (for planetary waves, energy and enstrophy conservation) which must be satisfied by all free waves.

Forced waves

We discarded the forced wave response of (38.15) under the assumption that the resonance interaction would quickly grow to an amplitude of the order of that of the primary waves and would thus completely swamp the forced waves, which remain at all times of order ϵ^2. This argument is indeed valid for small ϵ, but we now recall that the primary planetary waves (38.12) are *exact* solutions of (38.11), so that ϵ is not limited in value. If ϵ is 0(1) or greater, the forced wave will also be of 0(1) or greater and may surpass the resonant wave in amplitude. The solution to (38.15) in the non-resonant case ($-\boldsymbol{k}_c' = \boldsymbol{k}_a - \boldsymbol{k}_b; -\omega_c' = \omega_a - \omega_b$) is

$$\psi_c' = C_c' \sin \theta_c', \tag{38.32}$$

$$\text{with} \quad C_e' = \frac{DC_aC_b}{\kappa_c'^2\omega_c' + \beta k_c'}. \tag{38.33}$$

If we take for a representative amplitude of the resonant wave that of a primary wave (assuming C_a and C_b nearly equal), then the ratio of the forced wave amplitude to that of the resonant wave is

$$\frac{C_c'}{C_b} = \frac{DC_a}{\kappa_c'^2\omega_c' + \beta k_c'}$$

$$= \frac{(k_a m_b - k_b m_a)(\kappa_a^2 - \kappa_b^2)C_a}{2\beta\left[\kappa_c'^2\left(\dfrac{k_a}{\kappa_a^2} - \dfrac{k_b}{\kappa_b^2}\right) - (k_a - k_b)\right]}. \tag{38.34}$$

Most simply, in terms of a characteristic magnitude of the primary wavenumber κ, say,

$$C_c'/C_b \sim 0(\kappa^3/\beta)C_a. \tag{38.35}$$

Since a typical wave frequency is $|\omega| \sim \beta/\kappa$, the forced wave amplitude as given by (38.35) is much less than that of the resonant wave only if

$$\kappa C_a \ll \omega/\kappa. \tag{38.36}$$

This inequality is satisfied if the particle velocity of the primary wave (proportional to κC_a) is much smaller than the phase velocity ω/κ. This is a common criterion for linearity in wave motion as noted earlier. We may then identify the unspecified small parameter ϵ with the ratio $\kappa^2 C_a/\omega$ as a characteristic value of nondimensionalized wave amplitude. Weak wave–wave interactions are thus restricted to $\epsilon \ll 1$; the predominantly forced wave interactions which dominate for $\epsilon \sim 0(1)$ are more properly viewed as a type of turbulent energy cascade, vigorously broadening the narrow spectrum of primary waves through a succession of energy interchanges between the primary waves and a hierarchy of their interaction products. These nonlinear interactions are probably of great relevance to ocean dynamics (typical values of ϵ in the ocean easily reach unity). They have been discussed by Pedlosky (1962) and more recently by Rhines (1975, 1977). Their importance has already been discussed in Section 19.

Feynman diagrams

A useful symbolic representation of wave–wave interaction processes may be obtained by exploiting the analogy between plane waves and particles, and adapting the formalism of quantum mechanical Feynman diagrams (Schweber, 1961) to resonant wave–wave interactions (cf. Hasselmann, 1966, 1967a). Each interacting wave (as well as non-propagating Fourier components of the inhomogeneity of the medium of propagation) is represented by an arrow of a length equal to the magnitude of its wavenumber and pointing in the direction of propagation of the wave. Wave components with negative frequencies are labelled "anti-waves" (just like anti-particles) and denoted by cross-stroked arrows. The interaction occurs at the vertex where the arrows meet and out of which issues the interaction product, also represented by a vector according to the same conventions. Free waves, satisfying the dispersion relation as well as the resonance conditions (and thus a set of conservation laws, to stress the quantum mechanical analogy), are drawn as full lines; forced waves, satisfying the resonance equations, but not the dispersion relation, are drawn as broken lines and identified with virtual particles. Examples are shown in Fig. 38.4. Note that in high-order interactions it is possible, as shown in Fig. 38.4d, for a virtual component to interact with a free wave to yield a free component. The use of Feynman diagrams is not restricted to interactions between progressive waves. Any Fourier component of the inhomogeneity of the medium of propagation may be represented as a wave of zero frequency and incorporated within the same symbolism.

Feynman diagrams give a useful shorthand picture of the waves entering the interaction; unless supplemented with labels giving the strength of the coupling coefficients, they give no indication of the efficiency of the interaction process. These coefficients may be calculated by the perturbation analysis applied above to planetary waves; they may also be found by variational methods applied to the averaged wave Lagrangian density (cf. Section 6), as shown by Simmons (1969).

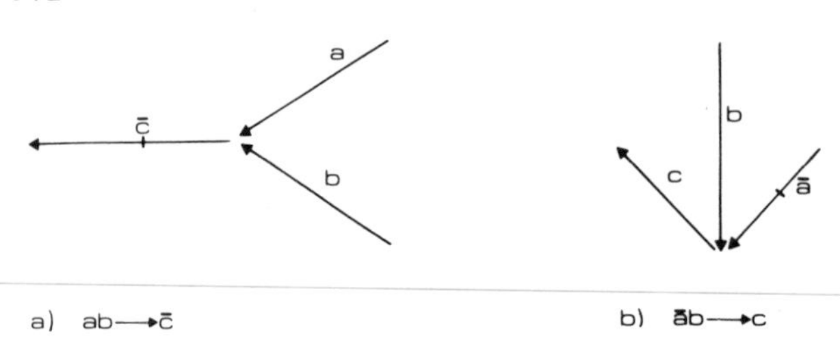

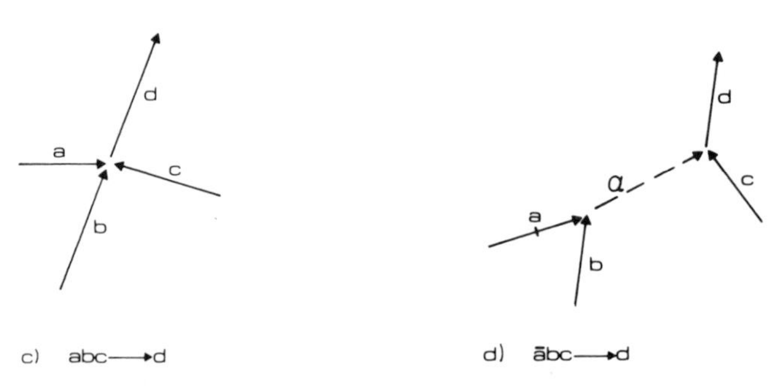

Fig. 38.4. Feynman diagrams for second-order (triplet) and third-order (quartet) interactions. The triplets (a), (b) represent the interactions shown in Fig. 38.1; the quartets are fictional.

Surface gravity wave interactions

The relative algebraic simplicity of the planetary wave interaction problem outlined above is rather atypical of the wave interaction problems encountered in oceanography. For surface gravity waves in deep water, the nonlinearity stems from the surface boundary conditions. In the absence of surface tension ($B = 0$), the dispersion relation (11.16) reduces, for $kH \gg 1$, to

$$\omega = \sqrt{gk}. \tag{38.37}$$

Resonance between a triplet of waves is not possible (Philips, 1960; see also Exercise 38.6). Three waves must interact to produce a resonant fourth wave; each wave must satisfy (38.37), and the conditions

$$\boldsymbol{k}_a + \boldsymbol{k}_b + \boldsymbol{k}_c + \boldsymbol{k}_d = \boldsymbol{0}, \tag{38.38a}$$

$$\omega_a + \omega_b + \omega_c + \omega_d = 0 \tag{38.38b}$$

must hold at resonance. The interaction between surface gravity waves is thus much weaker than between planetary waves, since the interaction product is now of third order in a small parameter characterizing the wave amplitude. The conditions (38.38) hold provided two of both the wavenumbers and frequencies are negative. Thus (38.38) may be rewritten as

$$\boldsymbol{k}_a + \boldsymbol{k}_b = \boldsymbol{k}_c + \boldsymbol{k}_d, \tag{38.39}$$

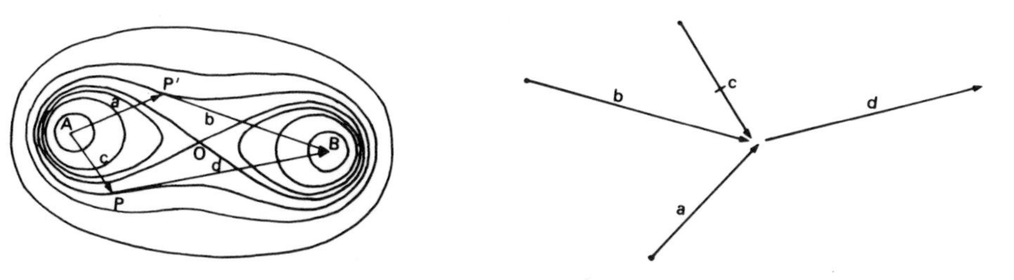

Fig. 38.5. The family of resonance loci for four interacting short surface gravity waves. If P and P' are any pair of points on one of the family of curves, the vectors AP', AP, $P'B$ and PB (which obviously satisfy $\boldsymbol{k}_a + \boldsymbol{k}_b - \boldsymbol{k}_c = \boldsymbol{k}_d$) form a resonant quartet. The Feynman diagram of the interaction is shown on the right.

where all wavenumbers are positive. The curve on which the wavenumbers must lie to obey the resonance conditions is found by a geometrical analysis similar to that given above for planetary waves. The resonance locus is shown in Fig. 38.5, together with a Feynman diagram representation. The coupling coefficients and the energy interchanges between resonant quartets of surface gravity waves have been calculated by Benney (1962) and Bretherton (1964a). It was found in particular that the total energy and momentum of the set of four waves are conserved: the resonance conditions may again be viewed as conditions for the satisfaction of these conservation rules.

A simpler class of interactions is found when two of the wavenumbers in (38.39) are indetical; thus $\boldsymbol{k}_c = \boldsymbol{k}_d$ and $\omega_c = \omega_d$ so that the resonance conditions become

$$\boldsymbol{k}_a + \boldsymbol{k}_b = 2\boldsymbol{k}_c ; \quad \omega_a + \omega_b = 2\omega_c . \tag{38.40}$$

With reference to Fig. 38.5, the vector $2\boldsymbol{k}_c$ goes from the point A through O to B. The other wavenumbers are defined by the point of contact P' on any one of the family of curves in Fig. 38.5. Experimental confirmation of the resonance process has been obtained by McGoldrick et al. (1966) for the special case where $\boldsymbol{k}_b$ and $\boldsymbol{k}_c$ are mutually perpendicular. The primary wave trains, here $\boldsymbol{k}_b$ and $\boldsymbol{k}_c$, were generated mechanically on contiguous walls of a square wave tank. At resonance, the interaction product $\boldsymbol{k}_a$ was observed to grow linearly with distance across the tank. The appearance of the resonant frequency $\omega_a = 2\omega_c - \omega_b$ is evident in the periodogram shown in Fig. 38.6.

The ideas of wave–wave interaction are also valuable in understanding the instability which Benjamin and Feir (1967) discovered in surface gravity waves, and already mentioned in Section 12. It was found by these authors that nearly monochromatic wave pulses generated at one end of a long channel gradually disintegrated as they propagated along the channel; an example of this phenomenon is shown in Fig. 38.7. The nonlinear interaction which leads to the degradation of a monochromatic group may be represented in terms of a resonant interaction between the second harmonic of the fundamental wave and two side bands. The three waves entering the interaction are:

(1) the upper side band:

$$\eta_a = C_a \operatorname{Re}\left[\exp\{i[k(1+\mu)x - \omega(1+\delta)t - \alpha_a]\}\right];$$

(2) the lower side band:

$$\eta_b = C_b \operatorname{Re}\left[\exp\{i[k(1-\mu)x - \omega(1-\delta)t - \alpha_b]\}\right],$$

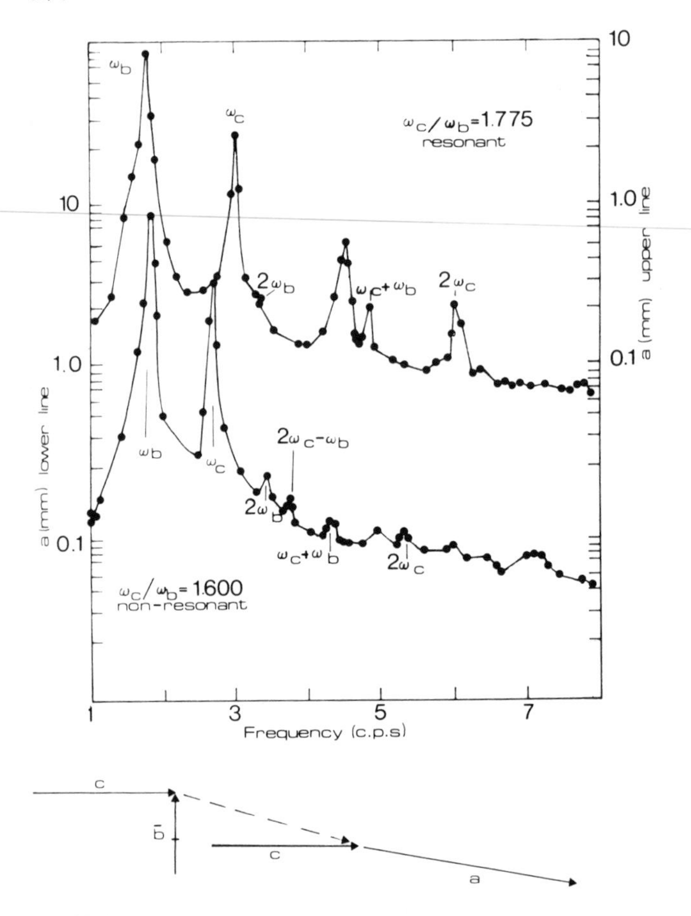

Fig. 38.6. Resonant wave interactions for the process $\boldsymbol{k}_a = 2\boldsymbol{k}_c - \boldsymbol{k}_b$, when $\boldsymbol{k}_b$ and $\boldsymbol{k}_c$ are perpendicular. The primary waves have frequencies ω_b, ω_c. The upper and lower curves are periodograms of surface amplitude under resonant and nonresonant conditions, respectively. Note the forced wave at $2\omega_c$ (dotted line in Feynman diagram) and the resonant wave at $\omega_a = 2\omega_c - \omega_b$. The Feynman diagram is also shown. (From McGoldrick et al., 1966.)

(3) the fundamental wave:

$$\eta_c \;=\; C_c \text{ Re } [\exp \{i(kx - \omega t)\}].$$

The second harmonic of η_c has phase $2(kx - \omega t)$. Both μ and δ are much less than unity; whenever the phases α_a, α_b satisfy $\alpha_a + \alpha_b =$ constant, a resonant interaction is possible between the second harmonic of η_c and the two side bands. The resonance is a special case of (38.40) in which $\boldsymbol{k}_a = (1 + \mu)\boldsymbol{k}$ and $\boldsymbol{k}_b = (1 - \mu)\boldsymbol{k}$ are collinear with the fundamental wavenumber $\boldsymbol{k}_c$ and lie along AO in Fig. 38.5, Actually, for $0 < \mu \leqslant 1$, the inter action is only approximately resonant since the tips of $\boldsymbol{k}_a$ and $\boldsymbol{k}_b$ do not lie exactly on the resonance curve at the point O. In this approximate resonance [the wavenumber relation in (38.40) being satisfied exactly, but the frequency relation only approximately], the rate of growth of each side band is proportional to the amplitude of the other and each grows exponentially at the expense of the primary. The Feynman diagram

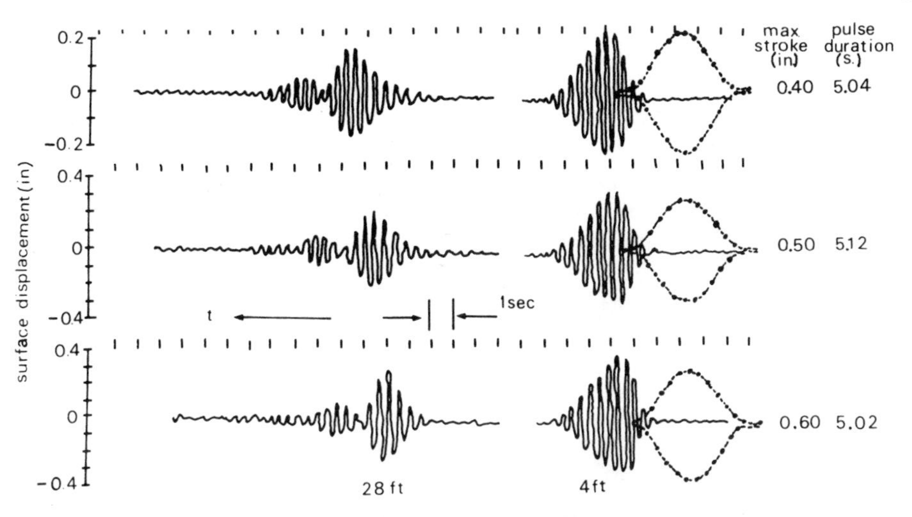

Fig. 38.7. The disintegration of wave pulses in deep water. The envelope of the wave pulse at the wavemaker is shown in dotted lines on the right. The oscillations of the surface after propagating 4 ft and 28 ft from the wavemaker are shown on the left, as a function of time, increasing towards the left. Three cases are shown, with nearly equal pulse durations but increasing wave amplitudes (note the different amplitude scales). (From Feir, 1967.)

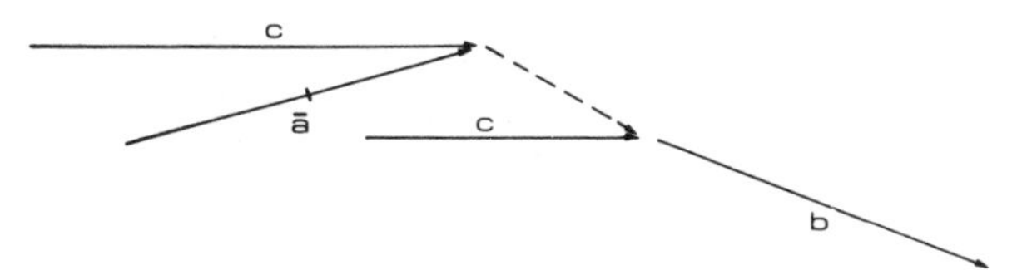

Fig. 38.8. Feynman diagram for the approximate resonance between two side bands and the second harmonic of a primary wave, leading to a disintegration of the primary in favour of the side bands.

for side bands which propagate at a small angle to the primary is shown in Fig. 38.8. The Benjamin-Feir instability was shown by Hasselmann (1967b) to be a special case of an instability which occurs through nonlinear coupling of two wave components of infinitesimal amplitude (a) and (b) with a finite-amplitude wave (o) under resonance conditions. Hasselmann showed that for a conservative triplet of waves satisfying a certain dispersion relation, and the resonance conditions

$$\boldsymbol{k}_a \pm \boldsymbol{k}_b = \boldsymbol{k}_0; \quad \omega_a \pm \omega_b = \omega_0,$$

the interaction is unstable (i.e., the small waves grow at the expense of the large one) for the sum interaction (the plus sign) and neutrally stable for the difference interaction. Although the Benjamin-Feir instability occurs at the third order and involves four waves, the finite-amplitude wave occurs twice and Hasselmann's result applies with $\boldsymbol{k}_0$ and ω_0 replaced by $2\boldsymbol{k}_0$ and $2\omega_0$.

The existence of the instability found by Benjamin and Feir came as a surprise to a generation of fluid dynamicists who had been raised with the convergence proofs of Levi-Civita (1925) and Struik (1926) for the Stokes expansion of small-amplitude waves

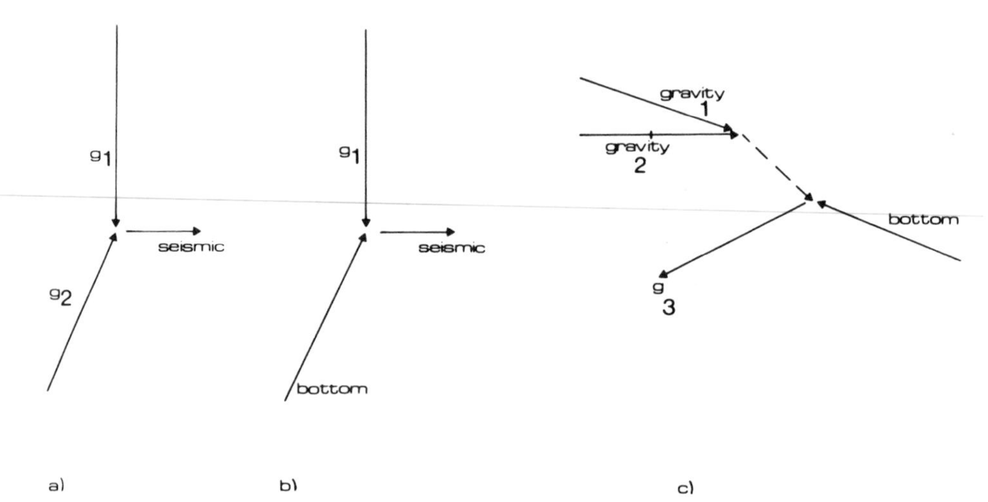

Fig. 38.9. Feynman diagrams for wave interactions near the shore: (a) and (b), two mechanisms for the generation of microseisms; (c), "surf-beat" resonance.

in terms of a wave-slope parameter. The discovery of this instability shows in a rather striking manner that a proof of convergence of a series solution does not prove the stability of that solution.

Surface wave interactions are also of interest in the generation of microseisms (high-frequency, small-amplitude seismic waves observed mainly in coastal areas). In that case, the result of the interaction of two surface gravity waves is an elastic wave. The Feynman diagram of the interaction is shown in Fig. 38.9a. This generation mechanism for microseisms was proposed by Longuet-Higgins (1950), who showed that the pressure field associated with standing surface waves had a component which does not decay with depth and could couple with elastic deformations of the bottom (Exercise 38.4). Another generation mechanism attributes the origin of microseisms to interactions of gravity waves with the sea bottom (Fig. 38.9b). The generation theories have been reviewed by Hasselmann (1963b). Comparisons of microseism and swell spectra may be found in Haubrich et al. (1963).

Other relevant interaction processes in the coastal zone involve edge waves (Kenyon, 1970) and their coupling with the surf and bottom and shoreline features (Guza and Davis, 1974; Guza and Inman, 1975) which lead to the formation of rip currents (cf. Section 37). The long period (0.5–5 min) observed near the shore and described as "surf-beats" (Munk, 1949b) were described theoretically by Gallagher (1971) in terms of nonlinear interaction between incoming waves. Recent observations (Huntley, 1976) shows that "surf-beats" have all the characteristics of longshore propagating edge waves.

For short enough surface waves, the influence of capillarity becomes important, and in deep water ($kH \gg 1$) the dispersion relation (11.16) has the form

$$\omega^2 = gk + \sigma k^3/\rho_*. \tag{38.41}$$

There then exist two real wavenumber magnitudes for a given ω. Resonance between triplets becomes possible, as was first examined by McGoldrick (1965), and later by Simmons (1969), who also studied third-order interactions (between wave quartets).

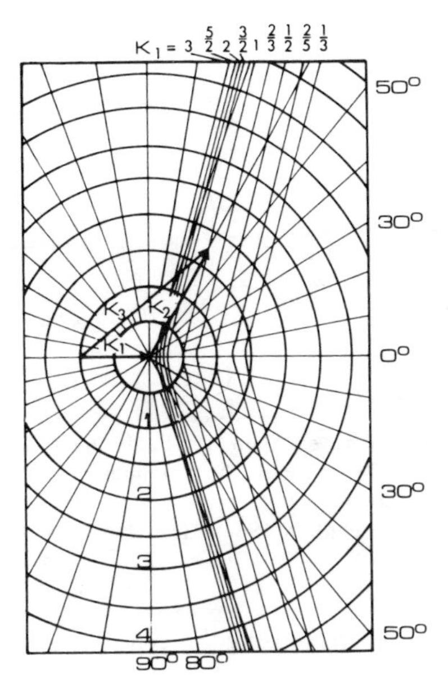

Fig. 38.10. Solution curves of the resonance conditions for capillary–gravity waves. Choosing a wavenumber K_1, terminating at the origin, and a wavenumber K_3 starting at the same point as K_1 and ending anywhere on the curve labelled with the appropriate magnitude of K_1, the third wavenumber K_2 completes the triangle. (From McGoldrick, 1965.)

The solution of the resonance conditions, as given by McGoldrick (1965), is illustrated in Fig. 38.10. The existence of second-order interactions between capillary–gravity waves was also demonstrated experimentally by McGoldrick (1970).

Internal gravity wave interactions

We have shown that plane internal gravity waves in an unbounded nonrotating Boussinesq fluid with uniform Brunt-Väisälä frequency satisfy (8.29):

$$\omega = N\cos\theta, \tag{38.42}$$

where θ is the angle between the wavenumber vector and the horizontal (see Fig. 8.3b). Resonance is possible at the second order; the resonance conditions may be written

$$\boldsymbol{k}_a \pm \boldsymbol{k}_b = \boldsymbol{k}_c \tag{38.43a}$$

$$\cos\theta_a \pm \cos\theta_b = \cos\theta_c. \tag{38.43b}$$

Some resonant interaction diagrams are shown in Fig. 38.11. As the plane waves with dispersion relation (38.42) are transverse, they are exact solutions to the nonlinear momentum, mass and density conservation equations. The amplitude of the primary interacting waves is not restricted to be small and, as discussed earlier in connection with planetary waves, the forced wave products of the interaction may, at high energy levels, be more important than the resonant terms. Thus, a turbulent energy cascade, strongly affected by buoyancy effects, will arise through interaction of large amplitude

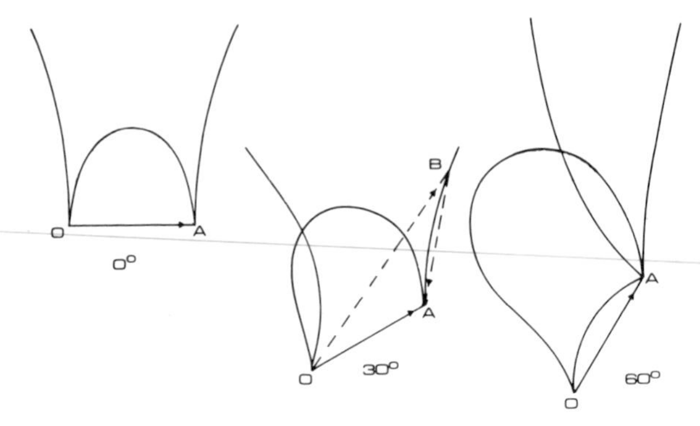

Fig. 38.11. Resonant interaction diagrams for short internal gravity waves, when $\theta_c = 0$, 30° and 60°. Any point B on a branch of the curve specifies a resonant triad OB, BA, OA; the wavenumber with the least slope is always on the vector sum of the other two. Only wavenumbers in the vertical plane are shown. The wavenumber scale is arbitrary. (Reproduced from Phillips, 1966, "The Dynamics of the Upper Ocean", by permission of the Cambridge University Press.)

internal waves. The "buoyancy subrange" of turbulence characterizing this type of interaction has been explored by Lumley (1964).

A plane wave representation of internal gravity waves will be useful in the oceans only for very short waves. Longer waves will be affected by the presence of the surface and bottom boundaries and may be analysed in terms of vertical modes propagating horizontally (see Section 10). Resonant triplet interactions between internal wave modes are possible for any stable stratification (Thorpe, 1966) provided the three waves do not all belong to the same mode. The simplest case is that of uniform Brunt-Väisälä frequency N, explored in theory and experiment by Martin et al. (1972). In that case, the dispersion relation is obtained from (10.45) and (10.72b) as

$$\omega^2/N^2 = k_h^2/(k_h^2 + n^2\pi^2/H^2), \tag{38.44}$$

with k_h the magnitude of the horizontal wavenumber, n the vertical mode number and H the depth of the fluid. All waves in a resonant triplet satisfy (38.44), the conditions (38.8) for ω and the horizontal wavenumber k_h, as well as the condition

$$n_a + n_b + n_c = 0 \tag{38.45}$$

for the vertical mode numbers. Kenyon (1968) has discussed internal wave mode interactions in a non-Boussinesq fluid. Internal wave interactions have also been examined experimentally by McEwan (1971) and McEwan et al. (1972).

Resonant interactions between a pair of surface gravity waves and an internal mode are also possible for any stable stratification (see again Thorpe, 1966). Since the surface waves always travel much faster than the internal mode, their wavenumbers must be nearly equal and opposite and also much larger than that of the internal wave. The simplest case, that occurring in a two-layer fluid, was examined by Ball (1964), and experiments have been performed by Lewis et al. (1974). The resonant interaction between internal and surface gravity waves in a uniformly stratified fluid have been examined by Brekhovskikh et al. (1972). Interactions between standing waves have been studied by Joyce (1974), both theoretically and experimentally.

Only the general nature and the simplified mechanics of weak resonant wave interactions between isolated multiplets have been discussed here. In nature, of course, waves present themselves in continuous spectra, from which it is impossible to extract isolated triplets of quartets of mutually interacting waves. The interactions take place in a continuous fashion and resonant and forced energy transfers act continuously over wavenumber and frequency domains. This problem of energy transfer by weak interactions will be discussed in the next section.

Parametric oscillations

When a mechanical system is forced by externally imposed variations of one of its parameters rather than by the excitation of one or more of its generalized coordinates, its response takes the form of parametric oscillations. Parametric excitation may be achieved in an electrical circuit by varying the impedance of one of the components; in a fluid system, by varying the depth of the fluid; in an elastic system, by modulating an elastic modulus. A simple example from classical mechanics is the parametric oscillation of a pendulum by alternating changes in its length. The reader is referred to Bogoliubov and Mitropolsky (1961) for examples and an account of the mathematical treatment of parametric oscillations.

In their most elementary form, parametric oscillations of a physical system with natural frequency of oscillation ω_0 are described by Mathieu's equation:

$$\phi_{tt} + \omega_0^2(1 + \epsilon \cos \omega t)\phi = 0. \qquad (38.46)$$

One of the geometrical or constitutive parameters entering the natural frequency ω_0 [a parameter λ, in the notation of (6.8)] is modulated at a frequency ω. The behaviour of the solutions of Mathieu's equation is well known (McLachlan, 1947). For small values of ϵ, unstable (or resonant) oscillations of a frequency $\sigma \simeq M\omega/2$ arise whenever $\omega \simeq 2\omega_0/M$, where M is a positive integer. For $M = 1$, the response is subharmonic, at a frequency equal to half the excitation frequency. Within the framework of wave interaction theory, parametric oscillations are a special case with a resonance condition of the form $\omega - 2\omega = -\omega$.

A well studied example of parametric resonance is that of "cross-waves". A homogeneous fluid of uniform depth lying in a channel of rectangular cross-section is driven by an oscillating paddle at one end of the channel. The standing waves which appear with crests at right angles to the wavemaker are called cross-waves. Their theory has been examined by many authors (Garrett, 1970b, Mahony, 1972) and recent experimental results have been presented by Barnard and Pritchard (1972). The interaction of two wave fields of very different scales may also be represented as a problem of parametric excitation, the large-scale wave field modulating in time and space the properties of the medium on which the short waves propagate. This problem has been studied for internal gravity wave interactions by McEwan and Robinson (1975).

A spatially periodic variation in a parameter also leads to Mathieu's equation. This problem has been studied extensively in solid-state physics (Brillouin, 1966) and applied to wave propagation in an ocean of varying depth by Rhines (1970b). Over a sinusoidal bottom, gaps appear in the dispersion relation and no energy can propagate at frequencies contained in these stopping bands.

Exercises Section 38

1. After the resonant interaction product of a pair of primary planetary waves has grown to be of a magnitude comparable to that of the primaries, all three waves of the resonant triplet are of the same order, and one may seek solutions of (38.11) of the form

$$\psi = C_a(t)\cos S_a + C_b(t)\cos S_b + C_c(t)\cos S_c.$$

Show that under the resonance conditions (38.16), the amplitudes C_a, C_b, C_c are governed by three equations of the form (38.19). It will be noticed that these three equations could be written directly from (38.19) by permutation of the indices *a, b, c*.

2. Show, using the results of the previous problem, that the total energy and the total vorticity squared (the enstrophy) of a planetary wave triplet are conserved.

3. Using (38.31), and the definition of ϕ' in (38.27) and $\boldsymbol{k}_0$ in (38.24), determine the range of values of ϕ' (and thus the portions of the resonance curve) for which a pair of planetary waves $\boldsymbol{k}_a$, $\boldsymbol{k}_b$ will interact resonantly with and feed energy into a third wave $\boldsymbol{k}_c$ for which $\phi_c = -60°$.

4. Show that the pressure field of standing surface gravity waves in deep water has a component which does not decay with depth. This pressure contribution on the bottom forms the basis of Longuet-Higgins' (1950) theory of microseism generation by ocean waves.

5. For pure capillary waves [high Bond number, see (11.16)], the dispersion relation (38.41) reduces to $\omega^2 = \sigma k^3/\rho_*$. Obtain an equation for the angle between two wavenumbers $\boldsymbol{k}_a$ and $\boldsymbol{k}_b$ in resonance with a third wave $\boldsymbol{k}_c$, and show that this equation has a unique positive real solution (McGoldrick, 1965). The resonance conditions may be taken as

$$\boldsymbol{k}_a + \boldsymbol{k}_b = \boldsymbol{k}_c; \quad \omega_a + \omega_b = \omega_c.$$

6. Show that the resonance conditions between two primary surface gravity waves and their interaction product are satisfied only for the trivial cases where the coupling coefficient vanishes (Phillips, 1960).

39. SPECTRAL MODIFICATIONS BY REFRACTION AND WAVE–WAVE INTERACTIONS

The mechanics of wave refraction and of wave–wave interactions have been explored in Sections 37 and 38 in terms of individual Fourier components and multiplets thereof, i.e., for line spectra. We now broaden our point of view to include continuous wave spectra.

We first recall our notation for the spatial densities of mean wave energy $\langle E \rangle$ and action $\langle A \rangle$, and their spectral densities S and n:

$$\langle E(\boldsymbol{x}, t) \rangle = \int S(\boldsymbol{k}; \boldsymbol{x}, t)\, \mathrm{d}\boldsymbol{k}, \tag{39.1}$$

$$\langle A(\boldsymbol{x}, t) \rangle = \int n(\boldsymbol{k}; \boldsymbol{x}, t)\, \mathrm{d}\boldsymbol{k}, \tag{39.2}$$

$$\text{with} \quad n(\boldsymbol{k}; \boldsymbol{x}, t) = S(\boldsymbol{k}; \boldsymbol{x}, t)/\omega_0(\boldsymbol{k}; \boldsymbol{x}, t). \tag{39.3}$$

Over a wave band δk,

$$S(\boldsymbol{k}; \boldsymbol{x}, t)\delta k = \langle \eta\eta^* \rangle, \tag{39.4}$$

where η is the wave amplitude and δk is an element of *volume* of wavenumber space, identical to the $\mathrm{d}\boldsymbol{k}$ used in the integrals, and defined by (33.4). We note again that this definition of S includes only the significant part of the energy and may not have the correct dimensions of energy per unit volume, or area, as the case may be. For surface gravity waves, for example, the complete energy density per unit area is $\frac{1}{2}\rho_* g$ times $\langle E(\boldsymbol{x}, t) \rangle$ as obtained from (39.1) and (39.4). As earlier, $\omega_0(\boldsymbol{k}; \boldsymbol{x}, t)$ is the intrinsic (or Doppler-shifted) frequency, as seen by an observer who moves with the fluid.

In purely refractive situations, there are no sources of wave action and (33.8) holds along a ray:

$$\mathrm{d}n/\mathrm{d}t = 0. \tag{39.5}$$

If the spectral energy density S and frequency ω_0 are known at a point $(\boldsymbol{x}, t)$, their values S' and ω_0' at some other point $(\boldsymbol{x}', t')$ down a ray satisfy the relation

$$S'(\boldsymbol{k}'; \boldsymbol{x}', t) = \frac{\omega_0'(\boldsymbol{k}'; \boldsymbol{x}', t')}{\omega_0(\boldsymbol{k}; \boldsymbol{x}, t)} S(\boldsymbol{k}; \boldsymbol{x}, t), \tag{39.6}$$

where ω_0' is determined from (36.2) and $\boldsymbol{k}'$ from (6.15b); the ray path itself (i.e., the position of $\boldsymbol{x}', t)$ is found by integration of (6.15a). As the wavenumber changes along a ray from $\boldsymbol{k}$ to $\boldsymbol{k}'$, the spectral energy density associated with $\boldsymbol{k}$ shifts to a different wavenumber $\boldsymbol{k}'$. The spatial energy density $\langle E \rangle$ may then increase or decrease upon refraction according to the variation in the bandwidth δk. For a sharply peaked wave spectrum, we may approximate (39.1) by

$$\langle E \rangle \simeq S\delta k, \tag{39.7}$$

where (when ω_0 is constant) S is a constant on a ray and δk varies according to (33.7). The changes in bandwidth may also be deduced from the ray equations, which, like (33.7), are a consequence of the equation for the conservation of wave crests (6.4) (cf. Longuet-Higgins, 1957). The wave energy may also be represented in terms of a spectral density over frequency and direction of propagation, as specified (in three dimensions) by the angles θ and ϕ:

$$\langle E(\boldsymbol{x}, t)\rangle = \iiint S(\omega, \theta, \phi; \boldsymbol{x}, t)\, \mathrm{d}\omega\, \mathrm{d}\theta\, \mathrm{d}\phi. \tag{39.8}$$

Averaging over all angles of propagation,

$$\langle E(\boldsymbol{x}, t)\rangle = \int S(\omega; \boldsymbol{x}, t)\, \mathrm{d}\omega. \tag{39.9}$$

The same symbol S is used for all spectral densities; no confusion should arise when the explicit functional dependence is given for each spectral function. A similar averaging may be used to write a one-dimensional spectrum with respect to the magnitude of the wavenumber, k

$$\langle E(\boldsymbol{x}, t)\rangle = \int S(k; \boldsymbol{x}, t)\, \mathrm{d}k. \tag{39.10}$$

In comparing wave spectra at different points along a ray, it is often more convenient to use the frequency spectrum $S(\omega; \boldsymbol{x}, t)$ than the one-dimensional wavenumber spectrum $S(k; \boldsymbol{x}, t)$. While k changes along a ray, ω does not (at least in time-independent media, which is the usual case), and successive spectral peaks of a narrow band event, such as swell entering shallow water, will not be displaced with respect to each other in an ω-spectrum, as they would in a k-spectrum. Moreover, refraction effects which are nearly invisible in the wavenumber spectrum may be obvious in the frequency spectrum. Consider, for example, a long, narrow bandwidth swell approaching a beach normally; in the absence of currents, $S(\boldsymbol{k}; \boldsymbol{x}, t)$ is invariant along a ray, and since the approach is normal and the wavenumber one-dimensional, so is $S(k; \boldsymbol{x}, t)$.

From (39.1) and (39.9),

$$S(k; \boldsymbol{x}, t)\, \mathrm{d}k = S(\omega; \boldsymbol{x}, t) \frac{\mathrm{d}\omega}{\mathrm{d}k}\, \mathrm{d}k. \tag{39.11}$$

Although $S(k; \boldsymbol{x}, t)$ is invariant along a ray, $S(\omega; \boldsymbol{x}, t)$ is not, since $\mathrm{d}\omega/\mathrm{d}k = c_g$ varies. Thus, the peak in the frequency spectrum of swell entering into shallow water should increase in amplitude as the depth H, and hence c_g, decreases. This effect is visible in Fig. 51.11, which shows a series of spectra taken from JONSWAP data (Hasselman et al., 1973), where most of the refraction takes place between curves labelled 1 and 2. The general decay of the spectral levels in Fig. 51.11 with decreasing distance from shore (as indicated by the curve numbers) is due to dissipation and will be discussed later.

Another interesting refractive effect may be observed in the JONSWAP records. In Fig. 39.1a, we notice a semi-diurnal modulation of wave frequency, in addition to the increasing linear trend in frequency characterizing swells arriving from a distant source, a feature already discussed in Section 11. This phenomenon was first noted by Barber and Ursell (1948) who attributed it to Doppler shifting by tidal currents. From (6.19b), one should expect the frequency of a wave packet in a spatially uniform but temporally variable medium to vary along a ray according to

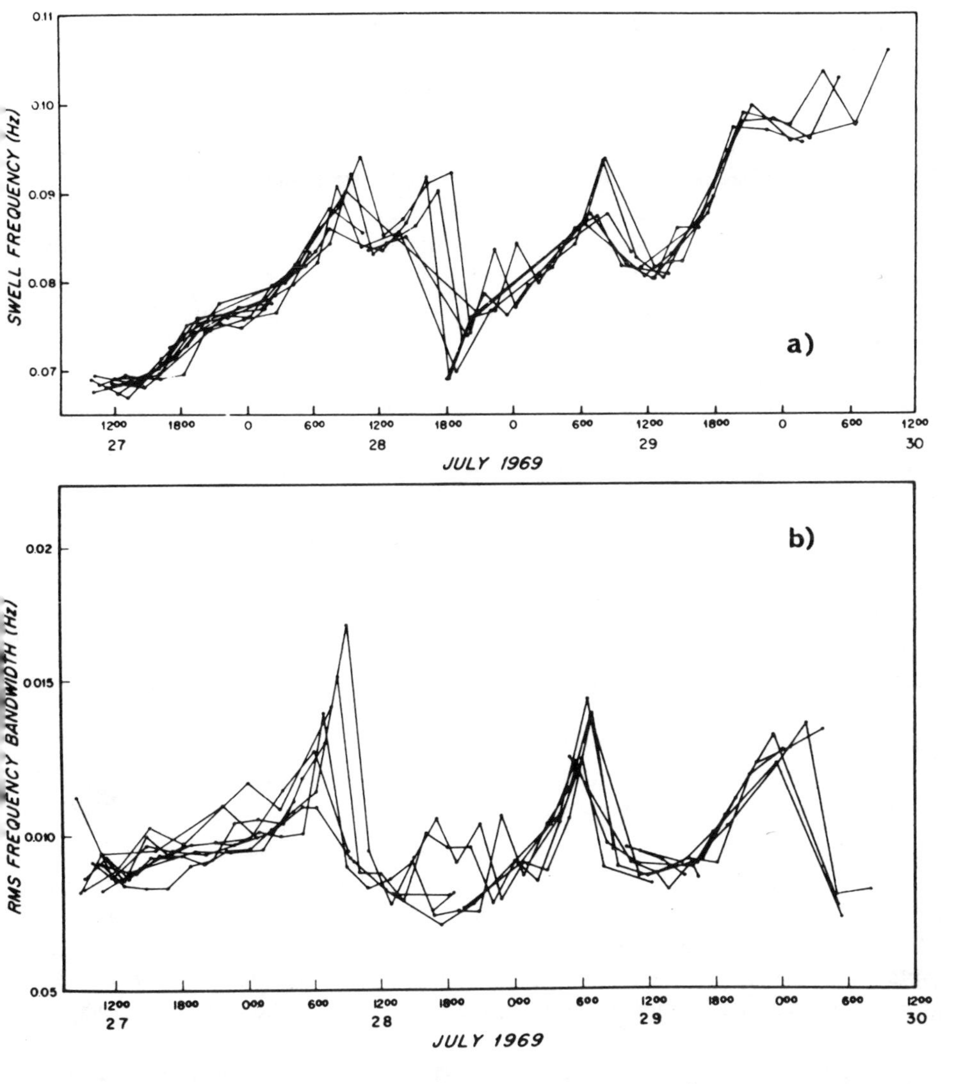

Fig. 39.1. The variation with time of swell frequency in hertz ($f = \omega/2\pi$) (above) and frequency bandwidth (below) for a JONSWAP swell event. Different curves correspond to different stations along the observation array. (From Hasselmann et al., 1973.)

$$\frac{\mathrm{d}\omega}{\mathrm{d}t} = \frac{\partial}{\partial t}(\omega_0 + \boldsymbol{k} \cdot \boldsymbol{U}) = \boldsymbol{k} \cdot \frac{\partial \boldsymbol{U}}{\partial t}, \tag{39.12}$$

and thus mirror the time variation of the current $\boldsymbol{U}$. The frequency bandwidth $\delta\omega$ is even more noticeably modulated (Fig. 39.1b): the frequency shows a steady trend in time but $\delta\omega$ does not. Thus, in one dimension for simplicity,

$$\delta\omega = (U + c_g)\delta k \tag{39.13}$$

and $\delta\omega$ shows the same time modulation as $U(t)$.

Spectral modifications are considerably more complicated for the wave–wave interaction process. In a continuous wave spectrum, no resonant multiplet may be considered isolated, since one or more of its components may be a member of one or more other resonant multiplets. Hence, in a continuous spectrum, resonant wave–wave interactions lead to a redistribution of wave action between all possible interacting wave multiplets. The total action density of any one resonant multiplet is no longer conserved: wave action becomes a conservative property of the whole spectrum. In general, one should expect some tendency towards equipartition of wave action between all those waves which participate in the lowest order, and hence the strongest resonant interaction. However, in actual wave spectra other sources of action density due to generating forces or dissipation will be at work and the observed spectral shape will be a result of a balance between the input terms and the redistributing nonlinear interaction terms.

For a second-order interaction the radiation balance equation may be written as

$$\frac{\mathrm{d}}{\mathrm{d}t}n_3(\boldsymbol{k}_3) = \iint T_+(n_1n_2 - n_3n_1 - n_3n_2)\delta(\boldsymbol{k}_1 + \boldsymbol{k}_2 - \boldsymbol{k}_3)\delta(\omega_1 + \omega_2 - \omega_3)$$
$$+ 2T_-(n_1n_2 + n_3n_1 - n_3n_2)\delta(\boldsymbol{k}_1 - \boldsymbol{k}_2 - \boldsymbol{k}_3)\delta(\omega_1 - \omega_2 - \omega_3)\,\mathrm{d}\boldsymbol{k}_1\,\mathrm{d}\boldsymbol{k}_2, \qquad (39.14)$$

where the factors T_+ and T_- are quadratic functions of the three resonant wavenumbers. The change in action density of the interaction product is seen as resulting from the effect of all possible resonant triads in which it enters (hence the two δ-functions). The direction of the wave action flow between the pair $\boldsymbol{k}_1$ and $\boldsymbol{k}_2$ and their resonant interaction product $\boldsymbol{k}_3$ depends on the sign of the integrand in (39.14). However, over the whole spectrum, as was shown by Hasselmann (1963a),

$$\int \frac{\mathrm{d}n_3(\boldsymbol{k}_3)}{\mathrm{d}t}\,\mathrm{d}\boldsymbol{k}_3 = 0, \qquad (39.15)$$

and the total (integrated) action density is conserved. Equation (39.14) was first derived by Peierls (1929) for interacting lattice vibrations; it is also discussed in Hasselmann (1967a). For third-order interactions, between wave quartets, the action transfer expression is similar in structure to that appearing in (39.14) but involves cubic products of n_1, n_2, n_3, n_4, and δ-functions representing resonance conditions between four wavenumbers and frequencies.

It is only for surface gravity waves that sufficient observational evidence is available to compare theory and observations. Pure surface gravity waves interact resonantly only at the fourth order, according to (38.39). The analogue of (39.14) was derived by Hasselmann (1962, 1968) and has the compact form

$$\frac{\mathrm{d}n_4(\boldsymbol{k}_4)}{\mathrm{d}t} = \iiint Q\delta(\boldsymbol{k}_1 + \boldsymbol{k}_2 - \boldsymbol{k}_3 - \boldsymbol{k}_4)\delta(\omega_1 + \omega_2 - \omega_3 - \omega_4)\,\mathrm{d}\boldsymbol{k}_1\,\mathrm{d}\boldsymbol{k}_2\,\mathrm{d}\boldsymbol{k}_3, \qquad (39.16)$$

where Q is a homogeneous cubic function of n_1, n_2, n_3, n_4 and a homogeneous quadratic function of the wavenumbers $\boldsymbol{k}_1, \boldsymbol{k}_2, \boldsymbol{k}_3, \boldsymbol{k}_4$. Again, the total action density of the spectrum is conserved. Further, we have already noted that the energy and the momentum of a resonant quartet of surface gravity waves is conserved; for a continuous spectrum, as shown by Hasselmann (1963), these quantities are conserved over the entire spectrum.

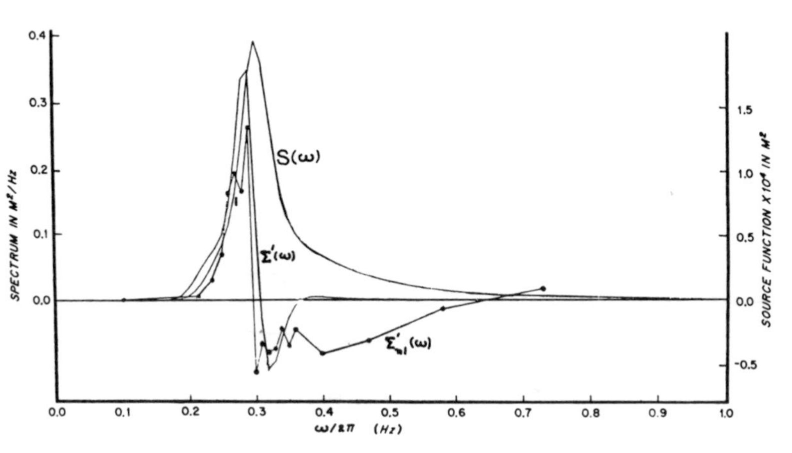

Fig. 39.2. The mean JONSWAP spectrum $S(\omega)$ and source functions $\Sigma'(\omega)$ together with the computed nonlinear energy transfer Σ'_{nl}. (From Hasselmann et al., 1973.)

Experiments by Snyder and Cox (1966), Barnett and Wilkerson (1967) and Hasselmann et al. (1973) have provided some knowledge of the evolution of travelling surface wave spectra. In the JONSWAP experiment (Hasselmann et al. 1973), a series of instruments was deployed along a 160 km long array normal to the coast of the island of Sylt, in the North Sea. By comparing energy spectra at successive sampling points it was possible to construct an energy source function Σ' accounting for the change in form and amplitude of the spectrum of waves approaching the shore. The mean energy and source function spectra for JONSWAP are shown in Fig. 39.2, together with a computed nonlinear energy transfer rate, Σ'_{nl}. The frequency spectra plotted in Fig. 39.2 have been averaged over all directions of propagation, according to (39.9) (but in two dimensions only, i.e., over only one angle); the energy spectra plotted are thus $S(\omega)$.

It is clear from Fig. 39.2 that the nonlinear interaction term does not account for the complete estimated energy source. One may split the energy source Σ' into three parts:

$$\Sigma' = \Sigma'_{in} + \Sigma'_{nl} + \Sigma'_{ds}, \tag{39.17}$$

where Σ'_{in} represents the input from the atmosphere and Σ'_{ds} is the dissipation term. These two terms do not integrate to zero over the whole spectrum; they will be discussed in Sections 49 and 51. A schematic representation of the energy balance is shown in Fig. 39.3. The nonlinear wave–wave interaction process redistributes energy from the centre of the spectrum, where it is put in by the atmosphere, to the low- and high-frequency ends of the spectrum. The nonlinear energy flux to high frequency is cancelled by dissipative effects and the main role of the wave–wave interactions appears to be to feed energy towards the low-frequency forward face of the spectrum. Indeed, Hasselmann et al. (1973) conclude that: "The shape of the spectrum is determined primarily by the nonlinear energy transfer from the central region of the spectrum to both shorter and longer wave components. In particular, the pronounced peak and steep forward face can be explained by a self-stabilizing feature of this process. The rapid wave growth observed for waves on the forward face of the spectrum can also

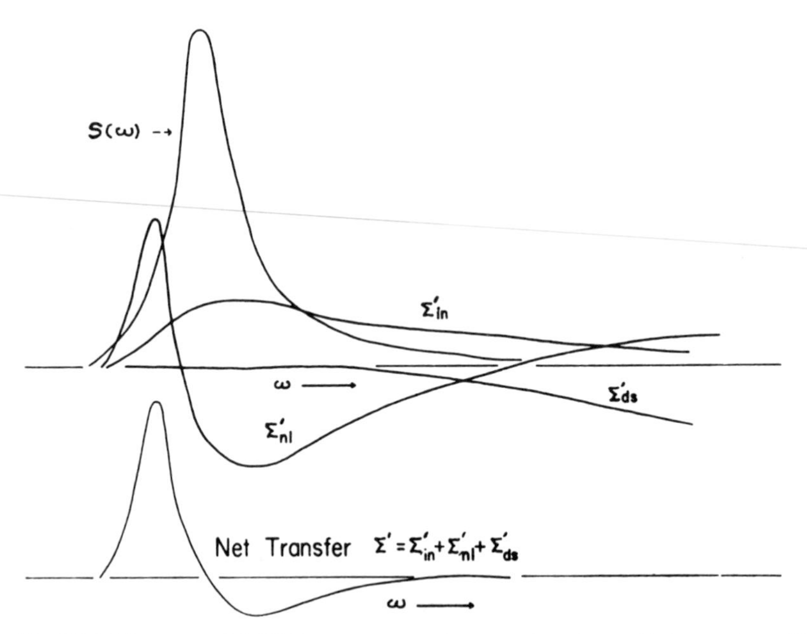

Fig. 39.2. The mean JONSWAP spectrum and source functions $\Sigma'(\omega)$ together with the computed nonlinear energy transfer Σ'_{nl}. (From Hasselmann et al., 1973.)

be largely attributed to the nonlinear transfer to longer waves." More detailed calculations of nonlinear energy transfer near the peak of a wind–wave spectrum by Longuet-Higgins (1976) and Fox (1976) have confirmed this assessment.

The role played by resonant wave–wave interactions in surface gravity waves is of surprising importance considering that these interactions are mere third-order effects in wave slope! An even stronger contribution should be expected in other waves, such as internal waves or planetary waves, where second-order interactions are possible. The energy balance of an internal wave field characterized by a Garrett and Munk (1972b) spectrum (GM 72) has been examined by Olbers (1974) and Müller and Olbers (1975). These results are shown in Fig. 53.3, in which wave–wave interactions are seen to play a significant role in the transfer of energy from the centre of the spectrum to high and low vertical wavenumbers. The influence of wave–wave interactions in planetary wave spectra has been discussed by Kenyon (1967) and Rhines (1975, 1977), and has already been discussed in Section 19 in relation to nonlinear effects in these waves. The observational evidence for both internal and planetary waves is still, however, well below the quality of that available for surface waves, and the role of nonlinear interactions for these waves cannot yet be verified by good data.

CHAPTER 7

WAVE–CURRENT INTERACTIONS: CRITICAL LAYER ABSORPTION AND STABILITY OF PARALLEL FLOWS

40. INTRODUCTION

In the previous chapter we discussed examples of *weak* wave–current and wave–wave interactions. In such phenomena as wave refraction by currents and nonlinear resonant interactions, the waves slowly exchange energy with large-scale mean flows or with other waves. The space and time scales characterizing these processes greatly exceed the characteristic wavelength and period of the waves involved, which enabled us to use the transport equation (33.11), valid for action-conserving interactions. In this chapter we discuss two types of *strong* wave–current interactions that are generally not action-conserving: critical layer absorption and inertial stability of parallel flow. In both of these phenomena significant energy exchanges occur over scales comparable with the wavelength and the period.

In the phenomenon of critical layer absorption of internal gravity waves in a nonrotating stratified fluid, a wave is severely attenuated and loses momentum to a vertically sheared flow as it propagates across a layer at which the intrinsic (or Doppler-shifted) frequency is equal to zero. At such a layer (or level) the vertical component of the group velocity also vanishes. Critical layer absorption can also occur for planetary waves propagating through a large-scale, laterally sheared flow. However, most of our discussion of critical layer absorption (Section 41) will be devoted to the internal wave case.

The topic of stability of parallel flows has had a long history in fluid dynamics. The fundamental problem is to determine whether a given shear flow is stable to travelling wave perturbations. It is beyond the scope of this book to catalogue the stability properties of all the types of flows that arise in the ocean. Therefore, we shall concentrate on the stability of two representative cases: (1) relatively small-scale, vertically sheared flows of a nonrotating stratified fluid (Sections 42 and 43); and (2) large-scale flows with lateral and/or vertical shears on the β-plane (Sections 44–46). In case (1) the wave perturbations are modified internal waves which grow exponentially with time if the flow is unstable, leading to overturning and hence vertical mixing over scales of meters to hundreds of meters. In case (2) the wave perturbations are modified planetary waves; for unstable flows, the waves develop into large eddy-like motions with horizontal scales of hundreds of kilometers. Thus we shall examine how simple linear internal waves and planetary waves are significantly affected by the two classes of mean flow. Further, in their modified form we shall see how these waves play an important role in the energy balance of the ocean.

41. CRITICAL LAYER ABSORPTION

In Section 36 we showed that long surface gravity waves always undergo total internal reflection in the presence of a mean current that is sufficiently horizontally sheared; no critical layer is ever reached. In the case of internal gravity waves, however, an upward propagating wave (with respect to wave energy transport) can have its amplitude severely attenuated as it crosses that *critical layer or level* $z = z_c$ at which the (vertically sheared) mean current $U(z)$ is equal to the horizontal phase speed. At the critical layer, the vertical component of $\boldsymbol{c}_g$ vanishes and there is a substantial transfer of momentum to the mean flow.

Using the WKB method, which in this problem is valid for a large Richardson number (defined below), Bretherton (1966) showed that an upward-propagating internal wave packet† would approach the critical layer for the dominant frequency and wavenumber of the packet, but that it would not reach the critical layer in any finite time. This is because along a ray path, $dz/dt \propto (z - z_c)^2$ as $z \to z_c$, which gives $t - t_0 \propto 1/(z - z_c)$ as $z \to z_c$. Thus he inferred that it would neither be reflected nor transmitted, but effectively absorbed. Further, as the critical layer is approached, Bretherton showed that the vertical wavenumber becomes large and the perturbation velocity, being transverse to the wavenumber vector, becomes more and more horizontally oriented (see Fig. 41.1). However, in the vicinity of the critical layer, the WKB approximation is not valid; in order to determine the behaviour of a wave crossing the critical layer, the governing wave equation has to be solved in another way (by the method of Frobenius). This was done by Booker and Bretherton (1967); it is in this paper that an explicit expression was obtained for the attenuation factor referred to above.

These and other early studies were motivated by a desire to understand the behaviour of internal gravity waves propagating into the upper atmosphere in the presence of the mean wind shear. There is still considerable interest in critical layers and their relevance to the dynamics of the atmosphere (for example, see Geller et al., 1975). However, the possible importance of critical layers in the ocean has only been considered quite recently. For a brief discussion of critical layer absorption (of internal waves) in the ocean we refer the reader to Thorpe (1975).

Although most meteorologists and oceanographers usually think of critical layers in relation to internal waves, there also exist critical layers for planetary waves. In the atmospheric context the latter were first discussed by Dickinson (1968, 1970); however, it was only recently that they were studied in the oceanic context (Geisler and Dickinson, 1975; Yamagata, 1976a). Critical layers for continental shelf waves have recently been discussed by McKee (1977).

We wish to add here a cautionary note on the definition of a critical layer. In the internal wave case, the intrinsic frequency does *not* vanish at the critical layer *if* rotation is included; however, the vertical component of the group velocity does vanish there (as it also does when there is no rotation). On these grounds it has been advocated by some authors that a critical layer be defined as that location where the component of the group velocity normal to the mean flow vanishes (a definition which is, however, not applicable to shear flows in a homogeneous fluid).

---

† For the case of a flow with constant shear, Hartman (1975) showed that the wave-packet concept is also meaningful for a Richardson number of order unity.

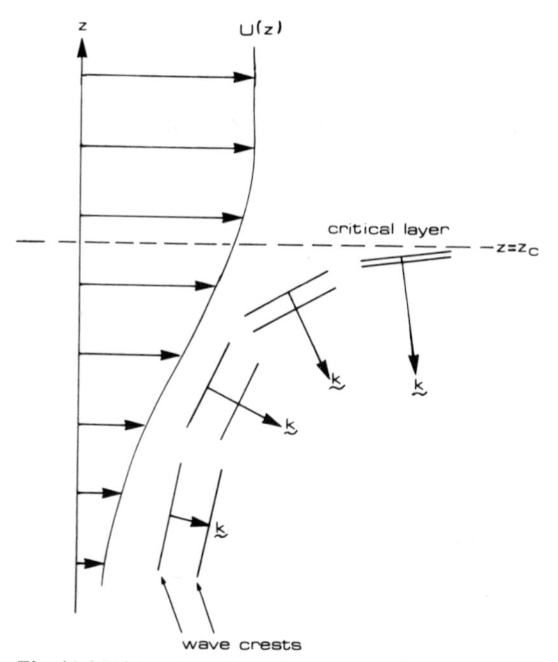

Fig. 41.1. The progression of a wave packet upward toward a critical layer at $z = z_c$. The particle motions are parallel to the wave crests.

Critical layers for internal gravity waves

Let us now proceed with the analysis of critical layers for internal gravity waves. We consider two-dimensional motions ($v \equiv 0$ and $\partial_y \equiv 0$) of a stably stratified, nonrotating inviscid fluid in which there is a steady mean shear flow $U(z)$ in the x-direction. If $p_0(z)$ and $\rho_0(z)$ denote the hydrostatic pressure and density as usual, then the basic state p_0, ρ_0 and $\boldsymbol{U} = [U(z), 0, 0]$ exactly satisfies the two-dimensional (nonlinear) adiabatic equations (3.9)–(3.11) in which $\boldsymbol{\Omega} = \boldsymbol{0}$. We now suppose there are small perturbations of this basic state. Thus for the total field we let

$$\boldsymbol{u} = (u' + U, 0, w'), \quad p = p_0 + p', \quad \rho = \rho_0 + \rho', \tag{41.1}$$

where the perturbation (primed) quantities are functions of x, z and t and are assumed to be small compared to their basic-state counterparts. Substituting (41.1) into (3.9)–(3.11), using the hydrostatic relation for p_0 and ρ_0 and then neglecting the products of all perturbation quantities, we obtain, upon dropping the primes, the following linearized equations:

$$\frac{D_0 u}{Dt} + wU_z + \frac{1}{\rho_0} p_x = 0, \text{ and } \frac{D_0 w}{Dt} + \frac{1}{\rho_0} p_z = -\frac{\rho}{\rho_0} g, \tag{41.2}$$

$$\frac{D_0 \rho}{Dt} + \rho_{0z} w = 0, \tag{41.3}$$

$$u_x + w_z = 0, \tag{41.4}$$

where $\dfrac{D_0}{Dt} = \dfrac{\partial}{\partial t} + U \dfrac{\partial}{\partial x}$.

We note that implicit in the linearization approximation is the requirement that

$$|(u\partial_x + w\partial_z)u| \ll |(\partial_t + U\partial_x)u|, \tag{41.5}$$

and similarly for operations on w. The validity of (41.5) will be discussed later.

Cross-differentiating (41.2) to eliminate p and using (41.3) and (41.4) to eliminate ρ and u, we obtain, under the Boussinesq approximation [$\rho_0 = \rho_* =$ constant in (41.2)],

$$\frac{D_0^2}{Dt^2}(w_{xx} + w_{zz}) - \frac{D_0}{Dt}(U_{zz}w_x) + N^2 w_{xx} = 0, \tag{41.6}$$

where $N^2 = -g\rho_{0z}/\rho_* > 0$ (for static stability). In the absence of a mean flow, (41.6) reduces to the governing equation for internal gravity waves [cf. (8.22) in which the Boussinesq approximation is applied]. In a shear flow, however, there is now a new restoring effect proportional to U_{zz}, which is due to the vertical variations in U_z, the mean vorticity. We now consider travelling wave solutions of (41.6) of the form

$$w(x, z, t) = w(z)\,e^{ik(x-ct)}, \tag{41.7}$$

where $k > 0$. Substituting (41.7) into (41.6) gives

$$(U-c)^2 w'' + [N^2 - (U-c)U'' - (U-c)^2 k^2]w = 0 \tag{41.8}$$

as the equation for the vertical structure; here the prime denotes differentiation with respect to z. Unless otherwise specified, in this chapter k will always denote the wave-number component in the $\boldsymbol{x}$-direction.

Equation (41.8) is a simplified version of the Taylor-Goldstein equation (see Section 42) used to study the stability of stratified shear flows contained between two parallel boundaries (one or both of which may be at infinity). When $N^2 = 0$ (homogeneous fluid), (41.8) reduces to the Rayleigh equation, which was first investigated in the nineteenth century; for a modern account of the stability of homogeneous shear flows, see Drazin and Howard (1966). Along with appropriate homogeneous boundary conditions on $w(z)$, (41.8) represents an eigenvalue problem for the eigenfunction $w = w(z; c)$, with corresponding eigenvalue c. If any of the eigenvalues c are complex with $\mathrm{Im}(c) > 0$, the perturbations grow exponentially with time (see 41.7) and the flow is said to be *dynamically unstable.* More precisely, such flows are *absolutely unstable* since the perturbations grow with time at all positions in the fluid. A *convective instability*, on the other hand, is one in which exponential growth is seen only by an observer moving with a nonzero velocity that lies in a certain range. The two types of instability can be distinguished in terms of the group velocity, $\boldsymbol{c}_g$. An absolutely unstable flow is one for which there is an unstable wave that has $\boldsymbol{c}_g = \boldsymbol{0}$; a convectively unstable flow on the other hand has no unstable waves with $\boldsymbol{c}_g = \boldsymbol{0}$ [see Lee (1975) for a further discussion]. The distinction is important when considering flows in regions of limited horizontal extent. For, if a certain flow is convectively unstable only, then the instability may in fact never be observed in a finite region and hence is of little concern in practice. However, in most elementary accounts of hydrodynamic stability (Drazin and Howard, 1966) the distinction is not made and it is implicitly assumed that the flows, if unstable, are absolutely unstable.

Miles (1961) showed that if the Richardson number Ri, defined as the square of the ratio of the buoyancy frequency to the mean shear, is everywhere greater than 0.25, viz.,

$$Ri \equiv N^2/U_z^2 > 1/4, \tag{41.9}$$

then the eigenvalues c of (41.8) are real. A proof of this result will be given in Section 43. The inequality in (41.9) will clearly hold for sufficiently weak shears and/or strong stratification. We shall assume in the remainder of Section 41 that $Ri > 1/4$ everywhere and hence that in (41.8) c is real. The case of $Ri < 1/4$ and hence the possibility of unstable waves will be considered in Sections 42 and 43.

For a given real c the critical layer (or level), if it exists, occurs at $z = z_c$ where

$$U(z_c) - c = 0. \tag{41.10}$$

At $z = z_c$, (41.8) has a singular point, and care must be exercised in connecting the solution across the singularity. However, it is precisely the nature of the solution near $z = z_c$ that gives rise to the phenomenon of critical layer absorption. Thus our next step will be to determine the solution of (41.8) near $z = z_c$ by the method of Frobenius.

We assume that U and N can be expanded in power series about the point $z = z_c$:

$$U = c + U'(z_c)(z - z_c) + \ldots,$$

$$N = N(z_c) + N'(z_c)(z - z_c) + \ldots, \tag{41.11}$$

where (41.10) has been invoked. For convenience we denote $N(z_c)$ by N_c and similarly for $U'(z_c)$ and $N'(z_c)$. We also assume that $U_c' \neq 0$ so that $z = z_c$ is a regular singularity. Hence we seek a solution of (41.8) of the form

$$w(z) = \sum_{n=0}^{\infty} a_n (z - z_c)^{n+\alpha}, \quad a_0 \neq 0. \tag{41.12}$$

Substitution of (41.11) and (41.12) into (41.8) leads to the indicial equation

$$\alpha^2 - \alpha + Ri_c = 0, \tag{41.13}$$

where $Ri_c = N_c^2/(U_c')^2$. Since we have assumed that $Ri > 1/4$ for all z, (41.13) has roots of the form

$$\alpha = 1/2 \pm i\mu, \tag{41.14}$$

where $\quad \mu = (Ri_c - 1/4)^{1/2} > 0.$

Thus, using (41.14) in (41.12), we see that near $z = z_c$,

$$\begin{aligned} w(z) &\simeq A(z - z_c)^{1/2+i\mu} + B(z - z_c)^{1/2-i\mu} \\ &= A \exp\left[(1/2 + i\mu)\{\log|z - z_c| + i\arg(z - z_c)\}\right] \\ &\quad + B \exp\left[(1/2 - i\mu)\{\log|z - z_c| + i\arg(z - z_c)\}\right] \\ &\equiv w_A + w_B. \end{aligned} \tag{41.15}$$

We have neglected the purely analytic part of the solution [which is just a power series in $(z - z_c)$, see (41.12)] since it is single-valued at $z = z_c$ and therefore does not give rise to any discontinuity at the critical level. From (41.15) we see that both w_A and w_B have a branch point at $z = z_c$. For the sake of definiteness we choose that branch of the log function for which $\arg(z - z_c) = 0$ when $z > z_c$ and introduce the branch cut from $z = z_c$ along the negative z-axis. Thus we have

$$\left.\begin{aligned} w_A &= A(z-z_c)^{1/2} \exp\,[i\mu \log\,(z-z_c)] \\ w_B &= B(z-z_c)^{1/2} \exp\,[-i\mu \log\,(z-z_c)] \end{aligned}\right\} z > z_c. \tag{41.16}$$

As $z-z_c$ decreases continuously from positive to negative values, $\arg(z-z_c)$ can change continuously from 0 to π or from 0 to $-\pi$, depending on whether near $z = z_c$ we pass around the singularity along a small semicircle in the upper half plane or lower half plane. The first choice takes us to a point above the cut, the second to a point below the cut. The problem thus arises as to which is the proper choice for $\arg(z-z_c)$ when $z < z_c$. One way to resolve this difficulty is to solve an initial value problem by transform methods and invoke causality (Miles, 1961). Alternatively, we can introduce a small amount of linear damping (Jones, 1967) into the original equations (41.2) and (41.3) which, for an $e^{-i\omega t}$ time dependence, amounts to replacing ω by $\omega + i\omega_i$ where ω_i is small and positive. Since $\omega = kc$ and $k > 0$, this is equivalent to replacing c by $c + ic_i$, where $c_i > 0$ and $|c_i/c| \ll 1$. Thus in the governing equation (41.8), $U-c$ is replaced by $U-c-ic_i$ and the singular point now occurs at some complex number $z = z_c'$. Thus we now formally treat z as a complex variable and introduce power series expansions in powers of $z-z_c'$. After completing the analysis we then take the limit $c_i \downarrow 0$. We assume that z_c' is close to z_c and hence for z near z_c' we have

$$\begin{aligned} U-c-ic_i &\simeq U'(z_c')(z-z_c') \\ &\simeq U_c'(z-z_c'), \end{aligned}$$

where, we recall, $U_c' = U'(z_c)$. Solving for z_c' and using $U-c \simeq U_c'(z-z_c)$, we obtain

$$z_c' = z_c + ic_i/U_c'.$$

Thus with linear damping included, the solution for $w(z)$ analogous to (41.15) now has a branch point at $z = z_c'$ which lies above (below) the Re(z)-axis according to whether $U_c' > 0$ (< 0) (see Fig. 41.2). We observe that as $\mathrm{Re}(z) - z_c$ decreases from positive to negative values (i.e., as point A moves towards point B in Fig. 41.2), $\arg(z-z_c')$ changes from $-\delta$ to $-\pi+\epsilon$ if $U_c' > 0$ and from δ to $\pi - \epsilon$ if $U_c' < 0$. Hence in the limit $c_i \downarrow 0$, we see that we must choose $\arg(z-z_c) = -\pi \operatorname{sgn} U_c'$ when $z < z_c$. Using this result in (41.15), we obtain

$$\left.\begin{aligned} w_A &= A|z-z_c|^{1/2} \exp\,[i\mu \log|z-z_c| - (1/2)\pi i \operatorname{sgn} U_c' + \mu\pi \operatorname{sgn} U_c'] \\ w_B &= B|z-z_c|^{1/2} \exp\,[-i\mu \log|z-z_c| - (1/2)\pi i \operatorname{sgn} U_c' - \mu\pi \operatorname{sgn} U_c'] \end{aligned}\right\} z < z_c. \tag{41.17}$$

For future reference we let $w^<$ and $w^>$ denote $w(z)$ for $z < z_c$ and $z > z_c$, respectively.

From (41.16) and (41.17) we now obtain the following ratios:

$$|w_A^>|/|w_A^<| = \exp\,(-\mu\pi \operatorname{sgn} U_c'), \quad |w_B^>|/|w_B^<| = \exp\,(\mu\pi \operatorname{sgn} U_c'). \tag{41.18}$$

That is for $U_c' > 0$ say, the solution w_A is attenuated by a factor $e^{-\mu\pi}$ as the critical layer is crossed from below, whereas the solution w_B is amplified by a factor $e^{\mu\pi}$ in the same process. We now show that for the case $U_c' > 0$, w_A represents a wave whose energy is propagating upward for both $z < z_c$ and $z > z_c$, and analogously that w_B represents a downward propagating wave. Thus when the direction of energy propagation is taken into account, both waves are attenuated as they cross the critical layer. When $U_c' < 0$, the roles

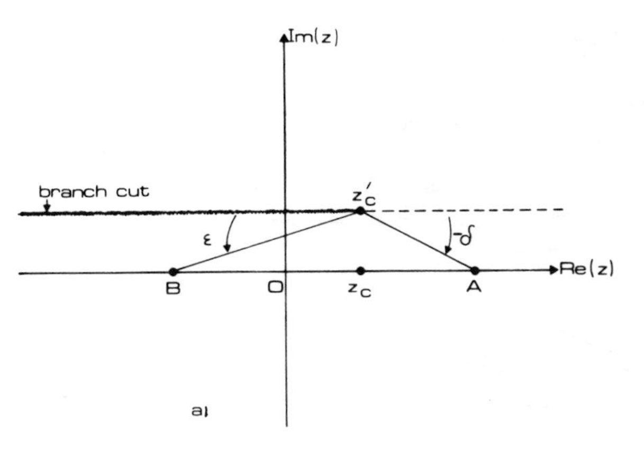

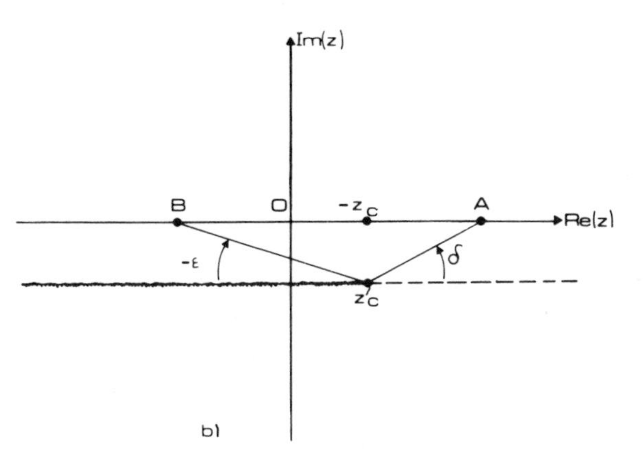

Fig. 41.2. Location of branch point with linear damping included, (a) $U_c' > 0$; (b) $U_c' < 0$. For illustrative purposes we have taken $z_c > 0$.

of the two solutions are reversed. The most direct way to establish this result concerning energy propagation is to compute the *Reynolds stress* and hence the vertical energy flux. The Reynolds stress represents the momentum transferred across a unit area by the perturbation velocity. More precisely, if $\langle\cdot\rangle$ denotes the average over one wave cycle (see 8.60), then the Reynolds stress in the x-direction across a unit area in the z-direction is defined as $\tau_{13} = -\rho_0\langle uw\rangle$ (with u and w real), this being the rate at which the x-momentum, $\rho_0 u$, is transported vertically by the vertical velocity, w. From (41.4) and (41.7) we find

$$\langle \tfrac{1}{2}(u+u^*)\tfrac{1}{2}(w+w^*)\rangle \equiv \langle u_r w_r\rangle = -\frac{1}{4ik}[w^*(z)w'(z) - w(z)w^{*\prime}(z)]$$

$$= \frac{\mu}{2k}|A|^2 \begin{cases} -1, z > z_c \\ e^{2\mu\pi}, z < z_c \end{cases} \tag{41.19}$$

for w_A, upon using (41.16) and (41.17). Similarly, for w_B

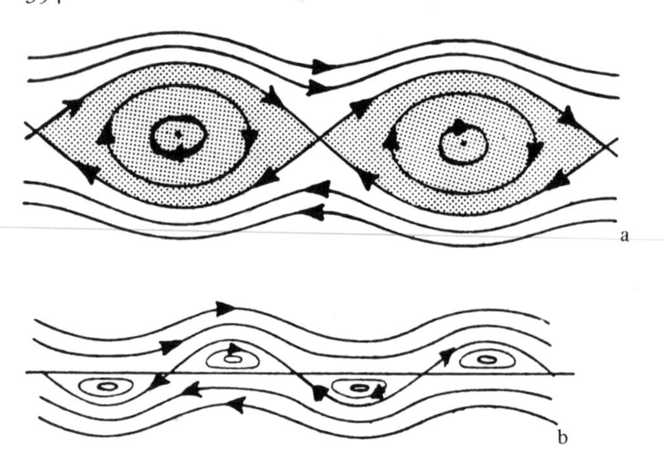

Fig. 41.3. (a) Kelvin's cat's-eye streamline pattern near a critical layer in a homogeneous fluid. (b) Taylor's cockeyed cat's-eye streamline pattern near a critical layer in a stratified fluid.

$$\langle u_r w_r \rangle = \frac{\mu}{2k}|B|^2 \begin{cases} 1, z > z_c \\ -e^{-2\mu\pi}, z < z_c. \end{cases} \tag{41.20}$$

Hence the vertical energy flux has the signature

$$-\rho_0(U-c)\langle u_r w_r \rangle \begin{cases} >0 \text{ for all } z \text{ for } w_A \\ <0 \text{ for all } z \text{ for } w_B \end{cases} \tag{41.21}$$

since $U-c \lessgtr 0$ for $z \lessgtr z_c$ for the case under consideration ($U_c' > 0$). We conclude that for $U_c' > 0$ the solution w_A is associated with an upward propagation of energy. We note that the flux is discontinuous across the critical layer, with the attenuation factor being $e^{-2\mu\pi}$. Thus, although the vertical flux is everywhere positive, the attenuation factor suggests there may be a significant transfer of horizontal momentum to the mean flow at the critical layer. The solution w_B, on the other hand, represents downward propagation of energy, with the wave losing energy as it passes through the critical layer.

Before commenting on a few refinements in the theory of critical layers for internal waves, it is worthwhile to show some typical streamline patterns of the fluid motions near a critical layer. In particular, it is instructive to look at patterns for both the homogeneous and stratified fluid cases. When $N=0$ (homogeneous fluid), the w'' term in (41.8) is multiplied by $U-c$, rather than by $(U-c)^2$, as in the stratified case; this fact results in quite different streamline patterns in the two cases. For $U-c \propto z-z_c$, Kelvin (1880b) showed that the resulting streamline pattern in a homogeneous fluid is of the form of a "cat's eye" near the critical layer (Fig. 41.3a). However, in a stratified fluid, this pattern can become asymmetric about the critical layer. For the case of a three-layer fluid with $U-c \propto z-z_c$ in the middle layer, Taylor (1931a) obtained the "cockeyed cat's eye" pattern as one of the possible solutions for $w(z)$ (Fig. 41.3b).

Refinements of the theory

From either (41.16) or (41.17) we note that $w(z) \to 0$ like $|z-z_c|^{1/2}$ as $z \to z_c$ either from above or below. Since $u(z) = -w'(z)/ik$, it follows that the horizontal velocity perturbation becomes infinitely large there. Thus clearly the linearization approximation (41.5) breaks down near $z = z_c$. Hence one must include other effects which will not eliminate critical layer absorption but which will lead to a solution that is consistent with all the approximations in the model. Hazel (1967) introduced viscosity and heat conduction into the linearized equations (41.2)–(41.4), which then lead to a sixth-order differential equation for $w(z)$ which is not singular at $z = z_c$. He solved this equation numerically and found precisely the *same* attenuation factor across the layer as derived above on the basis of the nondissipative theory! Kelly and Maslowe (1970) and Maslowe (1972) have also examined the influence of nonlinearity on the structure of a stratified shear flow in the vicinity of a critical layer. For a small Richardson number, Kelly and Maslowe (1970) showed that velocity and thermal boundary layers must exist within the nonlinear critical layer in order to make the solution continuous. The basic streamline pattern in the limit of a zero Richardson number is the Kelvin cat's-eye configuration shown in Fig. 41.3a. Maslowe (1972) extended the analysis to arbitrary values of the Richardson number and examined the influence of nonlinearity on the occurrence of instability at critical layers. His results indicate that Clear Air Turbulence in the atmosphere is likely to owe its origin to such an instability mechanism. These results may also apply to oceanic critical layers and to the billow-like disturbances observed by Woods (1968) (see below, Fig. 42.4). Breeding (1971) also studied the nonlinear and dissipative problem, and solved the perturbation equations by finite-difference methods. For waves incident from above he found that when $Ri_c > 2.0$ the nonlinear results agreed very well with those predicted by the linear theory. For $0.25 < Ri_c < 2.0$ a significant fraction of an incident wave is now reflected (in the linear theory only a small fraction is reflected). Nevertheless, energy is still transferred to the mean flow at the critical layer. This result has also been confirmed by Grimshaw (1975a) on the basis of a WKB solution of the nonlinear equations.

In all the above calculations the Coriolis force has been neglected. In the absence of mean flows f can be neglected provided $\omega \gg f$. However, in the presence of a mean flow the intrinsic frequency $\omega_0 = \omega - kU$ may well be comparable to f and therefore it is questionable whether the results described in this section can be applied to the ocean or atmosphere. This question was raised by Jones (1967) who showed that for a geostrophically balanced mean flow, the singularities of the equation for $w(z)$ in a rotating system differ from those of (41.8), both in number and form: there are critical layers where $U - c = \pm f/k$† as well as at $z = z_c$ given by (41.10). For the case of a constant shear Jones has shown that sufficiently far on either side of the critical layers the Booker-Bretherton solution is adequate; however, near the critical layers the solutions are quite different. Jones also showed that away from the critical layers the vertical flux of horizontal momentum is not conserved unless $f = 0$, but that the vertical flux of angular momentum (about the vertical axis) is conserved and is a more suitable measure of the wave intensity in a rotating system. Grimshaw (1975b) has generalized Jones' analysis to include $\tilde{f}$ (the horizontal component of the Coriolis parameter) as well as f. Grimshaw showed that the vertical component of the *wave-action flux*, $F_3 \equiv \langle E \rangle c_{g3}/\omega_0$, is now the conserved quantity away from the critical layers, which in this case also occur at $U - c = \pm f/k$, as well as at $z = z_c$ as given by (41.10). When there is no rotation, F_3 is proportional to the vertical

† Thus the Doppler-shifted frequency is not zero at these layers! The vertical component of the group velocity is, however, zero there.

flux of horizontal momentum; when $\tilde{f}=0$ only, F_3 can be related to the vertical flux of angular momentum. The effect of $\tilde{f}$ is to act as a valve at each of the critical layers: for a certain range of values of ω_0, the vertical propagation speed, $\tilde{f}$ and the horizontal wave-number vector, an upward propagating wave will pass through the critical layer without attenuation!

Oceanic applications

Bell (1975) has recently studied the generation of internal waves in the deep ocean by the interaction of the barotropic tide and quasi-steady geostrophic currents with bottom topography. He found that the relatively low-frequency waves [$\omega = 0(f)$] generated by the first process are quite likely to pass through any critical layers. However, waves generated by the second process are effectively dissipated by critical layer absorption in the lowest kilometer or so of the ocean. In this region $N \sim 7 \cdot 10^{-4}$ rad s^{-1} and $U_z \sim 10^{-4}$ s^{-1} and hence $Ri \sim 50$, which makes the factor $e^{-\mu\pi}$ very small. Mied and Dugan (1975), on the other hand, numerically integrated (41.8) for the reflection of upward propagating internal waves for models of variable N and U that describe oceanic conditions near the thermocline. They found that significant portions of the internal wave spectrum can be reflected back downward by the variations in N and U.

As mentioned at the beginning of this section, planetary waves can also be absorbed by mean shear flows in the ocean. The particular case of planetary wave absorption in a barotropic fluid with a mean north–south geostrophic flow $V(x)$ (x = eastward) was considered by Geisler and Dickinson (1975) and also by Yamagata (1976a). Geisler and Dickinson showed that for $V_c' > 0$, a short wave whose energy propagates eastward is attenuated by a factor $\exp(-\beta\pi/k_2 V_c')$ across the critical layer. However, in contrast to the internal wave case, both perturbation velocity components remain finite at the critical layer. In barotropic ocean circulation models (Pedlosky, 1965b; N. Phillips, 1966) the time-dependent behaviour of the motion near the western boundary is often interpreted in terms of reflecting planetary waves. Geisler and Dickinson's results show that mean currents can significantly alter the western boundary reflection coefficient used in such models. A nonlinear theory of critical layer absorption of planetary waves has recently been presented by Redekopp (1977).

Exercises Section 41

1. For a two-dimensional mean shear flow $[U(z), V(z), 0]$, show that the vertical perturbation velocity $w(x, y, z, t)$ satisfies

$$\frac{D_0^2}{Dt^2}(w_{xx}+w_{yy}+w_{zz})-\frac{D_0}{Dt}(U_{zz}w_x+V_{zz}w_y)+N^2(w_{xx}+w_{yy}) = 0,$$

where $D_0/Dt = \partial_t + U\partial_x + V\partial_y$.

2. From (41.2)–(41.4) derive the analogue of (41.6) for a non-Boussinesq fluid.

3. Suppose we introduce, on the right-hand sides of (41.2) and (41.3), the linear damping terms $-Ku$, $-Kw$ and $-K\rho (K > 0)$. Show that with an $e^{-i\omega t}$ time-dependence this is equivalent to replacing ω by $\omega + iK$. (See also Chapter 8.)

4. Work out all the steps in (41.19).

5. Show that the vertical component of wave-action flux is conserved on each side of a critical layer.

6. Let $V(x)$ be a geostrophic mean flow in a homogeneous fluid of constant depth and rigid top; then,

$$-fV + \frac{1}{\rho_*} p_{0x} = 0,$$

where p_0 is the basic pressure. Suppose now the total horizontal velocity and pressure have the form

$$u_h = (u', V + v'), \quad p = p_0 + p'.$$

Show that the linearized vorticity equation for the perturbation stream function is given by

$$(\partial_t + V\partial_y)\nabla_h^2 \Psi + \beta\Psi_x - V_{xx}\Psi_y = 0. \tag{41.22}$$

If Ψ has the form

$$\Psi = \psi(x) \exp [ik_2(y - ct)],$$

show that $\psi(x)$ satisfies

$$(V - c)\psi'' - (i\beta/k_2)\psi' - V''\psi = 0. \tag{41.23}$$

If $V =$ constant, discuss the propagating solutions of (41.27). The more ambitious reader might like to examine the behaviour of these solutions near the critical layer!

7. By differentiating the phase in the (upward propagating) solution w_A as given by (41.16), show that the local vertical wavenumber is given by

$$k_3 = \mu/(z - z_c). \tag{41.24}$$

Hence show that the directions of phase propagation and particle motions are consistent with those shown in Fig. 41.1. Which way does the wave packet move above the critical layer in this figure?

8. Suppose in (41.8) U and N are slowly varying, so that U_{zz} can be neglected. Then for $w(z) = \exp(ik_3 z)$, we obtain the local dispersion relation

$$k^2 + k_3^2 = N^2k^2/(Uk - \omega)^2. \tag{41.25}$$

From (41.25), show that

$$c_{g3} = \frac{kN^2}{(k^2 + k_3^2)} \frac{k_3}{(U - c)}. \tag{41.26}$$

Hence show that $c_{g3} > 0$ for all $z \neq z_c$ for an upward propagating wave.

42. INTRODUCTION TO STABILITY OF STRATIFIED SHEAR FLOWS

We now begin the investigation of a phenomenon in which internal waves play an important role in the production of small-scale turbulence in the ocean: inertial instability of stratified shear flows. If a mean flow with vertical shear is unstable with respect to internal wave perturbations, then the latter may grow into finite-amplitude billows which in turn break and produce vertical mixing (turbulence) on a scale that completely dominates molecular diffusion (cf. Turner, 1973). Clearly, a full understanding of a given stratified shear flow ultimately depends on a numerical integration of a nonlinear system, which is beyond the scope of this book. Here we shall only discuss the *onset* of shear-flow instabilities, as predicted by linear analysis. Our analysis is based on the Taylor-Goldstein equation for the vertical structure $w(z)$, this equation being a non-Boussinesq version of (41.8). In this section we discuss the solution of the Taylor-Goldstein equation (cf. 42.6) for a simple shear flow of a two-layer fluid with a free surface. Then, in Section 43 we shall derive some general results concerning the stability characteristics of shear flows of any continuously stratified fluid bounded above and below by rigid surfaces.

The Taylor-Goldstein equation

We start again from the two-dimensional linearized perturbation equations (41.2)–(41.4) for a nonrotating stratified fluid and assume that each dependent variable is of the plane-wave form

$$\phi(x, z, t) = \phi(z)\, e^{ik(x-ct)}, \quad k > 0. \tag{42.1}$$

Under this assumption (41.2)–(41.4) reduce to

$$\rho_0[ik(U-c)u + wU'] + ikp = 0, \tag{42.2}$$

$$\rho_0 ik(U-c)w + p' = -\rho g, \tag{42.3}$$

$$ik(U-c)\rho + \rho_0' w = 0, \tag{42.4}$$

$$iku + w' = 0, \tag{42.5}$$

where u, w, p, ρ, U and ρ_0 are functions of z alone and a prime denotes differentiation with respect to z. We now use (42.5) to eliminate u in (42.2) and substitute the resulting expression for p into (42.3). Then, upon eliminating ρ by means of (42.4), (42.3) yields the following equation for $w(z)$:

$$[\rho_0(U-c)w']' - (\rho_0 U' w)' - \left[\frac{\rho_0' g}{U-c} + \rho_0 k^2 (U-c)\right] w = 0. \tag{42.6}$$

This is known as the *Taylor-Goldstein equation*; it was first derived by Taylor (1931a) and Goldstein (1931) in their studies of the stability of stratified shear flow. When ρ_0 is held constant in the first two terms, we readily retrieve the Boussinesq form (41.8).

We note here two fundamental restrictions associated with (42.6). The first concerns the two-dimensional nature of the disturbances, i.e., waves travelling only parallel to the mean flow were introduced. This simplification has its origin in *Squire's theorem* (Squire, 1933), which for a homogeneous fluid states that: "For each unstable three-dimensional wave there is always a more unstable two-dimensional one travelling parallel to the flow."

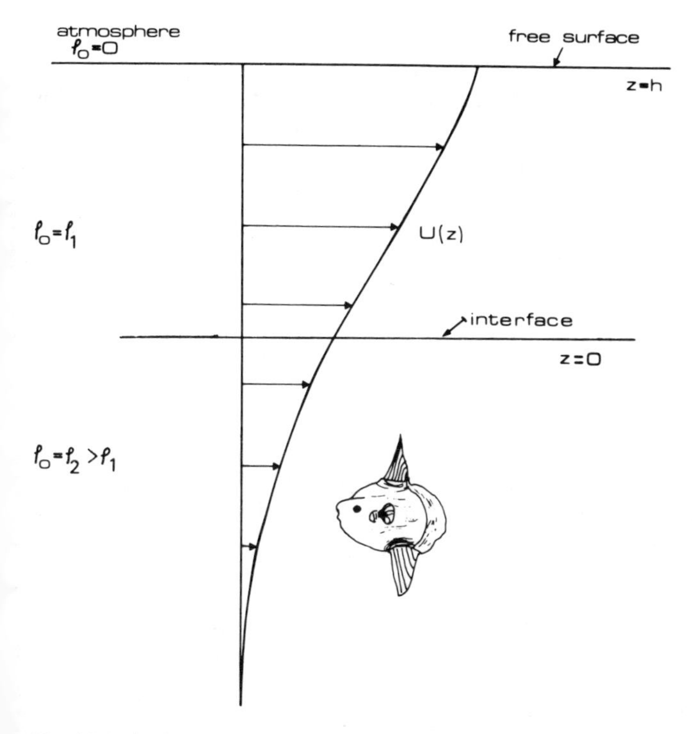

Fig. 42.1 The basic state for a shear flow in a deep, two-layer ocean with free surface. The eigenvalue equation is derived explicitly for the special vortex sheet profile (42.13).

For a simple proof of this, see Drazin and Howard (1966). Yih (1955) extended this theorem to the case of stratified fluids, which provides the motivation for our consideration of two-dimensional disturbances only since these are always the fastest growing waves. On the other hand, waves travelling strictly normal to the flow are unaffected by the current (Yih, 1965). Even with rotation included ($f \neq 0$), the interaction between a baroclinic mean flow (Section 44) and internal waves propagating normally to it is passive: the dispersion relation, as modified by the current, never yields growing or decaying wave solutions (cf. Healey and LeBlond, 1969; Mooers, 1975b). The second restriction concerns the plane-wave decomposition (42.1); this approach a priori eliminates the transient solution (a continuous spectrum) that would arise in any initial value calculation. However, for large times the solution will be dominated by the growing unstable plane-wave modes and the latter thus deserve first consideration [see Drazin and Howard (1966) for a further discussion].

Shear flow of a two-layer fluid with a free surface

We now consider the stability of the flow shown in Fig. 42.1 in which the basic density stratification is modelled by a two-layer structure:

$$\begin{aligned}\rho_0(z) &= \lim_{d \downarrow 0} [\rho_1 \Theta(z-d) + \rho_2 \Theta(-z-d)] \\ &= \rho_1 \Theta(z) + \rho_2 \Theta(-z), \quad -\infty < z < h,\end{aligned} \tag{42.7}$$

where $\rho_2 > \rho_1$ and $\Theta(z)$ denotes the unit step function. This flow can be thought of as a simple model of a shear flow concentrated in the upper mixed layer of the open ocean. It may also be interpreted as a crude model of estuarine flow (Esch, 1962). We consider this model because it contains three interesting special cases: (1) $U(z) = 0$, in which case the travelling wave perturbations reduce to pure sinusoidal surface and interfacial gravity waves; (2) $\rho_1 = \rho_2$, in which case the perturbations are modified surface gravity waves; and (3) $h \to \infty$, which corresponds to the classical Kelvin-Helmholtz stability problem of an unbounded, two-layer fluid. For future reference we describe the surface and interfacial perturbations by the equations

$$z = h + \eta(x, t), \quad z = \eta_i(x, t). \tag{42.8}$$

Since ρ_0 is discontinuous, we must solve (42.6) in each layer separately and then impose two matching conditions across the interface at $z = \eta_i$. The first condition is obtained by requiring continuity of the interfacial displacement. To first order in small quantities we have $w(x, 0\pm, t) = \eta_{it} + U(0\pm)\eta_{ix}$, which for travelling wave solutions of the form (42.1) reduces to $w(0\pm) = ik[U(0\pm) - c]\eta_i$. Hence we obtain

$$\frac{w(0+)}{U(0+)-c} = ik\eta_i = \frac{w(0-)}{U(0-)-c},$$

or equivalently, in the usual notation (cf. 23.29),

$$\left[\frac{w}{U-c}\right] = 0 \quad \text{at} \quad z = 0 \tag{42.9}$$

as our first matching condition. When $U = 0$, (42.9) reduces to (10.52), derived earlier for interfacial waves. The second condition is obtained by requiring the continuity of total pressure at $z = \eta_i$. The linearized version of this condition is most easily obtained mathematically by integrating (42.6) over the small interval $(-\epsilon, \epsilon)$ and then taking the limit $\epsilon \downarrow 0$. However, before doing so it is convenient to approximate ρ_0' in (42.6) by the expression (cf. 42.7)

$$\rho_0' = \rho_1\delta(z-d) - \rho_2\delta(-z-d),$$

where $d < \epsilon \ll 1$, and then take the limit $d \downarrow 0$ after performing the integration in the term involving ρ_0'. Thus (42.6) yields

$$\rho_0\{(U-c)w' - U'w\}\Big|_{-\epsilon}^{\epsilon} - g\left[\frac{\rho_0 w}{U-c}\right] - k^2\int_{-\epsilon}^{\epsilon}\rho_0(U-c)w\,dz = 0.$$

In the limit $\epsilon \downarrow 0$, the last term vanishes since the integrand is bounded, and we finally obtain

$$[\rho_0\{(U-c)w' - U'w - gw/(U-c)\}] = 0 \quad \text{at} \quad z = 0 \tag{42.10}$$

as our second matching condition. This condition reduces to (10.54) when $U = 0$ and $c^2 = gh_n$. It is important to note that the continuity conditions (42.9) and (42.10) also apply at points where U or U' is discontinuous, independent of whether ρ_0 is continuous or discontinuous at these points.

As boundary conditions we require that

$$w \to 0 \quad \text{as} \quad z \to -\infty, \tag{42.11}$$

and that at $z = h + \eta$, the usual kinematic and pressure conditions hold. Following a procedure similar to that in Section 36, we readily find that for $p_a =$ constant, these conditions, when linearized, can be combined into a single one for w:

$$(U-c)^2 w' - (U-c)U'w - gw = 0 \text{ at } z = h. \tag{42.12}$$

This relation also follows from (42.10) applied at $z = h$ since $\rho_0 = 0$ for $z > h$.

Solution for a vortex sheet mean current

The simplest shear profile is the so-called vortex sheet in which each layer is moving with a different uniform velocity. By using coordinates moving with the lower layer velocity, we can simply consider $U(z)$ of the form

$$U(z) = \begin{cases} U_1, 0 < z \leqslant h \\ 0, -\infty < z < 0. \end{cases} \tag{42.13}$$

The term "vortex sheet" is used to describe (42.13) because all the vorticity of the flow is concentrated at the level $z = 0$. The eigenvalue equation for perturbations on this profile was derived by Esch (1962) using potential theory (see Section 11), which is applicable in this case since in each layer itself the motion is irrotational. However, we shall derive the eigenvalue equation by solving (42.6) in each region and then applying the above matching and boundary conditions. This is usually the approach now taken in the literature since it is not restricted to flows which are irrotational in each layer. For example, the method described here can also be used for the shear-layer profile in which the vorticity is uniformly distributed over the upper layer (cf. Esch, 1962):

$$U(z) = \begin{cases} U_1 z/h, 0 < z \leqslant h \\ 0, -\infty < z < 0. \end{cases} \tag{42.14}$$

The graphical solution for this case will be presented below and compared with the solution for the profile (42.13).

For the profile (42.13) [and for (42.14) also] and the density structure (42.7), equation (42.6) reduces to

$$w'' - k^2 w = 0 \tag{42.15}$$

in each layer provided $U - c \neq 0$. Invoking (42.11), we write the solution to (42.15) as

$$w = \begin{cases} A\, e^{kz} + B\, e^{-kz}, & 0 < z \leqslant h \\ C\, e^{kz}, & -\infty < z < 0. \end{cases} \tag{42.16}$$

Application of (42.9), (42.10) and (42.12) yields three homogeneous equations for A, B, C. For a nontrivial solution, we require the determinant of the coefficients to be zero, which gives the following eigenvalue equation for c:

$$\{(U_1 - c)^2 k - g\}\{\Delta(c^2 k - g) + (U_1 - c)^2 k + g\}$$
$$+ e^{-2kh}\{(U_1 - c)^2 k + g\}\{\Delta(c^2 k - g) - [(U_1 - c)^2 k - g]\} = 0, \tag{42.17}$$

where $\Delta = \rho_2/\rho_1 > 1$. We note that (42.17) is a quartic in c. Esch (1962) has shown that for given values of U_1, h, Δ and k, there are either four real roots or two real roots and a complex conjugate pair of roots. In the first situation each eigenwave is purely sinusoidal and the flow is stable. In the second situation one of the eigenwaves associated with the complex roots will be a growing unstable wave and the flow is unstable. It is of particular interest in stability theory to determine, as a function of the relevant physical parameters, the range of wavenumbers for which the flow is unstable. This question leads to the concept of a *neutral stability curve* or *stability boundary*, which will be explored later.

Special cases of the eigenvalue equation

At this point it is illuminating to discuss the three special cases [mentioned before (42.8)] that arise from (42.17). For the general case (i.e., $U_1 \neq 0$, $\Delta \neq 1$ and kh finite), there are three mechanisms in operation in the problem:

(1) The destabilizing influence of the shear flow, which is called Helmholtz instability. Over a century ago Helmholtz (1868) considered the stability of the flow $U = U_1\Theta(z)$ of an unbounded, homogeneous fluid.

(2) The weak stabilizing influence of the small density change at the interface.

(3) The strong stabilizing influence of the large density change at the free surface.

The existence of the complex solutions to (42.17) thus expresses the fact that the stabilizing stratification cannot always prevent the overturning tendency of the shear flow. When $U_1 = 0$, however, the destabilizing mechanism is absent and therefore we would expect strictly sinusoidal disturbances.

(1) *Case* $U_1 = 0$

Upon setting $U_1 = 0$ in (42.17), we find that the quartic can be factored into two quadratics:

$$(c^2k - g)\{c^2k(\Delta + 1) - g' + [c^2k(\Delta - 1) + g']\,\mathrm{e}^{-2kh}\} = 0, \tag{42.18}$$

where $g' = (\Delta - 1)g$ is the reduced gravity. Setting the first factor equal to zero gives the dispersion relation for deep-water surface gravity waves (cf. 11.22b):

$$c^2 = g/k. \tag{42.19}$$

The second factor in (42.18) gives

$$c^2 = \frac{g'}{k}\left[\frac{1 - \mathrm{e}^{-2kh}}{\Delta + 1 + (\Delta - 1)\,\mathrm{e}^{-2kh}}\right], \tag{42.20}$$

which is the dispersion relation for interfacial waves; since $g' \ll g$, these waves move much slower than the surface waves. For short waves ($kh \gg 1$), (42.20) gives

$$c^2 = \frac{g}{k}\frac{\rho_2 - \rho_1}{\rho_2 + \rho_1} \equiv c_{\mathrm{I}}^2, \tag{42.21}$$

which agrees with (10.66) upon setting $f = 0$ and taking the short wave limit $1/gh_n = k^2/\omega^2 \to \infty$.

(2) *Case* $\Delta = 1$ ($\rho_1 = \rho_2$)

This case eliminates the weak stabilizing influence of the interface. The importance of that influence is exhibited by its removal: (42.17) always has a pair of complex conjugate roots for $\Delta = 1$. When $\Delta = 1$, (42.17) can be written as

$$[(U_1 - c)^2 - g/k][(U_1 - c)^2 + c^2] - e^{-2kh}[(U_1 - c)^2 + g/k][(U_1 - c)^2 - c^2] = 0. \tag{42.22}$$

When kh is large (short waves), (42.22) gives

$$(U_1 - c)^2[(U_1 - c)^2 + c^2] = 0,$$

which implies $c = (U_1 \pm iU_1)/2$ since we have excluded the root $U_1 = c$ earlier in the analysis. These unstable waves grow rapidly, with an e-folding time comparable to the period. When kh is small (long waves), (42.22) gives

$$khc^2 + (U_1 - c)^2(1 - kh) = 0,$$

which implies $c = U_1(1 - kh) \pm iU_1[(1 - kh)kh]^{1/2}$. In this case $\text{Im}(c) \ll U_1$ and the growth rate is relatively small. We leave it as an exercise for the reader to show that (42.22) also has a complex conjugate pair of roots for intermediate wavenumbers. It turns out, however, that the addition of a rigid bottom is a strongly stabilizing influence. Fenton (1973) has recently considered various vertical shear models for surface gravity waves in a wide channel of finite depth. Depending on the wavenumber and other factors, he finds that the wave perturbations are either marginally stable or decay with position as one moves with a wave down the channel.

(3) *Case* $h \to \infty$

This last limiting case leads to the classical Kelvin-Helmholtz stability problem which is usually the starting point for most elementary studies of stratified shear flow. Kelvin (1871) generalized Helmholtz's original calculation for a homogeneous fluid by including a two-layer model for the density stratification. Taking the limit $h \to \infty$ in (42.17) gives

$$[(U_1 - c)^2 k - g][\Delta c^2 k + (U_1 - c)^2 k - g'] = 0.$$

The first factor gives $c = U_1 \pm (g/k)^{1/2}$, corresponding to (surface) gravity waves being advected with the flow. These roots arise because of the presence of the free surface at $z = h + \eta$. Setting the second factor equal to zero, we obtain the classical Kelvin-Helmholtz result:

$$c = \frac{\rho_1 U_1}{\rho_1 + \rho_2} \pm \left[\frac{g}{k}\frac{\rho_2 - \rho_1}{\rho_2 + \rho_1} - \frac{\rho_1 \rho_2}{(\rho_1 + \rho_2)^2} U_1^2\right]^{1/2}. \tag{42.23}$$

In the absence of mean flow this result reduces to the expression for the interfacial wave speed c_I as given by (42.21). With $U_1 \neq 0$, however, the disturbances move relative to the weighted mean flow speed $\bar{U} = \rho_1 U_1/(\rho_1 + \rho_2)$ with speed $\pm s$, where

$$s^2 = c_I^2 - \rho_1 \rho_2 U_1^2/(\rho_1 + \rho_2)^2, \tag{42.24}$$

provided of course $s^2 > 0$. If $s^2 < 0$, on the other hand, c is complex and the disturbances grow but remain stationary with respect to $\bar{U}$. Thus the flow is stable or unstable according to whether $s^2 > 0$ or $s^2 < 0$. When $s^2 = 0$, the transitional case, we say the flow is *marginally* or *neutrally stable*.

It is clear that $s^2 < 0$ will hold for all k and U_1 when $\rho_2 < \rho_1$, which is expected since this corresponds to a statically unstable configuration. For $\rho_2 > \rho_1$ and a given U_1, it also follows from (42.21) and (42.24) that $s^2 < 0$ for sufficiently large k. However, for these very short waves, the surface tension becomes important and indeed acts as a stabilizing mechanism [see Yih (1965) for details].

The conditions $s^2 > 0$ (< 0) are known as *stability* (*instability*) *criteria*. When written in the form (valid for small density differences)

$$\mathcal{R} = g\frac{\rho_2 - \rho_1}{\rho_1 U_1^2} > \tfrac{1}{2}k \text{ for stability,}$$

$$< \tfrac{1}{2}k \text{ for instability,} \qquad (42.25)$$

it is clear that the criteria define regions of stability and instability in the $(k, \mathcal{R})$-plane. The curve separating these two regions, namely the straight line $\mathcal{R} = k/2$, is known as a *neutral stability curve* or a *stability boundary*. Thus in this plane the region above (below) the stability boundary corresponds to stable (unstable) flows. If (42.25) is divided by k, then $\mathcal{R}k^{-1}$ represents a Richardson number (cf. 42.9) based on the wavenumber of the disturbance (which is the only length scale in the problem in this limiting case). Thus the first statement in (42.25) is analogous to Miles' sufficiency criterion $Ri > 1/4$ for stability of shear flow of a *continuously* stratified, rigidly bounded fluid, a result already quoted in Section 41 and which will be proved in Section 43.

Neutral stability curves for the complete eigenvalue equation

We now return to the full equation (42.17), which was analyzed numerically by Esch (1962). In order to introduce the stability curves it is first convenient to nondimensionalize (42.17) by multiplying the equation by h^2/U_1^4; this gives

$$[(1-c')^2k' - F^{-2}][\Delta(c'^2k' - F^{-2}) + (1-c')^2k' + F^{-2}]$$
$$+ e^{-2k'}[(1-c')^2k' + F^{-2}][\Delta(c'^2k' - F^{-2}) - (1-c')^2k' + F^{-2}] = 0, \qquad (42.26)$$

where $c' = c/U_1$, $k' = kh$ and $F = U_1/(gh)^{1/2}$. F is known as a *Froude number*; for a given stratification, we note from (42.25) that F^{-1} is proportional to the square root of the Richardson number. We observe that the roots of (42.26) can be determined as a function of the nondimensional wavenumber k', with F and Δ as nondimensional parameters, viz.,

$$c' = c'(k'; F, \Delta).$$

The roots in this functional form were computed by Esch and then used to plot the neutral stability curves in (k', F^{-1})-space for various values of the stratification parameter Δ. The results are shown in Fig. 42.2. For each stability curve the imaginary part of c' is exactly zero. The parabolic shape of these curves is not unexpected since $1/F \propto \sqrt{\mathcal{R}} \propto \sqrt{k}$ for marginal stability in the limiting case $h \to \infty$ discussed above. For each unstable solution, Esch found that $0 < \mathrm{Re}(c') < 1$ and hence the disturbance is convected along with the flow, as in the classical Kelvin-Helmholtz case ($h \to \infty$).

Other velocity and density profiles

One of the unsatisfactory consequences of the vortex sheet profile is the unlimited instability at high wavenumbers. One way of stabilizing the situation in this range is to

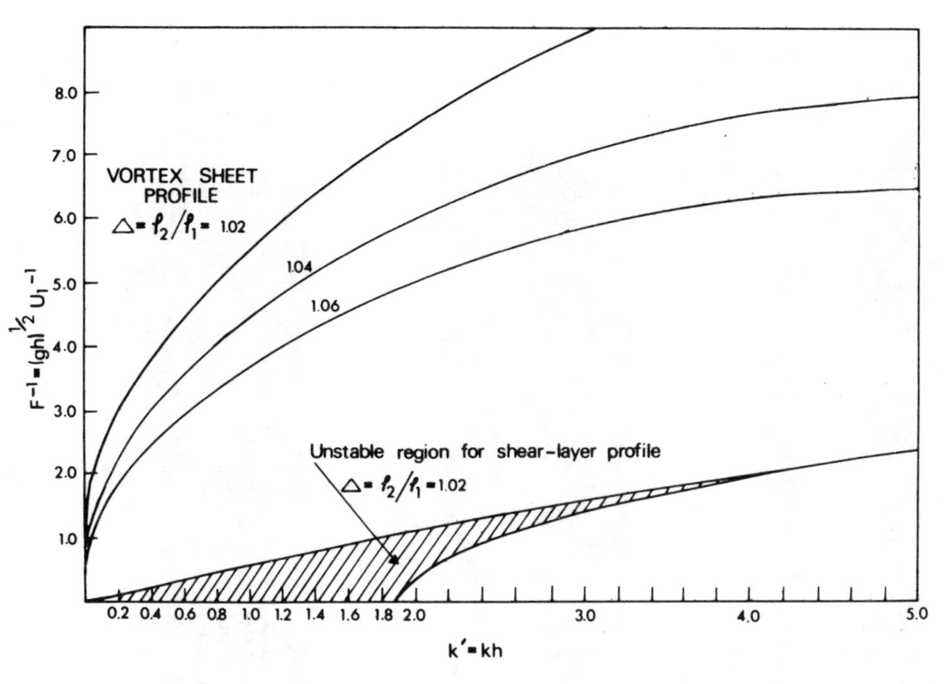

Fig. 42.2. Neutral stability curves. For the vortex sheet the region of instability is below each stability curve, as computed from (42.26). For the shear-layer profile (42.14), the region of instability is represented by the cross-hatched area. (Adapted from Esch, 1962.)

distribute the vorticity of the flow across a layer of finite thickness. The shear-layer profile (42.14) is a good example of such a distributed vorticity model. The solution for this case can also be obtained in terms of elementary functions since (42.15) again holds in each layer. The new contributions to the eigenvalue equation come from the now nonzero term $-\rho_0 U'w$ in (42.10) and (42.12). The calculation of the eigenvalue equation for the shear-layer profile is left as an exercise. From it one obtains, for each Δ, two stability boundaries which enclose a finite region in the (k', F^{-1})-plane (see Fig. 42.2). The flow is thus stable at high wavenumbers. Moreover, for a fixed F^{-1}, the flow is only unstable for a small band of wavenumbers.

These properties are also found in the (k', Ri)-plane for shear-layer profiles in unbounded or rigidly bounded fluids with discontinuous density profiles. If the density varies continuously across the shear layer, however, the stability boundary is approximately parabolic in shape, with the parabola concave downward. As generalizations of the classical Kelvin-Helmholtz instability problem, these cases are also known as Kelvin-Helmholtz (K-H) instabilities, or alternatively, as stratified shear flow instabilities.

For flows with a smoothly varying shear, $U'' \neq 0$, (42.6) no longer has elementary functions as solutions, independently of whether ρ_0 is continuous or discontinuous. While some analytical solutions of (42.6) are possible in terms of special functions (e.g., see Drazin and Howard, 1966; Thorpe, 1969), much effort has been expended on seeking numerical solutions [see Turner (1973) for a discussion of some of these].

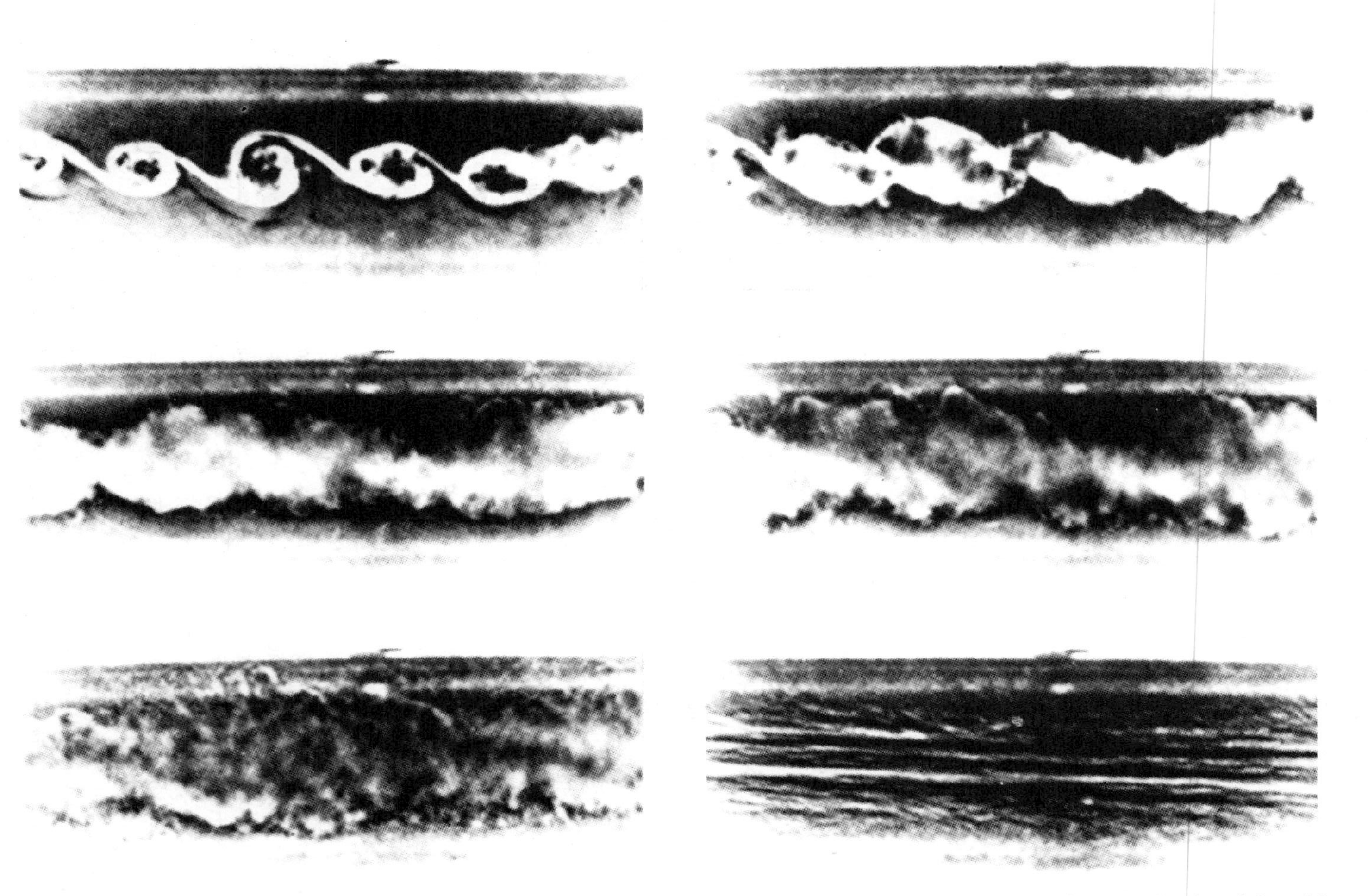

Fig. 42.3 A sequence of shadowgraph pictures showing the evolution of a density interface in the presence of a steady shear. Reading from left to right, top to bottom, an instability of the *K-H* form grows, turbulence is produced, the interface thickens and finally the turbulence is suppressed. (From Thorpe, 1971.)

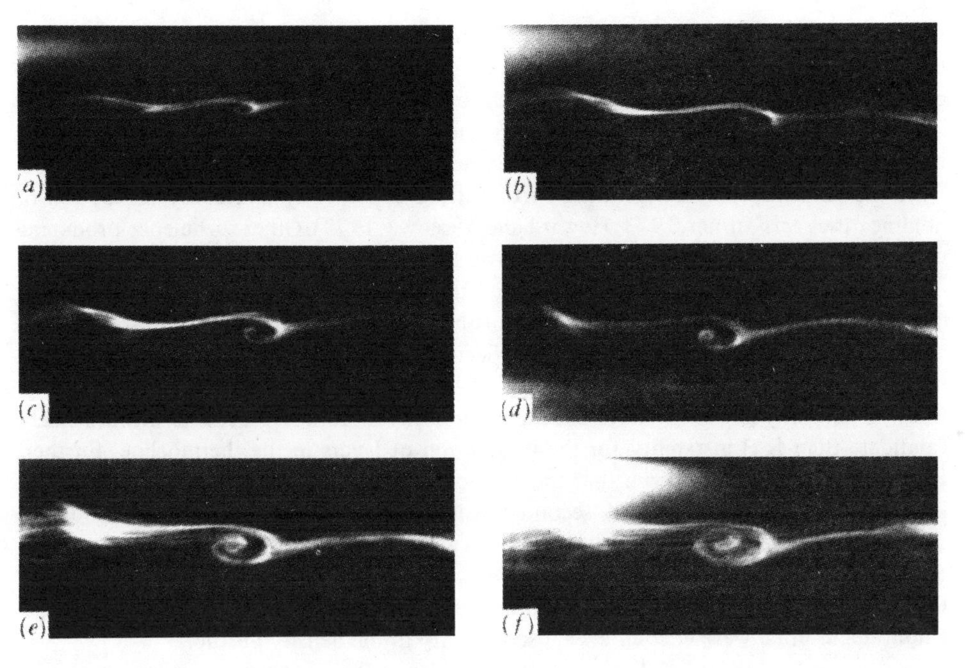

Fig. 42.4. Stages in the development of a billow produced by a long wave travelling along an interface in the thermocline. (From Woods, 1968.)

Oceanic applications

As mentioned earlier, Esch (1962) considered the basic state in Fig. 42.1 as a possible model of estuarine flow. While no data from observed flows were presented, Esch did find that his theoretical results agreed qualitatively with those obtained in a simple laboratory experiment. The application of the (Kelvin-Helmholtz) shear instability mechanism, on the other hand, has had a long history in geophysical fluid dynamics. Kelvin (1871) suggested that it might be an initiating mechanism for the wind generation of surface gravity waves (with ρ_1 and ρ_2 being the densities of the air and water, respectively). However, this mechanism is now largely out of favour since instability would occur only at very high wind speeds (nearly 7 m/s); even a more realistic profile does not alter the situation (Miles, 1959b). However, Miles (1959b) has established that K-H instability might occur at an air–oil interface. Secondly, we remark that clear-air turbulence in the atmosphere is now commonly believed to be generated by K-H instability, especially at its nonlinear stages (Atlas et al., 1970). This observation in part has motivated many laboratory studies of this instability and of its subsequent evolution into small-scale turbulence (see Fig. 42.3 and Thorpe, 1973). Thirdly, it has recently been proposed (Woods, 1968; Woods and Wiley, 1972; Munk and Garrett, 1973; Turner 1973) that K-H instability in the ocean may be an important mechanism that leads to billowing and hence to the breaking of internal waves which in turn produces vertical mixing and horizontal layering. As a consequence of this mixing, the mean flow profile will be modified and therefore will give different growth rates for any new wave perturbations. The observed development of an unstable interfacial wave at the thermocline into a large billow due to K-H instability is shown in Fig. 42.4. Finally, Gargett (1975) has suggested that random occurrences of large-scale K-H instabilities may be a possible explanation of the observed low horizontal correlation of the temperature field at depths between 100 and 300 m.

We conclude ths section by emphasizing that stratified shear flow instability may not be the most effective turbulence-producing mechanism if the growth rates are very slow or if the amplitudes of the growing perturbations are severely limited by dissipation or nonlinearities (cf. Turner, 1973; Howard and Maslowe, 1973). Other turbulence-producing mechanisms involving breaking internal waves are as follows. First, there is the possibility of dynamic gravitational instability in which the vertical density gradient associated with large-amplitude internal waves becomes positive (we recall that $\rho_{0z} < 0$ for static stability). This is a nonlinear instability that occurs when the particle velocity of the wave motion exceeds the wave phase velocity, with the result being convective overturning. Orlanski and Bryan (1969) have shown numerically that this process is a more likely candidate than K-H instability for the production of layers in the thermocline. Further, Frankignoul (1972) has shown that this instability is also likely to develop before shear instability in the deep ocean. A second possibility (McEwan, 1973) is that turbulence develops from very localized intensifications ('traumata') of the density gradient that occur abruptly before an internal wave breaks (due to steepening). Finally, Delisi and Orlanski (1975) have shown experimentally that local overturning occurs when a large-amplitude internal wave encounters a discontinuity in the density profile.

Exercises Section 42

1. Derive (42.12) from the kinematic and pressure conditions at the free surface.

2. Show that (42.21) has a complex conjugate pair of roots for intermediate wavenumbers.

3. Derive the analogue of (42.23) for the case in which both layers are moving with uniform velocities U_1 and U_2.

$$\left(\text{Ans.}\quad c = \frac{\rho_1 U_1 + \rho_2 U_2}{\rho_1 + \rho_2} \pm \left[\frac{g}{k}\frac{\rho_2 - \rho_1}{\rho_2 + \rho_1} - \frac{\rho_1 \rho_2}{(\rho_1 + \rho_2)^2}(U_1 - U_2)^2\right]^{1/2}.\right)$$

4. Derive the eigenvalue equation for the shear-layer profile (42.14).

5. Show that the inclusion of a lower solid boundary has a destabilizing effect on Kelvin-Helmholtz waves with long wavelengths. Also show that this boundary introduces additional families of modes. (Lalas et al., 1976; Lindzen and Rosenthal, 1976.)

43. GENERAL RESULTS ON THE STABILITY OF STRATIFIED SHEAR FLOWS

In this section we first examine the energy transfer equation implied by the perturbation equations (41.2)–(41.4), valid for a nonrotating system. We then prove three theorems concerning stability criteria from the Taylor-Goldstein equation (42.6). Most of these theorems are fairly straightforward extensions of those which are well known in the stability theory of shear flow of a homogeneous fluid (cf. Drazin and Howard, 1966).

Energy transfer equation

From the linearized perturbation equations (41.2)–(41.4), we first form an energy equation which is a generalization of (8.58). Taking the scalar product of (41.2) written in vector form with (u, w), using (41.3) to eliminate w on the right-hand side, and invoking (41.4), we readily obtain

$$\frac{\mathrm{D}_0}{\mathrm{D}t}\left\{\frac{1}{2}\rho_0(u^2+w^2)-\frac{g\rho^2}{2\rho_0'}\right\} = -(pu)_x-(pw)_z-\rho_0 uwU', \tag{43.1}$$

where $\mathrm{D}_0/\mathrm{D}t = \partial_t + U\partial_x$ and the prime denotes d/dz. The total density ρ_T is given by

$$\rho_\mathrm{T} = \rho_0 + \rho;$$

but we can also write ρ_T as (Miles, 1961)

$$\rho_\mathrm{T} = \rho_0 - \rho_0'\xi,$$

where $\xi(x, t)$ is the vertical displacement of a particle from its initial position at depth z. Comparison of these two expressions gives the relation

$$\rho = -\rho_0'\xi. \tag{43.2}$$

Using (43.2) we thus find

$$-\frac{g\rho^2}{2\rho_0'} = \tfrac{1}{2}\rho_0 N^2\xi^2, \tag{43.3}$$

which is the usual expression for the perturbation potential energy density in a stratified fluid. When $U = 0$ we note that (43.1) reduces to the energy conservation law (8.58).

We now assume that all the perturbation quantities are periodic in x with wavelength L, and integrate (43.1) over one wavelength in the x-direction and over the vertical domain of interest, (z_1, z_2) say. Denoting the x-integration $\left[\int_x^{x+L}(\cdot)\,\mathrm{d}x\right]$ by an overbar, this operation gives the *energy transfer equation*

$$\frac{\partial}{\partial t}(T+V) = P+Q, \tag{43.4}$$

where $T = \dfrac{1}{2}\displaystyle\int_{z_1}^{z_2}\rho_0\overline{(u^2+w^2)}\,\mathrm{d}z$

denotes the perturbation kinetic energy (per unit width in the y-direction),

$$V = \frac{1}{2}\int_{z_1}^{z_2} \rho_0 N^2\, \overline{\xi^2}\, dz$$

denotes the perturbation potential energy (per unit width in the y-direction),

$$P = -pw\Big|_{z_2}^{z_1}$$

represents the rate at which work is done on the perturbation flow by the external pressure at the boundaries, and finally

$$Q = \int_{z_1}^{z_2} \tau_{13} U' dz$$

is the rate at which energy is transferred from the mean flow to the perturbation flow by the *Reynolds stress*

$$\tau_{13} = -\rho_0 \overline{uw}.$$

As mentioned in Section 41, τ_{13} represents the rate at which horizontal momentum, $\rho_0 u$, is transported vertically by w. If the boundaries at z_1 and z_2 are rigid (or the boundaries are at infinity) then $P = 0$, and examination of (43.4) reveals that when instability occurs $[(T+V)_t > 0]$, energy is transferred from the shear flow to the perturbation flow, resulting in a net loss of kinetic energy of the basic state.

Synge's generalized Rayleigh criterion

Nearly a century ago Rayleigh (1880) showed that for flow of a homogeneous fluid with rigid boundaries (or boundaries at infinity), a necessary condition for instability is that the profile $U(z)$ should have at least one point of inflection. An analogous but more complicated necessary condition for instability was obtained by Synge (1933) for the case of a stratified fluid. However, his paper was overlooked for several decades and the same result was proved independently by Yih (1957) and Drazin (1958).

To prove Synge's necessary condition, we start with the Taylor-Goldstein equation (42.6) written in the following form:

$$(\rho_0 W w')' - (\rho_0 U' w)' + \rho_0 (N^2 W^{-1} - k^2 W) w = 0, \quad z_1 \leqslant z \leqslant z_2, \tag{43.5}$$

where $W = U - c$. (Note that W is *not* a mean vertical velocity!) Since $W' = U'$ the first two terms can be written as $W(\rho_0 w')' - (\rho_0 U')' w$; hence (43.5) becomes

$$(\rho_0 w')' - [(\rho_0 U')' W^{-1} - \rho_0 N^2 W^{-2} + \rho_0 k^2] w = 0. \tag{43.6}$$

Next we multiply (43.6) by w^* and its complex conjugate by w, subtract and then integrate from z_1 to z_2:

$$\int [w^*(\rho_0 w')' - w(\rho_0 w^{*\prime})']$$

$$- \int |w|^2 \{(\rho_0 U')'(W^{-1} - W^{*-1}) - \rho_0 N^2 (W^{-2} - W^{*-2})\} = 0, \tag{43.7}$$

where $\int \equiv \int_{z_1}^{z_2} dz$ and $|w|^2 = ww^*$. If the boundaries are rigid or at infinity, w and w^* vanish there and the first integral in (43.7) drops out after an integration by parts. Putting $c = c_r + ic_i$ in the second integral and simplifying yields

$$c_i \int \frac{|w|^2}{[(U-c_r)^2 + c_i^2]^2} \{(\rho_0 U')' [(U-c_r)^2 + c_i^2] - 2\rho_0 N^2 (U-c_r)\} = 0. \tag{43.8}$$

Hence if $c_i > 0$ (unstable waves), the expression in the curly brackets must change sign and hence must vanish for some $z \in (z_1, z_2)$. Thus formally we obtain *Synge's theorem*: "A necessary condition for a stratified shear flow to be unstable is that

$$(\rho_0 U')' [(U-c_r)^2 + c_i^2] - 2\rho_0 N^2 (U-c_r) = 0 \tag{43.9}$$

for at least one value of $z \in (z_1, z_2)$." When ρ_0 = constant (homogeneous fluid), (43.9) reduces to $U'' = 0$, which is Rayleigh's well-known necessary condition for instability. The condition (43.9) has no simple interpretation, however, since it involves the unknown eigenvalue c.

Miles' sufficiency condition for stability

Of the many stability properties Miles (1961) established for stratified shear flow, the most celebrated is undoubtedly the stability criteria involving the Richardson number, a result which we have already referred to several times in this chapter. On the assumption of analyticity of U' and ρ_0, Miles showed that a sufficient condition for stability is that $Ri \geqslant 1/4$ everywhere in the flow. Here we shall present Howard's proof (Howard, 1961) of this theorem, which is simpler and does not require the analyticity assumption.

To prove Miles' result we first make two transformations in the Taylor-Goldstein equation. First let $w = FW$, where $W = U - c$ and $F = F(z)$ represents a new dependent variable. Then (43.7) becomes

$$[\rho_0 W(WF' + W'F)]' - (\rho_0 WW'F)' + \rho_0(N^2 - k^2 W^2)F = 0, \quad z_1 \leqslant z \leqslant z_2. \tag{43.10}$$

The rigid-wall boundary conditions imply that $F(z_1) = F(z_2) = 0$ provided $W \neq 0$ at $z = z_1$ and z_2, which we will assume to be the case. Carrying out the differentiation in (43.10) gives

$$(\rho_0 W^2 F')' + \rho_0(N^2 - k^2 W^2)F = 0, \quad z_1 \leqslant z \leqslant z_2. \tag{43.11}$$

Suppose now that F is an unstable solution. Then c is complex and $W = U - c \neq 0$ for any z, and we can choose one branch of $w^{1/2}$ for all (z_1, z_2) which will be as differentiable as is U. Now set $F = G/W^{1/2}$ in (43.11). After a little algebra it follows that

$$(\rho_0 WG')' - [\tfrac{1}{2}(\rho_0 U')' + \rho_0 k^2 W + \rho_0 W^{-1}(\tfrac{1}{4}U^{-1} - N^2)]G = 0, \quad z_1 \leqslant z \leqslant z_2, \tag{43.12}$$

with $G(z_1) = G(z_2) = 0$. Multiplying by G^* and integrating over (z_1, z_2), (43.12) yields, after integrating by parts,

$$\int \rho_0 W(|G'|^2 + k^2|G|^2) + \frac{1}{2}\int (\rho_0 U')'|G|^2 + \int \rho_0 W^* \left|\frac{G}{W}\right|^2 (\tfrac{1}{4}U'^2 - N^2) = 0. \tag{43.13}$$

Equating the imaginary part of (43.13) to zero, we obtain

$$c_i\left\{\int\rho_0(|G'|^2+k^2|G|^2)+\int\rho_0\left|\frac{G}{W}\right|^2(-\tfrac{1}{4}U'^2+N^2)\right\} = 0. \tag{43.14}$$

Hence if $c_i > 0$ (unstable waves), (43.14) implies that $-\frac{1}{4}U'^2+N^2<0$ for some range of z. Thus, as Howard put it, a necessary condition for instability is that $-\frac{1}{4}U'^2+N^2$ be somewhere negative. On the other hand if $-\frac{1}{4}U'^2+N^2\geqslant 0$ everywhere, then (43.14) implies that $c_i = 0$. Thus we obtain (for $U'\neq 0$) *Miles' theorem*: "A sufficient condition for stability of a stratified shear flow is that

$$Ri = N^2/U'^2 \geqslant \tfrac{1}{4} \tag{43.15}$$

everywhere in the flow."

Unlike Synge's necessary condition for instability, Miles' sufficient condition for stability has a simple physical interpretation. Since the Richardson number represents the ratio of buoyancy to inertia, Miles' therorem effectively states that if the stabilizing influence of stratification dominates the destabilizing influence of the nonlinear terms, then the flow is stable. The simplicity of Miles' sufficiency condition also makes it easy to check in the laboratory. Scotti and Corcos (1969) measured small-amplitude perturbations on shear flows in a wind tunnel and found that whenever Ri was everywhere greater than 1/4, the flows were stable to small disturbances. They also found that for unstable flows the waves grew at rates which were predicted by the linear theory. A further discussion of recent experimental results can be found in Thorpe (1973).

Howard's semicircle theorem

This theorem defines a semicircular region in the complex c-plane in which the eigenvalue c for an unstable wave is located. This result was first established by Howard (1961).

Our starting point is (43.11), the equation for F, where $w = FW = F(U-c)$. If this equation is multipled by F^* and the result integrated over (z_1, z_2), we find (again, for rigid boundaries or boundaries at infinity)

$$\int\rho_0 W^2(|F'|^2+k^2|F|^2)-\int\rho_0 N^2|F|^2 = 0. \tag{43.16}$$

Setting the real and imaginary parts of (43.16) equal to zero gives

$$\int\rho_0[(U-c_r)^2-c_i^2](|F'|^2+k^2|F|^2)-\int\rho_0 N^2|F|^2 = 0, \tag{43.17}$$

$$c_i\int\rho_0(U-c_r)(|F'|^2+k^2|F|^2) = 0. \tag{43.18}$$

The second equation immediately gives another of Synge's results (Synge, 1933): "If $c_i > 0$, then there exists some point z_r such that $c_r = U(z_r)$." This is sometimes rephrased to read: "Instability implies that c_r must lie in the range of U." Now let

$$Q = \rho_0(|F'|^2+k^2|F|^2) > 0. \tag{43.19}$$

Then (43.18) becomes, for $c_i > 0$,

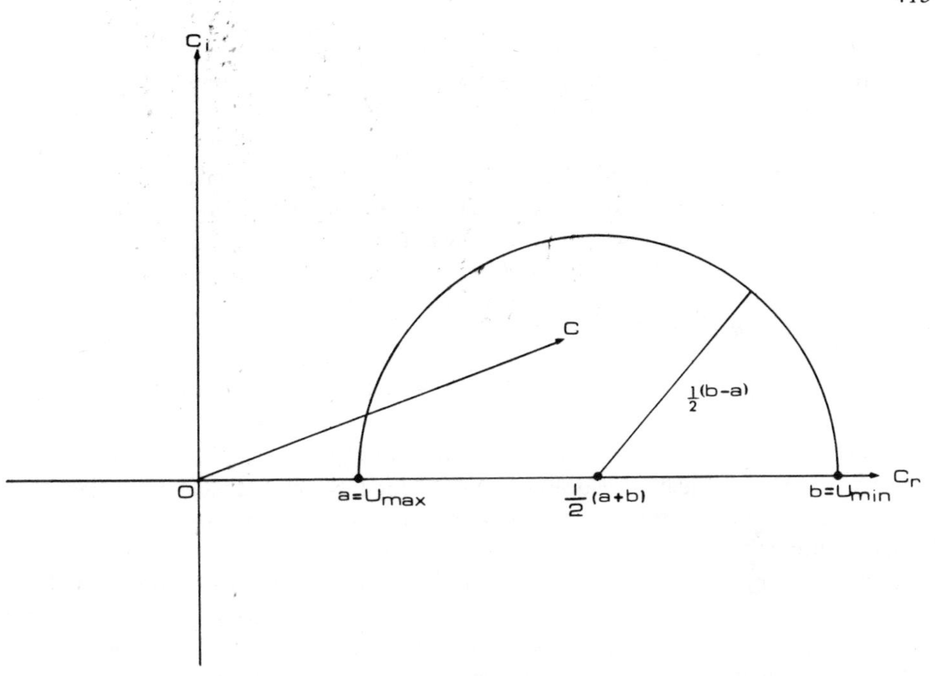

Fig. 43.1. Semicircular region in the complex c-plane defined by (43.23) which contains any unstable mode in a stratified shear flow.

$$\int UQ = c_r \int Q, \tag{43.20}$$

and by virtue of (43.20), (43.17) reduces to

$$\int U^2 Q = (c_r^2 + c_i^2) \int Q + \int \rho_0 N^2 |F|^2. \tag{43.21}$$

We now suppose that $U_{\min} \equiv a \leqslant U(z) \leqslant b \equiv U_{\max}$. Then

$$0 \geqslant \int Q(U-a)(U-b) = \int U^2 Q - (a+b) \int UQ + ab \int Q$$

$$= [c_r^2 + c_i^2 - (a+b)c_r + ab] \int Q + \int \rho_0 N^2 |F|^2 \tag{43.22}$$

using (43.20) and (43.21). But $\int Q > 0$ [see (43.19)] and $\int \rho_0 N^2 |F|^2 > 0$; hence the inequality (43.22) implies that $c_r^2 + c_i^2 - (a+b)c_r + ab \leqslant 0$, or equivalently,

$$[c_r - \tfrac{1}{2}(a+b)]^2 + c_i^2 \leqslant [\tfrac{1}{2}(a-b)]^2. \tag{43.23}$$

Thus we have (see also Fig. 43.1) *Howard's semicircle theorem*: "The complex wave velocity c for any unstable mode ($c_i > 0$) must lie on or inside the semicircle in the upper half c-plane with center at $[(a+b)/2, 0]$ and diameter equal to the range of U."

Howard's theorem is particularly useful when the Taylor-Goldstein equation is numerically integrated as a two-point boundary-value problem because it can be used to find first approximations to the eigenvalues. Nevertheless, because the eigensolutions often oscillate rapidly, the numerical techniques for boundary-value problems are often complicated (see, for example, Betchov and Criminale, 1967). A promising way out of these difficulties may be provided by invariant embedding techniques in which the original *linear* equation is replaced by a system of *nonlinear Riccati* equations, giving an initial-value problem. The nonlinearity gives no difficulty on the computer, while the initial-value formulation eliminates the guessing of unknown derivatives usually encountered in two-point methods.

Shear flows with lateral variations

Thus far we have only treated the problem of two-dimensional shear flow instability in the sense that the basic current is a function of the depth alone. By restricting $U = U(z)$ we were able to formulate a simple two-point boundary-value problem in terms of an ordinary differential equation. However, a model for U which is a much better representation of actual mean currents in the ocean and atmosphere is one in which U also has lateral variations: $U = U(y, z)$. Such a model in general leads to a nonseparable eigenvalue problem involving a partial differential equation and to date only special cases have been treated. For example, Blumen (1971) has studied wave perturbations in a parallel shear flow $U(y, z)$ under the hydrostatic approximation. Drazin (1974), on the other hand, used the WKB method to find approximate eigenvalues and eigenfunctions for a slowly varying flow of the form

$$U(y, z) = U(\epsilon y)\Theta(z), \quad 0 < \epsilon \ll 1,$$

of a two-layer fluid, which is a generalization of the classical, Kelvin-Helmholtz instability problem. Finally, we mention that some progress has also been made on the problem of stability of *nonplanar* shear flow (Drazin and Howard, 1966; Blumen, 1975). However, it is beyond the scope of this book to pursue these interesting topics here.

Exercises Section 43

1. If $(\rho_0 U')'$ is of the same sign for all $z \in (z_1, z_2)$, use (43.8) to show that (Synge, 1933)

$$|c_i| < \max |\rho_0 N^2/(\rho_0 U')'|.$$

2. Derive (43.12)

3. Upon observing that $|W|^{-2} \leqslant c_i^{-2}$, show that (43.14) implies the following bound for the growth rate (Howard, 1961):

$$k^2 c_i^2 \leqslant \max\left(\tfrac{1}{4}U'^2 - N^2\right).$$

4. Assuming $c_i > 0$, show that the substitution $F = W^{-n}H$, with some definite branch being selected if n is not an integer, into (43.11) yields (Howard, 1961)

$$(\rho_0 W^{2(1-n)} H')' - \{k^2 \rho_0 W^{2(1-n)} + n W^{1-2n} (\rho_0 U')' + \rho_0 W^{-2n}[n(1-n)U'^2 - N^2]\}H = 0. \qquad (43.24)$$

5. Show that multiplication of (43.24) by H^* and integration over (z_1, z_2) yields the integral relation

$$\int \rho_0 W^{2(1-n)}[|H'|^2 + k^2|H|^2] + n \int W^{1-2n}(\rho_0 U')'|H|^2$$

$$+ \int \rho_0 W^{-2n}[n(1-n)U'^2 - N^2]\,|H|^2 = 0. \tag{43.25}$$

Hence show that taking $n = 0$ in (43.25) yields Howard's semicircle theorem, $n = 1/2$ gives Miles' theorem, and $n = 1$ gives Synge's theorem.

6. Show that Miles' theorem and Howard's semicircle theorem also hold if the upper rigid surface at $z = z_2$ is replaced by a free surface. [Hint: Use (43.24) for the appropriate values of n together with the new boundary condition at $z = z_2$.]

7. Show that for solutions of the form (42.1) with $c = c_r + ic_i$,

$$\tau_{13} = -\rho_0 \overline{uw} \propto \exp(2kc_i t).$$

8. Find an expression for $\partial\tau_{13}/\partial y$ that is proportional to $\overline{w^2}$. Use this expression together with the condition $\tau_{13} = 0$ at $z = z_1, z_2$ to establish Synge's theorem.

44. INTRODUCTION TO BAROTROPIC AND BAROCLINIC INSTABILITY

In Sections 42 and 43 we discussed the stability of stratified flows with vertical shear in the presence of horizontally propagating wave disturbances with wavelengths comparable to the depth of the ocean. The wave motions considered were two-dimensional ($v \equiv 0$, $\partial/\partial y \equiv 0$), and it was intimated that such motions, when unstable, could be an initiating mechanism for relatively small-scale, vertical mixing (turbulence) in the ocean. Here, and in the remaining sections of this chapter, we shall discuss the stability of large-scale flows on the β-plane that are horizontally and/or vertically sheared. The wave perturbations in this case are hydrostatic, vertically constrained, and have wavelengths comparable to the internal (Rossby) radius of deformation (or order 100 km in the open ocean). It is conceivable that these unstable waves may be an initiating mechanism for the meso-scale eddies that occur in different parts of the ocean. They have been observed, for example, in the vicinity of intense boundary currents such as the Gulf Stream (see Fig. 2.4). As mentioned in Section 18, these eddies have also been observed in the open ocean in the 1970 Polygon experiment (Brekhovskikh et al., 1971) conducted off the west coast of Africa, and in the 1973 MODE experiment off the southeast coast of the United States. Similar motions are currently being investigated in the international POLYMODE experiment which is taking place in a large mid-Atlantic region. In the same way that small-scale, vertically oriented eddies arising from stratified shear flow instabilities can transport heat, salt, and energy in the vertical direction, these meso-scale eddies can transport such quantities over large horizontal distances in the ocean. It seems likely that they play a central role in the mean circulation of the ocean since it has been known for some time that such large-scale eddies play a very important role in the mean circulation of the atmosphere (Starr, 1968).

Governing equations and the basic state

Before discussing the basic-state flow field it is convenient to rewrite here the governing nonlinear adiabatic equations for large-scale, time-dependent hydrostatic motions on the β-plane. For a Boussinesq incompressible fluid, the appropriate equations are, from (5.31), (5.32) and (5.36),

$$\frac{\mathrm{D}u}{\mathrm{D}t} - fv + \rho_*^{-1} p_x = 0, \tag{44.1}$$

$$\frac{\mathrm{D}v}{\mathrm{D}t} + fu + \rho_*^{-1} p_y = 0, \tag{44.2}$$

$$p_z = -\rho g, \tag{44.3}$$

$$\frac{\mathrm{D}\rho}{\mathrm{D}t} = 0, \tag{44.4}$$

$$\nabla \cdot \boldsymbol{u} = 0, \tag{44.5}$$

where ρ_* is a constant reference density.

As our basic-state density and pressure fields we take

$$\rho(y,z) = \rho_0(z) + \bar{\rho}(y,z), \quad \text{and} \quad p(y,z) = p_0(z) + \bar{p}(y,z), \tag{44.6}$$

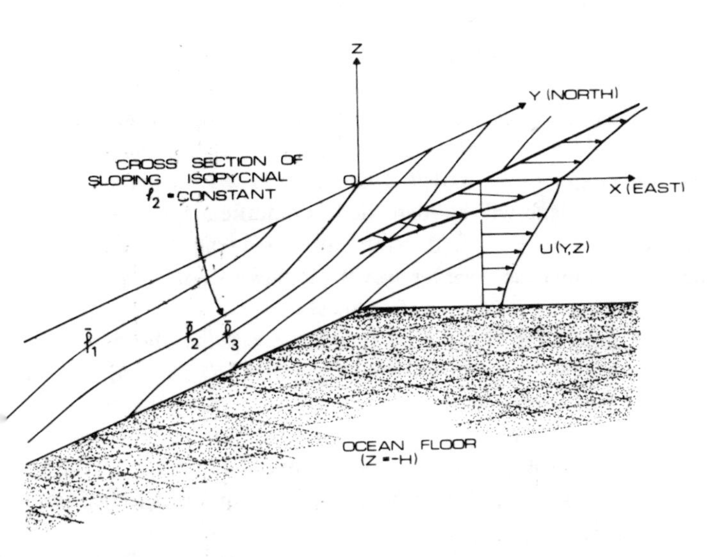

Fig. 44.1. The basic state for the combined barotropic–baroclinic instability problem. For $U_z > 0$ the isopycnal surfaces $\bar{\rho}$ = constant have a positive northward gradient ($\bar{\rho}_1 < \bar{\rho}_2 < \bar{\rho}_3 < \ldots$), so that at a given depth the ocean density increases as one moves northward (in the Northern Hemisphere). In addition to $\bar{\rho}$ there is also a stably stratified vertical density gradient $\partial\rho_0/\partial z < 0$. The ocean surface (assumed rigid) is located at $z = 0$ and the ocean floor (assumed flat) is located at $z = -H$.

where ρ_0 and p_0 are the usual stably stratified hydrostatic density and pressure (thus $\partial p_0/\partial z = -\rho_0 g$ and $\partial\rho_0/\partial z < 0$) and $\bar{\rho}$ corresponds to some mean density field with a north–south gradient (see Fig. 44.1). Substituting (44.6) into (44.3) gives

$$\bar{p}_z = -\bar{\rho}g. \tag{44.7}$$

As the basic flow field, we take a steady zonal current with both vertical and horizontal shear (see Fig. 44.1):

$$u = [U(y,z), 0, 0]. \tag{44.8}$$

Note that equations (44.6) and (44.8) identically satisfy (44.1), (44.4) and (44.5). Substitution of (44.6) and (44.8) into (44.2) yields the *geostrophic relation*

$$fU + \rho_*^{-1}\bar{p}_y = 0. \tag{44.9}$$

Eliminating $\bar{p}$ from (44.7) and (44.9) gives the *thermal wind relation*

$$fU_z = \rho_*^{-1}\bar{\rho}_y g, \tag{44.10}$$

a term commonly used in the meteorological literature. Equation (44.10) shows that in the presence of a horizontal density gradient, the basic current in a rotating fluid must possess a vertical shear. Such geostrophically maintained currents are called *baroclinic*. When $U_z = 0$, on the other hand, $\bar{\rho}_y = 0$ and the current is said to be *barotropic*. When $U_z > 0$ everywhere (44.10) implies that $\bar{\rho}_y > 0$ everywhere also. Since in the ocean the density gradient is of opposite sign to the temperature gradient (see Section 3), this simple configuration implies that at a given depth in the Northern Hemisphere the temperature decreases monotonically to the north. This situation of sloping isopycnals is a state of higher gravitational potential energy than that of flat isopycnals ($\bar{\rho}_y = 0$), a result which will be of utmost importance later when we consider the energy transfer equation.

In the subsequent development we shall consider the stability of the above basic state with respect to zonally propagating perturbations. Thus the (x, t)-dependence of the perturbations will have the form $\exp[ik(x-ct)]$, $k>0$. If the flow is unstable, $\operatorname{Im} c > 0$ and the unstable waves on the β-plane will be identified as modified zonally propagating planetary waves. These unstable waves will develop into large-scale lateral meanders which in the atmosphere are known as *weather waves* or *cyclone waves*. In the ocean these waves may lead to the formation of mesoscale eddies.

For future reference we note that when $U_z \neq 0$ and $U_y = 0$ the above instability phenomenon is known as *baroclinic instability*. When $U_z = 0$ and $U_y \neq 0$, on the other hand, the instability is called *barotropic*. In this latter case, the perturbation density and vertical velocity fields, ρ and w, are identically zero. The combined case ($U_z \neq 0$ and $U_y \neq 0$) is known as baroclinic–barotropic instability. An excellent review of the basic state of geostrophic motion and the wave motions which it may support is to be found in N. Phillips' review article (1963).

Energy transfer equation

To examine the energetics of the perturbation field we proceed as in Section 43. We assume that in (44.1)–(44.5) the dependent variables have the form

$$\rho = \rho_0 + \bar{\rho} + \rho', \quad p = p_0 + \bar{p} + p', \quad \text{and} \quad \boldsymbol{u} = (U + u', v', w'), \tag{44.11}$$

where the basic state takes the form described above and the primed (perturbation) quantities are functions of x, y, z and t. Substituting (44.11) into (44.1)–(44.5), neglecting products of the primed quantities and invoking (44.7) and (44.9), we obtain, upon dropping the primes, the following linearized equations (see also next paragraph):

$$u_t + Uu_x + vU_y - fv + \rho_*^{-1} p_x = 0, \tag{44.12}$$

$$v_t + Uv_x + fu + \rho_*^{-1} p_y = 0, \tag{44.13}$$

$$p_z = -\rho g, \tag{44.14}$$

$$\rho_t + U\rho_x + v\bar{\rho}_y + w\rho_{0z} = 0, \tag{44.15}$$

$$\nabla \cdot \boldsymbol{u} = 0. \tag{44.16}$$

In deriving the above set of equations, we have also neglected the terms wU_z and $w\bar{\rho}_z$ in (44.12) and (44.15), respectively (see Pedlosky, 1971b). If the wU_z term were retained, it would lead to a vertical Reynolds stress contribution in the energy equation [as in (43.4)] that, for large-scale motions, is much smaller than the horizontal Reynolds stress contribution associated with the lateral shear U_y. Since the vertical density gradient is assumed to be largely determined by ρ_0, $w\rho_{0z} \gg w\bar{\rho}_z$, which explains why the latter term can be dropped in (44.15). In the case that $U_{zz} = 0$, the term $w\bar{\rho}_z$ is identically zero [see (44.10)]. These approximations will be verified later by means of a scale analysis of the original nonlinear equations (44.1)–(44.5).

Taking the scalar product of (44.12) and (44.13) with (u, v) and integrating over the meridional plane $0 \leqslant y \leqslant l$, $-H \leqslant z \leqslant 0$ and over one wavelength L in the x-direction (a region we shall call R_w), we obtain

$$\frac{\partial}{\partial t} \int_{R_w} \frac{1}{2} \rho_*(u^2 + v^2) = \int_0^l \int_{-H}^0 \tau_{12} U_y \, dy dz - \int_{R_w} (up_x + vp_y), \tag{44.17}$$

where $\int_{R_w} (\cdot) = \int_x^{x+L} \int_0^l \int_{-H}^0 (\cdot) \, dx dy dz$

and $\tau_{12} = -\rho_* \int_x^{x+L} uv \, dx$

is the horizontal Reynolds stress. In arriving at (44.17) we have assumed that $\boldsymbol{u}$ is periodic in x with period L. (We shall henceforth assume that p and ρ are also periodic in x with period L.) Using (44.16), we find that

$$\begin{aligned} -\int_{R_w} (up_x + vp_y) &= -\int_{R_w} \nabla \cdot (p\boldsymbol{u}) + \int_{R_w} wp_z \\ &= \int_{R_w} wp_z \end{aligned} \tag{44.18}$$

upon integrating the divergence term and using the periodicity condition on u and p and the boundary conditions $v = 0$ at $y = 0, l$ and $w = 0$ at $z = 0, -H$. Hence combining (44.18) with (44.17) and using (44.14) to eliminate p_z, we obtain the *energy transfer equation*

$$\frac{\partial}{\partial t} KE_h = \int_0^l \int_{-H}^0 \tau_{12} U_y dy dz - g \int_{R_w} w\rho, \tag{44.19}$$

where $KE_h = \frac{1}{2} \int_{R_w} \rho_*(u^2 + v^2)$

is the horizontal perturbation kinetic energy. Hence we conclude that for barotropic instability ($\rho = w = 0$), kinetic energy must be transferred from the mean, horizontally sheared flow through the horizontal Reynolds stress τ_{12} in order for the perturbation kinetic energy to increase with time. If $U_y = 0$, on the other hand (baroclinic instability), (44.19) implies that the perturbation vertical velocity must be negatively correlated with the perturbation density for the perturbation energy to grow; that is, the integrated vertical buoyancy flux must be positive. This condition means that on the average, lighter (warm) fluid must rise and heavier (cold) fluid must sink for the system to be unstable. Of course for the general case, the sum of these two different contributions must be positive for barotropic–baroclinic instability to occur. Let us now continue with the case of baroclinic instability alone ($U_y = 0$). If (44.15) is used to eliminate w, (44.19) (with $U_y = 0$) can be written as

$$\frac{\partial}{\partial t} (KE_h + PE) = \int_{R_w} v\rho(\bar{\rho}_y / \rho_{0z}), \tag{44.20}$$

where $PE = -\frac{1}{2} \int\limits_{R_w} g\rho^2/\rho_{0z}$

is the perturbation potential energy [see also (43.3)]. Now $\rho_{0z} < 0$ (stable stratification) and usually $\bar{\rho}_y > 0$ in the oceans (see Fig. 44.1). Therefore, for baroclinic instability to occur (44.20) implies that $v\rho < 0$ on the average. That is, lighter (warm) fluid must move northward and heavier (cold) fluid must move southward. Combining this with the earlier condition $w\rho < 0$ on the average, we see that for instability warm fluid must rise and move northward and cold fluid must sink and move southward. Clearly such a motion will tend to tip the basic-state isopycnals towards the horizontal, i.e., to reduce the potential energy due to stratification. Hence we conclude that when baroclinic instability occurs, the potential energy of the sloping-isopycnal mean state (called the *available potential energy* since it represents potential energy in excess of that due to the purely gravitational stratification) is converted into kinetic and potential energy of the perturbation flow. We also note that baroclinic instability is a mechanism whereby heat from the low latitudes is transported to northern latitudes. For a further discussion of the energetics of this fascinating topic, we refer the reader to Pedlosky (1971a) and Holton (1972).

Derivation of the potential vorticity equation

From the linearized equations (44.12)–(44.16) it is possible to derive a linearized potential vorticity equation (see Exercises). However it is more instructive to return to the original nonlinear equations (44.1)–(44.5) and show by means of a scale analysis and a regular perturbation expansion in a small parameter that the geostrophic pressure satisfies a nonlinear potential vorticity equation. Also, in carrying out this analysis we shall obtain a justification for dropping the two terms in (44.12) and (44.15).

We now define a set of nondimensional (primed) quantities by the equations

$$(x,y) = l(x',y'), \quad z = Hz', \quad t = (l/U_0)t',$$

$$(u,v) = U_0(u',v'), \quad w = (HU_0/l)w', \tag{44.21}$$

where l and H are characteristic horizontal and vertical length scales respectively, and U_0 is a characteristic horizontal velocity scale. Further, we write the density and pressure fields as follows:

$$\rho = \rho_0(z) + (\rho_* f_0 U_0 l/gH)\rho', \quad \text{and} \quad p = p_0(z) + (\rho_* f_0 U_0 l)p', \tag{44.22}$$

where $p_{0z} = -\rho_0 g$ and $f_0 = 2\Omega \sin \phi_0$. Substituting (44.21) and (44.22) into (44.1)–(44.5) we obtain, upon dropping all the primes,

$$Ro\frac{Du}{Dt} - \left(1 + \frac{l}{R}\cot \phi_0 y\right) v + p_x = 0, \tag{44.23}$$

$$Ro\frac{Dv}{Dt} + \left(1 + \frac{l}{R}\cot \phi_0 y\right) u + p_y = 0, \tag{44.24}$$

$$p_z = -\rho, \tag{44.25}$$

$$Ro\,\frac{\mathrm{D}\rho}{\mathrm{D}t} - B(z)w = 0, \tag{44.26}$$

$$\nabla \cdot \boldsymbol{u} = 0, \tag{44.27}$$

where R is the radius of the Earth and

$$Ro = U_0/f_0 l \quad \text{and} \quad B(z) = N^2H^2/f_0^2l^2 \tag{44.28}$$

are the *Rossby* and *Burger numbers*, respectively. From (44.23), (44.24) and (44.26) it is clear that Ro is a measure of the departure of the motion from geostrophy. In barotropic–baroclinic instability theory it is assumed that $Ro \ll 1$. For weak mid-ocean flows, $U_0 \sim 0.1\ \mathrm{m\,s^{-1}}$; therefore, with $l \sim 10^5\ \mathrm{m}$ and $f_0 \sim 10^{-4}\ \mathrm{rad\,s^{-1}}$, we find $Ro = 0(10^{-2})$. Even for strong flows ($U_0 \sim 1\ \mathrm{m\,s^{-1}}$), Ro is still much less than unity. The Burger number can be written in the form

$$B(z) = (r_i/l)^2,$$

that is, as the square of the ratio of the internal (Rossby) radius of deformation to the horizontal length scale (cf. 20.33). Another fundamental assumption in barotropic–baroclinic instability theory is that these two horizontal length scales are comparable, and hence $B(z)$ is of order unity. For $N^2 \sim 4 \cdot 10^{-6}\ \mathrm{rad\,s^{-1}}$, $H \sim 5 \cdot 10^3\ \mathrm{m}$ and the above scales for f_0 and l, we indeed see that $B = 0(1)$. Finally, we assume that $l \cot \phi_0/R = 0(Ro)$ (in fact typically, $l \cot \phi_0/R \sim 2 \cdot 10^{-2}$) and hence define a nondimensional β^* by the relation

$$\frac{l}{R}\cot\phi_0 = Ro\left(\frac{l\cot\phi_0}{RRo}\right) \equiv Ro\beta^*. \tag{44.29}$$

The quantity β^* is sometimes called the *planetary vorticity factor.*

The astute reader will recognize the similarity between the nondimensional equations obtained here and those that were obtained in Section 20 in connection with topographic planetary waves in a stratified fluid [cf. (20.27)–(20.31)]. In the latter, the Burger number and a planetary vorticity factor (cf. 20.32) also appeared as 0(1) quantities in the nondimensional equations, whereas the fractional change in depth, δ_s, appeared as the small parameter. In Section 20 we obtained an asymptotic solution (in terms of a regular perturbation expansion) in powers of δ_s; here we proceed analogously, and expand the solution in powers of the small Rossby number Ro.

Under the condition of small Ro, we therefore write the dependent variable u in the form

$$u = u^{(0)} + Ro\,u^{(1)} + \ldots$$

and similarly for v, w, p and ρ. Thus, using (44.29), equations (44.23)–(44.27) yield the following zeroth-order equations:

$$-v^{(0)} + p_x^{(0)} = 0, \tag{44.30a}$$

$$u^{(0)} + p_y^{(0)} = 0, \tag{44.30b}$$

$$p_z^{(0)} = -\rho^{(0)}, \tag{44.31}$$

$$w^{(0)} = 0, \tag{44.32}$$

$$u_x^{(0)} + v_y^{(0)} = 0. \tag{44.33}$$

We note that (44.30) implies that $p^{(0)}$ can be identified as a stream function and hence (44.33) is identically satisfied; $p^{(0)}$ is known as the *geostrophic stream function* and (44.30)–(44.33) are known as the *geostrophic equations*. However, as we found in Section 20, the system (44.30), (44.31) is not closed: we have three equations for four unknowns, and to determine $p^{(0)}$ say, we must go to the next order. Taking into account (44.32), the first-order equations in Ro are given by

$$\frac{\mathrm{D}_h u^{(0)}}{\mathrm{D}t} - v^{(1)} - \beta^* y v^{(0)} + p_x^{(1)} = 0, \tag{44.34a}$$

$$\frac{\mathrm{D}_h v^{(0)}}{\mathrm{D}t} + u^{(1)} + \beta^* y u^{(0)} + p_y^{(1)} = 0, \tag{44.34b}$$

$$p_z^{(1)} = -\rho^{(1)}, \tag{44.35}$$

$$\frac{\mathrm{D}_h \rho^{(0)}}{\mathrm{D}t} - B(z) w^{(1)} = 0, \tag{44.36}$$

$$\nabla \cdot \boldsymbol{u}^{(1)} = 0, \tag{44.37}$$

where $$\frac{\mathrm{D}_h}{\mathrm{D}t} = \frac{\partial}{\partial t} + u^{(0)} \frac{\partial}{\partial x} + v^{(0)} \frac{\partial}{\partial y}. \tag{44.38}$$

These are known as the *quasi-geostrophic equations*.

We note that the vertical velocity does not appear in (44.34), in keeping with our neglect of the term wU_z in the analogous linearized dimensional equation (44.12). Also, in (44.36) there is no vertical gradient of $\rho^{(1)}$, in keeping with the omission of $w\bar{\rho}_z$ in (44.15).

To derive a single equation for $p^{(0)} \equiv \psi$, we eliminate $p^{(1)}$ from (44.34) by cross-differentiation, then use (44.37) to eliminate $u_x^{(1)} + v_y^{(1)}$, and finally use (44.36) to eliminate $w_z^{(1)}$. Writing $\rho^{(0)}$, $u^{(0)}$ and $v^{(0)}$ in terms of ψ by using (44.30) and (44.31), we finally obtain

$$\frac{\partial}{\partial y}\left(\frac{\mathrm{D}_h \psi_y}{\mathrm{D}t}\right) + \frac{\partial}{\partial x}\left(\frac{\mathrm{D}_h \psi_x}{\mathrm{D}t}\right) + \beta^* \frac{\partial \psi}{\partial x} + \frac{\partial}{\partial z}\left(\frac{1}{B}\frac{\mathrm{D}_h \psi_z}{\mathrm{D}t}\right) = 0, \tag{44.39}$$

where now (cf. 44.38)

$$\frac{\mathrm{D}_h}{\mathrm{D}t} = \frac{\partial}{\partial t} + \frac{\partial \psi}{\partial x}\frac{\partial}{\partial y} - \frac{\partial \psi}{\partial y}\frac{\partial}{\partial x}.$$

It is easy to show that

$$\frac{\partial}{\partial y}\frac{\mathrm{D}_h \psi_y}{\mathrm{D}t} = \frac{\mathrm{D}_h}{\mathrm{D}t}\frac{\partial^2 \psi}{\partial y^2}$$

and similarly for the other terms involving $\mathrm{D}_h/\mathrm{D}t$. Hence (44.39) takes the form

$$\frac{D_h}{Dt}[\nabla_h^2\psi + (B^{-1}\psi_z)_z + \beta^* y] = 0, \tag{44.40}$$

which expresses the conservation of potential vorticity for the zeroth-order flow. We shall see below that the linearized version of (44.40) is equivalent to the usual planetary wave equation for a stratified fluid.

We now assume that the motions are confined to the rectangular channel $0 \leqslant y \leqslant 1$, $-1 \leqslant z \leqslant 0$. On the walls of the channel, ψ must obey the boundary conditions

$$v^{(0)} = \psi_x = 0 \quad \text{at} \quad y = 0, 1 \tag{44.41}$$

and $w^{(1)} = 0$ at $z = 0, -1$, which, from (44.31) and (44.36), is equivalent to

$$\frac{D_h \psi_z}{Dt} = 0 \quad \text{at} \quad z = 0, -1. \tag{44.42}$$

The linearized vorticity equation

We now suppose that $p^{(0)} = \psi$ has the form

$$\psi = \Psi_s(y,z) + \Psi(x,y,z,t), \tag{44.43}$$

where Ψ_s represents a zonally uniform steady flow field and Ψ is the small perturbation field: $|\Psi| \ll |\Psi_s|$. From (44.30) and (44.31) we note that the basic-state velocity and density are given by

$$U(y,z) = -\Psi_{sy} \quad \text{and} \quad \rho = -\Psi_{sz}. \tag{44.44}$$

In terms of the basic state discussed at the beginning of this section, these quantities can be identified with the (dimensional) U and $\bar{\rho}$ in (44.8) and (44.7), respectively. We note that any function $\Psi_s = \Psi_s(y,z)$ satisfies (44.40) exactly. Substituting (44.43) into (44.40), using (44.44) and neglecting the quadratic terms in Ψ, we obtain the following linearized vorticity equation:

$$(\partial_{t'} + U'\partial_{x'})[\Psi'_{x'x'} + \Psi'_{y'y'} + (B^{-1}\Psi'_{z'})_{z'}] + [\beta^* - U'_{y'y'} - (B^{-1}U'_{z'})_{z'}]\Psi'_{x'} = 0, \tag{44.45}$$

where the primes have been reintroduced to emphasize that all the quantities are dimensionless. Substituting (44.43) into (44.41) and (44.42) and linearizing, we obtain

$$\Psi'_{x'} = 0 \quad \text{at} \quad y' = 0, 1, \tag{44.46a}$$

$$(\partial_{t'} + U'\partial_{x'})\Psi'_{z'} - U'_{z'}\Psi'_{x'} = 0 \quad \text{at} \quad z' = 0, -1. \tag{44.46b}$$

To rewrite these in dimensional form, put $U' = U/U_0$, $\Psi' = \Psi/\rho_* f_0 U_0 l$, and use (44.21), (44.28) and (44.29); thus (44.45) and (44.46) yield the following dimensional potential vorticity equation and boundary conditions:

$$(\partial_t + U\partial_x)[\nabla_h^2 + f_0^2\partial_z(N^{-2}\partial_z)]\Psi + q_y\Psi_x = 0, \tag{44.47}$$

$$\Psi_x = 0 \quad \text{at} \quad y = 0, l, \tag{44.48}$$

$$(\partial_t + U\partial_x)\Psi_z - U_z\Psi_x = 0 \quad \text{at} \quad z = 0, -H, \tag{44.49}$$

$$\text{where} \quad q_y = \beta - U_{yy} - f_0^2(N^{-2}U_z)_z \tag{44.50}$$

is the cross-stream gradient of the basic-state potential vorticity.

When $U = 0$ (no basic flow), (44.47) reduces to the low-frequency approximation ($\omega \ll f$) of (15.21), the usual planetary wave equation for a Boussinesq fluid. For the barotropic case ($\partial_z = 0$), (44.47)–(44.50) reduce to

$$(\partial_t + U\partial_x)\nabla_h^2\Psi + (\beta - U_{yy})\Psi_x = 0, \tag{44.51}$$

with $\Psi_x = 0$ at $y = 0, l$.

Equation (44.51) is also easily derived directly from the perturbation equations (44.12)–(44.16), which for the barotropic case ($\partial_z = 0$, $\rho = w = \bar{\rho}_y = 0$) reduce to (44.12) and (44.13) and $u_x + v_y = 0$. The latter implies there exists a stream function Ψ such that $u = -\Psi_y$, $v = \Psi_x$; substituting these expressions into (44.12) and (44.13) and eliminating p we immediately arrive at (44.51). From (44.51) one can also derive the energy-transfer equation (44.19) with $\rho w = 0$ upon multiplying by $\rho_*\Psi$ and integrating over one wavelength (L) and over the width of the channel (l):

$$\frac{\partial}{\partial t}\int_x^{x+L}\int_0^l \frac{1}{2}\rho_*(\Psi_x^2 + \Psi_y^2)\,\mathrm{d}x\mathrm{d}y = -\int_x^{x+L}\int_0^l \rho_* uvU_y\,\mathrm{d}x\mathrm{d}y.$$

With a little more effort, the combined barotropic–baroclinic energy transfer equation can also be obtained by integrating (44.47) times $\rho_*\Psi$ over R_w.

Necessary condition for instability

We now suppose that the perturbation geostrophic stream function consists of zonally propagating waves:

$$\Psi = \Psi(y,z)\,\mathrm{e}^{i(kx-ct)}, \quad k > 0. \tag{44.52}$$

Substituting (44.52) into (44.47)–(44.49) yields the following boundary-value problem:

$$(U-c)[\partial_{yy} - k^2 + f_0^2\partial_z(N^{-2}\partial_z)]\,\Psi + q_y\Psi = 0,$$
$$0 \leqslant y \leqslant l, \quad -H \leqslant z \leqslant 0, \tag{44.53}$$

with boundary conditions

$$\Psi = 0 \quad \text{at} \quad y = 0, l, \tag{44.54}$$

$$(U-c)\,\Psi_z - U_z\Psi = 0 \quad \text{at} \quad z = 0, -H. \tag{44.55}$$

Since the system (44.53)–(44.55) is in general a nonseparable problem, it is extremely difficult to find exact solutions corresponding to a given mean flow with both horizontal and vertical shear. One recourse is to model the continuous stratification by a two-layer model and hence work with two coupled equations with y-derivatives only. This approach has been taken by Pedlosky (1964a, b), among others, and will be discussed briefly later. However, it is possible to obtain many qualitative stability results for arbitrary flows by using techniques similar to those employed in Section 43. Here we shall prove one of the simplest of these results, a necessary condition for instability akin to Synge's necessary condition (43.9) for instability of stratified shear flows. This condition represents an extension of the baroclinic instability criteria obtained by Fjortoft (1950) and Charney and Stern (1962).

Let us suppose that (44.53) has unstable wave solutions with $c = c_r + ic_i$ and $c_i > 0$. We can then divide (44.53) by $U - c$ since $U - c \neq 0$. Upon multiplying by Ψ^*, integrating over the domain of the equation and using the boundary conditions (44.54) and (44.55), we obtain

$$\int_0^l \int_{-H}^0 (|\Psi_y|^2 + k^2|\Psi|^2 + |\Psi_z|^2 f_0^2 N^{-2})\, dy dz$$

$$= \int_0^l \int_{-H}^0 \frac{|\Psi|^2 q_y}{U - c}\, dy dz + f_0^2 \int_0^l \frac{N^{-2}|\Psi|^2 U_z}{U - c}\Bigg|_{z=-H}^0 dy. \tag{44.56}$$

Next we put $U - c = U - c_r - ic_i$ into the right-hand side of (44.56) and rearrange the whole equation into the form $A + iB = 0$, where A and B are real. Thus $A = 0$ and $B = 0$; the latter equation takes the form

$$c_i \left\{ \int_0^l \int_{-H}^0 \frac{|\Psi|^2 q_y}{|U - c|^2}\, dy dz + f_0^2 \int_0^l \frac{N^{-2}|\Psi|^2 U_z}{|U - c|^2}\Bigg|_{z=-H}^0 dy \right\} = 0. \tag{44.57}$$

Hence if $c_i > 0$, the expression inside the curly brackets must vanish. Thus a necessary condition for instability is that

$$\int_0^l \int_{-H}^0 \frac{|\Psi|^2 q_y}{|U - c|^2}\, dy dz + f_0^2 \int_0^l \frac{N^{-2}|\Psi|^2 U_z}{|U - c|^2}\Bigg|_{z=-H}^0 dy = 0. \tag{44.58}$$

Another necessary condition for instability is obtained from the equation $A = 0$ (Pedlosky, 1964a); however, we shall leave the derivation of this as an exercise for the reader.

The condition (44.58) is rather complicated for the combined barotropic–baroclinic instability case. In the case of baroclinic instability alone for which y-variations are ignored, (44.58) reduces to

$$\int_{-H}^0 A(z)[\beta - f_0^2(N^{-2}U_z)_z]\, dz + f_0^2 N^{-2} U_z|_{-H}^0 = 0,$$

where $A(z) = |\Psi|^2/|U - c|^2 > 0$. Suppose $U_z(0) - U_z(-H) > 0$. Then the last term as well as the first is positive, and instability can only occur if the interior vortex-stretching term (the middle term, proportional to f_0^2) is sufficiently negative. For barotropic instability, there is no z-dependence in any of the quantities and (44.58) reduces to

$$\int_0^l \frac{|\Psi|^2}{|U - c|^2} q_y\, dy = 0, \tag{44.59}$$

where now (cf. 44.50) $\quad q_y = \beta - U_{yy}. \quad$ (44.60)

Therefore $\beta - U_{yy}$ must vanish at least once in the interval $(0, l)$. For sufficiently large β, q_y may not vanish and hence we can say that β is a stabilizing influence. On the f-plane, the necessary condition (44.59) reduces to Rayleigh's criterion for a homogeneous, nonrotating and nonhydrostatic flow (cf. Drazin and Howard, 1966).

The above results are of course valid only for oceans of constant depth, a rather unrealistic assumption. For generalizations of the above results when $H = H(x, y)$, we refer the reader to Pedlosky (1964a). Some of the consequences of a variable-depth ocean will be explored in Section 45. We also refer the reader to Pedlosky (1964a) for derivations of bounds on the phase speed and growth rate of unstable waves, including, for example, the analogue of Howard's semicircle theorem proved in Section 43 for stratified shear flows.

A two-layer model

We saw in Chapters 2 and 3 and in Section 42 that a two-layer model for the stratification is a good approximation for an ocean with a sharp thermocline just below the mixed layer. A good overview of the theory of barotropic–baroclinic instability for such a model is contained in the papers of Pedlosky (1964a, b). Before consulting these references, the reader is encouraged to derive the governing potential vorticity equations for ψ_1 and ψ_2, the geostrophic stream functions for each layer (Exercise 44.5). Here we shall merely quote the two-layer analogue of (44.58), the necessary condition for instability in a continuously stratified ocean. The result (44.61) quoted below is readily derived from the linearized version of the two-layer potential vorticity equations.

Let ρ_1, H_1 and $U_1(y)$ be, respectively, the density, the mean depth in the absence of motion and the mean velocity in the upper layer, and similarly for ρ_2, H_2 and $U_2(y)$ in the lower layer. Let $H = H_1 + H_2$, $g' = (\rho_2 - \rho_1)g/\rho_2$, $r_e = \sqrt{gH}/f_0$, $r_1 = \sqrt{g'H_1}/f_0$ and $r_2 = \sqrt{g'H_2}/f_0$. The length r_e is the usual external deformation radius, and r_1, r_2 are of the order of r_{i2}, the internal deformation radius for a two-layer fluid $[r_{i2} = (g'H_1H_2/Hf_0^2)^{1/2}]$. For mesoscale motions of horizontal scale l such that

$$l^2/r_e^2 \ll 1, \quad l^2/r_1^2 = 0(1), \quad l^2/r_2^2 = 0(1),$$

it can be shown that a necessary condition for instability in the presence of zonally propagating disturbances of the form $\Psi_n(y)\,e^{ik(x-ct)}$ is given by

$$\int_0^l \sum_{n=1}^{2} H_n \frac{|\Psi_n(y)|^2}{|U_n(y) - c|^2} \frac{dq_n}{dy}\, dy = 0, \tag{44.61}$$

where Ψ_n is the perturbation stream function in layer n and where

$$\frac{dq_1}{dy} = \beta - \frac{d^2U_1}{dy^2} + \frac{U_1 - U_2}{r_1^2}, \tag{44.62a}$$

$$\frac{dq_2}{dy} = \beta - \frac{d^2U_2}{dy^2} + \frac{U_2 - U_1}{r_2^2}. \tag{44.62b}$$

Examination of (44.61) and (44.62) reveals that if the flow is unstable, the potential vorticity gradient must be somewhere positive and somewhere negative. We note also that a sufficiently large β will stabilize the flow since then the potential vorticity gradient will be positive in both layers. In particular, consider the simple example $U_1 = \text{constant} > 0$ and $U_2 \ll U_1$. Then (44.62) implies that $dq_1/dy > 0$ and $dq_2/dy \simeq \beta - U_1/r_2^2$. Therefore when instability occurs we must have $\beta - U_1/r_2^2 < 0$ or $U_1 > r_2^2\beta$. For the open ocean, H_2 and hence r_2 is relatively large and hence U_1 would have to be relatively large for this inequality to hold. Thus on the basis of this simple argument we can say that strong shears are necessary for baroclinic instability in the open ocean.

Exercises Section 44

1. Show that a homogeneous fluid in a container with a sloping surface has more gravitational potential energy for the same total volume of fluid than when the surface is level.

2. By eliminating the pressure from (44.12) and (44.13), derive the linearized perturbation vorticity equation.

3. From (44.51) derive the energy transfer equation for barotropic instability.

4. By considering the real part of (44.56) equal to zero along with (44.57), derive another necessary condition for instability involving the product of U and q_y.

5. Derive the following (dimensional) potential vorticity equations for the geostrophic stream functions ψ_1 and ψ_2 that describe mesoscale motions in a two-layer fluid (Pedlosky, 1964a):

$$\frac{\mathrm{D}_1}{\mathrm{D}t}\left[\nabla_h^2\psi_1 + \beta y + \frac{1}{r_1^2}(\psi_2 - \psi_1)\right] = 0, \tag{44.63a}$$

$$\frac{\mathrm{D}_2}{\mathrm{D}t}\left[\nabla_h^2\psi_2 + \beta y + \frac{1}{r_2^2}(\psi_1 - \psi_2)\right] = 0, \tag{44.63b}$$

where $r_n^2 = g'H_n/f_0^2 \quad (n = 1, 2)$

and $\dfrac{\mathrm{D}_n}{\mathrm{D}t} = \dfrac{\partial}{\partial t} + \dfrac{\partial\psi_n}{\partial x}\dfrac{\partial}{\partial y} - \dfrac{\partial\psi_n}{\partial y}\dfrac{\partial}{\partial x} \quad (n = 1, 2).$

Governing equations

In the preceding section we found that the stream function $\Psi(x,y,t)$ for a small perturbation flow superimposed on a barotropic zonal current $U(y)$ satisfies (44.51). For solutions of the usual travelling wave form

$$\Psi(x,y,t) = \Psi(y)\,e^{ik(x-ct)}, \quad k>0, \tag{45.1}$$

(44.51) reduces to

$$(U-c)(\Psi''-k^2\Psi)+(\beta-U'')\Psi = 0, \tag{45.2}$$

which also follows from (44.53) upon setting $\partial_z = 0$ (the barotropic case). In addition to investigating a few solutions of (45.2) in this section, we shall also consider zonal flows over variable topography with north–south variation: $H=H(y)$. In this case (and also under the rigid-lid approximation) the conservation of mass equation is (cf. 20.9)

$$Hu_x+(Hv)_y = 0. \tag{45.3}$$

We now introduce the mass transport stream function ψ [not to be confused with the geostrophic stream function introduced following (44.33)] by the equations

$$Hu = -\psi_y, \quad Hv = \psi_x. \tag{45.4}$$

Equation (45.3) is then identically satisfied, and substitution of (45.4) into (44.12) and (44.13) yields, after cross-differentiation, the following vorticity balance equation:

$$(\partial_t+U\partial_x)\left[\left(\frac{1}{H}\psi_y\right)_y+\frac{1}{H}\psi_{xx}\right]+\left(\frac{f-U_y}{H}\right)_y\psi_x = 0. \tag{45.5}$$

We note that $(f-U_y)/H$ is now the potential vorticity of the basic state. For solutions $\psi(x,y,t)$ of the form (45.1), (45.5) implies that the amplitude $\psi(y)$ satisfies

$$(U-c)\left[\left(\frac{1}{H}\psi'\right)'-\frac{k^2}{H}\psi\right]+\left(\frac{f-U'}{H}\right)'\psi = 0, \tag{45.6}$$

which clearly reduces to (45.2) when $H=$ constant.

At a vertical wall ($y=y_0$, say) the normal flux Hv must vanish, which implies that (cf. 45.4)

$$\psi(y_0) = 0 \tag{45.7}$$

for zonally propagating waves. For an ocean of constant depth, the same condition applies to $\Psi(y)$. If the basic flow extends to $y=\infty$ say, ψ and Ψ must be bounded there and satisfy the Sommerfeld radiation condition that there be no inward flux of energy from $y=\infty$. If U, U' or H is discontinuous at $y=0$ say, then we need two matching conditions at $y=0$. These are most easily derived by integrating the differential equation for the amplitude function over the interval $(-\epsilon, \epsilon)$ and then taking the limit $\epsilon\to 0$. Thus integrating (45.6) in this manner and then integrating by parts, we obtain

$$\left(\frac{U-c}{H}\right)\psi'|_{-\epsilon}^{\epsilon}+\left(\frac{f-U'}{H}\right)\psi|_{-\epsilon}^{\epsilon}-k^2\int\limits_{-\epsilon}^{\epsilon}\left(\frac{U-c}{H}\right)\psi\mathrm{d}y-\int\limits_{-\epsilon}^{\epsilon}\frac{f}{H}\psi'\mathrm{d}y = 0. \tag{45.8}$$

Assuming the integrands in the two integrals are bounded in $(-\epsilon, \epsilon)$, (45.8) thus yields the matching condition

$$\left[\left(\frac{U-c}{H}\right)\psi'+\left(\frac{f-U'}{H}\right)\psi\right] = 0 \quad \text{at} \quad y = 0, \tag{45.9}$$

where the bold square brackets have the usual meaning (cf. 23.29). It can be shown that (45.9) ensures the continuity of pressure at $y = 0$. Performing a similar analysis on (45.2) gives

$$[(U-c)\Psi' - U'\Psi] = 0 \quad \text{at} \quad y = 0. \tag{45.10}$$

To obtain a second matching condition for ψ say, we first replace the upper limit ϵ in (45.8) by y, multiply by $H/(U-c)^2$ and then integrate over $(-\delta, \delta)$; this gives

$$\int\limits_{-\delta}^{\delta}\left(\frac{\psi}{U-c}\right)'\mathrm{d}y+\int\limits_{-\delta}^{\delta}\frac{f}{(U-c)^2}\psi\mathrm{d}y+\int\limits_{-\delta}^{\delta}\frac{H}{(U-c)^2}C(-\epsilon)\,\mathrm{d}y-k^2\int\limits_{-\delta}^{\delta}\left\{\frac{H}{(U-c)^2}\right.$$

$$\left.\times\int\limits_{-\epsilon}^{y}\left(\frac{U-c}{H}\right)\psi(y')\,\mathrm{d}y'\right\}\mathrm{d}y-\int\limits_{-\delta}^{\delta}\left\{\frac{H}{(U-c)^2}\int\limits_{-\epsilon}^{y}\frac{f}{H}\psi'(y')\,\mathrm{d}y'\right\}\mathrm{d}y = 0, \tag{45.11}$$

where $C(-\epsilon)$ denotes all the quantities in (45.8) that are evaluated at $y = -\epsilon$. Hence integrating the first term and then proceeding to the limit $\delta \to 0$, (45.11) yields

$$\left[\frac{\psi}{U-c}\right] = 0 \quad \text{at} \quad y = 0. \tag{45.12}$$

In physical terms, (45.12) ensures that the normal mass transport at the material interface centered at $y = 0$ is continuous. It is easy to show that (45.12) also applies to $\Psi(y)$.

Early work

Equation (45.2) was first derived by Rossby et al. (1939) with reference to the zonal circulation in the atmosphere, but a formal investigation of its qualitative theory and its solutions came much later (for example see Kuo, 1949; Lipps, 1962, Howard and Drazin, 1964). In particular, Kuo (1949) showed that a necessary condition for instability is that $\beta - U'' = 0$ for some y, a result we derived in Section 44 [cf. (44.59) and (44.60)]. Lipps (1962) obtained the solution for the symmetric jet

$$U(y) = U_0\,\mathrm{sech}^2 y, \quad -\infty < y < \infty, \tag{45.13}$$

for which case (45.2) can be transformed into Legendre's equation. He showed that the perturbations are divided into symmetric and antisymmetric modes, the former having their maximum amplitude at $y = 0$ (where the jet has a maximum) and the latter having zero amplitude at $y = 0$. Lipps also found that the antisymmetric modes are more stable than the symmetric ones. Solutions for several other flows are presented in Howard and Drazin (1964), including the solution for a vortex sheet (see below).

The topographic stability equation (45.6) for the case $f = f_0$ (a constant) was analyzed by Niiler and Mysak (1971) with reference to the Gulf Stream meanders. Its qualitative theory has recently been investigated by Grimshaw (1976). The first attempt to explain the existence of the Gulf Stream meanders in terms of an unstable barotropic flow was given by Haurwitz and Panofsky (1950) who based their calculations on the constant-depth equation (45.2) with $\beta = 0$.

Stability of a vortex sheet on the β-plane

Examination of (45.2) reveals that the only example of a shear flow on the β-plane which leads to a constant coefficient equation is that of a vortex sheet:

$$U(y) = U_0 \operatorname{sgn} y, \quad -\infty < y < \infty \; (U_0 > 0). \tag{45.14}$$

Substitution of (45.14) into (45.2) yields

$$\left.\begin{aligned} \Psi'' - [k^2 + \beta/(c - U_0)]\Psi &= 0, \quad y > 0 \\ \Psi'' - [k^2 + \beta/(c + U_0)]\Psi &= 0, \quad y < 0 \end{aligned}\right\} \tag{45.15}$$

provided that $c \neq \pm U_0$ (i.e., that there are no critical layers). The solution of (45.15) which is bounded at $y = \pm\infty$ and which satisfies (45.12) is given by

$$\Psi = \begin{cases} (c - U_0)\Psi_0 \exp\{-[k^2 + \beta/(c - U_0)]^{1/2} y\}, & y > 0 \\ (c + U_0)\Psi_0 \exp\{[k^2 + \beta/(c + U_0)]^{1/2} y\}, & y < 0, \end{cases} \tag{45.16}$$

where the positive square root is implied. Application of the remaining matching condition (45.10) gives the following equation for c:

$$(c - U_0)^2[k^2 + \beta/(c - U_0)]^{1/2} + (c + U_0)^2[k^2 + \beta/(c + U_0)]^{1/2} = 0. \tag{45.17}$$

When $U_0 = 0$, this yields $k^2 + \beta/c = 0$, or equivalently,

$$\omega = kc = -\beta/k,$$

which is the dispersion relation for a barotropic, westward-propagating planetary wave (see 18.28). On the other hand, when $\beta = 0$, (45.17) has the solution

$$c = \pm i U_0,$$

which are the nonpropagating solutions found for the case of Helmholtz instability of a vortex sheet in a homogeneous fluid (see answer of Exercise 42.3, with $\rho_1 = \rho_2$ and $U_2 = -U_1$).

To gain further information about the general case, we take the second term in (45.17) to the right-hand side and square both sides; this gives the following cubic for c:

$$c(c^2 + U_0^2) + \frac{\beta}{4k^2}(3c^2 + U_0^2) = 0. \tag{45.18}$$

It can be shown that the discriminant of (45.18) is positive for all real β/k^2, which implies that there is a complex conjugate pair of roots and that the third root is real; it can also be shown that the latter is admissible only for $0 \leqslant \beta/k^2 < \infty$ and $-3\beta/4k^2 \leqslant c \leqslant -U_0$ (Drazin and Howard, 1966). Therefore, for all U_0 and all wavenumbers, the vortex sheet

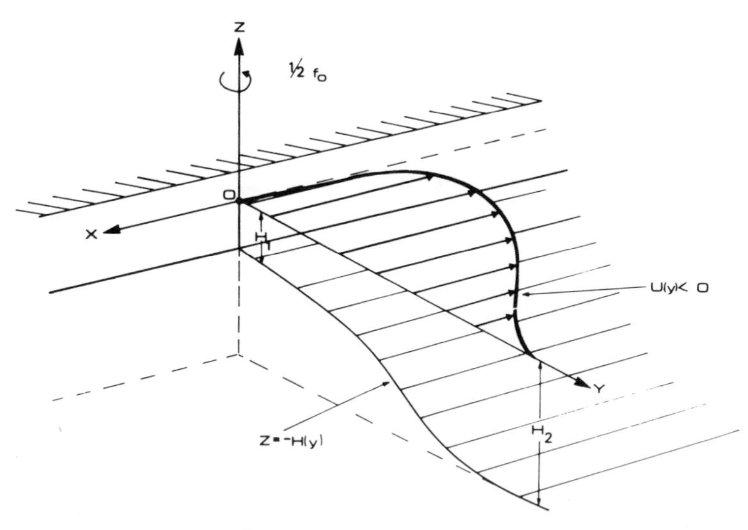

Fig. 45.1. Schematic diagram of a barotropic, laterally sheared mean flow along a continental shelf.

is unstable. Further, it can be shown that the growth rate of the unstable waves decreases as β increases, which illustrates the stabilizing effect of β. However, even as $\beta \to \infty$, the waves remain unstable since in this limit (45.18) implies that $c \to \pm iU_0/\sqrt{3}$.

Stability of a shear flow along the continental shelf

Orlanski (1969) and Nikitin and Tareyev (1972) have suggested that the meandering behaviour of the Gulf Stream and other strong western boundary currents may originate from an instability of a baroclinic jet which flows over a bottom topography which has a strong variation across the jet. On the basis of numerical integrations of the governing equations for a two-layer model, Orlanski concluded that unstable waves on the Gulf Stream are likely to have wavelengths of a few hundred kilometers, which is similar to the spatial scale of the observed meanders (for example, see Hansen, 1970). Orlanski's analysis is rather complicated, however, since the basic flow has both lateral and vertical shear and since the calculations are done for two models of the topography (one representing the coastal topography south of Cape Hatteras, and the other the continental rise beyond this Cape). We shall discuss a simple but related barotropic instability problem first solved by Niiler and Mysak (1971) in which a laterally sheared current flows along a continental shelf region (see Fig. 45.1). We shall assume that topographic variations are relatively large so that we can approximate f by f_0 (that is, ignore the restoring force due to β). Hence the governing equation for the amplitude of a travelling wave disturbance is given by (45.6) in which $f = f_0$. Since β is neglected, the problem has rotational symmetry about the vertical axis and applies to a flow of arbitrary orientation.

To motivate our choice of a model for $U(y)$ and $H(y)$, we show in Fig. 45.2a the topography and vertically averaged mean axial current of the Gulf Stream across the Blake Plateau, at latitude 30°N. We note that the current profile is approximately V-shaped (a symmetric jet) and has its maximum speed where the depth rapidly changes from a small shelf region to a constant-depth region (the Blake Plateau). Thus a simple model for $U(y)$ and $H(y)$ is given by

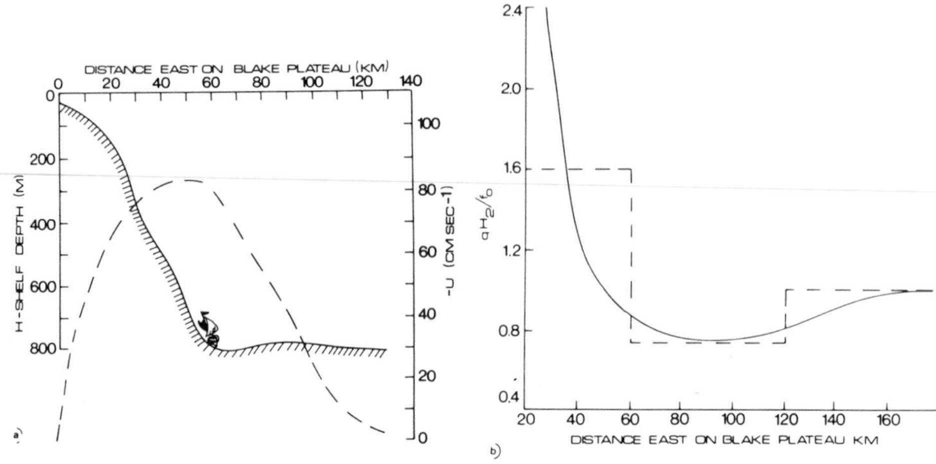

Fig. 45.2. (a) Depth profile and vertically averaged mean axial current of the Gulf Stream at latitude 30°N on the Blake Plateau. (b) The nondimensional potential vorticity (qH_2/f_0) for the profiles shown in (a). The dashed line is the mathematical model used in the analysis. (From Niiler and Mysak, 1971.)

$$U(y) = \begin{cases} -U_0 y/l & , \quad 0 \leqslant y \leqslant l \\ -U_0(2-y/l), & l \leqslant y \leqslant 2l \\ 0 & , \quad 2l \leqslant y < \infty \end{cases} \tag{45.19}$$

and $$H(y) = \begin{cases} H_1, & 0 \leqslant y < l \\ H_2, & l < y < \infty. \end{cases} \tag{45.20}$$

For a discussion of solutions of this barotropic instability problem with smoothly varying profiles for $U(y)$ and $H(y)$, we refer the reader to Tareyev (1971). For the profiles (45.19) and (45.20) the potential vorticity $q(y) = (f_0 - U')/H$ is piecewise constant, and for the Blake Plateau region this model of q is shown in Fig. 45.2b; in this figure q as computed from the data is also shown. We note that for the data, $q' = 0$ for one value of y; hence Kuo's necessary condition for instability is satisfied by the observations. The attractive feature of the model (45.19), (45.20) is that in regions 1, 2, and 3, defined by $0 \leqslant y \leqslant l$, $l \leqslant y \leqslant 2l$ and $2l \leqslant y < \infty$ respectively, (45.6) reduces to the simple constant-coefficient differential equation:

$$\psi_i'' - k^2\psi_i = 0 \quad (i = 1, 2, 3), \tag{45.21}$$

provided $U - c \neq 0$ (no critical layers). (As mentioned in Section 41, the critical layer case has been examined by McKee, 1977.)

The solution of (45.21), which vanishes at $y = 0$ (see 45.7), is bounded at $y = \infty$ and is continuous at $y = l$ and $2l$ (see 45.12), is

$$\psi = \begin{pmatrix} \psi_1 \\ \psi_2 \\ \psi_3 \end{pmatrix} = \left\{ \begin{array}{l} A \sinh ky \\ A \sinh ky + B \sinh k(y-l) \\ [A \sinh 2kl + B \sinh kl] \exp[-k(y-2l)] \end{array} \right\}. \tag{45.22}$$

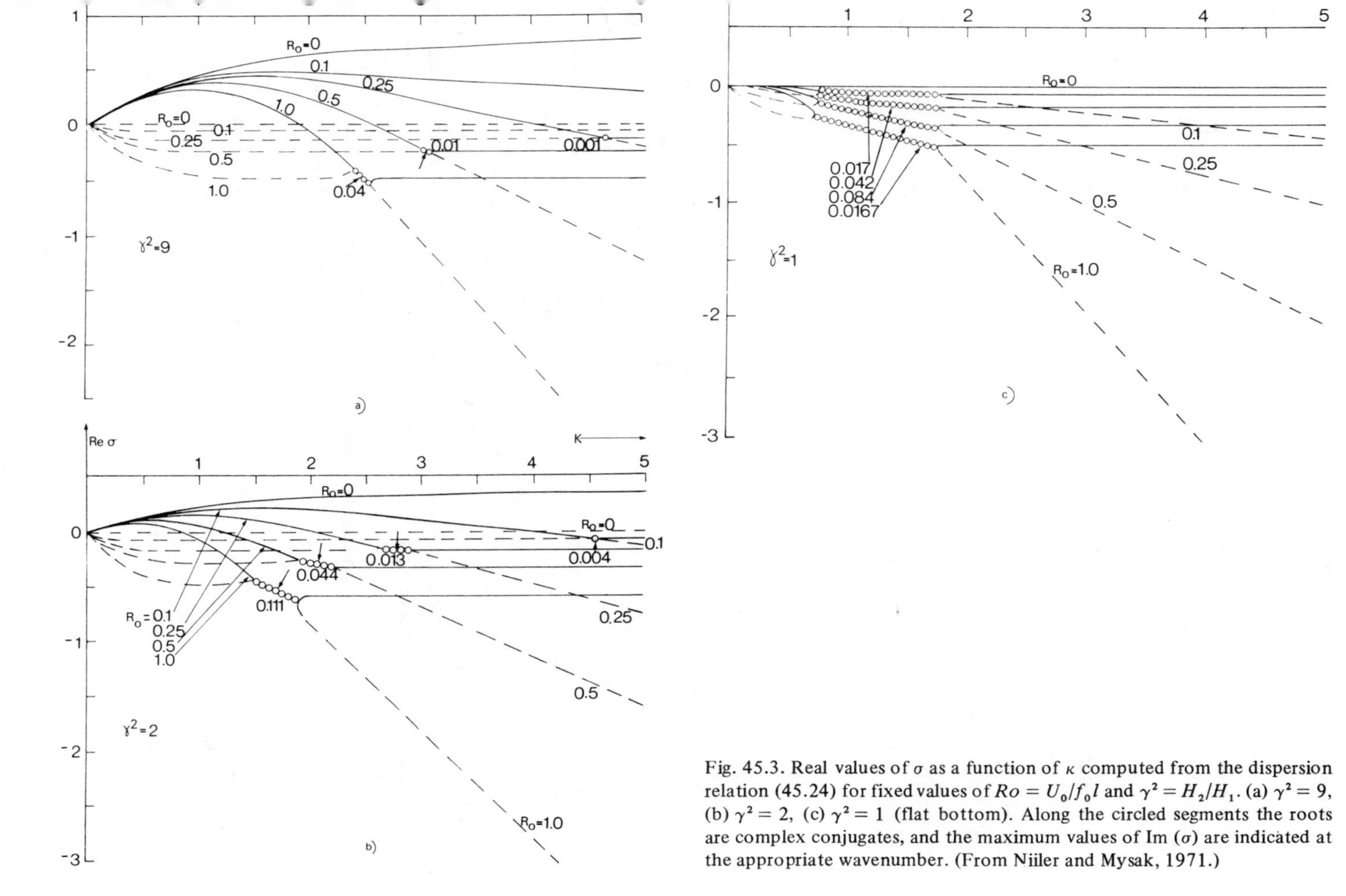

Fig. 45.3. Real values of σ as a function of κ computed from the dispersion relation (45.24) for fixed values of $Ro = U_0/f_0 l$ and $\gamma^2 = H_2/H_1$. (a) $\gamma^2 = 9$, (b) $\gamma^2 = 2$, (c) $\gamma^2 = 1$ (flat bottom). Along the circled segments the roots are complex conjugates, and the maximum values of Im (σ) are indicated at the appropriate wavenumber. (From Niiler and Mysak, 1971.)

The application of the other matching condition (45.9) at $y = l$ and $2l$ yields two homogeneous equations for A and B:

$$\begin{pmatrix} k(U_0+c)(\gamma^2-1)\cosh kl - \left[f_0(\gamma^2-1)+\dfrac{U_0}{l}(\gamma^2+1)\right]\sinh kl, & -k(U_0+c) \\ ck\cosh 2kl + \left(\dfrac{U_0}{l}+ck\right)\sinh 2kl, & ck\cosh kl + \left(\dfrac{U_0}{l}+ck\right)\sinh kl \end{pmatrix}\begin{pmatrix} A \\ B \end{pmatrix} = \begin{pmatrix} 0 \\ 0 \end{pmatrix} \tag{45.23}$$

For a nontrivial solution the determinant of the matrix in (45.23) must vanish, which yields the following dispersion relation:

$$\begin{aligned} &\sigma^2[(\gamma^2-1)\cosh\kappa + e^{\kappa}]\,e^{\kappa} + \sigma\{Ro\,[(\gamma^2-1)\cosh\kappa(\sinh\kappa+\kappa e^{\kappa}) - e^{\kappa}(\gamma^2+1)\sinh\kappa \\ &\quad + \kappa e^{2\kappa} + \sinh 2\kappa] - e^{\kappa}(\gamma^2-1)\sinh\kappa\} + Ro\sinh\kappa\{Ro\,[\kappa(\gamma^2-1)\cosh\kappa \\ &\quad - (\gamma^2+1)\sinh\kappa + 2\kappa\cosh\kappa] - (\gamma^2-1)\sinh\kappa\} = 0, \end{aligned} \tag{45.24}$$

where $\sigma = ck/f_0 \equiv \omega/f_0$, the nondimensional frequency

$\kappa = kl$, the nondimensional wavenumber

$Ro = U_0/f_0 l$, the Rossby number

$\gamma^2 = H_2/H_1 > 1$, the topographic parameter.

For a given κ, Ro, and γ^2, (45.24) gives two roots for σ. In the Northern Hemisphere ($f_0 > 0$), positive (negative) frequency corresponds to waves whose phase travels in the positive (negative) x-direction, that is against (with) the mean current (see Fig. 45.1). Figure 45.3 shows the real part of the roots of (45.24) as a function of κ for various values of the parameters Ro and γ^2. When the two curves coalesce, the roots are complex conjugates and there is an unstable wave whose phase propagates with the current. For any given Ro and γ^2 this occurs for a small range of intermediate wavenumbers. We also note that one root is always negative (the "lower branch"), and hence this wave's phase, for all κ, moves with the current. The wave associated with the upper branch, which is a modified shelf wave (cf. Section 25), propagates against the current for small wavenumbers but with the current for large wavenumbers. When this branch crosses the wavenumber axis, corresponding to a zero frequency, a stationary Rossby lee wave is produced. For a given γ^2, Fig. 45.3 shows that instability increases with increasing Ro in two ways: both the maximum growth rate and the wavenumber band of instability increase with increasing Ro. More specifically, since Ro is proportional to the magnitude of the shear of the mean flow (U_0/l), we conclude that the shear exerts a destabilizing influence. Alternatively, for a fixed shear, we can say that rotation has a stabilizing influence since Ro is proportional to f_0^{-1}. Finally we note that topography is a stabilizing influence, which, as we saw earlier, is the case for β in a flat-bottomed ocean. These different destabilizing and stabilizing mechanisms are most clearly illustrated in Fig. 45.4 where the stability boundaries in the (κ, Ro)-plane are plotted for the values of γ^2 used in Fig. 45.3.

When $Ro = 0$ (no basic current), (45.24) reduces to $\sigma = 0$ (steady motions) or to

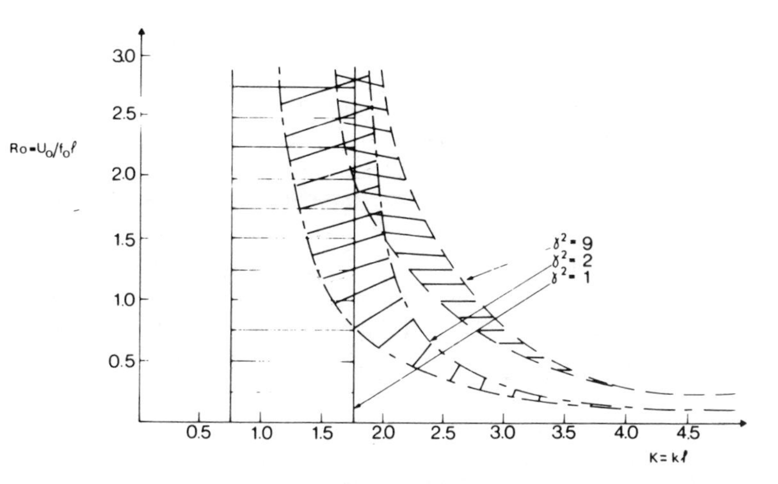

Fig. 45.4. The neutral stability curves in the (κ, Ro)-plane computed from the dispersion relation (45.24) for various values of γ^2. The region of instability for each γ^2 is shaded.

$$\sigma = \frac{\gamma^2 - 1}{1 + \gamma^2 \coth \kappa} > 0, \tag{45.25}$$

which is the dispersion relation for a right-bounded shelf wave on a single-step topography [cf. (25.61) and note the different orientation of axes used in Section 25, which accounts for the minus sign in (25.61)]. From Fig. 45.3 we see that $\sigma(\kappa)$ as given by (45.25) is a monotonic increasing function of κ. The other interesting limiting case is $\gamma^2 = 1$ (flat bottom), when waves exist only because of the current and always propagate with the current. We shall see an analogous situation in Section 46 where we examine Eady's model for baroclinic instability on the f-plane.

For the Gulf Stream over the Blake Plateau, $R_0 \simeq 0.25$ and $\gamma^2 \simeq 2$; the dispersion relation for this case is shown in Fig. 45.3b. The most unstable wave has a wavelength of 140 km and a period of 10 days; its phase and group velocities (both in the direction of the current) are 14 and 31 km/day, respectively. In practice, if such a growing wave is excited over the Blake Plateau, it will propagate in about two weeks into the deep water on the continental rise, which is to the northeast of Cape Hatteras. However, in this new region a wave with a period of 10 days will be stable according to another dispersion relation which is valid for a deep continental rise region. Further, the wave in the latter region travels slower and hence is longer: the phase velocity is 11 km/day and the wavelength is 180 km (see Niiler and Mysak, 1971). These theoretical values are consistent with some of the measurements made by Hansen (1970) who found that the Gulf Stream meanders travel northeastward from Cape Hatteras with a velocity of 4–9 km/day and have a wavelength of 200–400 km.

We close this topic by remarking that Lee (1975) has recently studied the wind-stress generation of these unstable waves. He found that the response is always larger at the Stream's outer edge and that a wind system moving slowly with the Stream is most efficient at generating the unstable waves.

Other considerations

In the solutions presented in this section we have assumed that no critical levels occur ($U \neq c$ for any y). Geisler and Dickinson (1974) numerically solved the initial-value problem of a forced planetary wave encountering a critical level in a zonal flow which initially has the profile $U(y) = U_m \tanh(y/L)$. The main result of their calculation is that the shape of the profile $U(y)$ changes with time in such a way that the potential vorticity gradient $q' = \beta - U''$ is reduced to zero at the critical level. Yamagata (1976b), on the other hand, solved the initial-value problem for planetary waves in a Couette shear flow ($U' =$ constant) by using convected coordinates (Phillips, 1966, p. 180) and Fourier transforms. He found that the ray trajectories of a planetary wave packet are similar to the observed trajectories of mesoscale eddies found to the southeast of the Gulf Stream (Richardson et al., 1973).

In keeping with much of the literature on barotropic (and baroclinic) instability, we have focused our attention on basic currents that are steady and uniform in the direction of the flow. Recently, there has been considerable interest in the theory of barotropic instability for situations in which the basic flow is neither steady nor zonally uniform. For example, Lorenz (1972) considered the stability of a barotropic flow whose stream function has the form

$$\Psi = -U_0 y + \psi_0 \sin\,[k_0(x - ct)]. \tag{45.26}$$

where $c = U_0 - \beta/k_0^2$ is the phase speed of a zonally propagating planetary wave embedded in a westerly current of speed U_0. Lorenz found that the flow (45.26) may be unstable with respect to further zonally propagating disturbances. For sufficiently large ψ_0 or large wavenumber, Lorenz showed that the flow is unstable, but that for small ψ_0 or small wavenumber the β-effect may render the flow stable. Gill (1974) and others have generalized Lorenz' stability conditions, and Duffy (1975) has examined the effect of horizontal divergence on the earlier nondivergent results. In particular, Duffy showed that in addition to the unstable and neutral planetary waves found by Lorenz, there exist modes corresponding to stable inertia–gravity waves.

We conclude this section on barotropic instability on a slightly philosophical note. We recall that our objective in most of this chapter has been to determine whether a given mean (zonal) flow is stable with respect to travelling wave disturbances. If the flow is unstable we found from the linearized governing equations that the disturbances grow exponentially with time. From this behaviour we inferred that stratified shear flow instability or barotropic–baroclinic instability may be an initiating mechanism for small vertical, or large horizontal eddy-like motions (turbulence) in the ocean. It is natural to ask two questions at this stage. What happens to these eddies after they are formed? And second, where does the mean flow come from in the first place? For some of the answers to these questions in connection with motions in a nonrotating stratified fluid, we refer the reader to the recent book by Turner (1973). For a discussion of these questions for motions in a homogeneous fluid on the β-plane, we refer the reader to Rhines (1975) and the references quoted therein. It has long been known that two-dimensional eddies that are closely packed in a homogeneous fluid at high Reynolds number evolve towards larger scales. On the β-plane, this cascade towards larger scales produces nonlinear planetary waves (cf. Section 19), and the turbulent migration from large to small wavenumbers ceases at $k_\beta = (\beta/2U_{rms})^{1/2}$, independent of initial conditions other than U_{rms}, the root-mean-square particle speed. For typical mid-latitude flows ($U_{rms} = 5$ cm s^{-1}), $k_\beta^{-1} =$ 70 km. After passing through a state of propagating waves, the cascade, when homogeneous in space, tends towards a flow of alternating zonal jets, which are almost perfectly steady (cf. Fig. 14.4). Hence we are led back to the beginning of barotropic stability theory: the stability of zonal flows.

Exercises Section 45

1. Derive the matching condition (45.10).
2. Show that for the jet profile (45.13), (45.2) can be transformed into a Legendre equation after an appropriate change of the independent variable.
3. Derive (45.18) and show that this equation always has two complex conjugate roots.
4. Compare the solution obtained here for the vortex-sheet basic flow on the β-plane with that obtained in Exercise 42.3, in which $U_2 = -U_1$.
5. From (45.6) show that a necessary condition for instability is that $[(f - U')/H]' = 0$ for at least one value of y in the domain of the differential equation.

In the presence of a basic baroclinic current without lateral shear, the amplitude of the perturbation stream function for zonally propagating waves (cf. 44.52) satisfies (cf. 44.53)

$$(U-c)[\partial_{yy}-k^2+f_0^2\partial_z(N^{-2}\partial_z)]\Psi+q_y\Psi = 0, \tag{46.1}$$

where here $q_y=\beta-f_0^2(N^{-2}U_z)_z$. The appropriate boundary conditions for a rectangular channel of width l and depth H are

$$\Psi = 0 \quad \text{at} \quad y = 0, l, \tag{46.2}$$

$$(U-c)\Psi_z-U_z\Psi = 0 \quad \text{at} \quad z = 0, -H. \tag{46.3}$$

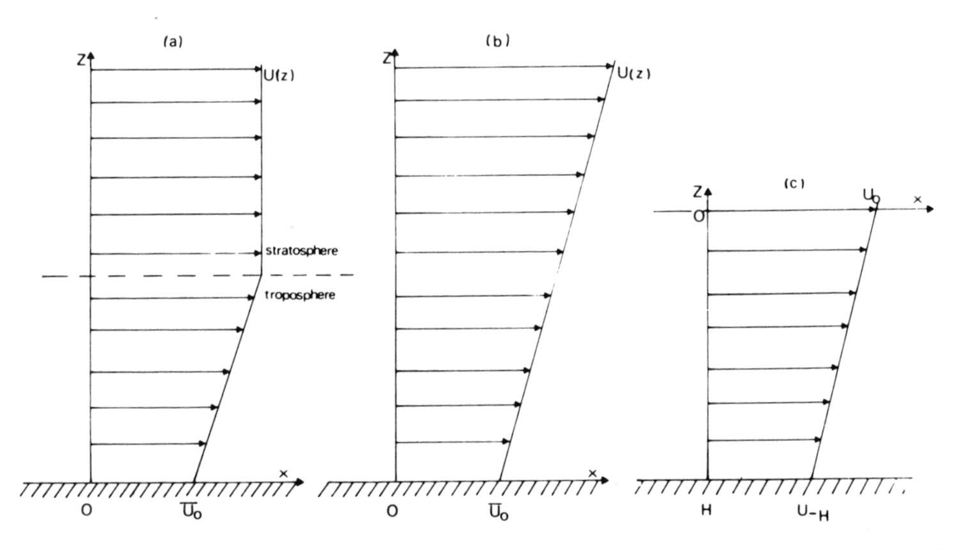

Fig. 46.1. Profiles of zonal flows used in the early literature of baroclinic instability. (a and b) Charney (1947); (c) Eady (1949).

The first thorough study of the solutions of (46.1) was made by Charney (1947) for typical baroclinic westerly flows in the atmosphere. However, the first clear physical explanation of baroclinic instability in the atmosphere was given by Bjerknes and Holmboe (1944); but their results are mathematically incomplete, being based on semi-empirical considerations of the thermal wind relation (44.10) rather than on the complete dynamical equations. Charney (1947) constructed y-independent solutions of (46.1) for the mean flow profiles (a) and (b) shown in Fig. 46.1. Examination of (46.1) reveals that for constant N and for either of these flows on the β-plane, the differential equation has variable coefficients. The solutions can be written in terms of confluent hypergeometric functions, and we refer the reader to Charney (1947) for details. Charney found that the stability boundaries for profiles (a) and (b) were very similar, from which he concluded that motions in the upper atmosphere had little influence on those in the lower atmosphere. The stability boundary (neutral stability curve) for the current profile (b) is shown in Fig. 46.2 for an atmosphere with constant lapse rate (i.e., constant vertical temperature

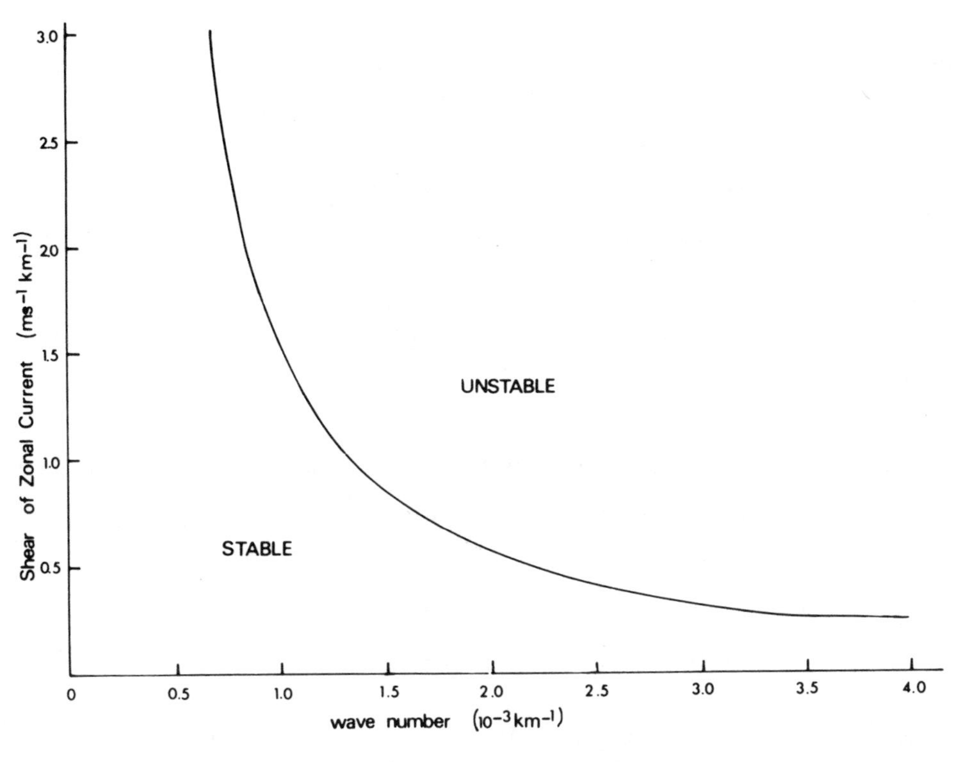

Fig. 46.2. Neutral stability curve for baroclinic waves on a basic current with constant shear in an unbounded, continuously stratified atmosphere on the β-plane. (Adapted from Charney, 1947.)

gradient). The result shown in this figure indicates that instability increases with shear and decreases with wavelength.

Charney's pioneering study provoked further work on baroclinic instability in different models (cf. Eady's model, below), and eventually a re-examination of Charney's work itself. The instability problem for a constant shear profile of the form (b) in Fig. 46.1 was attacked anew by Green (1960) and by Burger (1962), who discovered that the constant shear model actually implies instability at almost all wavelengths and speeds. Miles (1964a) conjectured that the neutral curve found for that model may generally be neither unique nor single-valued, so that the curve shown in Fig. 46.2 may represent a transition between different unstable modes rather than a transition from a stable to an unstable mode. A closer analysis of Charney's model (Miles, 1964b) revealed the existence of two types of instability: a "weak" instability, which owes its existence to the presence of a critical level and derives its energy from processes occurring at that level, and a "strong" instability, which is not related to the presence of a critical level but derives its energy from the processes described in Section 44.

Eady's model

For the ocean, of course, the above results are not directly relevant since they were obtained for a fluid of semi-infinite depth. A more realistic flow which is likely to occur in the ocean is shown in Fig. 46.1c. The solution of (46.1) for this model was first obtained by Eady (1949) for the case of constant N and uniform rotation ($\beta = 0$). Under these conditions $q_y \equiv 0$, and provided $U - c \neq 0$, (46.1) reduces to

$$[\partial_{yy} - k^2 + (f_0^2/N^2)\partial_{zz}]\Psi = 0. \tag{46.4}$$

The solution of (46.4) and (46.2) can be written in the form

$$\Psi(y,z) = \sin(n\pi y/l)\{A\cosh[\kappa(z/H + 1/2)] + B\sinh[\kappa(z/H + 1/2)]\} \tag{46.5}$$

$$\text{where} \quad \kappa = r(k^2 + n^2\pi^2/l^2)^{1/2} \tag{46.6}$$

in which $r = NH/f_0$ is the usual internal deformation radius. For simplicity we chose $U_{-H} = 0$, so that the basic flow has the representation

$$U(z) = U_0(z + H)/H. \tag{46.7}$$

Application of the boundary condition (46.3) to (46.5) for the profile (46.7) yields

$$\begin{aligned} A[\kappa(U_0 - c)\sinh\tfrac{1}{2}\kappa - U_0\cosh\tfrac{1}{2}\kappa] + B[\kappa(U_0 - c)\cosh\tfrac{1}{2}\kappa - U_0\sinh\tfrac{1}{2}\kappa] &= 0, \\ A[\kappa c\sinh\tfrac{1}{2}\kappa - U_0\cosh\tfrac{1}{2}\kappa] + B[-\kappa c\cosh\tfrac{1}{2}\kappa + U_0\sinh\tfrac{1}{2}\kappa] &= 0. \end{aligned} \tag{46.8}$$

Thus for a nontrivial solution, (46.8) implies that c must satisfy the following quadratic:

$$\kappa^2 c^2 - \kappa^2 U_0 c + U_0^2[\tfrac{1}{2}\kappa(\tanh\tfrac{1}{2}\kappa + \coth\tfrac{1}{2}\kappa) - 1] = 0. \tag{46.9}$$

We immediately see from (46.9) that when there is no mean flow ($U_0 = 0$), $c = 0$ and hence there are no propagating waves. This is because we have taken $\beta = 0$ so that the restoring force for planetary wave propagation is absent from the model. This is analogous to the barotropic stability problem considered in Section 45 for flow along a continental shelf. When $\gamma^2 = H_2/H_1 = 1$ (no topographic restoring force), the dispersion relation (45.24) also has only the trivial solution $c = 0$ when $U_0 = 0$. Solving for c from (46.9) we find

$$c = \frac{U_0}{2} \pm \frac{U_0}{\kappa}[(\tfrac{1}{2}\kappa - \tanh\tfrac{1}{2}\kappa)(\tfrac{1}{2}\kappa - \coth\tfrac{1}{2}\kappa)]^{1/2}. \tag{46.10}$$

From (46.10) we see that the wave speed depends on the vertical average of the mean flow ($U_0/2$), the shear (U_0/H, since $\kappa \propto H$) and the parameter κ, which is defined by (46.6). The waves are stable whenever the radicand is positive. which is determined only by the magnitude of κ, and not by the strength of the shear! Since $\tanh\frac{1}{2}\kappa \leqslant \frac{1}{2}\kappa$ for all κ, we have stability when $\coth\frac{1}{2}\kappa < \frac{1}{2}\kappa$, which holds for relatively large κ, and instability when $\coth\frac{1}{2}\kappa > \frac{1}{2}\kappa$, which holds for small κ. The transitional case is discussed in the next paragraph. In the unstable case the waves propagate with the vertically averaged current speed $U_0/2$ and grow like $\exp(kc_i t)$, where

$$kc_i = (U_0 k/\kappa)[(\tfrac{1}{2}\kappa - \tanh\tfrac{1}{2}\kappa)(\coth\tfrac{1}{2}\kappa - \tfrac{1}{2}\kappa)]^{1/2},$$

which is proportional to the shear.

The critical value of κ where c changes from real to complex values is given by the solution of

$$\coth\tfrac{1}{2}\kappa = \tfrac{1}{2}\kappa:$$

approximately, $\kappa = 2.4$. Thus a necessary condition for instability is given by $\kappa < 2.4$; that is (cf. 46.6),

$$r(k^2 + n^2\pi^2/l^2)^{1/2} < 2.4,$$

which can be rewritten as

$$k^2 < 5.8/r^2 - n^2\pi^2/l^2. \tag{46.11}$$

For r somewhat smaller than l the right-hand side of (46.11) will be positive for low cross-channel mode numbers. Therefore, for any constant shear we have instability when the waves are sufficiently long. In the limit of very wide channels ($l \to \infty$), which is equivalent to neglecting y-variations in Ψ, (46.11) gives

$$k < 2.4/r \equiv k_c \text{ for instability.} \tag{46.12}$$

Since this holds for any constant shear, the stability boundary in the (wavenumber, shear)-plane for the Eady model simply consists of a vertical line through $k = k_c$. The unstable region thus consists of the rectangular strip in the first quadrant defined by $0 < k < k_c$, $0 < U_0/H < \infty$. That is, only relatively long waves are unstable.

Two-level models

We have already commented in Section 44 on the usefulness of two-layer models for the stratification and mean flow. A related approach is that given by Holton (1972), who presents an account of zonally propagating waves in a *two-level model* for a baroclinic fluid on the β-plane with constant flows U_1 and U_2 in an upper and lower level, respectively. More specifically, a two-level model is derived from the equations for a continuously stratified fluid by replacing the vertical derivatives with their linear finite difference forms. Further, in an ocean of depth H with the surface at $z = 0$, the geostrophic stream function is determined at the levels $z = -H/4$ (upper level) and $z = -3H/4$ (lower level), and the vertical velocity at $z = -H/2$ (middle level). At $z = 0, -H$ the vertical velocity is set equal to zero. The stability boundary implied by Holton's solution is shown in Fig. 46.3. The stabilizing influence of β is also clearly manifested in this figure: instability occurs only for sufficiently strong shears $[\frac{1}{2}(U_1 - U_2) > \beta r'^2]$.

When $\beta = 0$, Holton's solution for the wave speed reduces to

$$c = \tfrac{1}{2}(U_1 + U_2) \pm \tfrac{1}{2}(U_1 - U_2)\left(\frac{k^2 r'^2 - 1}{k^2 r'^2 + 1}\right)^{1/2}, \tag{46.13}$$

which is clearly the analogue of Eady's solution (46.10) for continuous stratification. Thus now we have

$$k < 1/r' \text{ for instability,} \tag{46.14}$$

which is the same as (46.12) except for the constant of proportionality. From these two solutions for baroclinic flows on the f-plane, it is clear that the internal deformation radius plays a fundamental role in determining the length scale of the unstable waves.

The above two-level model (with $\beta \neq 0$) for baroclinic instability has recently been applied to the Gulf Stream meanders east of Cape Hatteras by Saltzman and Tang (1975).

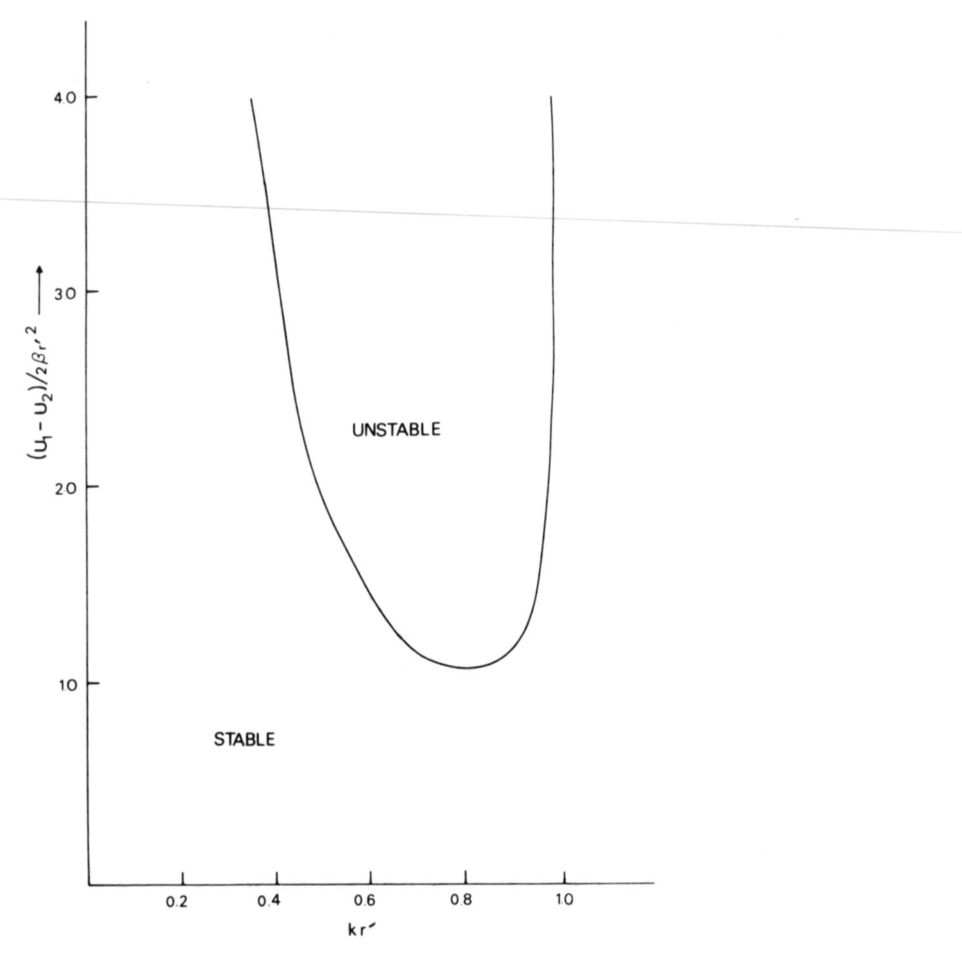

Fig. 46.3. Neutral stability curve for baroclinic waves in a two-level fluid, where U_1 and U_2 are the upper and lower level velocities and r' is the internal deformation radius. (Adapted from Holton, 1972.)

The results of this first-order instability theory (as opposed to second-order instability theory – see below) for the most unstable wave after 24 days of growth are shown in Fig. 46.4. Here we have plotted the sum of the basic-state and the usual perturbation fields of the current patterns at the surface (that is, upper level). The thermal field (which is proportional to the shear) and the vertical velocity field at the middle level are also shown in Fig. 46.4.

The meandering pattern of the current and the alternating cold and warm pools in the thermal field are clearly visible. The most unstable wave (which occurs when $\beta = 0$) has a length of 300 km, which is in general agreement with the observations. However, more detailed observations (e.g., Saunders, 1971) indicate that there are asymmetric departures from the pure sinusoidal wave form predicted by the first-order theory. For example, observations suggest the shedding of cyclonic cold pools to the right-hand side of the axis of the mean current looking downstream (to the east), and of somewhat weaker anticyclonic

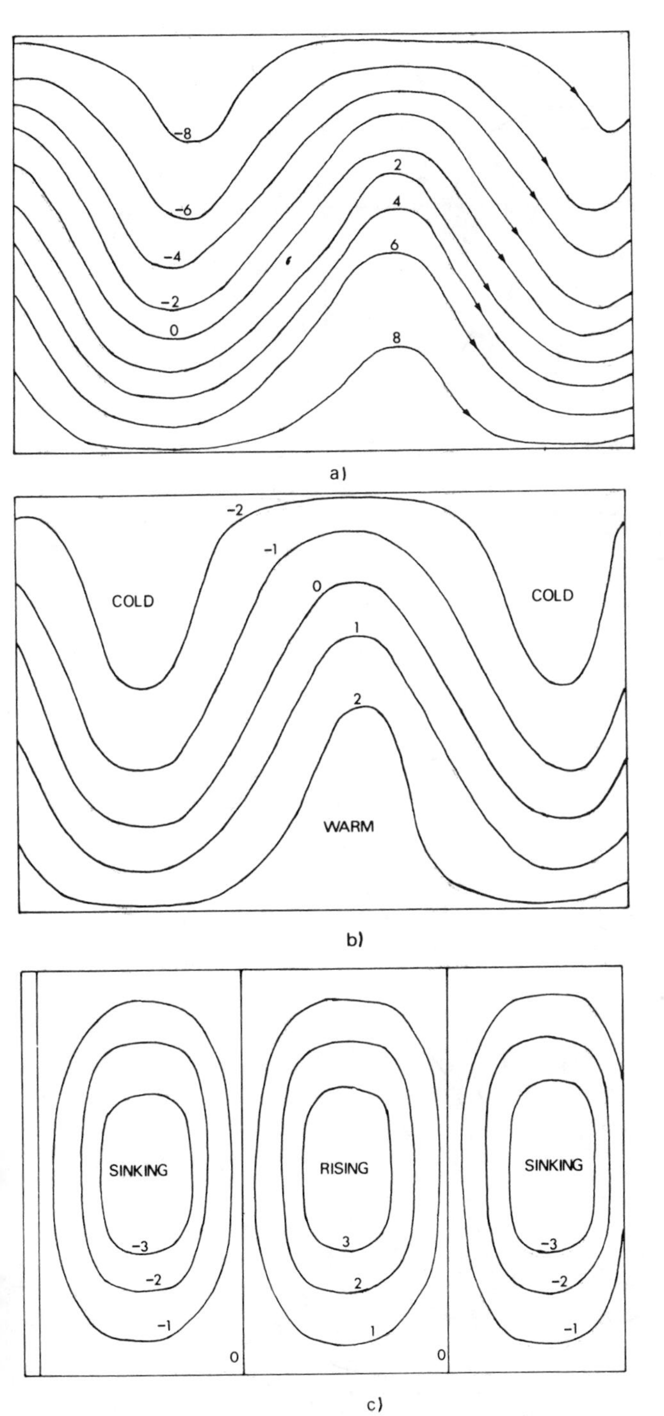

Fig. 46.4. Contours of (a) the current pattern, (b) the thermal field, and (c) the vertical velocity field, based on first-order baroclinic instability theory. The units in (a) and (b) are 10^8 cm^2 s^{-1}; in (c), 10^{-2} cm s^{-1}. (From Saltzman and Tang, 1975.)

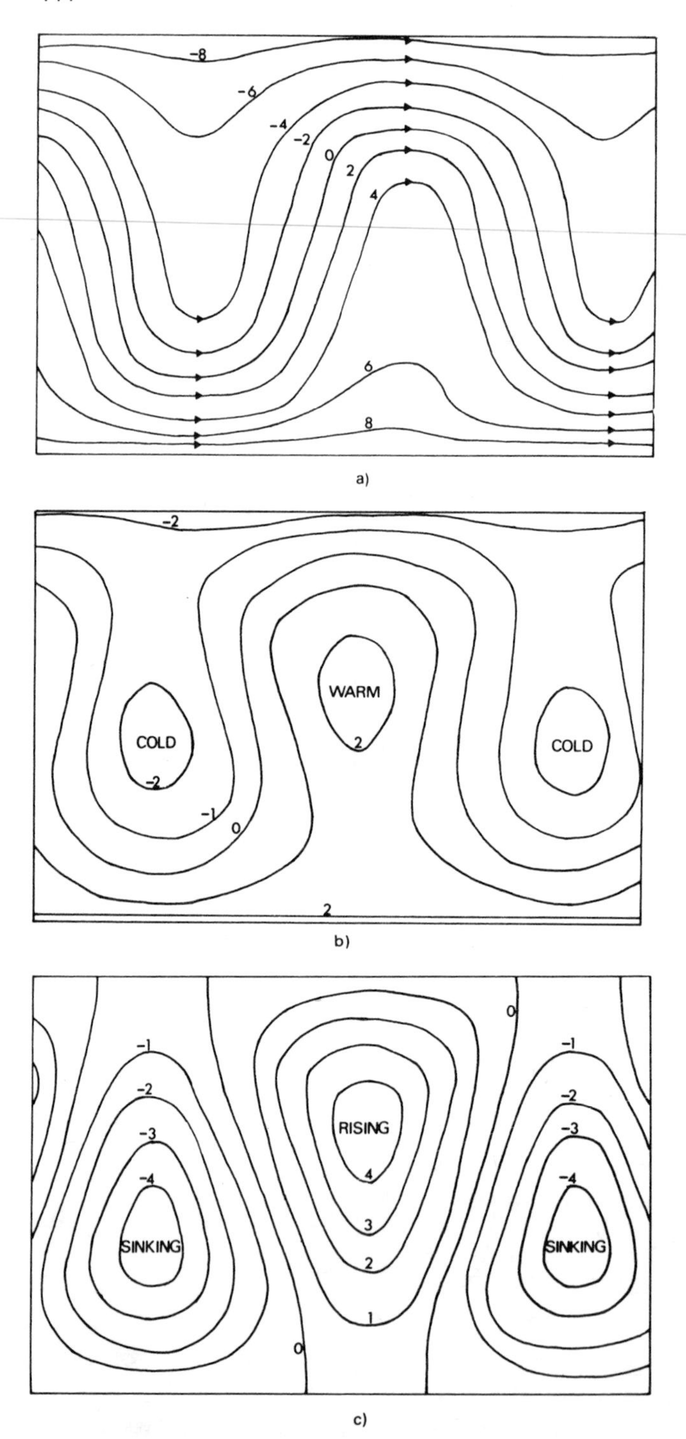

Fig. 46.5. Same contours as in Fig. 46.4 based on second-order baroclinic instability theory. (From Saltzman and Tang, 1975.)

warm pools to the left side. To explain these features theoretically, Saltzman and Tang showed that a second-order baroclinic instability theory must be used. That is, in the analysis each dependent variable ψ say, is resolved into the usual basic-state ψ_0, a zonally propagating primary or first-order disturbance ψ_1 (the quantity found in our examples using linear theory), and a further second-order eddy disturbance ψ_2. The field ψ_2 represents the response to the amplifying field ψ_1 caused by nongeostrophic effects. The results of the superposition of these three quantities after 24 days of growth for each of the fields shown in Fig. 46.4 are shown in Fig. 46.5. Thus with the second-order effects included, the cold cyclonic pools are displaced to the south and the warm anticyclonic pools to the north (see Fig. 46.5b and c), in agreement with the observations. Further, Fig. 46.5a and b shows that the second-order effects cause the current pattern to narrow into a meandering jet with an associated frontal temperature structure. For a more mathematical discussion of the second-order (nongeostrophic) theory of baroclinic instability we refer the reader to Tareyev (1968), Blumen (1973), Pedlosky (1975b) and the references quoted therein.

Other extensions of the theory of baroclinic instability

Because of its wide range of applicability to the ocean and the atmosphere (and also to other planetary atmospheres), there has recently been considerable effort devoted to extending and refining the rudimentary theory of baroclinic instability as presented above. Here we shall give a brief outline of some of the topics which have been examined in the literature.

In connection with Gulf Stream meanders we have already mentioned in Section 45 the papers of Orlanski (1969) and Nikitin and Tareyev (1972) in which the effects of topographic variations across a baroclinic mean flow have been studied. Several qualitative results of the effects of cross-stream topography on the stability of zonal flows on the β-plane have been established by Pedlosky (1964a). Blumsack and Gierasch (1972) considered the influence of a uniformly sloping bottom on the stability of the Eady model. For flows across slopes which characterize the topography of Mars, they found that the wavelength and growth rate of the most unstable waves are decreased. Also, the long wave cutoff for instability in the Eady model [$L_c = 2\pi/k_c$, where $k_c = 2.4/r$, as in (46.12)] is decreased. More recent papers which deal with the baroclinic instability of oceanic flows over a constant slope are those of Hart (1975a, b), Smith (1976) and Mysak and Schott (1977). In each case two-layer models were used to study the stability of the following current systems: the Arctic Ocean circulation, the Denmark Strait Overflow and the Norwegian Current. The consequences of cross-stream topographic variations (e.g., parabolic) on the stability of the Eady model in a channel were discussed by De Szoeke (1975). Finally, Robinson and McWilliams (1974) have determined the effects of slowly varying topography on the stability of a two-layer, β-plane model in which the mean flows have arbitrary orientation. (That is, the horizontal mean velocities in the two layers, U_1 and U_2, are not necessarily parallel.) Optimization techniques were used to determine the most unstable waves under various configurations.

The determination of solutions of baroclinic instability for cases which include lateral shear in the mean flow has proceeded along two lines. Several authors (for example, Stone, 1969; Gent, 1974, 1975) have considered Eady's model in which the flow has a *weak* lateral shear, viz.,

$$U(y,z) = zF(\epsilon y), \quad 0 < \epsilon \ll 1.$$

Because of the introduction of the small parameter ϵ, they were able to use the WKB method or multiple-scale perturbation techniques. For flows with strong lateral shear, on the other hand, simple solutions have been obtained only for two-layer models (Pedlosky, 1964b and Abramov et al., 1972).

To study the stability of flows with respect to zonal waves with lengths comparable to the circumference of the Earth, the β-plane must be abandoned and a true spherical geometry used. The baroclinic instability problem for such long waves has been examined by Miles (1965), and more recently by Moura and Stone (1976). On the other hand, the stability of barotropic flows with respect to very long wave disturbances has recently been explored in depth by Baines (1976).

The limiting of the growth of the unstable waves by eddy viscosity (see Chapter 8) and perturbation nonlinearities has also been a topic of much current activity. Tareyev (1965) has shown that eddy viscosity alone does not generally damp out unstable waves. Pedlosky (1970), on the other hand, showed that with sufficient dissipation and the inclusion of the nonlinear terms, a baroclinic wave reaches a final state in which the amplitude is steady. Related nonlinear studies have been conducted by Drazin (1970), Pedlosky (1971b) and Simons and Rao (1972).

The effects of the gravity-wave modes on the stability of a two-layer Eady model have been investigated by Rao and Simons (1970). They showed that an instability can occur as a result of a coupling between an internal gravitational mode and a rotational (Eady) mode. Stone (1971), on the other hand, presented a theory that incorporated nonhydrostatic terms into Eady's model. He found that these added terms decreased the growth rates of the unstable waves.

Another interesting variation of the usual baroclinic instability theory has been presented by Schulman (1967), who investigated the stability of a flow which models the oceanic thermocline. The basic flow has both meridional and vertical velocity components, and the wavelike perturbations have two horizontal wavenumber components. For typical thermocline parameters, the unstable waves grow very slowly, with e-folding times of the order of one year.

Among the more recent work, Pedlosky (1976) has developed a theory for the propagation of inviscid, finite-amplitude baroclinic wave packets in a baroclinic current whose properties vary slowly in the downstream direction. In this theory the stability of a wave travelling with the stream is determined by both local and upstream conditions.

The generation of mid-ocean eddies by baroclinic instability

There are perhaps four ways in which mid-ocean eddies may be generated. They may be forced by travelling atmospheric disturbances. They may be cast off or radiated by an intense boundary current (such as the Gulf Stream) which starts to meander away from the coast because of baroclinic instability. Thirdly, they may be the result of open-ocean baroclinic instability of the broad, slow mean currents near the region where the eddies are formed. Finally, they may be produced by wave–wave interactions.

While the first mechanism has not received much attention in the literature (see

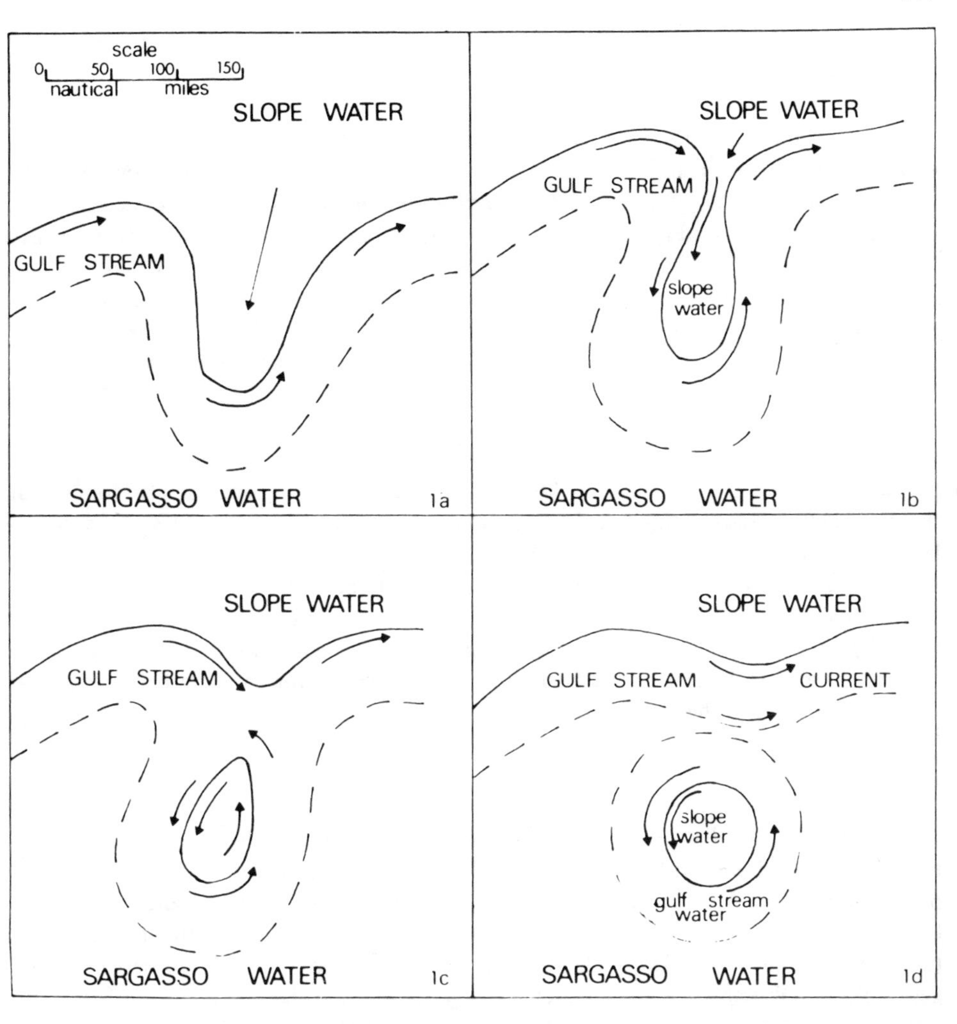

Fig. 46.6. The formation of a cyclonic eddy or ring from Gulf Stream meander development (1a), to separation from the Stream (1d). The solid line represents the 15° isotherm at 200 m. Dashed lines represent the approximate limit of the Sargasso side of the Gulf Stream. (From Parker, 1974.)

Hunkins, 1974, for a brief discussion in this topic), the second has been explored, for example, in the aforementioned papers of Orlanski (1969), Nikitin and Tareyev (1972) and Saltzman and Tang (1975). Their results have been extended further by the more lengthy studies of Orlanski and Cox (1973) and Robinson et al. (1975) which respectively present a three-dimensional numerical model for unstable waves on the Gulf Stream and a spatial initial-value problem for the Gulf Stream path. A schematic representation of the formation of a cyclonic eddy in the meandering Gulf Stream is shown in Fig. 46.6. Because such eddies contain three distinct water types with a circular configuration (cold, less saline slope water enclosed by warm Gulf Stream water – the ring – which is in turn surrounded by Sargasso Sea water), they are now often referred to in the literature as Gulf Stream rings (Parker, 1974). That the Gulf Stream is indeed a source of eddies has now been confirmed by observations. For example, Dantzler (1974) and Schmitz (1976, 1977) have found that the eddy kinetic energy rapidly decays away from the Gulf Stream region.

The third mechanism, in-situ baroclinic instability, has recently been explored by Gill et al. (1974), Robinson and McWilliams (1974) and others. Gill et al. have shown that the available potential energy in the large-scale mean circulation of a two-layer ocean model can be converted to eddy (unstable wave) energy by baroclinic instability (see also Pedlosky, 1975a). Maximum growth rates were obtained for eddies with wavelengths of 200 km, typical e-folding times being about 80 days. These eddies have significant velocities only in the surface layers. Secondary maxima in the growth rate curves gave longer wavelengths and e-folding times. These larger eddies tend to be confined to the deeper water. These basic results have also been confirmed numerically by Holland and Lin (1975). In particular they showed that in a two-layer, closed-basin model, eddies spontaneously appear due to baroclinic instability and propagate westward a few kilometers per day, with wavelengths of a few hundred kilometers. Holland and Lin also found that the Reynolds stresses computed from the eddy motions tend to slow down the upper-layer mean flow, but effectively drive the lower-layer flow.

The fourth mechanism – resonant wave–wave interactions – has only recently been proposed (Pedlosky, 1975c). Pedlosky examined the evolution of a resonant triad of baroclinic waves that are superimposed on a slightly unstable baroclinic current. Under certain initial conditions an equilibrium triad can exist, with characteristic horizontal scales of the order of the internal radius of deformation.

Eddy observations

We conclude this discussion with a brief survey of the observations of oceanic eddies, some of which have already been mentioned in Section 18. They were first observed in the western North Atlantic in the 1960's (Crease, 1962; Swallow, 1971) and the progress of one particular eddy in the southeastern North Atlantic was documented by the USSR Polygon experiment (Brekhovskikh et al., 1971). An eddy was also observed in the USA–UK MODE experiment. However, it did not appear to be due to in-situ baroclinic instability; as mentioned in Section 18 the best theoretical fit to the data was obtained by the superposition of two barotropic and two first-mode baroclinic planetary waves with no mean currents present (McWilliams and Flierl, 1976). Evidence of large baroclinic eddies in the eastern half of the North Pacific has been presented by Bernstein and White (1974). A simple theory based on westward-propagating planetary waves with no mean flow gives excellent agreement with the observations. Large, warm eddies off the coast of Japan have recently been observed by Kitano (1975). In the Arctic, on the other hand, smaller eddies of 5–20 km diameter have been observed (Newton et al., 1974; Hunkins, 1974), and some of them are believed to have their origin in baroclinic instability (Hunkins, 1974). Finally, we remark that small eddies and related low-frequency current motions have also been observed in relatively small bodies of water, such as gulfs, straits and lakes. In particular such motions have been found in the Gulf of Suez (Otterman, 1974), the Great Lakes of North America (Simons, 1976), the Strait of Georgia, British Columbia (Chang et al., 1976) and the Baltic Sea (A. Aitsam, personal communication, 1976).

Exercises Section 46

1. Consider a two-layer ocean on the f-plane in which each layer has a constant Brunt–Väisälä frequency. Suppose the basic zonal current has the form

$$U(z) = \begin{cases} U_1(H_1 + z)/H, & -H_1 \leqslant z \leqslant 0 \\ 0, & -H \leqslant z \leqslant -H_1. \end{cases}$$

Find the dispersion relation for this generalized Eady model of baroclinic instability. (Tang, 1975.)

2. Compare the stability boundaries of the Eady model with continuous stratification and the Niiler–Mysak model for barotropic instability in a constant-depth ocean (Section 45).

CHAPTER 8

THE GENERATION AND DISSIPATION OF WAVES

47. INTRODUCTION

In this final chapter, we abandon at last our adiabatic view of oceanic waves and enquire into the sources and sinks of energy for the many types of waves which fill the sea. Whereas the previous chapter dealt with the problems of continuous wave growth in the interior of the ocean – by examining the wavelike perturbations of oceanic flow fields – in this chapter we now open the whole ocean to energy inputs from its surroundings. Before looking into the details of generation mechanisms, it is pertinent to examine the general nature of sources and sinks of energy in the ocean: this will be done in the next Section (48). We follow in Section 49 with an elementary account of the influence of turbulent diffusion on waves and of the rationale behind the introduction of eddy coefficients to parameterize this influence. Some general remarks on wave generation, viewed as an initial-value problem, form the content of Section 50. The remaining Sections (51 to 55) present a review of generation theories and mechanisms for the various types of waves encountered in the earlier chapters.

48. THE GENERAL ENERGETICS OF THE OCEAN

Mechanical energy fluxes into the ocean arise from the action of stresses on its boundaries and of body forces through its bulk. Meteorological influences, seismic disturbances and tide-generating forces are foremost examples of such forces. Electromagnetic radiation enters the ocean from the top; geothermal heat flow penetrates it from the bottom. As we are primarily interested in energy inputs into time-dependent propagating motions, we shall omit consideration of seasonal or long-term heat fluxes into the ocean. We have already discussed the adiabatic energy balance in Section 4; we now add diffusive effects (in the form of friction and heat conduction) to this energy equation.

Momentum diffusion is hidden in our original momentum equation (3.3) in the unspecified force term denoted by F. For molecular friction this force may be written for an incompressible fluid in terms of a stress τ_{ij}; thus,

$$F_i = \frac{\partial}{\partial x_j} \tau_{ij}, \tag{48.1a}$$

$$\text{where} \quad \tau_{ij} = \mu(\partial u_i/\partial x_j + \partial u_j/\partial x_i) \tag{48.1b}$$

and μ is the molecular coefficient of viscosity. The ratio μ/ρ (usually written as ν) is called the kinematic viscosity. Other types of frictional forces are used, including turbulent eddy viscosities, as will be discussed in Section 49. The frictional force per unit volume exerted by a rigid boundary on a current flowing over it is often parameterized in terms of a drag coefficient C_{D}:

$$F = \rho_* C_\mathrm{D} \boldsymbol{u}|\boldsymbol{u}|, \tag{48.2}$$

where C_D may vary with position according, among other factors, to the roughness of the boundary. The velocity $\boldsymbol{u}$ used in (48.2) is that in the vicinity of the boundary and cannot be replaced by a mean velocity (over the depth, for example) without additional assumptions. Linear frictional forces, of the form

$$\boldsymbol{F} \propto \boldsymbol{u} \tag{48.3}$$

are occasionally used. The only justification of (48.3) lies in its analytic simplicity. The appropriateness of a linear friction law of this form to the damping of surface gravity waves has been explored by Pite (1973), who found that (48.3) gives good results for long waves ($kH \ll 1$) of small amplitude.

Putting F_i in the form (48.1a), we may rewrite the momentum equation (3.3) as

$$\frac{\partial}{\partial t}\rho u_i + 2\rho\epsilon_{ijk}\Omega_j u_k + \frac{\partial}{\partial x_j}[\rho u_i u_j + p\delta_{ij} - \tau_{ij}] = -\rho g \delta_{i3}. \tag{48.4}$$

The term including the centrifugal force due to the Earth's rotation has been neglected. The third-order antisymmetric tensor ϵ_{ijk} is equal to zero if any of the indices are equal, one if the indices are in cyclic order (1, 2, 3; 2, 3, 1; etc.), and minus one if the indices are in anticyclic order (such as 1, 3, 2).

Taking the scalar product of (48.4) with the velocity vector u_i gives, with the help of the continuity equation (3.10), the energy equation

$$\frac{\partial}{\partial t}\left(\frac{\rho u_i u_i}{2} + \rho g z\right) + \frac{\partial}{\partial x_j}\left[\left(\frac{\rho u_i u_i}{2} + \rho g z + p\right) u_j - u_i \tau_{ij}\right] = -\tau_{ij}\frac{\partial u_i}{\partial x_j}. \tag{48.5}$$

This equation differs from (4.10) only in the presence of the stresses τ_{ij}, which contribute to the energy flux and to a net source/sink term appearing on the right-hand side.

The influence of heat diffusion is already included in the heat equation (3.6), which may be expanded as

$$\frac{\partial}{\partial t}(\rho c_v T) + \frac{\partial}{\partial x_j}\left(\rho c_v T u_j - k_T \frac{\partial T}{\partial x_j}\right) = Q_T. \tag{48.6}$$

Local departures (T) from the mean temperature field ($\langle iT \rangle$) are eroded by diffusion at a rate which is determined by the thermal diffusion coefficient: $K_T = k_T/\rho c_v$. The influence of thermal diffusion on wave attenuation will be particularly important in regions of strong temperature gradients. The smoothing of temperature (and hence density) gradients by conduction produces an increase in the gravitational potential energy of stratification by raising cold heavy fluid and lowering warmer and lighter fluid. By contributing to this mixing process, internal waves transfer a part of their energy of motion to the potential energy of the mean stratification. Salt diffusion plays a similar role and will not be brought into the discussion. For an exposé of the various double-diffusion phenomena arising from the difference between K_T and K_S, the reader is referred to Turner (1973).

We may now add (48.5) and (48.6) to obtain an equation for the total energy of a parcel of water. Let us denote, for brevity, the total energy density per unit volume as E, its flux as $\boldsymbol{F}$, and the source/sink terms as $-\Sigma$; we then find

$$\frac{\partial E}{\partial t} + \frac{\partial}{\partial x_j} F_j = -\Sigma. \tag{48.7}$$

Integrating over a volume V of fixed geometry and using the divergence theorem, (48.7) becomes

$$\frac{\partial}{\partial t} \int_V E \, \mathrm{d}V = -\int_A F_j n_j \, \mathrm{d}A - \int_V \Sigma \, \mathrm{d}V, \tag{48.8}$$

where A is the closed surface with outward normal n_j enclosing the volume V. Except for the exclusion of salinity effects, this equation is quite general and describes the energetics of the mean state as well as that of any time-dependent motion superposed upon it. It is possible, by splitting all variables into a steady and a perturbation or fluctuating part as we did in Section 37 for gravity waves in a mean horizontal flow, to derive a pair of coupled energy equations, one for the mean state, the other for the fluctuations. This splitting will be performed again below (Section 49) for a more general situation to illustrate the effects of fluctuations (be they waves or turbulence) on the mean state.

The surface integral of the energy flux represents the rate of doing work by the chosen volume of fluid on its surroundings. The energy input from the surroundings is the negative of that term; explicitly,

$$-\int_A F_j n_j \, \mathrm{d}A = -\int_A \left\{ \rho \left(\frac{u_i u_i}{2} + gz + c_v T \right) u_j - k_T \frac{\partial T}{\partial x_j} + p u_j - \tau_{ij} u_i \right\} n_j \, \mathrm{d}A. \tag{48.9}$$

The terms $\frac{1}{2}(\rho u_i u_i) u_j$, $\rho g z u_j$ and $\rho c_v T u_j$ represent bodily injections of fluid carrying kinetic, gravitational potential and thermal energy through the surface of the volume. If the ocean is to remain of constant volume (as it is assumed incompressible), any net mass flux into it must be balanced by an efflux somewhere else: for example, net river inflow must be balanced by evaporation. Nevertheless, a net energy flux into or out of oceanic boundaries remains possible, as the injected mass may contain more internal, gravitational potential or kinetic energy than the exciting flow. The term $-k_T \partial T/\partial x_j$ represents heat diffusion through the boundaries. The last two terms in (48.9) represent the rate of doing work by normal and tangential stresses and dominate the budget of oceanic energy variance. The stresses due to the wind generate the whole spectrum of surface gravity waves and, on a larger time scale, planetary wave oscillations and the mean wind-driven circulation. Rapid deformations of the ocean bottom during earth tremors lead to variable bottom pressures giving rise to tsunamis. We note that energy injection at the sea surface, or at the bottom, will in general excite the whole spectrum of vertical normal modes and thus generate both surface and internal waves. The body force associated with tidal generation belongs into the source term Σ; as we shall see in Section 52 it is sometimes possible and convenient to replace this force by an equivalent surface pressure which will produce the same accelerations.

If, instead of considering the total energy E, which contains contributions from the mean flow, from turbulence, and from all the possible oceanic wave types, we wish to focus our attention on some particular type of waves, we should then use a restricted form of energy balance equation rather than (48.7). Under the conditions of ray theory, i.e., when the space and time scales of the chosen type of wave are much smaller than those of the other phenomena which coexist with it in the ocean, the energy balance is expressed by the action conservation equation (6.44). The influence of phenomena which occur on scales which are much *shorter* than those of the wave field under consideration and which contribute to gains or losses of wave action may be represented by the inclusion of inhomogeneous terms on the right-hand side of (6.44). We have already discussed a spectral form of the inhomogeneous energy balance in the form of the "radiation balance" equation (33.11).

Just how much energy is there in oceanic waves? It is interesting to compare estimates of energy density per unit area of the sea surface for various kinds of waves and for the mean oceanic circulation. For surface gravity waves (neglecting the surface-tension effect), the energy per unit area is given from (13.24) as $\frac{1}{2}\rho_* g a^2$. With $a = 1$ m, we find $5 \cdot 10^3$ J m^{-2}; for 3-meter waves, $4.5 \cdot 10^4$ J m^{-2}. Garrett and Munk (1972b) estimate that a total energy density of $4 \cdot 10^3$ J m^{-2} gives the best fit to their empirically fitted spectra of high-frequency ($f < \omega < N$) internal waves in the deep ocean. On the basis of a numerical integration of Laplace's tidal equation (29.23), Hendershott (1972) estimates the energy contained in the principal semi-diurnal lunar tide (M2) as $7 \cdot 10^{17}$ Joules. Taking the oceanic area as 70% of the Earth's surface (R = radius of the Earth = $6.4 \cdot 10^6$ m), the ocean's area is about $3.6 \cdot 10^{14}$ m^2 and the energy density for the M2 tide becomes $2 \cdot 10^3$ J m^{-2}. This may be doubled to include the other tidal constituents (cf. Section 52). Energy estimates for long-period (planetary wave) motions range from about $0.5 \cdot 10^3$ J m^{-2} in the numerical experiments of Holland and Lin (1975), to about 10^5 J m^{-2} based on a generous interpretation of MODE data. Calculations of eddy energies by Wyrtki et al. (1967a) based on ship drift observations (see Fig. 48.1) indicate surface speed variances $\langle u_s^2 \rangle$ of $4 \cdot 10^{-2}$ m^2 s^{-2} over most of the oceans and higher values in regions of intense flows. To express that number in terms of energy per unit area of the sea surface, some assumptions must be made about the vertical structure of the horizontal velocity, i.e., about its partition between barotropic and baroclinic modes. If we take the vertically averaged horizontal velocity as having one half of the surface value u_s, the energy per unit area becomes $\frac{1}{8}\rho_* H \langle u_s^2 \rangle$; for a depth of 5000 m and the value of $\langle u_s^2 \rangle$ quoted above, this gives $2.5 \cdot 10^4$ J m^{-2}. Chang et al. (1976) have found low-frequency ($T > 4$ days) energy levels of about 10^3 J m^{-2} in the semi-enclosed Strait of Georgia, British Columbia. For comparison, Gill et al. (1974) estimate energy densities in a two-layer model of the North Atlantic circulation as 10^2 J m^{-2} for the kinetic energy and 10^5 J m^{-2} for the potential energy associated with the mean tilt of isopycnals. All these estimates are grouped together in Table 48.I. It is remarkable, and perhaps also entirely fortuitous, that the four categories of waves contain similar amounts of energy!

TABLE 48.I

Estimates of energy levels in various kinds of wave motions and in the mean circulation (see text for references)

Surface gravity waves ($a = 2$ m)	$2 \cdot 10^4$ J m^{-2}
Internal gravity waves ($f < \omega < N$)	$4 \cdot 10^3$ J m^{-2}
Tides [$\omega = 0(f)$]	$4 \cdot 10^3$ J m^{-2}
Long period waves ($\omega \ll f$)	$10^3 - 10^5$ J m^{-2}
Mean circulation: kinetic energy	10^2 J m^{-2}
potential energy	10^5 J m^{-2}

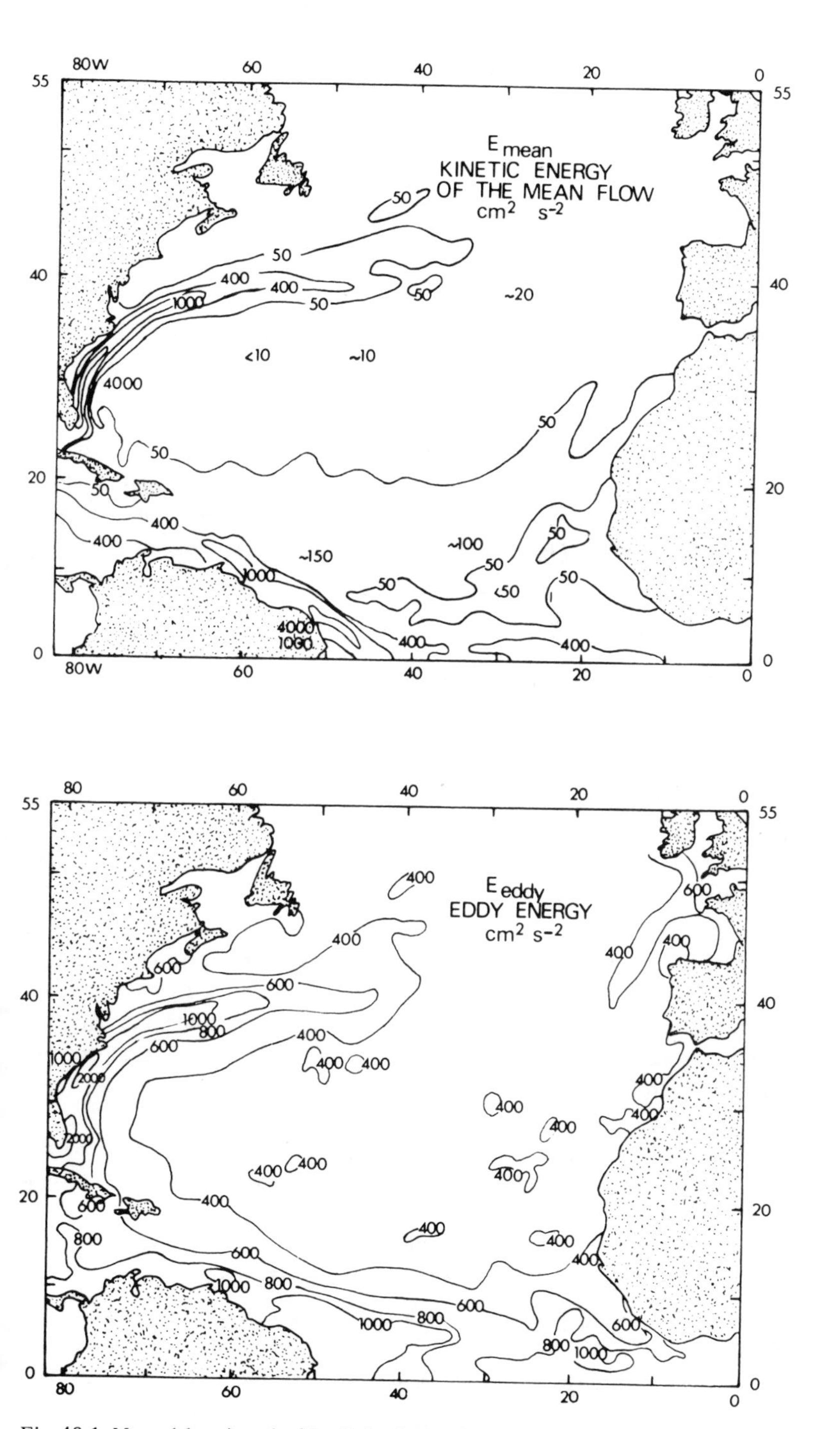

Fig. 48.1. Mean (above) and eddy (below) kinetic energy per unit area of the sea surface in the North Atlantic. (From Wyrtki et al., 1976a. Copyrighted by American Geophysical Union.)

49. REYNOLDS STRESSES AND EDDY COEFFICIENTS

In the ocean, and especially near its boundaries, the transfer of momentum, vorticity, heat and salt between fluid elements takes place mostly by means of turbulent processes. Turbulence is also responsible for mixing and for the dissipation of energy in the ocean, and must necessarily be included in any nonadiabatic description of ocean waves. In view of the complexity of turbulent flow, a practical representation of turbulent stresses and mixing effects on larger scale mean or wave motions is usually based on a simplified parameterization procedure, a method which sacrifices accuracy of description to formal simplicity. The simplest parameterization assumes that turbulent transfer of momentum and scalar properties is analogous to, but much more vigorous than, molecular diffusion. In that case, the effect of turbulence is simply modelled by replacing molecular diffusion coefficients by much larger "eddy diffusion" coefficients. The purpose of this section is to give a short account of the role of turbulence on waves and of the reasoning which leads to the introduction of eddy coefficients. For more details, the reader should refer to texts on turbulence, such as Hinze (1975) or Chorin (1975).

Equations for the mean and fluctuating fields

Let us suppose that the velocity, density and pressure (with the hydrostatic part removed) can be decomposed into ensemble mean and fluctuating parts [cf. (31.2) and (31.6)]:

$$u_i = \bar{u}_i + u_i', \quad p = \bar{p} + p', \quad \rho = \bar{\rho} + \rho'. \tag{49.1}$$

The primed quantities have zero (ensemble) mean and represent fluctuating contributions to the field variables. They can be thought of as a combination of random (turbulent) motions and waves. The ensemble average is found as the arithmetic mean of the variables over a large number of realizations. In order for this averaging process to be useful, the estimates of $\bar{u}_i$ and of the other mean quantities must converge after a reasonable number of realizations. Ensemble averaging is applicable to laboratory situations, in which a series of identical experiments can be carried out under controlled conditions. Geophysical data, on the other hand, usually consist of time or space series taken under circumstances which are beyond the control of the observer. Any meaningful averaging process must then be over time or space and an ergodic hypothesis (cf. Section 31) must be invoked to render this process statistically equivalent to an ensemble average. The condition on the number of realizations necessary to define an ensemble average becomes, for a time average, a condition on the interval of averaging, which should be shorter than the duration of the total time series. A separation of the variables into two paths, such as in (49.1) will thus be possible only if the fluctuations die out for periods greater than a certain value. This condition is met if there exists a spectral gap in the frequency distribution of the observed quantities: everything above the gap can then be identified as a fluctuation, everything below it, fluctuating on a much longer time scale, is defined as the average.

Substitution of (49.1) into (3.9)–(3.11) gives, for a vertically stratified fluid

$$\frac{\partial}{\partial x_i}(\bar{u}_i + u_i') = 0, \tag{49.2}$$

$$\frac{\partial}{\partial t}(\bar{\rho} + \rho') + (\bar{u}_i + u_i')\frac{\partial}{\partial x_i}(\bar{\rho} + \rho') + (\bar{w} + w')\frac{\partial \rho_0}{\partial z} = 0, \tag{49.3}$$

$$\frac{\partial}{\partial t}(\bar{u}_i + u_i') + (\bar{u}_j + u_j')\frac{\partial}{\partial x_j}(\bar{u}_i + u_i') + 2\epsilon_{ijk}\Omega_j(\bar{u}_k + u_k')$$

$$+ \frac{1}{\rho_*}\frac{\partial}{\partial x_i}(\bar{p} + p') + \frac{1}{\rho_*}(\bar{\rho} + \rho')g\delta_{i3} = 0, \tag{49.4}$$

where in (49.4) we have introduced the Boussinesq approximation, with ρ_* being a constant representative density. Taking the ensemble average of (49.2) gives

$$\partial \bar{u}_i/\partial x_i = 0, \tag{49.5}$$

since the operations of ensemble average and differentiation commute and $\overline{u_i'} = 0$. Subtracting (49.5) from (49.2) we then obtain

$$\partial u_i'/\partial x_i = 0. \tag{49.6}$$

Thus both $\bar{u}_i$ and u_i' satisfy the same equation independently, a result due to the linearity of (3.10). However, because of the nonlinear terms in (3.9) and (3.11), the averaging of (49.3) and (49.4) does not give a set of equations for $\bar{u}_i, \bar{p}$ and $\bar{\rho}$ alone; thus:

$$\frac{\partial \bar{\rho}}{\partial t} + \bar{u}_j\frac{\partial \bar{\rho}}{\partial x_j} + \bar{w}\frac{\partial \rho_0}{\partial z} = -\frac{\partial}{\partial x_j}\overline{\rho' u_j'}, \tag{49.7}$$

where (49.6) has been used to incorporate u_j' into the derivative on the right-hand side. Similarly, and again with the help of (49.6),

$$\frac{\partial \bar{u}_i}{\partial t} + \bar{u}_j\frac{\partial \bar{u}_i}{\partial x_j} + 2\epsilon_{ijk}\Omega_j\bar{u}_k + \frac{1}{\rho_*}\frac{\partial \bar{p}}{\partial x_i} + \frac{\bar{\rho}}{\rho_*}g\delta_{i3} = -\frac{\partial}{\partial x_j}\overline{u_i' u_j'}. \tag{49.8}$$

Equations (49.7) and (49.8) are known as Reynolds' equations (Reynolds, 1895). To solve for $\bar{u}_i$, $\bar{p}$, and $\bar{\rho}$ from (49.5), (49.7) and (49.8) we first need to know u_i', p' and ρ' which in turn will enable us to find the correlations $\overline{\rho' u_j'}$ and $\overline{u_i' u_j'}$. To find equations for u_i', p' and ρ' [in addition to (49.6)] it is logical to substract (49.7) and (49.8) from (49.3) and (49.4) respectively. This gives

$$\frac{\partial \rho'}{\partial t} + u_j'\frac{\partial \bar{\rho}}{\partial x_j} + \bar{u}_j\frac{\partial \rho'}{\partial x_j} + u_j'\frac{\partial \rho'}{\partial x_j} + w'\frac{\partial \rho_0}{\partial z} = \frac{\partial}{\partial x_j}\overline{\rho' u_j'}, \tag{49.9}$$

$$\frac{\partial u_i'}{\partial t} + u_j'\frac{\partial \bar{u}_i}{\partial x_j} + \bar{u}_j\frac{\partial u_i'}{\partial x_j} + u_j'\frac{\partial u_i'}{\partial x_j} + 2\epsilon_{ijk}\Omega_j u_k' + \frac{1}{\rho_*}\frac{\partial p'}{\partial x_i} + \frac{\rho'}{\rho_*}g\delta_{i3} = \frac{\partial}{\partial x_j}\overline{u_i' u_j'}. \tag{49.10}$$

The two pairs of momentum and density equations, for the mean and fluctuating variables, respectively, are inextricably coupled. If the fluctuations were to consist entirely of small-amplitude waves, the correlations $\overline{\rho' u_j'}$ and $\overline{u_i' u_j'}$ could be found from (49.6) and linearized forms of (49.9) and (49.10). Their substitution into (49.7) and (49.8) would then give expressions for the buoyancy and momentum sources of the mean state attributable to the wave fluctuations. This is the procedure which we have already exploited in our discussion of the interaction of gravity waves and currents in Section 37. For large-amplitude waves and a fortiori for turbulent fluctuations this linearization is out of the question. The description of turbulent flow is indeed one of the most challenging problems of classical physics and has taxed the physical intuition and mathematical ingenuity of two generations of fluid dynamicists. Even in terms of a statistical representation, the equations (49.6), (49.9) and (49.10) can be solved only approximately. The accuracy of the statistical description depends on the stage at which some closure hypothesis is invoked to truncate the approximation scheme. We note for example that the determination of u_i', p' and ρ' involves the second-order correlations $\overline{u_i'u_j'}$ and $\overline{\rho' u_j'}$; equations for these quantities in turn involve higher correlations such as

$$\overline{u_i'u_j'\frac{\partial u_i'}{\partial x_j}}.$$

Each correlation depends on a higher order correlation, and the series of equations for successively higher correlations can be terminated only if, at some stage, the higher correlations are related to the lower ones by some closure assumption. The accuracy of the description is usually accompanied by an increase in its complexity, so that for practical evaluation of the effect of the turbulence on the mean flow a simple rough approximation may be as useful as a more refined but more complicated representation.

Eddy coefficients

The simplest parameterization of turbulence is one which is based on a general analogy with molecular diffusion and represents the correlations $\overline{u_i'u_j'}$ and $\overline{\rho' u_j'}$ in terms of eddy diffusion coefficients without any effort at solving the turbulent equations (49.6), (49.9) and (49.10). The introduction of such coefficients must be viewed as a "lazy man's closure", applied before any attempt at describing turbulent flow.

The *Reynolds stress* τ_{ij} is defined as

$$\tau_{ij} = -\rho\overline{u_i'u_j'}; \tag{49.11}$$

its divergence, on the right-hand side of (49.8), represents the contribution of the fluctuations on the momentum balance of the mean flow. We have already encountered this stress in Chapter 7 in connection with our study of the stability of parallel flows with respect to wave perturbations. By analogy to the molecular stress, as given in (48.1), we write

$$-\overline{u_i'u_j'} = K_{ijlm}\left(\frac{\partial \bar{u}_l}{\partial x_m} + \frac{\partial \bar{u}_m}{\partial x_l}\right), \tag{49.12}$$

where the fourth-order tensor K_{ijlm} is symmetric ($K_{ijlm} = K_{jilm}$), in order to preserve the symmetry of the Reynolds stress. K_{ijlm} is an eddy diffusion coefficient for momentum and may depend on position.

A number of simplifications of (49.12) are used. The simplest, applicable to a homogeneous isotropic turbulent field, was proposed by Boussinesq (1877), and represents K_{ijlm} by

$$K_{ijlm} = K_M\delta_{il}\delta_{jm}, \tag{49.13}$$

where K_M is a constant momentum exchange coefficient. In this case, and with the help of (49.5), the right-hand side of (49.8) takes the form

$$K_M \frac{\partial^2 \bar{u}_i}{\partial x_j \partial x_j},$$

which is formally identical to the viscous force due to molecular diffusion. Another form in common use, chosen to represent the anisotropy of turbulent mixing in a stratified fluid is

$$K_{ijlm} = K_{Mj} \delta_{il} \delta_{jm}, \tag{49.14}$$

where $K_{Mj} = K_{Mh}$ for $j = 1, 2$ (a horizontal eddy coefficient), and $K_{Mj} = K_{Mv}$ for $j = 3$ (a vertical eddy coefficient); in general, $K_{Mh} \gg K_{Mv}$. For homogeneous conditions, these coefficients are independent of position, and the right-hand side of (49.8) takes the form

$$-\frac{\partial}{\partial x_j} \overline{u_i' u_j'} = K_{Mh} \left(\frac{\partial^2 \bar{u}_i}{\partial x^2} + \frac{\partial^2 \bar{u}_i}{\partial y^2} \right) + K_{Mv} \left(\frac{\partial^2 \bar{u}_i}{\partial z^2} \right). \tag{49.15}$$

More sophisticated parameterizations of the Reynolds stresses in terms of eddy viscosities are discussed by Pope (1975) and by Hinze (1975).

The molecular value of the kinematic viscosity $\nu = \mu/\rho$ for water is about $0.01\ \text{cm}^2\ \text{s}^{-1}$. Eddy viscosities are assigned much larger values. In analytical (Veronis, 1973) and numerical (Pond and Bryan, 1976) models of the large-scale ocean circulation, for example, the values of the coefficients are chosen so that the solutions will give best fits to observed flow patterns, while preserving computational stability. Typically, K_{Mh} and K_{Mv} lie in the ranges $10^6\ \text{cm}^2\text{s}^{-1} < K_{Mh} < 10^{10}\ \text{cm}^2\text{s}^{-1}$ and $3 \cdot 10^{-1}\ \text{cm}^2\text{s}^{-1} < K_{Mv} < 2 \cdot 10^2\ \text{cm}^2\text{s}^{-1}$; the values used generally increase with the scale of the phenomena considered. This variability illustrates the shortcomings of this simple-minded representation of turbulent stresses.

The analogy between turbulent and molecular mixing suggests that the role of the Reynolds stresses due to turbulence or to wave motion should be to transfer momentum out of the mean flow into the fluctuations. In that case, the eddy coefficients take positive values. However, Webster (1961), among others, has shown that K_{Mh} could take negative values. From an analysis of data taken south of Cape Hatteras, he established that there is in fact a transfer of momentum from the fluctuations to the mean flow, rather than vice-versa. If large-scale ocean turbulence is assumed to be nearly two-dimensional, energy transfer from small scales to larger scales of motion is indeed expected as we have noted in Section 19, and the negative eddy viscosity associated with this apparent reverse diffusion represents a normal state of affairs, rather than an anomaly. The relevance of the reverse energy cascade, and its interpretation in terms of a negative eddy viscosity, on the atmospheric circulation has been discussed by Starr (1968). The role played by two-dimensional turbulence in oceanic circulation is still subject to debate (cf. Rüdiger, 1976; Rhines, 1977). We now discuss a parameterization for the buoyancy flux term $\overline{\rho' u_j'}$ in (49.7). In analogy to (49.14) we define the eddy buoyancy coefficients by the relations

$$\overline{\rho' u_j'} = -K_{Bh} \frac{\partial \bar{\rho}}{\partial x_j} \quad (j = 1, 2); \quad \overline{\rho' w'} = -K_{Bv} \frac{\partial \bar{\rho}}{\partial z}. \tag{49.16}$$

Heat and salt transfer by turbulence are usually assumed to proceed at similar rates (Turner, 1973). The eddy values of K_T and K_S to be used in (3.6) and (3.7) are essentially the same, so that the effect of turbulent diffusion on the density field may be described adequately by a single equation such as (49.16). Estimates of turbulent diffusion coefficients may be obtained from dye experiments (Csanady, 1973b), inferred from microstructure measurements (Gregg et al., 1973), or found from internal wave–current interactions (Frankignoul, 1976). The ranges of values taken by K_{Th} and K_{Tv} are similar to those quoted earlier for K_{Mh} and K_{Mv}. Values for the salinity eddy coefficients K_{Sh} and K_{Sv} may be found by similar methods. For example, by fitting a simple model including advection and three-dimensional diffusion to data on the Mediterranean high-salinity tongue in the North Atlantic, Needler and Heath (1975) obtained values of $1.5–3 \cdot 10^7$ cm^2 s^{-1} for K_{Sh} and $0.3-0.7$ cm^2 s^{-1} for K_{Sv}. With such values of eddy coefficients, turbulent diffusion completely swamps molecular effects ($K_T \simeq 10^{-3}$ cm^2 s^{-1} and $K_S \simeq 10^{-5}$ cm^2 s^{-1} for molecular processes).

50. WAVE GENERATION AS AN INITIAL-VALUE PROBLEM

Given a forcing function applied initially at $t = 0$, say, together with appropriate initial conditions, a wave generation problem may be treated as an initial boundary-value problem for a forced wave equation, the solution of which reveals the form of the radiated wave field and its repartition into the various allowed spatial modes. While much of the physics of wave generation consists in the proper specification of the forcing function and initial conditions, a matter which will be addressed in more detail in the following sections, we now bypass such matters and leap to a discussion of wave propagation from specified sources. The advantage of this step is that it sets the problem of the origin of oceanic waves in the applied mathematical framework already developed for other kinds of waves. In view of the availability of excellent reference works (Morse and Feshbach, 1953; Duff and Naylor, 1966; Whitham, 1974) on the solution of forced wave equations, we feel that there is no need to belabour the methods of solution. Only a few examples illustrating some specific properties of oceanic waves will be treated.

The Cauchy-Poisson problem

Gravity wave propagation from an initial surface disturbance was first analyzed by Poisson (1816) and Cauchy (1827). In spite of Scott Russell's opinion that "... the greater part of the investigations of M. Poisson and M. Cauchy under the name of wave theory is rather to be regarded as mathematical exercises than a physical investigation" (Russell, 1844), the solution of the initial-value problem for small-amplitude waves is instructive in that it allows us to distinguish between the roles of phase dispersion and initial wave form in the shape of a travelling wave group. The theory finds current application in the prediction of wave forms produced by earthquakes (tsunamis) (Kajiura, 1963, 1970), and by explosions (Kranzer and Keller, 1959; Le Méhauté, 1971).

Consider the simple case of a radially symmetric disturbance of the sea surface (the general case of a submerged arbitrary disturbance is treated by Kajiura, 1963). In terms of the velocity potential ϕ defined in (11.12), the radially symmetric, linearized surface wave problem in water of uniform depth is stated, in cylindrical coordinates (r, θ, z), as

$$\nabla^2\phi = \frac{1}{r}\frac{\partial}{\partial r}\left(\frac{r\partial\phi}{\partial r}\right) + \frac{\partial^2\phi}{\partial z^2} = 0, \quad -H \leqslant z \leqslant 0, \tag{50.1}$$

$$\text{with} \quad \phi_z = 0 \quad \text{on} \quad z = -H \tag{50.2}$$

$$\phi_z = \eta_t \quad \text{on} \quad z = 0 \tag{50.3a}$$

$$\phi_t = -g\eta \quad \text{on} \quad z = 0. \tag{50.3b}$$

A separated solution of (50.1), harmonic in time, satisfying the boundary condition (50.2) and bounded everywhere, is readily obtained as

$$\phi(r, z, t) = J_0(kr) \cosh\,[k(z+H)][A \sin \omega t + B \cos \omega t], \tag{50.4}$$

where k is a separation constant (essentially a wavenumber), ω the angular frequency and A, B are constants determined by the initial form of the disturbance. Combining the two parts of (50.3) into $\phi_{tt} + g\phi_z = 0$, we find, upon substitution of (50.4) into this equation, the familiar dispersion relation (11.16), without surface tension effects:

$$\omega^2 = gk \tanh kH. \tag{50.5}$$

The surface displacement is obtained from (50.3a) and (50.4) as

$$\eta = -\frac{k}{\omega} J_0(kr) \sinh(kH)[A \cos \omega t - B \sin \omega t]. \tag{50.6}$$

Integrating over all values of the separation constant, the complete solution for the surface displacement becomes

$$\eta(r,t) = -\int_0^\infty \frac{k}{\omega} J_0(kr) \sinh(kH)[A(k) \cos \omega t - B(k) \sin \omega t]\, \mathrm{d}k. \tag{50.7}$$

The dependence of A and B on k is acknowledged explicitly: these functions are to be determined from the initial conditions.

Appropriate initial conditions involve the initial displacement and the initial vertical velocity:

$$\eta(r,0) = \eta_0(r), \tag{50.8a}$$

$$\eta_t(r,0) = w_0(r). \tag{50.8b}$$

From (50.7)

$$\eta_0(r) = -\int_0^\infty \frac{k}{\omega} J_0(kr) \sinh(kH) A(k)\, \mathrm{d}k, \tag{50.9a}$$

$$w_0(r) = \int_0^\infty k J_0(kr) \sinh(kH) B(k)\, \mathrm{d}k. \tag{50.9b}$$

Using the Hankel transform relations (Duff and Naylor, 1966)

$$f(r) = \int_0^\infty \bar{f}(k) J_0(kr) k\, \mathrm{d}k, \tag{50.10a}$$

$$\bar{f}(k) = \int_0^\infty f(r) J_0(kr) r\, \mathrm{d}r, \tag{50.10b}$$

the functions $A(k)$ and $B(k)$ have the integral representations

$$A(k) = -\frac{\omega}{\sinh(kH)} \int_0^\infty \eta_0(r) J_0(kr) r\, \mathrm{d}r, \tag{50.11a}$$

$$B(k) = \frac{1}{\sinh(kH)} \int_0^\infty w_0(r) J_0(kr) r\, \mathrm{d}r. \tag{50.11b}$$

Combining (50.11) with (50.7), the surface displacement $\eta(r, t)$ is then given in terms of the initial elevation and velocity as

$$\eta(r,t) = \int_0^\infty J_0(kr)\left\{\int_0^\infty \left[\eta_0(r')\cos\omega t + \frac{w_0(r')}{\omega}\sin\omega t\right] J_0(kr')r'\mathrm{d}r'\right\} k\mathrm{d}k. \tag{50.12}$$

The influence of the initial displacement manifests itself through $\eta_0(r')$, that of the initial velocity, through $w_0(r')$; since the problem is linear the two effects may be separated and treated individually.

Let us put $w_0(r') = 0$ for simplicity, and abbreviate the source dependence through

$$\int_0^\infty \eta_0(r')J_0(kr')r'\mathrm{d}r' = F(k). \tag{50.13}$$

Concentrating our attention on the far field ($kr \gg 1$), we make use of the asymptotic form

$$J_0(kr) \sim \sqrt{\frac{2}{\pi kr}}\sin(kr + \pi/4) \quad \text{as} \quad kr \to \infty \tag{50.14}$$

to reduce (50.12) [with $w_0(r') = 0$] to

$$\eta(r,t) \sim \sqrt{\frac{2}{\pi r}}\int_0^\infty F(k)\sin(kr+\pi/4)\cos\omega t\,\sqrt{k}\mathrm{d}k. \tag{50.15}$$

Note that in order that the asymptotic representation of $J_0(kr)$ be valid down to the lower limit of integration of (50.15), it is necessary that kr be large even as $k \to 0$, an assumption which implies that infinitely long waves have an infinite speed of propagation! The above analysis may still be correct, however, if $\eta_0(r)$ is such that the contribution to the integral in (50.15) from small k is negligible [this requires that $F(k)$ be "small" near $k = 0$]. The product of trigonometric functions in the integrand in (50.15) may be expanded as

$$\sin(kr+\pi/4)\cos\omega t = \tfrac{1}{2}[\sin(kr+\pi/4+\omega t) + \sin(kr+\pi/4-\omega t)]. \tag{50.16}$$

Upon taking $\omega = (gk\tanh kH)^{1/2} > 0$, the first term in the right-hand side of (50.16) represents waves propagating towards the origin, and will not contribute to the solution in the far field, well beyond the range of the initial displacement. Keeping only the second term, (50.15) may be rewritten as [assuming $F(k)$ is real]

$$\eta(r,t) = \frac{1}{\sqrt{2\pi r}}\,\mathrm{Im}\left\{\int_0^\infty F(k)\exp[i(kr+\pi/4-\omega t)]\sqrt{k}\mathrm{d}k\right\}. \tag{50.17}$$

This integral is of the form (6.21) and, for large r and t, may be evaluated by the method of stationary phase. Let us first pause, however, to recognize the $r^{-1/2}$ spatial decay preceding the integral in (50.17) as due to geometrical spreading in two dimensions. Additional spatial attenuation may also arise because of dispersion and will appear upon integration.

The wavenumber of stationary phase, k_0, is found from (6.22):

$$r - \omega'(k_0)t = 0. \tag{50.18}$$

From (6.23), and noting that $\omega''(k_0) < 0$, the integral (50.17) is approximately given by

$$\eta(r,t) \simeq \frac{1}{\sqrt{r}}\,\frac{F(k_0)\sqrt{k_0}}{\sqrt{-t\omega''(k_0)}}\cos[k_0 r - \omega(k_0)t]. \tag{50.19}$$

Substituting for the time t from (50.18), (50.19) becomes

$$\eta(r,t) \simeq \frac{1}{r}\left(\frac{k_0\omega'(k_0)}{-\omega''(k_0)}\right)^{1/2} F(k_0)\cos\left[k_0 r - \omega(k_0)t\right]. \tag{50.20}$$

The local oscillations (the cosine term) are at a wavenumber and frequency corresponding to the point of stationary phase, i.e., to the wave having travelled to the point of observation at the group velocity. Phase dispersion affects the amplitude through the square root term and through an additional $r^{-1/2}$ spreading term. The initial wave form continues to manifest itself [through $F(k_0)$] in the envelope of the wave train.

The Airy phase

We must now remind ourselves that the method of stationary phase, as used above, is valid only when $\omega''(k_0) \neq 0$. Near extrema of $\omega(k)$, an additional term must be kept in the approximation to the phase of the exponential function in order to obtain meaningful results. From (50.5), $\omega''(k) = 0$ when $k = 0$, i.e., at the maximum wave speed $c_0 = \sqrt{gH}$. Considering again a prototype equation of the form (6.21), i.e.,

$$\eta(r,t) = \int_0^\infty g(k)\exp\left[if\{k,\omega(k);x,t\}\right]\mathrm{d}k, \tag{50.21}$$

we now expand the phase $f(k)[\equiv f(k,\omega;x,t)]$ about a certain value of k (k_0, say):

$$f(k) = f(k_0) + (k-k_0)f'(k_0) + \tfrac{1}{2}(k-k_0)^2 f''(k_0) + \tfrac{1}{6}(k-k_0)^3 f'''(k_0) + \ldots, \tag{50.22}$$

where the primes indicate derivatives with respect to k. Making the substitutions

$$\alpha = f'(k_0), \quad \beta = f''(k_0), \quad \gamma = f'''(k_0), \quad \xi = (k-k_0) + \beta/\gamma, \tag{50.23}$$

and assuming that $f(k)$ is slowly varying near k_0, (50.21) becomes

$$\eta(r,t) = g(k_0)\exp\left\{i\left[f(k_0) + \frac{\beta}{\gamma}\left(\frac{\beta^2}{3\gamma} - \alpha\right)\right]\right\}\int_{-k_0+\beta/\gamma}^{\infty} \exp\left\{i\left[\frac{\gamma\xi^3}{6} + \left(\alpha - \frac{\beta^2}{2\gamma}\right)\xi\right]\right\}\mathrm{d}\xi. \tag{50.24}$$

If it is assumed that the main contribution to the integral comes from the vicinity of the wavenumber k_0, the lower limit of integration may be changed to $-\infty$ without substantially affecting the estimate of the integral. Since the argument of the exponential is odd, the integral may be rewritten as

$$2\int_0^\infty \cos\left[\frac{\gamma\xi^3}{6} + \left(\alpha - \frac{\beta^2}{2\gamma}\right)\xi\right]\mathrm{d}\xi.$$

Using a result from Abramowitz and Stegun (1965, p. 447), this integral can be evaluated as

$$2\pi\left(\frac{\gamma}{2}\right)^{-1/3} Ai\left[\left(\alpha - \frac{\beta^2}{2\gamma}\right)\left(\frac{\gamma}{2}\right)^{-1/3}\right], \tag{50.25}$$

where Ai is the Airy function encountered earlier (Section 36).

Returning to our gravity wave example, we now focus our attention on the small wavenumber limit of (50.5). For $kH \ll 1$,

$$\omega^2 \simeq gk[kH - k^3H^3/3],$$

$$\omega \simeq c_0[k - k^3H^2/6],$$

$$\omega' \simeq c_0[1 - k^2H^2/2],$$

$$\omega'' \simeq -c_0kH^2,$$

$$\omega''' \simeq -c_0H^2, \tag{50.26}$$

where $c_0 = \sqrt{gH}$. At the head of the wavefront, $k_0 = 0$, and $\omega(0) = 0$, $\omega'(0) = c_0$ and $\omega''(0)[= -\beta$ (in 50.23)$] = 0$. The integral (50.17) becomes, using (50.24)–(50.26) and $f(k) = kr - \omega t$,

$$\eta(r,t) = \frac{\pi}{r}\left\{\lim_{k_0 \to 0}[F(k_0)\sqrt{k_0}]\right\}\left(\frac{H^2c_0t}{2}\right)^{-1/3} Ai\left[(r - c_0t)\left(\frac{H^2c_0t}{2}\right)^{-1/3}\right]. \tag{50.27}$$

The transition from the irradiated region ($r < c_0t$) to the domain which the waves have not yet penetrated ($r > c_0t$) occurs, as for the caustic studied in Section 36, through an Airy function. Near the wavefront, both frequency and wavenumber vanish, so that the oscillations of the form exp $[if(k_0)]$ disappear. The oscillation seen there is that associated with the spatial dependence of the Airy function. At the point $r = c_0t$, $Ai(0) = 0.355$ and the spatial decay associated with wave dispersion is proportional to $r^{-1/3}$. The total decay there is like $r^{-5/6}$ rather than r^{-1} as found in (50.20), well behind the wavefront. As a wave group advances, the wavefront should then become progressively more pronounced with respect to the waves following it. This region of a wave group is often called the Airy phase. The appellation is particularly favoured in seismology, where the Airy phase occurs at a minimum of the group velocity of seismic waves in layered media. In the above example, the Airy phase occurs at a maximum of the group velocity; had we included surface tension effects, an Airy phase could also occur at the minimum group velocity which occurs at short wavelengths (cf. Section 11).

Comparison of the observed radiated pattern to that predicted by theory is possible whenever the source function $\eta_0(r)$ is known. Taking an initial perturbation of the form shown in Fig. 50.1a:

$$\eta_0(r) = \begin{cases} \eta_{0\max}\left[2\left(\dfrac{r}{r_0}\right)^2 - 1\right], & r \leqslant r_0 \\ 0, & r > r_0 \end{cases} \tag{50.28}$$

as resulting from a localized explosion, Le Méhauté (1971) obtained from (50.20) the wave form

$$\eta(r,t) = -\frac{\eta_{0\max}r_0}{k_0r}J_3(k_0r)\left(\frac{\omega'}{-\omega''}\right)^{1/2}\cos[k_0r - \omega(k_0)t]. \tag{50.29}$$

In Fig. 50.1b, the envelope of this curve [the terms preceding cos $\{k_0r - \omega(k_0)t\}$] is compared to the observed wave in an experimental situation.

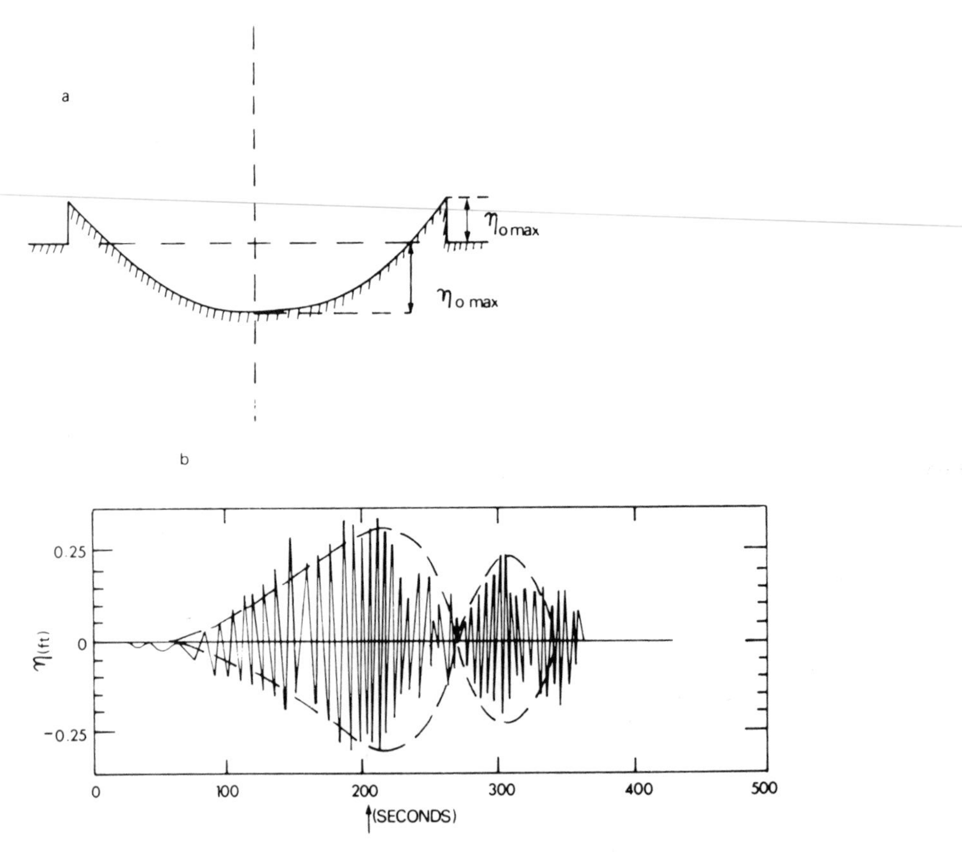

Fig. 50.1. (a) The initial surface displacement described by (50.28). (b) A comparison between the wave envelope predicted by (50.29) (dashed curve) and the observed wave train (high-frequency continuous curve). (From Le Méhauté, 1971.)

Some idea of the influence of the shape of the initial perturbation may be obtained from the tsunami simulations of Carrier (1971). Assuming a bottom velocity distribution (a formulation which differs from that adopted here, but which leads to similar results, as the reader may verify in Exercise 50.2) of the form

$$w_0(x, -H, t) = (4\pi\alpha)^{-1/2} \exp\left[-x^2/4\alpha H^2\right]\delta(t), \tag{50.30}$$

where α is a width factor and $\delta(t)$ the Dirac delta function, Carrier calculates the wave form in shallow water ($kH \ll 1$) from an expression equivalent to (50.27) for a series of values of α at a distance $x = 1000H$. The results are shown in Fig. 50.2. Note the considerable importance of the width of the initial perturbation on the amount of dispersion seen in the wave field. The double Fourier transform of $w_0(x, -H, t)$ is given by

$$\bar{\bar{w}}_0(k, -H, \omega) = \int_{-\infty}^{\infty}\int_{-\infty}^{\infty} \exp\left[-i\omega t - ikx\right] w_0(x, -H, t)\,\mathrm{d}x\mathrm{d}t = H\exp\left(-\alpha k^2 H^2\right). \tag{50.31}$$

For large α, the wavenumber spectrum decays rapidly with increasing k and the initial disturbance consists mainly of very long waves, for which there is little dispersion. For small α, shorter waves are important contributors to the initial profile, and a significant amount of dispersion occurs. Real tsunamis are rather more complicated; examples are shown in Fig. 50.3.

Solutions of the Cauchy–Poisson problem for viscous fluids have been presented by Nikitin and Potetyunko (1967) and by Miles (1968). The nonlinear Cauchy–Poisson problem has recently been tackled by Shinbrot (1976).

Green's functions

Wave propagation from localized sources may also be investigated by using the Green's function appropriate to the problem at hand. Green's function techniques are standard in wave propagation problems and details may be found in Morse and Feshbach (1953, Chapter 7) and in Duff and Naylor (1966). Basically, given an inhomogeneous equation of the form

$$\mathcal{L}(\psi) = f(\boldsymbol{x}, t), \tag{50.32}$$

with given boundary and initial conditions and where $\mathcal{L}$ is a self-adjoint differential operator, the Green's function is that solution of

$$\mathcal{L}(G) = \delta(\boldsymbol{x}-\boldsymbol{x}_0)\delta(t-t_0), \tag{50.33}$$

where G satisfies the same auxiliary conditions. Multiplying (50.32) by G and (50.33) by ψ, subtracting the resulting equations, then integrating with respect to $\boldsymbol{x}$ and t, and finally interchanging $(\boldsymbol{x}_0, t_0)$ and $(\boldsymbol{x}, t)$, we find that

$$\psi(\boldsymbol{x},t) = \iint G(\boldsymbol{x},t;\boldsymbol{x}_0,t_0)\, f(\boldsymbol{x}_0,t_0)\,\mathrm{d}\boldsymbol{x}_0\mathrm{d}t_0 - \iint G(\boldsymbol{x},t;\boldsymbol{x}_0,t_0)\,\mathcal{L}[\psi(\boldsymbol{x}_0,t_0)]\,\mathrm{d}\boldsymbol{x}_0\mathrm{d}t_0$$

$$+ \iint \psi(\boldsymbol{x}_0,t_0)\,\mathcal{L}[G(\boldsymbol{x},t;\boldsymbol{x}_0,t_0)]\,\mathrm{d}\boldsymbol{x}_0\mathrm{d}t_0. \tag{50.34}$$

The solution thus consists of an integral of the contributions over all points in the source at all times of generation of the waves which travelled from the source (the first term) plus contributions (the second and third integrals) from sources on the boundaries due to inhomogeneous boundary conditions; the latter result follows after the appropriate integrations by parts.

The Green's function for surface gravity waves in infinitely deep water is given by Stoker (1957); Kajiura (1963) derived the analogous function for a fluid of finite depth. The Cauchy–Poisson problem discussed above may also be solved through the Green's function, as was done by Kajiura (1963).

The surface elevation of long waves in a homogeneous rotating fluid (on the f-plane) obeys (17.27); for a flat bottom and harmonic time-dependence ($\mathrm{e}^{i\omega t}$),† this equation reduces to Helmholtz's equation:

---

† Note the departure from the usual case of $\mathrm{e}^{-i\omega t}$ time-dependence used in this book. We use $\mathrm{e}^{i\omega t}$ here as this is what is also used in much of the literature quoted below.

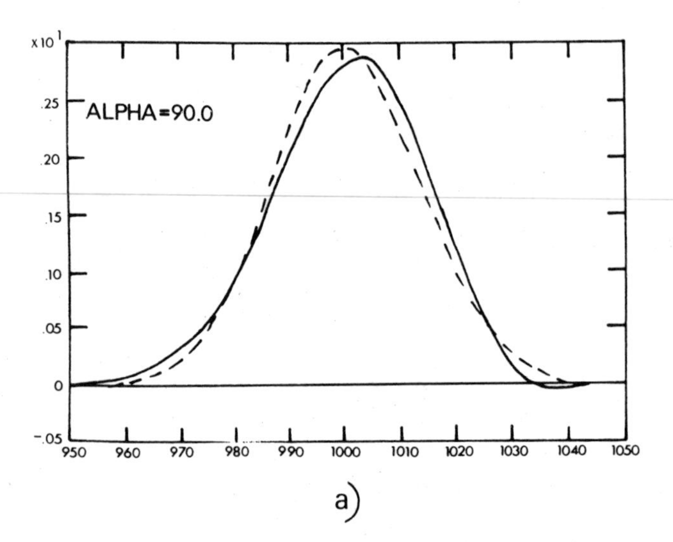

a)

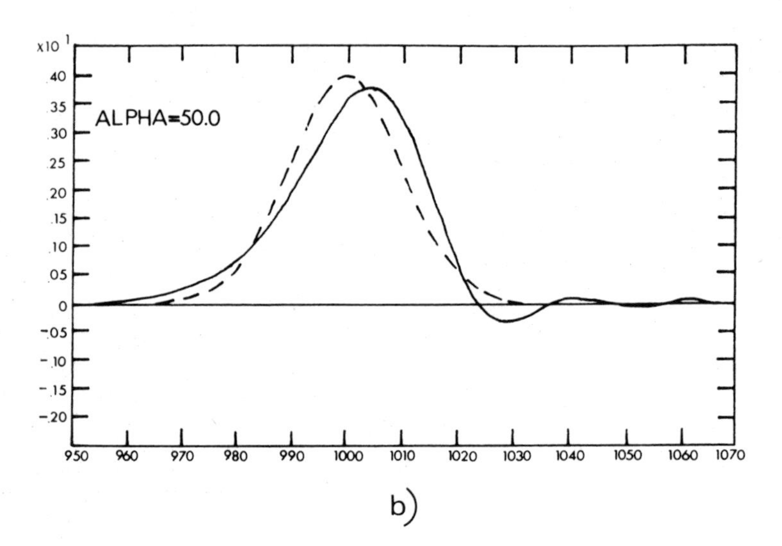

b)

$$(\nabla_h^2 + k^2)\eta = 0, \tag{50.35}$$

$$\text{with} \quad k^2 = (\omega^2 - f^2)/gH. \tag{50.36}$$

The Green's function of Helmholtz's equation representing cylindrical wave propagation into an unbounded plane away from a two-dimensional source at $\boldsymbol{x}_0$ is simply (Morse and Feshbach, 1953)

$$G(\boldsymbol{x},\boldsymbol{x}_0) = -\frac{i}{4} H_0^{(2)}(kr), \tag{50.37}$$

where $r = |\boldsymbol{x} - \boldsymbol{x}_0|$. The solution for a specified harmonic source $f(\boldsymbol{x})$ is then, from (50.34),

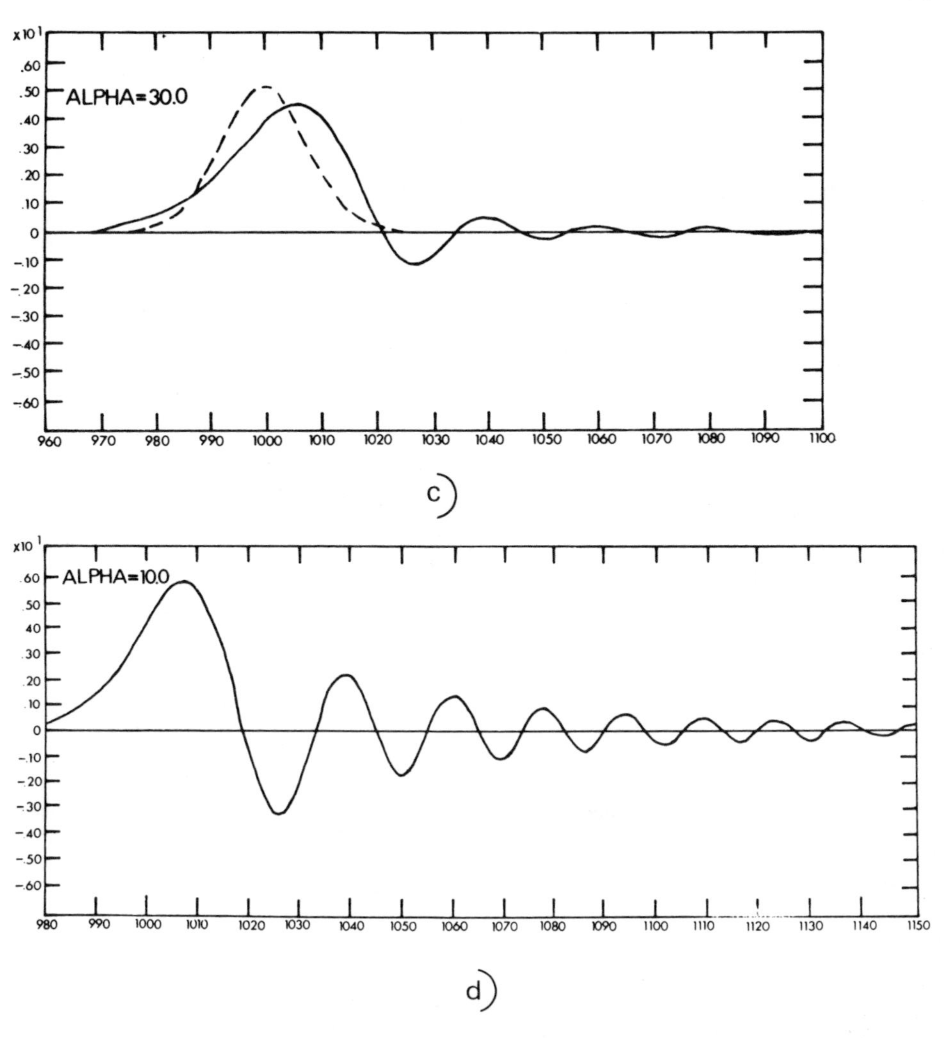

Fig. 50.2. The influence of source width (α) on the amount of dispersion occurring in a wave group issuing from a source of the form (50.30). The shape of the source is shown by the dashed curve. From (a) to (d), the width parameter α decreases from 90 to 10. In all cases, the wave group is shown at a time $t = 1000$ (in units of $\sqrt{H/k}$) following its creation at the source. The distance x and the elevation η are also measured in units of the depth H. (From Carrier, 1971.) (Reprinted with permission of the publisher, American Mathematical Society, from 'Lectures in Applied Mathematics'. Copyright © 1971, Volume 13, No. 1, pp. 164–168.)

$$\eta(x, t) = -\frac{i}{4} \int H_0^{(2)}(kr)\, f(x_0)\, \mathrm{d}x_0 \mathrm{e}^{i\omega t}. \tag{50.38}$$

Of greater oceanographic relevance is the problem discussed by Voit (1958) and Buchwald (1971), wherein waves originate from a pulsating mass flux at the mouth of a narrow channel in the wall of a semi-infinite ocean. Consider an ocean of uniform depth H on the f-plane, bounded at $x = 0$ by a rigid vertical wall. In $x \geqslant 0$, η satisfies (50.35); on the wall, the boundary condition is given by

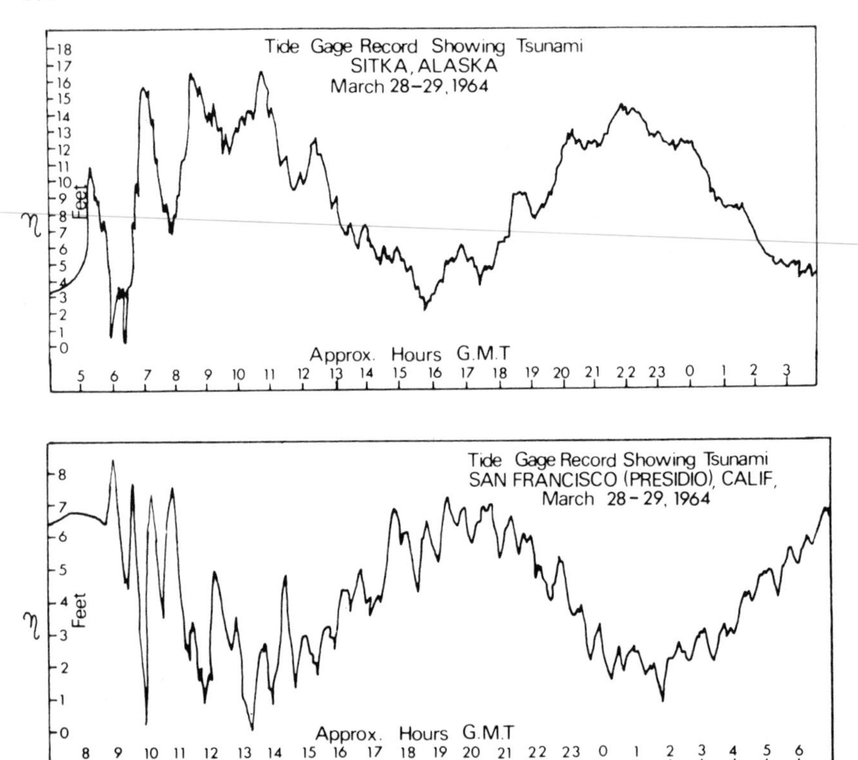

Fig. 50.3. Sea-level variation at two locations showing the tsunami generated by the Alaska earthquake of May 28, 1964. Great circle distances from the source (60°N, 147°W) to Sitka and San Francisco are 405 and 1626 miles, respectively. The shorter periods at San Francisco are indicative of the amount of dispersion along the path of propagation. (From Spaeth and Berkman, 1967.)

$$u(0,y) = \begin{cases} u_0(y), & |y| \leqslant b \\ 0 \quad , & |y| > b. \end{cases} \tag{50.39}$$

From (15.19), the u velocity component is related to η through

$$u = g(i\omega\eta_x + f\eta_y)/(\omega^2 - f^2), \tag{50.40}$$

which we abbreviate as

$$u = \mho\eta, \tag{50.41}$$

where $\mho$ is the appropriate differential operator. The solution for η in $x \geqslant 0$ may be found from the Green's function $G(x,y)$ satisfying

$$(\nabla_h^2 + k^2)G = 0, \tag{50.42}$$

$$\mho G = \delta(y) \quad \text{on} \quad x = 0. \tag{50.43}$$

It is readily verified that the solution to (50.35), subject to the boundary condition (50.39) is

$$\eta(x,y) = \int_{-b}^{b} u_0(y')G(x,y-y')\,\mathrm{d}y'. \tag{50.44}$$

We now find the Green's function satisfying (50.42) and (50.43). Let us write G in terms of a Fourier integral:

$$G(x,y) = \int_{-\infty}^{\infty} \Lambda(k_2)\exp\left[-ik_2y - sx\right]\mathrm{d}k_2. \tag{50.45}$$

In order to satisfy (50.42), s and k_2 must be related to k through

$$s = (k_2^2 - k^2)^{1/2}. \tag{50.46}$$

For convergence of the integral in (50.45), it is necessary to choose that branch for which $s \to |k_2|$ as $k_2 \to \pm\infty$ (cf. Buchwald, 1971). Substituting (50.45) into the boundary condition (50.43) and using the Fourier inversion theorem, we find

$$\Lambda(k_2) = \frac{i(\omega^2 - f^2)}{2\pi g(\omega s + fk_2)}, \tag{50.47}$$

which is to be used in (50.45) to construct $G(x, y)$. Let us rewrite $\Lambda(k_2)$ in a different form to aid in the interpretation of the Green's function. Multiplying top and bottom of $(\omega^2 - f^2)/(\omega s + fk_2)$ by $(\omega s - fk_2)$ and using (50.36), we find

$$\frac{\omega^2 - f^2}{\omega s + fk_2} = \frac{(\omega s - fk_2)(\omega^2 - f^2)}{(\omega^2 s^2 - f^2 k_2^2)} = \frac{\omega s}{(k_2^2 - \omega^2/gH)} - \frac{fk_2}{(k_2^2 - \omega^2/gH)}. \tag{50.48}$$

Multiplying the first term of the latter pair of terms by s/s and again using (50.36), we can finally write (50.45) as

$$G(x,y) = \frac{i}{2\pi g}\int_{-\infty}^{\infty}\left[\frac{\omega}{s} + \frac{\omega f^2}{gHs(k_2^2 - \omega^2/gH)} - \frac{fk_2}{(k_2^2 - \omega^2/gH)}\right]\exp\left(-ik_2y - sx\right)\mathrm{d}k_2. \tag{50.49}$$

Transformations of standard forms (Campbell and Foster, 1942, pairs 444, 445 and 867, 868) can be used to show that

$$G(x,y) = \sum_{j=1}^{3} B_j(x,y), \tag{50.50}$$

where
$$B_1(x,y) = \frac{\omega}{2g}H_0^{(2)}(kr), \tag{50.51}$$

$$B_2(x,y) = \frac{-if^2}{4g\sqrt{gH}}\int_{-\infty}^{\infty} H_0^{(2)}\left[k(x^2 + \xi^2)^{1/2}\right]\exp\left[-i(\omega|y - \xi|)/\sqrt{gH}\right]\mathrm{d}\xi, \tag{50.52}$$

$$B_3(x,y) = \frac{ifkx}{4g}\int_{-\infty}^{\infty}\frac{H_1^{(2)}\left[k(x^2 + \xi^2)^{1/2}\right]}{(x^2 + \xi^2)^{1/2}}\exp\left[-i(\omega|y - \xi|)/\sqrt{gH}\right]\mathrm{sgn}\,(y - \xi)\,\mathrm{d}\xi, \tag{50.53}$$

$$\text{and} \quad B_3(0,y) = \frac{-f}{2g} \exp\left[-i\omega|y|/\sqrt{gH}\right] \operatorname{sgn}(y). \tag{50.54}$$

This is the solution for $\omega/f > 1$; the modifications for $\omega/f < 1$ are given by Buchwald (1971). The first term is of the same form as the Green's function derived for an isolated point source in the absence of boundaries: it represents the radiation in the form of (cylindrical) Poincaré waves (cf. Section 23). For a nonrotating fluid ($f = 0$) the Green's function consists of this first term only. The terms B_2 and B_3 each contain one half of the Kelvin wave which propagates along the coast in the direction $y < 0$.

When $\omega/f < 1$, there is no Poincaré wave propagation, and only the Kelvin wave remains. The evaluation of specialized forms of the above solution for G for far and near fields goes beyond the scope of this book; results are available in Buchwald (1971) (see also Exercise 50.5).

Finally, we note that wave forcing problems may also be approached formally by a combination of Laplace and Fourier transform techniques (Exercises 50.1 and 50.2). Numerous examples are found in the literature and will be referred to in the discussion of the generation of specific types of waves below.

Modal partition

So far in this section, we have discussed the radiation of surface gravity waves. The presence of density stratification leads to the possibility of a sequence of vertical internal modes (Section 10); that of lateral boundaries, to a quantization of the horizontal dependence into a double sequence of eigenmodes (Sections 10, 28 and 29). Let us first consider each case separately.

A localized source in a stratified and laterally unbounded fluid may be specified through the spatial and temporal dependence of an imposed vertical forcing velocity w_0. For a flat-bottom ocean, a complete set of eigenfunctions is given by the surface and internal wave solutions of (10.41) and (10.42). Hence, the vertical dependence of the forcing velocity $w_0(z)$ may be expanded in terms of the eigenfunctions $Z_n(z)$:

$$w_0(z) = \sum_n a_n Z_n(z), \quad -H \leqslant z \leqslant 0. \tag{50.55}$$

For a linear system, the energy radiated into the nth vertical mode from the source will be proportional to a_n^2. Each mode may then be treated through the methods sketched above (Green's functions, transform methods, . . .) and the solution synthesized at any horizontal position by summing the vertical modes. This method has been used by Miropolsky (1975) to study the impulsive generation of waves in a stratified fluid. Because of dispersion and attenuation, the vertical modal structure changes with position. The higher modes are attenuated faster, roughly as $\exp[-n^2\kappa]$, where κ is an appropriate eddy coefficient (Section 53), and are also left behind because of their lower group velocity (Section 10). For any given horizontal wavenumber, the vertical structure will simplify with distance of propagation from the source until only a few lower modes are present. This smoothing of the vertical structure with distance of propagation from the source may account in part for the success of the Garrett and Munk (1972, 1975) model (cf. Section 53) in reproducing the spectral characteristics of deep ocean internal waves by using only the first twenty internal wave modes.

In the presence of bottom topography, modal decomposition is no longer possible and wave propagation must be considered in terms of the characteristics; nevertheless the arguments concerning attenuation and dispersion remain applicable.

The situation in a horizontally bounded fluid is analogous to that described above. The spatial distribution of any source function may be expressed in terms of a double sum of horizontal eigenfunctions to give a modal decomposition of the forcing. To each mode there now corresponds a specific eigenfrequency, so that the time structure of the response of the basin will reflect not only the temporal, but also the spatial structure of the forcing function, the response being generally enhanced at resonant frequencies. The advisability of a two-fold modal decomposition depends in practice on the role played by friction in the basin. Should frictional effects be large, waves originating at some point within the basin will be damped out before bouncing back and forth from the walls a sufficient number of times to establish a discernible standing wave system. Such seems to be the case, for example, for internal waves of tidal and higher frequencies in the ocean (Rattray, 1957; LeBlond, 1966). In the case of the barotropic tides, on the other hand, where the forcing is continuous in time and spread over the whole ocean area (Section 52), a knowledge of the oceanic eigenfunctions is essential to an understanding of the response to the applied forces.

In a bounded and stratified ocean, each one of the double infinity of horizontal modes is further split in a vertical sequence. The vertical structure of the source function determines the vertical energy distribution in any horizontal mode; the horizontal form of the forcing is reflected in the distribution of energy among horizontal eigenmodes for each vertical mode. Our remarks on the influence of friction still hold and in many instances the finite dimensions of the ocean (although not of smaller basins, such as lakes or estuaries) may be of little relevance to the propagation of the internal modes.

Exercises Section 50

1. Derive a Fourier integral representation of the surface elevation for surface gravity waves in one horizontal dimension due to prescribed initial displacement and velocity distributions.

2. Obtain, in integral form, the surface perturbation generated in one horizontal dimension by a bottom displacement of the form (50.30). (Carrier, 1971.)

3. Show that, if instead of a perturbation of the form (50.28), a parabolic depression without lips, of the form

$$\eta_0(r) = \begin{cases} \eta_{0\,\max}[(r/r_0)^2 - 1], & r \leqslant r_0 \\ 0 & , \quad r > r_0, \end{cases}$$

is assumed, the envelope modulation factor preceding the cosine term in (50.29) becomes (Le Méhauté, 1971)

$$\frac{2\eta_{0\,\max}}{rk^2} J_2(kr)(-\omega'/\omega'')^{1/2}.$$

4. Let $A(k, r)$ denote the envelope function preceding the cosine term in (50.29). The peaks of this envelope will be found at those points $k = k_{\max}$ where $\partial A/\partial k = 0$. Show

that if kH is assumed large enough for the deep water approximation to hold for surface gravity waves, the first envelope maximum occurs at (Le Méhauté, 1971)

$$r_0 k_{\text{max}} = 4.20.$$

5. For $\omega/f > 1$, Buchwald's results (50.46)–(50.50) give, in the far field, in plane polar coordinates taken from the point source, $G(r, \theta) = K_p + K_k \Theta(\theta_0 - \theta)$, where K_p represents the Poincaré wave contribution

$$K_p = kH \left(\frac{k}{2\pi r}\right)^{1/2} \frac{\cos\theta}{(\omega\cos\theta - if\sin\theta)} e^{-i(kr - \pi/4)} + 0(r^{-3/2}), \tag{50.56}$$

propagating into the open ocean from the source, and K_k is the Kelvin wave

$$K_k = \frac{f}{g} \exp\left(i\omega y - fx/\sqrt{gH}\right) \tag{50.57}$$

travelling along the wall. $\Theta(z)$ is the Heaviside unit function and $\theta_0 = -\sin^{-1}(k\sqrt{gH}/\omega)$. Sketch the domains in which the Poincaré and Kelvin waves contribute to the solution and compute the energy flux associated with the two types of waves as a function of ω/f.

6. Gravity waves are excited in a stratified rotating fluid by a very thin horizontal pulsating jet (of frequency ω) at the upper surface. Assuming that the Brunt–Väisälä frequency is uniform with depth, calculate the initial amplitude distribution between the various vertical modes.

Surface gravity waves are the most obvious oceanic waves. Long the bane of seafarers, their relation to the wind which generates them is obvious to the most casual landlubber. It also does not require special perception to note that heavy objects dropped into the sea, or ships travelling through it also generate waves. We have already discussed the topic of tsunami generation (by landslides or bottom displacements) in the previous section. Ship waves will not be considered in this book; although they may be of some relevance to the surface waviness of localized areas, their role in the overall oceanic energy budget is minor. The reader is referred to Stoker (1957) for some of the fundamental principles of ship wave generation and to Wehausen (1973) for a fuller discussion.

In this section, we shall concern ourselves with the principal mechanism of surface wave generation: the momentum transfer from the wind to the sea surface. In spite of the clear connection between wind and waves and a long history of theoretical work, it is only recently that some degree of understanding of the mechanisms of wind-wave generation has been acquired. Following a short description of the wind field over the sea (which specifies the forcing conditions), we shall review the development of wind–wave generation theories, with emphasis on the more recent results. Our account of the subject is rather sketchy compared to that found in Kinsman's (1965) book, which the reader avid for a very detailed presentation of the early works would do well to consult. An excellent review of recent progress in the field of wind-wave generation has been published by Barnett and Kenyon (1975).

The air flow over waves

The air flow over the sea is invariably turbulent, and the characteristics of this turbulence are of profound relevance to the air–sea interaction process. Low-frequency spectra of mid-latitude wind speed and air pressure, based on 3-hourly observations, are shown in Fig. 51.1 (Fissel et al., 1976). Spectral peaks occur at time scales of about 3.1 and 15 days for wind and pressure fluctuations, respectively, reflecting the passage of synoptic disturbances of horizontal scales of hundreds of kilometers. Higher frequency spectra taken very near the sea surface (Fig. 51.2, from Elliott, 1972) show a general decay with increasing frequency and, at elevations above the sea surface comparable to the wavelength, a contamination of the pressure fluctuations by surface wave activity. At greater heights the influence of the sea motion is no longer visible and the decays are similar to those observed over a flat plate. On the large scale, the air flow may be considered as a fully developed turbulent boundary layer; the small-scale structure is, however, influenced by the presence of wave motion at the lower boundary.

Consider a turbulent boundary layer flowing over a rigid surface lying in the (x, y)-plane. Under statistically steady $(\partial/\partial t = 0)$ and laterally homogeneous conditions $(\partial/\partial x = 0, \partial/\partial y = 0)$, the horizontal momentum equations reduce to

$$\frac{\partial}{\partial z}\left[\tau_{i3} + \mu\frac{\partial U_i}{\partial z}\right] = 0, \tag{51.1}$$

where τ_{i3} is the Reynolds stress (cf. 49.11) representing downward transfer of horizontal momentum in the x_i-direction, and $\mu(\partial U_i/\partial z)$ is the molecular stress (cf. 48.1b). Under the

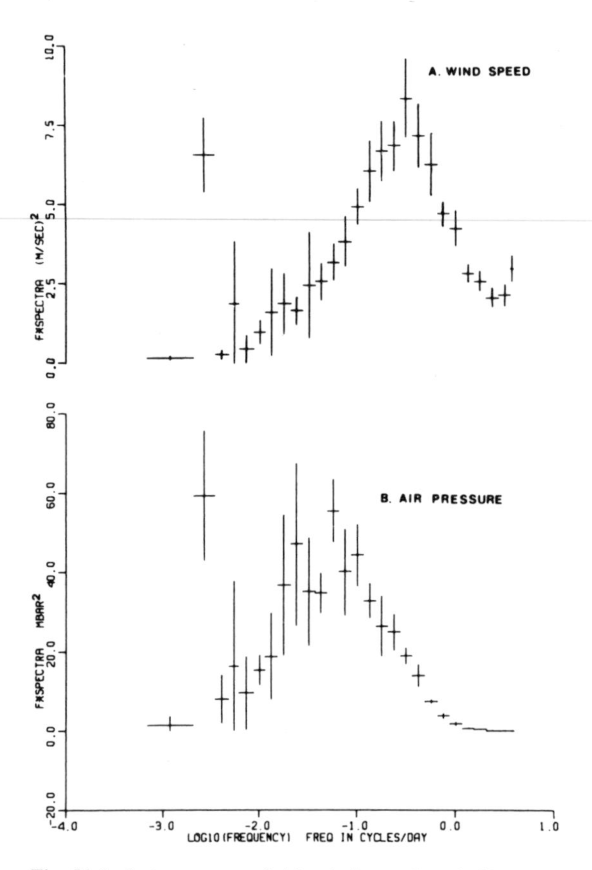

Fig. 51.1. Auto spectra of (a) wind speed and (b) air pressure at Station P. The vertical error bars represent approximate 95%-confidence intervals about the mean of each spectral estimate. (From Fissel et al., 1976.)

idealized conditions envisaged, the total stress is thus independent of position. Except perhaps very near the rigid boundary, where the turbulent stress must vanish since $w' = 0$ at that boundary, the Reynolds stress greatly exceeds the molecular term. Choosing $U_i = U_1$, (51.1) then reduces to

$$\tau_{13} = -\rho_a \overline{u'w'} \equiv \tau, \tag{51.2}$$

where τ is a constant, ρ_a the air density, and u' and w' the x and z components of the turbulent velocity fluctuations; the overbar denotes a suitable averaging procedure (cf. Section 49). It is convenient to introduce a turbulent velocity scale u_*, defined through

$$u_*^2 = -\overline{u'w'}. \tag{51.3}$$

In terms of this "friction" velocity, the stress becomes

$$\tau = \rho_a u_*^2. \tag{51.4}$$

The stress may also be related to the wind velocity at some reference level (10 m, say)

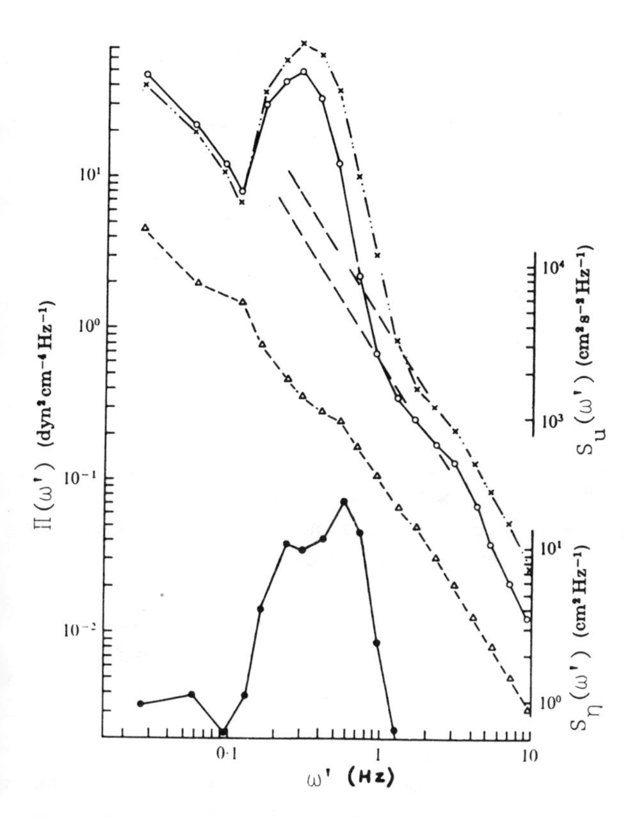

Fig. 51.2. Air pressure spectra $\Pi(\omega')$, downstream wind velocity spectra $S_u(\omega')$ near the sea surface, and a wave amplitude spectrum $S_\eta(\omega')$, with $\omega' = \omega/2\pi$. The wave spectrum is given by the curve $-\cdot-\cdot-$; the wind velocity spectrum by the curve $--\triangle--$. Two air-pressure spectra are shown; the curve $--\circ--$ is the spectrum of pressure at 30 cm above the mean water level; the curve $--\times--$ is a pressure spectrum at 80 cm. Note the hump in the air-pressure spectra in the frequency range where surface waves are found. The monotonically sloping spectra of the type found over a flat surface (– – – –) are shown for comparison. (From Elliott, 1972.)

by introducing a "drag coefficient" c_D, defined through

$$\tau = \rho_a c_D U_{10}^2, \tag{51.5}$$

where U_{10} is the wind speed measured at a height of ten metres above the mean sea level. For $U_{10} < 10\,\mathrm{m\,s^{-1}}$, c_D does not appear to depend on wind speed and has observed values of $c_D = (1.5 \pm 0.3) \cdot 10^{-3}$ (Pond, 1975).

Another feature of the steady turbulent boundary layer is that, on the average, the rate of production of turbulent energy by the mean flow must be balanced locally by the rate at which turbulent energy is lost by molecular friction. Thus, by a straightforward extension of the results of Section 49 to the energy equation,

$$\rho_a \overline{u'w'} \frac{\partial U_1}{\partial z} = \mu \overline{u_i' \frac{\partial u_i'}{\partial x_j \partial x_j}}. \tag{51.6}$$

The molecular dissipation rate on the right-hand side cannot be estimated precisely without some knowledge of the spatial structure of the turbulence. It is clear, however, that this term must have the same dimensions as $\partial(\rho_a u_i' u_i')/\partial t$. Using u_* as a velocity scale and z/u_* as a time scale, the dissipation term becomes proportional to $-\rho_a u_*^3/z$ (with a minus sign since this is a rate of loss of energy). Using (51.2) and (51.3), (51.6) becomes

$$\rho_a u_*^2 \frac{\partial U_1}{\partial z} = \frac{1}{K}\rho_a u_*^3/z, \tag{51.7}$$

where K is a constant of proportionality (von Karman's constant: $K \simeq 0.4$). Integrating (51.7) we obtain the logarithmic velocity profile

$$U(z) = \frac{u_*}{K} \ln\left(\frac{z}{z_0}\right), \tag{51.8}$$

where z_0 is a constant of integration. The constant z_0 is a virtual origin for the logarithmic profile, determined either by the roughness scale of the surface or by the thickness of the laminar sublayer for a smooth surface. The reader interested in a more critical justification of (51.8), or in other properties of the atmospheric turbulent shear layer should consult Lumley and Panofsky (1964).

Over a wavy surface, such as that of the sea, the profile (51.8) must be modified to conform to the shape of the surface; the relevant modifications have been discussed by Benjamin (1959). Furthermore, since the sea surface is not rigid, the air flow over it is not expected to be identical to that over a similar rigid surface. Nevertheless, as established by a number of observers (Brocks, 1959; Badgley et al., 1972, for example), the air flow well above the wave crests conforms closely to (51.8). Departures from the logarithmic velocity profile are to be expected in the vicinity of the free air–sea interface, at levels where measurements are difficult to make. Such deviations, as we shall see below, are critical to the wave generation process.

Energy and momentum are transferred from the air flow to the sea through the action of the air pressure and shear stress at the sea surface. Both components of the stress are of course continuous across the air–sea interface. Furthermore, since the total stress exerted by the wind on the sea surface is $\rho_a u_*^2$, the sum of the horizontal pressure force and the shear stress, averaged over a wave length, must satisfy

$$\langle \text{pressure force}_a \rangle + \langle \text{shear}_a \rangle = \rho_a u_*^2 \tag{51.9}$$

at any level from the sea surface up. Since surface gravity waves are irrotational, their direct generation by the wind can only be through the action of the pressure force at the sea surface. The net rate of energy input into wave motion is thus equal to $-\langle p_a(\eta)\eta_t \rangle$, and thus depends critically on the relative phase of the surface air pressure field to the wave profile.

Continuity of the shear stress across the air–sea interface implies that a shear layer should also be expected to be present at the sea surface. This so-called "wind-drift" layer is commonly observed to have a maximum surface velocity (u_d) of about 3% of the wind speed U_{10} (Banner and Phillips, 1974). The influence of the presence of the drift layer on the initiation of wave breaking has already been mentioned in Section 12. Its relevance on the wave generation process will also be noted below.

It is extremely convenient to consider the flow field over travelling waves in a

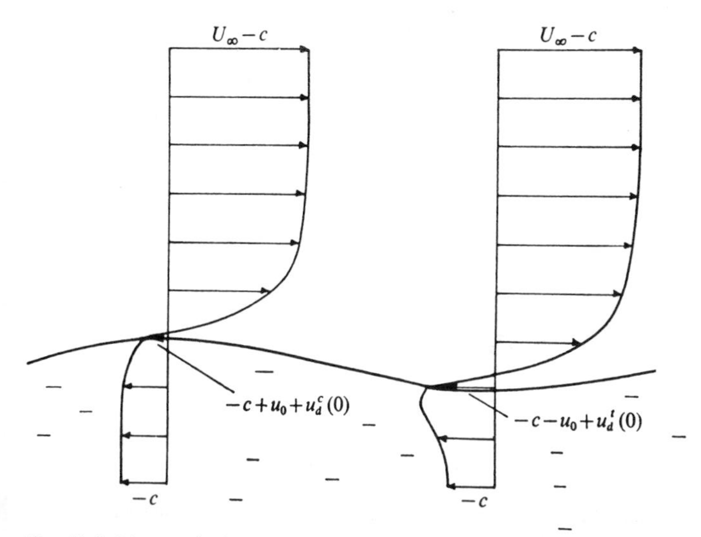

Fig. 51.3. Mean velocity profiles in the air and in the water at the crest and at the trough of an unbroken wave, in a coordinate system travelling with the wave profile. $u_d^c(0)$ and $u_d^t(0)$ are the values of the wind draft velocity at the crest and at the trough, respectively; u_0 is the maximum value of the wave orbital velocity. The critical level occurs where the wind profile changes sign. (Adapted from Banner and Melville, 1976.)

coordinate system moving with the phase speed of the waves, since the wave profile is time-independent in that reference frame. The velocity profiles at the crest and at the trough of a wave in this moving frame of reference are illustrated in Fig. 51.3. We note that in that coordinate system a critical layer occurs at $z = z_c$, where $U(z_c) - c = 0$. From our previous encounter with critical layers in Section 41, we may expect phenomena of interest to occur at that level. The role of the critical layer is particularly important in the wave generation theory of Miles, discussed below.

When a boundary layer proceeds against an adverse pressure gradient, it may be slowed down until it separates from the supporting wall (Batchelor, 1967, p. 325). At the point of separation the flow along the boundary changes sign. Variations in air pressure along the wave profile may then, if they are of the right sign, oppose the flow in the viscous sublayer and, if strong enough, lead to flow separation over the waves. The possibility of flow separation was the central point of an early model of wind-wave generation (Jeffreys, 1925, 1926) and has received renewed attention in recent years (Longuet-Higgins, 1973b; Banner and Melville, 1976; Gent and Taylor, 1977). Looking at the wind flow over a wave profile in a coordinate system moving at the phase speed, it is clear that separation occurs at a point where the component of air flow along the surface vanishes and changes sign. The occurrence of separation may be described by the fact that a streamline leaves the water surface at this point (Fig. 51.4). But since the water velocity is continuous with the air velocity across the air–sea interface, the former also vanishes at the point of separation. In a frame of reference moving with the wave profile, this means that the water particles at the surface at that point move at the phase speed. As we have seen earlier (Section 12), this is precisely the criterion for breaking to occur. Hence, flow separation is to be found only over breaking waves, a conclusion confirmed experimentally by Banner and Melville (1976).

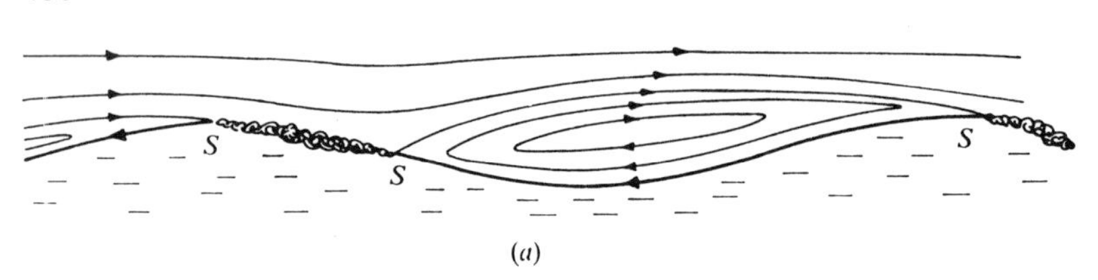

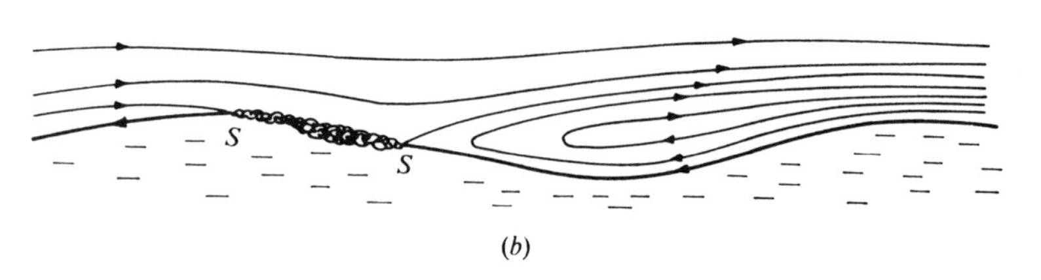

Fig. 51.4. Conjectured mean streamlines, in a coordinate system travelling at the phase speed, for the air flow over a broken wave if the downstream wave is (a) broken and (b) unbroken. The points marked S are stagnation points. (From Banner and Melville, 1976.)

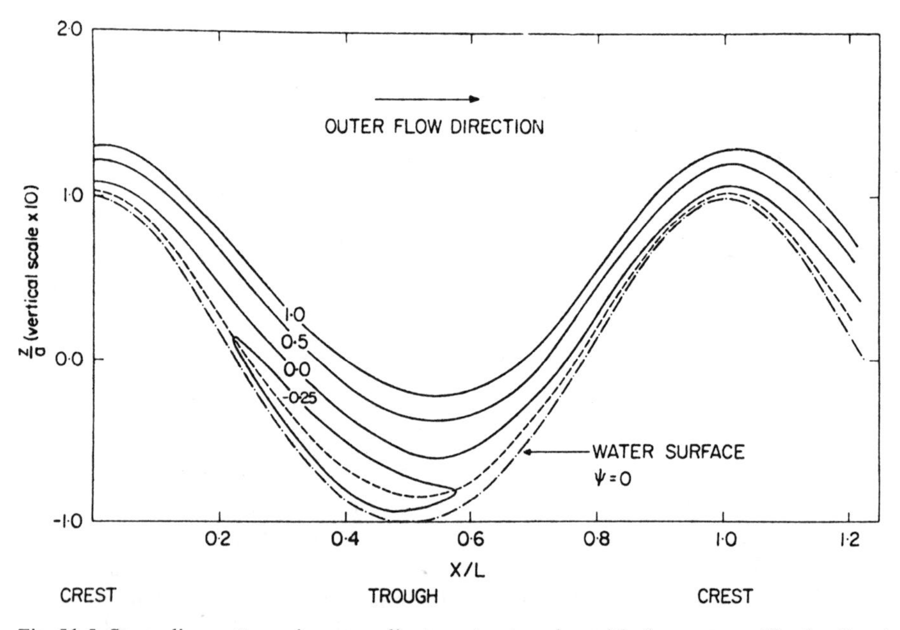

Fig. 51.5. Streamline pattern, in a coordinate system moving with the wave profile, for the air flow over a sinusoidal wave of steepness $ak = 0.157$. Other parameters: $R = -\ln(kz_0) = 8; c = 8u_*$. Water surface: — · — · — · —; critical level: — — — — — —. The vertical scale is exaggerated 10 times and the streamlines are labelled with the values of the stream-function ψ, scaled with respect to au_*. (From Gent and Taylor, 1977.)

Even though separation does not occur over unbroken waves, the flow field in their trough shows some interesting and important features. We recall from Section 41 that closed streamline flow patterns (Fig. 41.3) appear in a shear flow in the neighbourhood of a critical level. As was argued by Lighthill (1962b) similar "cat's eye" patterns must

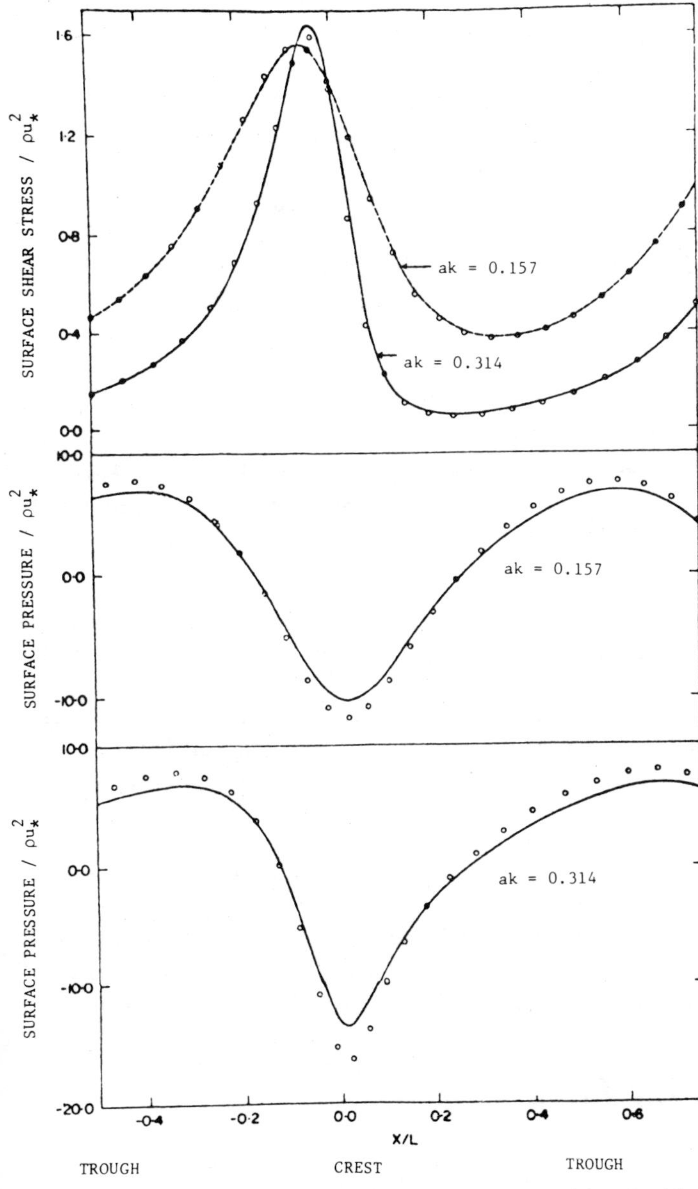

Fig. 51.6. Surface shear stress and pressure distributions with and without surface drift u_d, for two values of wave steepness $ak = 0.157$ and $ak = 0.314$. $R = -\ln(kz_0) = 5.07$, $c = 1.5u_*$ throughout. The case $u_d = 0$ corresponds to the continuous curves; $u_d = 0.5u_*$ to the circles. (From Gent and Taylor, 1977.)

exist, in a coordinate system moving with the wave profile, in the vicinity of the critical level $U(z_c) = c$. Should this critical level be very near the water surface, the position of the closed streamline pattern would be expected to be affected by the form of the wave profile, influencing in turn the distribution of air pressure and wind shear stress over the

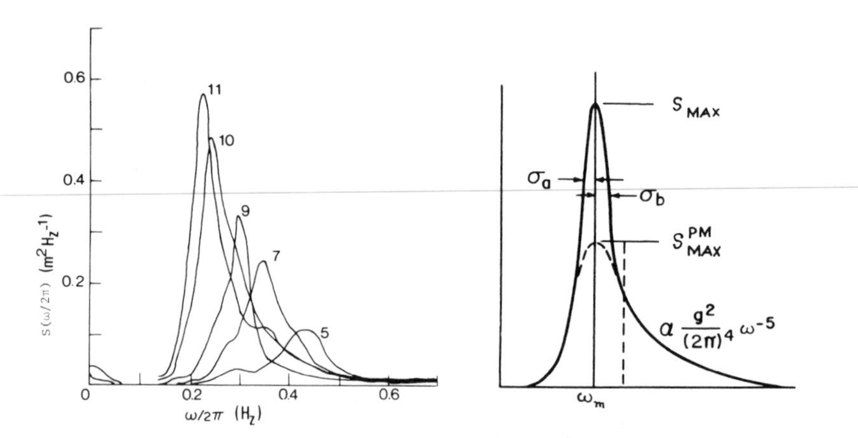

Fig. 51.7. The evolution of surface wave spectra with fetch, from the JONSWAP observations. The fetch increases with the number labelling the spectral peak. (From Hasselmann et al., 1973.)

Fig. 51.8. Definition sketch for the parameters entering the JONSWAP spectrum (51.10) and (51.11). S_{MAX}^{PM} denotes the height of the spectral peak of the Pierson-Moskowitz spectrum; S_{MAX} is the height of the peak of the JONSWAP spectrum. The ratio S_{MAX}/S_{MAX}^{PM} is denoted by γ (cf. 51.10).

water surface. Wind–water tunnel measurements by Shemdin (1969) and Shemdin and Lai (1973), as well as numerical computations by Gent and Taylor (1976) confirm the existence of an eddy above the forward face of the wave profile (Fig. 51.5). The exact form and position of the eddy depend on the wave steepness (*ka*) as well as on the magnitude and on the sign of the ratio of wave speed to wind speed (as measured at some reference level well above the waves, say 10 metres) c/U_{10}, and also on the strength of the wind drift velocity u_d at the water surface (cf. Gent and Taylor, 1977).

The distribution of pressure and shear stress associated with the air flow over the wave profile determines the normal and tangential forces exerted by the wind on the water surface and hence the rate of momentum transfer. Computed distributions of pressure and shear stress over a wave similar (but not identical) to that shown in Fig. 51.5 are illustrated in Fig. 51.6. We shall delay the discussion of the effects of these stresses, i.e., the actual wave generation process, and review first the observational evidence concerning wind waves.

Review of surface gravity wave observations

The evolution of surface gravity waves with increasing wind fetch is illustrated in Fig. 51.7 by a sequence of frequency spectra. The sea state is characterized by a sharp spectral peak at a frequency ω_m, say, where ω_m decreases with increasing fetch. The forward face (at $\omega < \omega_m$) of the peak is very steep: very little energy is to be found in waves of longer period than that of the peak. The high-frequency tail of the spectrum follows the ω^{-5} form appropriate to a fully developed sea, already described in Section 32. Note also in Fig. 51.7 that the spectral peak at short fetches reaches energy levels which exceed that reached at the same frequency for longer fetches. This overshooting phenomenon has already been discussed in Section 36.

A number of analytical expressions have been proposed to describe observed surface wave spectra $S(\omega)$. (Neumann, 1953; Burling, 1959; Bretschneider, 1959; Pierson and Moskowitz, 1964; Liu, 1971). The latest proposed form, based on the JONSWAP data (Hasselmann et al., 1973), is sketched in Fig. 51.8 and has the analytical representation

$$S(\omega) = \alpha g^2 (2\pi)^{-4} \omega^{-5} \exp\left[-\frac{5}{4}\left(\frac{\omega}{\omega_m}\right)^{-4}\right] \gamma^q, \tag{51.10}$$

where $S(\omega)$ is, apart from the dimensional factor $g\rho_w/2$, the energy density as a function of frequency [i.e., $\langle\eta\eta^*\rangle = \int S(\omega) d\omega$], ω_m the frequency of the spectral peak, g the acceleration of gravity and α an amplitude constant. The spectrum (51.10) differs from the earlier Pierson-Moskowitz (1964) spectrum through the presence of the asymmetrical peak enhancement factor γ^q, where γ is the ratio of the JONSWAP peak to the Pearson-Moskowitz peak (Fig. 51.8) and q is given by

$$q = \exp\left[-(\omega - \omega_m)^2 / 2\sigma^2\right], \tag{51.11}$$

$$\sigma = \begin{cases} \sigma_a & \text{for} \quad \omega \leqslant \omega_m \\ \sigma_b & \text{for} \quad \omega > \omega_m. \end{cases}$$

The JONSWAP spectrum described by (51.10) and (51.11) reduces to (32.5) for $\omega \gg \omega_m$. We have already noted in Section 32 that a spectrum of the form $S(\omega) \propto \omega^{-5}$ arises where breaking dominates the energetics of the wave field. We can thus conclude that actual wind wave spectra are dominated by wave breaking at high frequencies. For $\omega < \omega_m$, $S(\omega)$ is a very sharply increasing function of frequency (see Exercise 51.2). The JONSWAP spectrum is specified by the five parameters ω_m, α, γ, σ_a and σ_b. The dependence of these parameters on wind fetch x is best expressed in terms of dimensionless frequency and fetch variables:

$$\tilde{\omega}_m = \omega_m U_{10}/g, \tag{51.12a}$$

$$\tilde{x} = xg/U_{10}^2, \tag{51.12b}$$

where U_{10} is the wind speed measured at 10 meters, as defined above. The peak frequency is related to the fetch through (Hasselmann et al., 1973)

$$\tilde{\omega}_m = 3.5\tilde{x}^{-0.33}. \tag{51.13}$$

We note that in deep water ($kH \gg 1$), $c = g/\omega$, so that the wind speed $U_{10} \geqslant c_m$ for $\tilde{\omega}_m \geqslant 1$. Waves for which $\tilde{\omega}_m \leqslant 1$ travel faster than the wind.

A reasonable fit to field and wave tank observations of wave growth for the amplitude constant α is given by

$$\alpha = 0.076(2\pi)^5 \tilde{x}^{-0.22}; \tag{51.14}$$

slightly larger values of the exponent have, however, been reported (Hasselmann et al., 1973). The same authors observed that the shape parameters γ, σ_a and σ_b do not show a discernible trend with fetch within the scatter of the data. The mean values $\gamma = 3.3$, $\sigma_a = 0.07$ and $\sigma_b = 0.09$ are used in (51.10) and (51.11), indicating that the spectral shape factor $\{S(\omega)/\alpha g^2 (2\pi)^{-4} \omega^{-5}\}$ is self-preserving with fetch. Other shape factors have been proposed and two of them are compared with the JONSWAP spectrum in Fig. 51.9.

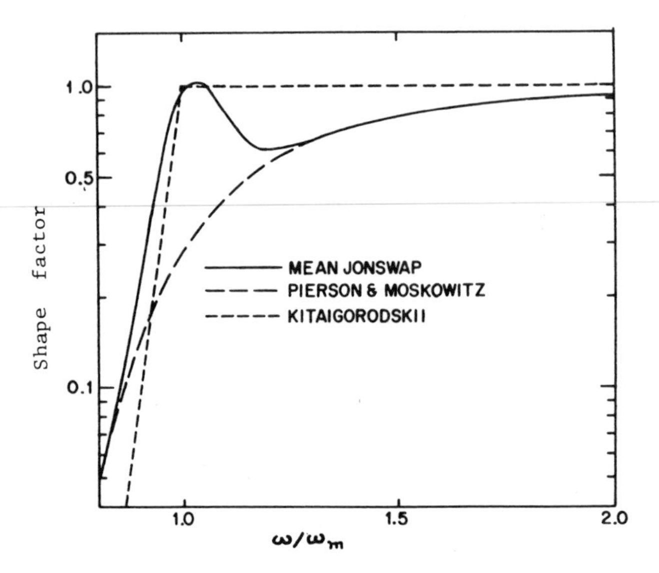

Fig. 51.9. Comparison of the shape factor, defined as $S(\omega)/\alpha g^2(2\pi)^{-4}\,\omega^{-5}$, of the mean JONSWAP spectrum (51.10) to other shape factors, as proposed by Pierson and Moskowitz (1964) and Kitaigorodskii (1962).

With the fetch dependences (51.13) and (51.14) for α and ω_m, it follows from (51.10) that the energy density per unit frequency bandwidth at the spectral peak varies with fetch as

$$S(\omega_m) \propto \tilde{x}^{1.43}. \qquad (51.15)$$

Since the spectral shape is self-preserving,

$$\int_0^\infty S(\omega)\,\mathrm{d}\omega \propto S(\omega_m)\,\omega_m \tilde{x}^{1.10}. \qquad (51.16)$$

A plot of energy against fetch (Fig. 51.10) shows a closer fit to a linear $\tilde{x}$-dependence. The discrepancy with (51.16) is within the error bounds of the estimates of the exponents expressing the ω_m and α dependence on fetch. The simple proportionality between the wave energy per unit area and wind fetch implies that the difference between the rate of energy input into the gravity wave spectrum and the dissipation rate of these waves is very nearly constant over the range of fetches for which the spectrum is described by the shape-preserving form (51.10). The observations of Donelan (1976) show that the rate of momentum transfer from the wind to the water decreases with wave age (as measured by $\tilde{\omega}$) and hence also with fetch. One must thus also conclude that the rate of wave dissipation decreases in the same fashion with fetch.

A relation of the form (51.16) cannot hold indefinitely; once the wave speed at the peak of the spectrum exceeds the wind speed, an equilibrium is reached and the wave energy no longer grows with fetch. This flattening off of wave growth is visible in Fig. 51.11 at very long fetches, and is reflected in empirical wave growth formulae (Snyder and Cox, 1966; Dobson, 1971) expressing the rate of growth per radian (in a time-independent situation) as

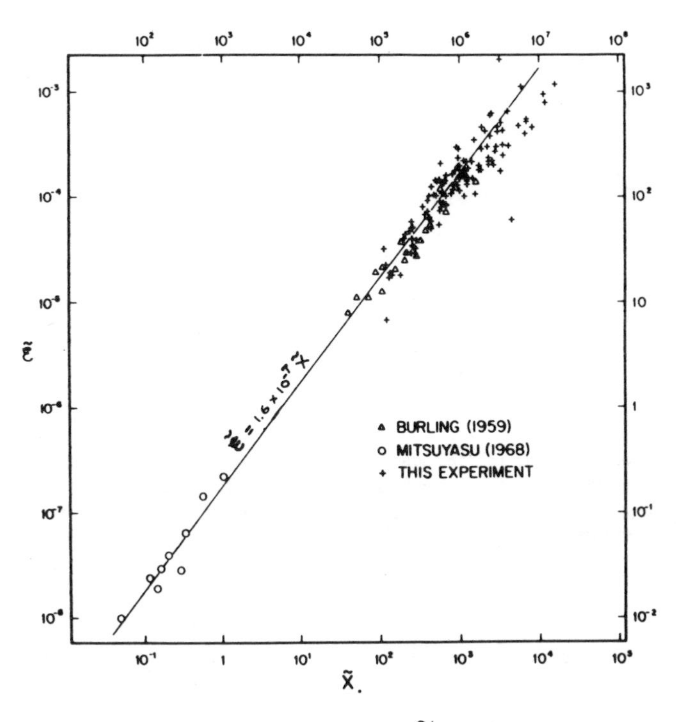

Fig. 51.10. Total wave energy against fetch $\tilde{\mathcal{E}} = g^2\langle E\rangle/U_{10}^4$ (data from Burling, 1959; Mitsuyasu, 1968, 1969; and Hasselmann et al., 1973). (From Hasselmann et al., 1973.)

$$\frac{1}{\omega S}\frac{\mathrm{D}S}{\mathrm{D}t} = \frac{c_g}{\omega S}\frac{\partial S}{\partial x} = \frac{\rho_a}{\rho_w}\left(\frac{U_\lambda}{c} - 1\right), \tag{51.17}$$

where U_λ is the wind speed at a height of one wavelength above the mean water surface. The relation (51.17) was found by the authors quoted, to hold for $U_\lambda/c \leqslant 1$.

In the engineering literature (Wiegel, 1970) one encounters the concept of the "significant wave-height", $H_{1/3}$, defined as the average height of the highest one-third of the waves (the wave-height is twice the amplitude). Similarly, the average period, T_a, of the whole wave field and the average period of the highest one-third class of waves, $T_{H_{1/3}}$, are used to characterize the time-dependence of the wave field. These quantities are related to the spectral parameters [such as those in (51.10)] through a description of the frequency distribution of sea-level displacements, with regard to which the reader may refer to Kinsman (1965, Chapter 7) and Longuet-Higgins (1963b). Although this type of characterization of the wind field antedates the spectral representation, its continued use requires some mention. Empirical best-fit formulae for wave speed and significant height are given by Wilson (1966) as

$$\frac{c}{U_{10}} = 1.37[1-(1+0.008\tilde{x}^{1/3})^{-5}], \tag{51.18}$$

$$\frac{gH_{1/3}}{U_{10}^2} = 0.30[1-(1+0.004\tilde{x}^{1/2})^{-2}], \tag{51.19}$$

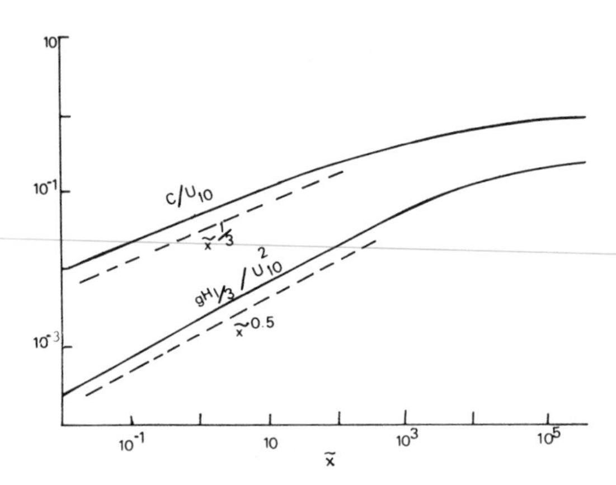

Fig. 51.11. Wilson's (1966) empirical formulae (51.18) and (51.19) for significant wave height and period as a function of fetch. The exact provenance of the data points may be found in Wiegel (1970). The curves $H_{1/3} \propto \tilde{x}^{0.5}$ and $T_m \propto \tilde{x}^{0.33}$ ($T_m = \omega_m^{-1}$) are also shown as a comparison with the fetch dependence inferred from the JONSWAP spectrum.

where c, the phase velocity, is related to the representative period ($T_{H_{1/3}}$, say) in deep water through $c = gT_{H_{1/3}}/2\pi$. A linear regression due to Scott (1965) relates the frequency of the spectral peak to $H_{1/3}$ through

$$\frac{1}{\omega_m} = 0.03H_{1/3} + 1.35, \tag{51.20}$$

where ω_m is in radians s^{-1} and $H_{1/3}$ in feet. The relations (51.18) and (51.19) are shown graphically in Fig. 51.11. A slightly refined version of (51.18) and (51.19) has also been presented by Bretschneider (1970).

We now have two sets of observational relations for the fetch-dependence of wave height and period: the JONSWAP spectrum: (51.13) and (51.14), and the formulae of Wilson: (51.18) and (51.19). Are the two sets consistent with each other? Let us associate $H_{1/3}$ with that part of the JONSWAP spectrum over which $S(\omega) \geqslant S(\omega_m)/\sqrt{3}$. Since the spectral shape is self-preserving with fetch, any partial integral of the spectrum over a finite energy band is also self-preserving [(51.16) is thus only a special case] and the energy associated with the highest one-third waves, $S_{1/3}$, say, also varies as $\tilde{x}^{1.10}$ (or, perhaps more fittingly as $\tilde{x}$, as indicated by Fig. 51.10). Thus $H_{1/3}$, according to the spectral representation (51.10) should vary as $\tilde{x}^{0.5}$. A comparison with Wilson's formula (51.19) shows excellent agreement, except at very large fetches, where Wilson's formula has been adjusted to the limitation of wave growth as $\tilde{\omega}_m \to 1$. Similarly, (51.13) gives $T_m \propto \tilde{x}^{0.33}$ which is again seen to provide as good a fit as (51.18) over all but the largest fetches.

Finally, although wind-generated surface waves run mainly with the wind, there is a degree of angular spread in energy density about the wind direction. Measurements from a floating buoy (Longuet-Higgins et al., 1963, Cartwright and Smith, 1964) and by radio back-scatter from the sea surface (Tyler et al., 1972) suggest an angular spread function of the form

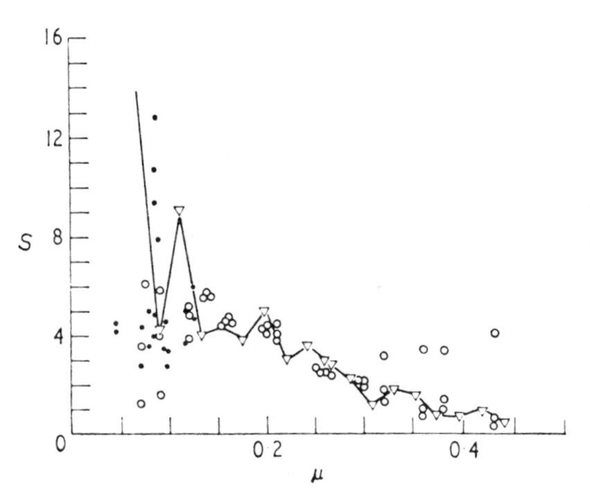

Fig. 51.12. The angular spread of wind generated waves. The angular exponent s is defined in (51.21). The parameter μ is equal to U/cK, where U is the wind speed, c the wave phase velocity and K is von Karman's constant. (From Barnett and Kenyon, 1975, copyrighted by the Institute of Physics.)

$$A(\theta) = \cos^s(\theta/2), \tag{51.21}$$

where $-\pi \leqslant \theta \leqslant \pi$ and $\theta = 0$ is the wind direction. The exponent s is a function of frequency, varying from 1 at high frequencies (and hence short fetches) to about 10 at low frequencies (long fetches). The directional spectrum $S(\omega, \theta)$, given by $S(\omega)A(\theta)$, thus becomes increasingly concentrated about the wind direction with increasing fetch. A summary of the variation of the exponent s with wave frequency is shown in Fig. 51.12.

Early theories of wind-wave generation

The problem of wind-wave generation consists in explaining the observed features of wind waves, as reviewed above, in terms of the properties of the air flow over a liquid surface. A good theory should explain the original appearance of waves over a previously unruffled surface and account for their subsequent rate of growth and for the shape of their energy spectrum and of its angular distribution. As the observational evidence on the nature of wind waves has unfolded over the years, the demands on the quality of wind observations and on the completeness of the theoretical explanation have increased accordingly. The complexity of the modern theoretical framework makes it difficult to accommodate within the confines of a book such as this one. Our review of wind-wave generation theory will thus present the highlights of the various mechanisms proposed to transfer momentum between the air into water wave motion, without going into intimate details.

It is convenient to view the problem of wind-wave generation from the point of view of the radiation balance equation (33.11). In a medium at rest, the frequency is constant along a ray, and we find it more useful to work in terms of the spectral energy density S, rather than that of wave action n. With Σ now representing energy sources, the radiation balance equation takes the form

$$\frac{\mathfrak{D}S}{\mathfrak{D}t} = \frac{\partial S}{\partial t} + \left(\frac{\mathrm{d}\boldsymbol{k}}{\mathrm{d}t}\right) \cdot \left(\frac{\partial S}{\partial \boldsymbol{k}}\right) + \left(\frac{\mathrm{d}\boldsymbol{x}}{\mathrm{d}t}\right) \cdot \left(\frac{\partial S}{\partial \boldsymbol{x}}\right) = \Sigma, \tag{51.22}$$

where $S = S(\boldsymbol{k}; \boldsymbol{x}, t)$. The energy spectrum in terms of the wavenumber vector embodies the directional properties of the wave field. One may pass from the wavenumber spectrum to the frequency spectrum through the relations presented in Section 39. Thus,

$$\int S(\boldsymbol{k}; \boldsymbol{x}, t)\, \mathrm{d}\boldsymbol{k} = \iint S(\omega; \boldsymbol{x}, t) A(\theta, \omega; \boldsymbol{x}, t)\, \mathrm{d}\omega \mathrm{d}\theta. \tag{51.23}$$

The physics of the wave generation process lies of course in the specification of the various parts of the source term Σ. In the notation of Hasselmann (1968), the following types of sources occur: Σ_1, a constant, such as is found in Phillips' (1957) generation theory in which the turbulent atmospheric pressure fluctuations are in resonance with the waves; Σ_2, a constant times $S(k)$, a term which arises from a coupling of the air flow with the wave profile in the theories of Jeffreys and Miles (see below). Hasselmann (1968) includes other terms, representing nonlinear corrections to the above, wave–wave interactions (Section 39) and frictional dissipation. The main contributions are now thought to arise from the terms Σ_1 and Σ_2, although nonlinear interactions between waves play an important role in determining the shape of the spectrum and must also be the energy source for waves which propagate faster than the wind (i.e., for which $\tilde{\omega} < \tilde{\omega}_m \simeq 1$ at long fetches).

The first surface wave generation mechanism was proposed by Lord Kelvin (1871) and has already been discussed in Section 42 as the Kelvin–Helmholtz instability mechanism. An excellent presentation is also to be found in Chandrasekhar (1961). The assumption used that the air flow over the sea is uniform with height was obviously a poor model of the wind distribution over a wavy water surface (cf. Fig. 51.5). For the conditions considered by Kelvin, the stability criterion may be found from (42.24). The main result of the analysis, that instability and hence wave generation would not occur at wind speeds less than $6.5\,\mathrm{m\,s^{-1}}$, was recognized by Kelvin himself as a major obstacle to the application of his work to wind-wave generation. The fact that the instability occurs at wavelengths of about 3 cm is also a strong limitation of this model.

A different approach was suggested by Jeffreys (1925, 1926). He assumed that as the air flows over the wavy water surface, boundary-layer separation occurs on the leeward of the wave crests, as would be the case over a solid surface. As we have seen above, flow separation implies wave breaking, a fact which was not recognized for a long time. Jeffreys then argued that the low-pressure area associated with the separated eddy would, because of its location over a part of the wave profile where the particle velocity is upwards, produce a negative correlation $\langle p\eta_t \rangle$ (cf. 48.9) corresponding to a positive rate of momentum transfer from the pressure field of the mean wind to the waves. With a monochromatic wave of the form $\eta = a \sin(kx - \omega t)$, Jeffreys assumed that the out-of-phase part of the pressure would take the form

$$p = s\rho_a (U - c)^2 \eta_x, \tag{51.24}$$

where s is a "sheltering coefficient", the free parameter of the theory. The rate of energy input into the wave is then given by

$$-\langle p\eta_t \rangle = \pm \tfrac{1}{2} s\rho_a k^2 c (U - c)^2 a^2, \tag{51.25}$$

where the plus sign is used when $U > c$ and the minus sign (corresponding to a net damping of waves which travel faster than the wind) for $U < c$. In terms of energy density

($\langle E \rangle = \frac{1}{2}\rho_w g a^2$, neglecting surface tension),

$$-\langle p\eta_t \rangle = \pm \left(\frac{s\rho_a}{\rho_w g}\right) k^2 c(U-c)^2 \langle E \rangle. \tag{51.26}$$

As the rate of energy input is proportional to the wave energy itself, we recognize this source term as being of the form Σ_2, in Hasselmann's (1968) notation. Jeffreys estimated the value of the sheltering coefficient s through an energy balance between pressure forces and molecular viscosity, equating the least wind which could sustain the waves against dissipation with observed least winds which seemed just capable of generating waves (the reader may retrace his argument in Exercise 51.4). The high value of the sheltering coefficient ($s = 0.3$) deduced by Jeffreys was not substantiated by pressure measurements carried out over solid models of wavy surfaces (see Ursell, 1956, for a critique), which indicated (with values of s of the order of 0.05) that the pressure forces produced were much too small for Jeffreys' mechanism to be effective in wave generation. Because of the different boundary conditions which apply over a moving surface, flow measurements performed over a solid boundary are not relevant to the study of the air flow over water waves. It is only recently that good measurements of the air flow over waves have been obtained (Shemdin, 1969; Shemdin and Lai, 1973; Banner and Melville, 1976) and that evidence for flow separation over breaking waves has been found. It is now clear that Jeffreys' mechanism, at least in its original formulation, cannot account for the growth of unbroken waves, since flow separation does not occur over the latter. In a fully developed sea, however, there are many breaking waves: the high-frequency part of the spectrum (the ω^{-5} region) is dominated by wave breaking. As we have discussed in Section 36, short waves have steeper slopes and hence are more likely to break near the crests of the longer waves on which they are superimposed. Flow separation due to the breaking of short waves will then occur preferentially over the forward face of longer waves. Banner and Melville (1976) observed a sharp increase in momentum transfer to the sea surface over breaking waves, and it is quite likely that the separation induced by the breaking of short waves would contribute to the growth of longer waves in a well developed sea.

The sheltering mechanism cannot, however, explain the appearance of undulations on an initially smooth water surface. A model pertinent to wave growth on an unperturbed surface was proposed by Eckart (1953). He calculated the effect of a random distribution of pressure gusts associated with a circular storm. The atmospheric pressure field was assumed undisturbed by the waves: no sheltering effect leading to form drag was included. Eckart concluded that this mechanism was not effective in producing surface waves, the pressure field being too weak to bring about the observed response. We shall see below, however, how a resonant response of the sea surface, as suggested by Phillips (1957), may indeed account for the initial stages of wind-wave generation.

At more or less the same time, Wüest (1949) and Lock (1954) extended the early Kelvin–Helmholtz instability theory to obtain a better estimate of the conditions under which small oscillations may grow. Their results were found hard to relate to any field or laboratory observations of actual waves and were soon superseded by the works of Miles (see below).

The applicability of the early theories to actual wind-wave generation was critically reviewed by Ursell (1956), who was forced to conclude that all available theories were grossly inadequate to account for observations. Ursell's criticism stimulated renewed efforts leading to the independent and complementary works of Phillips (1957) and Miles (1957), the cornerstones on which now rests much of the edifice of our theoretical understanding of wind-wave generation.

Initial wave growth: Phillips' theory

Phillips' model of wind-wave generation rests on a simple but crucial extension of Eckart's ideas. Turbulent pressure fluctuations are advected over the sea surface at some velocity U related to the wind speed. This pressure field may be represented as a two-dimensional superposition of Fourier components, each component being of the form

$$p_a = \Pi \exp\left[i(k_1 x + k_2 y - \boldsymbol{k}\cdot \boldsymbol{U}t)\right]. \tag{51.27}$$

Note that the advection speed may vary with eddy size: $\boldsymbol{U} = \boldsymbol{U}(\boldsymbol{k})$ in general; indeed, $\boldsymbol{U}(\boldsymbol{k})$ should be expected to be a decreasing function of k since high wavenumber pressure variations will be generated by smaller eddies, closer to the sea surface, and [according to (51.8)], will travel more slowly. If, instead of assuming that the surface pressure p_a is constant, as we did in deriving the properties of free waves, we represent p_a by (51.27), we find, from Section 11, that in a homogeneous nonrotating fluid, the pressure must satisfy (in the linearized case)

$$\nabla^2 p = 0, \quad -H \leqslant z \leqslant 0; \tag{51.28}$$

$$p_z = 0, \quad z = -H; \tag{51.29}$$

$$p_{tt} + g p_z = (p_a)_t, \quad z = 0. \tag{51.30}$$

In deep water, (51.29) may be replaced by $p \to 0$ as $z \to -\infty$ and a solution of (51.28) is

$$p = B \exp\left[kz + i(k_1 x + k_2 y - \boldsymbol{k}\cdot \boldsymbol{U}t)\right], \tag{51.31}$$

where, from (51.30),

$$B = \Pi(\boldsymbol{k}\cdot \boldsymbol{U})^2/[(\boldsymbol{k}\cdot \boldsymbol{U})^2 - gk]. \tag{51.32}$$

Resonance is thus possible between the advected pressure field and those waves which travel at the right speed to keep in step with the forcing. Letting θ be the angle between $\boldsymbol{k}$ and $\boldsymbol{U}$, resonance is observed at phase speeds

$$c = \pm U(\boldsymbol{k}) \cos\theta. \tag{51.33}$$

This simple explanation is somewhat misleading, as it implies a growth in amplitude proportional to time (as in usual resonant responses), and of energy as t^2. In practice, the energy transfer is complicated by the fact that the atmospheric pressure pattern is turbulent in nature and gradually loses coherence with its former self as it is advected: resonant forcing does not take place continuously, but rather in a random series of linearly independent impulses. Furthermore, even if the atmospheric pressure pattern were rigidly advected by the wind, the waves generated would continuously be left behind since their group velocity is less than the phase velocity [a point discussed by Stewart and Manton (1971)]. These dephasing effects concur to limit the growth of the wave amplitude to a $\sqrt{t}$ dependence, as shown by a more careful analysis.

Instead of looking at the monochromatic response to a single air-pressure Fourier component (51.27), let us instead examine the growth of a spectrum. We now write

$$\Pi(\boldsymbol{k}, t) = \int_{-\infty}^{\infty} e^{i\boldsymbol{k}\cdot\boldsymbol{x}} p_a(\boldsymbol{x}, t)\, d\boldsymbol{x} \tag{51.34}$$

as the spatial Fourier transform of the air-pressure field. Similarly, we may define the transforms $B(\boldsymbol{k}, t)$ for the water-pressure field and $A(\boldsymbol{k}, t)$ for the sea-level displacement. Combining the vertical momentum equation (8.69) and the surface kinematic condition (9.18b), we find

$$\eta_{tt} = -\frac{1}{\rho_w} p_z(0). \tag{51.35}$$

Hence $$A_{tt} = -\frac{k}{\rho_w} B, \tag{51.36}$$

and the surface boundary condition (51.30) becomes

$$A_{tt} + gkA = -k\Pi/\rho_w. \tag{51.37}$$

The solution of (51.37) which satisfies the initial conditions $A(\boldsymbol{k}, 0) = 0 = A_t(\boldsymbol{k}, 0)$, is

$$A(\boldsymbol{k}, t) = -\frac{1}{\rho_w c} \int_0^t \Pi(\boldsymbol{k}, \tau) \sin\,[\omega(t-\tau)]\, d\tau, \tag{51.38}$$

where $\omega = \sqrt{gk}$, and $c = \sqrt{g/k}$ is the deep-water phase speed.

In order to relate the spectrum of atmospheric pressure fluctuations to that of surface elevation, let us define a time lag covariance function $\Omega(\boldsymbol{k}, \tau)$ for the pressure fluctuations by the equation

$$\langle \Pi(\boldsymbol{k}, t)\Pi^*(\boldsymbol{k}', t+\tau)\rangle = 4\pi^2 \Omega(\boldsymbol{k}, \tau)\delta(\boldsymbol{k}' - \boldsymbol{k}) \tag{51.39}$$

and a spectral energy density of wave amplitude $\Psi(\boldsymbol{k}, t)$ by the equation

$$\langle A(\boldsymbol{k}, t)A^*(\boldsymbol{k}', t)\rangle = 4\pi^2 \Psi(\boldsymbol{k}, t)\delta(\boldsymbol{k}' - \boldsymbol{k}). \tag{51.40}$$

Then, using (51.38) in (51.40) and the relation (51.39), we can relate Ψ and Ω through

$$\Psi(\boldsymbol{k}, t) = \frac{1}{2\rho_w^2 c^2} \int_0^t d\tau_1 \int_0^t d\tau_2 \Omega(\boldsymbol{k}, \tau_2 - \tau_1)\{\cos\,[\omega(\tau_2 - \tau_1)] - \cos\,[\omega(2t - \tau_1 - \tau_2)]\}. \tag{51.41}$$

Since $\Omega(\boldsymbol{k}, \tau_2 - \tau_1)$ depends only on the difference $\tau_2 - \tau_1$, it is advantageous to change the integration variables to $r = \tau_2 - \tau_1$, $s = (\tau_2 + \tau_1)/2$. Taking proper care of the limits of integration, we can rewrite (51.41) as

$$\Psi(\boldsymbol{k}, t) = \frac{1}{2\rho_w^2 c^2} \left\{ \int_0^t dr \int_{r/2}^{t-r/2} ds\, \Omega(\boldsymbol{k}, r)\{\cos \omega r - \cos\,[2\omega(s-t)]\} \right.$$

$$+\int_{-t}^{0} \mathrm{d}r \int_{-r/2}^{t+r/2} \mathrm{d}s\,\Omega(\boldsymbol{k},r)\{\cos\omega r - \cos[2\omega(s-t)]\}\Bigg\}, \tag{51.42}$$

which is readily reduced to

$$\Psi(\boldsymbol{k},t) = \frac{1}{2\rho_w^2c^2}\int_0^t (t-r)\cos\omega r\,[\Omega(\boldsymbol{k},r)+\Omega(\boldsymbol{k},-r)]\,\mathrm{d}r$$

$$+\frac{1}{4\rho_w^2c^2\omega}\int_0^t [\Omega(\boldsymbol{k},r)+\Omega(\boldsymbol{k},-r)][\sin\omega r - \sin\{\omega(2t-r)\}]\,\mathrm{d}r. \tag{51.43}$$

For $\omega t \gg 1$, the presence of the linear factor $(t-r)$ suffices to establish the dominance of the first integral in (51.43); thus, at times well in excess of one wave period,

$$\Psi(\boldsymbol{k},t) \simeq \frac{1}{2\rho_w^2c^2}\int_0^t (t-r)\cos\omega r\,[\Omega(\boldsymbol{k},r)+\Omega(\boldsymbol{k},-r)]\,\mathrm{d}r. \tag{51.44}$$

In the idealized case in which atmospheric pressure variations are merely advected by the wind without losing any of their coherence, we can write

$$\Omega(\boldsymbol{k},r) = \Omega(\boldsymbol{k})\exp(i\boldsymbol{k}\cdot\boldsymbol{U}r). \tag{51.45}$$

Upon substitution into (51.44), this "frozen-field" covariance function yields, at the resonant frequencies $\omega = \pm\boldsymbol{k}\cdot\boldsymbol{U}$, $\Psi = 0(t^2)$ as $t\to\infty$, as expected for a resonant process (see Exercise 51.5). The "frozen-field" hypothesis is, however, unrealistic at $\omega t \gg 1$, since turbulent pressure fluctuations gradually lose coherence. To characterize the loss of coherence, Phillips (1957) defined the integral time scale $\hat{T}(\boldsymbol{k})$ as

$$\hat{T}(\boldsymbol{k}) = \int_{-\infty}^{\infty} \exp[i\boldsymbol{k}\cdot\boldsymbol{U}\tau]\,\frac{\Omega(\boldsymbol{k},\tau)}{\Omega(\boldsymbol{k},0)}\,\mathrm{d}\tau. \tag{51.46}$$

Introducing the Fourier transform of $\Omega(\boldsymbol{k},\tau)$ (which is also the spectral density of the pressure fluctuation) as $F(\boldsymbol{k},\omega)$, viz.,

$$F(\boldsymbol{k},\omega) = \frac{1}{2\pi}\int_{-\infty}^{\infty}\exp(i\omega\tau)\,\Omega(\boldsymbol{k},\tau)\,\mathrm{d}\tau, \tag{51.47}$$

we find that

$$\hat{T}(\boldsymbol{k}) = 2\pi F(\boldsymbol{k},\boldsymbol{k}\cdot\boldsymbol{U})/\Omega(\boldsymbol{k},0). \tag{51.48}$$

The determination of $\hat{T}(\boldsymbol{k})$ is one of the major difficulties of Phillips' theory. We note that if the decay rate $\hat{T}(\boldsymbol{k})$ is slow enough, i.e., if the pressure fluctuations remain coherent over many cycles, as established by the measurements of Willmarth and Wooldridge (1962), then $\boldsymbol{k}\cdot\boldsymbol{U}\hat{T}(\boldsymbol{k}) \gg 1$, and $\Omega(\boldsymbol{k},r)$ may be written in the form (51.45) with the addition of a decay factor $\exp\{-[r/\hat{T}(\boldsymbol{k})]\}$. Thus the covariance function $\Omega(\boldsymbol{k},\tau)$ has amplitude $\Omega(\boldsymbol{k},0)$, oscillates with frequency $\boldsymbol{k}\cdot\boldsymbol{U}$ and decays with a time scale $\hat{T}(\boldsymbol{k})$. From the basic considerations of Section 32, the transform of $\Omega(\boldsymbol{k},\tau)$, namely $F(\boldsymbol{k},\omega)$, must be peaked at $\omega = \boldsymbol{k}\cdot\boldsymbol{U}$, with amplitude of the order of $\Omega(\boldsymbol{k},0)\hat{T}(\boldsymbol{k})/2\pi$ and bandwidth $\hat{T}(\boldsymbol{k})$. Since the decay is slow, $F(\boldsymbol{k}, \omega)$ must be sharply peaked and may be approximated by (Stewart and Manton, 1971)

$$F(\boldsymbol{k}, \omega) = \Omega(\boldsymbol{k}, 0)\delta_s(\omega - \boldsymbol{k} \cdot \boldsymbol{U}), \tag{51.49}$$

$$\text{where} \quad \delta_s(\omega) = (\Delta\omega\sqrt{\pi})^{-1} \exp\left[-(\omega/\Delta\omega)^2\right] \tag{51.50}$$

is a "smudged" delta function with bandwidth $\Delta\omega = 2\pi/\hat{T}(\boldsymbol{k})$. Substituting (51.47) into (51.44), we can express the spectrum of sea surface displacement in terms of that of the air pressure fluctuations:

$$\Psi(\boldsymbol{k}, t) \simeq \frac{\pi t}{2\rho_w^2 c^2}\left[F(\boldsymbol{k}, \omega) + F(\boldsymbol{k}, -\omega)\right]. \tag{51.51}$$

Using (51.49),

$$\Psi(\boldsymbol{k}, t) \simeq \frac{\Omega(\boldsymbol{k}, 0)\pi t}{2\rho_w^2 c^2}\left[\delta_s(\omega - \boldsymbol{k} \cdot \boldsymbol{U}) + \delta_s(\omega + \boldsymbol{k} \cdot \boldsymbol{U})\right]. \tag{51.52}$$

For a delta function, $\delta[f(k)] = \delta(k - k_0)/|f'(k_0)|$, where k_0 is a simple zero of $f(k)$ (Duff and Naylor, 1966). Since δ_s is nearly a delta function, we write

$$\delta_s(\omega - kU\cos\theta) \simeq \delta_s(k - k_+) \Big/ \left|\left(U\cos\theta - \frac{\partial\omega}{\partial k}\right)\right|_{k=k_+}, \tag{51.53}$$

where θ is the angle between the wind and the waves as given by (51.33), and k_+ satisfies the resonance condition (51.33) with the plus sign. At resonance, $U\cos\theta = c(k_+)$; thus

$$\delta_s(\omega - kU\cos\theta) \simeq \delta_s(k - k_+)/|c - c_g|_{k=k_+}. \tag{51.54}$$

Substituting (51.54) into (51.52), we obtain for the spectral energy density per unit area of the sea surface,

$$\Psi(\boldsymbol{k}, t) \simeq \frac{\Omega(\boldsymbol{k}, 0)\pi t\delta_s(k - k_+)}{2\rho_w^2 c^2 |c - c_g|_{k=k_+}}. \tag{51.55}$$

The energy is found at resonance with the wind, as expected from (51.33), but because of the dispersive nature of surface gravity waves, it increases only as the first power of the time. Phillips' mechanism thus contributes a term of the form Σ_1 to the energy source in (51.22). The total energy density per unit area in a given direction θ is given by

$$\langle E(\theta, t)\rangle = \rho_w g \int_0^\infty k\Psi(\boldsymbol{k}, t)\,\mathrm{d}k.$$

Treating the function δ_s as an exact δ-function, we find, after substitution of $\Psi(\boldsymbol{k}, t)$ from (51.55),

$$\langle E(\theta, t)\rangle = \frac{k_+^2 \pi t\Omega[\boldsymbol{k}(k_+, \theta), 0]}{2\rho_w |c - c_g|_{k=k_+}}, \tag{51.56}$$

which is independent of the turbulent time scale $\hat{T}(\boldsymbol{k})$. The relevant time scale is now associated with the dephasing effect of the difference between c and c_g. According to the arguments of Stewart and Manton (1971), continuous forcing near resonance does tend to make the wave amplitude go up linearly with time; the bandwidth over which the waves appear, however, decreases with time as the energy is focused ever closer to the resonant frequency. The net energy density (proportional to the amplitude squared times the bandwidth) thus increases only linearly with time.

The resonance mechanism introduced by Phillips and briefly presented above accounts for the excitation and initial growth of waves on an undisturbed water surface; it predicts a bimodal angular spectral distribution of energy in the resonant directions given by (51.33). This angular spread has been observed by Gilchrist (1966) and merges with (51.21) at longer fetches. The resonance mechanism is, however, too weak to account for the continued rate of growth of wind waves, and must be supplemented by a momentum transfer mechanism which takes over once the waves have become large enough to modify the air flow over them.

Continued wave growth: Miles' theory

Once waves have appeared on the sea surface, their presence modifies the air flow, as shown in Fig. 51.5. If the wave-induced fluctuations in the air flow are treated as small perturbations of a mean shear flow, their effect may be analyzed through linear stability theory (Drazin and Howard, 1966). In its simplest inviscid form (Miles, 1957) the coupled air–water motion may be investigated by solving the unstratified version of the Taylor-Goldstein equation (41.8) in the air and Laplace's equation in the water, with suitable matching and boundary conditions.

Following Miles' (1957) original approach, let us consider a monochromatic deep water wave with surface displacement given by

$$\eta(x,t) = a \exp\left[ik(x-ct)\right], \tag{51.57}$$

Then the solution of Laplace's equation for a velocity potential ϕ (see 11.12) which vanishes at depth is

$$\phi = -ica \exp\left[kz + ik(x-ct)\right], \quad -\infty < z \leqslant 0. \tag{51.58}$$

The surface displacement and the potential are related through the surface kinematic condition (11.39a). The pressure associated with the waves is found from Bernoulli's equation (11.13):

$$p = -\rho_w \phi_t = ak\rho_w c^2 \exp\left[kz + ik(x-ct)\right], \quad -\infty < z \leqslant 0. \tag{51.59}$$

In the air, the flow field is composed of a mean wind $U(z)$ and a two-dimensional perturbation related to the surface wave. Thus,

$$u(x,z,t) = U(z) + \psi_z, \quad 0 \leqslant z < \infty, \tag{51.60a}$$

$$w(x,z,t) = -\psi_x, \quad 0 \leqslant z < \infty, \tag{51.60b}$$

where $\psi(x,z,t)$ is a stream function which, to first order in the wave slope, ka, has the form

$$\psi = kaF(z) \exp\left[ik(x-ct)\right], \quad 0 \leqslant z < \infty. \tag{51.61}$$

The air pressure perturbation will be denoted by $p_a(x,z,t)$, where

$$p_a(x,z,t) = \rho_a ka\Pi(z) \exp[ik(x-ct)]. \tag{51.62}$$

The functions $F(z)$ and $\Pi(z)$ are related through the vertical momentum equation in the air domain. The vertical dependence $F(z)$ must satisfy the unstratified form of (41.8):

$$(U-c)(F''-k^2F)-U''F = 0, \quad 0 \leqslant z < \infty, \tag{51.63}$$

$$F \to 0 \quad \text{as} \quad z \to \infty. \tag{51.64}$$

The second boundary condition follows from the identity of interfacial displacements as seen from either the air or the water side of the boundary (Exercise 51.10). From (42.9), we already know that $w/(U-c)$ must be continuous in this type of shear flow. Applying this condition, using (51.58) on the wet side of the interface and (51.60b) on the dry side, we obtain

$$F(0) = [c-U(0)]/k. \tag{51.65}$$

Further progress necessitates solution of the boundary-value problem (51.63)–(51.65). One solution has been obtained by Conte and Miles (1959) for a logarithmic wind profile. The salient features of the flow may, however, be examined without reference to the exact solution.

We note that (51.63) has a regular singular point at the critical level $z=z_c$, where $U(z_c)=c$. In the inviscid case, the horizontal component of velocity has a singularity at that level; in a real fluid, the critical level will be a zone of strong shears. The behaviour of the wavy secondary flow (51.60) at $z=z_c$ is analogous to that of internal waves at a critical level, discussed in detail in Section 41. In particular, as we found in that section [equations (49.19) and (49.20)], the shear stress $\tau_{13}=-\rho_a\langle u'w'\rangle$ has a discontinuity at the critical level, passing from a very small value (essentially zero) to a constant value $\tau(0)$ across $z=z_c$. For the problem considered here, the stress is equal to $\tau(0)$ below the critical level. The stress $\tau(0)$ depends on the form of the wind profile at z_c; it may also be related to the pressure field at the free surface, part of which does work on the waves. The effectiveness of this wave generation mechanism will then depend (rather critically) on the velocity profile and streamline configuration around the critical level.

At the free surface, continuity of normal stress dictates that

$$p(\eta) = p_a(\eta),$$

$$\text{i.e.,} \quad p(0)-\rho_w g\eta = p_a(0)-\rho_a g\eta. \tag{51.66}$$

Using (51.57), (51.59) and (51.62), this relation becomes

$$\rho_w(kc^2-g) = \rho_a(k\Pi(0)-g),$$

which we solve for c:

$$c = \left[\frac{g}{k}\left(1-\frac{\rho_a}{\rho_w}\right)+\frac{\rho_a}{\rho_w}\,\Pi(0)\right]^{1/2}. \tag{51.67}$$

In general c is complex, because the air pressure at $z=0$ need not be in phase with η, and

$\Pi(0)$ may be complex. Since $\rho_a/\rho_w \ll 1$, and provided $\Pi(0) = 0(g/k)$, Re (c) departs very little from the free wave speed $c_0 = \sqrt{g/k}$. The imaginary part of c is approximated by

$$\frac{\text{Im}\,(c)}{c_0} = \frac{\rho_a}{2\rho_w}\frac{\text{Im}\,\{\Pi(0)\}}{c_0^2} + 0\left[\left(\frac{\rho_a}{\rho_w}\right)^2\right]. \tag{51.68}$$

With $\langle E\rangle = \rho_w g\langle\eta\eta^*\rangle/2$ for the wave energy density, and η as given by (51.57), the rate of energy input from the air to the sea is equal to

$$\frac{\partial\langle E\rangle}{\partial t} = 2k\,\text{Im}\,(c)\langle E\rangle = \frac{k\langle E\rangle}{\rho_w c_0^2}\frac{\rho_a}{\rho_w}\,\text{Im}\,\{\Pi(0)\}. \tag{51.69}$$

The rate of energy input per unit area into the waves is proportional to that part of the air pressure which is out of phase with the surface displacement (see also Exercise 51.6). The forcing term provided by Miles' theory is thus of the same form (Σ_2) as that found by Jeffreys using the sheltering hypothesis.

The form of the air flow at the critical level is related to the wind stress exerted below that level, and hence onto the sea. The stress $\tau(0)$ at the sea surface may be related (see Exercise 51.7), through the definitions of u and w in terms of the stream function, to the rate of increase of wave energy:

$$\frac{\partial\langle E\rangle}{\partial t} = c\tau(0). \tag{51.70}$$

The shear stress below the critical level all the way down to the wave surface is determined by the wind profile at $z = z_c$ (Miles, 1957):

$$\tau(z) = \tfrac{1}{2}\pi\rho_a k^3|\eta|^2|F^2(z_c)|[-U''(z_c)/U'(z_c)], \quad 0 < z < z_c. \tag{51.71}$$

Thus, in this rather idealized model, in which the turbulence does not participate explicitly in the momentum transfer, wind conditions at the critical level $z = z_c$ entirely determine the energy input into the waves. Since the model is inviscid, the momentum transfer to the sea surface is entirely through pressure forces. The origin of the net pressure force over the wave profile is to be attributed to the position of the closed circulation pattern (in a coordinate system moving with the wave profile) around the critical level above the forward face of the waves. Lighthill (1962b) advanced a physical interpretation of the wave generation process in terms of the vortex force due to the modified pressure field in the "cat's eye" circulation pattern (see Exercise 51.8). The exact position of this eddy pattern as a function of wavelength and amplitude remained rather uncertain until the detailed numerical calculations of Gent and Taylor (1976, 1977) became available.

After Miles (1960) combined Phillips' resonance mechanism with the above shear flow perturbation mechanism, it was felt that, at last, a suitable theory of wave generation was at hand. It did not take long before measurements (Snyder and Cox, 1966; Barnett and Wilkerson, 1967), showed that although Phillips' theory accounted reasonably well for the initial wavegrowth, the major portion of the spectral development appeared to occur at a rate exceeding that predicted by Miles by an order of magnitude! Furthermore, even after a number of refinements (Miles, 1959a, 1959b, 1962), a number of uncertainties concerning the details of the pressure force acting on the waves in Miles' shear flow mechanism remained unresolved.

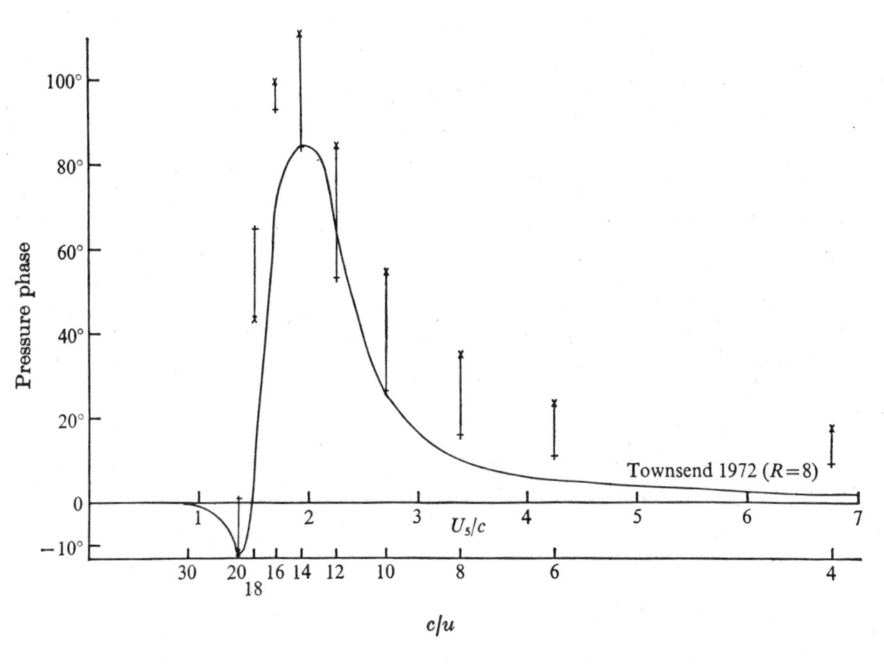

Fig. 51.13. Phase difference of the surface pressure maximum measured from the wave trough as a function of U_5/c and c/u_*. Results for z_0 = constant, for wave slopes between $ak = 0.01$ (+) and $ak = 0.157$ (X). The continuous curve is that obtained by Townsend (1972) for $R = -\ln(kz_0) = 8$. (From Gent and Taylor, 1976.)

Refinements of the wave generation theories

Conditions at the critical level play a crucial role in the growth rate predicted by Miles' shear flow stability theory. A glance at Fig. 51.5 shows that the height of the critical level varies considerably over the wave profile. In order to de-emphasize the importance of the proper specification of the mean flow at the critical level, Townsend (1972) re-evaluated the rate of momentum transfer to the sea surface in a fully turbulent boundary layer over a moving profile. In such a formulation, the critical level becomes an unimportant part of an equilibrium layer (in which turbulent energy production and dissipation balance, on the average). The pressure forces calculated by Townsend turned out to be of the same magnitude as those given by Miles (1959a), and the calculated rate of momentum transfer could not be reconciled in this way with the observed rates of wave growth.

The recent numerical calculations of Gent and Taylor (1976, 1977) have provided a more detailed picture of the air flow over water waves and have shown in particular how the air flow pattern is affected by the wave steepness and by the ratio of wind speed to wave phase velocity. The distributions of surface pressure and surface shear stress calculated by Gent and Taylor (1977) have already been illustrated in Fig. 51.6. The stresses acting on the water surface do not have the same sinusoidal symmetry as the wave profile itself. The shear stress shows a strong peak near the wave crest, a fact to which we shall return below in connection with the generation of short waves by such a stress, and which is relevant to the discussion of the ideas advanced by Garrett and Smith (1976) on the

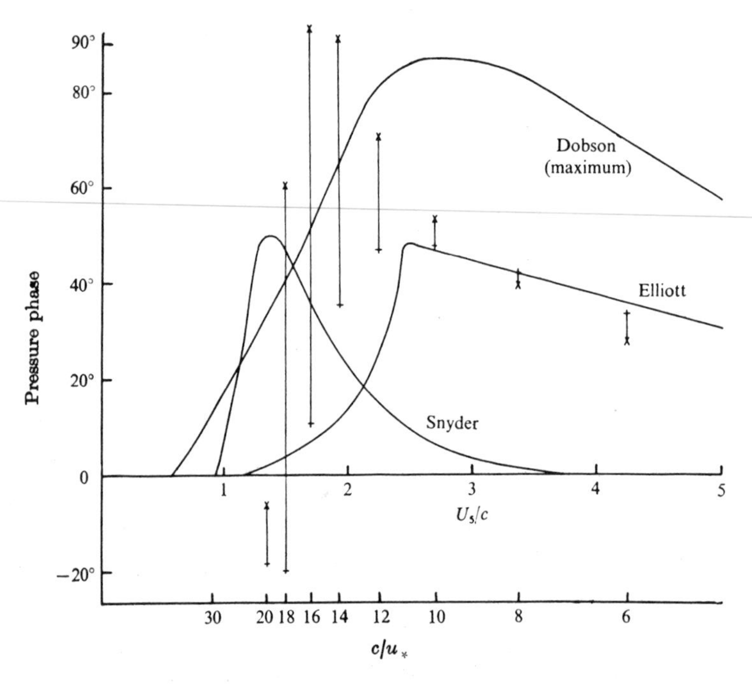

Fig. 51.14. A comparison of the phase difference of the pressure maximum as calculated by Gent and Taylor (1976) with values measured by Dobson (1971), Elliott (1972) and Snyder (1974). The calculated results are for a variable roughness length of the form $z_0 = \langle z_0 \rangle [1 - \gamma \cos (k\xi - \delta)]$ (where ξ is the distance along the wave profile). Parameter values: $\delta = \pi/4$; (+), $ak = 0.01$ and $\gamma = 0.5$; (×), $ak = 0.157$ and $\gamma = 0.75$. (From Gent and Taylor, 1976.)

interaction between short and long waves (Section 36). The pressure distribution at the water surface shows a minimum just forward of the wave crest for all values of wave slope. The pressure maximum, on the other hand, starts at a position which is almost exactly 180° out of phase with the minimum when the wave slope (ka) is small, and migrates forward from the wave trough as the wave slope and the wind speed increase. The angle by which the maximum of the pressure distribution leads the wave trough is shown in Fig. 51.13. Averaged over a wave length, a net rate of momentum transfer to wave motion by the pressure force begins when the phase angle of the pressure maximum becomes positive, i.e., from Fig. 51.13, at $U_5/c \simeq 1.6$, where U_5 is the mean wind speed at $z = 5$ m. A comparison of the position of the pressure maximum at the surface, as calculated by Gent and Taylor (1976), with observed pressure distributions over actual wind waves is shown in Fig. 51.14. The calculated values for small wave slopes ($ak = 0.157$, labelled +) agree rather well with Elliott's (1972) measurements. At large wave slopes ($ak = 0.314$, labelled ×), pressure phase angles as high as those measured by Dobson (1971) are calculated for a range of c/u_*.

Although the surface pressure exceeds the surface shear stress by about an order of magnitude (Fig. 51.6), it is the average value over a wavelength which contributes to the net wave generation. From the upper panel of Fig. 51.6, we note that the surface shear stress averages to about $0.8\rho_a u_*^2$ for small wave slopes ($ak = 0.157$) and to about $0.4\rho_a u_*^2$ for steeper waves ($ak = 0.314$). From (51.9), the momentum transfer to wave motion by

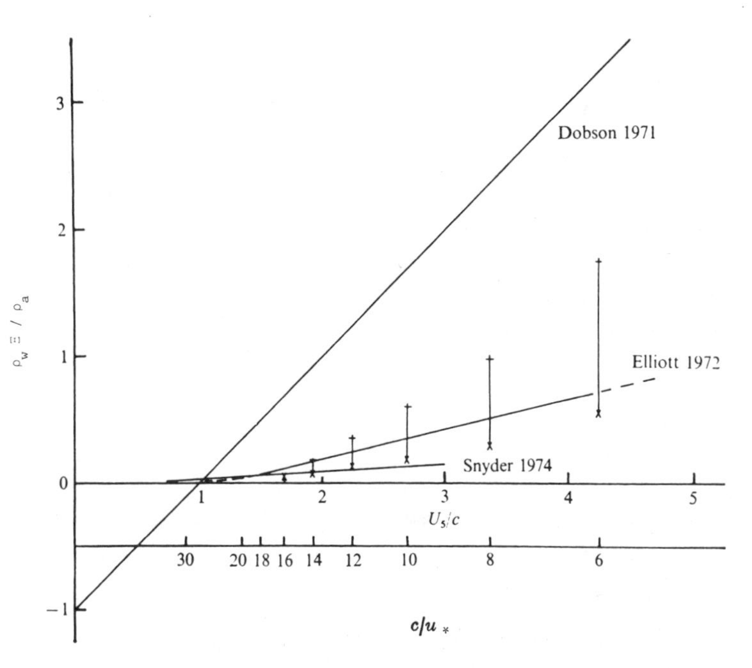

Fig. 51.15. The fractional rate of energy gained by the waves per radian advance in phase, Ξ, against U_5/c. Same notation as in Fig. 51.14. (From Gent and Taylor, 1976.)

the average pressure then lies between 0.2 and $0.6\rho_a u_*^2$. The fractional rate of energy input into the waves per unit radian (cf. 51.17), as calculated by Gent and Taylor (1976), is plotted in Fig. 51.15 and compared to that obtained from field pressure measurements by Dobson (1971), Elliott (1972) and Snyder (1974). Of the observed rates of energy input shown in Fig. 51.15, only that found by Dobson (1971) is large enough to match the observed rates of wave growth measured by Snyder and Cox (1966) and Barnett and Wilkerson (1967). Subsequent observations of the rate of work by the pressure forces, as well as the computed rates of Gent and Taylor, all give lower values. The rate of work measured by Snyder (1974) is of the same order as that predicted by the linear theories of Miles and Townsend; that found by Elliott (1972) and the calculated values of Gent and Taylor (1976) are about twice that given by linear theories, but still smaller than the rate required to account for the wave growth observed by Snyder and Cox (1966) and Barnett and Wilkerson (1967).

If the rate at which work is done by the wind, through the action of pressure forces, on the sea surface is too small, at a given surface wavenumber, to explain the rate of growth of such a wave, where does the additional energy flux come from? Except for very short gravity–capillary waves, the contribution of the surface shear stresses to the rate of energy transfer is too small to account for the difference. As the main difficulty lies in explaining the large rate of growth of the spectral peak [at $\omega = \omega_m$ in (51.10)], a number of authors have attempted to account for part of its growth through nonlinear wave–wave interaction mechanisms. The "maser" mechanism proposed by Longuet-Higgins (1969a), whereby short waves, amplified near long wave crests by radiation stress interactions, would lose their momentum to the long waves and hence contribute to their growth has already been discussed in Section 36. More recent work by Garrett and Smith (1976) has shown that such an interaction process may be effective in transferring energy from short to long waves provided short wave generation occurs most actively on the crests of the long waves. Judging from the strong peak in surface shear stress over the wave crest evident in Fig. 51.6, this condition is likely to be satisfied. The wave–wave interaction terms calculated for the JONSWAP spectra by Hasselmann et al. (1973) strongly indicate that nonlinear interactions within the wave spectrum are responsible for the presence of a sharp peak in the wind-wave spectrum, and for the rapid rate of growth of this peak with fetch (cf. Section 39). Theoretical calculations by Longuet-Higgins (1976) and Fox (1976) also lead to this conclusion.

Finally, as we have noted earlier in this section when discussing Jeffreys' sheltering mechanism, flow separation over short breaking waves (Banner and Melville, 1976) is associated with much larger momentum and energy transfer rates than over unbroken waves, a fact which may account in part for the high growth rates observed in a fully developed sea.

One should also note that all the above theories of wind-wave generation assume the air flow over the sea to be two-dimensional. As all sailors well know, this is a gross idealization of the actual wind field. The need for considering the role of the three-dimensional structure of the wind field in wave generation has been emphasized by Dorman and Mollo-Christensen (1971). Stewart (1974) has presented some preliminary arguments which show that the presence of downward motion over part of the sea surface could lead to a significant contribution to the wind generating force.

The generation of capillary and short gravity waves

As noted above, wave generation results mainly from pressure forces induced by the distortion of the air flow over a wavy profile. For very short waves of moderate slopes, however, the displacement of the sea surface is so small that the wave profile is found mainly within the laminar sublayer. For such waves one would expect the mean flow distortion to be minimal and the shear stresses to dominate over the pressure forces. Laminar shear flow instability of the viscous sublayer was first suggested by Miles (1962) as the principal mechanism responsible for the generation of very short gravity waves and of capillary waves. Valenzuela (1976) has extended the analysis of Miles to take into account the presence of a drift current at the water surface. The rate of growth of short waves predicted by these theories is exponential (a term of the form Σ_2) in (51.22) and in very good agreement up to wavenumbers $k = 8.67\,\mathrm{cm}^{-1}$ with the measurements of Larson and Wright (1975).

The Phillips' resonance mechanism discussed above for the generation of gravity waves also operates to produce capillary waves. Phillips (1957) has shown that the strongest response occurs for waves of minimum phase velocity, as given by (11.19). The shear-flow instability mechanism is usually dominant and obscures the resonant response to air pressure fluctuations.

The wavenumber range of capillary waves is strongly limited at the short wavelength end by viscous dissipation, a point which will be taken up again below. For longer waves, there is a range of wavenumbers for which dissipation is not crucial and over which the characteristics of the capillary wave spectrum are influenced only by the restoring force of surface tension (σ), the density of the fluid (ρ_*) and the wave properties (ω and k). Over this equilibrium range, the frequency spectrum $S(\omega)$ may be related to σ, ρ_*, ω and k by dimensional analysis, following arguments similar to those used to obtain the gravity wave equilibrium spectrum (32.5) (cf. Exercise 51.10). Thus,

$$S(\omega) = A\left(\frac{\sigma}{\rho_*}\right)^{2/3} \omega^{-7/3}, \tag{51.72}$$

where A is a constant. This spectral form has been discussed by Phillips (1958, 1966). Observations of capillary waves have focused on optical measurements of the frequency spectrum of the surface slope, $R(\omega)$, defined as

$$\langle \nabla_h \eta \cdot \nabla_h \eta \rangle = \int_0^\infty R(\omega)\, d\omega. \tag{51.73}$$

Again from dimensional arguments, in the equilibrium range,

$$R(\omega) = C\omega^{-1}, \tag{51.74}$$

where C is a constant. Recent measurements by Long and Huang (1976a, b) have confirmed the existence of a range of frequencies over which (51.74) is satisfied.

In his search for complementary momentum transfer mechanisms which might reconcile the Phillips-Miles theory with observed rates of wave growth, Longuet-Higgins (1969b) showed how tangential stresses can contribute to the growth of already existing waves. The shear stress at the sea surface is strongly correlated with the amplitude of the wave profile (Fig. 51.6). The surface wind drift driven in the water by the shear stress must then also vary in speed and thickness along the wave profile. The mass flux associated with a wind drift current u_d along the sea surface is

$$M_d = \int \rho_w u_d\, dz. \tag{51.75}$$

Ignoring the tangential stress underneath the drift layer, conservation of momentum parallel to the sea surface gives

$$\partial M_d/\partial t = \tau, \tag{51.76}$$

where τ is the surface shear stress. Letting Δ be the thickness of the drift layer,

$$\frac{\partial \Delta}{\partial t} = w_d(\eta) - w_d(\eta - \Delta) = \int \frac{\partial w_d}{\partial z}\, dz.$$

which, from the continuity equation, gives

$$\frac{\partial \Delta}{\partial t} = -\int \frac{\partial u_d}{\partial x}\, dz. \tag{51.77}$$

In a coordinate system moving with the wave profile, $\partial/\partial x = -(1/c)\partial/\partial t$; substituting into (51.77) and using (51.75) and (51.76), we find that

$$\partial \Delta/\partial t = \tau/\rho_w c. \tag{51.78}$$

The thickness of the wind drift layer, and hence also of the hydrostatic pressure field associated with it, is out of phase with the surface shear stress and with the wave profile. This pressure field can then contribute a nonzero correlation $\langle p\eta_t \rangle$ and add to wave generation at a rate proportional to the wave amplitude, adding another term of the form Σ_2 to (51.22). This additional contribution is, however, insufficient to reconcile the theory with the observed rates of wave growth.

The dissipation of surface waves

Although surface waves can prove astonishingly long-lived, as shown by the observations of Snodgrass et al. (1966) of waves having propagated across the whole Pacific Ocean (cf. Section 11), they eventually succumb to dissipative processes.

Let us first examine the influence of molecular viscosity on short gravity–capillary waves. From (48.5), the rate of energy dissipation per unit volume by viscous stresses is $\tau_{ij}\partial u_i/\partial x_j$. With τ_{ij} given by (48.1b), and in view of the symmetry of the stress tensor, we may rewrite the dissipation rate as

$$\tau_{ij}\frac{\partial u_i}{\partial x_j} = \frac{\mu}{2}\left(\frac{\partial u_i}{\partial x_j} + \frac{\partial u_j}{\partial x_i}\right)^2. \tag{51.79}$$

Integrating over the depth of the fluid, we find for the dissipation rate per unit area of the sea surface

$$\frac{\partial \langle E \rangle}{\partial t} = -\frac{\mu}{2}\int_{-H}^{\eta}\left(\frac{\partial u_i}{\partial x_j} + \frac{\partial u_j}{\partial x_i}\right)^2 \mathrm{d}z. \tag{51.80}$$

The presence of viscosity leads to the formation of boundary layers near the top and bottom boundaries. Assuming a simple vorticity balance of the form

$$\zeta_t = \nu\zeta_{zz} \tag{51.81}$$

near the solid boundaries (with $\nu = \mu/\rho_w$), boundary layers of thickness of the order $\delta \sim (\nu/\omega)^{1/2}$ form near these boundaries. With $\nu = 10^{-2}\,\mathrm{cm^2\,s^{-1}}$ and a lower limit for the frequency ω of $10^{-1}\,\mathrm{rad\,s^{-1}}$ (60 second waves!), $\delta \sim 0.3$ cm. Thus $\delta \ll H$ in usual circumstances and we may assume that the flow in the bulk of the fluid remains irrotational.

Let us first compute the dissipation in the bulk of the fluid, using the deep-water ($kH \gg 1$) approximation to the gravity–capillary wave results of Section 11. The velocity components u and w are given by (11.28) and (11.30); letting $H \to \infty$ ($kH \gg 1$) and substituting for u and w into (51.80), we find that on the average,

$$\frac{\partial \langle E \rangle}{\partial t} = -2\mu\omega^2 a^2 k(1+B), \tag{51.82}$$

where B is the Bond number (cf. Section 11). This result is valid for small-amplitude deep-water waves damped by molecular viscosity only. From (11.16), we find $\omega^2 = gk(1+B)$, for $kH \gg 1$; the mean energy density, as given by (13.24), then becomes

$$\langle E \rangle = \rho_w\omega^2 a^2/2k. \tag{51.83}$$

Combining (51.80) and (51.81),

$$\frac{1}{\langle E\rangle}\frac{\partial\langle E\rangle}{\partial t} = -4\nu k^2(1+B). \tag{51.84}$$

If we assume that $\eta = a \exp[i(kx-\omega t)]$, where ω is complex ($\omega = \omega_r + i\omega_i$), then, since

$$\langle E\rangle \propto \langle\eta\eta^*\rangle,$$

(51.84) gives

$$\omega_i = -2\nu k^2(1+B). \tag{51.85}$$

Taking for the inviscid frequency the real part of ω, we find that the Q of the waves $[Q = \mathrm{Re}(\omega)/2\,\mathrm{Im}(\omega)]$ is given by

$$Q = [gk(1+B)]^{1/2}/4\nu k^2(1+B). \tag{51.86}$$

For normal values of g and ν, Q is small only for very large values of k, i.e., for very short capillary waves (see Exercise 51.11). Dissipation by molecular friction in the bulk of the fluid is thus of negligible consequence to all but extremely short gravity–capillary waves.

Contributions to energy dissipation by relatively large gradients in the boundary layers are left to the reader to explore (Exercise 51.12). The presence of a very thin layer of contaminants (a slick) at the sea surface makes the upper surface behave very much like a rigid wall as far as small-scale motions are concerned. To a first approximation (a more precise analysis may be found in Phillips, 1966), the result of Exercise 51.12 may be used to find the attenuation rate in the surface boundary layer in the presence of a slick:

$$\frac{\partial\langle E\rangle}{\partial t} = -\mu\frac{\langle u^2(0)\rangle}{\delta}, \tag{51.87}$$

where $u(0)$ is the velocity at the surface (from 11.28) and $\delta = (\nu/\omega)^{1/2}$ is the boundary layer thickness. In deep water ($kH \gg 1$), (51.87) reduces to

$$\frac{1}{\langle E\rangle}\frac{\partial\langle E\rangle}{\partial t} \simeq -\frac{\nu k}{\delta}, \tag{51.88}$$

which is also small, unless k becomes quite large. Surface films are thus particularly effective in damping out very short capillary waves (Exercise 51.13). Another phenomenon which is of even greater importance in the damping of short surface waves by a surface film is the alternate compression and extension to which the film is subjected by the wave motion. In a series of experiments with various types of surface contaminants Davies and Vose (1965) have shown that the compressional modulus of the contaminant is of prime importance in damping capillary waves, and that quite low values of this modulus can cause wave damping which considerably exceeds that at a rigid surface.

In addition to the small attenuation rates discussed above, the presence of viscosity induces secondary streaming effects, tending to increase the mean Stokes flow seen in Section 13. These motions have been analyzed by Longuet-Higgins (1953, 1960) and Ünlüata and Mei (1970), among others.

On the whole, then, before breaking occurs, viscous effects are of little importance in

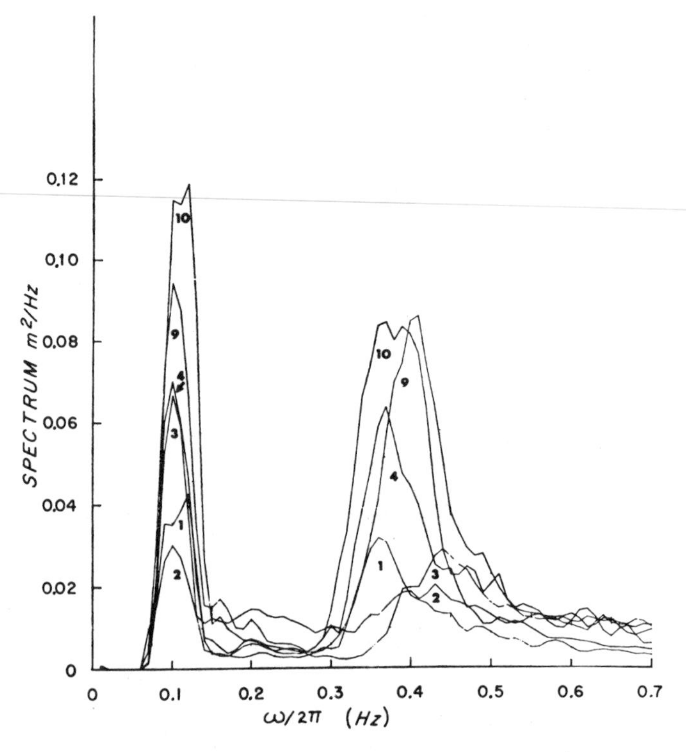

Fig. 51.16. Surface wave spectra $S(\omega/2\pi)$ from the JONSWAP experiment, showing the attenuation of swell (low frequency peaks) entering shallow water. Numbers refer to different points of observation, the highest numbers referring to deeper water locations. The spectral peak of the swell decreases as it advances into shallow water, except between locations *2* and *1*, where refraction effects mask the attenuation. (From Hasselmann et al., 1973.)

the dissipation of wind waves, a fact well borne out by the observations of Snodgrass et al. (1966) who observed only moderate attentuation of Pacific swell over distances of thousands of kilometers! In shallow water, however, stronger, turbulent boundary layers may develop before the waves actually break. The quadratic friction law mentioned in (48.2), of the form

$$\tau_b = c_b u|u| \tag{51.89}$$

has been used (Hasselmann et al., 1973) to represent the turbulent bottom tangential stress on swell propagating in shallow water. The JONSWAP data indicate a value of the drag coefficient of $c_b = 0.015$, which is consistent with other similar estimates. The influence of bottom friction is visible in the decrease in magnitude of the frequency spectra shown in Fig. 51.16 for swell approaching the shore.

Other processes involving backscattering through interactions with bottom irregularities (of the form studied in Section 38) may account for part of the attenuation seen in records such as that of Fig. 51.16 and have been proposed by Long (1973) and Bowen (1976).

One of the principal methods of energy extraction from wind waves is through wave breaking. Breaking occurs at sea, in the generation area, where it serves to limit the wave amplitude, and in the near-shore area in shallow water. We have already reproduced in Section 32 the dimensional argument of Phillips (1958) concerning the shape of the spectrum in a sea dominated by wave breaking. The expected ω^{-5} dependence is found in measured spectra (cf. 51.10).

Although whitecapping at sea and breaking on the shore are without doubt one of the most fascinating features of surface gravity waves, they occur with such violence as to make their study and analysis extremely difficult. That wave breaking is not amenable to the weak interaction approach followed so far goes without saying! Although some attempts at analytical description of breaking waves and of surf are available (Biesel, 1951; Ho et al., 1963; Longuet-Higgins and Turner, 1974), the most successful description of breaking waves is based on the "Particle-in-Cell" numerical method developed by Harlow and his collaborators (Harlow et al., 1965; Harlow and Shannon, 1967). The probability distribution of breaking wave heights has been examined by Nath and Ramsey (1976), on the basis of the JONSWAP spectrum. Observations of whitecap coverage at sea have been reported by Monahan (1971), and a striking periodicity in whitecap distribution has been noted by Donelan et al. (1972). For a recent report on laboratory investigations of wave breaking, the reader may consult Van Dorn and Pazan (1975).

Exercises Section 51

1. Show how the turbulent scale velocity u_*, and hence the Reynolds stress $\rho_a u_*^2$ may be evaluated from measurements of the wind profile (51.8). This is the basis of the "profile method" for estimating the wind stress on the ocean (Pond, 1975).

2. Calculate the slope $\partial S/\partial\omega$ of the JONSWAP spectrum at the frequency $\omega_m - \sigma_a$ on the steep low-frequency face of the spectral peak.

3. Show that for a shape-preserving spectrum of the form (51.10), the integrated energy over any frequency band has the same fetch dependence as $S(\omega_m)\omega_m$.

4. With an energy input as given by Jeffreys' expression (51.26) and dissipation of the form (51.82), show that, in deep water ($c^2 = g/k$), the minimum rate of wave growth for a given gravity wave speed occurs at $U = 3c$. With $\nu = 0.01\ \text{cm}^2\,\text{s}^{-1}$, $g = 981\ \text{cm}\,\text{s}^{-2}$, $\rho_a/\rho_w = 1.3 \cdot 10^{-3}$ and an observed minimum wind speed of $110\ \text{cm}\,\text{s}^{-1}$, find the value of the sheltering coefficient s.

5. Using the "frozen-field" hypothesis (51.45) for advected pressure fluctuations, show by integrating (51.44) that at resonance, $\Psi(\boldsymbol{k}, t) \propto t^2$ for large t.

6. Show, using the vertical momentum equation to relate the vertical velocity to the pressure, that the rate of energy input as calculated from $\partial\langle E\rangle/\partial t = -\frac{1}{2}\,\text{Re}\,(pw^*)$ is also given by (51.69).

7. From the formulation of the boundary-value problem for the air flow perturbation (51.63)–(51.65), and the definition of τ (51.66) (into which u and w are replaced by their equivalents in terms of F and F'), show that

$$\tau(0) = \tfrac{1}{2}\rho_a k^2 |\eta|^2\ \text{Im}\,[\Pi(0)]\,,$$

and hence that (51.70) holds.

8. In an homogeneous inviscid fluid, one may write the momentum equation as

$$\frac{\partial \boldsymbol{u}}{\partial t} = -(\nabla \times \boldsymbol{u}) \times \boldsymbol{u} - \nabla \left[\frac{p}{\rho_*} + \frac{\boldsymbol{u} \cdot \boldsymbol{u}}{2} \right].$$

The first term on the right-hand side may be called the "vortex force": it is analogous to the Coriolis force. Assuming that the "total" pressure field $p/\rho_* + (\boldsymbol{u} \cdot \boldsymbol{u})/2$ is periodic along the sea surface, we may eliminate its effect by averaging over a wave-length. In the x-direction then,

$$\partial \langle u \rangle / \partial t = \langle w \zeta \rangle,$$

where ζ is the horizontal (y-directed) vorticity at some level z. Since the air flow is perturbed by the wave motion, the vorticity at z is that belonging to the unperturbed flow at some lower height $z - h$, above which the flow has risen. Hence

$$\frac{\partial \langle u \rangle}{\partial t} = - \left\langle w \frac{\partial U(z-h)}{\partial z} \right\rangle = \langle wh \rangle \frac{\partial^2 U}{\partial z^2}.$$

By considering the flow pattern around a "cat's eye" eddy such as that shown in Fig. 51.5, show that with $U(z)$ as given by (51.8), a net forward force is exerted by the fluid above the critical height on the fluid lying below that level, and hence onto the sea surface (Lighthill, 1962b).

9. Use the surface kinematic boundary condition $\mathrm{D}\eta/\mathrm{D}t = w$ to derive (51.65).

10. Derive the spectral form (51.72) on dimensional grounds, assuming that the frequency spectrum $S(\omega)$, defined by $\langle \eta \eta^* \rangle = \int S(\omega)\, \mathrm{d}\omega$, depends only on σ, ρ_* and ω.

11. Given the values $g = 980\,\mathrm{cm\,s^{-2}}$, $\nu = 10^{-2}\,\mathrm{cm^2\,s^{-1}}$, $\sigma = 74\,\mathrm{dyn\,cm^{-1}}$ and $\rho_* = 1\,\mathrm{g\,cm^{-3}}$, find the value of k for which Q, as given by (51.86), is equal to unity.

(Ans. $k \simeq 24\,\mathrm{cm}^{-1}$.)

12. Consider a thin viscous boundary layer at $z = -H$ in the presence of surface waves. Show that by assuming that the shear may be approximated as $u(-H)/\delta$, where $u(-H)$ is the horizontal potential flow at the bottom, the energy dissipation rate [from (51.80)] is given by

$$\partial \langle E \rangle / \partial t = -\mu \langle u^2(-H) \rangle / \delta.$$

Calculate this attenuation rate explicitly and compare it to that prevailing over the bulk of the fluid, as given by (51.84).

13. Calculate the ratio of the dissipation rate in the upper boundary layer in the presence of a slick (51.88) to that in the bulk of the fluid (51.84) for pure capillary waves with dispersion relation (11.21a).

(Ans. ratio $= [\sigma/k\rho_w \nu^2]^{1/4}/4(1+B)$.)

Wind waves represent the high-frequency, small-scale response of the ocean surface to atmospheric forcing. As seen in Fig. 51.1, atmospheric pressures and wind speeds show a wide spectrum of fluctuations, peaked at synoptic scales associated with storm dimensions. The response of the ocean at low frequencies is less obvious to the casual observer, as it extends over longer times, but the currents and surges of water level associated with the passage of storms can be clearly noticed on low-lying coasts, where they may lead to serious flooding. Storm surges, or wind tides, were long confused with astronomical tides which occur on similar time scales, but with much greater regularity. In this section, we discuss both types of "tides" which make up the dominant oceanic motions on a time scale of the order of a day.

Wave motions on time scales of a day, spatial scales of tens to hundreds of kilometers and of sufficiently small amplitude may be described by the linearized long wave equations for the f-plane:

$$u_t - fv + g\eta_x = -\frac{1}{\rho_*}\Pi_x + (K_{Mv}u_z)_z, \tag{52.1}$$

$$v_t + fu + g\eta_y = -\frac{1}{\rho_*}\Pi_y + (K_{Mv}v_z)_z. \tag{52.2}$$

The forcing pressure field $\Pi = \Pi(x, y, t)$ includes variations in the atmospheric pressure as well as the astronomical tidal forces; K_{Mv} is the vertical eddy viscosity.

Let us integrate (52.1) and (52.2), defining vertically averaged horizontal velocities $\bar{u}$ and $\bar{v}$ by

$$(\bar{u}, \bar{v}) = \frac{1}{H}\int_{-H}^{0} (u, v)\,dz. \tag{52.3}$$

In addition to the usual conditions (9.18) concerning the pressure and the displacement at the free surface, we also have

$$\rho_* K_{Mv}\mathbf{u}_z|_{z=0} = \boldsymbol{\tau}_s; \quad \rho_* K_{Mv}\mathbf{u}_z|_{z=-H} = \boldsymbol{\tau}_b, \tag{52.4}$$

where $\boldsymbol{\tau}_s$ and $\boldsymbol{\tau}_b$ are surface and bottom tangential stresses. Thus, after integration, (52.1) and (52.2) become

$$\bar{u}_t - f\bar{v} + g\eta_x = -\frac{1}{\rho_*}\Pi_x + \frac{1}{\rho_* H}(\tau_{1s} - \tau_{1b}), \tag{52.5}$$

$$\bar{v}_t + f\bar{u} + g\eta_y = -\frac{1}{\rho_*}\Pi_y + \frac{1}{\rho_* H}(\tau_{2s} - \tau_{2b}). \tag{52.6}$$

The surface stress $\boldsymbol{\tau}_s$ is that due to the wind (as perhaps hampered by an ice sheet if ice should be present). The bottom stress is usually expressed in the form (51.89) as

$$\boldsymbol{\tau}_b = \rho_* c_b \bar{\mathbf{u}}|\bar{\mathbf{u}}|, \tag{52.7}$$

where c_b is $0(10^{-3})$.

Equations (52.5) and (52.6) apply to the study of long wave generation by atmospheric or astronomical forcing over an area of horizontal scale $L \ll R$, the radius of the Earth. For global studies, one must fall back on the full spherical-coordinate equations [Laplace's tidal equations, cf. (29.23)].

Storm surges

Storm surges are the direct response of the ocean to large-scale atmospheric pressure and wind fields. Taking $\Pi(x, y, t)$ and $\boldsymbol{\tau}_s$ from meteorological sources, equations (52.5) and (52.6) together with the integrated continuity equation (15.49) are to be integrated over the area of interest, given appropriate initial and boundary conditions. For realistic forcing distributions and bottom bathymetry such an integration is achieved only by machine computation.

As an example of an idealized storm surge problem which may be approached by analytical methods, we consider the case treated by Crease (1956a). Let us assume that conditions are uniform in the y-direction and that the forcing terms act only in the x-direction. Then (52.5), (52.6) and the linearized form of the continuity equation (15.49) reduce, for a flat bottom, to

$$\bar{u}_t - f\bar{v} + g\eta_x = -gK(x, t), \tag{52.8}$$

$$\bar{v}_t + f\bar{u} = 0, \tag{52.9}$$

$$H\bar{u}_x + \eta_t = 0, \tag{52.10}$$

where $K(x, t) = (\Pi_x - H^{-1}\tau_{1s})/\rho_* g$. The bottom stress is neglected (remember, this is an early model!). Eliminating $\bar{u}$ and $\bar{v}$ from the above in favour of η and integrating once with respect to time, we obtain

$$[\partial_{xx} - c^{-2}(\partial_{tt} + f^2)]\eta = -K_x(x, t) \tag{52.11}$$

as the governing equation for the surface elevation, with $c^2 = gH$. This equation is the forced Klein-Gordon equation, which has been extensively studied in quantum-mechanical problems (Morse and Feshbach, 1953). For zero initial conditions ($\eta = 0 = \eta_t$ at $t = 0$), the solution of (52.11) can be expressed, for arbitrary K_x, in terms of the appropriate Green's function:

$$\eta = \frac{1}{2} c \int_0^t \mathrm{d}t' \int_{-\infty}^{\infty} \mathrm{d}x' \, K_{x'}(x', t') J_0 \left\{ f \left[(t - t')^2 - \frac{(x - x')^2}{c^2} \right]^{1/2} \right\} \Theta \left[(t - t') - \frac{|x - x'|}{c^2} \right], \tag{52.12}$$

where $\Theta(x)$ is the unit step function.

Choosing a wind stress applied uniformly from $t = 0$ to the positive half-space $x < 0$, and neglecting the forcing pressure Π, K takes the form

$$K(x, t) = -K_0 \Theta(t) \Theta(-x). \tag{52.13}$$

Substituting (52.13) into (52.12) and integrating with respect to x', we find that

$$\eta = \frac{1}{2} c K_0 \int_0^t \Theta(t - \tau) J_0 \left[f \left(\tau^2 - \frac{x^2}{c^2} \right)^{1/2} \right] \Theta \left(\tau^2 - \frac{|x|}{c} \right) \mathrm{d}\tau. \tag{52.14}$$

We note first that the response is an even function of x. Secondly, the numerical integration of (52.14) reveals that the surface elevation approaches, through a series of

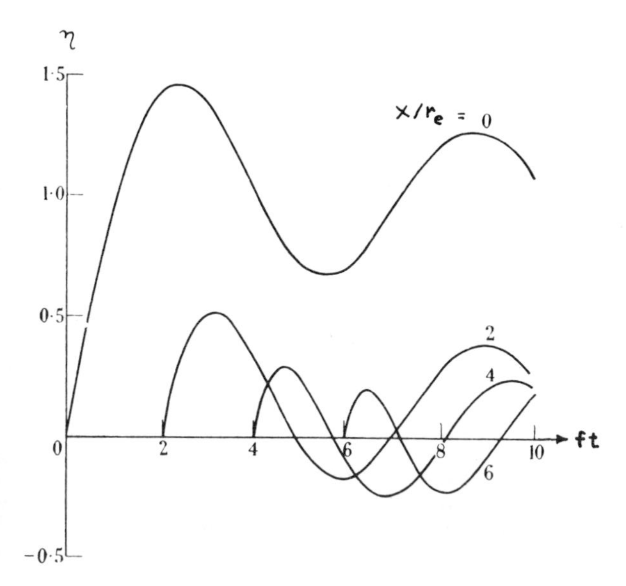

Fig. 52.1. The surface elevation η, as given by (52.14), as a function of time for three different distances from the edge of the generating area. $r_e = c/f$ is the barotropic Rossby radius of deformation. (From Crease, 1956b.)

near-inertial oscillations, a steady-state amplitude which decays with distance from the line $x = 0$, the edge of the generating area (see Fig. 52.1). In the absence of rotation, all waves would propagate with speed c and the response due to (52.13) would consist of an ever-growing perturbation travelling at a speed $c = \sqrt{gH}$. However, in the presence of rotation, long gravity waves are dispersive and the surge eventually tends towards a steady-state configuration in which the Coriolis forces are balanced by applied stresses and/or pressure gradients.

Crease (1956a) also examined the response to a moving stress field of the form

$$K(x, t) = \Theta(t)\Theta(Vt - x). \tag{52.15}$$

In the "resonant" case $V = c$, η is proportional to t at the leading edge $x - Vt = 0$, and a discontinuity develops.

The pioneering work of Crease has been extended by others (Platzman, 1963; Heaps, 1969; Henry and Heaps, 1976, to name but a few) to include bottom friction and topography as well as realistic meteorological forcing. Since the integrations become possible only by machine calculations, no advantage is to be gained from approximations designed towards the application of analytical methods. Hence approximations, such as linearity or the use of the f-plane, have largely been dropped. Readers are referred to the authors quoted above for descriptions of numerical techniques and integration methods.

As an example of the results achieved, one may compare the reasonable fit between calculated and observed sea-level variations at one location (Fig. 52.2). Numerical models of storm surges will surely, someday, be useful as predictive tools of sea-level variations to be expected under storm conditions. This stage has not yet been reached, and much work of a hindcasting nature remains to be done to refine the techniques used in parameterizing

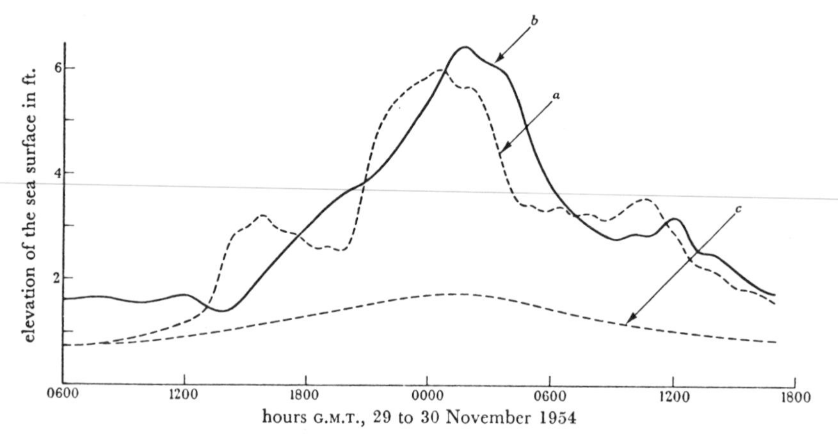

Fig. 52.2. A comparison of the storm surge calculated at Milford Haven (England), (*a*), with the observed sea-level displacement at the same location, (*b*). Curve (*c*) consists of that part of the computed surge which is due to changes in barometric pressure, i.e., the local inverse barometric response. (From Heaps, 1965.)

frictional effects and in dealing with radiation through open boundaries. (On the latter, see Bennett, 1976.)

A summary of tidal observations

The origin of tidal forces, resulting from minor gravitational imbalances due to the attraction of the Moon and the Sun, has become clear only to post-Newtonian man. The curious reader should consult Darwin (1898, Chapter 4) and Needham (1959, Vol. 3, Chapter 21) on antique western and Chinese ideas on the origin of tides. After a review of salient tidal data, this section continues with an exposé of the origin of tidal forces and of the oceanic response to their action.

The term "tide" is commonly used to denote regular sea-level changes and currents of primarily diurnal and semi-diurnal periods. Variations in sea level are best documented; they range from a purely semi-diurnal type to a purely diurnal type, passing through various mixed types (Fig. 52.3). These types are rather irregularly distributed throughout the oceans, even in fairly localized areas (Fig. 52.4). The temporal and spatial patterns in these two figures suggest an interference phenomenon between diurnal and semi-diurnal waves, with phases and amplitudes determined largely by local topography. Tides of mixed type are usually characterized by irregular oscillations such as shown in Fig. 52.5, with diurnal inequalities between the two high (HHW and LHW) and the two low (LLW and HLW) waters and unequal intervals between successive high waters.

It is also clear from Fig. 52.3 that the range of water levels varies with a longer, approximately fortnightly, period. Three types of variations are detectable:

(1) The *synodic* type, in which the largest amplitudes (SPRING tides) occur near full and new Moon, and lowest amplitudes (NEAP tides) are found at first and third quarters. The tides at Immingham (Fig. 52.3) are of this type.

(2) The *declinational* type, in which the magnitude of the diurnal inequality depends

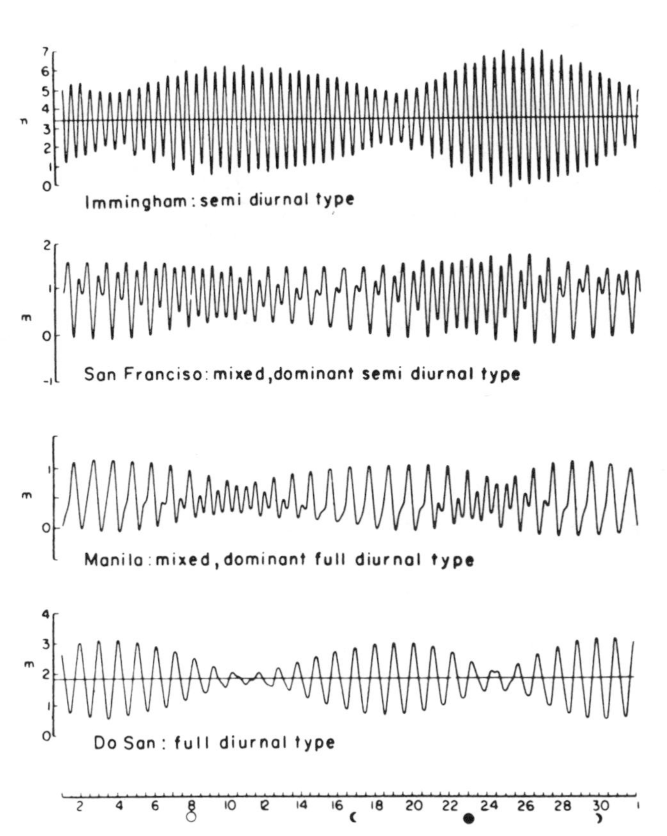

Fig. 52.3. Sea-level variations at selected ports, for March 1936. (From German Tide Tables for the year 1940, Vol. II, Berlin 1939, as shown by Defant, 1958.)

on the declination of the Moon.† One then speaks of tropic and equatorial tides, the Moon being at maximum declination and over the equator, respectively. An example is shown in Fig. 52.6.

(3) A third type of semi-monthly variation, called *anomalistic*, is discernible in some areas. Variations in tidal range are associated in that case with changes in the Earth–Moon distance.

All three types of fortnightly variation usually occur together at any one location, although one type may sometimes clearly dominate. Longer period variations can also be detected; they are associated with various astronomical and geophysical phenomena.

Tidal currents in the open sea are weak (typically 5–10 cm s^{-1}) and as a rule, rotary in nature: the tip of the velocity vector describes a horizontal ellipse, the magnitude, ellipticity and sense of tracing of which varies with position (cf. Fig. 17.2). Near the shore, current ellipses flatten and ultimately reduce to rectilinear motions of ingoing (flood) and outgoing (ebb) nature. "Slack" tide occurs when the current changes sign.

† The declination is the angle between the equatorial plane of the Earth and the line joining the Earth and the Moon.

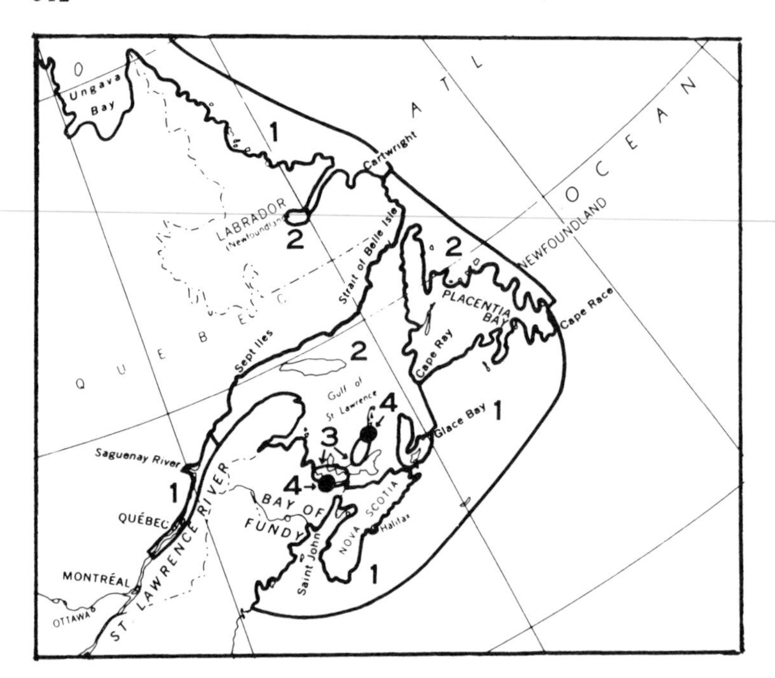

Fig. 52.4. The character of the tide in eastern Canadian waters. *1* = semi-diurnal; *2* = mixed, mainly semi-diurnal; *3* = mixed, mainly diurnal; *4* = diurnal. (From Dohler, 1964.)

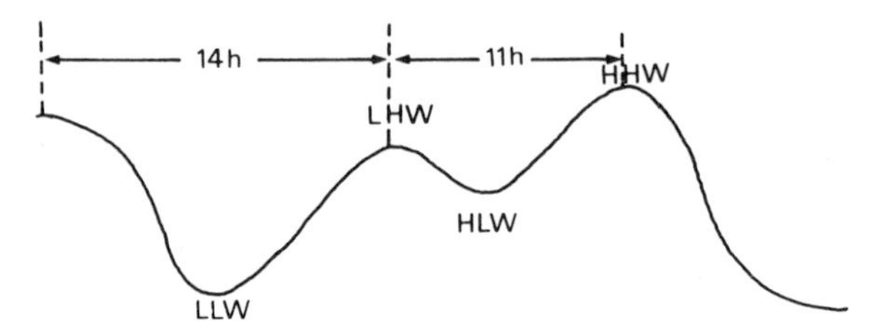

Fig. 52.5. Sea-level variation in mixed tides. The abbreviations stand for low-low water, low-high water, high-low water and high-high water. The diurnal range is HHW–LLW; the diurnal inequality in high waters is HHW–LHW; that in low waters, HLW–LLW.

Depending on the nature of the wave making up the tide, the currents will be in phase with the sea level (progressive wave) or out of phase (standing wave), or somewhere between the two. Nonlinear effects in constricted embayments and river mouths also affect the phase relationship. Coastal tidal currents can be very strong, frequently reaching up to five knots and more. In shallow rivers, tidal currents may reach supercritical values ($u > \sqrt{gH}$ and tidal bores appear. Well-known examples have been described in the Seine, the Severn, the Petitcodiac and the Chien Tang Kiang (Tricker, 1965; Clancy, 1968); the rising tide then takes the form of a hydraulic shock wave and obeys the nonlinear hydraulic equations (12.95) and (12.96). An example of a tidal bore has already been shown in Fig. 19.2.

In order to describe the geographical distribution of the tide, one plots lines of equal

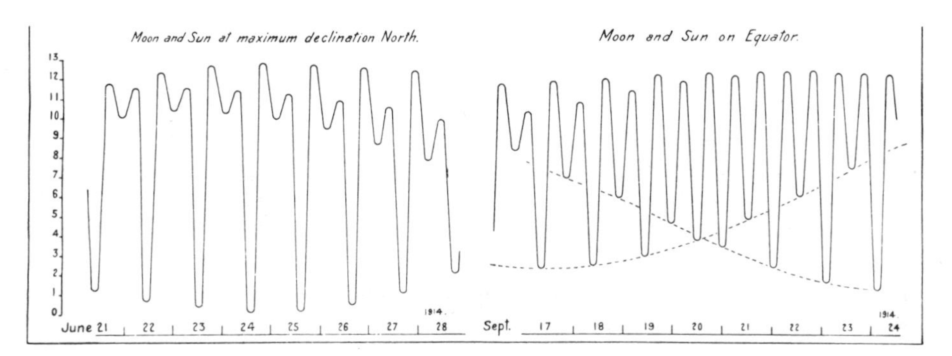

Fig. 52.6. An example of declinational variation: tides at Sand Heads, British Columbia, Canada. (a) Moon and Sun at maximum declination north. (b) Moon and Sun on the equator. Elevations are in feet. (From Dawson, 1920.)

amplitude (co-amplitude, or co-range, lines, cf. Sections 25 and 28) and equal phase (co-phase or co-tidal lines) for a given frequency. An example is shown in Fig. 52.7, where the pattern of amphidromic points is strongly reminiscent (as it should be) of that seen in Fig. 28.2 for oscillations in open and closed basins. Another example has been seen in Fig. 24.2.

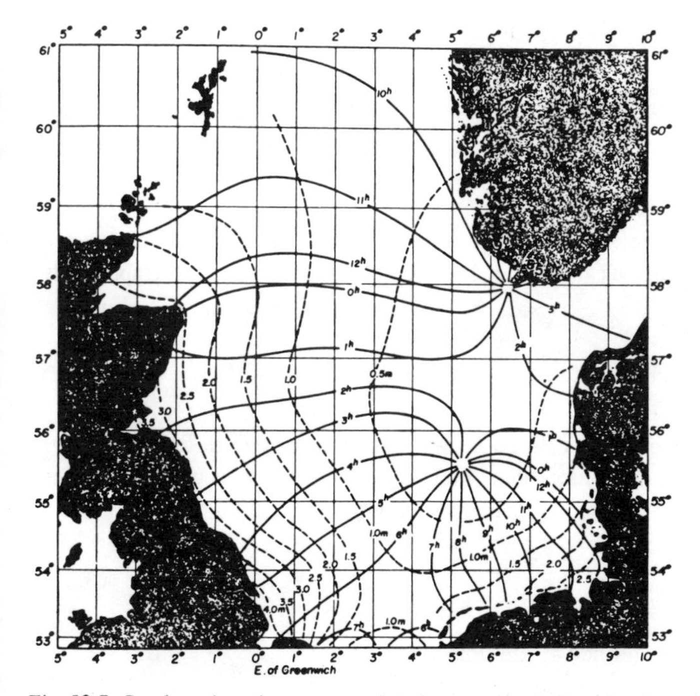

Fig. 52.7. Co-phase (continuous curves) and co-amplitude lines for the principal lunar semi-diurnal tide M_2 in the North Sea. Times of high water in hours after the Moon's transit over Greenwich; tidal amplitudes in meters. Note the strong Kelvin wave behaviour of the tide on the east coast of Great Britain. (From Defant, 1958.)

Tide generating forces

Tidal forces arise on any body of finite size moving in a spatially variable force field. Consider a spherical body, such as shown in Fig. 52.8, whose center of mass is in free fall under the influence of a gravitational field. Let one of the point masses contributing to that field be denoted by M, at a distance D (which may vary with time) from the centre of mass of the body of interest. Because of the finite size of this body, different points on its surface will in general be subject to different gravitational attractions by M. It is the differences between local values of attraction which constitute the tidal force. The gravitational acceleration at a point P due to the mass M is $\boldsymbol{a}_p$, a vector pointing towards the centre of gravity of M, and of magnitude

$$|\boldsymbol{a}_p| = \frac{GM}{r^2}, \tag{52.16}$$

where G is the universal gravitational constant ($G = 6.67 \cdot 10^{-11}\ \mathrm{m^3\ kg^{-1}\ s^{-1}}$).

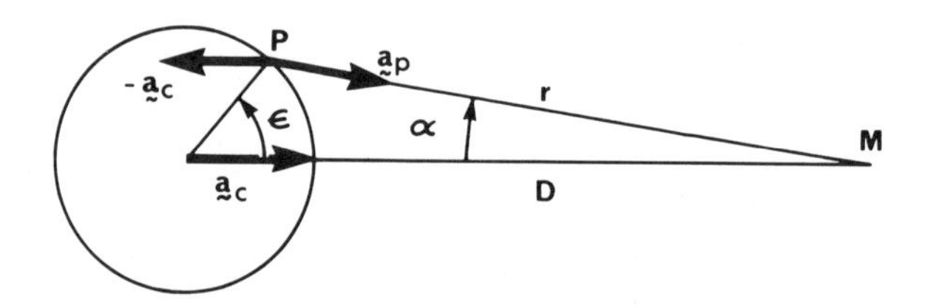

Fig. 52.8. The geometry of gravitational attraction by an external source M on a point P at the surface of a sphere of radius R. The vector $\boldsymbol{a}_p$ is the direct gravitational acceleration due to M at P; the vector $\boldsymbol{a}_c$, to which $\boldsymbol{a}_p$ is compared, is the gravitational acceleration at the centre of the sphere.

This acceleration can be resolved into components normal and parallel to the surface of the sphere (i.e., along its radius and in the direction of increasing angle ϵ, respectively) as

$$\boldsymbol{a}_p = \frac{GM}{r^2}\left[\cos(\alpha+\epsilon), -\sin(\alpha+\epsilon)\right]. \tag{52.17}$$

In order to estimate differences in acceleration, we must compare $\boldsymbol{a}_p$ to the acceleration at some other point. A natural choice of a comparison value is the acceleration at the center of the sphere $\boldsymbol{a}_c$; in the same components as (52.17),

$$\boldsymbol{a}_c = \frac{GM}{D^2}(\cos\epsilon, -\sin\epsilon). \tag{52.18}$$

The difference $\boldsymbol{a}_p - \boldsymbol{a}_c$ is then defined as the tidal acceleration:

$$\boldsymbol{a}_p - \boldsymbol{a}_c = \frac{GM}{D^2}\left[\frac{D^2}{r^2}\cos(\alpha+\epsilon) - \cos\epsilon, -\frac{D^2}{r^2}\sin(\alpha+\epsilon) + \sin\epsilon\right]. \tag{52.19}$$

From Fig. 52.8, $r^2 = D^2 + R^2 - 2RD\cos\epsilon$, and hence

$$\left(\frac{r}{D}\right)^2 = 1 - 2\frac{R}{D}\cos\epsilon + \frac{R^2}{D^2}. \tag{52.20}$$

The mean Earth–Moon distance is $D_m = 3.8 \cdot 10^8$ m; the mean Earth–Sun distance is $D_s = 1.49 \cdot 10^{11}$ m. Since R (the radius of the Earth) $= 6.37 \cdot 10^6$ m, the quantities $R/D_m = 0.017$ and $R/D_s = 4.28 \cdot 10^{-5}$ are both much smaller than unity. Thus, for $R/D \ll 1$,

$$\tan\alpha = \frac{R}{D}\frac{\sin\epsilon}{\left[1 - \frac{R}{D}\cos\epsilon\right]}$$

$$= \frac{R}{D}\sin\epsilon + O\left(\frac{R^2}{D^2}\right), \tag{52.21}$$

so that
$$\sin\alpha = \frac{R}{D}\sin\epsilon + O\left(\frac{R^2}{D^2}\right), \tag{52.22a}$$

$$\cos\alpha = 1 + O\left(\frac{R^2}{D^2}\right). \tag{52.22b}$$

Using (52.20) and (52.22), the difference $\boldsymbol{a}_p - \boldsymbol{a}_c$ as expressed by (52.19) becomes, to leading order in R/D:

$$\boldsymbol{a}_p - \boldsymbol{a}_c = \frac{GMR}{D^3}\left[3\cos^2\epsilon - 1, -\tfrac{3}{2}\sin 2\epsilon\right]. \tag{52.23}$$

The tidal acceleration is very small compared to that of the Earth's own gravity field ($9.8\ \mathrm{m\,s^{-2}}$). With the mass of the Moon $M_m = 7.33 \cdot 10^{22}$ kg and that of the Sun $M_s = 1.98 \cdot 10^{30}$ kg, the factor GMR/D^3, which we may take as representing the magnitude of the tidal acceleration, takes the value $5.67 \cdot 10^{-7}\ \mathrm{m\,s^{-2}}$ for lunar values, and $2.73 \cdot 10^{-7}\ \mathrm{m\,s^{-2}}$ for solar values. Solar tides are thus to be expected to be about half as strong as lunar tides. The radial component of the tidal acceleration merely leads to a minute change in local gravity. The tangential acceleration is just as small but is essentially unopposed and provides the "tractive" force which propels the water along the Earth's surface. Because of the relative thinness of the ocean, the tidal force is practically uniform with depth and acts as a homogeneous body force, to be inserted into the right-hand side of (52.1) and (52.2). Definition of a potential function (see Exercise 52.1) allows one to represent this force as part of the generalized forcing pressure Π in these equations.

The tidal accelerations affect all parts of the Earth, not just the ocean. There are thus measurable tides in the atmosphere and in the solid Earth. Only the oceanic response is discussed here. The reader is referred to Chapman and Lindzen (1970) for a description of atmospheric tides, and to Kuo (1975) and to Melchior (1966) for a discussion of Earth tides.

The equilibrium tide

The simplest theory of the oceanic response to tidal accelerations of the form (52.23) is that which assumes the ocean to take a shape which keeps its surface in instantaneous gravitational equilibrium with the combination of tidal accelerations and the Earth's gravity. In this "equilibrium theory", the surface of the ocean is one of uniform total gravitational potential. The tidal potential Π_t is given by (52.36); that of the Earth's gravity by

$$\Pi_0 \simeq -g(R + \eta^*), \tag{52.24}$$

where η^* is the displacement of the oceanic free surface from the spherical form. The condition $\Pi_t + \Pi_0 = \text{constant}$ is then expressed as

$$-g(R + \eta^*) + \frac{GMR^2}{2D^3}(3 \cos^2 \epsilon - 1) = \text{constant}, \tag{52.25}$$

from which one may deduce the equilibrium shape of the surface under the attraction of a single tide-producing body. The constant must be equal to $-gR$ if η^* is to vanish in the absence of tide-producing forces. Solving (52.25) for η^*, we find

$$\eta^* = \frac{GMR^2}{2gD^3}(3 \cos^2 \epsilon - 1), \tag{52.26}$$

which represents a distortion of the sphere into an ellipsoid of revolution with its long axis pointing towards M. The distortion in the direction of M ($\epsilon = 0$) corresponds to a bulge in "sea level" which is twice as large as the trough at $\epsilon = \pm \pi/2$. For lunar tides, $GMR^2/2gD^3 = 18.2$ cm; for solar tides, 8.4 cm. The equilibrium tidal model is clearly inadequate to represent the observed tidal sea-level amplitudes, reaching up to 10 m in some coastal areas. An equilibrium theory ignores the inertia of the ocean waters (while allowing them to preserve their mass and be subjected to gravitational forces) as well as the presence of continental barriers and is thus based on shaky physical and geometrical grounds! It nevertheless provides a useful image of the nature of the tidal forces: since $\eta^* = \Pi_t/g$, we see that the effect of the tidal forces is to attempt to distort the shape of the ocean surface into that of the equilibrium ellipsoid, and we can replace $\nabla\Pi_t$ by $g\nabla\eta^*$ as the forcing term in the momentum equations (52.1) and (52.2).

The dual influence of the Moon and Sun may be combined in a pair of ellipsoidal forcing terms, each with its long axis pointing towards the relevant attracting body. The ellipsoids sweep around the Earth as they follow the apparent course of the Moon and the Sun in the sky. At any one point on the Earth's surface, the equilibrium tide η^* is a time-dependent function incorporating all the periodicities of the Earth's motion about its own axis, and in its orbit around the Sun, as well as those of the orbital motion of the Moon. It is useful to separate all these time variations in a line spectrum of forcing influences and to calculate the strength of tidal forces associated with the various astronomical changes. Such a decomposition was first attempted by Laplace (1799) and carried to its ultimate completion by Doodson (1921). The proposed decomposition implies a representation of the varables ϵ and D in (52.26) in terms of orbital parameters of the Sun and the Moon and of motions of the Earth about its own axis of rotation. Let us introduce a Cartesian coordinate system, fixed in space, with the x-axis pointing towards the Point of

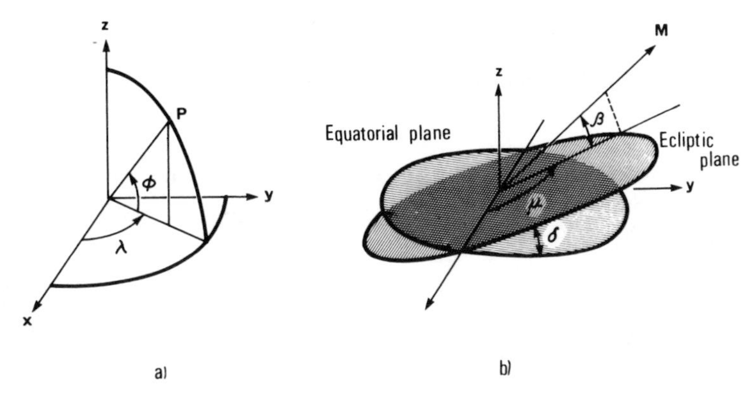

Fig. 52.9. The coordinates x, y, z, fixed in space, with respect to which are defined (a) the terrestrial latitude θ and longitude ϕ; (b) the celestial longitude λ and latitude β, as referred to the ecliptic plane.

Aries (a point on the celestial sphere which is the projection, in the direction of the ascending node, of the intersection of the ecliptic and equatorial planes) and the z-axis, in the direction of the mean polar axis (Fig. 52.9). The position of a point P on the earth's surface is then specified by its longitude ϕ and latitude θ. As is readily seen from Fig. 52.9a, a unit vector in the direction OP has components

$$\boldsymbol{OP} = (\cos\theta \cos\phi, \cos\theta \sin\phi, \sin\theta), \qquad (52.27)$$

referred to the chosen axes.

The position of an astral body is most conveniently defined in terms of angles about the plane of the ecliptic (on which lies the path of the Sun in the sky). The x-axis was chosen so as to lie along the intersection of the equatorial and ecliptic planes; the angle δ made by the two planes is equal to $23°7'$. The position of the astral body is specified in terms of its celestial longitude λ and latitude β. The unit vector $\boldsymbol{OM}$ has components (see Exercise 52.2)

$$\boldsymbol{OM} = (\cos\beta \cos\lambda, \cos\beta \sin\lambda \cos\delta - \sin\beta \sin\delta, \cos\beta \sin\delta + \cos\delta \cos\beta). \qquad (52.28)$$

The angle ϵ, to be expressed in terms of angles $(\theta, \phi; \lambda, \beta, \delta)$, is the angle $\widehat{MOP}$, so that its cosine is simply the scalar product of the two unit vectors $\boldsymbol{OP}$ and $\boldsymbol{OM}$; from (52.27) and (52.28),

$$\cos\epsilon = \cos\theta\,[\cos\phi \cos\beta \cos\lambda + \sin\phi(\cos\beta \sin\lambda \cos\delta - \sin\beta \sin\delta)$$
$$+ \sin\theta\,(\cos\beta \sin\lambda \sin\delta + \cos\delta \cos\beta)]. \qquad (52.29)$$

The angles θ and ϕ are given for any point on the Earth's surface. In order to take into account the Earth's rotation, ϕ, as referred to the fixed coordinate system, changes by 2π/day. Since the reference coordinates are oriented towards a fixed point in the heavens, the day in question is the SIDEREAL day, i.e., the time interval between successive passages of a fixed star overhead. This differs from the solar day commonly used. There are 365.256 mean solar days per year, but in that time the Earth goes around the Sun once, which makes up an extra revolution per year. The rate of change of the longitude ϕ is then $2\pi \times 366.256/365.256$ radians per solar day. This is also written as

$$d\phi/dt = 15°/\text{mean solar hour} + dL_s/dt,$$

where L_s is the mean solar longitude, and increases by 2π/solar year. Thus,

$$\phi = 15°t + L_s, \tag{52.30}$$

where t is measured in mean solar hours. The angles λ and β take one value for the Sun, another for the Moon. The Sun's apparent motion is simple and may be described to a good approximation by (Dronkers, 1964):

$$\beta_s = 0,$$
$$\lambda_s = L_s + 0.034 \sin (L_s - p_s), \tag{52.31}$$
$$D_s = \bar{D}_s[1 + e_s \cos (L_s - p_s)]^{-1},$$

where $\bar{D}_s$ is the mean Earth–Sun distance, p_s the mean longitude of the perihelion, $e_s = 0.017$ is the eccentricity of the Earth's orbit, and L_s has been defined above.

More complicated expressions hold for the lunar motion and involve trigonometric functions of linear combinations of $\bar{L}_s$, $\bar{L}_m$ (the mean lunar longitude, which increases by 2π/month), p_m the mean longitude of the Moon's perigee, the eccentricity of the lunar orbit, e_m, and K, the mean longitude of the ascending node where the lunar orbit intersects the ecliptic.

The rates of change of the orbital parameters $\bar{L}_s, p_s, \bar{L}_m, p_m$ and K are shown in Table 52.I, in degrees per mean solar day. The period associated with each of these motions is also given. The equilibrium tide η^* varies with a combination of the rates of change of all the parameters of Table 52.I and sums and combinations thereof. To make the dependence explicit one must substitute for $\cos \epsilon$ from (52.29) into (52.26) and then for λ, β, D from astronomical descriptions such as (52.31) for both lunar and solar effects. The results consists of a series of terms of the form

$$\cos [A\ 15°t + BL_m + CL_s + Dp_m - EK + Fp_s + G]. \tag{52.32}$$

A is an integer which may take the values 0, 1, or 2 and gives the "species" of the tide, a term introduced by Laplace to distinguish semi-diurnal ($A = 2$), diurnal ($A = 1$) and long-period ($A = 0$) tides; $B, C, \ldots, F$ are also integers ranging over $0, \pm 1, \pm 2 \ldots$, while G is a phase factor.

The frequency of the forcing terms in (52.32) depends on the particular values of $A, B, \ldots, F$. For example, if $A = 1, C = 1$ and $B = D = E = F = G = 0$, the period is just one sidereal day: 23.93 hours. To completely specify the tidal constituent in a simple fashion, Doodson introduced numbers which are now called after him. The Doodson number for a particular tidal forcing term is defined as

$$A(B + A + 5)(C - A + 5) \quad (D + 5)(E + 5)(F + 5).$$

When one of $B, C, \ldots, F$ is smaller than -4 or greater than $+4$ a special notation is used which we will not bother going into; this does not occur for the more important tidal constituents. Clearly, since the frequency associated with the F terms is so small (Table 52.I), the value of F will affect the total frequency very little. The main variation in the tidal forcing frequency comes from the variation in the values of A, B, C. As a matter of fact, tides with different values of D, E, F cannot be distinguished from each other in one year of data, since the difference in frequencies between them is less than a cycle per year.

TABLE 52.I

Frequencies (in degrees/mean solar day) and periods of astronomical periodic variables

Degrees/mean solar day		Description	Period
ϕ	361.054	(day)	23.93 hours
L_m	13.17644	(lunar longitude)	27.321 days
L_s	0.98565	(solar longitude)	365.256 days
p_m	0.11140	(perigee)	3231.597 days
p_s	0.00005	(perihelion)	1971.2 years
K	−0.052955	(nodal variation of Moon)	18.61 years

The simplest classification of tidal frequencies, in terms of tidal species, distributes the forcing terms into three classes: semi-diurnal, diurnal and long-period tides. The equilibrium tide may then be represented as the sum of three series:

$$\begin{aligned}\eta^* = {} & \sum_i Q_i \cos^2\theta \cos(\omega_i t + 2\phi + \gamma_i) \\ & + \sum_j Q_j \sin 2\theta \cos(\omega_j t + \phi + \gamma_j) \\ & + \sum_k Q_k (\cos^2\theta - \tfrac{2}{3}) \cos(\omega_k t + \gamma_k),\end{aligned} \tag{52.33}$$

where Q, ω and γ are the amplitude, frequency (incorporating combinations of the frequency variations of the orbital parameters and of the Earth's rotation) and phase (with respect to some arbitrarily chosen time origin) of each constituent. In the development of the equilibrium tide, Doodson (1921) identified and calculated up to 300 terms. A few constituents are sufficient to specify the tidal forcing to a reasonable degree of accuracy; a short list is shown in Table 52.II. More detailed tables may be found in Dietrich (1963) and in Neumann and Pierson (1966). An example of the fine structure of the line spectrum of the tidal force is shown below in Fig. 52.12.

Tidal response

The response of the ocean waters to the tidal body forces must be in the form of long barotropic waves of the types seen in Chapters 3 and 4. The irregularity of the continental boundaries and of the oceanic bathymetry is, however, such as to preclude any analytical solution of Laplace's tidal equations (29.23) or, for small basins, of the f-plane long wave equations (52.1) and (52.2). Machine computations have been performed by a number of authors; examples are shown for the Atlantic Ocean in Fig. 52.10. The similarity between these computations and empirical maps [such as that of Dietrich (1944)] suggests that Laplace's tidal equations are probably a suitable model of the oceanic response. One must bear in mind, however, that empirical tidal maps have until recently, been drawn exclusively from coastal data, relying on a few island observations to guide the extrapolation of co-range and co-phase lines in the deep ocean. It is only in a spirit of optimism that such maps may be said to show the "real" tide in the open ocean. A comparison of the numerical results from various authors, as in Fig. 52.10, shows significant differences in the

TABLE 52.II

The main harmonics of the tidal force. In practice, only those of relative magnitude > 10.0 are of practical importance. The abbreviations D. and S.D. stand for diurnal and semi-diurnal

Darwin name	Tidal harmonic	Period (mean solar hours)		Magnitude (percent of M_2)	Doodson number
Semi-diurnal					
M_2	Principal lunar S.D.	12.42		100	255 555
S_2	Principal solar S.D.	12.00		46.6	273 555
N_2	Longer lunar elliptic S.D.	12.66		19.2	245 655
K_2	Luni-solar declinational S.D.	11.97		12.7	275 555
ν_2	Evectional term to M_2	12.62		3.8	247 455
L_2	Smaller lunar elliptic S.D.	12.19		2.8	265 455
T_2	Principal solar elliptic S.D.	12.01		2.7	272 556
$2N_2$	Elliptic correction to M_2	12.90		2.5	235 755
μ_2	Variational correction to M_2	12.87		2.4	237 555
Diurnal					
K_1	Soli-lunar declinational D.	23.93		58.4	165 555
O_1	Main lunar D.	25.82		41.5	145 555
P_1	Main solar D.	24.07		19.3	163 555
Q_1	Lunar elliptic D. terms	26.87		8.0	135 655
NO_1	Lunar elliptic D. terms	24.83		3.3	155 655
J_1	Lunar elliptic D. terms	23.09		3.3	175 655
OO_1	Smaller lunar declinational D.	22.31		1.8	185 555
π_1	Solar elliptic D.	24.13		1.1	162 556
Long period		hours	days		
M_f	Lunar fortnightly	327.8	(13.6)	17.2	075 555
Mm	Lunar monthly	661.3	(27.5)	9.1	065 455
Ssa	Solar semi-annual	4383.3	(182.6)	8.0	057 555
Mtm	Lunar tri-monthly	219.2	(9.1)	3.3	085 455
Sa	Solar annual	8767.6	(365.2)	1.3	056 554
Msm	Luni-solar monthly	763.5	(31.8)	1.3	063 655

positions, and sometimes in the number, of amphidromic points and in the computed tidal amplitudes. These differences are related to the size of the finite-difference mesh used and its effect on the form of the boundaries, to the form of the lateral boundary conditions, to the amount and form of dissipation included and to the role assigned to Earth tides (the yield of the solid Earth to the tidal potential). Tidal computations have been critically discussed by Hendershott and Munk (1970) and by Hendershott (1973). The role of Earth tides on the oceanic response is discussed by Bogdanov and Sebekin (1976).

The possibility of resonance of the tidal forcing with some of the free modes of oscillation of the ocean has often been invoked to account for the tidal amplitudes being well in excess of those predicted by the equilibrium tide (1–2 m as compared to ~ 20 cm). A comparison between computations of the forced response of the Atlantic Ocean (Fig. 52.10) and the 12.8-hour free mode (Fig. 52.11), computed by Platzman (1975) for the combined Atlantic and Indian Oceans, shows a great similarity and is strongly suggestive of a near resonant response of the Atlantic Ocean at semi-diurnal frequencies. When the

Fig. 52.10. Solutions of Laplace's tidal equation for the M_2 tide in the Atlantic Ocean; co-phase lines are solid, co-range lines broken. From left to right, top to bottom: Bogdanov and Magarik (1967), Hendershott (1972), Zahel (1970), Pekeris and Accad (1969) (frictionless, 2° mesh), Dietrich (1944) (an empirical map), Pekeris and Accad (1969) (dissipative, 1° mesh). (From Hendershott, 1973. Copyrighted by American Geophysical Union.)

free modes of oscillation of the world ocean become well documented, it may prove worthwhile to seek an expansion of the tidal forces in terms of these modes, and to analyze the response of the ocean as a complex mechanical system, in a series of resonant and nonresonant normal modes. However, the oceanic normal modes have a very complex spatial structure and it is very unlikely that the spatial dependence of the tidal potential (cf. 52.33) should be orthogonal to any one of the free oceanic modes. The occurrence of resonance will in that case depend primarily on the forcing frequency rather than on its spatial structure.

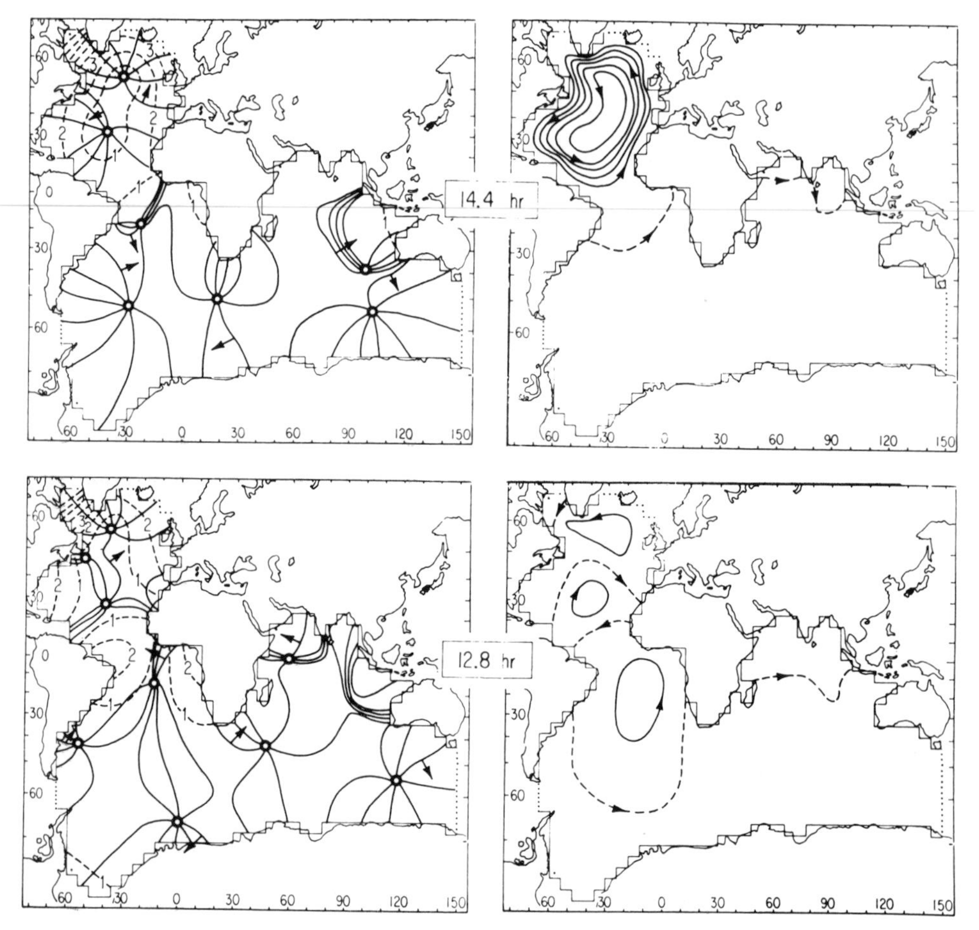

Fig. 52.11. Free gravity modes of the combined Atlantic and Indian oceans. (From Platzman, 1975.)

The description and prediction of the tide at a given point may, fortunately, be carried out without reference to the global response. Since tidal forces act at precisely known frequencies, the observed local sea-level and current fluctuations may be decomposed into a "harmonic" series of components at these same frequencies. With the addition of some higher harmonics to account for the nonlinearity of the local response, such an analysis successfully accounts for most of the tide. Having determined the amplitude and the phase appropriate to each of a series of tidal constituents, prediction proceeds from a straightforward time extrapolation of the series beyond the time interval over which its coefficients were obtained by fitting it to the observations. It is somewhat ironical that the foremost predictive triumph of oceanography is based solely on a form of time series analysis rather than on an understanding of the actual time-dependent behaviour of the ocean.

Another method of interpretation and prediction of tidal sea-level variations, introduced by Munk and Cartwright (1966), is based on a comparison of the observed tide to the equilibrium tide in terms of an admittance function, $Z(\omega)$, defined as the ratio of the Fourier transforms of the actual astronomical tide to that of the equilibrium tide:

$$Z(\omega) = H(\omega)/G(\omega), \tag{52.34}$$

where
$$H(\omega) = \int_{-\infty}^{\infty} \eta(t)e^{i\omega t}\,dt,$$
$$G(\omega) = \int_{-\infty}^{\infty} \eta^*(t)e^{i\omega t}\,dt. \tag{52.35}$$

The admittance embodies an amplitude response, $|Z(\omega)|$, and phase angle, arg $[Z(\omega)]$. Sea-level variations also occur because of meteorological forcing, or in response to fluctuations in ocean currents, in a broad spectrum which overlaps the tides. In order to estimate a proper "tidal" response to astronomical forcing, it is necessary to extract that part of the sea-level variation which is coherent with the tidal potential. This coherence analysis may be performed for long enough records [see Munk and Cartwright (1966) for details].

Figure 52.12 shows the oceanic response to semi-diurnal tides at Honolulu, based on about 50 years of data. The same groups of frequencies are prominent in the recorded and in the equilibrium spectra, but there are no gaps in the recorded spectrum. A non-tidal continuum, of energy level of about 1 cm^2/cycle/day, underlies the tidal lines. The admittance function $Z(\omega)$ is shown for that part of the sea-level energy which is coherent with the tidal potential.

Part of the meteorological forcing may occur as a line spectrum at precisely given frequencies. The diurnal alternation of land and sea breezes near a coast, for example, would produce effects which would superimpose themselves exactly upon the S_1 component (a small 24.00 hr period solar tidal component). Any semi-diurnal harmonic of the same type of breezes would fall with the S_2 12.00 hr component. These contributions would be lumped together with the astronomical tide in any harmonic analysis. An examination of the admittance function at sufficient resolution (1 cycle/year, say), however, allows identification of these so-called "radiational" tides (as they are associated with the alternation of sunlight and darkness). Assuming that the admittance is a slowly varying function of frequency (Munk and Cartwright's "credo of smoothness"), one should expect the admittance at S_1 to be equal to that at the neighbouring lines P_1 and K_1. Similarly (Fig. 52.12), the admittance at S_2 should not differ much from that at T_2 and R_2. Any difference may be attributed to "radiational" effects. Zetler (1971) has used this method to estimate the effect of radiational tides at U.S. coastal stations.

In simple mechanical systems (see Exercise 52.3), resonance to an applied force is detectable by the presence of a pronounced peak in the amplitude of the response at the resonant frequency and by a rapid change of phase in the vicinity of that frequency. While one should not expect such simple ideas to apply directly to the ocean, a strong frequency dependence of the admittance may still be taken as an indication of resonance at a nearby frequency. The admittances calculated by Wunsch (1972) from the sea-level spectrum at Bermuda (Fig. 52.13) are best fitted by an admittance function with a strong peak at 14.8 hr and secondary maxima at 9.3 and 36.6 hr. It is remarkable that Platzman's (1975) results show free modes of the combined Atlantic and Indian Ocean at 14.4 hr and also at 9.20 and 9.05 hr. The significance of the 36.6 hr peak is unknown (Wunsch suspects that it might be an artifact of the theoretical fit to the measured admittances). One may also wonder why the 12.8 hr free mode obtained by Platzman (Fig. 52.12) does not manifest itself as a peak in the ocean's response at Bermuda.

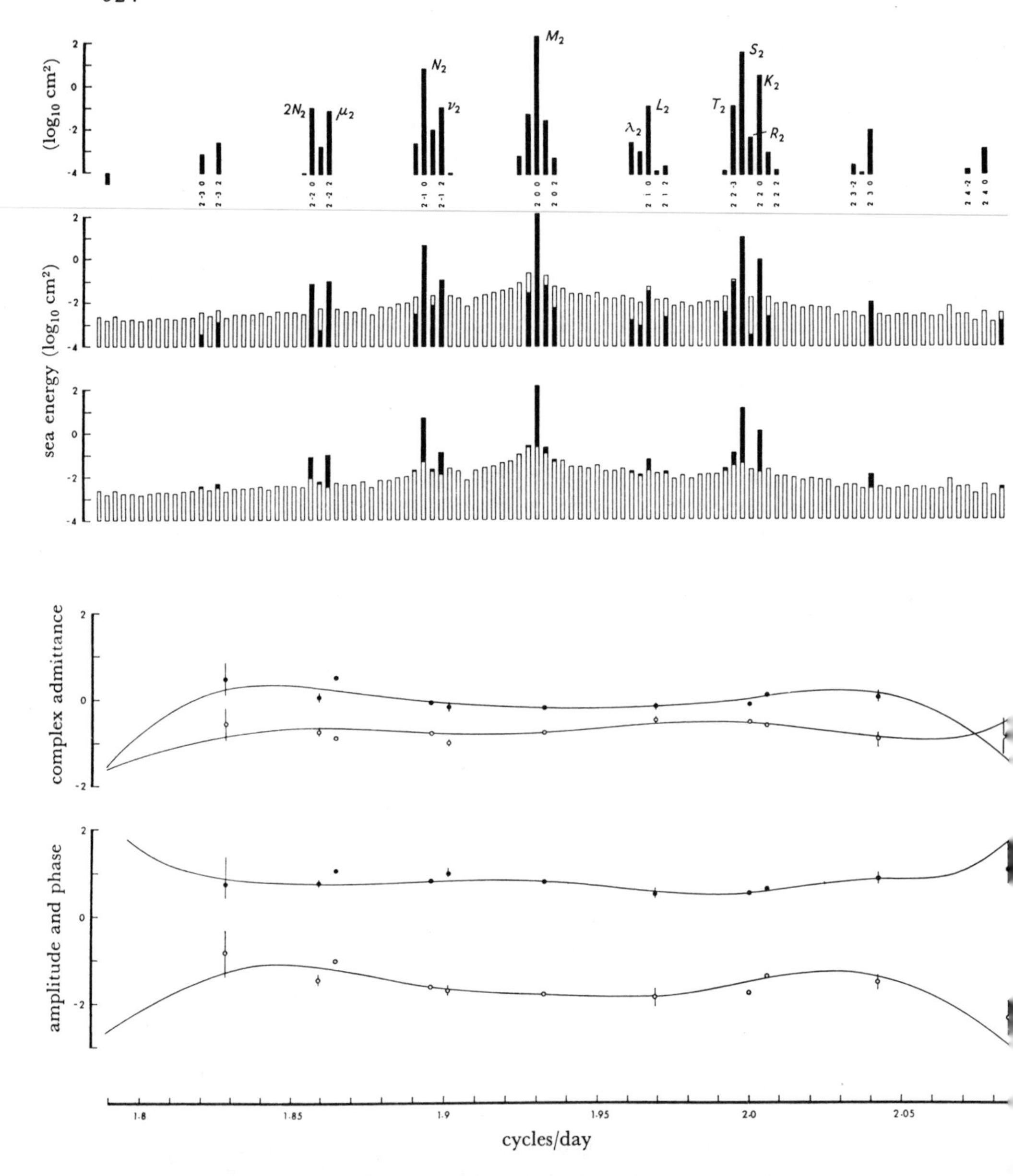

Fig. 52.12. Honolulu tide spectra at 1 cycle/year resolution for the semi-diurnal tide. The upper panel shows the energy of the gravitational equilibrium tide at various frequencies [i.e., $G(\omega)$] relative to $10^{-4}\,cm^2$. In the next two panels the energy of the observed sea spectrum is designated by the total height of the columns. In the upper of the two, the height of the filled portion designates the energy coherent with the equilibrium tide; in the lower, the height of the unfilled portion designates the non-coherent energy. The left scale refers to energy per column, the right scale to energy per cycle/day. In the fourth panel, the filled circles refer to the real part, the open circles the imaginary part of the admittance. In the fifth panel, they refer to |admittance| and phase level in radians, respectively. The vertical lines show the 95% confidence limits of the circles. The curves represent the calculated admittance functions. (From Munk and Cartwright, 1966.)

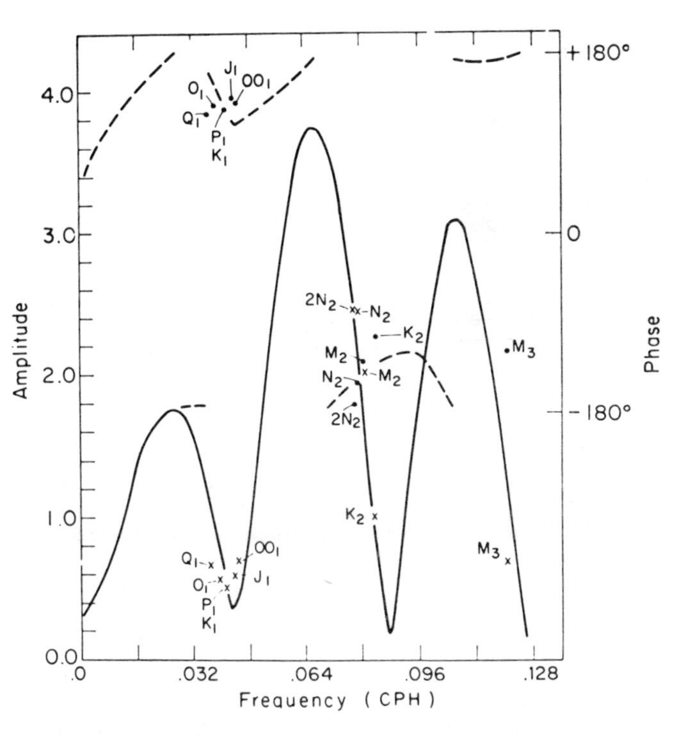

Fig. 52.13. The tidal admittance at Bermuda. The tidal lines are identified by conventional symbols (see Table 52.II), a cross denoting the amplitude of the admittance, a dot, its phase. The solid line is a theoretical fit of the amplitude of the admittance; the broken line is a fit to the phase. (From Wunsch, 1972. Copyrighted by American Geophysical Union.)

Tidal dissipation

The phase lag between the actual tide and the equilibrium forcing function produces a torque which tends to slow the Earth's rotation. Historical and paleontological data (Munk and MacDonald, 1960) on the increase of the length of the day give a value of about $3 \cdot 10^{12}$ watts for the rate of work done by this torque. A more recent calculation based on the study of satellite orbits (Rochester, 1973) yields a higher value: $4.6–6.4 \cdot 10^{12}$ watts. This energy is all supposed to be dissipated by atmospheric, oceanic and solid-earth tides. The main question concerning tidal dissipation is then not how much is lost, but where and how.

Wunsch (1975b) estimates that the Earth's mantle can only account for at most $2.4 \cdot 10^{10}$ watts and the core for $1.3 \cdot 10^{11}$ watts. Atmospheric tidal dissipation is thought to be negligible (Hendershott, 1973), so that the bulk of the total energy dissipation must occur in the oceans. A flow diagram of the possible dissipation mechanisms (adapted from Schott, 1977) is shown in Fig. 52.14. The main energy sink has long been considered to be frictional dissipation in shallow water, through bottom friction of the form (52.7). Early calculations by Jeffreys (1921) and Heiskanen (1921) gave frictional dissipation values of $1.1 \cdot 10^{12}$ and $1.9 \cdot 10^{12}$ watts, respectively. More recently, Miller (1966) has computed the energy flux into shallow seas at $1.7 \cdot 10^{12}$ watts ± 20%. These rates of

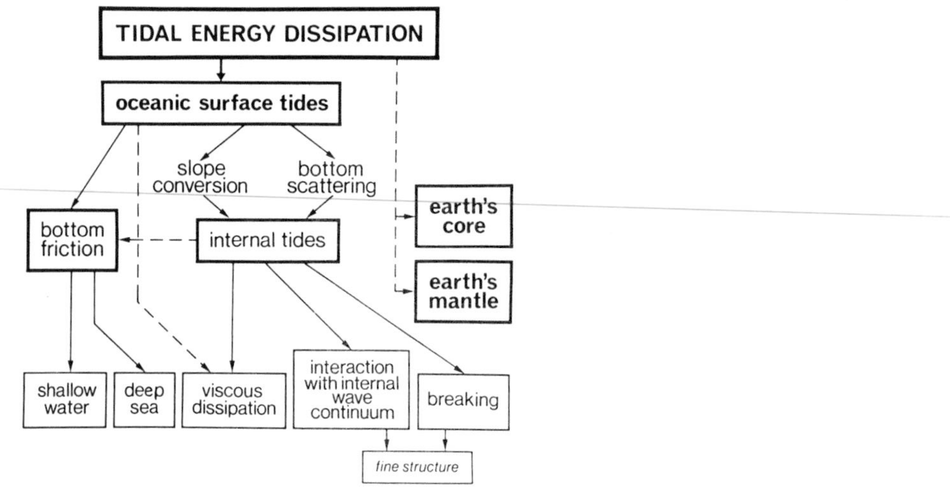

Fig. 52.14. A flow diagram of the energy dissipation processes of the barotropic tide. (From Schott, 1977.)

energy loss are insufficient to account for the astronomically based estimates. Deep-sea friction cannot close the gap and amounts to only about 1% of the dissipation in shallow seas.

It was believed for some years that conversion of barotropic tidal energy to baroclinic waves at the continental shelf margins could account for the remainder of the energy dissipation. Recent measurements by Wunsch (1975b) and Schott (1977) indicate that this is not so: this mechanism would remove energy from the barotropic tide at a rate of only about $1.0 \cdot 10^9$ watts. An estimate of tidal dissipation by turbulence in the deep ocean (Gordeyev et al., 1974) yields a rate of about $3 \cdot 10^9$ watts. Internal wave generation by scattering from small-scale topographic features in the deep ocean, based on calculations by Munk (1968) and Bell (1975), would be more important than energy conversion at continental slopes, withdrawing barotropic tidal energy at a rate of 2.5–$5 \cdot 10^{11}$ watts. The total transfer of energy from barotropic into baroclinic tides nevertheless seems to be only of the order of about 10% of the total required dissipation.

The elucidation of the proper mechanisms of tidal energy dissipation thus remains in suspense. A new approach to the problem has been attempted by Brosche and Sündermann (1971), who calculated the torques exerted by tidal currents, and hence their effect on the rotational energy of the Earth, rather than the scalar energy dissipation. In that case, the direction of the tidal currents and the latitude at which they occur become relevant. Brosche and Sündermann found that for a numerical world ocean model with a coarse 10° grid the integrated torques yielded a considerably larger energy loss than the scalar dissipation rate for the same current field. It thus appears probable that oceanic frictional torques may yet account for the observed retardation of the Earth's rotation.

An indirect measure of dissipation of an oscillating system may be found from an estimate of its Q factor, defined in Section 51. For an oscillation of the form

$$\eta = A \exp(i\omega t),$$

where ω is complex,

$$Q = \text{Re}(\omega)/2\text{Im}(\omega).$$

The Q factor may also be determined from the width of the resonance peaks as the ratio of the resonant frequency to the bandwidth at the half power points. Thus, from his interpretation of the data in Fig. 52.13, Wunsch (1972) estimates a $Q > 5$ for the 14.8 hr peak of the Bermuda admittance curve. Garrett and Munk (1971b) used measurements of the "age of the tide" (the phase lag between the maximum forcing potential and the highest observed tide at a point) to obtain a $Q > 25$ for the world's ocean. Webb (1973), on the other hand, reinterpreted the information on the age of the tide to find a much lower value of $Q \sim 8$, corresponding to a decay time of about 30 hours. From his estimate of the total energy content of the M_2 tide ($7 \cdot 10^{17}$ joules), Hendershott (1972) calculates a Q of 34 (see Exercise 52.4).

Tidal power

Energy can, in principle, be extracted from the tide by placing submerged windmills in the path of tidal currents or exploiting the potential energy made available by sea-level fluctuations. Numerous projects have been put forward, applicable to regions of high tidal energy density (the Bay of Fundy, for example), but only a pair of tidal power generating stations are in operation, one at the mouth of the Rance River, on the French coast of the English Channel, the other in the USSR, near Murmansk. Reviews of the technical problems involved in assessing the available tidal energy from a given area and in optimizing the operation of a power station have been presented by Godin (1969), Gibrat (1966), and Gray and Gashus (1972). The construction of structures for harnessing tidal energy modifies the tide itself. The resulting modifications have been most thoroughly discussed by Garrett and Greenberg (1977).

Exercises Section 52

1. The tidal acceleration is often expressed in terms of a potential function Π_t such that

$$\boldsymbol{a}_p - \boldsymbol{a}_c = \left(\frac{\partial}{\partial R}, \frac{1}{R}\frac{\partial}{\partial \epsilon}\right)\Pi_t.$$

Show from (52.23) that

$$\Pi_t = GMR^2(3\cos^2\epsilon - 1)/2D^3. \tag{52.36}$$

2. Derive the expression (52.28) for the unit vector in the direction of a tide-producing gravity source in the coordinates illustrated in Fig. 52.9b.

3. Consider a simple damped oscillator under forced motion as described by

$$\eta_{tt} + \gamma\eta_t + \omega_0^2\eta = F\,e^{i\omega t}.$$

Assuming a response of the form $\eta = A\exp[i(\omega t - K)]$, solve for and plot the admittance function $Z = A\,e^{-iK}/F$ in the vicinity of the resonant frequency ω_0.

4. Hendershott (1972) estimates the total energy content of the M_2 tide at $7 \cdot 10^{17}$ joules. Assume that the total energy dissipation (estimated at $3 \cdot 10^{12}$ watts) is due to that tidal component to compute a Q factor for the ocean. Use $Q = \langle E \rangle / \omega \langle E \rangle_t$.

53. THE GENERATION AND DISSIPATION OF INTERNAL WAVES

The first scientific observations of the generation and propagation of internal waves were made by the Norwegian explorer Fridtjof Nansen (1902) in the last decade of the nineteenth century. During his passage across the Barents Sea, he observed that the progress of his ship (the *Fram*) was considerably impeded when sailing through a thin layer of fresh water overlying saltier water. This irksome phenomenon, commonly called "dead-water", had long been known to sailors of estuarine waters.

The significance of Nansen's observation lies in that he attracted Ekman's attention to the problem. Ekman (1904) confirmed theoretically that the passage of the ship was generating interfacial waves and that the momentum of the ship was reduced by the transfer of its momentum to the waves which it generated. This type of wave drag is also well known for surface vessels and for aircraft (the so-called sound barrier). Since then, of course, many observations have established the ubiquity of internal waves in stratified waters and the theoretical aspects of their generation have been extensively studied; recent reviews are presented by Briscoe (1975a, b), Thorpe (1975) and Müller and Olbers (1975). However, despite intensive efforts we still lack a complete understanding of the overall energetics of internal waves in the ocean. That is, we still cannot identify the relative roles played by all the sources, sinks and interactions of the internal wave field. Or, as put succinctly by Briscoe (1975a), we cannot yet answer satisfactorily the questions "where does the internal wave energy come from, where does it go, and what happens to it along the way?"

The Garrett-Munk spectrum

An important step in establishing the time and length scales and the energy levels of internal waves in the deep ocean was taken by Garrett and Munk (1972b) who contrived, by synthesizing existing internal wave data, a universal, semi-empirical frequency-wavenumber spectrum (GM 72) for the internal wave field.

This spectrum assumes horizontal isotropy and takes the general form

$$S(k_h, \omega) \propto \mu^{-1}\omega^{-p+1}(\omega^2 - f^2)^{-1/2} \quad \text{for} \quad k_h \leqslant \mu, \tag{53.1}$$

where $k_h = (k_1^2 + k_2^2)^{1/2}$, μ is a cut-off wavenumber of the form $\mu(\omega) \propto \omega^{r-1}(\omega^2 - f^2)^{1/2}$, and $f < \omega < N$. The indices p and r are determined empirically. In particular, for relatively high frequencies ($\omega^2 \gg f^2$), the model simplifies to

$$S(k_h, \omega) \propto \omega^{-p-r} \quad \text{for} \quad k_h \leqslant \mu, \tag{53.2}$$

where $\mu \propto \omega^r$. Integration of the latter with respect to wavenumber k_h, and frequency and wavenumber k_2, respectively, yields frequency and wavenumber spectra of the form

$$S(\omega) \propto \omega^{-p}, \quad S(k_1) \propto k_1^{-q}, \tag{53.3}$$

where $q = (p + r - 1)/r$. Observed moored and towed spectra suggest that p and q both lie between 5/3 and 2 yielding values of r between 2/3 and 3/2. After examining several data sets, Garrett and Munk conclude that the choice $(p, q, r) = (2, 2, 1)$ is not inconsistent with existing evidence.

An example of a measured internal wave spectrum is shown in Fig. 53.1. The spectrum

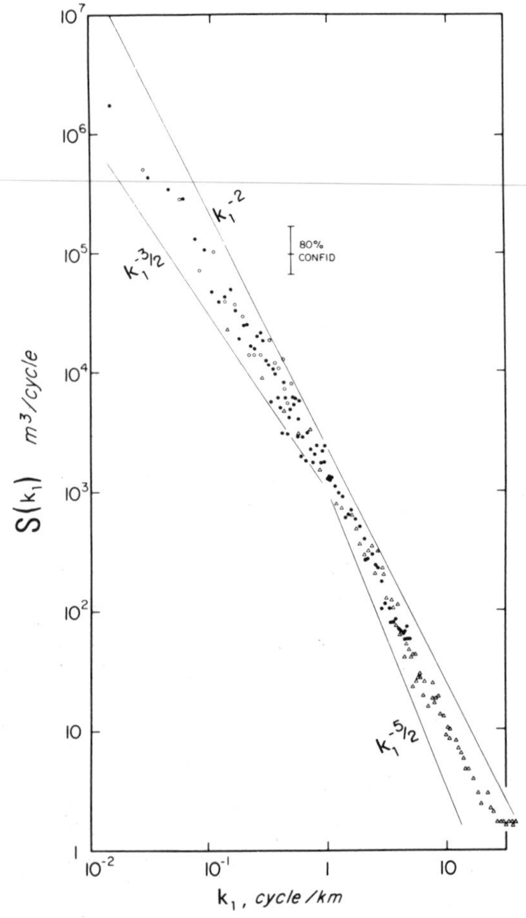

Fig. 53.1. Vertical displacement spectrum $S_\xi(k_1)$ about the 12° isotherm in the MODE area. Power law slopes of the form k_1^{-q} are also shown for comparison. (From Katz, 1975. Copyrighted by American Geophysical Union.)

illustrated (due to Katz, 1975) is that of the vertical displacement ξ as a function of wavenumber [i.e., only part of $S(k_1)$]. Thus the quantity plotted is $S_\xi(k_1)$, defined by

$$\langle \xi\xi^* \rangle = \int_0^\infty S_\xi(k_1)\, \mathrm{d}k_1 .$$

Despite the many simplifying and ad-hoc assumptions introduced in order to obtain GM 72, it nevertheless provided a unifying framework for the design and conduct of subsequent internal wave experiments, the results of which were then used to produce a modified internal wave spectrum, GM 75 (Garrett and Munk, 1975). (A further refinement of the GM 75 has been recently proposed by Desaubies, 1976.) The main revision by Garrett and Munk involved a reduction of the wavenumber bandwidth associated with moored horizontal and vertical coherences and stemmed from accepting the hypothesis (which does not always hold in practice) that microstructure (step-like structure in the

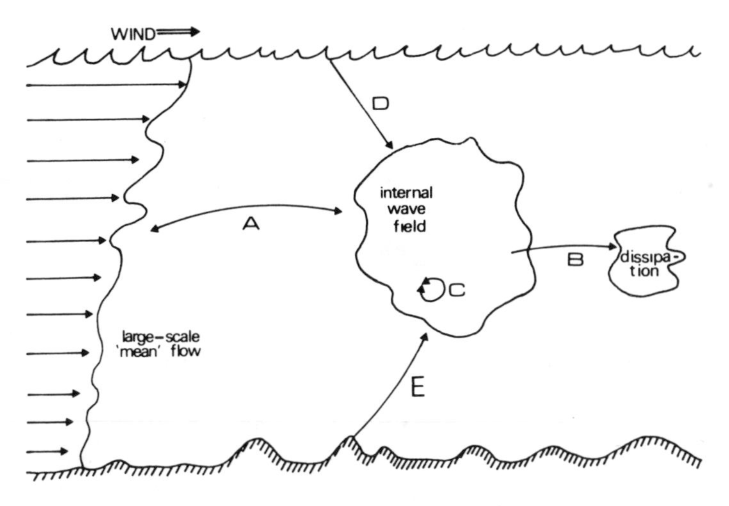

Fig. 53.2. Schematic dynamic balance of the internal wave field with the mean flow (*A*), with losses to dissipation (*B*), with self-interactions (*C*), with surface generation (D), and with topographic generation (*E*) (From Briscoe, 1975b. Copyrighted by American Geophysical Union.)

temperature and salinity profiles) is due to internal waves distorting a smooth profile rather than due to other mechanisms. However, it must be emphasized that GM 72 and GM 75 are phenomenological descriptions of the kinematic structure of the observed internal wave field and give only indirect information on the dynamic balance of the field. By attempting to prescribe a universal, isotropic spectrum, information on the generation, propagation, dissipation and interaction of internal waves is immediately ruled out. As pointed out by Wunsch (1975a), instead of always looking for a confirmation of the GM spectra by fiddling the parameters in these spectral models, observationalists should be looking for the deviations from the base models since such deviations provide us with valuable clues on the energy sources and sinks of the internal wave field. However, a recent search for such sources and sinks in the deeper part of the North Atlantic was only moderately successful (Wunsch, 1976). The underlying assumptions of stationarity and isotropy are, of course, also open to question. Indeed, Brekhovskikh et al. (1975) and Frankignoul (1974, 1976) have presented convincing evidence that in some regions, the internal wave field is nonstationary and/or anisotropic.

The dynamic balance in the deep ocean

To arrive at a complete picture of the dynamic balance of the internal wave field in the ocean we must determine: (1) the amount of energy being fed into the waves by various input (source) mechanisms; (2) the rates of energy transfer between waves of different lengths through wave–wave interactions; and (3) the amount of energy being dissipated by various output (sink) mechanisms. A schematic diagram showing examples of each of these processes is shown in Figure 53.2: *A* is the system of wave interactions with the mean shear flow, *B* the losses due to dissipation, *C* the wave–wave interactions, *D* the input from the surface wind field due to stress, pressure and buoyancy fluxes, and *E* the input resulting from flow over the bottom topography. In Chapters 6 and 7 we briefly

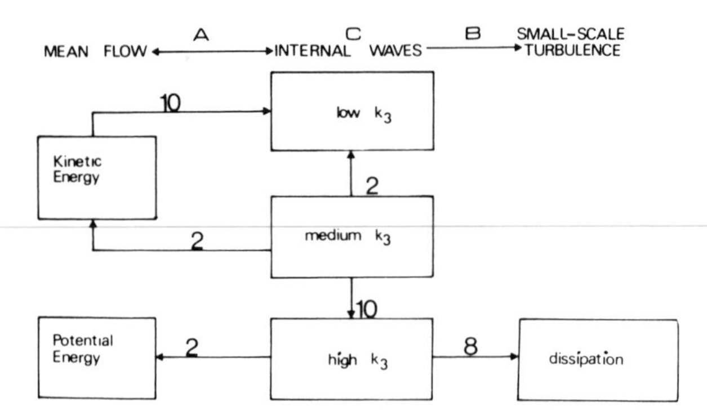

Fig. 53.3. Energy transfers (with directions indicated by arrows) according to Müller and Olbers (1975) that correspond to processes A, B and C in Fig. 53.2. The numbers are transfer rates per unit volume in units of 10^{-6} erg cm^{-3} s^{-1}.

discussed the processes A and C without, however, making any attempt to estimate actual transfer rates likely to be found in the ocean. Processes B, D and E, together with other source and sink mechanisms not shown in Fig. 53.2, have been discussed individually in the literature and have been reviewed by Thorpe (1975). We shall discuss some of these later. Generally speaking, most of the past studies on internal wave generation have been based on a linear analysis and have involved inputs at specific frequencies and wavenumbers. These approaches are evidently too simplified to be of much use in practice since the observed internal wave spectra are broad-banded (see Fig. 53.1). Thus it appears that one must take into account nonlinear wave–wave interactions, since they would tend to fill out the whole $(\omega, \boldsymbol{k})$-space. Alternatively, or in addition, the broad-band nature of the observed spectra suggest that we could profitably employ the statistical methods discussed in Chapter 5 to determine the spectral response to broad-band spectral inputs. Although the energy balance for surface gravity waves has been treated by these statistical methods for some time (see Sections 39 and 51), work along these lines for internal waves has only begun recently.

A pioneering effort on the spectral dynamic balance of internal waves has been made by Müller and Olbers (1975). The analysis is based on (33.11), the radiative transfer equation which describes the evolution of the wave-action spectral density along wave group trajectories. Müller and Olbers assumed that the wave field can be treated in the WKB (short wave) approximation and presented a lengthy list of possible source terms in this equation. However, in the final analysis, they provided estimates only for the energy transfers involved in processes A, B and C (see Fig. 53.3). The estimates for A and B were determined from the WKB wave–current interaction study of Müller (1974) and the study of internal wave dissipation at high wavenumbers due to wave breaking (Olbers, 1974), 1976). During wave breaking most of the energy goes into small-scale turbulence, but part also increases the mean potential energy by mixing. We note that their balance involves a net transfer of eight units from the mean flow to the wave field at low wavenumbers and a loss of the same amount to turbulence (by wave breaking) at high wavenumbers. From the latter process Olbers estimated an effective eddy viscosity of $K_{Mv} = 0.3\,\text{cm}^2\text{s}^{-1}$. However, the picture is not complete. Resonant interactions mainly transfer energy away from intermediate wavenumbers (see Fig. 53.3) and hence the question remains as to how the internal wave energy gets into the intermediate k_3-box in the first place.

One obvious input missing in the balance proposed by Müller and Olbers is that due to meteorological forcing. Using the spectral response method discussed in Section 34, Käse and Tang (1976) have recently shown that energy can be fed into the intermediate k_3-box referred to above by a randomly varying isotropic wind stress applied at the sea surface. More specifically, the random internal wave field in the interior of the ocean is generated by the vertical displacement of the bottom of the upper mixed layer. The energy of the waves is then dissipated in the interior by transfer to small-scale turbulence which is parameterized by a vertical eddy coefficient K_{Mv} with values in the range 1–10 cm^2 s^{-1}. An Ekman boundary layer is inserted at the bottom of the ocean to model bottom friction, which also helps to dissipate the wave energy. A similar study, without dissipation and Ekman layers, is due to Miropolsky (1975a). Miropolsky (1975b) also considered the spectral response of stable and unstable internal waves in the presence of a basic two-layer shear flow, the waves being excited by random atmospheric pressure fluctuations. This work generalizes the results of an earlier paper by Leonov and Miropolsky (1973). As there are no dissipative mechanisms in Miropolsky's model, he found that under resonant conditions the response spectrum of the vertical displacement in the fluid continually increases with time, growing linearly at low wavenumbers and exponentially at high wavenumbers. It is thus conceivable that both wind stress and atmospheric pressure forcing could add broad-band energy to the intermediate k_3-box in Fig. 53.3. However, spectra of wind stress and pressure fluctuations over the ocean have only recently become available (Fissel et al., 1976), and estimates of their role in generating internal waves are still lacking.

The other main source term which has recently been treated from a statistical viewpoint is that process represented by E in Fig. 53.2. Bell (1975) considered the generation of internal waves in the open ocean by the WKB interaction of quasi-steady and barotropic tidal flows over randomly irregular bottom topography. For bottom currents of 4 cm s^{-1}, the upward propagating waves have lengths of between 400 and 4000 m. Bell also estimated that the energy flux generated by either tidal or steady flows is approximately 1 erg cm^{-2} s^{-1} (10^{-3} joules m^{-2} s^{-1}), which is roughly an order of magnitude smaller than the mean-flow input in Fig. 53.3. Thus it appears unlikely that topographically generated waves can feed very much energy into the intermediate k_3-box [at least in the open ocean; in some coastal regions topographic generation seems important (Halpern, 1971; Samuels and LeBlond, 1977)]. In fact Bell suggests that because of critical layer absorption (Section 41) in the bottom kilometer of the ocean, the waves generated by steady flows never reach the upper part of the ocean. From this process he estimates an equivalent vertical eddy coefficient of 10 cm^2 s^{-1}. The tidally generated waves are not affected by critical layers, however, and in the middle of the ocean Bell argues that these waves spread over the entire spectrum by weak wave–wave interactions. Although short internal waves generated by bottom scattering of the tide do not appear to be an important input into the deep ocean internal wave field, Bell estimates that they do represent an important means of dissipating about 10% of the total barotropic tidal energy in the world's oceans.

We now mention one last process which is not explicitly represented in Fig. 53.2 but which nevertheless has been dealt with statistically in the literature: the interaction between the internal wave field and microstructure. Assuming that the latter can be modelled in the form of a randomly varying Brunt-Väisälä frequency, viz.,

$$N^2 = N_0^2[1 + \mu(x, z)],$$

where μ is a zero-mean, stationary random process, Mysak and Howe (1976) have derived a transport equation describing the evolution of the wave action in the presence of microstructure. For horizontally uniform microstructure $[\mu = \mu(z)]$, the wave action decays due to Bragg scattering. Further, as shown by Müller and Olbers (1975), the principal effect of the microstructure on the internal wave spectrum is to smooth out any vertical asymmetries. However, there are as yet no reliable estimates of the relative importance of this process in the overall dynamical balance of the internal wave field.

The generation of internal waves

In the remainder of this section we review briefly all the mechanisms that are believed to provide sources and sinks of energy for the internal wave field in the ocean. Several of those which involve modern statistical methods have already been dealt with above; therefore, we shall concentrate mostly on the more classical studies involving the generation and dissipation of plane internal waves. Many of the papers that we refer to are discussed in more detail by Thorpe (1975), and therefore to aid the reader for future study we shall approximately follow his order of presentation.

We shall divide our discussion of generation into three parts, according to where the excited internal waves emanate from: (1) the surface; (2) bottom or lateral boundaries; and (3) the interior of the ocean.

The meteorological forces associated with the atmospheric pressure, the geostrophic wind and the sun and rain are the chief mechanisms for exciting downward propagating internal waves.

In the case of generation by the atmospheric pressure, large vertical motions result from a resonant coupling between a moving pressure field and the internal waves. The mechanism is analogous to that proposed by Phillips (1957) for the initial generation of surface gravity waves by moving atmospheric pressures (see Section 51). For details of this mechanism for internal wave generation, we refer the reader to Voit (1959), Krauss (1966), Dotsenko and Cherkesov (1971), Bukatov (1971) and Käse (1971). From the general statistical theory of Leonov and Miropolsky (1973), it can be shown that for a pressure fluctuation of 1 mbar moving at $0.37\,\mathrm{m\,s^{-1}}$ with a wavelength of 450 m, the amplitude of the generated internal wave would grow about 1 m/day.

By way of contrast we mention here that the generation of internal waves by a point-source moving pressure field, such as a moving ship, produces a steady, nongrowing pattern of waves. The problems for a two-layer and continuously stratified ocean have been respectively solved by Hudimac (1961) and Keller and Munk (1970). For the two-layer ocean, when the speed of the point source U is less than some critical value $U_0 > c$, (where c is the wave phase speed) both transverse and divergent (vee-shaped) wave systems are generated. However, when $U > U_0$, there are no waves that can keep up with the source and the wake consists only of a set of divergent waves that spread laterally, away from the line of motion. Figure 53.4, on the other hand, shows one of the possible phase surfaces that is generated in a continuously stratified ocean. In general, the pattern is

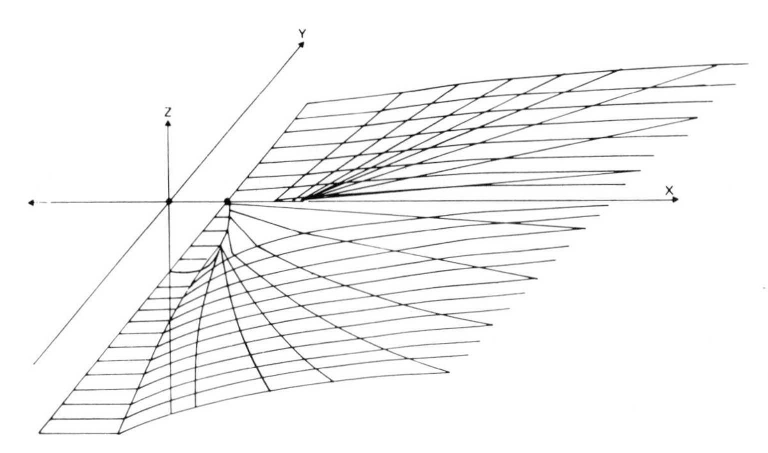

Fig. 53.4. One of the set of internal wave constant-phase surfaces belonging to the wake of a point disturbance moving in the negative x-direction on the surface of a nonrotating stratified fluid with constant N. (From Thorpe, 1975; copyrighted by American Geophysical Union.)

determined by the method of stationary phase (see Section 6). When rotation is present the pattern becomes much more complicated and we refer the reader to Abrashina (1966) and Thorpe (1975) for details.

The wind produces a horizontal momentum flux $\boldsymbol{\tau} = (\tau_1, \tau_2, 0)$, called the surface wind stress, which acts on the sea surface. For general space–time varying wind fields (including travelling disturbances), this applied stress results in a divergent current field in a thin, nearly homogeneous upper layer (the mixed layer). By continuity, there results a vertical velocity w_s (the Ekman suction velocity) which then excites internal waves at some depth in the fluid. Using a nondissipative, continuously stratified model, Käse (1972) worked out the mathematical details of the generated wave field for any given w_s. In earlier theories, two-layer models were used (with two layers of different constant density or of different constant N), and the wind stress was assumed to act as a body force, independent of depth, throughout the upper layer. In much of the recent work, the usual linearized internal wave equations are used with the addition of eddy coefficients as defined in (49.16) and (49.18) to simulate turbulent processes. The following boundary conditions are applied:

$$\text{At} \quad z = 0: \quad w = 0 \quad (= -\zeta_t \quad \text{if a free surface is included}), \tag{53.4}$$

$$\rho_* K_{Mv} \frac{\partial u}{\partial z} = -\tau_1, \quad \rho_* K_{Mv} \frac{\partial v}{\partial z} = -\tau_2; \tag{53.5}$$

$$\text{At} \quad z = -H\,(\text{constant}): \quad u = v = w = 0. \tag{53.6}$$

The special case of constant eddy coefficients, which is a good approximation for shallow seas, has been discussed in detail by Tomczak (1967) and Krauss (1972). For the case of variable coefficients one can use: (1) boundary-layer techniques in which the interior region under the mixed layer is assumed to be nondissipative (Magaard, 1973); (2) numerical models (Magaard, 1971); or (3) multi-layer models in which the coefficients take on different constant values in each layer (Käse and Tang, 1976). For a stress field of

1 dyn cm^{-2} [10^{-1} N m^{-2}] moving with a speed of 1 m s^{-1}, Käse estimated from a modified version of his earlier work (Käse, 1972) that an internal wave of frequency $2f$ will grow at a rate of 2.7 m/day, which is considerably larger than the growth rate of 1 m/day for waves generated by travelling atmospheric pressure fields.

Finally, we mention that Magaard (1973) has calculated the vertical velocity induced in the interior by a travelling buoyancy flux due to solar heating and precipitation. He then applied Käse's theory to solve the generation problem and estimated that the growth rate of first-mode internal waves in the deep ocean which are in resonance with a typical travelling periodic temperature perturbation is only about 0.2 m/day.

Thus, upon comparing the different growth rates for the pressure, wind and buoyancy generation of internal waves, we see that the wind stress appears to provide the largest meteorological input into the internal wave field. The strength of this conclusion remains to be confirmed through calculations based on actual oceanic spectra of air pressure and surface wind stress.

It is interesting to note here that the surface wind stress is commonly believed to be the most likely source of energy for the lower frequency, gyroscopic (inertial) waves that are observed in many parts of the ocean (e.g., see Fig. 2.3 and Leaman, 1976, for recent observations). Of the more recent papers on the wind-generation of inertial waves, we mention those of Tomczak (1968), Pollard (1970) and Kroll (1975), who, respectively, used models with constant eddy coefficients, two layers with N_1 and N_2 as constants, and a viscous boundary layer above an inviscid interior.

We conclude our discussion of the surface generation of internal waves with a comment on two other mechanisms: resonant interaction and Ekman layer instability. In Section 38 we mentioned that two relatively high-frequency surface gravity waves can interact resonantly to produce a relatively low-frequency internal wave. According to Brekhovskikh et al. (1972) the growth rate for such a generated wave is only a few centimeters per day in a constant N ocean; for a two-layer ocean, however, a typical growth rate is 2.6 m/day (Thorpe, 1975), which is comparable to that for a growing wind-generated wave. The second mechanism mentioned above has so far only been applied to the atmosphere. Kaylor and Faller (1972) have described how an unstable stratified Ekman layer can generate internal waves. It is conceivable that an instability of flow in the upper mixed layer can also generate internal waves.

The generation of internal waves by flow over topography has already been discussed briefly earlier. The pattern generated by a *steady* flow produces what are known as lee waves, and the governing theory and its application to atmospheric flow over mountains have been discussed in detail by meteorologists (Gossard and Hooke, 1975). An example of the streamline pattern for flow of a constant N fluid over a semi-elliptical ridge is shown in Fig. 53.5. The semicircular pattern of waves of constant phase that is evident here is identical with that seen in the plane $y = 0$ behind a moving pressure point (see Fig. 53.4). In the ocean, however, it is the *oscillatory* barotropic tidal flow that is of most interest. Cox and Sandstrom (1962) used a normal mode approach to determine the internal wave field of tidal frequency generated in the open ocean by tidal flow over arbitrary but small topographic variations. The problem for an isolated ridge has been considered by Baines (1973). The generation of the internal tide by barotropic tidal flow over the continental shelf and slope, on the other hand, has received wide attention and is a mechanism that is now well formulated. Although first conceived by Zeilon (1912,

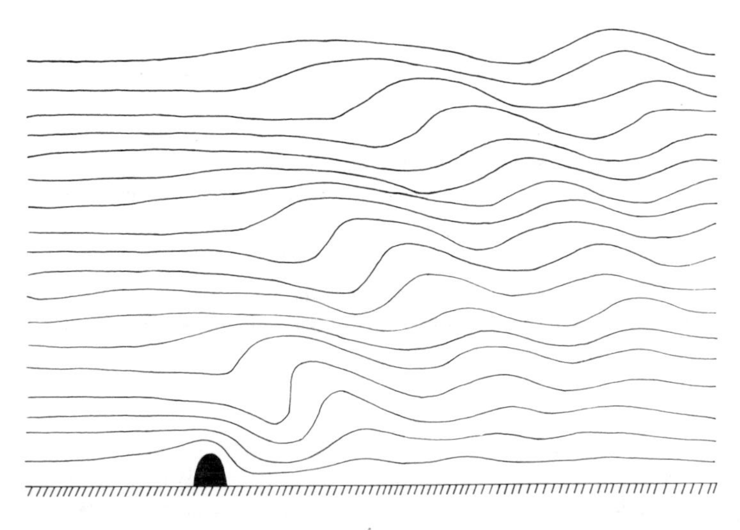

Fig. 53.5. The streamline pattern of stationary lee waves generated by a stratified shear flow over a two-dimensional semi-elliptical obstacle. (From Huppert and Miles, 1969.)

1934), it has been the work of Rattray (starting in 1960) and his collaborators and also of Baines (1974) that has given us our understanding of this mechanism. The motivation of these studies stems from the observations which indicate that the internal tides have very large amplitudes over continental slopes (Wunsch, 1975b; Barbee et al., 1975). When a long wave representing the surface tide passes over a step-like shelf, internal waves of the same frequency are generated to compensate for the mismatch in horizontal and vertical velocity fields. The generated waves progress both seaward and shoreward and the former form a beam (of rays; see Section 9) that emanates from the shelf break. The attempts to observe this beam off steep shelves have so far been inconclusive (Barbee et al., 1975; Schott, 1977). The shoreward propagating waves, however, are generally dissipated before they reflect back to the shelf break. For a discussion of the refinements of the theory we refer the reader to the several references quoted in Barbee et al. (1975).

As our last mechanism of bottom-generated internal waves, we mention the work of Cherkesov (1968) who discusses the generation of internal waves (internal tsunamis) by vertical motions of the sea bottom. Since large seismic motions occur infrequently, internal waves generated by this method are likely to make up a very minor part of the total wave field.

Very little is known about the generation of internal waves in the interior of the ocean by natural forces. It has been hypothesized by Konyaev and Sabinin (1973) that internal tidal modulation of the thickness of the thermocline creates a waveguide for high-frequency internal waves in the upper part of the ocean. It is also conceivable that two processes that generate internal waves in the atmosphere – penetrative convection and geostrophic adjustment – may also be operative in the ocean. In the first process internal waves are generated during stabilization following the buoyant rise of a miscible fluid parcel. This process has been observed in the laboratory (McLaren et al., 1973) and simulated numerically (Orlanski and Ross, 1973); it is further discussed by Gossard and Hooke (1975) and Stull (1976). With regard to the second process, it is known that as large-scale flows in the atmosphere adjust from a nongeostrophic condition to one that is nearly so, internal and gyroscopic waves are radiated. This is the Rossby adjustment problem, first discussed by Rossby (1938), and solved exactly by Mihaljan (1963) for a nongeostrophic jet flowing in a two-layer ocean. The atmospheric adjustment problem has been discussed by Blumen (1972). Geisler (1970) has shown that the response of the ocean to a moving hurricane is essentially one of geostrophic adjustment in that internal and gyroscopic waves propagate outward from the source, leaving a geostrophically balanced ridge along the storm track. The large gap in the frequency spectra of deep ocean currents between the internal wave band ($f < \omega < N$) and the planetary wave band (periods greater than 4 to 5 days) (see Fig. 15.1) indicates that there is little or no cascade of energy flow from the low-frequency motions to internal waves, as would be expected during geostrophic adjustment of large-scale flows.

Although not part of the usual oceanographic literature, we mention here the generation of internal waves by moving sources or bodies in a stratified fluid, a topic which has received wide attention by applied mathematicians and experimentalists. Wave generation by an oscillating localized source is shown in Fig. 8.4. The damping effects of molecular viscosity on waves generated this way have been analyzed by Thomas and Stevenson (1972). Related studies have been made by Hurley (1969) and Gordon and Stevenson (1972) for an oscillating cylinder, and by Larsen (1969b), for an oscillating sphere. The generation of internal waves by a steadily rising sphere was discussed by Mowbray and Rarity (1967b). The problem of wave generation by horizontally moving sources in the fluid has been discussed by Miles (1971b) for the case of no rotation, and by Redekopp (1975), who included rotation and allowed the source to oscillate as well. The latter paper employs a general technique developed by Lighthill for wave generation by travelling forcing effects; it will be discussed briefly in the next section. This technique was also used by Rao and Rao (1971) to determine the pattern of waves in a rotating stratified fluid produced by a vertically moving disturbance that also oscillates.

The dissipation of internal waves

The attenuation of long internal waves in a rotating two-layer ocean by a constant vertical eddy viscosity was investigated by Rattray (1957). Johns and Cross (1969, 1970) presented similar studies of internal wave attenuation in a multi-layered ocean in which different eddy coefficients for each layer were employed. The first thorough study of internal wave attenuation by eddy processes in a continuously stratified ocean was carried out by LeBlond (1966). He included rotation, allowed for different lateral and vertical viscosity coefficients and diffusion coefficients of heat and salt, and assumed that the effects of bottom friction are confined to a thin bottom boundary layer. For a reasonable range of values for the eddy coefficients he concluded that: (1) short internal waves (length < 100 m) are strongly damped; (2) long internal tides (length of about 200 km) will propagate for distances of 2000 km or more, but will nevertheless be damped sufficiently rapidly that cross-ocean baroclinic standing waves will not be established; (3) very long internal waves (length > 200 km) are damped primarily by bottom friction. It is instructive to see how the internal wave dispersion relation is altered when eddy coefficients are included in the basic equations. For the sake of simplicity we shall consider the case of waves in a Boussinesq, nonrotating ocean with constant eddy viscosity and diffusion coefficients: $K_{Mh} = K_{Mv} \equiv K_M =$ constant and $K_{Bh} = K_{Bv} \equiv K_B$. Thus from (49.5) and the combination of (49.7) with (49.18) and (49.8) with (49.16), we arrive at the following linearized equations (LeBlond, 1965):

$$\nabla \cdot \boldsymbol{u} = 0, \tag{53.7}$$

$$(\partial_t - K_B \nabla^2)\rho + \rho_{0z} w = 0, \tag{53.8}$$

$$(\partial_t - K_M \nabla^2)\boldsymbol{u} + \frac{1}{\rho_*}\nabla p + \frac{\rho}{\rho_*} g\hat{\boldsymbol{k}} = 0, \tag{53.9}$$

where in the above the overbar denoting the ensemble average has been dropped. As in the nondissipative case, we can combine (53.7)–(53.9) into one equation for w:

$$(\partial_t - K_B \nabla^2)(\partial_t - K_M \nabla^2)\nabla^2 w + N^2 \nabla_h^2 w = 0. \tag{53.10}$$

For time-dependent solutions of the form

$$w = w_0 \exp(\omega t + i\boldsymbol{k} \cdot \boldsymbol{x}) \tag{53.11}$$

where Im(ω) is the frequency and Re(ω) the damping rate, (53.10) gives the characteristic equation

$$\omega^2 + k^2(K_B + K_M)\omega + k^4 K_B K_M + k_h^2 N^2/k^2 = 0 \tag{53.12}$$

provided N is constant. Solving (53.12) for ω we obtain the complex dispersion relation

$$\omega = -\tfrac{1}{2}k^2(K_B + K_M) \pm i\{k_h^2 N^2 k^{-2}[1 - \tfrac{1}{4}k^6(K_B - K_M)^2 k_h^{-2} N^{-2}]\}^{1/2}. \tag{53.13}$$

Combining (53.13) with (53.11) we note that when

$$1 - \tfrac{1}{4}k^6(K_B - K_M)^2 k_h^{-2} N^{-2} > 0, \tag{53.14}$$

the waves are always attenuated [Re(ω) < 0] and propagate with a frequency Im(ω) less than that found earlier for a nondissipative ocean. The attenuation rate increases linearly with the eddy coefficients and quadratically with wavenumber, the latter property explaining why short waves are strongly damped. When the inequality in (53.14) is reversed, however, the system is overly damped and no waves propagate [Im(ω) = 0]. For fixed wavelength and eddy coefficients this would occur when N is sufficiently small. That is to say, the buoyancy restoring force is overcome by frictional forces. A similar result was also obtained by Thomas and Stevenson (1973) for waves in a viscous and diffusive ocean stratified by both temperature and salinity.

The dissipation of internal waves by turbulent friction in the bottom boundary layer has been examined by LeBlond (1966), as mentioned above, and was formally introduced by Müller and Olbers into their radiation balance equation for the wave action spectral density. However, a full understanding of the dissipative processes involved in this region is still lacking. Wunsch and Hendry (1972) made a series of current measurements along the continental slope south of Cape Cod with the aim of studying the interaction of the internal wave field with the bottom boundary layer. They found that large internal tides were generated where the bottom slope was "critical" (i.e., equalled the characteristic slope for internal waves, cf. Section 9); however, at higher frequencies, the energy appeared to be more associated with turbulence than waves.

The remaining known mechanisms for transferring or dissipating energy out of the

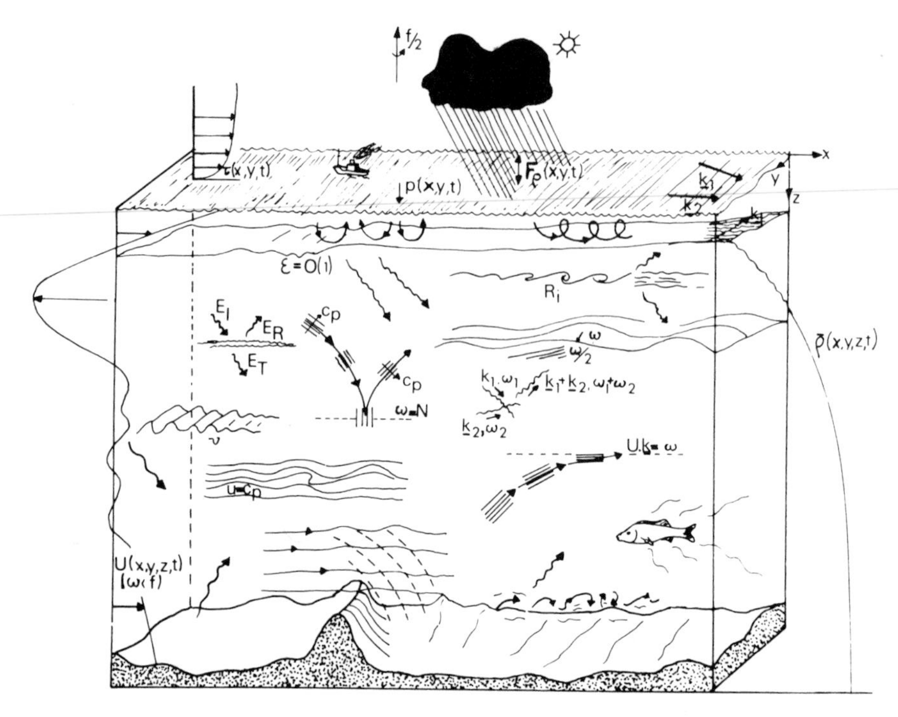

Fig. 53.6. Pictorial representation of the physical processes involved in the generation, interaction and dissipation of internal waves. (From Thorpe, 1975. Copyrighted by American Geophysical Union.)

internal wave field have nearly all been discussed in earlier parts of the book and thus will only be briefly touched upon here. The effects of a randomly varying Brunt-Väisälä frequency or horizontal velocity field on a coherent internal plane wave have been determined by Tang and Mysak (1976) [see also McGorman and Mysak (1973)] and Thomson (1976), respectively. It is shown in these studies that for short correlation scales, the coherent wave is attenuated in the direction of energy propagation ($\boldsymbol{c}_g$), with the energy being transferred to the random field (see also Section 34). A similar sort of energy transfer would take place when a coherent wave strikes the ocean bottom in regions where there are small topographic irregularities. This process has also been referred to as another possible source term in the radiation balance equation of Müller and Olbers (1975).

Absorption of internal wave energy at critical levels was discussed in Section 41. As mentioned earlier, Bell (1975) showed that upward propagating waves generated by quasi-geostrophic flows over topography are likely to be dissipated in the lower kilometer of the ocean by this mechanism. It is possible that critical layers may also exist in the thermocline region; however, observational evidence suggests that waves here lose their energy by breaking, a process which may be initiated by either convective instability or by Kelvin-Helmholtz instability (Section 42).

Thorpe (1975) has vividly summarized the various physical processes that affect internal waves in the deep ocean by means of a picture, which is reproduced here as Fig. 53.6. As well as generation and dissipation mechanisms, many interaction processes are also shown, all of which, we believe, have been discussed in this book. In ending this section we challenge the reader to decipher the different processes represented symbolically in Fig. 53.6!

Exercises Section 53

1. Consider a rotating stratified Boussinesq fluid with constant vertical eddy viscosity K_v and no other dissipation. For plane waves of the form $\exp\,[i(\mathbf{k}\cdot\mathbf{x}-\omega t)]$, show that the dispersion relation is given implicitly by (Tomczak, 1967):

$$K_v^2 k_3^6 - 2i\omega K_v k_3^4 - [\omega^2 - f^2 - iK_v k_h^2 (N^2-\omega^2)\omega^{-1}]\,k_3^2 + (N^2-\omega^2)k_h^2 = 0. \qquad (53.15)$$

2. By iteration about the solution for $K_v = 0$, which is the usual internal wave dispersion relation [see (8.56)], find two roots of (53.15) for $k_3 = k_3(\omega)$. The other four roots of (53.15) do not represent waves, but damped solutions confirmed to top and bottom Ekman layers.

3. After first invoking the hydrostatic approximation in (53.15) ($\omega^2 \ll N^2$), discuss the roots for the special case $N = 0$ and $K_v = 0$.

54. THE GENERATION OF PLANETARY WAVES

The main generation mechanisms for planetary waves arise from the meterorological forces acting on the sea surface. The generation of planetary waves by the wind stress has in particular received much attention in the literature. Very little is known about the dissipation of these waves; consequently, this topic will only be touched upon briefly at the end of the section.

Wind-stress generation of planetary waves

Early theories of the wind-driven circulation of mid-latitude oceans dealt with the response of the upper mixed layer to various steady distributions of the curl of the wind stress applied at the ocean surface (see Robinson, 1963). The deep ocean was assumed to be at rest, and all transient effects due to changes in the applied winds are ignored. In fact, the large-scale wind patterns at mid-latitudes are time-variable (with time scales of days to months – see Fig. 51.1), and the response of a stratified ocean must be time-dependent, consisting of a mixture of barotropic and baroclinic modes. Before one can confidently formulate steady-state theories of ocean circulation, it is necessary, therefore, to know the time scales in which such modes are excited by changing winds. It is considerations such as these which motivated Veronis and Stommel (1956) to study the wind-stress generation of barotropic and baroclinic planetary waves in a two-layer ocean of uniform depth.

Veronis and Stommel (1956) derived various solutions for special cases (see below) of the following *forced* two-layer equations:

$$u_{1t} - fv_1 + g\eta_{1x} = \tau_1/\rho_1 H_1, \tag{54.1a}$$

$$v_{1t} + fu_1 + g\eta_{1y} = \tau_2/\rho_1 H_1, \tag{54.1b}$$

$$H_1(u_{1x} + v_{1y}) + \phi_t = 0, \tag{54.2}$$

$$u_{2t} - fv_2 + g\eta_{1x} - g'\phi_x = 0, \tag{54.3a}$$

$$v_{2t} + fu_2 + g\eta_{1y} - g'\phi_y = 0, \tag{54.3b}$$

$$H_2(u_{2x} + v_{2y}) + (\eta_1 - \phi)_t = 0, \tag{54.4}$$

where $\phi = \eta_1 - \eta_2$, $g' = (1 - \rho_1/\rho_2)g$ and τ_1 and τ_2 are respectively the x- and y-components of the wind stress at the sea surface. These equations are a forced version of (16.8)–(16.11), which were analyzed in terms of normal modes in Section 16. Charney (1955) has shown that it is justifiable to treat the wind stress, which is actually applied at the sea surface, as a body force distributed evenly with depth over the upper layer.

From the system (54.1)–(54.4) it is easy to obtain the following vorticity balance equations:

$$\zeta_{1t} - \frac{f}{H_1}\phi_t + \beta v_1 = \frac{1}{\rho_1 H_1}(\tau_{2x} - \tau_{1y}) = \frac{1}{\rho_1 H_1}\nabla \times \boldsymbol{\tau} \cdot \hat{z}, \tag{54.5}$$

$$\zeta_{2t} - \frac{f}{H_2}(\eta_1 - \phi)_t + \beta v_2 = 0, \tag{54.6}$$

where $\zeta_1 = v_{1x} - u_{1y}$ and $\zeta_2 = v_{2x} - u_{2y}$ are the relative vorticities in the upper and lower layers, respectively. From (54.5) it is clear that only a wind stress with a nonzero curl can excite planetary wave motions in an ocean of uniform depth. It should also be noted that both layers will be in motion since (54.5) and (54.6) are coupled through ϕ.

As a simplifying hypothesis, Veronis and Stommel considered only zonally propagating wave solutions of (54.5), (54.6). In this case no y-derivatives appear in (54.5), (54.6) and therefore it is only the meridonal component of the wind stress, $\tau_2(x, t)$, which entered into their analysis. Further, they assumed τ_2 to be periodic in x and either periodic or step-like in t. For the latter type of time-dependence, Veronis and Stommel showed that when winds blow longer than about one day, most of the energy goes into a combination of barotropic and baroclinic planetary waves, rather than into inertia–gravity waves. [For a further discussion of the repartition between the two classes of waves, see also Veronis (1956).] For time-periodic winds, on the other hand, they found that as the forcing period increases, the baroclinic effects became more important. For forcing periods in the range of one to seven weeks, only the barotropic mode is generated. This is because the upper cut-off frequency for the baroclinic mode is very low. From (18.9) we find that

$$\omega_c = \beta(h^{(1)}/4f^2)^{1/2} = \beta(\delta H_1 H_2/4Hf^2)^{1/2} = O(10^{-7}\,\mathrm{rad\,s^{-1}})$$

for typical mid-ocean conditions, which corresponds to a period of O(1 year). For forcing periods of about one year, the response is partly barotropic and partly baroclinic. When the wind period is of the order of several decades a purely baroclinic motion is generated.

Longuet-Higgins (1965b) extended the work of Veronis and Stommel (1956) to include waves of both wavenumber components, generated by initial impulses localized in both space and time. Using the divergent long wave equations for the β-plane, he first showed that a concentrated initial disturbance gives rise to two systems of long-period waves, a longitudinal and a transverse system, in which the amplitude falls off like t^{-1} (see Fig. 54.1). The two systems of waves meet on a caustic, which expands outward from the origin at a constant rate. Note that this outward rate of propagation is determined by the magnitude of r_e, the external Rossby deformation radius. On the caustic the amplitude is large, whereas outside it the disturbance is exponentially small. Longuet-Higgins also considered the response to a rapidly moving storm path. In view of our earlier work on normal modes in a two-layer fluid (Sections 16 and 18), it follows that the same set of patterns are also obtained for the baroclinic mode, with r_e replaced by r_{i2}, the internal deformation radius for a two-layer fluid. Since $r_{i2} \ll r_e$ and ξ and η are proportional to $1/r_{i2}^2 t$ for the baroclinic mode, the pattern shown in Fig. 54.1 expands outward at a much slower rate in this case.

An important theory of generation of dispersive waves by travelling disturbances, that may be steady, oscillatory or transient in character, has been developed by Lighthill (1967). He applied the theory to nondivergent barotropic planetary waves excited by travelling patterns of wind stress. In contrast to the case treated by Longuet-Higgins (1965b), the speed of the moving disturbance can be comparable to the wave speed. He showed that a steady westward-moving wind-stress pattern generates semicircular waves of length $2\pi(U/\beta)^{1/2}$ in the wake ($U =$ speed of moving wind stress), and also signals directly ahead. An eastward-travelling pattern, however, only produces a wake-like disturbance.

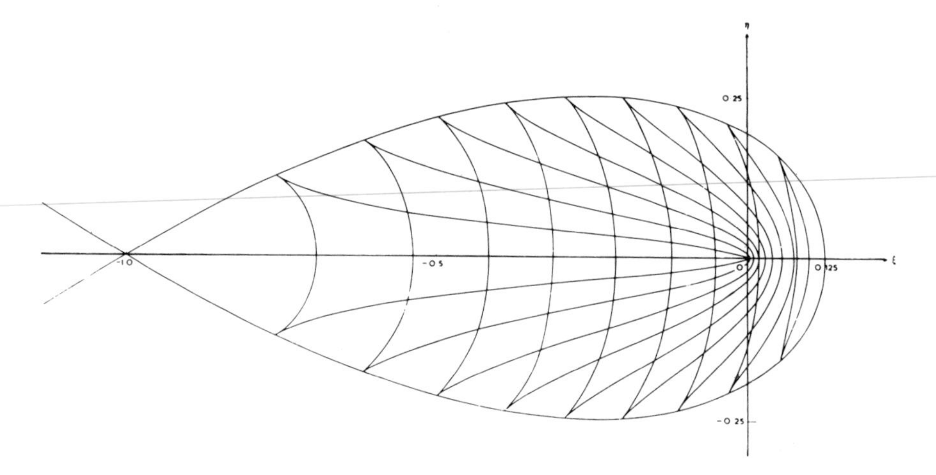

Fig. 54.1. The divergent planetary wave pattern (lines of constant phase) produced by a concentrated initial stress curl at the origin. The coordinate axes are $\xi = x/r_e^2\beta t$, $\eta = y/r_e^2\beta t$ ($r_e^2 = gH/f^2$, the square of the external deformation radius). (From Longuet-Higgins, 1965b.)

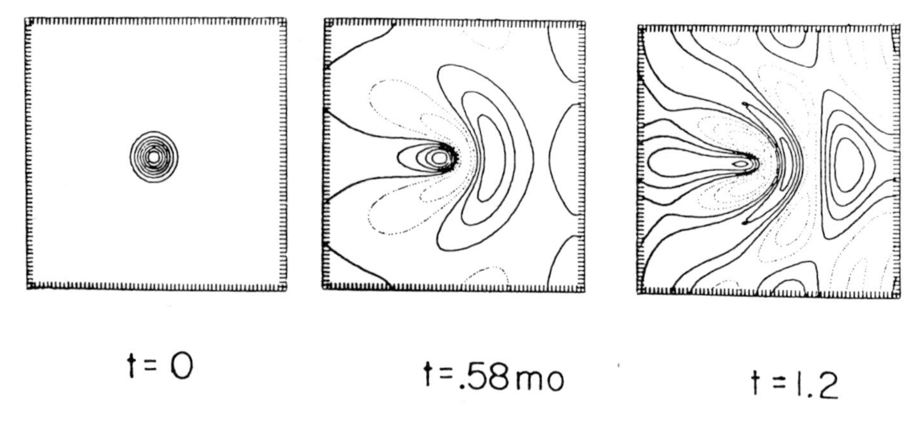

Fig. 54.2. Streamline patterns of nondivergent barotropic planetary waves dispersing from an initial Gaussian current pattern. Solid and dashed contours correspond to negative and positive ψ. The boundary conditions are periodic. (From Rhines, 1977.)

Some interesting initial-value calculations illustrating planetary wave dispersion have recently been computed by Rhines (1977). Figure 54.2 shows nondivergent barotropic planetary waves dispersing from an initial Gaussian current pattern, a configuration that may result from an applied wind stress that suddenly stops. This picture clearly shows the characteristic rapid penetration of long, east–west creasted waves to the west of the origin, since such waves have a nearly westward directed group velocity that is relatively large since, as noted earlier, $c_g \propto 1/k^2$. On the other hand, short waves, whose group velocity is relatively small and is directed eastward, appear east of the origin. Similar patterns based on asymptotic rather than numerical calculations, were also obtained by Longuet-Higgins (1965b).

Two noteworthy papers concerning the effects of one or two lateral boundaries on wind-stress generated planetary waves are those of Lighthill (1969) and Anderson and Gill (1975). As mentioned in Section 21, Lighthill (1969) proposed that the Somali current on the east coast of equatorial Africa may be due to the accumulation of energy from westward-propagating, equatorially trapped planetary waves that are generated by the onset of the Southwest Monsoon in the Indian Ocean. Both the barotropic and first-mode baroclinic waves travel westward fairly quickly (about 1 m/s) and the characteristic time for the build-up of the Somali current (which is a mixture of the barotropic and first baroclinic mode) is only about one month (of which two-thirds is for the propagation of the waves and one-third is for the concentration of the current). This response time is clearly much shorter than the "decades" predicted by the wind-stress generation of planetary waves at mid-latitudes, as discussed above.

Anderson and Gill (1975) were concerned with the spin-up of a two-layer ocean with either two boundaries (an eastern and western) or one eastern boundary. In the first problem, the response to an east–west wind stress was considered. In the interior they found that the zonal velocity increases linearly with time until a long planetary wave arrives from the eastern boundary. Then the interior flow stops growing and oscillates about a steady value. At the west, a boundary layer forms and gets progressively thinner and more intense. In the second problem, a comparison is made between upwelling at the eastern boundary induced by a north–south wind stress when there are no planetary waves (f-plane) and when there are (β-plane). They found that on the f-plane, upwelling is restricted to within about 30 km of the coast (i.e., to within one internal Rossby radius), whereas, on the β-plane, the width of the upwelling region increases with time. In both models, a Kelvin wave carries energy poleward along the boundary; however, in the β-plane model, the amplitude of the wave is attenuated as it propagates poleward.

Nonlinear and topographic effects

The above discussions of course only apply to oceans that are flat-bottomed and have weak mean currents. Only for flat-bottomed oceans can we separate the motions into barotropic and baroclinic modes. Also, a modal decomposition is permissible only when the dynamics are linear, which for wave motions means that U/c is small, were U and c are the current and phase speeds, respectively. We have seen earlier that gradual variations in topography can lead to new modes of oscillation [topographic planetary waves (Section 20), shelf waves (Section 25), bottom-trapped oscillations (Section 20), and so on]; the generation of these will be discussed below in this section, and in Section 55. The coupling of barotropic and baroclinic motions due to strong nonlinearities, on the other hand, cannot generally be determined by analytical methods and numerical techniques must be invoked. Rhines (1977) has recently made an important contribution on this topic. For example, he showed that an initial baroclinic flow in a two-layer fluid tends towards eddy-like motions of the internal Rossby deformation scale in each of the two layers. However, these eddies above and below the thermocline become phase-locked and a barotropic, wave-like motion results. But topography turns out to be a significant factor. From numerical experiments Rhines found that ocean-floor roughness appears to preserve the baroclinity of vertically sheared flows, with no cascade towards barotropic motions taking place.

Generation of topographic planetary waves

The generation of two types of long-period, open-ocean topographic waves has been considered in the literature: double Kelvin waves (Section 24) and bottom-intensified waves (Section 20). The double Kelvin wave is trapped along an escarpment in a homogeneous fluid. The bottom-intensified wave [also called a "fast baroclinic" planetary wave by Rhines (1977)] propagates in a stratified fluid with a uniformly sloping bottom. For all but very long wavelengths, its horizontal kinetic energy increases with depth; whence the term "bottom-intensified".

Mysak (1969) considered the response of an infinite homogeneous ocean with a discontinuity in depth (a model escarpment) to a variable wind stress. Fourier and Laplace transforms were used to obtain the sea surface elevation that results from either transient or time-periodic wind stresses that are applied to an initially calm sea surface. In each case it was found that, far from the forcing region, the response is dominated by a double Kelvin wave whose associated current speeds are $0(1 \text{ cm s}^{-1})$. Pinsent (1971) considered an alternative mechanism for double Kelvin wave generation involving diffraction (Section 27). He showed that part of the energy of a subinertial Kelvin wave propagating along a semi-infinite barrier that lies along a discontinuity in depth can be transmitted to a double Kelvin wave (see Fig. 27.4). Such a situation could be envisaged as taking place, for example, at the southern tip of South America. A Kelvin wave moving southward down the Pacific coast of this continent would give up some of its energy to a double Kelvin wave as it (the Kelvin wave) rounded Cape Horn and emerged into the shallower waters of the Atlantic Ocean.

Suarez (1971) considered a variety of mechanisms for the generation of bottom-intensified waves: (1) excitation by planetary waves incident upon a finite-width region of constant slope; (2) generation by initially imposed disturbances; and (3) generation by surface wind stresses. For typical mid-ocean conditions only the second of these can effectively generate bottom-intensified waves, and even then the scale of the initially imposed disturbance must be smaller than the internal deformation radius $r_i = NH/f = 0(10^2 \text{ m})$. Suarez also considered the interaction between small-scale topography and bottom-intensified waves; these waves are only strongly bottom-trapped when the topographic scale is smaller than r_i. This interaction was further explored by McWilliams (1974), who also examined the wind-stress forcing of these waves over complex topography of several length scales.

Other sources and possible sinks for planetary waves

We have discussed a few energy sources for planetary waves. The stability analyses of Chapter 7 are also relevant to the growth of such waves. In particular, for historical reasons, we have devoted a fair amount of time to the wind-stress generation of planetary waves. However, in practice, the winds may be very inefficient in generating low-frequency planetary waves since wind spectra are peaked at relatively high frequencies (around 3 days – see Fig. 51.1). Because of this mismatch, other mechanisms have to be seriously considered. One important possibility is the generation by surface buoyancy fluxes. Magaard (1977) has recently shown that such a mechanism may in fact be responsible for producing very long, low-frequency baroclinic waves, of the type recently observed by Emery and Magaard (1976) in the eastern Pacific. In the previous chapter we have discussed the generation of eddies (unstable planetary waves) by baroclinic instability of intense boundary currents or weaker, mid-ocean flows. Also, in Section 21 we mentioned that transient motions associated with El Niño may excite (equatorial) planetary waves. It has also been suggested (e.g., Rhines, 1977) that planetary waves and eddies can be generated by tidal and other flows over rough topography and irregular coastlines, by penetrative convection (the sinking of cold water), and by internal wave stresses. However, direct evidence of many of these generation processes is still lacking. The study of such processes will certainly attract more attention in the future.

Apart from possible energy losses due to scattering from rough topography (Rhines and Bretherton, 1973; Thomson, 1975a, Hall, 1976), very little is known about how planetary waves are attenuated. For example, the significance of critical layer absorption of planetary waves in the ocean has not been determined. Dissipation by lateral friction, as modelled by eddy coefficients, does not appear to be important. At the moment, it is conjectured that all dissipation takes place in the bottom Ekman layer, a few meters above the bottom, and that vertical mixing is negligible elsewhere (Bretherton, 1975). This is clearly a topic that needs further investigation.

Exercises Section 54

1. Show that for nondivergent barotropic planetary waves to be excited, the forcing frequency and (total) wavenumber k of the wind stress must satisfy the inequality (Lighthill, 1971)

$$\omega k < \beta.$$

Thus, show that inputs with time scales of weeks must have length scales of several hundred kilometers. (In other words, to generate motions uniform with depth, the winds must get a large "grip" on the ocean.)

2. By taking the usual combinations

$$u = u_1 + su_2,$$

$$v = v_1 + sv_2,$$

where s is a root of (16.25), show that the forced two-layer equations (54.1)–(54.4) (after eliminating η_1 and η_2) can be reduced to the *inhomogeneous* normal mode equations

$$u_{tt} - fv_t = gh^{(n)}(u_{xx} + v_{xy}) + (\tau_1/\rho_1 H_1)_t,$$

$$v_{tt} + fu_t = gh^{(n)}(u_{xy} + v_{yy}) + (\tau_2/\rho_1 H_1)_t,$$

where $h^{(n)}$ $(n = 1, 2)$ are the equivalent depths for a two-layer fluid [cf. (16.28), (16.29)].

55. THE GENERATION OF COASTAL TRAPPED WAVES

In this final section we discuss the generation of Kelvin, edge and continental shelf waves. A few of the generation mechanisms for these waves have already been touched upon in earlier sections, so our purpose here will be to introduce other forcing mechanisms and hence consolidate our knowledge of all the possible energy sources for coastal trapped waves.

Kelvin waves

In coastal regions Kelvin waves make up a significant part of the tidal motions (cf. Section 24). Kelvin waves can also be generated by storm surges diffracted by vertical barriers and scattered by irregular coastlines (cf. Section 27). Further, they can be generated by a point source located far from the coast that is suddenly switched on and allowed to vary with some aperiodic time behaviour, as in an earthquake (Voit and Sebekin, 1970). Finally, atmospheric disturbances represent another important energy source for Kelvin waves.

Kajiura (1962) and Thomson (1970) considered the response of a semi-infinite, uniformly rotating homogeneous ocean to variable winds and atmospheric pressure acting on the sea surface near the coast. They showed that Kelvin waves are generated only by the longshore wind component and pressure gradient. Thomson applied his results to the Oregon coast and predicted that for typical summer winds, forced Kelvin waves at subinertial frequencies have amplitudes of the order of 5 m. However, low-frequency Kelvin waves generated by moving pressure sources alone (which model intense cyclones or anticyclones) have calculated amplitudes of less than 1 cm.

Gill and Clarke (1974) considered the wind generation of internal Kelvin waves in a continuously stratified ocean. The purpose of their study was to show how wind-induced upwelling at the coast is modulated by the propagation of low-frequency internal Kelvin waves. They used the normal-mode technique described in Section 10 to find the response as a sum of vertical modes; the amplitude of each mode in the low-frequency limit ($\omega^2 \ll f^2$) satisfies a first-order wave equation. Such equations can be easily integrated by the method of characteristics and the results, they suggest, may be of use in forecasting coastal upwelling. Modifications of this theory to allow for slow variations in the coastline have been carried out by Clarke (1977a). In particular, he found that a Kelvin wave speeds up when travelling along a bay and slows down at capes.

Edge waves

In contrast to the case for Kelvin waves, atmospheric pressure disturbances generally appear to be more important than the wind stress in the generation of edge waves on a sloping beach. To illustrate this we examine the forced shallow-water equations for a uniformly rotating fluid of depth $H = H(x, y)$. If $p_a(x, y, t)$ denotes the atmospheric pressure at the sea surface and $\boldsymbol{\tau} = (\tau_1, \tau_2)$ the surface wind stress, the vertically integrated equations (52.5), (52.6) and the linearized form of (15.49) become (dropping overbars)

$$u_t - fu + g\eta_x = -\frac{1}{\rho_*} p_{ax} + \frac{1}{\rho_* H} \tau_1, \tag{55.1a}$$

$$v_t + fu + g\eta_y = -\frac{1}{\rho_*}p_{ay} + \frac{1}{\rho_* H}\tau_2, \tag{55.1b}$$

$$(Hu)_x + (Hv)_y + \eta_t = 0. \tag{55.2}$$

For f = constant it is easy to solve (55.1) for $\mathcal{L}u$ and $\mathcal{L}v$, where $\mathcal{L} = \partial_{tt} + f^2$, and then substitute these equations into (55.2) operated on by $\mathcal{L}$ to obtain an equation for η:

$$\nabla_h \cdot (H\nabla_h \eta_t) + f(\nabla_h H \times \nabla_h \eta) \cdot \hat{z} - \frac{1}{g}\mathcal{L}\eta_t$$

$$= \frac{1}{\rho_* g}[-\nabla_h \cdot (H\nabla_h p_{at}) - f(\nabla_h H \times \nabla_h p_a) \cdot \hat{z} + \nabla_h \cdot \boldsymbol{\tau}_t + f(\nabla \times \boldsymbol{\tau}) \cdot \hat{z}] \tag{55.3}$$

The unforced version of this equation was derived in Section 20 [cf. (20.13a) with $\beta = 0$].

For edge waves with periods of a few hours or less, the motion is very nearly irrotational (see Section 25) and the terms proportional to f can be neglected. Examination of the right-hand side of (55.3) reveals that such an approximation eliminates the vorticity inputs into the fluid. With $f = 0$, (55.3) can be integrated once with respect to t. Putting $H = \alpha x$, which represents a uniformly sloping beach with the mean shoreline located at $x = 0$, we obtain:

$$\alpha x \nabla_h^2 \eta + \alpha \eta_x - (1/g)\eta_{tt} = (1/\rho_* g)(-\alpha x \nabla_h^2 p_a - \alpha p_{ax} + \nabla_h \cdot \boldsymbol{\tau}). \tag{55.4}$$

The first observations of edge waves were reported by Munk et al. (1956). In particular, along the eastern U.S. coast, near Atlantic City, N.J., they found that first-mode edge waves were generated by low-pressure hurricanes travelling along the shelf, the latter having the average bottom slope $\alpha = 10^{-3}$ and maximum depth $H_m = 200$ m. These hurricanes were characterized by a sharp pressure drop, δp_a, of 70 mbar ($= 70 \cdot 10^3$ dyn/cm^2) over a horizontal distance of $L = 100$ km, and the associated winds U_w were of the order of 15 m s^{-1}. The wind stress can be calculated using the familiar formula (cf. 52.7)

$$\boldsymbol{\tau} = \rho_a c_D |U_w| U_w, \tag{55.5}$$

where $\rho_a = 1.25 \cdot 10^{-3}$ gm cm^{-3} is the air density and $c_D = 1.5 \cdot 10^{-3}$ is the drag coefficient. Using these values for U_w, ρ_a and c_D we find the wind stress has a magnitude of about $\tau_0 = 5$ dyn/cm^2. Thus the terms in the brackets on the right-hand side of (55.4) have the following typical values (with dimensions of dyn/cm^3):

$$\begin{aligned} \alpha x \nabla_h^2 p_a &\sim H_m \delta p_a / L^2 \sim 14 \cdot 10^{-6}, \\ \alpha p_{ax} &\sim \alpha \delta p_a / L \sim 7 \cdot 10^{-6}, \\ \nabla_h \cdot \boldsymbol{\tau} &\sim \tau_0 / L \sim 5 \cdot 10^{-7}, \end{aligned} \tag{55.6}$$

where we have assumed that the wind stress associated with hurricanes has the same length scale as the pressure field. From (55.6) it is clear that the pressure input completely dominates the wind-stress input in this case. On the basis of this result Greenspan (1956) calculated the sea-level response due to various model pressure systems moving parallel to the coast and found that in each case the response was always dominated by a first-mode edge wave travelling in the wake of the disturbance. The computed periods and amplitudes of this wave were about 6 hr and 1 m, respectively, which agreed quite well with the observations of Munk et al. (1956). However, for these fairly long-period edge waves, the Coriolis force could be important, which led Kajiura (1958) to consider its effect on the generation of edge waves. He found that in the Northern Hemisphere, edge waves are more efficiently generated by pressure systems moving with the coast to their right. Kajiura also studied the evolution of an initial deformation of the water surface in the presence of rotation. He found that the initial shape splits up into two edge waves propagating in opposite directions, with the right-bounded wave having the larger amplitude.

Under more normal atmospheric conditions, the pressure systems near coastal regions are characterized by fluctuations $\delta p_a \sim 5$ mbar, and by horizontal length scales $L \sim 10^3$ km. The corresponding winds are approximately in geostrophic balance with the pressure, and therefore

$$U_w \sim \frac{\delta p_a}{\rho_a f L} \sim 4\,\mathrm{ms}^{-1}.$$

With this value for U_w in (55.5) we obtain a wind-stress scale of $\tau_0 \sim 1$ dyn/cm^2 or less. In these situations the wind-stress forcing term can be comparable to the pressure term. Indeed, the ratio of the first and third terms on the right-hand side of (55.4) is characterized by $H\delta p_a/L\tau_0$, which is of order unity for $H \sim 200$ m and the scales given above. In corroboration of this result Buchwald and deSzoeke (1973) found that edge waves generated by pressure disturbances alone did not have large enough amplitudes to explain the edge-wave motions observed on the East Australian shelf. They concluded that the wind stress is also likely to be an important energy source for these oscillations.

The generation of edge wave by nonmeteorological forces has been receiving an increasing amount of attention in the literature. For example, Kajiura (1972) and Morris (1976) have discussed the generation of edge waves by transient and sinusoidal sources located near the coast. Kajiura (1972) presented detailed calculations for the response of a semi-infinite ocean bounded on one side by a single-step shelf due to a source that is located either on or just off the shelf. Such a forcing mechanism is characteristic of broad crustal deformations due to earthquakes near the coast. He found that the proportion of energy trapped on the shelf as edge waves relative to that radiated into the deep water increases with the longshore dimension of the source. Also, edge waves are generated more efficiently when the source is close to the shoreline.

Other recent edge-wave generation studies are based on the concept of resonant wave–wave interactions (Section 38). For example, Guza and Davis (1974) showed that a standing gravity wave normally incident on a sloping beach can transfer energy to edge waves through a weak resonant interaction resulting from the instability of the incident wave with respect to perturbation by edge waves. Their results suggest that subharmonic, lowest-mode edge waves are preferentially excited. Guza and Bowen (1975) extended this theory to the case of arbitrary incidence. Fuller and Mysak (1977), on the other hand, showed that when a gravity wave is incident upon a single-step whose coastline has an irregular shape, one or more edge waves can be generated on the shelf through a resonant interaction of the incident wave with the wavenumber spectrum of the coastal irregularities.

As an introduction to the study of shelf wave generation, let us recall some of their salient properties. A shelf wave is a type of topographic planetary wave and hence is highly rotational: its motion consists of a sequence of horizontal eddies of alternating sign propagating along and confined to the shelf/slope region (see Fig. 24.5 for an experimental reproduction of such eddy motions). In the Northern Hemisphere, the coast lies to the right of the direction of phase propagation. It is now generally accepted that these waves are generated by the large-scale weather systems that move across or along the shelf. They have periods of about one week, wavelengths of order 10^3 km and pressure and wind-stress fluctuations of order 5 mbar and 1 dyne/cm^2, respectively. Consequently, in the ocean, shelf waves have long wavelengths ($L \gg l$, where l is the width of the shelf/slope region), low frequencies ($\omega \gg f$) and small amplitudes (a few centimeters).

Observations of low-frequency sea-level oscillations at several coastal regions indicate a nonbarometric response of the sea surface to low-frequency atmospheric pressure fluctuations (e.g., see Hamon, 1966; Mooers and Smith, 1968; Mysak and Hamon, 1969). This is to say, corresponding to a pressure increase of 1 mbar, the sea surface is not necessarily lowered by 1 cm. Nevertheless, the sea level and air pressure fluctuations are highly correlated at low frequencies. Thus it was originally proposed that this somewhat anomalous sea-level behaviour might be due to a resonant response of the sea surface to large-scale atmospheric pressure systems, with long, nondispersive shelf waves being generated in the process (Robinson, 1964; Mysak, 1967a, b).† However, as we shall see below, such a mechanism is not very plausible, and current opinion now favours the generation of shelf waves by the wind stress. This mechanism was originally proposed by Adams and Buchwald (1969), who argued that rotational shelf waves need vorticity inputs to be effectively generated. Gill and Schumann (1974) have given a more extensive account of the theory of wind-generated shelf waves, and their formulation of the theory was recently used by Hamon (1976) to show that along the East Australian coast, shelf waves are most likely generated by the longshore component of the wind stress. The observed high correlation between sea level and pressure changes which motivated the earlier generation theory of Robinson and Mysak can be explained by the fact that the pressure changes are also highly correlated with changes in the wind stress (Hamon, 1966).

To elucidate the details of shelf wave generation we now derive a nondimensionalized version of (55.3) for a shelf/slope region of depth $H(x)$ with mean value H_0 and of width l. We introduce the adjusted sea level η^a by the equation

$$\eta^a = \eta + \frac{1}{\rho_* g} p \equiv \eta - \phi, \tag{55.7}$$

where ϕ is the negative of the atmospheric pressure measured in centimeters of water. When $\eta^a = 0$, (55.7) implies that the sea-level response is "inverse barometric" or "isostatic"; thus η^a represents the nonisostatic contribution to the sea-level fluctuations.

† Crépon (1976) has recently proposed that the anomalous sea-level response is merely due to the geostropic adjustment of the ocean to atmospheric pressure variations. However, his theory does not allow for longshore variations and therefore ignores the possibility of coastal trapped waves.

Let us assume that the pressure and wind-stress variations associated with large-scale weather systems have a horizontal scale L such that $L \gg l$. Thus ϕ and $\boldsymbol{\tau}$ are slowly varying functions of x for $0 \leqslant x \leqslant l$; hence over the shelf/slope region we may take $\phi = \phi(y, t)$ and $\boldsymbol{\tau} = \boldsymbol{\tau}(y, t)$ only. Under these conditions (55.3) can be written as

$$(H\eta^a_{tx})_x + H\eta^a_{tyy} + fH_x\eta^a_y - (\mathcal{L}/g)\eta^a_t = (\mathcal{L}/g)\phi_t + (1/\rho_* g)(\tau_{2\,yt} - f\tau_{1\,y}), \quad 0 \leqslant x \leqslant l. \tag{55.8}$$

We now introduce the following normalized variables:

$$x' = x/l, \quad y' = y/L, \quad t' = \omega_0 t, \quad H' = H/H_0$$

$$\phi' = \phi/\phi_0, \quad \boldsymbol{\tau}' = \boldsymbol{\tau}/\tau_0, \quad \eta^{a\prime} = \eta^a/\phi_0. \tag{55.9}$$

Substitution of (55.9) into (55.8) yields, after dropping the primes,

$$(H\eta^a_{tx})_x + KH_x\eta^a_y + \epsilon_1 H\eta^a_{tyy} - \epsilon_2(\epsilon_3\partial_{tt} + 1)\eta^a_t$$

$$= \epsilon_2(\epsilon_3\partial_{tt} + 1)\phi_t + K\epsilon_4(\epsilon_3^{1/2}\tau_{2yt} - \tau_{1y}), \quad 0 \leqslant x \leqslant 1, \tag{55.10}$$

where $K = fl/\omega_0 L, \quad \epsilon_1 = l^2/L^2, \quad \epsilon_2 = f^2l^2/gH_0,$

$\epsilon_3 = \omega_0^2/f^2 \quad$ and $\quad \epsilon_4 = l\tau_0/\rho_* g\phi_0 H_0.$

For low-frequency, long shelf waves on mid-latitude continental shelves, $l \sim 10^5$ m, $L \sim 10^6$ m, $\omega_0 \sim 10^{-5}$ rad s^{-2} and $H_0 \sim 2 \cdot 10^2$ m; further, for typical weather systems, $\phi_0 \sim 5 \cdot 10^{-2}$ m (eg., see Fig. 25.12) and $\tau_0 \sim 1$ dyn/cm^2 (10^{-1} N/m^2). For these characteristic values it follows that $K = 0(1)$, $\epsilon_1 \sim \epsilon_2 \sim \epsilon_3 = 0(10^{-2})$ and $\epsilon_4 = 0(10^{-1})$. The quantity ϵ_2 is the familiar horizontal divergence parameter that arises in planetary wave theory, and for topographic planetary waves the criterion $\epsilon_2 \ll 1$ [see (20.12)] expresses the fact that the waves are nearly nondivergent. Hence the term η^a_t in (55.2) can be neglected for shelf waves. Examination of the right-hand side of (55.10) reveals that the pressure forcing term is an order of magnitude smaller than the wind-stress terms, which explains why the pressure fluctuations alone are likely to excite only isostatic variations in the sea level: if $\boldsymbol{\tau} \equiv 0$, (55.10) implies that

$$\eta^a = 0(\epsilon_2\phi).$$

Only a highly tuned resonant response of the adjusted sea level to the pressure will produce shelf waves for $\boldsymbol{\tau} = 0$. However, even the wind-stress terms in (55.10) are small, which indicates that shelf waves are not efficiently generated by so-called Ekman pumping due to divergences in the wind-mixed layer. Indeed, we see that when each ϵ_i is set equal to zero in (55.10), we obtain an unforced equation for η^a:

$$(H\eta^a_{tx})_x + KH_x\eta^a_y = 0. \tag{55.11}$$

Equation (55.11), being derived by taking a time derivative of the forced momentum equations, shows that shelf waves are not efficiently generated by momentum inputs. To see how the waves can be generated efficiently we have to turn to the boundary condition at $x = 0$, where $u = 0$. Solving (55.1) for $\mathcal{L}u$ and using (55.7), we can write this condition as

$$\eta^a_{xt} + f\eta^a_y = (1/\rho_* gH)(\tau_{1t} + f\tau_2) \quad \text{at} \quad x = 0. \tag{55.12a}$$

In terms of the nondimensional variables (55.12a) takes the form

$$\sqrt{\epsilon_3}\eta^a_{xt} + \sqrt{\epsilon_1}\eta^a_y = (\epsilon_4/H)(\sqrt{\epsilon_3}\tau_{1t} + \tau_2) \quad \text{at} \quad x = 0. \tag{55.12b}$$

Since $\sqrt{\epsilon_3} \sim \sqrt{\epsilon_1} \sim \epsilon_4$, it follows from (55.12b) that the τ_{1t} term is a lower-order forcing term and that the longshore component of the wind stress evaluated at the coast is the primary forcing mechanism for low-frequency shelf waves.

In physical terms, we can describe the situation as follows (Gill and Schumann, 1974): if there were no coast (and viscosity and density variations were taken into account), the wind would only produce a horizontal Ekman flux in a thin layer near the surface. However, with a coast present, this flux is blocked (if $\tau_2 < 0$), and because the motion is effectively nondivergent ($\epsilon_2 \ll 1$), the only way mass can be conserved is by a return offshore flow in the deeper water. This flow moves across the rapidly changing shelf topography and therefore provides a significant vorticity input (by vortex stretching) into the fluid. When $\tau_2 > 0$, the flows are reversed. Therefore, a time periodic wind stress, say, can be seen as an effective way of exciting alternating vortices: a shelf wave.

To visualize these ideas more clearly, we now derive the forced vorticity equation for long, nondivergent shelf waves. With η_t neglected in (55.2), we can introduce the usual mass transport stream function ψ:

$$Hu = -\psi_y, \quad Hv = \psi_x. \tag{55.13}$$

Since the length scale for y is much greater than that chosen for x, (55.13) implies that $u \ll v$. Thus we can neglect u_t in (55.1a). Dropping this term, cross-differentiating (55.1) and then using (55.13), we obtain

$$(\psi_x/H)_{xt} - f\psi_y(1/H)_x = -\tau_{1y}/\rho_* H - \tau_2 H_x/\rho_* H^2, \quad 0 \leqslant x \leqslant l. \tag{55.14}$$

The first term on the right-hand side of (55.14) represents the vorticity input due to the wind-stress curl and is what we referred to above as Ekman pumping. It is much smaller than the second term, which represents the vorticity input due to the longshore component of wind stress over the sloping shelf and which (for $\tau_2 < 0$) corresponds to a mass flux away from the coast that balances the Ekman flux towards the coast. Indeed, the ratio of the first to the second term is

$$\frac{\tau_0/\rho_* H_0 L}{\tau_0 \alpha/\rho_* H_0^2} = \frac{H_0}{\alpha L} \sim 0(4 \cdot 10^{-2})$$

for $H_0 \sim 200$ m, $L \sim 10^6$ m, $\alpha \sim 5 \cdot 10^{-3}$. Thus henceforth we shall drop the first term in (55.14) and consider the forced equation

$$(\psi_x/H)_{xt} + f(H_x/H^2)\psi_y = -\tau_2 H_x/\rho_* H^2, \quad 0 \leqslant x \leqslant l, \tag{55.15}$$

in which $H = H(x)$ and $\tau_2 = \tau_2(y, t)$.

The boundary conditions associated with (55.15) are

$$\psi = 0 \quad \text{at} \quad x = 0, \tag{55.16}$$

implying no mass flux through the coast, and

$$\psi_x = 0 \quad \text{at} \quad x = l. \tag{55.17}$$

Equation (55.17) implies that $v = 0$ at $x = l$, a condition derived by Buchwald and Adams

(1968) for long, unforced shelf waves on an exponential shelf. It is questionable whether this condition should be applied for all shelf profiles [it is certainly not the right boundary condition for the Robinson shelf model (25.30)] and also for the forced problem. Allen (1976a) has argued that in general (55.17) is not valid and that both the shelf and the deep-sea motions ought to be considered together as one coupled system.

Before discussing the solution of (55.15)–(55.17) we mention here that the results concerning the generation of long barotropic shelf waves carry over to the stratified case. However, with a vertical coast and stratification over the shelf, the "shelf wave" is now really a blend of an internal Kelvin wave trapped at the coast and a topographic planetary wave which, at short wavelengths, is bottom-trapped. For details we refer the reader to Allen (1976b), for the case of a two-layer fluid, and to Wang (1976) and Clarke (1977b) for the case of continuous stratification. These authors show that the main generating mechanism is the longshore component of wind stress which induces vortex inputs across the sloping shelf. Clarke (1977b) also gives observational evidence of long coastal trapped waves in various regions.

The Gill-Schumann solution

Gill and Schumann (1974) showed that the simplest way to solve (55.15)–(55.17) is via eigenfunction expansions in terms of the shelf wave modes obtained from the unforced problem. When $\tau_2 = 0$, (55.15)–(55.17) can be solved by separation of variables to yield the eigenfunctions

$$\psi_n = \Phi_n(y, t)F_n(x), \tag{55.18}$$

where $$\Phi_{nt} - c_n\Phi_{ny} = 0, \tag{55.19}$$

and $$\frac{\mathrm{d}}{\mathrm{d}x}\left(\frac{1}{H}\frac{\mathrm{d}F_n}{\mathrm{d}x}\right) + \frac{1}{c_n}\frac{f}{H^2}\frac{\mathrm{d}H}{\mathrm{d}x}F_n = 0, \qquad 0 \leqslant x \leqslant l, \tag{55.20}$$

with $F_n = 0$ at $x = 0$,

$$\frac{\mathrm{d}F_n}{\mathrm{d}x} = 0 \quad \text{at} \quad x = l. \tag{55.21}$$

Equation (55.19) is a first-order wave equation having the general solution

$$\Phi_n(y, t) = \Phi_n(y + c_n t),$$

which for $c_n > 0 (< 0)$ represents a nondispersive wave travelling in the negative (positive) y-direction with speed (c_n), the speed of the nth mode shelf wave. The speeds c_n appear as the eigenvalues in the Sturm-Liouville problem (55.20), (55.21), which is readily solved once the topography $H(x)$ is specified. In particular, we note that for $f\mathrm{d}H/\mathrm{d}x > 0, c_n > 0$. When suitably normalized, the eigenfunctions satisfy the orthogonality relation

$$\int_0^l \frac{H_x}{H^2} \cdot F_n(x)F_m(x)\,\mathrm{d}x = \delta_{nm}, \tag{55.22}$$

where δ_{nm} is the usual Kronecker delta symbol.

To solve the forced problem we put

$$\psi = \sum_n \phi_n(y, t)F_n(x), \tag{55.23}$$

so that the boundary conditions (55.16), (55.17) are satisfied, and

$$\tau_1(y, t) = \tau_1(y, t) \sum_n a_n F_n(x). \tag{55.24}$$

The Fourier coefficients a_n are determined so that

$$\sum_n a_n F_n(x) = 1, \quad 0 \leqslant x \leqslant l, \tag{55.25}$$

since $\tau_{1x} = 0$. In view of (55.22), it follows that

$$a_n = \int_0^l \frac{1}{H^2} \frac{dH}{dx} F_n(x)\, dx. \tag{55.26}$$

Substituting (55.23), (55.24) into (55.15), invoking (55.20), and then multiplying by $F_m(x)$ and integrating with respect to x over $(0, l)$, we finally obtain the forced first-order wave equation

$$(1/c_n)\phi_{nt} - \phi_{ny} = \tau_2(y, t)b_n/\rho_* f. \tag{55.27}$$

Equation (55.27) can be solved by the method of characteristics (see Exercise 55.4). The complete solution is then a sum over all the shelf wave modes. The (adjusted) sea-level changes are obtained from (55.19) with u_t and τ_1 neglected:

$$-fv + g\eta_x^a = 0,$$

or, using (55.13) and integrating,

$$\eta^a(x, y, t) - \eta^a(L, y, t) = \frac{f}{g} \sum_n \phi_n(y, t) \int_L^x \frac{1}{H} \frac{dF_n}{dx}\, dx. \tag{55.28}$$

Equation (55.28) could be used in practice to determine the sea-level difference across the shelf at different locations along the coast, as a function of the meteorological forcing. This is effectively the formula used by Hamon (1976) to study the wind-stress generation of shelf waves on the East Australian coast.

One of the more interesting solutions of (55.27) is that corresponding to a moving wind stress of the form

$$\tau_2(y, t) = \tau_0 \cos(ky + \omega t). \tag{55.29}$$

If τ_2 as given by (55.29) is applied at some initial time $t = 0$, then the integration of (55.27) yields the following response:

$$\phi_n = A_n[\sin(ky + \omega t) - \sin k(y + c_n t)], \tag{55.30}$$

where $\quad A_n = -\tau_0 b_n/\rho_* f(k - \omega/c_n).$

Thus the response consists of a forced disturbance moving with the wind at frequency ω

and a shelf wave with frequency kc_n, which is in general different from ω. If $\omega = kc_n$, resonance occurs and the solution is $\propto t$. If $\omega/k < 0$, the disturbance moving with the wind propagates in the direction opposite to that of the shelf wave.

This solution qualitatively explains an interesting result observed by Mooers and Smith (1968). In their analysis of daily mean sea level and atmospheric pressure fluctuations of three Oregon coast stations (see Fig. 55.1), they observed significant peaks at frequencies of 0.1 and 0.3 cpd in summer (Fig. 55.2). The barometric factor, on the other hand, showed broad peaks at 0.1 and 0.35 cpd (Fig. 55.3). To Mooers and Smith these results suggested the possibility of some interesting dynamical events in the neighbourhood of these frequencies. To see if travelling waves were present, coherence and phase of the adjusted sea levels between Brookings and the two more northerly stations were computed and are shown in Fig. 55.4. The interesting result to be noted here is that the "wave" of 0.1 cpd travels southward (i.e., is left-bounded), whereas the "wave" at around 0.35 cpd travels northward, in the correct direction for a shelf wave. Mooers and Smith also found that the (adjusted) sea level at 0.1 cpd is strongly coupled with the geostrophic winds at North Bend, which were also observed to move southward. These observations are thus in accordance with (55.30): a travelling wind-stress disturbance produces a forced wave moving with the disturbance at the same frequency and also a shelf wave which moves in the opposite direction with a different frequency!

Another success of the Gill-Schumann theory has recently been reported by Hamon (1976). Gill and Schumann give an explicit formula whereby one can calculate the amplitude of a shelf wave, given observations of the longshore component of the wind stress at points along the coast in the direction from which the waves are approaching. Hamon applied this formula to observations on the southeast coast of Australia, and his results strongly support the contention that in the frequency range 0.04–0.25 cpd, shelf waves are generated by the longshore component of the wind stress.

Observations of wind-stress generated continental shelf waves

The above successes of the wind-generation theory of shelf waves has stimulated many recent observational studies of shelf waves using current measurements as well as sea level, pressure and wind data. To many physical oceanographers this represents a very exciting topic of research under the broad subject of mesoscale coastal dynamics.

Huyer et al. (1975) analyzed the low-frequency variations in current measurements over the Oregon–Washington shelf. The currents were found to be highly coherent over an alongshore separation of 200 km, with strong signals at 0.16, 0.3 and 0.44 cpd. The signal at 0.16 cpd is highly coherent with the wind and sea level and appears to represent a forced shelf wave generated by the local winds. The signal at 0.3 cpd on the other hand, has high coherence between current and sea level and probably represents a shelf wave generated by distant wind fields. A more extensive study of current oscillations off the Oregon coast was done by Kundu et al. (1975). They carried out a vertical mode decomposition of the data and found that most of the energy is contained in the barotropic mode. Further, both velocity components were highly correlated with the longshore wind-stress component at frequencies of 0.06 and 0.14 cpd. This result, together with observed phase relations, is consistent with a resonant condition between the wind stress and forced long shelf waves.

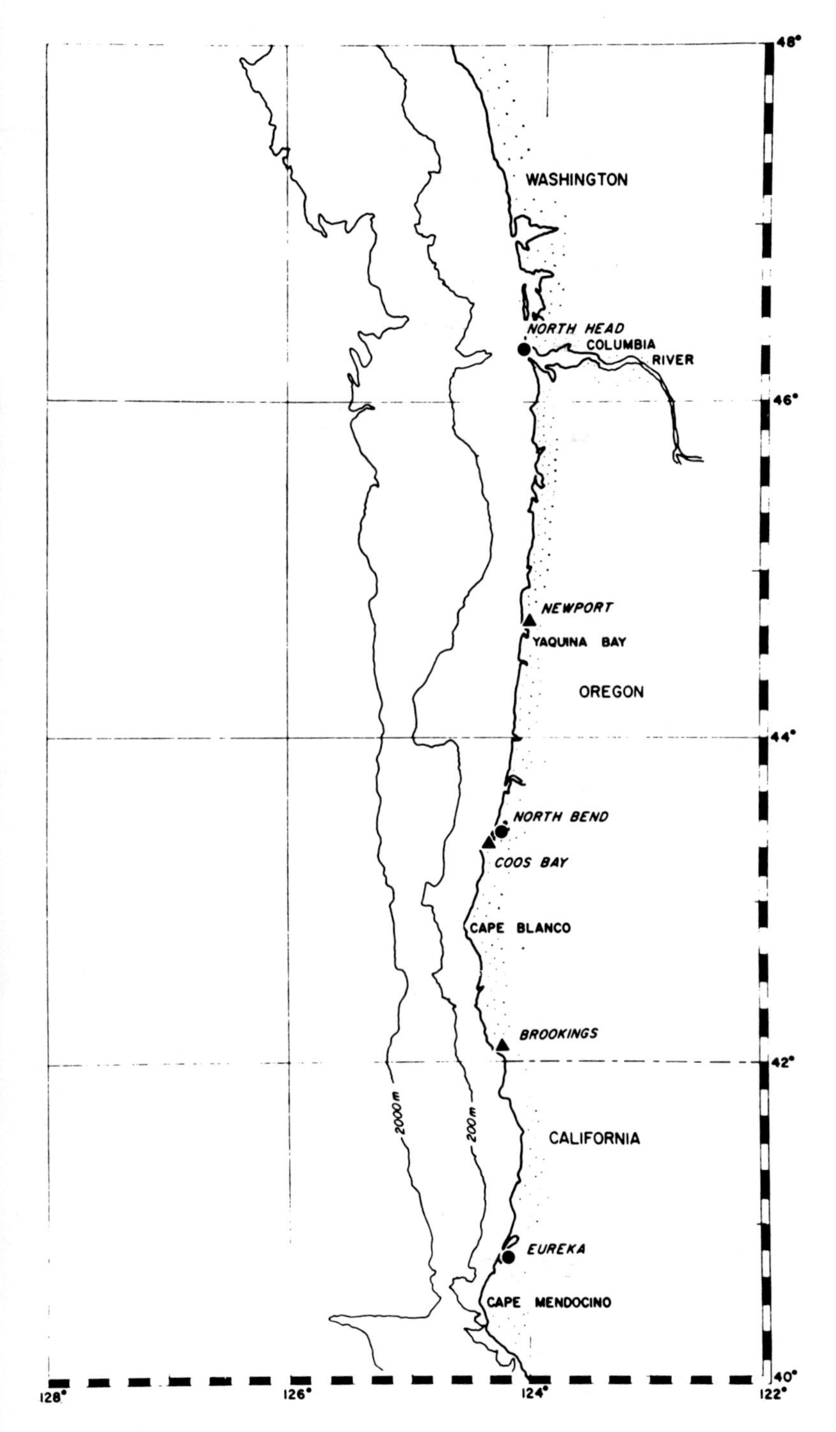

Fig. 55.1. Map of Oregon coast, showing the tide-guage stations (triangles) and weather stations (circles) used in the analysis by Mooers and Smith (1968). (Copyrighted by American Geophysical Union.)

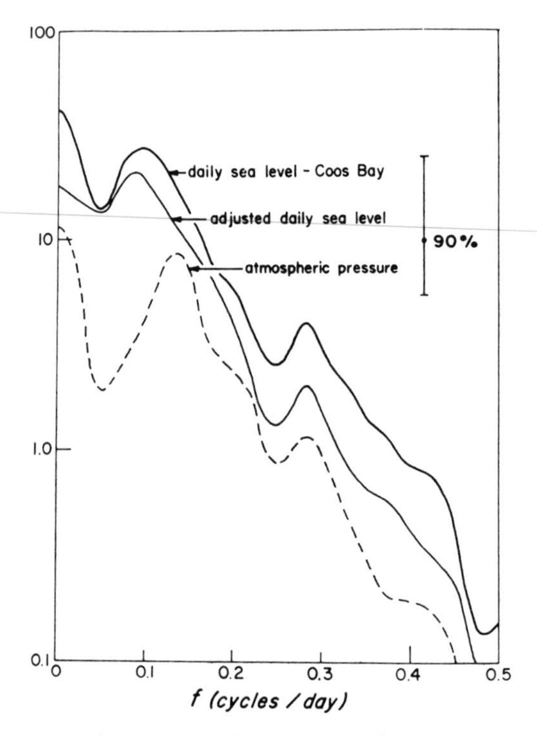

Fig. 55.2. Summer (1933) spectra (cm^2 or mb^2/cpd) of sea level and pressure of the Coos Bay–North Bend stations. The sea-level spectra at Newport and Brookings are similar to those shown here. (From Mooers and Smith, 1968.) (Copyrighted by American Geophysical Union.)

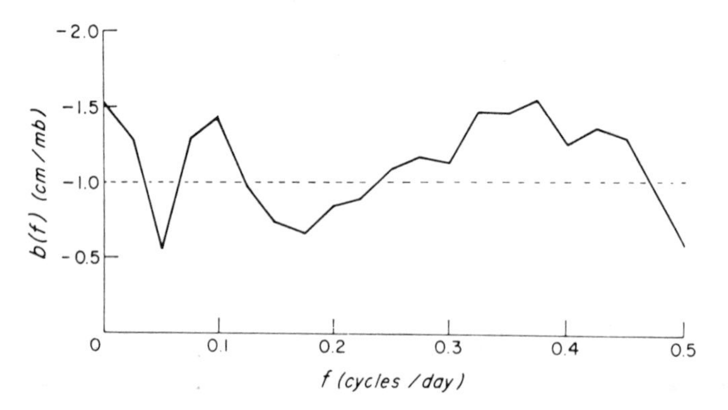

Fig. 55.3. The barometric factor $b(f)$ versus frequency for Coos Bay–North Bend stations. $b(f)$ is defined as the ratio of the sea level–atmospheric pressure co-spectrum to the auto-spectrum of the atmospheric pressure. For a sea-level response which is barometric at all frequencies, $b = -1$. (From Mooers and Smith, 1968.) (Copyrighted by American Geophysical Union.)

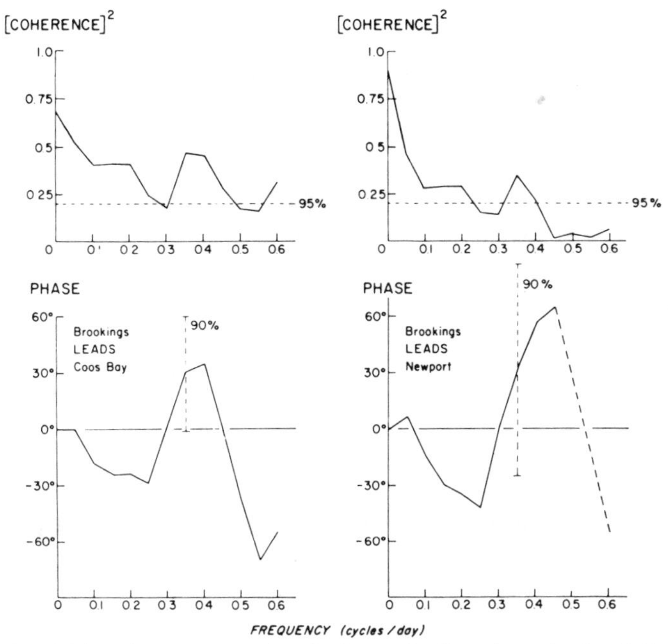

Fig. 55.4. Coherence-squared and phase of adjusted sea levels of Coos Bay versus Brookings and Newport versus Brookings. (From Mooers and Smith, 1968.) (Copyrighted by American Geophysical Union.)

Continental shelf waves have also recently been detected in the Florida Straits (Schott and Düing, 1976; Brooks and Mooers, 1977; Düing et al., 1977). These shelf waves appear to be responsible for much of the observed low-frequency variability in the Florida Current, and accordingly, the usual shelf wave theory discussed in this section has to be modified to include a basic flow that models the Florida current. Nevertheless, the modified theories predict stable, southward-propagating shelf waves, and these waves are indeed observed, with periods of around 10 days and wavelengths of 200 km. However, it appears that both the long-channel and cross-channel components of the wind stress play an important role in their generation.

Exercises Section 55

1. Estimate the relative importance of the forcing terms in (55.3) for the case of Kelvin wave generation off the Oregon coast. Take $H = 2000$ m and let $\delta p_a = 5$–50 mbar. Also, use the data in Fig. 55.5 for the wind stress. Consider both subinertial and super-inertial frequencies.

2. Give a heuristic argument based on scaling that supports the boundary condition (55.17) that $\psi_x = 0$ at $x = l$. (Hint: Consider also the motion in the deep-sea region where both x and y are scaled by $L \gg l$. See Allen, 1976a, for further details.)

3. Derive (55.27).

4. Use the method of characteristics to find an integral representation of the solution to (55.27).

5. Estimate the speed of propagation and wavelength of the shelf wave observed at

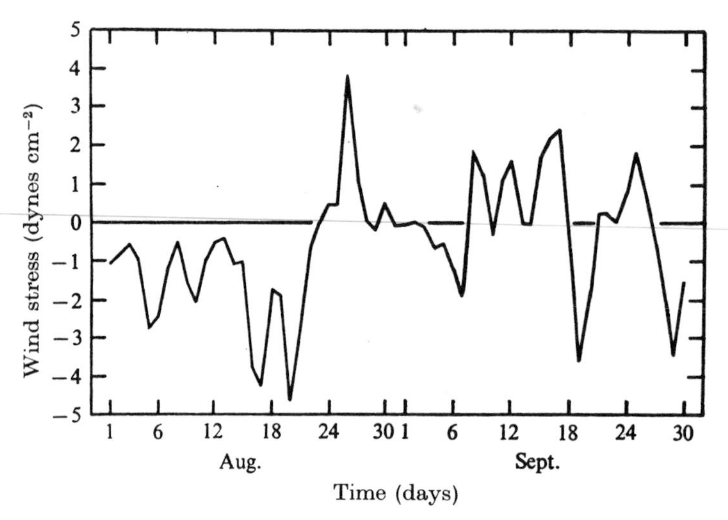

Fig. 55.5. Daily northward wind stress off the Oregon coast from August 1 to September 31, 1966. (From Thomson, 1970.)

0.35 cpd in the Oregon coast records shown in Figs. 55.2–55.4. How does your answer agree with theoretical speed(s) for a flat-shelf model and an exponential shelf model (see Section 25)? [It has recently been shown by Wright and Mysak (1977) that in a flat-shelf model with a two-layer stratification extending over the shelf, the theoretical phase speed differs from that of a purely barotropic (homogeneous fluid) model by at most a few percent.]

REFERENCES

Abramov. A.A., Tareyev, B.A. and Ul'yanova, V.I., 1972. Instability of a two-layer geostrophic current with asymmetric velocity profile in the top layer. Izv. Akad. Nauk S.S.S.R., Fiz., Atmos. Okeana, 8: 131–141 (in Russian).

Abramowitz, M. and Stegun, I.A. (Editors), 1965. Handbook of Mathematical Functions. Dover, New York, 1046 pp.

Abrashina, N.N., 1966. Long waves generated in an inhomogeneous liquid by a periodically moving pressure system. Izv. Akad. Nauk S.S.S.R., Fiz. Atmos. Okeana, 2: 191–197 (in Russian).

Adams, J.K. and Buchwald, V.T., 1969. The generation of continental shelf waves. J. Fluid Mech., 35: 815–826.

Aida, I., 1967. Water level oscillations on the continental shelf in the vicinity of Miyagi-Enoshima. Bull. Earthquake Res. Inst., 45: 61–78.

Aida, I., Matori, T., Koyama, M. and Kajiura, K., 1968. A model experiment of long-period waves travelling along a continental shelf. Bull. Earthquake Res. Inst., 46: 707–739 (in Japanese).

Allen, J.S., 1975. Coastal trapped waves in a stratified ocean. J. Phys. Oceanogr., 5: 300–325.

Allen, J.S., 1976a. On forced, long continental shelf waves on an f-plane. J. Phys. Oceanogr., 6: 426–431.

Allen, J.S., 1976b. Some aspects of the forced wave response of stratified coastal regions. J. Phys. Oceanogr., 6: 113–119.

Anderson, D.L.T. and Gill, A.E., 1975. Spin-up of a stratified ocean, with applications to upwelling. Deep-Sea Res., 22: 583–596.

Anderson, D.L.T. and Rowlands, P.B., 1976a. The role of inertia–gravity and planetary waves in the response of the tropical ocean to the incidence of an equatorial Kelvin wave on a meridional boundary. J. Mar. Res., 34: 295–312.

Anderson, D.L.T. and Rowlands, P.B., 1976b. The Somali current response of the southwest monsoon: the relative importance of local and remote forcing. J. Mar. Res., 34: 395–417.

Apel, J.R., Byrne, H.M., Proni, J.R. and Charnell, R.L., 1975. Observations of oceanic internal and surface waves from the Earth Resources Technology Satellite. J. Geophys. Res., 80: 865–881.

Apel, J.R., Byrne, H.M. Proni, J.R. and Sellers, R., 1976. A study of oceanic internal waves using satellite imagery and ship data. Remote Sensing Environ., 5: 125–135.

Arnold, L., 1974. Stochastic Differential Equations: Theory and Applications. Wiley-Interscience, New York, 228 pp.

Atlas, D., Metcalf, J.I., Richter, J.H. and Gossard, E.E., 1970. The birth of 'CAT' and microscale turbulence. J. Atmos. Sci., 27: 903–913.

Badgley, F.I., Paulson, C.A. and Miyake, M., 1972. Profiles of wind, temperature and humidity over the Arabian Sea. Int. Indian Ocean Exped. Meteorol. Monographs, No. 6, Univ. Press, Hawaii, 62 pp.

Baer, R.N. and Jacobson, M.T., 1974. Analysis of effect of a Rossby wave on sound speed in ocean. J. Acoust. Soc. Am., 55: 1178–1189.

Baines, P.G., 1973. The generation of internal tides by flat-bump topography. Deep-Sea Res., 20: 179–205.

Baines, P.G., 1974. The generation of internal tides over steep continental slopes. Philos. Trans. R. Soc. Lond., A, 277: 27–58.

Baines, P.G., 1976. The stability of planetary waves on a sphere. J. Fluid Mech., 73: 193–213.

Ball, F.K., 1964. Energy transfer between external and internal waves. J. Fluid Mech., 19: 465–480.

Ball, F.K., 1965. Second-class motions of a shallow liquid. J. Fluid Mech., 23: 545–561.

Ball, F.K., 1967. Edge waves in an ocean of finite depth. Deep-Sea Res., 14: 79–88.

Banner, M.L. and Melvillle, W.K., 1976. On the separation of air flow over water waves. J. Fluid Mech., 77: 825–842.

Banner, M.L. and Phillips, O.M., 1974. On small-scale breaking waves. J. Fluid Mech., 65: 647–657.

Barbee, W.B., Dworski, J.G., Irish, J.D., Larsen, L.H. and Rattray, M., Jr., 1975. Measurement of internal waves of tidal frequency near a continental boundary. J. Geophys. Res., 80: 1965–1974.

Barber, N.F. and Ursell, F., 1948. The generation and propagation of ocean waves and swell, 1. Philos. Trans. R. Soc. Lond., Ser. A, 240: 527–560.

Barnard, B.J.S. and Pritchard, W.G., 1972. Cross-waves. Part 2. Experiments. J. Fluid Mech., 55: 245–255.

Barnett, T.P. and Kenyon, K.E., 1975. Recent advances in the study of wind waves. Rep. Prog. Phys., 38: 667–729.

Barnett, T.P. and Wilkerson, J.C. 1967. On the generation of ocean wind waves as inferred from airborne radar measurements of fetch-limited spectra. J. Mar. Res., 25: 292–328.
Batchelor, G.K., 1967. An Introduction to Fluid Dynamics. Cambridge University Press, 615 pp.
Battjes, J.A., 1972. Radiation stresses in short-crested waves. J. Mar. Res., 30: 56–64.
Beardsley, R.C., 1969. A laboratory model of the wind-driven ocean circulation. J. Fluid Mech., 38: 255–271.
Beckerle, J.C. and Delnore, V., 1973. Interference of Rossby waves by reflection from Bahama Banks and Blake–Bahama Outer Ridge. J. Geophys. Res., 78: 6316–6324.
Beckman, R. and Spizzichino, A., 1963. The Scattering of Electro-Magnetic Waves from Rough Surfaces. Pergamon Press, New York, 503 pp.
Bell, T.H., Jr., 1975. Topographically generated internal waves in the open ocean. J. Geophys. Res., 80: 320–337.
Benjamin, T.B., 1959. Shearing flow over a wavy boundary. J. Fluid Mech., 6: 161–205.
Benjamin, T.B., 1966. Internal waves of finite amplitude and permanent form. J. Fluid Mech., 25: 241–270.
Benjamin, T.B., 1967a. Internal waves of permanent form in fluids of great depth. J. Fluid Mech., 29: 559–592.
Benjamin, T.B., 1967b. Instability of periodic wave trains in non-linear dispersive systems. Proc. R. Soc. Lond., A, 299: 59–75.
Benjamin, T.B. and Feir, J.E., 1967. The disintegration of wave trains on deep water. Part 1, Theory. J. Fluid Mech., 27: 417–430.
Benjamin, T.B. and Lighthill, M.J., 1954. On cnoidal waves and bores. Proc. R. Soc. Lond., A, 221: 448–460.
Bennett, A.F., 1975. Tides in the Bristol Channel. Geophys. J. R. Astron. Soc., 40: 37–43.
Bennett, A.F., 1976. Open boundary conditions for dispersive waves. J. Atmos. Sci., 33: 176–182.
Bennett, J.R., 1973. A theory of large-amplitude Kelvin waves. J. Phys. Oceanogr., 3: 57–60.
Benney, D.J., 1962. Nonlinear gravity wave interactions. J. Fluid Mech., 14: 557–584.
Bergsten, F., 1926. The seiches of Lake Vättern. Geogr. Ann., Stockh. 8: nos. 1, 2.
Bernstein, R.L. and White, W.B., 1974. Time and length scales of baroclinic eddies in the Central North Pacific Ocean. J. Phys. Oceanogr., 4: 613–624.
Betchov, R. and Criminale, W.O., Jr., 1967. Stability of Parallel Flows. Academic Press, New York, 330 pp.
Biesel, F., 1951. Study of wave propagation in water of gradually varying depth. In: Gravity waves. U.S. Natl. Bureau of Standards Circular 521, pp. 243–253.
Birkhoff, G., Bona, J. and Kampé de Fériet, J., 1973. Statistically well-set Cauchy problems. In: A.T. Bharucha-Reid (Editor), Probabilistic Methods in Applied Mathematics. Academic Press, New York, 3: 1–120.
Bjerknes, J. and Holmboe, J., 1944. On the theory of cyclones. J Meteorol., 1: 1–22.
Bjerknes, V., Bjerknes, J., Solberg, H. and Bergeron, H., 1933. Physikalische Hydrodynamik. Springer, Berlin, 797 pp.
Blackman, R.B. and Tukey, J.W., 1958. The Measurement of Power Spectra. Dover, New York, 190 pp.
Blandford, R., 1966. Mixed gravity–Rossby waves in the ocean. Deep-Sea Res., 13: 941–961.
Blumen, W., 1971. Hydrostatic neutral waves in a parallel shear flow of a stratified fluid. J. Atmos. Sci., 28: 340–344.
Blumen, W., 1972. Geostrophic adjustment. Rev. Geophys. Space Phys., 10: 485–528.
Blumen, W., 1973. Stability of two-layer fluid model to nongeostrophic disturbances. Tellus, 25: 12–19.
Blumen, W., 1975. Stability of non-planar shear flow of a stratified fluid. J. Fluid Mech., 68: 177–189.
Blumsack, S.L. and Gierasch, P.G., 1972. Mars: the effects of topography on baroclinic instability. J. Atmos. Sci., 29: 1081–1089.
Bogdanov, K.T. and Magarik, V., 1967. Numerical solutions to the problem of distribution of semi-diurnal tides M_2 and S_2 in the world ocean. Dokl. Akad. Nauk. S.S.S.R., 172: 1315–1317 (in Russian).
Bogdanov, K.T. and Sebekin, B.I., 1976. Generation of internal tidal waves and influence of Earth's tides on tidal motions in the ocean. Izv. Akad. Nauk. S.S.S.R., Fiz. Atmos. Okeana, 12: 539–545 (in Russian).
Bogoliubov, N.N. and Mitropolsky, Y.A., 1961. Asymptotic Methods in the Theory of Nonlinear Oscillations. Delhi, Hindustan.
Booker, J.R. and Bretherton, F.P., 1967. The critical layer for internal gravity waves in a shear flow. J. Fluid Mech., 27: 513–539.
Borisenko, Yu. D. and Miropolsky, Yu. Z., 1974. On the influence of non-linearities on statistical distributions of internal waves in the ocean. Okeanologiya, 14: 788–796 (in Russian).
Bourret, R.C., 1962. Stochastically perturbed fields, with application to wave propagation in random media. Nuovo Cimento, 26: 1–31.
Bourret, R.C., 1965. Ficton theory of dynamical systems with noisy parameters. Can. J. Phys., 43: 619–639.

Boussinesq, J., 1877. Mém. Pres. Acad. Sci., 3rd edition, Paris, XXIII, p. 46.
Boussinesq, J., 1903. Théorie analytique de la chaleur, Vol. 2. Gauthier-Villars, Paris.
Bowen, A.J., 1969a. The generation of longshore currents on a plane beach. J. Mar. Res., 27: 206–215.
Bowen, A.J. 1969b. Rip currents. 1. Theoretical investigations. J. Geophys. Res., 74: 5467–5478.
Bowen, A.J., 1976. Reflection and scattering of long waves by bottom topography. Can. Meteorol. Soc., 10th Ann. Congr., Québec, 1976 (Abstract).
Bowen, A.J. and Inman, D.L., 1969. Rip currents. 2. Laboratory and field observations. J. Geophys. Res., 74: 5479–5490.
Bowen, A.J. and Inman, D.L., 1971. Edge waves and crescentic bars. J. Geophys. Res., 76: 8662–8671.
Bowen, A.J., Inman, D.L. and Simmons, V.P., 1968. Wave 'set-down' and set-up. J. Geophys. Res., 73: 2569–2577.
Boyce, W.E., 1968. Random eigenvalue problems. In: A.T. Bharucha-Reid (Editor), Probabilistic Methods in Applied Mathematics. Academic Press, New York, 1: 1–73.
Boyce, W.E. and DiPrima, R.C., 1969. Elementary Differential Equations and Boundary Value Problems. Wiley, New York, 2nd Edit., 533 pp.
Boys, C.V. 1959. Soap Bubbles. Dover, New York, 156 pp.
Breeding, R.J., 1971. A non-linear investigation of critical levels for internal atmospheric gravity waves. J. Fluid Mech., 50: 545–563.
Brekhovskikh, L.M., 1960. Waves in Layered Media. Academic Press, New York, 561 pp.
Brekhovskikh, L.M., Fedorov, K.N., Fomin, L.M., Koshlyakov, M.N. and Yampolsky, A.D., 1971. Large-scale multi-buoy experiment in the tropical Atlantic. Deep-Sea Res., 18: 1189–1206.
Brekhovskikh, L.M., Goncharov, V.V. Kurtepov, V.M. and Naugol'nykh, K.A., 1972. Resonant excitation of internal waves by non-linear interaction of surface waves. Izv. Akad. Nauk. S.S.S.R., Fiz. Atmos. Okeana, 8: 192–203 (in Russian).
Brekhovskikh, L.M., Konyaev, K.V., Sabinin, K.D. and Serikov, A.N., 1975. Short-period internal waves in the sea. J. Geophys. Res., 80: 856–864.
Bretherton, F.P., 1964a. Resonant interactions between waves. J. Fluid Mech., 20: 457–480.
Bretherton, F.P., 1964b. Low-frequency oscillations trapped near the equator. Tellus, 16: 181–185.
Bretherton, F.P. 1966. The propagation of groups of internal gravity waves in a shear flow. Q.J.R. Meteorol. Soc., 92: 466–480.
Bretherton, F.P., 1967. The time dependent motion due to a cylinder moving in an unbounded rotating or stratified fluid. J. Fluid Mech., 28: 545–570.
Bretherton, F.P., 1969. On the mean motion induced by internal gravity waves. J. Fluid Mech., 36: 785–803.
Bretherton, F.P., 1971. The general linearized theory of wave propagation. In: W.H.Reid (Editor) Mathematical Problems in the Geophysical Sciences. Am. Math. Soc., Providence, R.I., 13: 61–102.
Bretherton, F.P., 1975. Recent developments in dynamical oceanography. Q.J.R. Meteorol. Soc., 101: 705–721.
Bretherton, F.P. and Garrett, C.J.R., 1968. Wavetrains in inhomogeneous moving media. Proc. R. Soc. Lond., A, 302: 529–554.
Bretschneider, C.L., 1959. Wave variability and wave spectra for wind generated gravity waves. Beach Erosion Board, U.S. Army Corps of Engineers, Tech. Mem. No. 118, 192 pp.
Bretschneider, C.L., 1970. Forecasting relations for wave generation. Univ. of Hawaii, Dept. Ocean Eng., JKK Look Lab. Quart., 1(3): 31–34.
Brillouin, L., 1966. Wave Propagation in Periodic Structures. Dover, New York, 255 pp.
Briscoe, M.G., 1975a. Introduction to collection of papers on oceanic internal waves. J. Geophys. Res., 80: 289–290.
Briscoe, M.G., 1975b. Internal waves in the ocean. Rev. Geophys. Space Phys., 13: 591–598.
Brocks, K., 1959. Ein neues Gerät für störungsfreie meteorologische Messungen auf dem Meer. Arch. Met. Geophys. Biokl., A, 11: 227–239.
Broer, L.J.F., 1964. On the interaction of non-linearity and dispersion in wave propagation, I. Boussinesq's equation. Appl. Sci. Res., B, 11: 273–285.
Brooks, D.A. and Mooers, C.N.K., 1977. Wind-forced continental shelf waves in the Florida current. J. Geophys. Res., 82: 2469–2576.
Brosche, P. and Sündermann, J., 1971. Die Gezeiten des Meeres und die Rotation der Erde. Pure Appl. Geophys., 86: 95–117.
Brown, P.J., 1973. Kelvin-wave reflection in a semi-infinite canal. J. Mar. Res., 31: 1–10.
Brown, R.C., 1974. The surface tension of liquids. Contemp. Phys., 15: 301–327.
Bryan, K. and Cox, M.D., 1972. An approximate equation of state for numerical models of ocean circulation. J. Phys. Oceanogr., 2: 510–514.
Buchwald, V.T., 1968a. Long waves on oceanic ridges. Proc. R. Soc. Lond., A, 308: 343–354.
Buchwald, V.T., 1968b. The diffraction of Kelvin waves at a corner. J. Fluid Mech., 31: 193–205.
Buchwald, V.T., 1971. The diffraction of tides by a narrow channel. J. Fluid Mech., 46: 501–511.

Buchwald, V.T., 1972. Energy and energy flux in planetary waves. Proc. R. Soc. Lond., A, 328: 37–48.
Buchwald, V.T., 1973a. Long-period divergent planetary waves. Geophys. Fluid Dyn., 5: 359–367.
Buchwald, V.T., 1973b. On divergent shelf waves. J. Mar. Res., 31: 105–115.
Buchwald, V.T., 1977. Diffraction of shelf waves by an irregular coastline. In: D. G. Provis and R. Radok (Editors), Lecture Notes in Physics. Australian Acad. Sci., Canberra, and Springer Verlag, Berlin, 64: 188–193.
Buchwald, V.T. and Adams, J.K., 1968. The propagation of continental shelf waves. Proc. R. Soc. Lond., A, 305: 235–250.
Buchwald, V.T. and Miles, J.W., 1974. Kelvin wave diffraction by a gap. J. Aust. Math. Soc., 17: 29–34.
Buchwald, V.T. and de Szoeke, R.A., 1973. The response of a continental shelf to travelling pressure disturbances. Aust. J. Mar. Freshwater. Res., 24: 143–158.
Buchwald, V.T. and Williams, N.V., 1975. Rectangular resonators on infinite and semi-infinite channels. J. Fluid Mech., 67: 497–511.
Bukatov, A., 1971. On internal waves in a continuously stratified ocean. Morsk. Gidrofiz. Issled., 56: 26–36 (in Russian).
Burger, A.P., 1962. On the non-existence of critical wavelengths in a continuous baroclinic stability problem. J. Atmos. Sci., 19: 31–38.
Burling, R.W., 1959. The spectrum of waves at short fetches. Dtsch. Hydrogr. Z., 12: 45–117.
Cabannes, H., 1970. Theoretical Magneto-Fluid Dynamics. Academic Press, New York, 233 pp.
Cacchione, D. and Wunsch, C., 1974. Experimental study of internal waves over a slope. J. Fluid Mech., 66: 233–239.
Cairns, J.L., 1967. Asymmetry of internal tidal waves in shallow coastal waters. J. Geophys. Res., 72: 3563–3565.
Caldwell, D.R. and Eide, S.A., 1976. Experiments on the resonance of long-period waves near islands. Proc. R. Soc. Lond., A, 348: 359–378.
Caldwell, D.R. and Longuet-Higgins, M.S., 1971. The experimental generation of double Kelvin waves. Proc. R. Soc. Lond., A, 326: 39–52.
Caldwell, D.R., Cutchin, D.L. and Longuet-Higgins, M.S., 1972. Some model experiments on continental shelf waves. J. Mar. Res., 30: 39–55.
Campbell, G.A. and Foster, R.M., 1942. Fourier Integrals for Practical Applications. Van Nostrand, New York, 177p.
Carrier, G.F., 1971. The dynamics of tsunamis. In: W.H. Reid (Editor), Mathematical Problems In the Geophysical Sciences. Am. Math. Soc. Providence, R. I., 13: 157–187.
Carrier, G.F., Krook, M. and Pearson, C.E., 1966. Functions of a Complex Variable. McGraw-Hill, New York, 438 pp.
Cartwright, D.E., 1962. Analysis and statistics. In: M.N. Hill (General Editor), The Sea. Wiley-Interscience, New York, 1: 567–589.
Cartwright, D.E., 1969. Extraordinary tidal currents near St. Kilda. Nature, 223: 928–932.
Cartwright, D.E. and Longuet-Higgins, M.S., 1956. The statistical distribution of the maxima of a random function. Proc. R. Soc. Lond., A, 237: 212–232.
Cartwright, D.E. and Smith, N.D., 1964. Buoy Technology. Mar. Tech. Soc., Washington, D.C., pp. 112–121.
Catchpole, J.P. and Fulford, G., 1966. Dimensionless groups. Ind. Eng. Chem., 58: 46–60.
Cauchy, A.L., 1827, Mémoire sur la théorie des ondes. Mém. Acad. R. Sci., Paris, lière serie, I, 38.
Chambers, Ll. G., 1964. Long waves on a rotating earth in the presence of a semi-infinite barrier. Proc. Edinburgh Math. Soc., 10: 92–99.
Chambers, Ll. G., 1965. On long waves on a rotating earth. J. Fluid Mech., 22: 209–216.
Chandrasekhar, S., 1961. Hydrodynamic and Hydromagnetic Stability. Clarendon Press, Oxford, 652 pp.
Chang, P., Pond, S. and Tabata, S., 1976. Subsurface currents in the Strait of Georgia, west of Sturgeon Bank. J. Fish. Res. Board. Can., 33: 2218–2241.
Chapman, S. and Lindzen, R.S., 1970. Atmospheric Tides. Reidel, Dordrecht, Holland, 200 pp.
Charney, J.G., 1947. The dynamics of long waves in a baroclinic westerly current. J. Meteorol., 4: 135–163.
Charney, J.G., 1955. The generation of oceanic currents by wind. J. Mar. Res., 14: 477–498.
Charney, J.G. and Stern, M.E., 1962. On the stability of internal baroclinic jets in a rotating atmosphere. J. Atmos. Sci., 19: 159–173.
Cherkesov, L.V., 1968. On the problem of tsunamis in a sea with continuously varying density. Izv. Akad. Nauk. S.S.S.R., Fiz. Atmos. Okeana, 4: 1101–1106 (in Russian).
Chorin, A.J., 1974. An Analysis of Turbulent Flow with Shear. Report FM–74–9. Department of Mathematics, Univ. of California, Berkeley.
Chorin, A.J., 1975. Lectures on Turbulence Theory. Publish or Perish, Inc., Boston, 159 pp.
Chow, P.L., 1975. Perturbation methods in stochastic wave propagation. SIAM Rev., 17: 57–81.
Christensen, N. Jr., 1973a. On the free modes of oscillation of a hemispherical basin centered on the equator, J. Mar. Res., 31: 168–174.
Christensen, N. Jr., 1973b. The effect of a coastal shelf on long waves in a rotating hemispherical basin. J. Mar. Res., 33: 175–187.

Christiansen, P.L., 1975. Comparative Studies of Diffraction Processes. Polyteknisk Forlag, Lyngby, 241 pp.
Clancy, E.P., 1968. The Tides: Pulse of the Earth. Doubleday-Anchor, Garden City, New York, 228 pp.
Clarke, A.J., 1977a. Wind-forced linear and nonlinear Kelvin waves along an irregular coastline. J. Fluid Mech., in press.
Clarke, A.J., 1977b. Observational and numerical evidence for wind-forced coastal trapped long waves. J. Phys. Oceanogr., 7: 231–247.
Clarke, D.J., 1971. Seiche motions for a basin of rectangular plan and of nonuniform depth. J. Mar. Res., 29: 47–59.
Clarke, D.J., 1974. Long edge waves over a continental shelf. Dtsch. Hydrogr. Z., 27: 1–8.
Clarke, R.A., 1971. Solitary and cnoidal planetary waves. Geophys. Fluid Dyn., 2: 343–354.
Conte, S.D. and Miles, J.W., 1959. On the numerical integration of the Orr-Sommerfeld equation. SIAM J. Appl. Math., 7: 361–366.
Corkan, R.H. and Doodson, A.T.. 1952. Free tidal oscillations in a rotating square sea. Proc. R. Soc. Lond., A, 215: 147–162.
Courant, R. and Friedrichs, K.O., 1948. Supersonic Flow and Shock Waves. Interscience, New York, 464 pp.
Courant, R. and Hilbert, D., 1962. Methods of Mathematical Physics, Vol. 2. Interscience, New York, 830 pp.
Cox, C.S. and Munk, W.H., 1954. Statistics of the sea surface derived from sun glitter. J. Mar. Res., 13: 198–227.
Cox, C.S. and Sandstrom, H., 1962. Coupling of surface and internal gravity waves in water of variable depth. J. Oceanogr. Soc. Japan, 20th Anniv. Vol. 2: 499–513.
Cox, M.D., 1976. Equatorially trapped waves and the generation of the Somali current. Deep-Sea Res., 23: 1139–1152.
Crapper, G.D., 1957. An exact solution for progressive capillary waves of arbitrary amplitude. J. Fluid Mech., 2: 532–540.
Crease, J., 1956a. Long waves on a rotating earth in the presence of a semi-infinite barrier. J. Fluid Mech., 1: 86–96.
Crease, J., 1956b. Propagation of long waves due to atmospheric disturbances in a rotating sea. Proc. R. Soc. Lond., A, 233: 556–569.
Crease, J., 1958. The propagation of long waves into a semi-infinite channel in a rotating system. J. Fluid Mech., 4: 306–320.
Crease, J., 1962. Velocity measurements in the deep water of the Western North Atlantic. J. Geophys. Res., 67: 3173–3176.
Crépon, M.R., 1976. Sea level, bottom pressure and geostrophic adjustment. Mém. Soc. R. Sci. Liège, 6e sérrie, tome X: 43–60.
Csanady, G.T., 1971a. On the equilibrium shape of the thermocline in a shore zone. J. Phys. Oceanogr., 1: 263–260.
Csanady, G.T., 1971b. Baroclinic boundary currents and longshore edge-waves in basins with sloping shores. J. Phys. Oceanogr., 1: 92–104.
Csanady, G.T., 1972. Response of large stratified lakes to wind. J. Phys. Oceanogr., 2: 3–13.
Csanady, G.T., 1973a. Transverse internal seiches in large oblong lakes and marginal seas. J. Phys. Oceanogr., 3: 439–447.
Csanady, G.T., 1973b. Turbulent Diffusion in the Environment. Reidel, Dordrecht, Holland, 248 pp.
Csanady, G.T., 1975. Hydrodynamics of large lakes. Ann. Rev. Fluid Mech., 7: 357–386.
Csanady, G.T., 1976. Topographic waves in Lake Ontario. J. Phys. Oceanogr., 6: 93–103.
Csanady, G.T. and Scott, J.T., 1974. Baroclinic coastal jets in Lake Ontario during IFYGL. J. Phys. Oceanogr., 4: 524–541.
Curtin, T.B. and Mooers, C.N.K., 1975. Observation and interpretation of a high-frequency internal wave packet and surface slick pattern. J. Geophys. Res., 80: 872–894.
Cutchin, D.L. and Smith, R.L., 1973. Continental shelf waves: Low frequency variations in sea level and currents over the Oregon continental shelf. J. Phys. Oceanogr., 3: 73–82.
Dalrymple, R.A., 1975. A mechanism for rip current generation on an open coast. J. Geophys. Res., 80: 3485–3487.
Dantzler, L., 1974. Main-thermocline depth fluctuations in the Central North Atlantic. MODE Hot Line News, No. 52 (Unpubl. Document).
Darwin, G.H., 1898. The Tides and Kindred Phenomena of the Solar System. Republished (1962) by Greeman and Cooper, San Francisco, 378 pp.
Davenport,W.B. and Root, W.L., 1958. An Introduction to the Theory of Random Signals and Noise. McGraw-Hill, New York, 393 pp.
Davies, J.T. and Vose, R.W., 1965. On the damping of capillary waves by surface films. Proc. R. Soc. Lond., A, 286: 218-234.
Davis, R.E. and Acrivos, A., 1967. Solitary internal waves in deep water. J. Fluid Mech., 29: 593–607.
Dawson, W.B., 1920. The Tides and Tidal Streams. Dept. Naval Service, Ottawa, Canada, 43 pp.
Defant, A., 1958. Ebb and Flow. Univ. of Michigan Press, 121 pp.

Defant, A., 1961. Physical Oceanography, Vol. II. Pergamon Press, New York, 598 pp.
Delisi, D.P. and Orlanski, I., 1975. On the role of density jumps in the reflexion and breaking of internal gravity waves. J. Fluid. Mech., 69: 445–464.
Dence, D. and Spence, J.E., 1973. Wave propagation in random anisotropic media. In: A.T. Bharucha-Reid (Editor), Probabilistic Methods in Applied Mathematics. Academic Press, New York, 3: 121–181.
Denman, K.L., 1975. Spectral Analysis: A Summary of the Theory and Techniques. Environment Canada, Fisheries and Marine Service Technical Report No. 539, 28 pp.
Desaubies, Y.J.F., 1976. Analytical representation of internal wave spectra. J. Phys. Oceanogr. 6: 976–981.
Dickinson, R.E., 1968. Planetary waves propagating vertically through weak westerly wind wave guides. J. Atmos. Sci., 25: 984–1002.
Dickinson, R.E., 1970. Development of a Rossby wave critical level. J. Atmos. Sci., 27: 627–633.
Dietrich, G., 1944. Die Schwingungssysteme der halb-und eintägigen Tiden in den Ozeanen. Veroeff. Inst. Meereskd. Univ. Berlin, A41: 7–68.
Dietrich, G., 1963. General Oceanography. Wiley, New York, 588 pp.
Dobson, F.W., 1971. Measurements of atmospheric pressure on wind-generated sea waves. J. Fluid Mech., 48: 91–127.
Dohler, G., 1964. Tides in Canadian Waters. Can. Hydrogr. Service, Dept. Mines and Tech. Surveys (now Dept., of the Environment), Ottawa, 14 pp.
Donelan, M.A., 1976. The drag coefficient versus wave age. Can. Meteorol. Soc., 10th Annual Congr. Québec, 1976 (abstract).
Donelan, M.A., Longuet-Higgins, M.S. and Turner, J.S., 1972. Periodicity in whitecaps. Nature, 239: 449–451.
Doodson, A.T., 1921. The harmonic development of the tide-generating potential. Proc. R. Soc. Lond., A, 100: 305–329.
Doodson, A.T., 1958. Oceanic tides. Adv. Geophys., 5: 117–152.
Dorman, C. and Mollo-Christensen, E., 1971. Reynolds stress and buoyancy fluctuations caused by moving gust patterns over the sea surface (Cat's paws). Conf. on the Interaction of the Sea and the Atmosphere, Fort Lauderdale, 1971.
Dotsenko, S.F. and Cherkesov, L.V., 1971. Generation of internal waves by a travelling field of pressure. Morsk. Gidrofiz. Issled., 56: 5–14 (in Russian).
Drazin, P.G., 1958. The stability of a shear layer in an unbounded heterogeneous inviscid fluid. J. Fluid Mech., 4: 214–224.
Drazin, P.G., 1970. Non-linear baroclinic instability of a continuous zonal flow. Q.J.R. Meteorol Soc., 96: 667–676.
Drazin, P.G., 1974. Kelvin-Helmholtz instability of a slowly varying flow. J. Fluid Mech., 65: 781–797.
Drazin, P.G. and Howard, L.N., 1966. Hydrodynamic stability of parallel flow of inviscid fluid. In: Advances in Applied Mechanics. Academic Press, New York, 9: 1–89.
Dronkers, J.J., 1964. Tidal Computations in Rivers and Coastal Waters. North-Holland, Amsterdam, 518 pp.
Duff, G.F.D. and Naylor, D., 1966. Differential Equations of Applied Mathematics. Wiley, New York, 423 pp.
Duffy, D.G., 1975. The barotropic instability of Rossby wave motion: a reexamination. J. Atmos. Sci., 32: 1271–1277.
Düing, W., Hisard, P., Katz, E., Meincke, J., Miller, J., Moroshkin, K.V., Philander, G., Ribnikov, A.A., Voigt, K. and Weisberg, R., 1975. Meanders and long waves in the equatorial Atlantic. Nature, 257: 280–284.
Düing, W., Mooers, C.N.K. and Lee, T.N., 1977. Low-frequency variability in the Florida Current and relations to atmospheric forcing from 1972 to 1974. J. Mar. Res., 35: 129–161.
Eady, E.T., 1949. Long waves and cyclone waves. Tellus, 1: 33–52.
Easton, A.K., 1971. Seiches of Sydney Harbour, N.S. Can. J. Earth Sci., 9: 857–862.
Eckart, C., 1951. Surface waves in water of variable depth. Marine Physical Laboratory of the Scripps Institute of Oceanography, Wave Rep. No. 100, S.I.O. Ref. 51–12, 99 pp. (unpubl. manuscr.).
Eckart, C., 1953. The generation of wind waves over a water surface. J. Appl. Phys., 24: 1485–1494.
Ekman, V.W., 1904. On Dead-Water. Norw. North Polar Exped. 1893–1896, 15. Part 1, 152.
Elliott, J.A., 1972. Microscale pressure fluctuations near waves being generated by the wind. J. Fluid Mech., 54: 427–448.
Elmore, W.C. and Heald, M.A., 1969. Physics of Waves. McGraw-Hill, New York, 477 pp.
Emery, W.J. and Magaard, L., 1976. Baroclinic Rossby waves as inferred from temperature fluctuations in the Eastern Pacific. J. Mar. Res., 34: 365–385.
Esch, R.E., 1962. Stability of the parallel flow of a fluid over a slightly heavier fluid. J. Fluid Mech., 12: 192–208.
Ewing, G., 1950. Slicks, surface films and internal waves. J. Mar. Res., 9: 161–187.
Ewing, W.M., Jardetzky, W.S. and Press, F., 1957. Elastic Waves in Layered Media. McGraw-Hill, New York, 380 pp.
Farmer, D.M. 1978. Observations of long nonlinear internal waves in a lake. J. Phys. Oceanogr., in press.

Feir, J.E., 1967. Some results from wave pulse experiments. Proc. R. Soc. Lond., A, 299: 54–58.
Fenton, J.D., 1973. Some results for surface gravity waves on shear flows. J. Inst. Maths. Appl., 12: 1–20.
Fissel, D., Pond, S. and Miyake, M., 1976. Spectra of surface atmospheric quantities at ocean weathership P. Atmosphere, 14: 77–97.
Fjeldstad, J.E., 1933. Wärmeleitung in Meere, Geofys. Publ. Oslo, 10: 20 pp.
Fjortoft, R., 1950. Application of integral theorems in deriving criteria of stability. Geophys. Publ., 17: 1–52.
Flattery, T.W., 1967. Hough Functions. Ph.D. thesis, Univ. of Chicago.
Fofonoff, N., 1962. The physical properties of sea water. In: M.N. Hill (General Editor), The Sea. Interscience, New York, 1: 3–30.
Fofonoff, N.P., 1969. Spectral characteristics of internal waves in the ocean. Deep-Sea Res., Suppl. to Vol. 16: 58–71.
Fofonoff, N.P. and Tabata, S., 1966. Variability of oceanographic conditions between Ocean Station P and Swiftsure Bank off the Pacific coast of Canada. J. Fish. Res. Board Can., 23: 825–868.
Forrester, W.D., 1974. Internal tides in St. Lawrence estuary. J. Mar. Res., 32: 55–66.
Fox, M.J.H., 1976. On the nonlinear transfer of energy in the peak of a gravity-wave spectrum. Proc. R. Soc. Lond., A, 349: 467–483.
Frankignoul, C.J., 1972. Stability of finite amplitude internal waves in a shear flow. Geophys. Fluid Dyn., 4: 91–99.
Frankignoul, C.J., 1974. Observed anisotropy of spectral characteristics of internal waves by low frequency currents. J. Phys. Oceanogr., 4: 625–634.
Frankignoul, C.J., 1976. Observed interaction between oceanic internal waves and mesoscale eddies. Deep-Sea Res., 23: 805–820.
Freeland, H.J., Rhines, P.B. and Rossby, T., 1975. Statistical observations of the trajectories of neutrally buoyant floats in the North Atlantic. J. Mar. Res., 33: 383–404.
French, A.P., 1968. Special Relativity. Norton, New York, 286 pp.
Frisch, U., 1968. Wave propagation in random media. In: A.T. Bharucha-Reid (Editor) Probabilistic Methods in Applied Mathematics. Academic Press, New York, 1: 75–198.
Fuller, J.D. and Mysalø, L.A., 1977. Edge Waves in the Presence of an Irregular Coastline. J. Phys. Oceanogr., 7, in press.
Gallagher, B., 1971. Generation of surf beat by non-linear wave interactions. J. Fluid Mech., 49: 1–20.
Galvin, C.J., 1965. Resonant edge waves on laboratory beaches. Trans. Am. Geophys. Union, 46. 112 (abstract)
Gargett, A.E., 1975. Horizontal coherence of oceanic temperature structure. Deep-Sea Res., 22: 767–776.
Gargett, A.E., 1976. Generation of internal waves in the Strait of Georgia, British Columbia. Deep-Sea Res., 23: 17–32.
Gargett, A.E. and Hughes, B.A., 1972. On the interaction of surface and internal waves. J. Fluid Mech., 52: 179–191.
Garrett, C.J.R., 1968. On the interaction between internal gravity waves and a shear flow. J. Fluid Mech., 34: 711–720.
Garrett, C.J.R., 1970a. Bottomless harbours. J. Fluid Mech., 43: 433–449.
Garrett, C.J.R., 1970b. On cross-waves. J. Fluid Mech., 41: 837–849.
Garrett, C.J.R., 1975. Tides in gulfs. Deep.Sea Res., 22: 23–35.
Garrett, C.J.R., 1976. Generation of Langmuir circulations by surface waves – a feedback mechanism. J. Mar. Res., 34: 117–130.
Garrett, C.J.R. and Greenberg, D., 1977. Predicting changes in tidal regime: the open boundary problem. J. Phys. Oceanogr., 7: 171–181.
Garrett, C.J.R. and Munk, W.H. 1971a. Internal wave spectra in the presence of fine-structure. J. Phys. Oceanogr., 1: 196–202.
Garrett, C.J.R. and Munk, W.H., 1971b. The age of the tide and the Q of the ocean. Deep-Sea Res., 18: 493–503.
Garrett, C.J.R. and Munk, W.H., 1972a. Oceanic mixing by breaking internal waves. Deep-Sea Res., 19: 823–832.
Garrett, C.J.R. and Munk, W.H., 1972b. Space–time scales of internal waves. Geophys. Fluid Dyn., 3: 225–264.
Garrett, C.J.R. and Munk, W.H., 1975. Space–time scales of internal waves: a progress report. J. Geophys. Res., 80: 291–297. (Also, Correction, Ibid., 3924.)
Garrett, C.J.R. and Smith, J., 1976. On the interaction between long and short surface waves. J. Phys. Oceanogr., 6: 925–930.
Gates, W.L., 1970. Effects of western coastal orientation on Rossby-wave reflection and the resulting large-scale oceanic circulation. J. Geophys. Res., 75: 4105–4120.
Geisler, J.E., 1970. Linear theory of the response of a two-layer ocean to a moving hurricane. Geophys. Fluid Dyn., 1: 249–272.
Geisler, J.E. and Dickinson, R.E., 1974. Numerical study of an interacting Rossby wave and barotropic zonal flow near a critical level. J. Atmos. Sci., 31: 946–955.

Geisler, J.E. and Dickinson, R.E., 1975. Critical level absorption of barotropic Rossby waves in a north–south flow. J. Geophys. Res., 80: 3805–3811.
Gelfand, I.M. and Fomin, S.V., 1963. Calculus of Variations. Prentice-Hall, Englewood Cliffs, New Jersey, 232 pp.
Geller, M.A., Tanaka, H. and Fritts, D.C., 1975. Production of turbulence in the vicinity of critical levels for internal gravity waves. J. Atmos. Sci., 32: 2125–2135.
Gent, P.R., 1974. Baroclinic instability of a slowly varying zonal flow. J. Atmos. Sci., 31: 1983–1994.
Gent, P.R., 1975. Baroclinic instability of a slowly varying zonal flow. Part 2. J. Atmos. Sci., 32: 2094–2102.
Gent, P.R. and Taylor, P.A., 1976. A numerical model of the air flow above water waves. J. Fluid Mech., 77: 105–128.
Gent, P.R. and Taylor, P.A., 1977. A note on "separation" over short wind waves. Boundary-Layer Meteorol., 11: 65–87.
Gibrat, R., 1966. L'énergie des marées. Presses Universitaires de France, Paris, 219 pp.
Gilchrist, A.W.R., 1966. The directional spectrum of ocean waves: an experimental investigation of certain predictions of the Miles-Phillips theory of wave generation. J. Fluid Mech., 25: 795–816.
Gill, A.E., 1974. The stability of planetary waves on an infinite beta-plane. Geophys. Fluid Dyn., 6: 29–47.
Gill, A.E. and Clarke, A.J., 1974. Wind-induced upwelling, coastal currents and sea-level changes. Deep-Sea Res., 21: 325–345.
Gill, A.E. and Schumann, E.H., 1974. The generation of long shelf waves by the wind. J. Phys. Oceanogr., 4: 83–90.
Gill, A.E., Green, J.S.A. and Simmons, A.J., 1974. Energy partition in the large-scale ocean circulation and the production of mid-ocean eddies. Deep-Sea Res., 21: 499–528.
Godin, G., 1965. Some remarks on the tidal motion in a narrow rectangular sea of constant depth. Deep-Sea Res., 12: 461–468.
Godin, G., 1969. Theory of the exploitation of tidal energy and its application to the Bay of Fundy. J. Fish. Res. Board Can., 26: 2887–2957.
Godin, G., 1972. The Analysis of Tides. Univ. Toronto Press, Toronto, 264 pp.
Goldstein, H., 1959. Classical Mechanics, Addison-Wesley, Reading, Mass., 399 pp.
Goldstein, S., 1931. On the stability of superposed streams of fluid of different densities. Proc. R. Soc. Lond., A, 132: 524–548.
Gordeyev, R.G., Kagan, B.A. and Rivkind, V.Y., 1974. Estimation of rate of dissipation of tidal energy in open ocean. Okeanologiya, 14: 226–229 (in Russian).
Gordon, D. and Stevenson, T.N., 1972. Viscous effects in a vertically propagating internal wave. J. Fluid Mech., 56: 629–639.
Görtler, H., 1944. Einige Bemerkungen über Strömungen in rotierenden Flüssigkeiten. Z. Angew. Mech., 24: 210–214.
Gossard, E.E. and Hooke, W.H., 1975. Waves in the Atmosphere. Elsevier, Amsterdam, 456 pp.
Gould, J. and McKee, W., 1973. Observation of the vertical structure of semi-diurnal tidal currents in the Bay of Biscay. Nature, 244: 88.
Gould, W.J., Schmitz, W.J. and Wunsch, C., 1974. Preliminary field results for a Mid-Ocean Dynamics Experiment (MODE–O). Deep-Sea Res., 21: 911–932.
Gray, T.J. and Gashus, O.K. (Editors), 1972. Tidal Power. Plenum Press, New York, 630 pp.
Green, G., 1837. On the motion of waves in a variable canal of small depth and width. Trans. Cambridge Philos. Soc., 6: 457–462.
Green, J.S.A., 1960. A problem in baroclinic stability. Q.J.R. Meteorol. Soc., 86: 237–251.
Greenspan, H.P., 1956. The generation of edge waves by moving pressure distributions. J. Fluid Mech., 1: 574–592.
Greenspan, H.P., 1968. The Theory of Rotating Fluids. Cambridge Univ. Press, 327 pp.
Greenspan, H.P., 1970. A note on edge waves in a stratified fluid. Stud. Appl. Math., 49: 381–388.
Gregg, M.S., Cox, C.S. and Hacker, P.W., 1973. Vertical microstructure measurements in the Central North Pacific. J. Phys. Oceanogr., 3: 458–469.
Grimshaw, R., 1974. Edge waves: a long-wave theory for oceans of finite depth. J. Fluid Mech., 62: 775–791.
Grimshaw, R., 1975a. Nonlinear internal gravity waves and their interaction with the mean wind. J. Atmos. Sci., 32: 1779–1793.
Grimshaw, R., 1975b. Internal gravity waves: critical layer absorption in a rotating fluid. J. Fluid Mech., 70: 287–304.
Grimshaw, R., 1975c. A note on the β-plane approximation. Tellus, 27: 351–357.
Grimshaw, R., 1976. The Stability of Continental Shelf Waves in the Presence of a Boundary Current Shear. School of Mathematical Sciences, Univ. of Melbourne, Res. Rep. No. 43, 19 pp.
Grimshaw, R., 1977a. Nonlinear aspects of long shelf waves. Geophys. Astrophys. Fluid Dyn., 8: 3–16.
Grimshaw, R., 1977b. The Stability of Continental Shelf Waves. I. Side band instability and long wave resonance. J. Austral. Math. Soc., Ser. B, in press.
Griscom, C.A., 1967. Application of a perturbation technique to the nonlinear equations of internal wave motion. J. Geophys. Res., 72: 5599–5611.

Groves, G.W. and Miyata, M., 1967. On weather-induced long waves in the equatorial Pacific. J. Mar. Res., 25: 115–128.
Guza, R.T. and Bowen, A.J., 1975. The resonant instabilities of long waves obliquely incident on a beach. J. Geophys. Res., 80: 4529–4534.
Guza, R.T. and Bowen, A.J., 1976. Finite amplitude edge waves. J. Mar. Res., 34: 269–293.
Guza, R.T. and Davis, R.E., 1974. Excitation of edge waves by waves incident on a beach. J. Geophys. Res., 79: 1285–1291.
Guza, R.T. and Inman, D.L., 1975. Edgewaves and beach cusps. J. Geophys. Res., 80: 2997–3012.
Hall, R.E., 1974. Diffraction of Rossby Waves by a Wedge-Shaped Lateral Boundary. Woods Hole Geophysical Fluid Dynamics Summer Study Program, 1974, Vol. II, pp. 73–94.
Hall, R.E., 1976. Scattering of Rossby waves by Topography in a Stratified Ocean. Ph.D. thesis, Univ. of California, San Diego, 116 pp.
Halpern, D., 1971. Observations on short-period internal waves in Massachusetts Bay. J. Mar. Res., 29: 116–132.
Hamon, B.V., 1962. The spectrums of mean sea level at Sydney, Coff's Harbour, and Lord Howe Island. J. Geophys. Res., 67: 5147–5155.
Hamon, B.V., 1963. Correction to "The spectrums of mean sea level at Sydney, Coff's Harbour, and Lord Howe Island". J. Geophys. Res., 68: 4635.
Hamon, B.V., 1965. The east Australian current, 1960–1964. Deep-Sea Res., 12: 899–922.
Hamon, B.V., 1966. Continental shelf waves and the effects of atmospheric pressure and wind stress on sea level. J. Geophys. Res., 71: 2883–2893.
Hamon, B.V., 1976. Generation of shelf waves on the east Australian coast by wind stress. Mém. Soc. R. Sci. Liège, 6e série, tome X: 359–367.
Hansen, D.V., 1970. Gulf stream meanders between Cape Hatteras and the Grand Banks. Deep-Sea Res., 17: 495–511.
Harlow, F.H. and Shannon, J.P., 1967. Distortion of a splashing liquid drop. Science, 157: 547–550.
Harlow, F.H., Shannon, J.P. and Welch, J.E., 1965. Liquid waves by computer. Science, 149: 1092–1093.
Hart, J.E., 1975a. Baroclinic instability over a slope. Part I: Linear Theory. J. Phys. Oceanogr., 5: 625–633.
Hart, J.E., 1975b. Baroclinic instability over a slope. Part II: Finite-amplitude theory. J. Phys. Oceanogr., 5: 634–641.
Hartman, R.J., 1975. Wave propagation in a stratified shear flow. J. Fluid Mech., 71: 89–104.
Harvey, R.R. and Patzert, W.C., 1976. Deep current measurements suggest long waves in the eastern equatorial Pacific. Science, 193: 883–885.
Hasselmann, K., 1962. On the nonlinear energy transfer in a gravity-wave spectrum. 1, General theory. J. Fluid Mech., 12: 481–500.
Hasselmann, K., 1963a. On the nonlinear energy transfer in a gravity-wave spectrum. 2, Conservation laws, wave–particle correspondence, irreversibility. J. Fluid Mech., 15: 273–281.
Hasselmann, K., 1963b. A statistical analysis of the generation of microseism. Rev. Geophys., 1: 177–210.
Hasselmann, K., 1966. Feynman diagrams and interaction rules of wave–wave scattering processes. Rev. Geophys., 4: 1–32.
Hasselmann, K., 1967a. Non-linear interactions treated by the methods of theoretical physics (with application to the generation of waves by wind). Proc. R. Soc. Lond., A, 299:77–100.
Hasselmann, K., 1967b. A criterion for nonlinear wave stability. J. Fluid Mech., 30: 737–739.
Hasselmann, K., 1968. Weak interaction theory of ocean waves. In: M.Holt (Editor), Basic Developments in Fluid Dynamics. Academic Press, New York, 2: 117–182.
Hasselmann, K., 1971. On the mass and momentum transfer between short gravity waves and larger-scale motions. J. Fluid Mech., 50: 189–205.
Hasselmann, K., Barnett, T.P., Bouws, E., Carlson, H., Cartwright, D.E., Enke, K., Ewing, J.A., Gienapp, H., Hasselmann, D.E., Kruseman, P., Meerburg, A., Müller, P., Olbers, D.J., Richter, K., Sell, W. and Walden, H., 1973. Measurements of wind-wave growth and swell decay during the joint North Sea wave project (JONSWAP). Dtsch. Hydrogr. Z., Suppl. A., 8(12).
Haubrich, R.A., Munk, W.H. and Snodgrass, F.E., 1963. Comparative spectra of microseisms and swell. Bull. Seismol. Soc. Am., 53: 27–37.
Haurwitz. B. and Panofsky, H., 1950. Stability and meandering of the Gulf Steam. Trans. Am. Geophys. Union, 31: 723–731.
Hayes, S.P. and Halpern, D., 1976a. Variability of the semidiurnal internal tide during coastal upwelling. Mém. Soc. R. Sci. Liège, 6e série, tome X: 175–186.
Hayes, S.P. and Halpern, D., 1976b. Observations of internal waves and coastal upwelling off the Oregon coast. J. Mar. Res., 34: 247–267.
Hazel, P., 1967. The effect of viscosity and heat conduction on internal gravity waves at a critical level. J. Fluid Mech., 30: 775–783.
Healey, D. and LeBlond, P.H., 1969. Internal wave propagation normal to a geostrophic current. J. Mar. Res., 27: 85–98.
Heaps, N.S., 1965. Storm surges on a continental shelf. Philos. Trans. R. Soc. Lond., A, 257: 351–383.

Heaps, N.S., 1969. A two-dimensional numerical sea model. Proc. R. Soc. Lond., A, 265: 93–137.
Heaps, N.S. and Ramsbottom, A.E., 1966. Wind effects on the water in a narrow two-layered lake. Philos. Trans. R. Soc. Lond., A, 259: 391–430.
Heiskanen, W., 1921. Über den Einfluss der Gezeiten auf die säkuläre Acceleration des Mondes. Ann. Acad. Sci. Fennicae, A, 18: 1–84.
Helbig, J.A. and Mysak, L.A., 1976. Strait of Georgia oscillations. Low-frequency currents and topographic planetary waves. J. Fish. Res. Board. Can., 33: 2329–2339.
Helmholtz, H., 1868. Über diskontinuierliche Flüssigkeitsbewegungen. Monatsber. Königl. Preuss. Akad. Wiss. Berlin, pp. 215–228; transl. by F. Guthrie, On discontinuous movements of fluids. Philos. Mag., 36: 337–348.
Hendershott, M.C., 1972. The effects of solid-earth deformation on global ocean tides. Geophys. J. R. Astron. Soc., 29: 389–403.
Hendershott, M.C., 1973. Ocean tides. EOS, Trans. Am. Geophys. Union, 54: 76–86.
Hendershott, M.C., 1977. Numerical models of ocean tides. In: E. Goldberg, I. McCave, J. O'Brien and J. Steele (Editors), The Sea. Wiley-Interscience, New York, 6: 47–95.
Hendershott, M.C. and Speranza, A., 1971. Co-oscillating tides in long narrow bays; the Taylor problem revisited. Deep-Sea Res., 18: 959–980.
Henry, R.F. and Heaps, N.S., 1976. Storm surges in the southern Beaufort Sea. J. Fish Res. Board Can., 33: 2362–2376.
Hidaka, K., 1932. Tidal oscillations in a rectangular basin of variable depth. Mem. Imper. Mar. Observ., Kobe, 5: 15–23.
Hino, M., 1974. Theory of formation of rip-current and cuspidal-coast. Proc. 14th Int. Conf. Coastal Eng., Copenhagen, 1974.
Hinze, J.O., 1975. Turbulence. McGraw-Hill, New York, 2nd edit., 790 pp.
Ho, D.V., Meyer, R.E. and Shen, M.C., 1963. Long surf. J. Mar. Res., 21: 219–232.
Hogg, N.G., 1971. Longshore current generation by obliquely incident internal waves. Geophys. Fluid Dyn., 2: 361–376.
Holland, W.R. and Lin, L.B., 1975. On the generation of mesoscale eddies and their contribution to the oceanic general circulation. J. Phys. Oceanogr., 5: 642–657 (Part I. A preliminary numerical experiment); 658–669 (Part II. A parameter study).
Holton, J.R., 1971. An experimental study of forced barotropic Rossby waves. Geophys. Fluid Dyn., 2: 323–341.
Holton, J.R., 1972. An Introduction to Dynamic Meteorology. Academic Press, New York, 319 pp.
Hoogstraten, H.W., 1972. Trapped shallow-water waves for continuous bottom topographies. Acta Mech., 14: 171–182.
Hough, S.S., 1897. On the application of harmonic analysis to the dynamical theory of the tides. Part I. On Laplace's 'oscillations of the first species', and on the dynamics of ocean currents. Philos. Trans. R. Soc. Lond., A, 189: 201–257.
Hough, S.S., 1898. On the application of harmonic analysis to the dynamical theory of the tides. Part II. On the general integration of Laplace's dynamical equations. Philos. Trans. R. Soc. Lond., A, 191: 139–185.
Howard, L.N., 1960. Lectures on Fluid Dynamics. In: E.A. Spiegel (Editor), Notes on the 1960 Summer Study Program in Geophysical Fluid Dynamics. Woods Hole, Mass., Vol. 1.
Howard, L.N., 1961. Note on a paper of John W. Miles. J. Fluid Mech., 10: 509–512.
Howard, L.N., and Drazin, P.G., 1964. On the instability of parallel flow of inviscid fluid in a rotating system with variable Coriolis parameter. J. Maths. Phys., 43: 83–99.
Howard, L.N. and Maslowe, S.A., 1973. Stability of stratified shear flow. Boundary-Layer Meteorol., 4: 511–523.
Howard, L.N. and Siegmann, W.L., 1969. On the initial-value problem for rotating stratified flow. Stud. Appl. Math., 48: 153–169.
Howe, M.S., 1971. Wave propagation in random media. J. Fluid Mech., 45: 769–783.
Howe, M.S., 1973. Conservation of energy in random media, with application to the theory of sound absorption by an inhomegeneous flexible plate. Proc. R. Soc. Lond., A, 331: 479–496.
Howe, M.S., 1974. The mean square stability of an inverted pendulum subject to random parametric excitation. J. Sound Vib., 32: 407–421.
Howe, M.S. and Mysak, L.A., 1973. Scattering of Poincaré waves by an irregular coastline. J. Fluid Mech., 57: 111–128.
Hudimac, A.A., 1961. Ship waves in a stratified fluid. J. Fluid Mech., 11: 229–243.
Hughes, B., 1964. Effect of rotation on internal gravity waves. Nature, 201: 798–801.
Hunkins, K.L., 1974. Subsurface eddies in the Arctic Ocean. Deep-Sea Res., 21: 1017–1033.
Huntley, D.A. 1976. Long-period waves on a natural beach. J. Geophys. Res., 81: 6441–6449.
Huntley, D.A. and Bowen, A.J., 1973. Field observations of edge waves. Nature, 243: 160–162.
Huppert, H.E. and Miles, J.W., 1969. Lee waves in a stratified flow, Part 3, Semi-elliptical obstacle. J. Fluid Mech., 35: 481–496.
Hurlburt, H.E., Kindle, J.C. and O'Brien, J.J., 1976. A numerical simulation of the onset of El Niño. J. Phys. Oceanogr., 6: 621–631.
Hurley, D.G., 1969. The emission of internal waves by vibrating cylinders. J. Fluid Mech., 36: 657–672.

Huthnance, J.M., 1975. On trapped waves over a continental shelf. J. Fluid Mech., 69: 689–704.
Huyer, A., Hickey, B.M., Smith, J.D., Smith, R.L. and Pillsbury, R.D., 1975. Alongshore coherence at low frequency in currents observed over the continental shelf off Oregon and Washington. J. Geophys. Res., 80: 3495–3505.
Ibbetson, A. and Phillips, N.A., 1967. Some laboratory experiments on Rossby waves in a rotating annulus. Tellus, 19: 81–87.
Iooss, G., 1972. Existence et stabilité de la solution périodique secondaire intervenant dans les problèmes d'évolution du type Navier-Stokes. Arch. Ration. Mech. Anal., 47: 301–329.
James, I.D., 1974a. Non-linear waves in the nearshore region: shoaling and set-up. Estuarine Coastal. Mar. Sci., 2: 207–234.
James, I.D., 1974b. A non-linear theory of longshore currents. Estuarine Costal. Mar. Sci., 2: 235–249.
Jeffreys, H., 1921. Tidal friction in shallow seas. Philos. Trans. R. Soc. Lond., A, 221: 239–264.
Jeffreys, H., 1924. The free oscillations of water in an elliptical lake. Proc. Lond. Math. Soc., Ser.2, 23: 455–476..
Jeffreys, H., 1925. On the formation of water waves by wind. Proc. R. Soc. Lond., A, 107: 189–206.
Jeffreys, H.. 1926. On the formation of water waves by wind. (Second paper) Proc. R. Soc. Lond., A, 110: 241–247.
Jenkins, G.M. and Watts, D.G., 1968. Spectral Analysis and its Applications. Holden-Day, San Fransisco, 525 pp.
Johns, B., 1965a. Fundamental mode edge waves over a steeply sloping shelf. J. Mar. Res., 23: 200–206.
Johns, B., 1965b. Inertia currents. Deep-Sea Res., 12: 825–830.
Johns, B. and Cross, M.J., 1969. The decay of internal wave modes in a multi-layered system. Deep-Sea Res., 16: 185–195.
Johns, B. and Cross, M.J., 1970. The decay and stability of internal wave modes in a multisheeted thermocline. J. Mar. Res., 28: 215–224.
Johns, B. and Hamzah, A.M.O., 1969. On the seiche motion in a curved lake. Proc. Cambridge Philos. Soc., 66: 607–615.
Jones, D.S., 1964. The Theory of Electromagnetism. Pergamon Press, Oxford, 807 pp.
Jones, W.L., 1967. Propagation of internal gravity waves in fluids with shear flow and rotation. J. Fluid Mech., 30: 439–448.
Jonsson, I.G. Skovgaard, O. and Brink-Kjaer, O., 1976. Diffraction and refraction calculations for waves incident on an island. J. Mar. Res., 34: 469–496.
Joyce, T.M., 1974. Nonlinear interactions among standing surface and internal gravity waves. J. Fluid Mech., 63: 801–825.
Kajiura, K., 1958. Effect of Coriolis force on edge waves. 2, Specific examples of free and forced waves. J. Mar. Res., 16: 145–157.
Kajiura, K., 1962. A note on the generation of boundary waves on the Kelvin type. J. Oceanogr. Soc. Japan, 18: 51–58.
Kajiura, K., 1963. The leading waves of tsunami. Bull. Earthquake Res. Inst., 41: 535–571.
Kajiura, K., 1970. Tsunami source, energy and the directivity of wave radiation. Bull. Earthquake Res. Inst., 48: 835–869.
Kajiura, K., 1972. The directivity of energy radiation of the tsunami generated in the vicinity of a continental shelf. J. Oceanogr. Soc. Japan, 28: 32–49.
Kajiura, K., 1974. Effect of stratification on long period trapped waves on the shelf. J. Oceanogr., Soc. Japan, 30: 271–281.
Kamenkovich, V.M., 1973. Principles of Ocean Dynamics. Gidromet. Leningrad, 240 pp. (in Russian).
Kanari, S., 1975. The long-period internal waves in Lake Biwa. Limnol. Oceanogr., 20: 544–553.
Karpman, V.I., 1973. Nonlinear Waves in Dispersive Media. Nauk, Moscow, 175 pp. (in Russian).
Kase, R.H., 1971. Uber zweidimensionale luftdruckbedingte interne Wellen im exponentiell geschichteten Meer. Dtsch. Hydrogr. Z., 24: 193–209.
Käse, R.H., 1972. Zur Erzeugung interner Wellen durch Verticalgeschwindigkeitsfelder an einer Grenzfläche. Ph.D. thesis, Universität Kiel, 84 pp.
Käse, R.H. and Tang, C.L., 1976. A dynamical model for the energy spectra and coherence of internal waves. J. Fish Res. Board Can., 33: 2323–2328.
Katz, E.J., 1975. Tow spectra from MODE. J. Geophys. Res., 80: 1163–1167.
Kaylor, R. and Faller, A.J. 1972. Instability of the stratified Ekman layer and generation of internal waves. J. Atmos. Sci., 29: 497–509.
Keller, J.B., 1962a. Wave propagation in random media. In: Proc. Symp. Appl. Maths., American Mathematical Society, Providence, R.I., 13: 227–246.
Keller, J.B., 1962b. Geometrical theory of diffraction. J. Opt. Soc. Am., 52: 116–130.
Keller, J.B., 1967. The velocity and attenuation of waves in random media. In: R.L. Rowell and R.S. Stein (Editors), Electromagnetic Scattering. Gordon and Breach, New York, pp. 823–834.
Keller, J.B. and Munk, W.H., 1970. Internal wave wakes of a body moving in a stratified fluid. Phys. Fluids, 13: 1425–1431.
Keller, J.B. and Veronis, G., 1969. Rossby waves in the presence of random currents. J. Geophys. Res., 74: 1941–1951.
Kelly, R.E. and Maslowe, S.A., 1970. The nonlinear critical layer in a slightly stratified shear flow. Stud. Appl. Math., 49: 301–326.

Kelvin, Lord W., 1871. The influence of wind on waves in water supposed frictionless. Philos. Mag., 42: 368–374.
Kelvin, Lord W., 1879. On gravitational oscillations of rotating water. Proc. R. Soc. Edinburgh, 10: 92–100.
Kelvin, Lord W., 1880a. Vibrations of a columnar vortex. Philos. Mag., 10: 155–168.
Kelvin, Lord W., 1880b. On a disturbing infinity in Lord Rayleigh's solution for waves in a plane vortex stratum. Nature, 23: 45–46.
Kenyon, K., 1968. Wave–wave interactions of surface and internal waves. J. Mar. Res., 26: 208–231.
Kenyon, K.E.. 1970. A note on conservative edgewave interactions. Deep-Sea Res., 17: 197–201.
Kinder, T.H., Coachman, L.K. and Galt, J.A., 1975. The Bering slope current system. J. Phys. Oceanogr., 5: 231–244.
Kinsman, B., 1965. Wind Waves. Prentice-Hall, Englewood Cliffs, N.J., 676 pp.
Kishi, M.J. and Suginohara, N., 1975. Effects of longshore variation of coastline geometry and bottom topography on coastal upwelling in a two-layer model. J. Oceanogr. Soc. Japan. 31: 48–50.
Kitaigorodskii, S.A., 1962. Applications of the theory of similarity to the analysis of wind-generated wave motion as a stochastic process. Bull. Acad. Sci. USSR Geophys. Ser., No. 1: 105–117 (in Russian).
Kitano, K., 1975. Some properties of the warm eddies generated in the confluence zone of the Kuroshio and Oyashio currents. J. Phys. Oceanogr., 5: 245–252.
Kittel, C., 1969. Thermal Physics. Wiley, New York, 418 pp.
Knauss, J.A., 1966. Further measurements and observations on the Cromwell current. J. Mar. Res., 24: 205–240.
Kolmogorov, A.N., 1950. Foundations of the Theory of Probability. Chelsea, New York, 71 pp.
Komar, P.D., 1973. Computer models of delta growth due to sediment input from rivers and longshore transport. Geol. Soc. Am. Bull., 84: 2217–2226.
Konyaev, K.V. and Sabinin, K.D., 1973. Resonant theory of generation of internal waves in the ocean. Dokl. Akad. Nauk. SSSR, 210: 1342–1345 (in Russian).
Kordzadze, A.A., 1974. On the uniqueness of the solution of a certain problem in ocean dynamics. Dokl. Akad. Nauk. SSSR., 219: 856–859 (in Russian).
Korteweg, D.J. and DeVries, G., 1895. On the change of form of long waves advancing in a rectangular canal, and on a new type of long stationary waves. Philos. Mag., Ser. 5, 39: 422–443.
Koshlyakov, M.N., 1973. Results obtained at the Atlantic Polygon test-site in 1970 compared to certain models of free Rossby waves. Okeanologiya, 13: 760–767 (in Russian).
Koshlyakov, M.N. and Grachev, Y.M., 1973. Meso-scale currents at a hydrophysical polygon in the tropical Atlantic. Deep-Sea Res., 20: 507–526.
Kramareva, L.K., 1973. The relation of Rossby waves to the large-scale characteristics of temperature waves in the NW Pacific. Okeanologiya, 13: 801–803 (in Russian).
Kranzer, H.C. and Keller, J.B., 1959. Water waves produced by explosions. J. Appl. Phys., 30: 398–407.
Kraus, E.B., 1974. Planetary waves. Earth-Sci. Rev., 10: 203–221.
Krauss, W., 1966. Methoden und Ergebnisse der theoretischen Ozeanographie. 2: Interne Wellen. Borntraeger, Berlin, 248 pp.
Krauss, W., 1972. On the response of a stratified ocean to wind and air pressure. Dtsch. Hydrogr. Z., 25: 49–61.
Kroll, J., 1975. The propagation of wind-generated inertial oscillations from the surface into the deep ocean. J. Mar. Res., 33: 15–51.
Kroll, J. and Niiler, P.P., 1976. The transmission and decay of barotropic Rossby waves incident on a continental shelf. J. Phys. Oceanogr., 6: 432–450.
Kundu, P.K., Allen, J.S. and Smith, R.L., 1975. Modal decomposition of the velocity field near the Oregon coast. J. Phys. Oceanogr., 5: 683–704.
Kuo, H.L., 1949. Dynamic instability of two-dimensional non-divergent flow in a barotropic atmosphere. J. Meteorol., 6: 105–122.
Kuo, J.T., 1975. Earth Tides. Rev. Geophys. Space Phys., 13: 260–263.
Lafond, E.C., 1951. Processing Oceanographic Data. U.S. Navy Hydrogr. Office, Publ. 614, 114 pp.
Lafond, E.C., 1962. Internal waves, Part I. In: H.M. Hill (General Editor), The Sea. Wiley-Interscience, New York, 1: 731–751.
Lalas, D.P., Finandi, F. and Fua, D., 1976. The destabilizing effect of the ground on Kelvin-Helmholtz waves in the atmosphere. J. Atmos. Sci., 33: 59–69.
Lamb, H., 1945. Hydrodynamics, Dover, New York, 6th edition, 738 pp.
Landau, L.D. and Lifschitz, E.M., 1959. Fluid Mechanics. Pergamon Press, London, 536 pp.
Landau, L.D. and Lifschitz, E.M., 1960. Mechanics. Pergamon Press, Oxford, 165 pp.
Laplace, P. S., 1775. Recherches sur plusieurs points du système du monde. Mem. Acad. R. Sci. Paris (1778); oeuvres 9: 9ff.
Laplace, P.S., 1799. Traité de mécanique céleste. Crapelet, Paris.
Larichev, V.D., 1974. Statement of an internal boundary-value problem for the Rossby-wave equation (registration of waves by coastal recorders). Izv. Akad. Nauk SSSR, Fiz. Atmos. Okeana, 10: 470–473 (in Russian).
Larichev, V.D. and Reznik, G.M., 1976a. Nonlinear Rossby waves on a large-scale current. Okeanologiya, 16: 200–206 (in Russian).

Larichev, V.D. and Reznik, G.M., 1976b. Highly nonlinear Rossby waves. Okeanologiya, 16: 381–388 (in Russian).
Larsen, J.C., 1969. Long waves along a single-step topography in a semi-infinite uniformly rotating ocean. J. Mar. Res., 27: 1–6.
Larsen, L.H., 1965. Comments on "Solitary waves in the Westerlies". J. Atmos. Sci., 22: 222–224.
Larsen, L.H., 1969a. Internal waves incident upon a knife edge barrier. Deep-Sea Res., 16: 411–419.
Larsen, L.H., 1969b. Oscillations of a vertically buoyant sphere in a stratified fluid. Deep-Sea Res., 16: 587–603.
Larson, T.R. and Wright, J.W., 1975. Wind-generated gravity–capillary waves: laboratory measurements of temporal growth rates using microwave backscatter. J. Fluid Mech., 70: 417–436.
Laura, P.A., 1970. Comments on the paper: Stress wave propagation in rectangular bars. Int. J. Solids Struct., 6: 693–694.
Lax, M., 1973. Wave propagation and conductivity in random media. In: J.B. Keller and W. McKean (Editors), Stochastic Differential Equations, SIAM-AMS Proceedings, American Mathematical Society, Providence, R.I., 6: 35–95.
Lax, P.D., 1968. Integrals of nonlinear equations of evolution and solitary waves. Comm. Pure Appl. Math., 21: 467–490.
Leaman, K.D., 1976. Observations of the vertical polarization and energy flux of near-inertial waves. J. Phys. Oceanogr., 6: 894–908.
Leaman, K.D. and Sanford, T.B., 1975. Vertical energy propagation of inertial waves: a vector spectral analysis of velocity profiles. J. Geophys. Res., 80: 1975–1978.
LeBlond, P.H., 1964. Planetary waves in a symmetrical polar basin. Tellus, 16: 503–512.
LeBlond, P.H., 1965. Über den Einfluss von Reibung und Vermischung auf interne Wellen. Kieler Meeresforsch, 21: 127–131.
LeBlond, P.H., 1966. On the damping of internal gravity waves in a continuously stratified ocean. J. Fluid Mech., 25: 121–142.
LeBlond, P.H. and Mysak, L.A., 1977. Trapped coastal waves and their role in shelf dynamics. In: E. Goldberg, I. McCave, J. O'Brien, and J. Steele (Editors), The Sea. Wiley-Interscience, New York, 6: 459–495.
LeBlond, P.H. and Tang, C.L., 1974. On energy coupling between waves and rip currents. J. Geophys. Res., 79: 811–816.
Lee, C.A., 1975. The Generation of Unstable Waves and the Generation of Transverse Upwelling: Two Problems in Geophysical Fluid Dynamics. Ph.D. thesis, The University of British Columbia, 157 pp.
Lee, C.Y. and Beardsley, R.C., 1974. The generation of long nonlinear internal waves in a weakly stratified shear flow. J. Geophys. Res., 79: 453–462.
Leibovich, S. and Seebass, A.R., 1974. Nonlinear Waves. Cornell Univ. Press, Ithaca, New York, 331 pp.
Le Méhauté, B., 1971. Theory of explosion-generated waves. In: V.T. Chow (Editor), Advances in Hydroscience. Academic Press, New York, 7: 1–79.
Lemon, D.D., 1975. Observations and Theory of Seiche Motions in San Juan Harbour, B.C., M.Sc. thesis, The University of British Columbia, 81 pp.
Leonov, A.I. and Miropolsky, Yu.Z., 1973. Resonant excitation of internal gravity waves in the ocean by atmospheric pressure fluctuations. Izv. Akad. Nauk SSSR, Fiz. Atmos. Okeana, 9: 480–485 (in Russian).
Levich, V.G., 1962. Physicochemical Hydrodynamics. Prentice-Hall, Englewood Cliffs, New Jersey, 700 pp.
Levi-Civita, T., 1924. Questioni di mecanica classica e relativista, 11, Zanichelli, Bologna, pp. 26–59.
Levi-Civita, T., 1925. Détermination rigoureuse des ondes permanentes d'ampleur finie. Math. Ann., 93: 264–314.
Lewis, J.E., Lake, B.M. and Ko, D.R.S., 1974. On the interaction of internal waves and surface gravity waves. J. Fluid Mech., 63: 773–800.
Lick, W., 1970. Nonlinear wave propagation in fluids. Ann. Rev. Fluid Mech., 2: 113–136.
Lighthill, M.J., 1962a. Fourier Series and Generalized Functions. Cambridge Univ. Press, 79 pp.
Lighthill, M.J., 1962b. Physical interpretation of the mathematical theory of wave generation by wind. J. Fluid Mech., 14: 385–398.
Lighthill, M.J., 1965. Group velocity. J. Inst. Maths. Appl., 1: 1–28.
Lighthill, M.J., 1967. On waves generated in dispersive systems by travelling forcing effects, with applications to the dynamics of rotating fluids. J. Fluid Mech., 27: 725–752.
Lighthill, M.J., 1969. Dynamic response of the Indian ocean to onset of the southwest monsoon. Philos. Trans. R. Soc. Lond., A, 265: 45–92.
Lighthill, M.J., 1971. Time-varying currents. Philos. Trans. R. Soc. Lond., A, 270: 371–390.
Lindzen, R.S. and Rosenthal, A.J., 1976. On the instability of Helmholtz velocity profiles in stably stratified fluids when a lower boundary is present. J. Geophys. Res., 81: 1561–1571.
Lipps, F.B., 1962. The barotropic stability of the mean winds in the atmosphere. J. Fluid Mech., 12: 397–497.
Liu, P., 1971. Normalized and equilibrium spectra of wind waves in Lake Michigan. J. Phys. Oceanogr., 1: 249–257.
Lock, R.C., 1954. Hydrodynamic stability of the flow in the laminar boundary layer between parallel streams, Proc. Cambridge Philos. Soc., 50: 105–124.

Lo Dato, V.A., 1973. Stochastic processes in heat and mass transport. In: A.T. Bharaucha-Reid (Editor). Probabilistic Methods in Applied Mathematics. Academic Press, New York, 3: 183–212.
Long, R., 1973. Scattering of surface waves by bottom irregularities. J. Geophys. Res., 78: 7861–7870.
Long, R.R., 1953. Some aspects of the flow of stratified fluids. I. A theoretical investigation. Tellus, 5: 42–57.
Long, R.R., 1956. Solitary waves in one- and two-fluid systems. Tellus, 8: 460-471.
Long, R.R., 1965. On the Boussinesq approximation and its role in the theory of internal waves. Tellus, 17: 46–52.
Long, S.R. and Huang, N.E., 1976a. On the variation and growth of wave-slope spectra in the capillary–gravity range with increasing wind. J. Fluid Mech., 77: 209–228.
Long, S.R. and Huang, N.E., 1976b. Observations of wind-generated waves on variable current. J. Phys. Oceanogr., 6: 962–968.
Longuet-Higgins, M.S., 1950. A theory of the origin of microseisms. Philos. Trans. R. Soc. Lond., A, 243: 1–35.
Longuet-Higgins, M.S., 1953. Mass transport in water waves. Philos. Trans. R. Soc. Lond., A, 245: 535–581.
Longuet-Higgins, M.S., 1957. On the transformation of a continuous spectrum by refraction. Proc. Cambridge Philos. Soc., 53: 226–229.
Longuet-Higgins, M.S., 1960. Mass transport in the boundary layer at a free oscillating surface. J. Fluid Mech., 8: 293–306.
Longuet-Higgins, M.S., 1963a. The generation of capillary waves by steep gravity waves. J. Fluid Mech., 16: 138–159.
Longuet-Higgins, M.S., 1963b. The effect of nonlinearities on statistical distributions in the theory of sea waves. J. Fluid Mech., 17: 459–480.
Longuet-Higgins, M.S., 1964a. Planetary waves on a rotating sphere. Proc. R. Soc. Lond., A, 279: 446–473.
Longuet-Higgins, M.S., 1964b. On the group velocity and energy flux in planetary wave motions. Deep-Sea Res., 11: 35–42.
Longuet-Higgins, M.S., 1965a. Planetary waves on a rotating sphere, II. Proc. R. Soc. Lond., A, 284: 40–68.
Longuet-Higgins, M.S., 1965b. The response of a stratified ocean to stationary or moving wind-systems. Deep-Sea Res., 12: 923–973.
Longuet-Higgins, M.S., 1966. Planetary waves on a sphere bounded by meridians of longitude. Philos. Trans. R. Soc. Lond., A., 260: 317–350.
Longuet-Higgins, M.S., 1967. On the trapping of wave energy round islands. J. Fluid Mech., 29: 781–821.
Longuet-Higgins, M.S., 1968a. On the trapping of waves along a discontinuity of depth in a rotating ocean. J. Fluid Mech., 31: 417–434.
Longuet-Higgins, M.S., 1968b. Double Kelvin waves with continuous depth profiles. J. Fluid Mech., 34: 49–80.
Longuet-Higgins, M.S., 1968c. The eigenfunctions of Laplace's tidal equations over a sphere. Philos. Trans. R. Soc. Lond., A, 262: 511–607.
Longuet-Higgins, M.S., 1969a. A nonlinear mechanism for the generation of sea waves. Proc. R. Soc. Lond., A, 311: 371–389.
Longuet-Higgins, M.S., 1969b. Action of a variable stress at the surface of water waves. Phys. Fluids, 12: 737-740.
Longuet-Higgins, M.S., 1969c. On the trapping of long-period waves round islands. J. Fluid Mech., 37: 773–784.
Louguet-Higgins, M.S., 1970a. Longshore currents generated by obliquely incident sea waves, 1. J. Geophys. Res., 75: 6778–6789.
Longuet-Higgins, M.S., 1970b. Longshore currents generated by obliquely incident sea waves, 2. J. Geophys. Res., 75: 6790–6801.
Longuet-Higgins, M.S., 1970c. Steady currents induced by oscillations round islands. J. Fluid Mech., 42: 701–720.
Longuet-Higgins, M.S., 1971. On the spectrum of sea level at Oahu. J. Geophys. Res., 76: 3517–3522.
Longuet-Higgins, M.S., 1972. Recent progress in the study of longshore currents. In: R.E. Meyer (Editor), Waves on Beaches and Resulting Sediment Transport. Academic Press, New York, 203–248.
Longuet-Higgins, M.S., 1973a. On the form of the highest progressive and standing waves in deep water. Proc. R. Soc. Lond., A, 331: 445–456.
Longuet-Higgins, M.S., 1973b. A model of flow separation at a free surface. J. Fluid Mech., 57: 129–148.
Longuet-Higgins, M.S., 1974. On the mass, momentum, energy and circulation of a solitary wave. Proc. R. Soc. Lond., A, 337: 1-13.
Longuet-Higgins, M.S., 1975a. Integral properties of periodic gravity waves of finite amplitude. Proc. R. Soc. Lond., A, 342: 157-174.
Longuet-Higgins, M.S., 1975b. On the joint distribution of the periods and amplitudes of sea waves. J. Geophys. Res., 80: 2688–2694.

Longuet-Higgins, M.S., 1976. On the nonlinear transfer of energy in the peak of a gravity-wave spectrum: a simplified model. Proc. R. Soc. Lond., A, 347: 311–328.
Longuet-Higgins, M.S. and Gill, A.E., 1967. Resonant interactions between planetary waves. Proc. R. Soc. Lond., A, 299: 120-140.
Longuet-Higgins, M.S. and Pond, G.S., 1970. The free oscillations of fluid on a hemisphere bounded by meridians of longitude. Philos. Trans. R. Soc. Lond., A, 266: 193–223.
Longuet-Higgins, M.S. and Stewart, R.W., 1960. Changes in the form of short gravity waves on long waves and tidal currents. J. Fluid Mech., 8: 565–583.
Longuet-Higgins, M.S. and Stewart, R.W., 1961. The changes in amplitude of short gravity waves on steady non-uniform currents. J. Fluid Mech., 10: 529–549.
Longuet-Higgins, M.S. and Stewart, R.W., 1962. Radiation stress and mass transport in gravity waves, with applications to "surf-beats". J. Fluid Mech., 13: 481–504.
Longuet-Higgins, M.S. and Stewart, R.W., 1964. Radiation stresses in water waves; a physical discussion, with applications. Deep-Sea Res., 11: 529–562.
Longuet-Higgins, M.S. and Turner, J.S., 1974. An "entraining plume" model of a spilling breaker. J. Fluid Mech., 63: 1–20.
Longuet-Higgins, M.S., Cartwright, D.E. and Smith, N.D., 1963. Observation of the directional wave spectrum of sea waves using the motions of a floating buoy. In: Ocean Wave Spectra. Natl. Acad. Sci., 111–136.
Lorenz, E.N., 1972. Barotropic instability of Rossby wave motion. J. Atmos. Sci., 29: 258–264.
Ludwig. D., 1966. Uniform asymptotic expansions at a caustic. Comm. Pure Appl. Math., 19: 215–250.
Lumley, J.L., 1964. The spectrum of nearly inertial turbulence in a stably stratified fluid. J. Atmos. Sci., 21: 99–102.
Lumley, J.L. and Panofsky, H.A., 1964. The Structure of Atmospheric Turbulence. Interscience, New York, 239 pp.
Luyten, J.R., 1974. Topographic Rossby waves, a cautionary tale. Mém. Soc. R. Sci. Liège, 63 série, tome VI: 167–177.
Lyne, A.G. and Rickett, B.J., 1968. Measurements of the pulse shape and spectra of the pulsating radio sources. Nature, 218: 326–330.
Magaard, L., 1962. Zur Berechnung interner Wellen in Meeresräumen mit nichtebenen Böden bei einer speziellen Dichteverteilung. Kieler Meeresforsch., 18: 161–183.
Magaard, L., 1965. Zur Theorie zweidimensionaler nichtlinearer interner Wellen in stetig geschichteten Medien. Kieler Meeresforsch., 221: 22–32.
Magaard, L., 1971. Zur Berechnung von luftdruck-und windbedingten Bewegungen eines stetig geschicheteten unbegrenzten Meeres. Dtsch. Hydrogr. Z., 24: 145–158.
Magaard, L., 1973. On the generation of internal gravity waves by a fluctuating buoyancy flux at the sea surface. Geophys. Fluid Dyn., 5: 101–111.
Magaard, L., 1977. On the generation of baroclinic Rossby waves in the ocean by meteorological forces. J. Phys. Oceanogr., 7: 359–364.
Magnus, W. and Oberhettinger, F., 1949. Formulas and Theorems for the Functions of Mathematical Physics. Chelsea, New York, 172 pp.
Mahony, J.J., 1972. Cross-waves. Part 1. Theory. J. Fluid Mech., 55: 229–244.
Mamayev, O.I., 1975. Temperature–salinity Analysis of World Ocean Waters. Elsevier, Amsterdam, 374 pp.
Manton, M.J. and Mysak, L.A., 1971. Construction of internal wave solutions via a certain functional equation. J. Math. Anal. Appl., 35: 237–248.
Manton, M.J., Mysak, L.A. and McGorman, R.E., 1970. The diffraction of internal waves by a semi-infinite barrier. J. Fluid Mech., 43: 165–176.
Marchuk, G.I., 1972. About formulation of the problems on the dynamics of the ocean. In: International Symposium on Stratified Flows, Novosibirsk, 1972. Am. Soc. Civil Eng., New York, pp. 69–85.
Martin, S., Simmons, W. and Wunsch. C., 1972. The excitation of resonant triads by single internal waves. J. Fluid Mech. 53: 17–44.
Maslowe, S.A., 1972. The generation of clear-air turbulence by nonlinear waves. Stud. Appl. Math. 51: 1–16.
Matsuno, T., 1966. Quasi-geostrophic motions in the equatorial area. J. Meterol. Soc. Japan, II, 44: 25–43.
Maxworthy, T. and Redekopp, L.G., 1976. A solitary wave theory of the Great Red Spot and other observed features of the Jovian atmosphere. Icarus, 29: 261–271.
McCreary, J., 1976. Eastern tropical ocean response to changing winds: with application to El Niño. J. Phys. Oceanogr., 6: 632–645.
McEwan, A.D., 1971. Degeneration of resonantly-excited internal gravity waves. J. Fluid Mech., 50: 431–438.
McEwan, A.D., 1973. Interactions between internal gravity waves and their traumatic effect on a continuous stratification. Boundary-Layer Meteorol., 5: 159–175.
McEwan, A.D. and Robinson, R.M., 1975. Parametric instability of internal gravity waves. J. Fluid Mech., 67: 667–687.
McEwan, A.D., Mander, D.W. and Smith, R.K., 1972. Forced resonant second-order interaction between damped internal waves. J. Fluid Mech., 55: 589–608.

McGoldrick L.F., 1965. Resonant interactions among capillary–gravity waves. J. Fluid Mech., 21: 305–331.
McGoldrick, L.F., 1970. An experiment on second-order capillary–gravity resonant wave interactions. J. Fluid Mech., 40: 251–271.
McGoldrick, L.F., Phillips, O.M., Huang, N.E. and Hodgson, T.H., 1966. Measurements of third-order resonant wave interactions. J. Fluid Mech., 25: 437–456.
McGorman, R.E. and Mysak, L.A., 1973. Internal waves in a randomly stratified fluid. Geophys. Fluid Dyn., 4: 243–266.
McKee, W.D., 1972. Scattering of Rossby waves by partial barriers. Geophys. Fluid Dyn., 4: 83–89.
McKee, W.D., 1974. Waves on a shearing current: a uniformly valid asymptotic solution. Proc. Cambridge Philos. Soc., 75: 295–301.
McKee, W.D., 1975. A two turning-point problem in fluid mechanics. Proc. Cambridge Philos. Soc., 97: 581–590.
McKee, W.D., 1977. Continental shelf waves in the presence of a sheared geostrophic current. In: D.G. Provis and R. Radok (Editors), Lecture Notes in Physics. Australian Acad. Sci., Canberra, and Springer Verlag, Berlin, 64: 212–219.
McLachlan, N.W., 1947. Theory and Application of Mathieu Functions. Clarendon Press, Oxford, 401 pp.
McLaren, T.I., Pierce, A.D., Fohl, T. and Murphy, B.L., 1973. An investigation of internal gravity waves generated by a buoyantly rising fluid in a stratified medium. J. Fluid Mech., 57: 229–240.
McLellan, H.J., 1965. Elements of Physical Oceanography. Pergamon Press, New York, 150 pp.
McWilliams, J.C., 1974. Forced transient flow and small-scale topography. Geophys. Fluid Dyn., 6: 49–79.
McWilliams, J.C., 1976. Large-scale inhomogeneities and mesoscale ocean waves: a single stable wave field. J. Mar. Res., 34: 423–456.
McWilliams, J.C. and Flierl, G.R., 1976. Optimal, quasi-geostrophic wave analyses of MODE array data. Deep-Sea Res., 23: 285–300.
McWilliams, J.C. and Robinson, A.R., 1974. A wave analysis of the polygon array in the tropical Atlantic. Deep-Sea Res., 21: 359–368.
Mei, C.C., 1973. A note on the averaged momentum balance in two-dimensional water waves. J. Mar. Res., 31: 97–104.
Melchior, P.J., 1966. The Earth Tides. Pergamon Press, New York, 458 pp.
Michell, J.H., 1893. The highest waves in water. Philos. Mag., Ser. 5, 36: 430–437.
Mied, R.P. and Dugan, J.P., 1975. Internal wave reflection by a velocity shear and density anomaly. J. Phys Oceanogr., 5: 279–287.
Mihaljan, J.M., 1963. The exact solution of the Rossby adjustment problem. Tellus, 15: 150–154.
Milder, M., 1976. A conservation law for internal gravity waves. J. Fluid Mech., 78: 209–216.
Miles, J.W., 1957. On the generation of surface waves by shear flows. J. Fluid Mech., 3: 185–204.
Miles, J.W., 1959a. On the generation of surface waves by shear flows. Part 2. J. Fluid Mech., 6: 568–582.
Miles, J.W., 1959b. On the generation of surface waves by shear flows. Part 3. Kelvin-Helmholtz instability. J. Fluid Mech., 6: 583–598.
Miles, J.W., 1960. On the generation of surface waves by turbulent shear flows. J. Fluid Mech., 7: 469–478.
Miles, J.W., 1961. On the stability of heterogeneous shear flows. J. Fluid Mech., 10: 496–508.
Miles, J.W., 1962. On the generation of surface waves by shear flows. Part 4. J. Fluid Mech., 13: 433–448.
Miles, J.W., 1964a. Baroclinic instability of the zonal wind. Rev. Geophys., 2: 155–176.
Miles, J.W., 1964b. A note on Charney's model of zonal-wind instability. J. Atmos. Sci., 21: 451–452.
Miles, J.W., 1965. Instability of very long waves in a zonal flow. Tellus, 17: 302–305.
Miles, J.W., 1968. The Cauchy-Poisson problem for a viscous liquid. J. Fluid Mech., 34: 359–370.
Miles, J.W., 1971a. Resonant response of harbors: An equivalent-circuit analysis. J. Fluid Mech., 46: 241–265.
Miles, J.W., 1971b. Internal waves generated by a horizontally moving source. Geophys. Fluid Dyn., 2: 63–87.
Miles, J.W., 1972. Kelvin waves on oceanic boundaries. J. Fluid Mech., 55: 113–127.
Miles, J.W., 1974a. Harbor seiching. Ann. Rev. Fluid Mech., 6: 17–35.
Miles, J.W., 1974b. Laplace's tidal equations revisited. In: Proceedings 7th U.S. National Congress on Applied Mechanics, Am. Soc. Mech. Eng., New York, pp. 27–38.
Miles, J.W., 1974c. On Laplace's tidal equations. J. Fluid Mech., 66: 241–260.
Miles, J.W. and Lee, Y.K., 1975. Helmholtz resonance of harbours. J. Fluid Mech., 67: 445–464.
Miles, J.W. and Munk, W.H., 1961. Harbor paradox. J. Waterways Harb. Div., Am. Soc. Civil Eng., 87: 111–130.
Miller, C.D. and Barcilon, A., 1976. The dynamics of the littoral zone. Rev. Geophys. Space Sci., 14: 81–91.
Miller, G.R., 1966. The flux of energy out of the deep oceans. J. Geophys. Res., 71: 2485–2489.

Minzoni, A.A. 1976. Nonlinear edge waves and shallow-water theory. J. Fluid Mech., 74: 369–374.
Miropolsky, Yu. Z., 1972. The effect of microstructure of the density field in the sea on the propagation of internal gravity waves. Izv. Akad. Nauk SSSR, Fiz. Atmos. Okeana, 8: 515–517 (in Russian).
Miropolsky, Yu. Z., 1973. On the probability distribution laws of some characteristics of internal waves in the ocean. Izv. Akad. Nauk SSSR, Fiz. Atmos. Okeana, 9: 411–419 (in Russian).
Miropolsky, Yu. Z., 1975a. On the generation of internal waves in the ocean by the wind field. Okeanologiya 15: 389–396 (in Russian).
Miropolsky, Yu. Z., 1975b. The influence of shear flow on the generation of short-period internal waves in the ocean. Izv. Akad. Nauk SSSR, Fiz. Atmos. Okeana, 11: 933–941 (in Russian).
Miropolsky, Yu. Z., 1975c. On the pulse propagation in the stratified rotating fluid. Izv. Akad. Nauk SSSR, Fiz. Atmos. Okeana, 11: 1314–1322 (in Russian).
Mitsuyasu, H., 1968. On the growth of the spectrum of wind-generated waves. 1. Rep. Res. Inst., Appl. Mech., Kyushu Univ., 16: 459.
Mitsuyasu, H., 1969. On the growth of the spectrum of wind-generated waves. 2. Rep. Res. Inst., Appl. Mech., Kyushu Univ., 17: 235.
Miura, R.M., 1976. The Korteweg-deVries equation: A survey of results. SIAM Rev., 18: 412–459.
Miyata, M. and Groves, W.G., 1968. Note on sea-level observations at two nearby stations. J. Geophys. Res., 73: 3965–3967.
Mofjeld, H.O. and Rattray, M. Jr., 1971. Free oscillations in a beta-plane ocean. J. Mar. Res., 29: 281–305.
Monahan, E.C., 1971. Oceanic whitecaps. J. Phys. Oceanogr., 1: 139–144.
Monin, A.S. and Yaglom, A.M., 1971. Statistical Fluid Dynamics. MIT Press, Cambridge, Mass., 769 pp.
Monin, A.S., Kamenkovich, V.M. and Kort, V.G., 1977. Variability of the Oceans. Wiley, New York, 241 p.
Mooers, C.N.K., 1975a. Sound-velocity perturbations due to low-frequency motions in the ocean. J. Acoust. Soc. Am., 57: 1067–1075.
Mooers, C.N.K., 1975b. Several effects of a baroclinic current on the cross-stream propagation of inertial-internal waves. Geophys. Fluid Dyn., 6: 245–275.
Mooers, C.N.K. and Smith, R.L., 1968. Continental shelf waves off Oregon. J. Geophys. Res., 73: 549–557.
Moore, D.W., 1963. Rossby waves in ocean circulation. Deep-Sea Res., 10: 735–747.
Moore, D.W., 1968. Planetary–Gravity Waves in an Equatorial Ocean. Ph.D. thesis, Harvard Univ., Cambridge, Mass., 207 pp.
Morris, C.A.N., 1976. The generation of surface waves over a sloping beach by an oscillating line source. III. The three-dimensional problem and the generation of edge waves. Math. Proc. Cambridge Philos. Soc., 79: 573–585.
Morrison, J.A. and McKenna, J., 1973. Analysis of some stochastic ordinary differential equations. SIAM–Am. Math. Soc. Proc., American Mathematics Society, Providence, Rhode Island, 6: 97–161.
Morse, P.M. and Feshbach, H., 1953. Methods of Theoretical Physics, Vol. I and II. McGraw-Hill, New York, 1978 pp.
Morse, P.M. and Ingard, K.U., 1968. Theoretical Acoustics. McGraw-Hill, New York, 927 pp.
Moura, A.D. and Stone, P.H., 1976. The effects of spherical geometry on baroclinic instability. J. Atmos. Sci., 33: 602–616.
Mowbray, D.E. and Rarity, B.S.H., 1967a. A theoretical and experimental investigation of the phase configuration of internal waves of small amplitude in a density stratified liquid. J. Fluid Mech., 28: 1–16.
Mowbray, D.E. and Rarity, B.S.H., 1967b. The internal wave pattern produced by a sphere moving vertically in a density stratified fluid. J. Fluid Mech., 30: 489–495.
Müller, P., 1974. On the interaction between short internal waves and larger scale motion in the ocean. Hamb. Geophys. Einzelschr., 23.
Müller, P. and Olbers, D.J., 1975. On the dynamics of internal waves in the deep ocean. J. Geophys. Res., 80: 3848–3860.
Müller, P. and Siedler, G., 1976. Consistency relations for internal waves. Deep-Sea Res., 23: 613–628.
Munk, W.H., 1949a. The solitary wave and its application to surf problems. Ann. New York Acad. Sci., 51: 376–442.
Munk, W.H., 1949b. Surf beats. Trans. Am. Geophys. Union, 30: 849–854.
Munk, W.H., 1968. Once again – tidal friction. Q.J.R. Astron. Soc., 9: 352–375.
Munk, W.H. and Cartwright, D.E., 1966. Tidal spectroscopy and prediction. Philos. Trans. R. Soc. Lond., A, 259: 533–581.
Munk, W.H. and Garrett, C.J.R., 1973. Breaking and microstructure (the chicken and the egg). Boundary-Layer Meteorol., 4: 37–45.
Munk, W.H. and MacDonald, G., 1960. The Rotation of the Earth. Cambridge University Press, London, 323 pp.
Munk, W.H. and Moore, D., 1968. Is the Cromwell current driven by equatorial Rossby waves? J. Fluid Mech., 33: 241–259.

Munk, W.H., Snodgrass, F.E. and Carrier, G., 1956. Edge waves on the continental shelf. Science, 123: 127–132.
Munk, W.H., Snodgrass, F.E. and Gilbert, F.J., 1964. Long waves on the continental shelf: An experiment to separate trapped and leaky modes. J. Fluid Mech., 20: 529–554.
Munk, W.H., Snodgrass, F.E. and Wimbush, M., 1970. Tides off-shore: Transition from California coastal to deep-sea waters. Geophys. Fluid Dyn., 1: 161–235.
Mysak, L.A., 1967a. On the theory of continental shelf waves. J. Mar. Res., 25: 205–227.
Mysak, L.A., 1967b. On the very low frequency spectrum of the sea level on a continental shelf. J. Geophys. Res., 72: 3043–3047.
Mysak, L.A., 1968a. Edgewaves on a gently sloping continental shelf of finite width. J. Mar. Res., 26: 24–33.
Mysak, L.A., 1968b. Effects of deep-sea stratification and current on edgewaves. J. Mar. Res., 26: 34–43.
Mysak, L.A., 1969. On the generation of double Kelvin waves. J. Fluid Mech., 37: 417–434.
Mysak, L.A. and Hamon, B.V., 1969. Low-frequency sea-level behavior and continental shelf waves off North Carolina. J. Geophys. Res. 74: 1397–1405.
Mysak, L.A. and Howe, M.S., 1976. A kinetic theory for internal waves in a randomly stratified fluid. Dyn. Atmos. Oceans, 1: 3–31.
Mysak, L.A. and Howe, M.S., 1978. Scattering of Poincaré waves by an irregular coastline. Part 2. Multiple scattering theory. J. Fluid Mech., in press.
Mysak, L.A. and LeBlond, P.H., 1972. The scattering of Rossby waves by a semi-infinite barrier. J. Phys. Oceanogr., 2: 108–114.
Mysak, L.A. and Schott, F., 1977. Evidence for baroclinic instability of the Norwegian current. J. Geophys. Res., 82: 2087–2095.
Mysak, L.A. and Tang, C.L., 1974. Kelvin wave propagation along an irregular coastline. J. Fluid Mech., 64: 241-261.
Nansen, F., 1902. The oceanography of the north polar basin. Norw. North Polar Exped. 1893–1896. Sci. Results, 3: 9.
Nath, J.H. and Ramsey, F.L., 1976. Probability distributions of breaking wave heights emphasizing the utilization of the JONSWAP spectrum. J. Phys. Oceanogr., 6: 316–323.
Needham, J., 1959. Science and Civilization in China, Vol. 3. Cambridge University Press, London, 877 pp.
Needler, G.T. and Heath, R.A., 1975. Diffusion coefficients calculated from the Mediterranean salinity anomaly in the North Atlantic Ocean. J. Phys. Oceanogr., 5: 173–182.
Needler, G.T. and LeBlond, P.H., 1973. On the influence of the horizontal component of the Earth's rotation on long period waves. Geophys. Fluid Dyn., 5: 23–46.
Neumann, G., 1953. On ocean wave spectra and a new method of forecasting wind-generated sea. Beach Erosion Board, U.S. Army Corps of Engineers, Tech. Mem. No. 43, 42 pp.
Neumann, G. and Pierson, W.J., 1966. Principles of Physical Oceanography. Prentice-Hall, Englewood Cliffs, New Jersey, 545 pp.
Newell, A.C., 1969. Rossby wave packet interactions. J. Fluid Mech., 35: 255–271.
Newman, J.N., 1965a. Propagation of water waves past long two-dimensional obstacles. J. Fluid Mech., 23: 23–29.
Newman, J.N., 1965b. Propagation of water waves over an infinite step. J. Fluid Mech., 23: 399–415.
Newton, J.L., Aagaard, K. and Coachman, L.K., 1974. Baroclinic eddies in the Arctic Ocean. Deep-Sea Res., 21: 707–719.
Niiler, P.P., 1968. On the internal tidal motions in the Florida Straits. Deep-Sea Res., 15: 113–123.
Niiler, P.P. and Mysak, L.A., 1971. Barotropic waves along an eastern continental shelf. Geophys. Fluid Dyn., 2: 273–288.
Nikitin, A.K. and Potetyunko, E.N., 1967. On the Cauchy-Poisson problem for waves propagating at the surface of a viscious liquid of finite depth. Dokl. Akad. Nauk SSSR, 174: 50–52 (in Russian).
Nikitin, O.P. and Tareyev, B.A., 1972. Meanders of the Gulf Stream interpreted as resultants of baroclinic instability predicted by a simple two-layer model. Izv. Akad. Nauk SSSR, Fiz. Atmos. Okeana, 8: 973–980 (in Russian).
Odulo, A.B., 1974. Edge waves in a rotating stratified fluid at an inclined shore. Izv. Akad. Nauk SSSR, Fiz. Atmos. Okeana, 10: 188–189 (in Russian).
Odulo, A.B., 1975a. Propagation of long waves in a rotating basin of variable depth. Okeanologiya, 15: 11–14 (in Russian).
Odulo, A.B., 1975b. Long-wave propagation in an infinite ocean of variable depth. Okeanologiya, 15: 531–533 (in Russian).
Olbers, D.J., 1974. On the energy balance of small-scale internal waves in the deep-sea. Hamb. Geophys. Einzelschr., 24.
Olbers, D.J., 1976. Nonlinear energy transfer and the energy balance of the internal wave field in the deep ocean. J. Fluid Mech., 74: 375–399.
Orlanski, I., 1969. The influence of bottom topography on the stability of jets in a baroclinic fluid. J. Atmos. Sci., 26: 1216–1232.
Orlanski, I. and Bryan, K., 1969. Formation of the thermocline step structure by large-amplitude internal gravity waves. J. Geophys. Res., 74: 6975–6983.

Orlanski, I. and Cox, M.D., 1973. Baroclinic instability in ocean currents. Geophys. Fluid Dyn., 4: 297–332.
Orlanski, I. and Ross, B.B., 1973. Numerical simulation of the generation and breaking of internal gravity waves. J. Geophys. Res., 78: 8808–8826.
Osborn, T.R. and LeBlond, P.H., 1974. Static stability in fresh water lakes. Limnol. Oceanogr., 19: 544–545.
Otterman, I., 1974. Oceanic eddy in the Gulf of Suez. Deep-Sea Res., 21: 163–165.
Packham, B.A. and Williams, W.E., 1968. Diffraction of Kelvin waves at a sharp bend. J. Fluid Mech., 34: 517–529.
Parker, C.E., 1974. Gulf Stream rings in the Sargasso Sea. Deep-Sea Res., 18: 981–993.
Parrish, D.F. and Niiler, P.P., 1971. Topographic generation of long internal waves in a channel. Geophys. Fluid Dyn., 2: 1–15.
Pearson, J.M., 1966. A Theory of Waves. Allyn and Bacon, 140 pp.
Pedlosky, J., 1962. Spectral considerations in two-dimensional incompressible flow. Tellus, 14: 125–132.
Pedlosky, J., 1964a. The stability of currents in the atmosphere and the ocean: Part I. J. Atmos. Sci., 21: 201–219.
Pedlosky, J., 1964b. The stability of currents in the atmosphere and the ocean: Part II. J. Atmos. Sci., 21: 342–353.
Pedlosky, J., 1965a. A study of time-dependent ocean circulation. J. Atmos. Sci., 22: 267–272.
Pedlosky, J., 1965b. A note on the western intensification of the oceanic circulation. J. Mar. Res., 23: 207–209.
Pedlosky, J., 1967. Fluctuating winds and the ocean circulation. Tellus, 19: 250–257.
Pedlosky, J., 1970. Finite-amplitude baroclinic waves. J. Atmos. Sci., 27: 15–30.
Pedlosky, J., 1971a. Geophysical fluid dynamics. In: W.M. Reid (Editor), Mathematical Problems in the Geophysical Sciences. Am. Math. Soc., Providence, Rhode Island, 13: 1–60.
Pedlosky, J., 1971b. Finite-amplitude baroclinic waves with small dissipation. J. Atmos. Sci., 28: 587–597.
Pedlosky, J., 1975a. A note on the amplitude of baroclinic waves in mid-ocean. Deep-Sea Res., 22: 575–576; Reply, 22: 577–578.
Pedlosky, J., 1975b. On secondary baroclinic instability and the meridional scale of motion in the ocean. J. Phys. Oceanogr., 5: 603–607.
Pedlosky, J., 1975c. The amplitude of baroclinic wave triads and meso-scale motion in the ocean. J. Phys. Oceanogr., 5: 608–614.
Pedlosky, J., 1976. Finite-amplitude baroclinic disturbances in down-stream varying currents. J. Phys. Oceanogr., 6: 335–344.
Peffley, M.B. and O'Brien, J.J., 1976. A three-dimensional simulation of coastal upwelling off Oregon. J. Phys. Oceanogr., 6: 164–180.
Peierls, R.E., 1929. Zur kinetischen Theorie der Wärmeleitung in Kristallen. Ann. Phys., 3: 1055–1101.
Pekeris, C.L., 1975. A derivation of Laplace's tidal equation from the theory of inertial oscillations. Proc. R. Soc. Lond., A, 344: 81–86.
Pekeris, C.L. and Accad, Y., 1969. Solution of Laplace's equation for the *M*2 tide in the world oceans. Philos. Trans. R. Soc. Lond., A, 265: 413–436.
Penney, W.G. and Price, A.T., 1952. Finite periodic stationary gravity waves in a perfect fluid. Philos. Trans. R. Soc. Lond., A, 244: 254–284.
Peregrine, D.H., 1972. Equations for water waves and the approximation behind them. In: R.E. Meyer (Editor), Waves on Beaches and Resulting Sediment Transport. Academic Press, New York, 95–121.
Philander, S.G.H., 1973. Equatorial undercurrent, measurements and theories. Rev. Geophys. Space Phys., 11: 513–570.
Philander, S.G.H., 1976. Instabilities of zonal equatorial currents. J. Geophys. Res., 81: 3725–3735.
Phillips, N.A., 1963. Geostrophic motion. Rev. Geophysics, 1: 123–176.
Phillips, N.A., 1965. Elementary Rossby waves. Tellus, 17: 295–301.
Phillips, N.A., 1966. Large-scale eddy motion in the western Atlantic. J. Geophys. Res., 71: 3883–3891.
Phillips, O.M., 1957. On the generation of waves by turbulent wind. J. Fluid Mech., 2: 417–445.
Phillips, O.M., 1958. The equilibrium range in the spectrum of wind-generated ocean waves. J. Fluid Mech., 4: 426–434.
Phillips, O.M., 1960. On the dynamics of unsteady gravity waves of finite amplitude, 1. The elementary interactions. J. Fluid Mech., 9: 193–217.
Phillips, O.M., 1963a. Energy transfer in rotating fluids by reflection of internal waves. Phys. Fluids, 6: 513–520.
Phillips, O.M., 1963b. On the attenuation of long gravity waves by short breaking waves. J. Fluid Mech., 16: 321–332.
Phillips, O.M., 1966. The Dynamics of the Upper Ocean. Cambridge University Press, 261 pp.
Pierson, W.J. and Moskowitz, L., 1964. A proposed spectral form for fully developed wind seas based on the similarity theory of S.A. Kitaigorodskii. J. Geophys. Res., 69: 5181–5190.
Pingree, R.D., 1969. Small-scale structure of temperature and salinity near Station Cavall. Deep-Sea Res., 16: 275–295.

Pinney, E., 1946. Laguerre functions in the mathematical foundations of the electromagnetic theory of the paraboloidal reflector. J. Maths. Phys., 25: 49–79.
Pinsent, H.G., 1971. The effect of a depth discontinuity on Kelvin wave diffraction. J. Fluid Mech., 45: 747–758.
Pinsent, H.G., 1972. Kelvin wave attenuation along nearly straight boundaries. J. Fluid Mech., 53: 273–286.
Pite, H.D., 1973. Studies in Frictionally Damped Waves. Ph.D. thesis, School of Civil Eng., Univ. of New South Wales.
Platzman, G.W., 1963. The dynamical prediction of wind tides on Lake Erie. Meteorol. Monogr., 4(26), 44 pp.
Platzman, G.W., 1968. The Rossby wave. Q.J.R. Meteorol. Soc., 94: 225–248.
Platzman, G.W., 1971. Ocean tides and related waves. In: W.H. Reid (Editor), Mathematical Problems in the Geophysical Sciences. Am. Math. Soc., Providence, Rhode Island, 14: 239–291.
Platzman, G.W., 1972. Two-dimensional free oscillations in natural basins. J. Phys. Oceanogr., 2: 117–128.
Platzman, G.W., 1975. Normal modes of the Atlantic and Indian oceans. J. Phys. Oceanogr., 5: 201–221.
Pnueli, A. and Pekeris, C.L., 1968. Free tidal oscillations in rotating flat basins of the form of rectangles and of sectors of circles. Philos. Trans. R. Soc. Lond., A, 263: 149–171.
Poincaré, H., 1910, Leçons de mécanique céleste. Vol. 3, Théorie des marées. Gauthier-Villars, Paris.
Poisson, S.D., 1816. Mémoire sur la théorie des ondes. Mém. Acad. R. Sci., 1e série, i.
Pollard, R.T., 1970. On the generation by winds of inertial waves in the ocean. Deep-Sea Res., 17: 795–812.
Pond, G.S., 1975. The exchanges of momentum, heat and moisture at the ocean–atmosphere interface. In: Numerical Models of Ocean Circulation. Natl. Acad. Sci. U.S.A., Washington, D.C., pp 26–38.
Pond, G.S. and Bryan, K., 1976. Numerical models of the ocean circulation. Rev. Geophys. Space Phys.. 14: 243–264.
Pope, S.B., 1975. A more general effective-viscosity hypothesis. J. Fluid Mech., 72: 331–340.
Porter, D., 1972. The transmission of surface waves through a gap in a vertical barrier. Proc. Cambridge Philos. Soc., 71: 411–421.
Porter, R.P., Spindel, R.C. and Jaffee, R.J., 1974. Acoustic internal wave interaction at long ranges in the ocean. J. Acoust. Soc. Am., 56: 1426–1536.
Prinsenberg, S.J. and Rattray, M. Jr., 1975. Effects of continental slope and variable Brunt-Väisälä frequency on the coastal generation of internal tides. Deep-Sea Res., 22: 251–263.
Prinsenberg, S.J., Wilmot, W.L. and Rattray, M. Jr., 1974. Generation and dissipation of coastal internal tides. Deep-Sea Res., 21: 263–281.
Putnam, J.A. and Arthur, R.S., 1948. Diffraction of water waves by breakwaters. Trans. Am. Geophys. Union, 29: 481–490.
Radok, R., 1964. Tide gauge on Macquarie Island. Aurora (Official Journal of A.N.A.R.E. Club), June, p. 17.
Rao, D.B., 1966. Free gravitational oscillations in rotating rectangular basins. J. Fluid Mech., 25: 523–555.
Rao, D.B. and Simons, T.J., 1970. Stability of a sloping interface in a rotating two-fluid system. Tellus, 22: 493–503.
Rao, D.B., Mortimer, C.H. and Schwab, D.J., 1976. Surface normal modes of Lake Michigan: Calculations compared with spectra of observed water level fluctuations. J. Phys. Oceanogr., 6: 575–588.
Rao, V.S. and Rao, G.V.P., 1971. On waves generated in rotating stratified liquids by travelling forcing effects. J. Fluid Mech., 46: 447–464.
Rattray, M. Jr., 1957. Propagation and dissipation of long internal waves. Trans. Am. Geophys. Union, 38: 495–500.
Rattray, M. Jr., 1960. On the coastal generation of internal tides. Tellus, 12: 54–62.
Rattray, M. Jr., 1964. Time-dependent motion in an ocean; a unified two-layer, beta-plane approximation. In: K. Yoshida (Editor), Studies on Oceanography. University of Tokyo Press, 568 pp.
Rattray, M. Jr. and Charnell, R.L., 1966. Quasigeostrophic free oscillations in enclosed basins. J. Mar. Res., 24: 82–102.
Rattray, M. Jr., Dworski, J.G. and Kovala, P.E., 1969. Generation of long internal waves at the continental slope. Deep-Sea Res., Suppl. to Vol. 16: 179–195.
Rayleigh, J.W.S., 1880. On the stability, or instability of certain fluid motions. Proc. Lond. Math. Soc., 9: 57–70.
Rayleigh, J.W.S., 1896. Theory of Sound, Vol. II. Republished (1945) by Dover, New York, 204 pp.
Rayner, J.N., 1971. An Introduction to Spectral Analysis. Pion, London, 174 pp.
Rea, C.C. and Komar, P.D., 1975. Computer simulation of a hooked beach shoreline configuration. J. Sediment. Petrol., 45: 866–872.
Redekopp, L.G., 1975. Wave patterns generated by disturbances travelling horizontally in rotating stratified fluids. Geophys. Fluid Dyn., 6: 289–313.
Redekopp, L.G., 1977. On the theory of solitary Rossby waves. J. Fluid Mech., 82: 725–746.

Reid, R.O., 1958. Effect of Coriolis force on edge waves, I. Investigation of the normal modes. J. Mar. Res., 16: 109–144.
Reynolds, O., 1895. On the dynamical theory of incompressible viscous fluids and the determination of the criterion. Philos. Trans. R. Soc. Lond., A, 186: 123.
Rhines, P.B., 1967. The Influence of Bottom Topography on Long-Period Waves in the Ocean. Ph.D. thesis, Univ. of Cambridge.
Rhines, P.B., 1969a. Slow oscillations in an ocean of varying depth. Part I. Abrupt topography. J. Fluid Mech., 37: 161–189.
Rhines, P.B., 1969b. Slow oscillations in an ocean of varying depth. Part 2. Islands and seamounts. J. Fluid Mech., 37: 191–205.
Rhines, P.B., 1970a. Edge-, bottom-, and Rossby waves in a rotating stratified fluid. Geophys. Fluid Dyn., 1: 273–302.
Rhines, P.B., 1970b. Wave propagation in a periodic medium with application to the ocean. Rev. Geophys. Space Physics, 8: 303–319.
Rhines, P.B., 1971a. A comment on the Aries observations. Philos. Trans. R. Soc. Lond., A, 270: 461–463.
Rhines, P.B., 1971b. A note on long-period motions at Site D. Deep-Sea Res., 18: 21–26.
Rhines, P.B., 1973. Observations of energy containing oceanic eddies, and theoretical models of waves and turbulence. Boundary-Layer Meteorol., 4: 345–360.
Rhines, P.B., 1975. Waves and turbulence on a beta-plane. J. Fluid Mech., 69: 417–443.
Rhines, P.B., 1977. The dynamics of unsteady currents. In: E. Goldberg, I. McCave, J. O'Brien and J. Steele (Editors), The Sea. Wiley-Interscience, New York, 6: 189–318.
Rhines, P.B. and Bretherton, F., 1973. Topographic Rossby waves in a rough-bottomed ocean. J. Fluid Mech., 61: 583–607.
Rice, R.O., 1945. Mathematical analysis of random noise. Bell System Tech. J., 24: 46–156.
Richardson, J.M., 1964. The application of truncated hierarchy techniques in the solution of a stochastic linear differential equation. In: Proc. Symp. Appl. Math. American Mathematics Society, Providence, R.I., 16: 290–302.
Richardson, P.L., Strong, E.A. and Knauss, J.A., 1973. Gulf Stream eddies: Recent observations in the western Sargasso Sea. J. Phys. Oceanogr., 3: 297–301.
Robinson, A.R. (Editor), 1963. Wind-Driven Ocean Circulation. Blaisdell, New York, 161 pp.
Robinson, A.R., 1964. Continental shelf waves and the response of sea level to weather systems. J. Geophys. Res., 69: 367–368.
Robinson, A.R. and McWilliams, J.C., 1974. The baroclinic instability of the open ocean. J. Phys. Oceanogr., 4: 281–294.
Robinson, A.R., Luyten, J.R. and Flierl, G., 1975. On the theory of thin rotating jets: a quasi-geostrophic time-dependent model. Geophys. Fluid Dyn., 6: 211–244.
Rochester, M.G., 1973. The Earth's rotation. EOS, Trans. Am. Geophys. Union, 54: 769–780.
Roseau, M., 1967. Diffraction by a wedge in an anisotropic medium. Arch. Ration. Mech. Anal., 26: 188–218.
Rossby, C.G., 1938. On the mutual adjustment of pressure and velocity distribution in certain simple current systems, II. J. Mar. Res., 1: 239–263.
Rossby, C.G. and collaborators, 1939. Relation between variations in the intensity of the zonal circulation of the atmosphere and the displacements of the semi-permanent centers of action. J. Mar. Res., 2: 38–55.
Rüdiger, G., 1976. Negative viscosity in the Gulf Stream? Tellus, 28: 183–189.
Russell, J.S., 1844. Report on Waves. Rep. 14th Meeting Brit. Assoc. Adv. Sci., Murray, London, pp. 311–390.
Saint-Gully, B., 1968. Ondes de frontière dans un bassin tournant dont le fond est incliné. C.R. Acad. Sci. Paris, Ser. A, 266: 1291–1293.
Saint-Guily, B., 1970. On internal waves. Effects of the horizontal component of the Earth's rotation and of a uniform current. Dtsch. Hydrogr. Z., 23: 61–73.
Saint-Guily, B., 1972. Oscillations propres dans un bassin tournant de profondeur variable. Modes de seconde classe. In: "Studi in honore di Guisepina Aliverti", Istituto Universitario Navale di Napoli, Ist. di Meteorologia e Oceanografia, Napoli, pp. 15–25.
Saint-Guily, B., 1976. Sur la propagation des ondes de seconde classe le long d'un talus continental. C.R. Acad. Sci. Paris, Ser. B, 282: 141–144.
Saint-Guily, B. et Rouault, C., 1971. Sur la présence d'ondes de seconde classe dans le golfe du Lion. C.R. Acad. Sci. Paris, Ser. D, 272: 2661–2663.
Saltzman, B. and Tang, C.M., 1975. Formation of meanders, fronts, cutoff thermal pools in a baroclinic ocean current. J. Phys. Oceanogr., 5: 86–92.
Samuels, G. and LeBlond, P.H., 1977. The energy of near-surface internal waves in the Strait of Georgia. Atmosphere, 15: 151–159.
Sandstrom, H., 1969. Effect of topography on propagation of waves in stratified fluids. Deep-Sea Res., 16: 405–410.
Sandstrom, H., 1976. On topographic generation and coupling of internal waves. Geophys. Fluid Dyn., 7: 231–270.

Saunders, P.M., 1971. Anticyclonic eddies formed from shoreward meanders of the Gulf Stream. Deep-Sea Res., 18: 1207–1219.
Saville, T., 1961. Experimental determination of wave set up. Proc. 2nd Tech. Conf. on Hurricanes. Miami Beach, Fla., pp. 242–252. U.S. Dept. of Commerce, Natl. Hurricane Res. Proj. Rep. No. 50.
Schiff, L.I., 1955. Quantum Mechanics. McGraw-Hill, New York, 2nd edition, 417 pp.
Schmitz, W.J., 1976. Eddy kinetic energy in the deep Western North Atlantic. J. Geophys. Res., 81: 4981–4982.
Schmitz, W.J., 1977. On the deep general circulation in the western North Atlantic. J. Mar. Res., 35: 21–28.
Schott, F., 1977. On the energetics of baroclinic tides in the North Atlantic. Ann. Géophys., 33: 41–62.
Schott, F. and Düing, W., 1976. Continental shelf waves in the Florida Straits. J. Phys. Oceanogr ., 6: 451–460.
Schulman, E.E., 1967. The baroclinic instability of mid-ocean circulation. Tellus, 19: 292–305.
Schweber, S.S., 1961. An Introduction to Relativistic Quantum Field Theory. Harper and Row, New York, 905 pp.
Scott, J.R., 1965. A sea spectrum for model tests and long-term ship prediction. J. Ship Res., 9: 145–152.
Scott, W.E., 1975. Refraction of plane inertial waves. Phys. Fluids, 18: 631–632.
Scotti, R.S. and Corcos, G.M., 1969. Measurements on the growth of small disturbances in a stratified shear layer. Radio Sci., 4: 1309–1313.
Shand, J.A., 1953. Internal waves in the Georgia Strait. Trans. Am. Geophys. Union, 34: 849–856.
Shemdin, O.H., 1969. Instantaneous velocity and pressure measurements above propagating waves. Coastal Eng. Lab., Univ. Florida, Tech. Rep. No. 4.
Shemdin, O.H. and Lai, R.J., 1973. Investigation of the velocity field over waves using a wave follower. Coastal Eng. Lab., Univ. Florida, Tech. Rep. No. 18.
Shen, M.C., 1975. Ray method for surface waves on fluid of variable depth. SIAM Rev., 17: 38–56.
Shen, M.C., Meyer, R.E. and Keller, J.B., 1968. Spectra of water waves in channels and around islands. Phys. Fluids, 14: 2289–2304.
Shinbrot, M., 1976. The initial value problem for surface waves under gravity. I: The simplest case. Indiana Univ. Math. J., 25: 281–300.
Siew, P.F. and Hurley, D.G., 1972. Diffraction of planetary waves by a semi-infinite plate. Bull. Aust. Math. Soc., 6: 145–156.
Simmons, W.F., 1969. A variational method for weak resonant wave interactions. Proc. R. Soc. Lond., A, 309: 551–575.
Simons, T.J., 1976. Continuous dynamical computations of water transports in Lake Erie for 1970. J. Fish Res. Board Can., 33: 371–384.
Simons, T.J. and Rao, D.B., 1972. Nonlinear interaction of waves and zonal current in a two-layer baroclinic model. Tellus, 24: 1–5.
Skjelbreia. L. and Hendrickson, J.A., 1961. Fifth-order gravity wave theory. Proc. 7th Conf. on Coastal Eng., 1, Chapter 10, p. 184.
Skovgaard, O., Jonsson, I.G. and Bertelsen, J.A., 1975. Computation of wave heights due to refraction and friction. Proc. Am. Soc. Civ. Eng., J. Waterways, Harbors and Coastal Engin. Div., 101 (WWL): 15–32.
Skovgaard, O., Jonsson, I.G. and Bertelsen, J.A., 1976. Closure to "Computation of wave heights due to refraction and friction". Proc. Am. Soc. Civ. Eng., J. Waterways, Harbors and Coastal Engin. Div., 102 (WWL): 100–105.
Smith, P.C., 1976. Baroclinic instability in the Denmark Strait overflow. J. Phys. Oceanogr., 6: 335–371.
Smith, R., 1970. Asymptotic solutions for high-frequency trapped wave propagation. Philos. Trans. R. Soc. Lond., A, 268: 289–324.
Smith, R., 1971. The ray paths of topographic Rossby waves. Deep-Sea Res., 18: 477–483.
Smith, R., 1972. Nonlinear Kelvin and continental shelf waves. J. Fluid Mech., 52, 379–391.
Smith, R., 1975. Second-order turning point problems in oceanography. Deep-Sea Res., 22: 837–852.
Snodgrass, F.E., Munk, W.H. and Miller, G.R., 1962. Long-period waves over California's continental borderland. Part I. Background spectra. J. Mar. Res., 20: 3–30.
Snodgrass, F.E., Groves, G.W., Hasselman, K.F., Miller, G.R., Munk, W.H. and Powers, W.M., 1966. Propagation of ocean swell across the Pacific. Philos Trans. R. Soc. Lond., A, 259: 431–497.
Snyder, R.L., 1974. A field study of wave-induced pressure fluctuations above surface gravity waves. J. Mar. Res., 32: 497–531.
Snyder, R.L. and Cox, C.S., 1966. A field study of the generation of ocean waves. J. Mar. Res., 24: 141–178.
Spaeth, M.G. and Berkman, S.C., 1967. The tsunami of March 28, 1964, as recorded at tide stations. U.S. Coast Geodet. Surv., Tech. Bull. 23, 86 pp.
Squire, H.B., 1933. On the stability of three-dimensional disturbances of viscous flow between parallel walls. Proc. R. Soc. Lond., A, 142: 621–628.
Starr, V.P., 1947a. A momentum integral for surface waves in deep water. J. Mar. Res., 6: 126–135.
Starr, V.P., 1947b. Momentum and energy integrals for gravity waves of finite height. J. Mar. Res., 6: 175–193.

Starr, V.P., 1968. Physics of Negative Viscosity Phenomena. McGraw-Hill, New York, 256 pp.
Stern, M.E., 1963. Trapping of low-frequency oscillations in an equatorial "boundary layer". Tellus, 15: 246–250.
Stewart, R.W., 1974. The air–sea momentum exchange. Boundary-Layer Meteorol., 6: 151–167.
Stewart, R.W. and Manton, M.J., 1971. Generation of waves by advected pressure fluctuations. Geophys. Fluid Dyn., 2: 263–272.
Stewartson, K. and Rickard, J.A., 1969. Pathological oscillations of a rotating fluid. J. Fluid Mech., 35: 759–773.
Stoker, J.J., 1957. Water Waves. Interscience, New York, 567 pp.
Stokes, G.G., 1846. Report on recent researches in hydrodynamics. Rep. 16th Meet. Brit. Assoc. Adv. Sci., Southampton, 1846. Murray, London, pp. 1–20. Also: Math. Phys. Pap., 1, p. 167.
Stokes, G.G., 1847. On the theory of oscillatory waves. Trans. Cambridge Philos. Soc., 8: 441–455.
Stokes, G.G., 1880. Mathematical and Physical Papers, Vol. 1. Cambridge University Press, London.
Stommel, H. and Fedorov, K.N., 1967. Small-scale structure in temperature and salinity near Timor and Mindanao. Tellus, 19: 306–325.
Stone, P.H., 1969. The meridional structure of baroclinic waves. J. Atmos. Sci., 26: 376–389.
Stone, P.H., 1971. Baroclinic instability under non-hydrostatic conditions. J. Fluid Mech., 45: 659–671.
Stratonovich, R.L., 1963. Topics in the Theory of Random Noise, Vol. 1. Gordon and Breach, New York, 292 pp.
Struik, D.J., 1926. Détermination rigoureuse des ondes irrotationelles périodiques dans un canal à profondeur finie. Math. Ann., 95: 595–634.
Stull, R.B., 1976. Internal gravity-waves generated by penetrative convection. J. Atmos. Sci., 33: 1279–1286.
Suarez, A.A., 1971. The Propagation and Generation of Topographic Oscillations in the Ocean. Ph.D. thesis, Massachusetts Institute of Technology, 196 pp.
Suginohara, H., 1974. Onset of coastal upwelling in a two-layer ocean by wind stress with longshore variation. J. Oceanogr., Soc. Japan, 30: 23–33.
Summerfield, W.C., 1967. On the six-minute period wave phenomenon, recorded at Macquarie Island. Horace Lamb Centre Oceanogr. Res., Flinders Univ. South Aust. Res. Pap., No. 16.
Summerfield, W.C., 1969. On the trapping of wave energy by bottom topography. Horace Lamb Centre Oceanogr. Res., Flinders Univ. South Aust., Res. Pap., No. 30.
Sverdrup, H.U., 1926. Dynamics of tides on the Northern Siberian shelf. Geophys. Publ., 4, No. 5, Oslo.
Swallow, J.C., 1971. The Aries current measurements in the Western North Atlantic. Philos. Trans. R. Soc. Lond., A, 270: 451–460.
Swallow, J.C. and Hamon, B.V., 1960. Some measurements of deep currents in the eastern North Atlantic. Deep-Sea Res., 6: 155–168.
Swanson, C.A., 1968. Comparison and Oscillation Theory of Linear Differential Equations. Academic Press, New York, 227 pp.
Synge, J.L., 1933. The stability of heterogeneous liquids. Trans. R. Soc. Can., 27 (III): 1–18.
Szoeke, R.A. de, 1971. Observation and Theory of Long Ocean Waves. M.Sc. thesis, Univ. New South Wales, 89 pp.
Szoeke, R.A. de, 1975. Some effects of bottom topography on baroclinic stability. J. Mar. Res., 33: 93–122.
Tabata, S. and Stickland, J.A., 1972. Summary of oceanographic records obtained from moored instruments in the Strait of Georgia – 1969–70. Current velocity and sea-water temperature from Station H-06. Marine Sciences Directorate, Pacific Region, Pac. Mar. Sci. Rep. No. 72–8: 132 pp.
Tam, C.K.W., 1973. Dynamics of rip currents. J. Geophys. Res., 78: 1937–1943.
Tang, C.L. and Mysak, L.A., 1976. A note on "Internal waves in a randomly stratified fluid". J. Phys. Oceanogr., 6: 243–246.
Tang, C.M., 1975. Baroclinic instability of stratified shear flows in the ocean and atmosphere. J. Geophys. Res., 80: 1168–1175.
Tareyev, B.A., 1965. Unstable Rossby waves and the instability of oceanic currents. Izv. Akad. Nauk SSSR, Fiz. Atmos. Okeana, 1: 426–438 (in Russian).
Tareyev, B.A., 1968. Nongeostrophic disturbances and baroclinic instability in two-layered oceanic flow. Izv. Akad. Nauk SSSR, Fiz. Atmos. Okeana, 4: 1275–1284 (in Russian).
Tareyev, B.A., 1971. Gradient-vorticity waves on the continental shelf. Izv. Akad. Nauk SSSR, Fiz. Atmos. Okeana, 7: 431–436 (in Russian)
Tatarski, V.I., 1961. Wave Propagation in a Turbulent Medium. McGraw-Hill, New York, 285 pp.
Taylor, G.I., 1920. Tidal oscillations in gulfs and rectangular basins. Proc. Lond. Math. Soc., 20: 148–181.
Taylor, G.I., 1921a. Experiments with rotating fluids. Proc. R. Soc. Lond., A, 100: 114–121.
Taylor, G.I., 1921b. Tides in the Bristol Channel. Proc. Cambridge Philos. Soc., 20: 320–325.
Taylor, G.I., 1931a. Effect of variation of density on the stability of superposed streams of fluid. Proc. R. Soc. Lond., A, 132: 499–523.
Taylor, G.I., 1931b. Internal waves and turbulence in a fluid of variable density. Rapp. P.-V. Reun., Cons. Perm. Int. Explor. Mer, 76: 35–42. Also: Sci. Pap., 2: 240–246, Cambridge University Press.
Taylor, G.I., 1936. Oscillations of the atmosphere. Proc. R. Soc. Lond., A, 156: 318–326.

Taylor, G.I., 1953. An experimental study of standing waves. Proc. R. Soc. Lond., A, 218: 44–59.
Thomas, N.H. and Stevenson, T.N., 1972. A similarity solution for viscous internal waves. J. Fluid Mech., 54: 495–506.
Thomas, N.H. and Stevenson, T.N., 1973. An internal wave in a viscous ocean stratified by both salt and heat. J. Fluid Mech., 61: 301–304.
Thompson, R.O.R.Y., 1971. Topographic Rossby waves at a site north of the Gulf Stream. Deep-Sea Res., 18: 11–19.
Thompson, R.O.R.Y. and Luyten, J.R., 1976. Evidence for bottom-trapped topographic Rossby waves from single moorings. Deep-Sea Res., 23: 629–635.
Thomson, R.E., 1970. On the generation of Kelvin-type waves by atmospheric disturbances. J. Fluid Mech., 42: 657–670.
Thomson, R.E., 1973. The energy and energy flux of planetary waves in an ocean of variable depth. Geophys. Fluid Dyn., 5: 385–399.
Thomson, R.E., 1975a. The propagation of planetary waves over a random topography. J. Fluid Mech., 70: 267–285.
Thomson, R.E., 1975b. Longshore current generation by internal waves in Georgia Strait. Can. J. Earth Sci., 12: 472–488.
Thomson, R.E., 1976. The attenuation of vertically propagating internal gravity waves by a randomly varying wind/current shear. J. Atmos Sci., 33: 52–58.
Thomson, R.E. and Stewart, R.W., 1977. The balance and redistribution of potential vorticity within the ocean. Dyn. Atmos. Oceans, 1: 299–321.
Thorpe, S.A., 1966. On wave interactions in a stratified fluid. J. Fluid Mech., 24: 737–751.
Thorpe, S.A., 1968. On the shape of progressive internal waves. Philos. Trans. R. Soc. Lond., A, 263: 563–614.
Thorpe, S.A., 1969. Neutral eigensolutions of the stability equation for stratified shear flow. J. Fluid Mech., 36: 673–683.
Thorpe, S.A., 1971. Experiments on the instability of stratified shear flows: Miscible fluids. J. Fluid Mech., 46: 299–319.
Thorpe, S.A., 1973. Turbulence in stably stratified fluids: A review of laboratory experiments. Boundary-Layer Meteorol., 5: 95–119.
Thorpe, S.A., 1975. The excitation, dissipation, and interaction of internal waves in the deep ocean. J. Geophys. Res., 80: 328–338.
Tolstoy, I., 1973, Wave Propagation. McGraw-Hill, New York, 466 pp.
Tolstoy, I. and Clay, C.S., 1966. Ocean Acoustics. McGraw-Hill, New York, 293 pp.
Tomczak, M., 1967. Uber den Einfluss fluktuierender Windfelder auf ein stetig geschichtetes Meer. Dtsch. Hydrogr. Z., 20: 101–129.
Tomczak, M., 1968. Über interne Wellen in der Nähe der Trägheitsperiode. Dtsch. Hydrogr. Z., 21: 145–151.
Townsend, A.A., 1972. Flow in a deep turbulent boundary layer over a surface distorted by water waves. J. Fluid Mech., 55: 719–735.
Trefethen, Ll., 1972. Surface tension in fluid mechanics. In: Illustrated Experiments in Fluid Mechanics. The NCFMF Book of Film Notes. MIT Press, Cambridge, Mass., 251 pp.
Tricker, R.A.R., 1965. Bore, Breakers, Waves and Wakes. Am. Elsevier, New York, 250 pp.
Turner, J.S., 1973. Buoyancy Effects in Fluids. Cambridge University Press, 367 pp.
Tyler, G.L., Faulkerson, W.E., Peterson, A.M. and Teague, C.C., 1972. Second-order scattering from the sea: Ten-meter radar observations of the Doppler continuum. Science, 177: 349–351.
Ünlüata, Ü. and Mei, C.C., 1970. Mass transport in water waves. J. Geophys. Res., 75: 7611–7618.
Ursell, F., 1952. Edge waves on a sloping beach. Proc. R. Soc. Lond., A, 214: 79–97.
Ursell, F., 1956. Wave generation by wind. In: G.K. Batchelor (Editor), Surveys in Mechanics. Cambridge University Press, London, pp. 216–249.
Valenzuela, G.R., 1976. The growth of gravity–capillary waves in a coupled shear flow. J. Fluid Mech., 76: 229–250.
Van Dantzig, D. and Lauwerier, H.A., 1960. The North Sea problem. IV. Free oscillations of a rotating rectangular sea. Proc. K. Ned. Akad. Wet. (Ser. A), 63: 339.
Van Dorn, W.G. and Pazan, S.E., 1975. Laboratory investigation of wave breaking. Part II: Deep water waves. Scripps Inst. Oceanogr., Univ. of California, San Diego. SIO Ref. No. 75–21.
Veronis, G., 1956. Partition of energy between geostrophic and non-geostrophic oceanic motions. Deep-Sea Res., 3: 157–177.
Veronis, G., 1963a. On the approximations in transforming the equations of motion from a spherical surface to the β-plane. I. Barotropic systems. J. Mar. Res., 21: 110–124.
Veronis, G., 1963b. On the approximations in transforming the equations of motion from a spherical surface to the β-plane. II. Baroclinic systems. J. Mar. Res., 21: 199–204.
Veronis, G., 1966. Rossby waves with bottom topography. J. Mar. Res., 24: 338–349.
Veronis, G., 1970. The analogy between rotating and stratified fluids. Ann. Rev. Fluid. Mech., 2: 37–66.
Veronis, G., 1973. Large-scale ocean circulation. In: C.-S. Yih (Editor), Advances in Applied Mechanics. Academic Press, New York, 13: 1–92.
Veronis, G. and Stommel, H., 1956. The action of variable wind stress on a stratified ocean. J. Mar. Res., 15: 43–75.

Voit, S.S., 1958. Propagation of tidal waves from a channel into an open basin. Izv. Akad. Nauk SSSR, Ser. Geofiz., 4: 486–496 (in Russian).
Voit, S.S., 1959. Waves developing at the interface of two fluids induced by a travelling and periodic pressure system. Trans. Mar. Hydrophys. Inst. Acad. Sci. USSR., 17: 117 (in Russian).
Voit, S.S. and Sebekin, B.I., 1970. Reflection of unsteady long waves in a rotating basin. Izv. Akad. Nauk. SSSR, Fiz. Atmos. Okeana, 6: 1022–1034 (in Russian).
Volosov, V.M., 1974. Asymptotic analysis of a certain type of gravitational–gyroscopic internal waves. Okeanologiya, 15: 589–594 (in Russian).
Voorhis, A.D., 1968. Measurements of vertical motion and the partition of energy in the New England slope water. Deep-Sea Res., 15: 599–608.
Wait, J.R., 1962. Electromagnetic Waves in Stratified Media. MacMillan, New York, 372 pp.
Wang, D.P., 1975. Coastal trapped waves in a baroclinic ocean. J. Phys. Oceanogr., 5: 326–333.
Wang. D.P., 1976. Coastal water response to the variable wind: theory and coastal upwelling experiment. Tech. Rep. TR 76-2, Rosenstiel School of Marine and Atmospheric Science, Univ. of Miami, 174 pp.
Wang, D.P. and Mooers, C.N.K., 1976. Coastal-trapped waves in a continuously stratified ocean. J. Phys. Oceanogr., 6: 853–863.
Webb, D.J., 1973. On the age of the semi-diurnal tide. Deep-Sea Res., 20: 847–852.
Webster, F., 1961. The effects of meanders on the kinetic energy balance of the Gulf Stream. Tellus, 13: 392–401.
Wehausen, J.V., 1973. The wave resistance of ships. Adv. Appl. Mech., 13: 93–245.
Welander, P., 1973. Lateral friction in the oceans as an effect of potential vorticity mixing. Geophys. Fluid Dyn., 5: 173–189.
Whitham, G.B., 1962. Mass, momentum and energy flux in water waves. J. Fluid Mech., 12: 135–147.
Whitham, G.B., 1965. A general approach to linear and non-linear waves using a Lagrangian. J. Fluid Mech., 22: 273–283.
Whitham, G.B., 1967. Non-linear dispersion of water waves. J. Fluid Mech., 27: 399–412.
Whitham, G.B., 1974. Linear and Nonlinear Waves. Wiley, New York, 636 pp.
Whitham, G.B., 1976. Nonlinear effects in edge waves. J. Fluid Mech., 74: 353–368.
Wiegel, R.L., 1964. Oceanographical Engineering. Prentice-Hall, Englewood Cliffs, New Jersey, 532 pp.
Wiegel, R.L., 1970. Ocean dynamics. In: H.E. Sheets and V.T. Boatwright (Editors), Hydronautics. Academic Press, New York, pp. 123–228.
Willebrand, J., 1975. Energy transport in a nonlinear and inhomogeneous random gravity field. J. Fluid Mech., 70: 113–126.
Willmarth, W.W. and Wooldridge, C.E., 1962. Measurements of the fluctuating pressure at a wall beneath a thick turbulent boundary layer. J. Fluid Mech., 14: 187–210.
Wilson, W.S., 1966. A method for calculating and plotting surface wave rays. U.S. Army Coastal Eng. Res. Center, Tech. Memo. 17.
Woods, J.D., 1968. Wave-induced shear instability in the summer thermocline. J. Fluid Mech., 32: 791–800.
Woods, J.D. and Wiley, R.L., 1972. Billow turbulence and ocean microstructure. Deep-Sea Res., 19: 87–121.
Wright, D.G. and Mysak, L.A., 1977. Coastal trapped waves, with application to the Northeast Pacific Ocean. Atmosphere, 15: 141–150.
Wüest, W., 1949. Beitrag zur Entstehung von Wasserwellen durch Wind. Z. Angew. Math. Mech., 29: 239–252.
Wunsch, C., 1968. On the propagation of internal waves up a slope. Deep-Sea Res., 15: 251–258.
Wunsch, C., 1972. Bermuda sea-level in relation to tides, weather and baroclinic fluctuations. Rev. Geophys. Space Phys., 10: 1–49.
Wunsch, C., 1975a. Deep ocean internal waves: What do we really know? J. Geophys. Res., 80: 339–343.
Wunsch, C., 1975b. Internal tides in the ocean. Rev. Geophys. Space Phys., 13: 167–182.
Wunsch, C., 1976. Geographical variability of the internal wave field: A search for sources and sinks. J. Phys. Oceanogr., 6: 471–485.
Wunsch, C. and Gill, A.E., 1976. Observations of equatorial trapped waves in Pacific sea-level variations. Deep-Sea Res., 23: 371–390.
Wunsch, C. and Hendry, R., 1972. Array measurements of the bottom boundary layer and the internal wave field on the continental slope. Geophys. Fluid Dyn., 4: 101–145.
Wyrtki, K., Magaard, L. and Hager, J., 1976a. Eddy energy in the oceans. J. Geophys. Res., 81: 2641–2646.
Wyrtki, K., Stroup, E., Patzert, W., Williams, R. and Quinn, W., 1976b. Predicting and observing El Niño. Science, 191: 343–346.
Yaglom, A.M. 1962. Introduction to the Theory of Stationary Random Processes. Prentice-Hall, Englewood Cliffs, New Jersey, 235 pp.
Yamagata, T., 1976a. On trajectories of Rossby wave-packets released in a lateral mean flow. J. Oceanogr., Soc. Japan, 31: 162–168.
Yamagata, T., 1976b. On the propagation of Rossby waves in a weak shear flow. J. Meteorol. Soc. Japan, 54: 126–128.
Yang, W.H. and Yih, C.S., 1976. Internal waves in a circular channel. J. Fluid Mech., 74: 183–192.

Yanowitch, M., 1962. Gravity waves in a heterogeneous incompressible fluid. Comm. Pure Appl. Math., 15: 45–61.
Yih, C.S., 1955. Stability of two-dimensional parallel flows for three-dimensional disturbances. Q. Appl. Math., 12: 434–435.
Yih, C.S., 1957. On stratified flows in a gravitational field. Tellus, 9: 220–227.
Yih, C.S., 1960. Exact solutions for steady two-dimensional flow of a stratified fluid. J. Fluid Mech., 9: 161–174.
Yih, C.S., 1965. Dynamics of Nonhomogeneous Fluids. MacMillan, New York, 306 pp.
Yih, C.S., 1974. Wave motion in stratified fluids. In: S. Leibovich and A.R. Seebass (Editors), Nonlinear Waves. Cornell University Press, Ithaca, pp. 263–290.
Yourgrau, W. and Mandelstam, S., 1968. Variational Principles in Dynamics and Quantum Theory. Saunders, Philadelphia, 201 pp.
Yuen, K.B., 1969. A numerical study of large-scale motions in a two-layer rectangular basin. Marine Sciences Branch, Ottawa, Manuscr. Rep. Ser. No. 14, 119 pp.
Zabusky, N.J. and Kruskal, M.S., 1965. Interaction of solitons in a collisionless plasma and the recurrence of initial states. Phys. Rev. Lett., 15: 240-243.
Zahel, W., 1970. Die Reproduktion Gezeitenbedingter Bewegungsvorgänge im Weltozean mittels des hydrodynamisch-numerischen Verfahrens. Mitt. Inst. Meereskd. Univ. Hamb., No. 17.
Zeilon, N., 1912. On tidal boundary waves. K. Svenska Vetens. Akad. Handl., 47: 4.
Zeilon, N., 1934. Experiments on boundary tides. Medd. Goteborgs. Högskolas, Oceanogr. Inst. No. 8.
Zetler, B.D., 1971. Radiational tides along the coasts of the United States. J. Phys. Oceanogr., 1: 34–38.

The following recent references came to our attention after completion of the manuscript and have been added in proofs.

Beland, M., 1976. Numerical study of nonlinear Rossby wave critical level development in a barotropic zonal flow. J. Atmos. Sci., 33:2066–2078.
Benny, D.J., 1977. General theory for interactions between short and long waves. Stud. Appl. Maths., 56:81–94.
Briscoe, M.G., 1977. Gaussianity of internal waves. J. Geophys. Res., 82:2117–2126.
Brooks, D.A. and Mooers, C.N.K., 1977. Free, stable continental shelf waves in a sheared, barotropic boundary current. J. Phys. Oceanogr., 7:380–388.
Cane, M.A. and Sarachik, E.S., 1976. Forced baroclinic ocean motions. I. The linear equatorial unbounded case. J. Mar. Res., 34:629–665.
Cane, M.A. and Sarachik, E.S., 1977. Forced baroclinic ocean motions: II. The linear equatorial bounded case. J. Mar. Res., 35:395–432.
Dickinson, R.E., 1978. Rossby waves–long-period oscillations of oceans and atmospheres. Ann. Rev. Fluid Mech., 10, in press.
Domaracki, A. and Loesch, A.Z., 1977. Nonlinear interactions among equatorial waves. J. Atmos. Sci., 34:486–498.
Fitz-Gerald, G.F., 1976. The reflection of plane gravity waves travelling in water of variable depth. Philos. Trans. R. Soc. Lond., A, 284:49–89.
Flierl, G.R., 1977. The application of linear quasigeostrophic dynamics to Gulf Stream rings. J. Phys. Oceanogr., 7:365–379.
Geisler, J.E. and Garcia, R.R., 1977. Baroclinic instability at long wavelengths in a β-plane. J. Atmos. Sci., 34:311–321.
Gent, R.P., 1977. A numerical model of airflow above water waves. Part 2. J. Fluid Mech., 82:349–369.
Gent, P.R. and Leach, H., 1976. Baroclinic instability in an eccentric annulus. J. Fluid Mech., 77: 769–788.
Gill, A.E., 1976. Adjustment under gravity in a rotating channel. J. Fluid Mech., 77: 603–621.
Goodman, L. and Levine, E.R., 1977. Generation of oceanic internal waves by advecting atmospheric fields. J. Geophys. Res., 82:1711–1717.
Grimshaw, R., 1976. The reflection of internal gravity waves from a shear layer. Q. J. Mech. Appl. Math., 29:511–525.
Grimshaw, R., 1977. The effects of a variable Coriolis parameter, coastline curvature and variable bottom topography on continental shelf waves. J. Phys. Oceanogr., 7:547–554.
Hogg, N.G., 1976. On spatially growing baroclinic waves in the ocean. J. Fluid Mech., 78:217–235.
Huntley, D.A., 1976. Long-period waves on a natural beach. J. Geophys. Res., 81:6441–6449.
Loesch, A.Z. and Domaracki, A., 1977. Dynamic of N resonantly interacting baroclinic waves. J. Atmos. Sci., 34:22–35.
McComas, C.H. and Bretherton, F.P., 1977. Resonant interactions of oceanic internal waves. J. Geophys. Res., 82:1397–1412.
Mied, R.P., 1976. The occurrence of parametric instabilities in finite-amplitude internal gravity waves. J. Fluid Mech., 78:763–784.
Miles, J.W., 1977. Asymptotic eigensolutions of Laplace's tidal equation. Proc. R. Soc. Lond., A, 353:377–400.
Minzoni, A.A. and Whitham, G.B., 1977. On the excitation of edge waves on beaches. J. Fluid Mech., 79:273–287.

Mooers, C.N.K. and Brooks, D.A., 1977. Fluctuations in the Florida current, summer 1970. Deep-Sea Res., 24:399–425.
Muller, P., 1976. On the diffusion of momentum and mass by internal gravity waves. J. Fluid Mech., 77: 789–823.
Mysak, L.A., 1977a. On the stability of the California Undercurrent off Vancouver Island. J. Phys. Oceanogr., 7, in press.
Mysak, L.A., 1977b. Trapped waves on a tropical beach. Appl. Math. Notes., 3, in press.
Mysak, L.A., Peregrine, D.H., Provis, D.G. and Smith, R., 1977. Waves on water of varying depth: a report on an IUTAM symposium. J. Fluid Mech., 79:499–462.
Pedlosky, J., 1976. On the dynamics of finite-amplitude baroclinic waves as a function of supercriticality. J. Fluid Mech., 78:621–637.
Peregrine, D.H., 1976. Interaction of water-waves and currents. Adv. Appl. Mech., 16:9–117.
Plumb, R.A., 1977. Stability of small amplitude Rossby waves in a channel. J. Fluid Mech., 80: 705–720.
Rao, S.T., Czapski, U. and Sedefian, L., 1977. Characteristics of internal oscillations in Lake Ontario. J. Geophys. Res., 82:1725–1734.
Smith, R.K., 1977. On a theory of amplitude vacillation in baroclinic waves. J. Fluid Mech., 79: 289–306.
Thorpe, S.A., Hall, A.J., Taylor, C. and Allen, J., 1977. Billows in Loch Ness. Deep-Sea Res., 24: 371–379.
Vermersch, J.A. and Beardsley, R.C., 1976. A note on the theory of low-frequency waves in a rotating stratified channel. Stud. Appl. Math., 55:281–292.
Warn, T. and Warn, J., 1976. On the development of a Rossby wave critical level. J. Atmos. Sci., 33:2021–2024.
Warren, F.W., 1976. On the method of Hermitian forms and its application to some problems of hydrodynamic stability. Proc. R. Soc. Lond., A, 350:213–237.

AUTHOR INDEX

SUBJECT INDEX